C++面向对象程序设计

[美] 贝赫鲁兹·A. 佛罗赞（Behrouz A. Forouzan） 理查德·F. 吉尔伯格（Richard F. Gilberg） 著

江红 余青松 余靖 译

C++ Programming

An Object-Oriented Approach

机械工业出版社
CHINA MACHINE PRESS

图书在版编目（CIP）数据

C++ 面向对象程序设计 /（美）贝赫鲁兹 · A. 佛罗赞（Behrouz A. Forouzan），（美）理查德 · F. 吉尔伯格（Richard F. Gilberg）著；江红，余青松，余靖译 . —北京：机械工业出版社，2020.6（2024.6 重印）
（计算机科学丛书）
书名原文：C++ Programming: An Object-Oriented Approach

ISBN 978-7-111-65670-8

I. C… II. ① 贝… ② 理… ③ 江… ④ 余… ⑤ 余… III. C++ 语言 – 程序设计 IV. TP312.8

中国版本图书馆 CIP 数据核字（2020）第 089568 号

北京市版权局著作权合同登记　图字：01-2019-7115 号。

Behrouz A. Forouzan, Richard F. Gilberg: *C++ Programming: An Object-Oriented Approach* (ISBN: 9780073523385).

本书采用 C++ 语言来讲解面向对象编程，针对复杂的 C++ 语言将教学内容划分为五个独立且关联的目标模块：讲授计算机程序设计；讲授 C++ 语言的语法；呈现 C++ 的新特点；讨论数据结构并介绍 STL；介绍面向对象程序设计的设计模式。读者可以根据课程设计的目标和课时安排，灵活有效地进行学习或组织教学内容。全书篇章结构精良、组织有序、概念清晰，围绕教学需求展开内容，程序文档形式一致，为学生日后在学术界和专业领域承担程序设计方面的工作打好了基础。

出版发行：机械工业出版社（北京市西城区百万庄大街 22 号　邮政编码 100037）

责任编辑：游　静	责任校对：李秋荣
印　　刷：北京捷迅佳彩印刷有限公司	版　　次：2024 年 6 月第 1 版第 2 次印刷
开　　本：185mm×260mm　1/16	印　　张：43.25
书　　号：ISBN 978-7-111-65670-8	定　　价：139.00 元

客服电话：（010）88361066　88379833　68326294

译者序

C++ 是从 C 语言和 B 语言派生出来的不断演进的程序设计语言，它是一种面向对象的程序设计语言，支持类、封装、继承、多态等特性。C++ 语言灵活，运算符的数据结构丰富，具有结构化控制语句，程序执行效率高，而且同时具有高级语言与汇编语言的优点。C++ 应用非常广泛，常用于系统开发、引擎开发等应用领域。C++ 是大学计算机编程语言以及课程设计的首选程序设计语言之一。

本书是美国麦格劳 - 希尔教育（McGraw-Hill Education）出版公司出版的，其特色是，针对复杂的 C++ 语言将教学内容划分为五个独立且关联的目标模块：讲授计算机程序设计；讲授 C++ 语言的语法；呈现 C++ 的新特点；讨论数据结构并介绍 STL；介绍面向对象程序设计的设计模式。读者可以根据课程设计的目标和课时安排，灵活有效地进行学习或组织教学内容。

系统、深入地掌握一门计算机程序设计语言是信息化时代的必由之路。本书在世界各地广泛使用，通过该书，读者可以快速有效地掌握 C++ 程序设计语言，开启信息化的大门。

本书由华东师范大学江红、余青松和余靖共同翻译。衷心感谢机械工业出版社的曲熠老师，她积极帮我们筹划翻译事宜并认真审阅译稿。翻译也是一种再创造，同样需要艰辛的付出，感谢朋友、家人以及同事的理解和支持。感谢我们的研究生刘映君、余嘉昊、刘康、钟善豪、方宇雄、唐文芳、许柯嘉等对本译稿的认真通读和指正。在本书翻译的过程中我们力求忠于原著，但由于时间和译者学识有限，故书中的不足之处在所难免，敬请诸位同行、专家和读者指正。

江红　余青松　余靖

2019 年 12 月

本书适用于使用 C++ 语言教授面向对象程序设计的课程。本书还为学生提供了高级的概念，如数据结构和设计模式。完成本书学习的学生将具备学习其他面向对象语言课程、数据结构课程或者设计模式课程的知识储备。

什么是 C++ 语言

C++ 是从 C 语言和 B 语言派生出来的不断演进的程序设计语言。C++ 语言将结构的思想扩展到类，可以从单个类的定义创建不同的对象，并赋予每个数据元素不同的值。

此外，C++ 语言采用面向对象语言的思想来模拟现实生活。在现实生活中，我们定义一个类型，然后创建该类型的对象。在 C++ 语言中，我们定义一个类，然后创建该类的对象。C++ 还包括继承的思想。在继承中，我们可以创建一个类，然后通过扩展定义来创建其他类，就像在现实生活中，可以扩展动物的概念以创建马、牛、狗等概念。

也许 C++ 语言最有意思的部分是多态性。多态性使我们能够编写多个具有相同名称的操作版本，供不同的对象使用。这种实践行为在现实生活中也存在，例如动词“开”（open）的使用。我们可以说开了一家公司，开了一盒罐头，开了一扇门，等等。尽管在这些场景下均使用了“开”这个字，但在不同的对象上引发了不同的动作。

C++ 的最新功能包括标准模板库（Standard Template Library，STL），它是预定义的复杂对象和可以应用于这些对象上的操作的集合，以及帮助用户更加高效和连贯地解决问题的设计模式。

为什么编写本书

本书包括五个独具特色的目标模块，分别阐述如下。

讲授计算机程序设计

本书可以适用于以 C++ 语言为载体的计算机程序设计的第一门课程。第 1~6 章就是基于上述目的而设计的。前六章讨论计算机系统和程序语言，同时还讨论 C++ 语法和程序控制的基本知识，例如选择结构和循环结构。第 1 章到第 6 章对于使用 C++ 语言学习程序设计而言是必不可少的组成部分。

讲授 C++ 语言的语法

第 7 ～ 12 章是研究面向对象程序设计的基础。虽然第 8 和 9 章与 C++ 的面向对象特性没有直接关系，但是我们认为这两章可以在学生理解了第 7 章讨论的面向对象程序设计的基础知识之后再进行讲授。

呈现 C++ 的新特点

第 13 ～ 17 章讨论第一门或者第二门程序设计课程中通常包含的其他主题，可以按任意顺序进行讲授。

讨论数据结构并介绍 STL

第 18 和 19 章（在线提供⊖）是数据结构入门知识，它们为学生进一步选修数据结构课程提供知识准备。

介绍设计模式

第 20 章（在线提供）给出了面向对象程序设计中一些典型问题的简单可行的解决方案，如果不使用设计模式，解决这些问题的方法将更加复杂。第 20 章通过一系列针对特定问题的标准解决方案，让学生更加深入地洞察面向对象程序设计。虽然设计模式通常在计算机图形学课程中讲授，但我们将其应用于非图形问题，以帮助没有图形程序设计经验的学生更好地掌握它。

课程大纲

本书可以按以下顺序讲授：

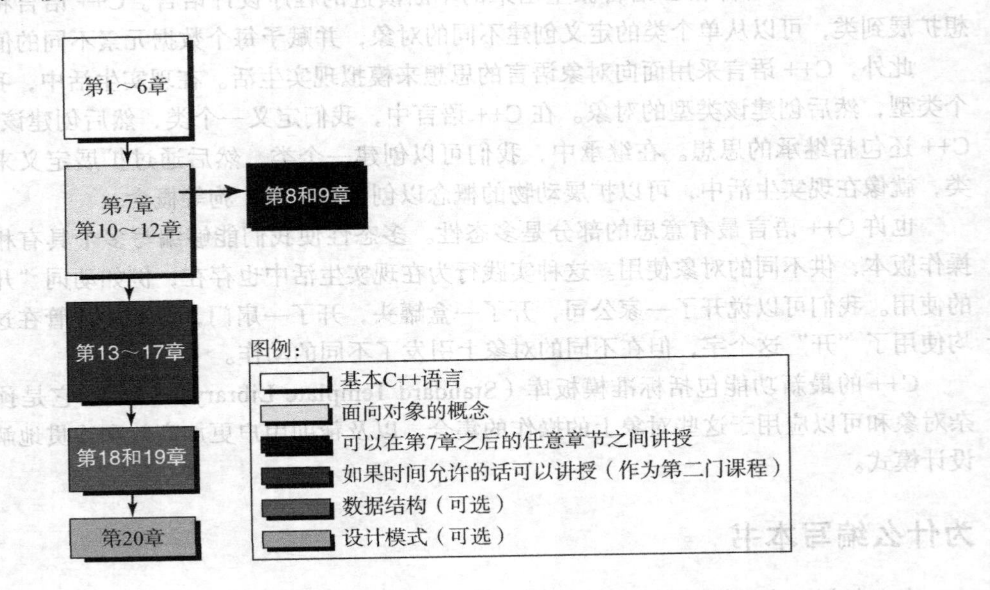

附录

本书配套网站提供了在线附录，包括六部分内容。

参考资料

附录 A 至附录 E 旨在为学生提供参考资料。学生在学习本书的相应章节时可能需要参考这些附录。

语言知识

附录 F 和附录 G 为学生提供有关 C++ 如何准备用于编译的源代码，以及如何在不同的节中处理名称的信息。

高级主题

附录 H 到附录 O 讨论了 C++ 语言中新增的一些高级主题。教师可以在课堂上讲授这些内容，或者学生可以将其作为附加信息的来源。

⊖ 在线资源请访问机工网站 www.course.cmpreading.com 下载。——编辑注

C++11 简介

附录 P 简要概述了在讨论高级主题的附录中没有涉及的有关 C++11 的主题。

UML 简介

本书使用了 UML 图。附录 Q 提供了作为面向对象项目设计工具的 UML 的一般性知识。

位集

当使用 C++ 进行网络编程时，位集（bitset）的概念变得越来越流行。我们在附录 R 中包含了这个主题。

教师资源[⊖]

本教程提供了若干附加资源，读者可以在 www.mhhe.com/forouzan1e 上找到。这些资源包括：知识点测验题，有助于指导教师检验学生对章节内容的理解；判断题和复习题，可以用于进一步测试学生对知识的掌握程度。此外，还提供了知识点测验题、判断题、复习题和思考题的完整答案。最后，提供了授课 PPT、文本图像文件和示例程序。

致谢

我们对本书的审阅者深表感谢。在过去几年中，他们的见解和建议对本书的出版影响巨大。按字母顺序，本书的审阅者如下：

Vicki H. Allan，犹他州立大学
Kanad Biswas，德里理工学院
Gary Dickerson，联合学院
Max I. Formitchev，马克西姆斯能源公司
Cynthia C. Fry，贝勒大学
Barbara Guillott，洛杉矶拉斐特 CGI 的 Q&A 分析师
Jon Hanrath，伊利诺伊理工学院
David Keathly，北得克萨斯大学
Robert Kramer，扬斯敦州立大学
Kami Makki，拉马尔大学
Christopher J. Mallery，微软首席软件工程主管
Michael L. Mick，加州普渡大学
Amar Raheja，加利福尼亚州立理工大学
Brendan Sheehan，内华达大学雷诺分校

我们还要感谢 McGraw-Hill 的编辑和制作人员：Thomas Scaife（高级投资组合经理）；Suzy Banbridge（执行投资组合经理）；Heather Ervolino（产品开发人员）；Shannon O'Donnell（营销经理）；Patrick Diller（业务项目经理）；Jane Mohr（内容项目经理）。

⊖ 关于本书教辅资源，只有使用本书作为教材的教师才可以申请，需要的教师可向麦格劳－希尔教育出版公司北京代表处申请，电话 010-57997618/7600，传真 010-59575582，电子邮件 instructorchina@mheducation.com。——编辑注

目　录

C++ Programming: An Object-Oriented Approach

在线章节[⊖]

⊖ 在线章节请访问机工网站 www.course.cmpreading.com 下载。其中，第 18 章和第 19 章为中文版，其余为英文原版资源。——编辑注

计算机与程序设计语言导论

在本章中，我们将描述计算机系统的组成部分，并讨论计算机语言背后的一般思想，以帮助读者为学习以后的章节做好准备。开始阅读本书时，读者也可以跳过本章，当对程序设计有了更好的理解后再回到这一章。

学习目标

阅读并学习本章后，读者应该能够

- 讨论计算机的两个主要组成部分：硬件和软件。
- 描述硬件的六个部分：中央处理器、主存储器、辅助存储器、输入系统、输出系统和通信系统。
- 描述软件的两大分类：系统软件和应用软件。
- 描述计算机语言从机器语言到汇编语言再到高级语言的演变。
- 讨论四种不同的计算机语言范式：面向过程的程序设计语言、面向对象的程序设计语言、函数式程序设计语言和逻辑式程序设计语言。
- 描述程序设计的两个步骤：理解问题和开发解决方案。
- 描述将 C++ 语言编写的程序转换为可执行程序的几个步骤。

1.1 计算机系统

计算机系统（computer system）由硬件和软件两大部分组成。**计算机硬件**（computer hardware）是物理设备。**软件**（software）是允许硬件完成其工作的程序（指令）的集合。

1.1.1 计算机硬件

计算机系统的**硬件**（hardware）由六部分组成：中央处理器（Central Processing Unit, CPU）、主存储器、辅助存储器、输入系统、输出系统和通信系统。这些组件通过所谓的总线连接在一起。图 1-1 显示了这六个组件及其连接。

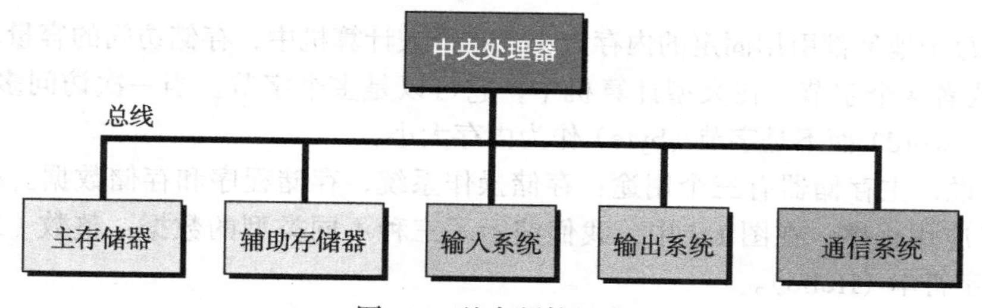

图 1-1 基本硬件组成

中央处理器（CPU）

中央处理器（CPU）由**算术逻辑单元**（Arithmetic-Logical Unit，ALU）、控制单元和一组用于在处理数据时暂时保存数据的寄存器组成。控制单元是系统的交通警察，它协调系统的所有操作。ALU 执行诸如数据的算术计算和比较等指令。图 1-2 显示了中央处理器的组成。

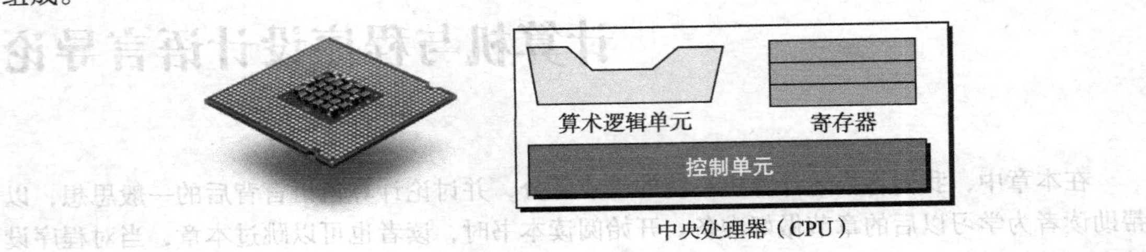

图 1-2　中央处理器

©leungchopan/Getty Images

主存储器

主存储器（primary memory，主内存，主存）是在处理过程中临时存储程序和数据的地方。当我们关闭计算机时，主存储器的内容会丢失。图 1-3 更详细地显示了主存储器。内存中的每个存储位置都有一个地址（很像街道地址），用于引用内存中的内容。图 1-3a 中的地址是左边显示的从 0 到 $n-1$ 的数值，其中 n 是内存的大小。在图 1-3b 中，地址以 x 符号表示。

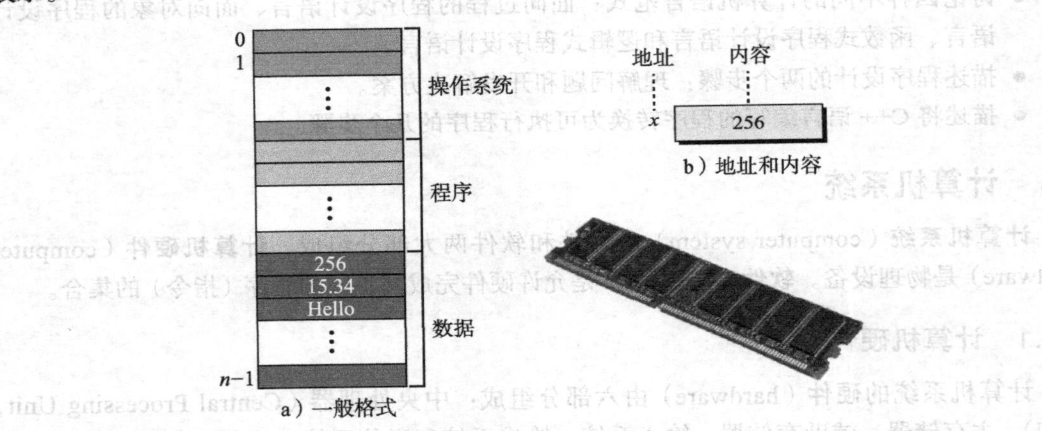

图 1-3　主存储器

©Simon Belcher/Alamy

通常，每个地址都引用固定的内存大小。在个人计算机中，存储访问的容量单位通常是 1 个、2 个或者 4 个字节。在大型计算机中，它可以是多个字节。当一次访问多个字节时，通常使用字（word）而不是字节（byte）作为内存大小。

一般来说，主存储器有三个用途：存储操作系统、存储程序和存储数据。存储的数据类型取决于应用程序。在图 1-3 中，我们演示了三种不同类型的数据：整数（256）、实数（15.34）和字符串（Hello）。

辅助存储器

程序和数据永久存储在**辅助存储器**（secondary storage，辅存或者外存）中。当我们关闭计算机时，程序和数据会保留在辅助存储器中，以便下次我们需要它们时使用。辅助存储器的示例包括硬盘、CD 和 DVD 以及闪存驱动器（图 1-4）。

图 1-4　常用的辅助存储器

©Shutterstock/PaulPaladin; ©David Arky/Getty Images; ©McGraw-Hill Education

输入系统

输入系统（input system）是将程序和数据输入到计算机中的设备，通常指键盘。其他输入设备的示例包括鼠标、笔或者手写笔、触摸屏或者音频输入设备（如图 1-5 所示）。

图 1-5　常用的输入系统

©JG Photography/Alamy; ©Keith Eng 2007

输出系统

输出系统（output system）是显示或者打印输出结果的设备，通常指显示器或者打印机。如果输出显示在监视器上，我们就称其为软拷贝。如果打印在打印机上，我们就称其为硬拷贝（如图 1-6 所示）。

图 1-6　常用的输出系统

©Roy Wylam/Alamy; ©Stephen VanHorn/Alamy

通信系统

我们可以通过多台计算机的连接来创建计算机网络。为此，需要在计算机系统上安装通信设备。图 1-7 显示了一些通信设备。

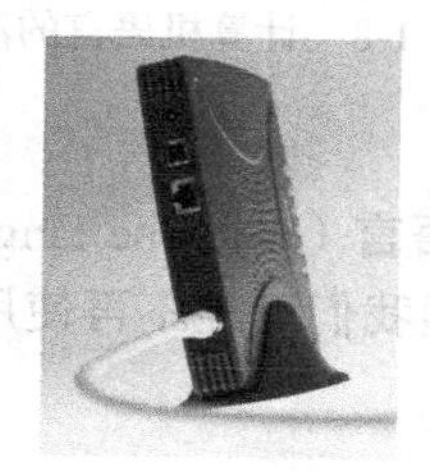

图 1-7　常用的通信设备

©Ingram Publishing; ©somnuek saelim/123RF

1.1.2 计算机软件

计算机软件（computer software）分为两大类：**系统软件**（system software）和**应用软件**（application software）。不管硬件系统架构如何，这种分类方法都是正确的。系统软件管理计算机资源，而应用软件则直接负责帮助用户解决问题。

系统软件

系统软件由管理计算机硬件资源并执行所需信息处理的程序组成。系统软件可以分为三种类型：**操作系统**（operating system）、**系统支持**（system support）**软件**和**系统开发**（system development）**软件**。

操作系统。操作系统提供诸如用户界面、文件和数据库访问以及通信系统的接口等服务。操作系统软件的主要目的是在允许用户访问系统的同时，以有效的方式操作系统。

系统支持软件。系统支持软件提供系统实用程序和其他操作服务。系统实用程序的示例包括排序程序和磁盘格式化程序。操作服务的示例包括为操作人员提供性能统计的程序以及保护系统和数据的安全监视器。

系统开发软件。系统开发软件包括将程序转换为可执行的机器语言的语言转换器、确保程序无错误的调试工具以及超出本书范围的计算机辅助软件工程（Computer-Assisted Software Engineering，CASE）系统。

应用软件

应用软件分为两类：**通用软件**（general-purpose software）和**专用软件**（application-specific software）。

通用软件。通用软件是从软件开发人员处购买，可以用于多种应用的软件。通用软件的示例包括文字处理器、数据库管理系统和计算机辅助设计系统。这些程序被称为通用程序，因为它们可以解决各种用户计算问题。

专用软件。专用软件只能用于特定的目标。会计人员使用的总账系统和工程师使用的物料需求规划系统是专用软件的示例。它们只能用于其所设计的任务，不能用于其他通用任务。

1.2 计算机语言

要为计算机编写程序，必须使用**计算机语言**（computer language）。多年来，计算机语言已经从机器语言发展到符号语言、高级语言等。计算机语言发展的时间线如图1-8所示。

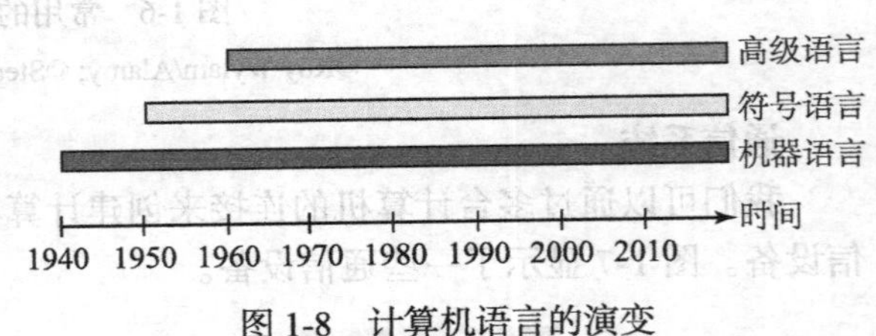

图1-8　计算机语言的演变

1.2.1 机器语言

在计算机的早期，唯一可用的程序设计语言是**机器语言**（machine language）。虽然每台计算机都有自己的机器语言（由0和1的序列组成），但我们已经不再使用机器语言编写程序。

计算机唯一能理解的语言是机器语言。

1.2.2 符号语言

很明显，如果程序员继续使用机器语言，就不会编写很多程序。20 世纪 50 年代初，格雷斯·霍珀（Grace Hopper，数学家、美国海军成员）提出了将程序转换为机器语言的特殊计算机程序的概念（见图 1-9）。

图 1-9 格雷斯·霍珀
©Cynthia Johnson/Getty Images

她的工作导致了程序设计语言的使用，这些语言只是使用符号或者助记符来表示各种机器语言指令。因为这些语言使用符号，所以被称为**符号语言**（symbolic language）。一个被称为**汇编程序**（assembler）的特殊程序用来把符号代码翻译成机器语言。因为符号语言必须被汇编成机器语言，所以它们很快就被称为**汇编语言**（assembly language）。这个名称至今仍然被用于那些与其计算机联系紧密的机器语言的符号语言。

> 符号语言使用助记符来表示机器语言指令。

1.2.3 高级语言

尽管符号语言大大提高了程序设计效率，但它们仍然要求程序员重点关注所使用的硬件。使用符号语言也非常烦琐，因为每个机器指令都必须单独编码。为了提高程序员的效率，并将重心从计算机转移到正在解决的问题上，**高级语言**（high-level language）应运而生。

高级语言可以移植到许多不同的计算机上，这样程序员就可以集中精力处理所面临的应用程序问题，而不是计算机的复杂问题。高级语言的设计目的是使程序员从汇编语言的细节中解脱出来。然而，高级语言与符号语言有一个共同点：必须转换为机器语言。这个过程称为编译（compilation）。

第一种广泛使用的高级语言 FORTRAN（Formula Translation，公式翻译）是由约翰·巴克斯（John Backus）和 IBM 团队于 1957 年发明的。紧接着在 FORTRAN 语言之后，又发明了 COBOL（Common Business-Oriented Language，面向业务的通用语言）。格雷斯·霍珀上将再次成为关键人物，这一次是因为她在开发 COBOL 业务语言上的贡献。

多年来，已经开发了若干高级语言，著名的包括 BASIC、Pascal、Ada、C、C++ 和 Java。今天，编写系统软件和新应用程序代码的流行高级语言之一是 C++，C++ 是本书讨论的对象。

1.3 计算机语言范式

计算机语言可以根据其解决问题的方法进行分类。范式是描述程序如何处理数据的模型或者框架。虽然存在几种不同的将当前计算机语言划分为一系列范式的分类法，但我们只讨论四种：面向过程的程序设计语言、面向对象的程序设计语言、函数式程序设计语言和逻辑式程序设计语言，如图 1-10 所示。图 1-10 还显示了根据我们的分类法各种语言所属的范式。

请注意，C++ 语言可以作为面向过程的程序设计语言范式，还可以作为面向对象的程序设计语言范式，我们将在后续章节讨论。

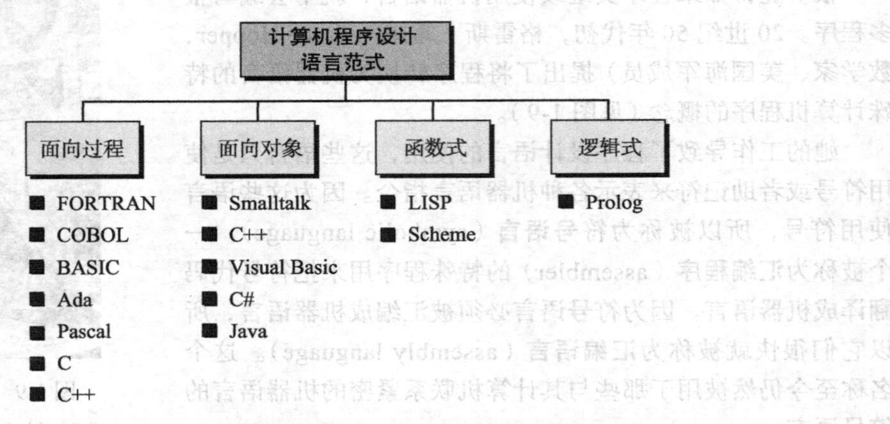

图 1-10　计算机程序设计语言范式

1.3.1　面向过程的程序设计语言范式

在面向过程（procedural，也称为 imperative）的程序设计语言范式中，程序是一组命令。每条命令的执行都会更改与该问题相关的内存状态。例如，假设我们想要求解任意两个值的和。我们保留了三个内存位置，分别称为 a、b 和 sum。在这种情况下，这三个内存位置的组合构成了一个状态。图 1-11 显示了一个面向过程的程序设计语言范式如何使用四条命令将内存状态更改四次。请注意，浅灰色的内存位置显示原始状态，这些位置是为程序保留的。

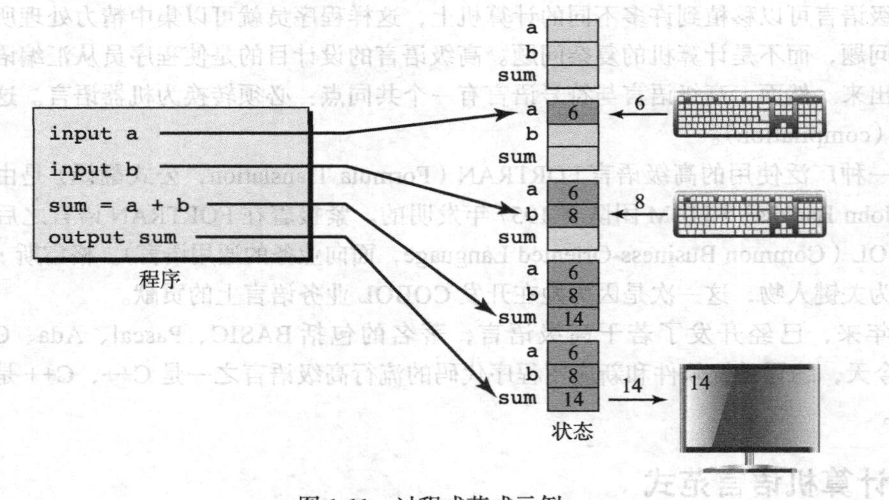

图 1-11　过程式范式示例

为了将第一个数值 a 的值存入内存，我们使用一条输入命令（input a）。执行此命令后，计算机等待我们在键盘上输入一个数值。我们键入：6。当我们按下键盘上的 Enter 键后，数值 6 存储在了第一个内存位置，内存状态改变。在第二条命令之后，内存的状态再次改变，现在 6 和 8 都存储在内存中。第三条命令通过把 a 和 b 的值相加并将结果存储到 sum 来更改内存状态。虽然最后一条命令（output sum）看起来不像是在改变内存状态，但认为它是一种改变，因为 sum 的值被输出了。

图 1-11 中编写代码的方式非常低效，原因有二：首先，它包含一些可以在同一程序或者其他程序中重复的命令；其次，如果要处理的数据集很大，需要逐个处理这些数据。为了解决这两个效率低下的问题，面向过程的程序设计语言范式允许封装命令和**数据项**。

1）如果为不同的程序编写代码，可以封装代码并创建所谓的**过程**（或者**函数**）。一个过程只需要编写一次，然后复制到不同的程序中。标准过程可以存储在语言库中，然后可以使用而不是重写。

2）如果处理的是一个大的数据项集（例如，成百上千个数值），需要将数据项集存储在一个包中（也可以使用不同的名称，例如数组或者记录），然后将这些数据全部一起输入，一起处理，一起输出。后台的操作仍然一次完成一个数据项，但程序可以将数据项视为包。

下面显示了面向过程的程序设计语言范式如何使用三行代码对任意大小的数值列表进行排序。当然，我们必须事先编写好了过程，并且已经将数据项封装到一个列表（list）中。

```
input (list);
sort (list);
output (list);
```

1.3.2　面向对象的程序设计语言范式

在面向过程的程序设计语言范式中，可以创建和保存常用的过程。然后，将这些过程的一个子集应用到相应的数据包中，以解决特定的问题。在面向过程的程序设计语言范式中，很明显过程集和数据包集之间没有明确的关系。当我们想要解决一个问题时，需要选择我们的数据包，然后去寻找合适的过程来处理数据。

面向对象的程序设计语言范式则更进一步，它规定应用于特定类型的数据包的过程集需要与数据封装在一起，封装的整体被称为一个对象（object）。换言之，对象是一个包含所有可能应用于特定类型的数据结构的操作的包。

这和我们在日常生活中发现的一些物理对象的概念是一样的。例如，我们把洗碗机看作一个对象，对洗碗机的最低要求包括清洗、漂洗和烘干。所有典型的洗碗机都包含这些操作。但是，每次我们给洗碗机加载一套不同的盘子（这与一组不同的数据相同）。需要注意的是，不能加载设计不支持的负载到机器上（例如，不能在洗碗机中洗衣服）。

在现实世界中，操作所需的所有硬件都包含在一个对象中。在面向对象的程序设计世界中，每个对象只保存数据，但是定义过程的代码是共享的。图 1-12 显示了面向对象的程序设计语言范式中过程和数据之间的关系。

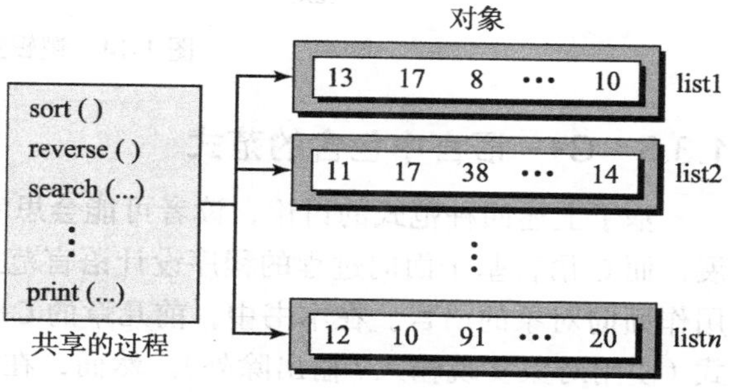

图 1-12　面向对象范式示例

1.3.3　函数式程序设计语言范式

在**函数式程序设计语言范式**中，程序是一个数学函数。在此上下文中，函数是一个黑盒子，它将输入列表映射到输出列表。例如，累加数值可以被视为一个函数，其中输入是要累加的数值列表，输出是只有一个项目的列表（即累加和）。换言之，函数式程序设计语言范式与数学函数的结果有关。在这种范式中，我们不使用命令，也不跟踪内存状态。基本思想是，我们有一些基本的函数，例如加、减、乘、除。同时，我们还有一些其他的基本函数，用于创建一个列表，或者从列表中提取第一个元素或其余元素。我们可以通过组合这些基本函数来编写程序或者新函数。图 1-13 显示了如何使用函数式程序设计语言范式来累加两个数值。代码使用伪代码符号，遵循函数式程序设计语言范式的每种语言都有其相应的函数定义。同时还要注意的是，在我们的代码中，把数值和列表区分开来。8 表示一个数值，而（8）则表示只包含一个数值的列表。函数 first 获取一个数值，而函数 rest 则获取一个列表。为了获取列表中的第二个数值，首先使用函数 rest 取得列表（8），然后使用函数 first 得到 8。

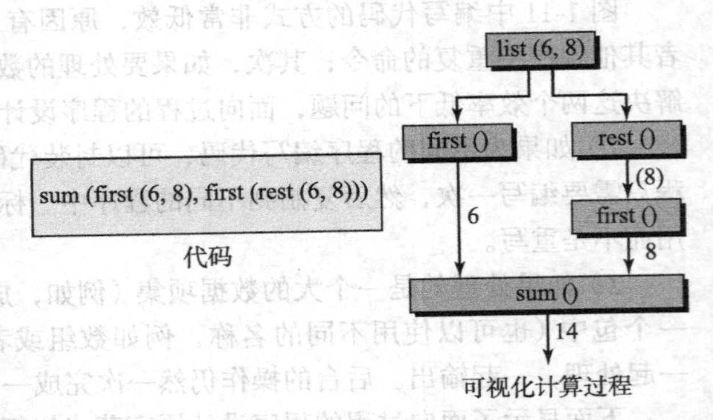

图 1-13　函数式范式示例

1.3.4　逻辑式程序设计语言范式

逻辑式程序设计语言范式使用一组事实和一组规则来回答查询。它以希腊数学家定义的形式逻辑为基础。在询问问题之前，我们首先向程序提供事实和规则。图 1-14 显示了逻辑式程序设计语言范式的简化伪代码符号版本。诸如"Parent(Fay, Tara)"的事实被理解为"Fay 是 Tara 的 parent（父母）"。

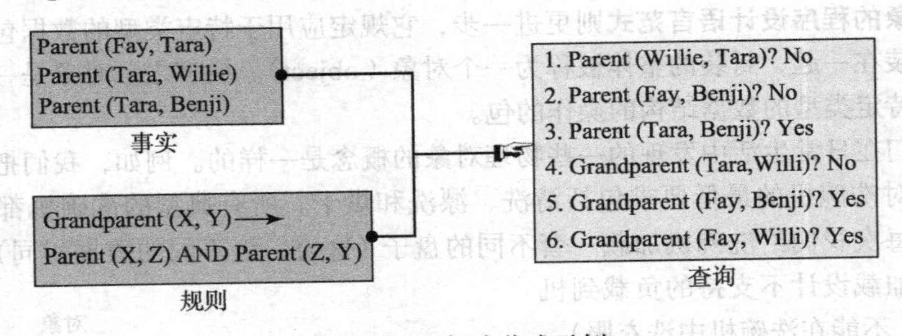

图 1-14　逻辑式范式示例

1.3.5　C++ 语言中包含的范式

基于上述四种范式的讨论，读者可能会思考 C++ 语言所属的范式。C++ 是 C 语言的扩展，而 C 语言基于面向过程的程序设计语言范式。然而，类和对象的存在允许将 C++ 语言用作面向对象的语言。在本书中，前几章的 C++ 语言主要用作面向过程的程序设计语言范式（使用对象实现输入 / 输出除外）。然而，在介绍性章节之后，C++ 语言将主要用作面向对象的程序设计语言范式。

1.4　程序设计

程序设计是包含如下两个步骤的过程：首先需要理解问题，然后开发解决方案。当我们被指派任务去开发一个程序时，会得到一个程序需求声明和有关程序接口的设计。换言之，我们被告知程序需要做什么。我们的工作是确定如何获取给定的输入，并将其转换为指定的输出。为了理解这个过程是如何工作的，我们来看一个简单的问题。

在一个数值列表中查找最大的数值。

我们该如何开始解决这个问题呢？

1.4.1 理解问题

程序设计的第一步是理解问题。我们首先仔细阅读需求说明书。当完全理解需求后，我们会与用户一起审查我们的理解。这通常包括提问以确认我们的理解是否正确。例如，在阅读了简单的需求说明书之后，我们应该提出如下几个需要澄清的问题。

需要处理什么类型的数值（包含小数部分，还是不包含小数部分）？数值是否按特殊顺序排列，例如从低到高排列？需要处理的数值的个数是多少？

如果不澄清问题，也就是说，如果只是对输入或者输出进行假设，我们可能会给出错误的答案。为了回答我们的问题，需要处理按任意顺序排列的整数，整数的数目没有限制。

如本例所示，即使是最简单的问题陈述也可能需要澄清。想象一下，对于一个包含成千上万行代码的程序，需要澄清的问题的数量将非常巨大。

1.4.2 开发解决方案

一旦完全理解了问题并且澄清了可能遇到的任何问题，我们就能够以算法的形式开发出一个解决方案。**算法**是解决问题所必需的一组逻辑步骤。算法有两个重要特点：第一，它们独立于计算机系统，这意味着它们可以被用来实现一个办公室中的手动系统，也可以被用来实现一个计算机上的程序；第二，算法接收数据作为输入，并将数据处理成输出。

要为问题编写算法，我们使用一种直观的方法，不仅使用问题的陈述，而且使用我们的知识和经验。我们首先从一个由五个数值组成的小集合开始：一旦为五个数值开发了一个解决方案，我们将把它扩展到任意数量的整数。

13	7	19	29	23

我们从一个简单的假设开始：算法一次只处理一个数值。我们将算法命名为 FindLargest。每个算法都有一个用于标识的名称。FindLargest 在并不知晓其他数值的具体值的前提下，依次查看每个数值。在处理每个数值时，它将该数值与目前已知的最大数值进行比较，并确定新数值是否更大。然后查看下一个数值以确认它是否更大，然后再查看下一个数值，直到所有的数值都被处理完毕。图 1-15 显示了确定五个整数中最大值的步骤。

该算法要求我们跟踪两个值：当前数值和已找到的最大值。我们使用以下步骤确定最大值。

- **步骤 1**：输入第 1 个数值 13。因为 *largest*（最大值）目前没有值，所以将 *largest*（最大值）设置为第 1 个数值的值。
- **步骤 2**：输入第 2 个数值 7。由于 7 小于最大值 13，因此不需要更改 *largest*（最大值）。
- **步骤 3**：输入第 3 个数值 19。当将 19 与最大值 13 进行比较时，发现 19 更大。因此，

将 *largest*（最大值）设置为19。

- **步骤4**：输入第4个数值29。当将29与最大值19进行比较时，发现29更大。将 *largest*（最大值）设置为29。
- **步骤5**：输入第5个数值23。因为它小于最大值29，所以不需要更改 *largest*（最大值）。因为没有更多的输入，所以任务完成，确定最大值是29。
- **步骤6**：输出 *largest*（最大值），即29。

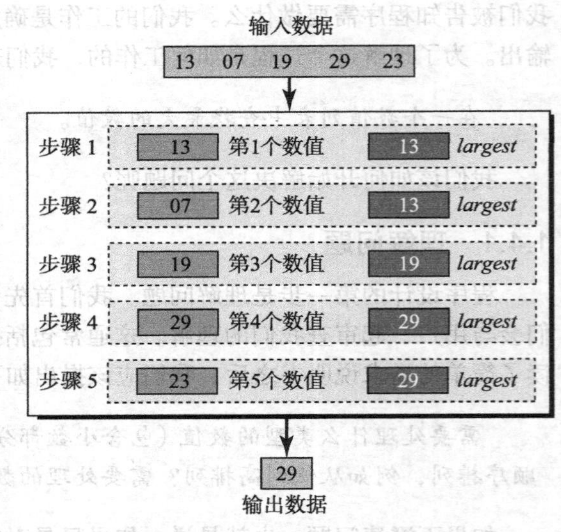

图1-15　在五个整数中查找最大值

算法泛化

图1-15所示的算法并不能完全解决我们原来的问题，因为它只能处理五个数值。为了使它适用于所有数值序列，我们需要替换步骤2到步骤5来处理一个未定数量的数值序列。这要求我们归纳这些语句，使它们相同。我们可以通过对如下所示的语句进行细微改写来实现这一点。

> 如果当前值大于 *largest*（最大值），则把 *largest*（最大值）设置为当前值。

然后，我们将重新措辞的语句包含在一个重复语句中，该语句执行这些步骤，直到处理完所有数值。结果算法如图1-16所示。

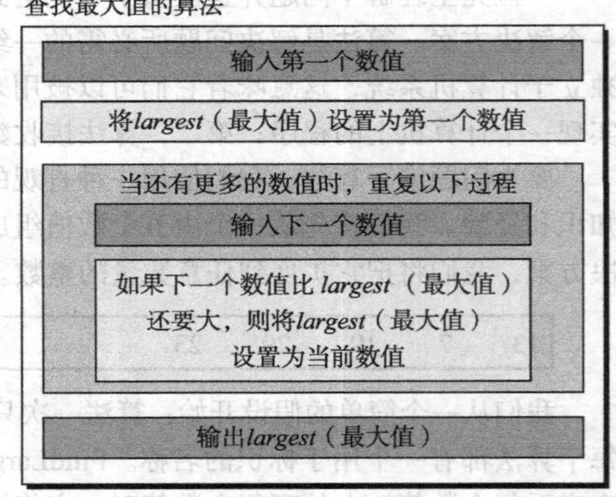

图1-16　在 *n* 个数值中查找最大值的算法

必须强调的是，设计是在编写程序之前完成的。从这一点而言，就像建筑的设计蓝图。如果没有一套详细的设计规划，没有人会开始建造一座房子。然而，无论是经验丰富的程序员还是新入门的程序员，最常见的错误之一就是在设计完成并全面文档化之前开始编写程序。

这种草率开始的部分原因是程序员认为他们已经完全理解了问题，还有一部分原因是程序员迫不及待想着手解决一个新问题。在第一种情况下，程序员会发现他们并没有完全理解问题。通过花时间设计程序，程序员会提出更多必须回答的问题，从而更好地理解问题。

程序员在完成设计之前进行编码的第二个原因恰恰是人的本性。程序设计是一项非常令人兴奋的任务。当看到你的设计开始成形，看到你的程序第一次运行时，都会带来一种自然而然的个人满足感。

统一建模语言（UML）

统一建模语言（UML）是用于设计、阐述和记录计算系统各个方面的标准工具。例如，它可以用于设计大型复杂系统、程序和程序中的对象。它也可以用于显示面向对象的语言（例如 C++）中对象之间的关系。我们将在后续章节中学习设计程序时讨论 UML。

1.5　程序开发

图 1-17 显示了将用任意语言编写的程序转换为机器语言的一般过程。C++ 程序的转换过程稍微有点复杂。这个过程是以一种直观明了的线性方式呈现的，但是我们需要认识到，在开发过程中，为了纠正错误并改进代码，这些步骤将重复多次。

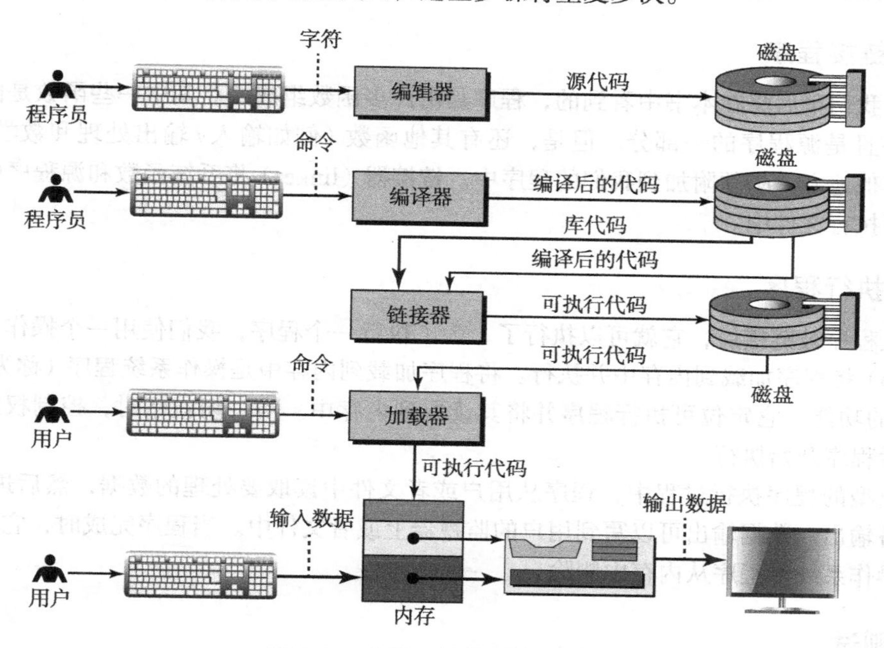

图 1-17　编写、编辑和执行程序

程序员的工作是编写程序，然后将其转换成**可执行文件**。这个过程包含如下四个步骤：

1）编写和编辑程序。

2）编译程序。

3）将程序与所需的库模块链接在一起（通常自动完成）。

4）执行程序。从我们的角度来看，执行程序是一个步骤。然而，从计算机的角度来看，它分为两个子步骤：加载程序和运行程序。

1.5.1　编写和编辑程序

用于编写程序的软件称为**文本编辑器**。文本编辑器帮助我们输入、修改和存储字符数据。使用系统提供的不同编辑器，我们可以撰写信函、创建报告或者编写程序。编写程序的文本处理器和其他形式的文本处理器的最大区别在于：程序面向代码行，而大多数文本处理则面向字符和段落。

文本编辑器可以是一个通用的文字处理器，但它通常是由提供编译器的公司提供的一个特殊的编辑器。用于编写程序的编辑器通常应该包含的功能包括：定位和替换语句的搜索命令、将语句从程序的一部分复制或者移动到另一部分的复制和粘贴命令、使用颜色显示程序关键代码的格式化命令以及将程序代码进行对齐和缩进的自动格式化功能。

完成一个程序的编写和编辑后，我们将文件保存到磁盘上，这个文件将成为编译器的输入，被称为**源文件**（source file）。

1.5.2　编译程序

存储在磁盘上的源文件中的信息必须翻译成机器语言，以便计算机能够理解。这是**编译器**（compiler）的工作。

1.5.3　链接程序

正如我们稍后将在本书中看到的，程序是由许多函数组成的。其中一些函数是由我们编写的，并且是源程序的一部分。但是，还有其他函数（例如输入/输出处理和数学库函数）存在于其他地方，必须附加到我们的程序中。**链接器**（linker）将系统函数和源程序中的函数组装到可执行文件中。

1.5.4　执行程序

一旦程序被链接后，它就可以执行了。为了执行一个程序，我们使用一个操作系统命令（例如 run）将程序加载到内存中并执行。将程序加载到内存中是操作系统程序（称为**加载器**，loader）的功能。它定位**可执行程序**并将其读取到内存中。当一切就绪时，控制权被交给程序，然后程序开始执行。

在典型的程序执行过程中，程序从用户或者文件中读取要处理的数据，然后进行处理，最后准备输出。数据输出可以写到用户的监视器上或者文件中。当程序完成时，它通知操作系统，操作系统将程序从内存中删除。

1.6　测试

在编写程序之后，我们必须对其进行测试。**程序测试**可能是程序开发中一个非常烦琐和耗时的部分。作为程序员，我们有责任对它进行全面的测试。我们必须确保每一条指令和每一种可能的情况都经过了测试。

1.6.1　设计测试数据

测试数据的开发应贯穿整个程序设计和开发过程。在设计程序时，我们创建测试用例来验证设计。在我们编写程序之后，这些测试用例就成为测试数据的一部分。

另外，当我们设计程序时，必须确定需要测试哪些情况（特别是特殊的情况），并记录下来。例如，在 FindLargest 算法中，如果只输入一个数值怎么办？类似地，如果数据是按顺序排列的或者完全相同的怎么办？当我们设计程序时，会着眼于测试用例来审查程序，并对所需的用例做额外的记录。最后，当我们编写程序时，会对测试用例做更多的记录。

当需要构造测试用例时，我们会审查所记录的笔记并将它们组织成逻辑集。除了非常简单的学生练习程序，一组测试数据永远无法完全验证程序。对于大型开发项目，可能需要运行 20、30 甚至更多的测试用例来验证程序。所有这些测试用例组成测试计划。

> 一组测试数据永远无法完全验证程序。

最后，当我们测试程序时，会发现更多的测试用例。同样，我们把它们记录下来，并将它们纳入测试计划中。当程序完成并投入生产运行时，我们仍然需要测试计划以验证程序的修改。针对修改进行的测试被称为**回归测试**，应该从编写程序时就开始开发测试计划。

那么，如何确定程序何时才算被完全测试了呢？事实上，没有办法确定，但我们可以做一些事情来帮助确定。虽然有些概念在我们学习后面的章节之前并不容易理解，但是为了完整起见，我们还是把这些概念罗列如下：

1）确保每行代码至少执行过一次。幸运的是，现在市场上有一些编程工具可以帮助我们做到这一点。

2）确保程序中的每个条件语句都执行了真假分支，即使其中一个分支为空。

3）对于每个有指定范围的条件，确保测试包括范围中的第一项和最后一项，以及第一项之前和最后一项之后的项。范围测试中最常见的错误往往发生在范围两端的界限处。

4）如果正在检测错误情况，确保测试所有的错误逻辑。这可能需要对程序进行临时修改以强制产生错误。例如，通常无法产生输入/输出错误，因此必须对其进行模拟。

1.6.2　程序错误

错误一般分三种：**规格错误**、**代码错误**和**逻辑错误**。

规格错误

当问题定义被错误地描述或者误解时，就会出现规格错误。通过与分析师和用户一起审查设计书，应该可以发现规格错误。

代码错误

代码错误通常会生成编译器错误信息。这些错误是最容易纠正的。有些代码错误会生成警告信息，这通常意味着编译器对代码产生了疑问，需要对其进行验证。警告信息表明代码可能是正确的，也可能是错误的。即使有警告消息程序也可以运行，但应该修改代码，以消除所有警告信息。

逻辑错误

最难发现和纠正的错误是逻辑错误。逻辑错误的示例包括除零错误，或者在FindLarget算法中忘记将第一个数值存储到largest（最大值）中。它们只能通过全面的测试来纠正。记住，在运行测试用例之前，我们应该知道正确的答案是什么。千万不要假定计算机的答案是正确的，如果存在逻辑错误，则答案将是错误的。

本章小结

计算机系统由两个主要部件组成：硬件（中央处理器、主存储器、辅助存储器、输入系统、输出系统和通信系统）和软件（系统软件和应用软件）。

计算机语言用于开发软件。计算机本身以机器语言方式运行。多年来，程序设计语言已经从符号语言发展到今天使用的诸多高级语言。

语言范式（面向过程的程序设计语言范式、面向对象的程序设计语言范式、函数式程序设计语言范式、逻辑式程序设计语言范式）描述了在计算机上解决问题的方法。C++基于面向过程和面向对象的程序设计语言范式。

程序设计是包含两个步骤的过程：首先需要理解问题，然后开发解决方案。

算法有两个重要的特点：它们独立于计算机系统；它们接收数据作为输入，并将数据处理成输出。

程序开发分为四个步骤：编写程序、编译程序、链接程序和执行程序。

测试程序需要验证每一条指令和每一种可能的情况。

思考题

1. 请描述以下面向过程的程序设计语言范式示例的内存状态（参见图 1-11）。

```
input a
input b
input c
sum = a + b + c
output sum
```

2. 请描述以下面向过程的程序设计语言范式示例的内存状态（参见图 1-11）。假设 length（长度）和 width（宽度）的值分别为 12 和 8，它们表示矩形的边长。

```
input length
input width
area = length × width
parameter = 2 × (length + width)
```

3. 假设我们需要使用面向对象的程序设计语言范式创建一个银行账户对象。请描述你认为需要用到的数据和需要与数据封装在一起的过程列表（参见图 1-12）。

4. 在函数式程序设计语言范式中，请描述以下函数的执行结果（参见图 1-13）。

```
first (rest (rest (a, b, c)))
```

5. 在函数式程序设计语言范式中，请描述以下函数的执行结果，假设 list(⋯) 创建给定元素的列表（参见图 1-13）。

```
list (first (rest (a, b)), first (a, b))
```

6. 根据图 1-14，以下查询的结果是什么？

```
Parent (Benji, Tara)?
GrandParent (Fay, Willi)?
```

7. 根据图 1-14，以下查询的结果是什么？

```
Parent (Fay, Tara)?
GrandParent (Tara, Willi)?
```

8. 请描述执行以下算法后 sum 的值。

```
sum = 0
sum = sum + 10
sum = sum × 10
sum = sum − 10
```

9. 请描述执行以下算法后 x 的值。

```
x = 5
x = x + 1
x = x − 10
```

10. 请描述执行以下算法后 x、y 和 z 的值。

```
x = 2
y = 5
x = x + 1
y = y - 10
z = 8
z = x + y
x = y + z
y = x + y + z
```

11. 请设计一个算法，使用以下公式将厘米值转换为英寸值：

1 英寸 = 2.54 厘米

12. 请设计一个算法，使用以下公式将英寸值转换为厘米值：

1 厘米 = 0.3937 英寸

13. 请设计一个算法，使用以下公式将华氏温度（F）值转换为摄氏温度（C）值：

C = (F − 32) × (100/180)

14. 请设计一个算法，计算购买如下商品的销售税和总销售价值：2 瓶软饮料（每瓶 1 美元）、3 瓶牛奶（每瓶 2 美元）和一罐咖啡（3 美元）。税率是 9%。
15. 请设计一个算法，查找一个数值列表中的最小值。
16. 请设计一个算法，计算一个数值列表中各数值的累加和。
17. 请设计一个算法，计算一个数值列表中各数值的累乘积。
18. 请设计一个算法，计算从 1 到 100 的累加和。

第 2 章

C++ Programming: An Object-Oriented Approach

C++ 程序设计基础

本章为使用 C++ 语言进行计算机程序设计奠定基础。每个 C++ 程序都包括输入、输出和赋值。首先，我们讨论如何使用两个对象和一个运算符来实现这些功能。接下来，介绍 C++ 中的基本数据类型。**基本数据类型**是基本的内置数据类型，可以不用声明直接使用。然后，我们讨论变量、值和常量，以及如何在程序中加以使用。最后，我们讨论程序的两个组件，即**标记符**（token）和**注释**（comment）。

学习目标

阅读并学习本章后，读者应该能够

- 讨论变量、值和常量，以及如何在 C++ 程序中加以使用。
- 讨论由标记符和注释组成的 C++ 程序的通用组件。
- 讨论 C++ 的基本数据类型及其所占内存大小。
- 描述如何在程序中使用整数类型。
- 描述如何在程序中使用字符类型。
- 描述如何在程序中使用布尔类型。
- 描述如何在程序中使用浮点类型。
- 讨论 void 数据类型及其用途。
- 介绍 C++ 字符串并简述其用途。
- 介绍和讨论一些简单的 C++ 程序。

2.1 C++ 程序

每个 C++ 程序由几节组成，每节又由几个部分组成。每一节或部分必须遵循 C++ 语言中定义的规则，就像用自然语言编写的文档必须遵循该语言的规则一样。

在正式深入研究 C++ 程序的结构和探索相关的规则之前，我们先研究一些示例程序。

2.1.1 第一个程序

程序 2-1 是一个简单的 C++ 示例程序，它将帮助我们对 C++ 程序有一个基本的了解。

程序 2-1 第一个简单程序

```
1  #include <iostream>
2
3  int main ()
4  {
5     std :: cout << "This is a simple program in C++ ";
6     std :: cout << "to show the main structure." << std :: endl;
7     std :: cout << "We learn more about the language ";
8     std :: cout << "in this chapter and the rest of the book.";
9     return 0;
10 }
```

```
运行结果：
This is a simple program in C++ to show the main structure.
We learn more about the language in this chapter and in the rest of the book.
```

程序代码与运行结果分离

我们将程序在垂直方向分成两个部分。顶部的两列显示我们编写的代码；底部显示程序运行的结果。请注意，第二部分中的第一行（运行结果：）并不是程序的一部分，也不是运行后的结果。它在此处用来说明这个部分是程序运行的结果。如果我们多次运行该程序，则会多次出现以（运行结果：）开始的内容。程序运行的结果通常显示在我们编译和运行程序的计算机屏幕上，但我们将它们显示在每个程序的底部以供快速参考。

> 我们将程序代码与运行结果分开进行演示；程序的运行结果通常显示在屏幕上。

区分大小写

每个 C++ 程序都是区分大小写的。我们必须严格使用语言中定义的术语。如果我们改变术语中一个字母的大小写，而这个术语是语言的一部分，我们就会得到一个编译错误。换言之，在程序的第一行中，我们不能使用 Include 或者 IOSTREAM 来代替 include 或者 iostream。在第三行中，我们不能用 Main 代替 main。类似的 std、cout、endl 和 return 等术语同样如此。

> C++ 语言是区分大小写的。必须按照定义使用命名实体的术语，不能改变字母的大小写。

程序代码行

虽然没有必要这样做，但是我们还是将程序分成几行来显示不同的部分，以提高其可读性。当我们讨论程序时，我们使用行号以便于参考。行号并不是程序的组成部分，在我们创建源代码时不应包括在程序中。请注意，我们还添加了空行来分隔程序的不同部分（例如，程序 2-1 中的第 2 行）。

> 行号不是程序的组成部分，不应包含在源代码中。在本书中，它们被用来引用代码行。

缩进

在程序 2-1 中，函数体中代码行进行了缩进。虽然没有必要这样做，但我们相信缩进可以提高程序的可读性。我们总是缩进属于封闭实体的行。第 5 行到第 9 行属于函数体，我们对它们进行了缩进，以表明它们位于两个花括号内。

> 行缩进可以提高程序的可读性，强烈推荐使用行缩进。

程序分析

让我们逐行简要分析程序的功能（除了第 2 行，这是空行，其目的是增加程序的可读性，表示分隔程序的两个部分）。

第 1 行：预处理指令（preprocessor directive）。预处理指令是编译器在编译程序之前采取某些操作的命令。程序的第 1 行（不包括行号）是一条预处理指令，如下所示：

```
#include <iostream>
```

C++ 程序需要一些不由我们编写的预定义代码行。这些代码行非常复杂，有时需要访问计算机的硬件，这意味着预处理指令的执行取决于我们使用的计算机类型。C++ 设计者创建了这些代码行，并将它们包含在被称为**头文件**（header file）的文件中。我们不必编写这些代码行，我们可以简单地将这些文件的内容复制到代码中。而要复制这些文件的内容，我们需要知道文件的名称，例如本示例中的 iostream，它代表输入 / 输出流。为了告诉编译器我们需要包含这个文件的内容，我们需要使用一个 include 行，如上所示。include 行以 # 符号开头。当程序中的代码行以 # 符号开头时，则指示编译器在编译程序之前需要进行预处理。在编译器开始编译代码之前，它运行另一个程序，叫作预处理器（preprocessor），它检查所有的预处理命令（include 指令就是其中之一）。编译器执行命令中需要的操作（例如包含头文件中的内容），然后删除预处理指令。在所有预处理指令都被处理好并被删除之后，程序就可以编译了。注意，每个预处理命令都需要独占一行，第一个字符不能是空白符，而且必须为 # 符号（磅符号）。在 include 指令中，文件名必须用两个尖括号括起来，例如 include 术语后的 <filename>。我们将在附录 F 中详细讨论预处理。

注意，所有的 include 指令后面都不能加分号，否则编译器将产生错误信息。

第 3 行：函数头。程序的第 3 行（不包括行号）如下所示：

```
int main ()
```

C++ 程序通常由多个函数组成。函数是一个将多行代码组织在一起的部分，我们将在后续章节阐述。C++ 程序由一个或者多个函数组成，但它必须包含一个名为 main 的函数。C++ 程序的执行从 main 函数开始，当 main 函数终止时程序终止。

每个 C++ 程序的执行都是从 main 函数开始的，这意味着每个程序必须有一个名为 main 的函数。

每个函数都有一个函数头（header）和一个函数体（body）。函数头定义函数的名称、传递给函数的参数内容（包含在括号内）以及函数返回的信息的类型（本示例中在函数名前面的 int）。第 3 行是 main 函数的函数头。它显示函数的名称是 main，没有任何传递给函数的参数内容（括号中是空的），函数返回一个整数值给操作系统（C++ 语言中，int 表示整数类型）。

第 4 行和第 10 行：开始花括号和结束花括号。第 4 行和第 10 行是一个左花括号和一个右花括号的组合，它们需要组合在一起以包含整个函数体。如果花括号不成对，则会导致编译错误。开始花括号和结束花括号组成的函数体显示如下：

```
{
   ...
}
```

C++ 中的每一个函数定义都必须是包含在左花括号和右花括号之间的函数体。

如果有左花括号但没有右花括号，或者有右花括号但没有左花括号，则是一个严重错误。

第 5 行：函数体的第 1 行代码。接下来从下面的第 5 行开始逐行讨论函数体的内容：

```
std :: cout << "This is a simple program in C++ ";
```

函数体中的大多数行都是命令，告诉计算机要执行什么操作。在自然语言中，命令（即指令句）由动词、直接宾语、间接宾语（目标或者接收者）和结束语（通常是句号）组成。这与我们在函数体中所看到的命令行一致，尽管顺序不同。在示例代码中，首先是目标对象（cout，它是监视器），其次是动词（写），最后是要写的内容（直接宾语，字符串）。命令以分号（而不是句点）结束。通过上述解释，程序 2-1 的第 5 行的含义是计算机应该在监视器上写对象字符串（This is a simple program in C++），然后移动到下一个命令。请注意，这两个引号是语言语法的一部分，它们不会输出到屏幕上。图 2-1 显示了第 5 行的图示化分析。

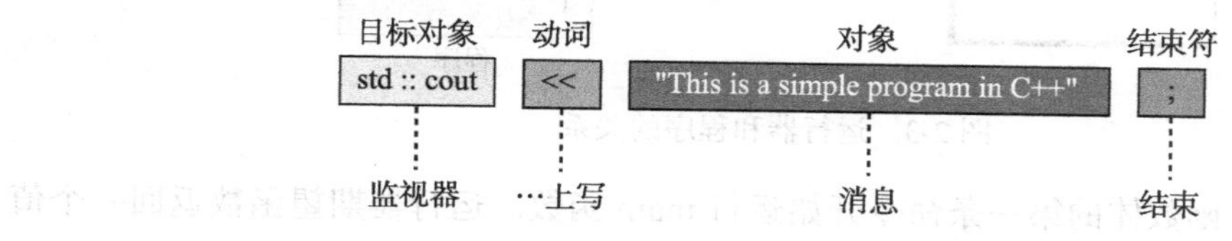

图 2-1　第 5 行代码的分析

我们将看到 C++（作为一种面向对象程序设计语言）中的每个对象都有一个名称。在 C++ 中，连接到计算机的监控器也有一个名称，被称为 std::cout。注意，此处的名称由两部分组成，std 和 cout，用两个冒号分隔。就像现实生活中的人有两个标识符（姓氏和名字），C++ 中的对象也有姓氏和名字。监视器的姓氏是 std（standard，标准的缩写），它的名字是 cout（console out，控制台输出的缩写）。姓氏定义对象所属的群组（家族），名字定义组中对象的实际名称。总之，第 5 行告诉计算机在屏幕上输出给定的消息。

第 6 行：函数体的第 2 行。示例程序的第 6 行如下所示：

```
std :: cout << "to show the main structure. " << std :: endl;
```

这一行与第 5 行非常相似，但略有不同。首先，消息的内容不同："to show the main structure."。其次，在消息被写入之后，动词（<<）被重复了一次。第三，没有定义新的消息，但是程序输出一个名为 std::endl 的预定义对象。这是另外一个对象，当发送到监视器时，该对象会在上一条消息的末尾添加新行，结果导致输出移动到下一行。此对象 endl 是结束行（end line）的缩写。下一个输出将打印在下一行。图 2-2 显示了目标对象是相同的（只提到一次），但是动词和宾语重复。在第一部分中，对象是一个字符串对象；在第二部分中，对象是 endl（注意最后一个字符是字母 L 的小写方式，而不是数字 1）。

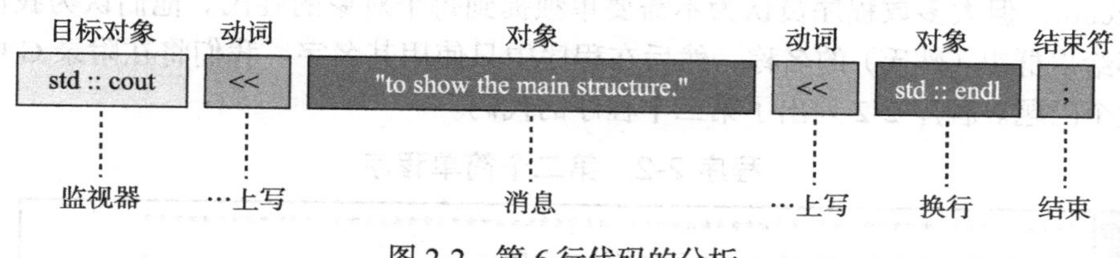

图 2-2　第 6 行代码的分析

第 7 行和第 8 行。第 7 行和第 8 行与第 5 行和第 6 行相似，但没有 endl 对象，因为第 8 行是文本的最后一行，故不需要 endl 对象。

第 9 行。函数体中的最后一行与其他行不同：

```
return 0;
```

这一行仍然是条命令，但没有明确的目标对象。此处目标对象是隐式的。它是 C++ 系统中的一个实体，被称为运行器（runner）。具体情况如图 2-3 所示。

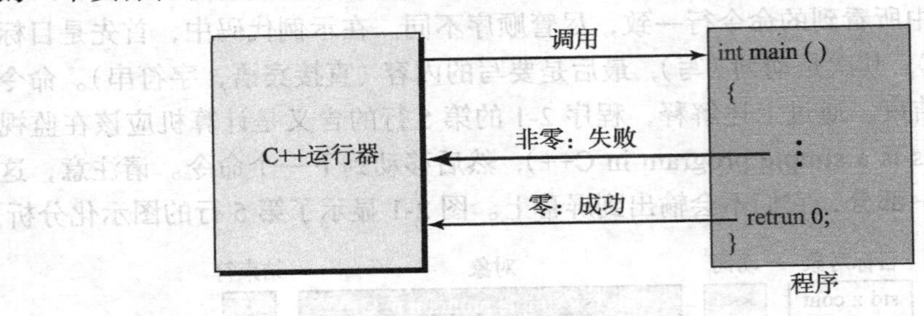

图 2-3　运行器和程序的关系

运行器从 main 函数体的第一条命令开始运行 main 函数。运行器期望函数返回一个值，该值显示程序是否成功。返回值定义成功或者失败。如果程序到达函数的最后一行并返回 0，则表示成功。图 2-3 显示了由运行器启动和停止一个程序的简单示例。

程序输出

执行目标对象为 cout 的命令的结果显示在连接到运行程序的计算机的监视器上。在示例程序中，我们将在监视器上显示两行文本。图 2-4 显示了监视器（控制台）窗口，其中显示了示例程序的输出结果（详细信息取决于平台）。

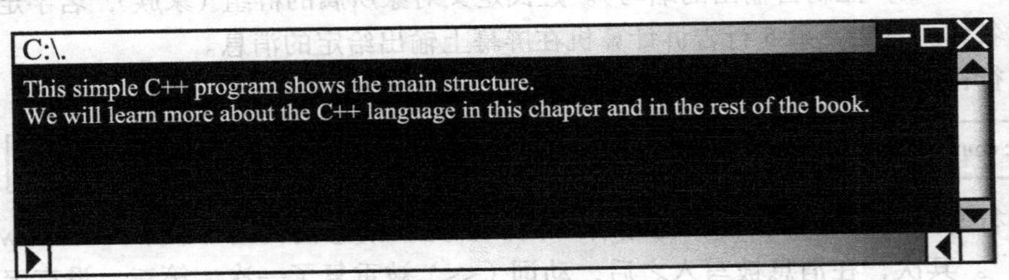

图 2-4　显示运行输出结果的控制台（cout 对象）

2.1.2　第二个程序

在本节中，我们编写了另一个与第一个程序类似的 C++ 程序，但是包含一些修改和添加。尽管许多纯粹主义者认为程序中的每个对象都应该用姓氏和名字来编码，例如 std::cout 或者 std::endl，但大多数程序员认为不需要单独提到每个对象的姓氏。他们认为我们应该在程序顶部给出群组（姓氏）的名称，然后在程序中只使用其名字。我们将在附录 G 中更详细地讨论这个问题。程序 2-2 列出了第二个程序的代码。

程序 2-2　第二个简单程序

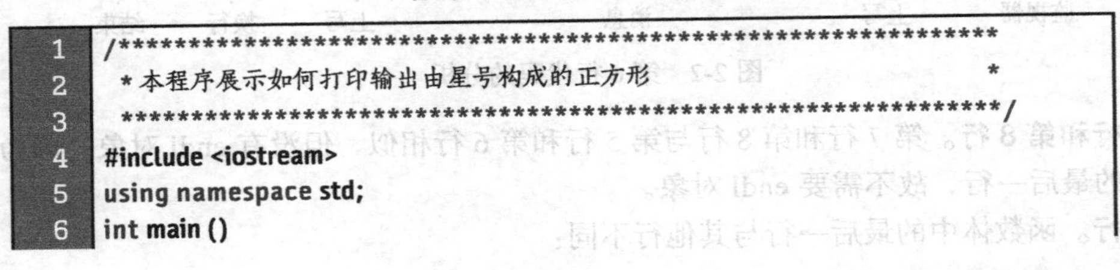

```
1  /*************************************************************
2   * 本程序展示如何打印输出由星号构成的正方形                  *
3   *************************************************************/
4  #include <iostream>
5  using namespace std;
6  int main ()
```

```
7  
8  {
9      // 打印输出由星号构成的正方形
10     cout << "******" << endl;
11     cout << "******" << endl;
12     cout << "******" << endl;
13     cout << "******" << endl;
14     cout << "******" << endl;
15     cout << "******";
16     return 0;
17 }
```

```
运行结果:
******
******
******
******
******
******
```

程序分析

下面简要分析程序 2-2 中不同于第一个程序的代码行。

第 1 行到第 3 行：块注释。第二个程序的第 1 行到第 3 行不同于第一个程序，它们被称为**块注释**（block comment）。块注释是展示给用户或者代码审查者的一行或者多行注释，被编译器完全忽视（块注释在程序被编译前被删除）。块注释的语法格式如下：

```
/* Text to be ignored
Text to be ignored
Text to be ignored */
```

块注释包括开始部分和结束部分。开始部分由两个字符（/*）组成，结束部分由两个同样的字符组成但顺序相反（*/）。二者之间包括的所有内容都是注释，将被忽略。注意，开始部分和结束部分之间可以包含任意字符（包括 * 和 /），但不能包括 * 和 / 的组合，否则会创建一个新的块注释的开始或者结束部分。注意，在第 1 行到第 3 行代码中，我们使用了一些星号（*）作为文本来创建注释外框，这仅仅是个人偏好。

第 5 行。该行也是程序中的新内容，如下所示：

```
using namespace std;
```

第 5 行告诉编译器，当对象没有姓氏时，在其前面插入 std::，以使其名称完整。换言之，它告诉编译器，我们在程序中使用的一些名称属于名称空间 std。通过包含这一行代码，我们可以直接使用 cout 和 endl 对象，而不必使用 std 名称空间限定它们。

第 9 行。该行也是注释，但不是块注释，而是**行注释**（line comment）。这种类型的注释以两个斜杠（//）开头，并以文本作为注释（一直到行尾）。整行被编译器忽略。

```
// 打印输出由星号构成的正方形
```

此类注释只能在一行。如果需要两行注释，则必须开始另一行注释或者使用块注释。

第 10 行到第 15 行。这些行与第一个程序中类似的行执行相同的操作，但是我们可以使用不带（std::）的cout和endl。每行在监视器上打印六个星号（*）组成的一行，以创建一个星号正方形。

2.2 变量、值和常量

在进一步编写程序之前，我们需要理解三个概念：变量、值和常量。

2.2.1 变量

用C++语言编写的程序接收输入数据，处理数据，并创建输出数据。由于输入数据和输出数据可能发生变化，我们必须能够在内存中存储输入数据、中间数据和输出数据。由于这个原因，与大多数程序设计语言一样，C++也使用变量的概念。计算机语言中的**变量**(variable)是一个具有名称和类型的内存块。因为其内容在程序执行过程中可能会发生变化，故称之为变量。变量的目的是存储和检索数据。变量必须有一个类型，因为不同的数据类型用于不同的目的。

变量是一个具有名称和类型的内存位置，用于在程序执行的每个时刻存储不同的值。

在使用变量之前，必须先定义变量。我们必须告诉编译器我们要使用具有给定名称和给定类型的内存位置。名称用于引用变量，类型用于告知存储在变量中的数据类型。

【例2-1】 图2-5显示了三个变量及其类型和名称。该图还显示了我们如何定义它们以确保编译器知道我们所引用的变量。

我们还没有讨论数据类型，但是我们指定所有三种数据类型都是整数类型（int），它们表示一个整数（没有小数部分），例如17、35、100等等，但不是23.67。

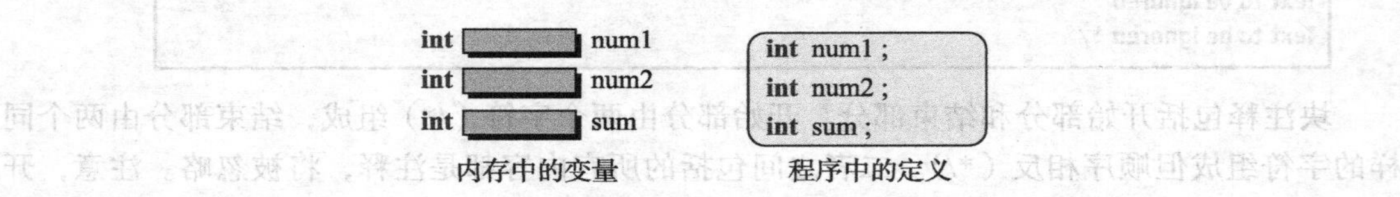

图2-5 内存中的变量以及变量在程序中的定义

2.2.2 值

变量的内容称为其**值**（value）。例如，如果此时num1的内容是整数17，则其值为17。当我们在程序中使用变量的名称时，有时我们指的是变量本身，有时指的是它所保存的值。下一个示例显示了其中的差异。

【例2-2】 假设我们想要编写一个程序：接收任意两个整数，将它们相加，并显示结果。虽然我们还没有学习数据类型，但是我们知道这两个数值以及求和结果都是整数类型（C++中的int）。为了存储这两个数值以及求和的结果，我们需要给它们命名。我们分别称它们为num1、num2和sum。然后，我们编写一个程序，每次运行时，都会计算在键盘上输入的两个数值的和并将求和结果打印出来（程序2-3）。

程序2-3 计算两个值的和的程序

```
1  /*************************************************************
2   *该程序从键盘上读取两个数值，                              *
```

```
3    *计算这两个数值的和，并在监视器上输出结果                    *
4    ***************************************************************/
5   #include <iostream>
6   using namespace std;
7
8   int main ()
9   {
10      // 定义
11      int num1;
12      int num2;
13      int sum;
14      // 获取输入
15      cout << "Enter the first number: ";
16      cin >> num1;
17      cout << "Enter the second number: ";
18      cin >> num2;
19      // 计算并存储结果
20      sum = num1 + num2;
21      // 显示结果
22      cout << "The sum is: " << sum;
23      return 0;
24  }
```

```
运行结果：
Enter the first number: 23
Enter the second number: 35
The sum is: 58
```

```
运行结果：
Enter the first number: 7
Enter the second number: 110
The sum is: 117
```

让我们讨论第 11 行到第 13 行代码的内容，如下所示：

```
int num1;
int num2;
int sum;
```

这三行声明变量。变量的声明意味着使用相应的名称保留内存。在上述示例中，三个声明的示意图如图 2-5 所示。请注意，此时程序中没有存储任何值。

接下来讨论第 15 行到第 18 行代码的内容，如下所示：

```
cout << "Enter the first number: ";
cin >> num1;
cout << "Enter the second number: ";
cin >> num2;
```

第 1 行代码我们是比较熟悉的。第 1 行代码在屏幕上打印一条消息，该消息被称为提示信息。其目的是告诉用户下一步要做什么。第 2 行代码是新的知识点，使用一个目标对象 cin（用于控制台输入，console in 的简写）和一个命令（>>，其含义是读取）。此命令等待程序用户在键盘上输入一个值（本例中为整数），并按 Enter 键。然后程序将输入的值存储在变量 num1 中。第 3 行与第 1 行相同。第 4 行与第 2 行相同，但它将用户输入的值存储在变量

num2 中。在这四行之后，两个值存储在两个变量中。假设我们输入了23和35，那么内存位置的示意图如图2-6所示。

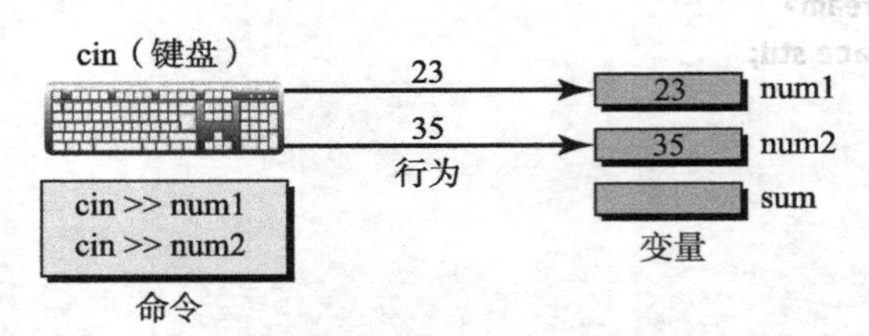

图2-6　两条输入命令之前和之后变量的内容

接下来分析第20行代码，如下所示：

```
sum = num1 + num2;
```

在这一行代码中，我们使用两个运算符：赋值运算符（=）和加法运算符（+）。我们将在下一章中详细地讨论运算符。现在，让我们先描述它们在这一行代码中的作用。赋值运算符（=）表示存储，其右侧为值，左侧为变量。换言之，其语法形式如下：

```
variable = value;
```

在本示例中，左侧的变量是sum，右侧的值是存储在num1中的值的副本与存储在num2中的值的副本之和，示意图如图2-7所示。

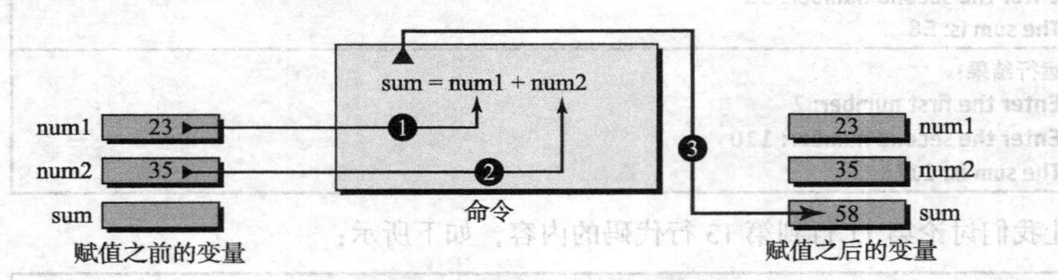

图2-7　赋值之前和之后变量的内容

第22行与我们之前接触的代码类似。如下所示：

```
cout << "The sum is: " << sum;
```

在本示例中，屏幕上会显示两个实体的内容。第一个是消息，第二个是变量sum的值，如图2-8所示。

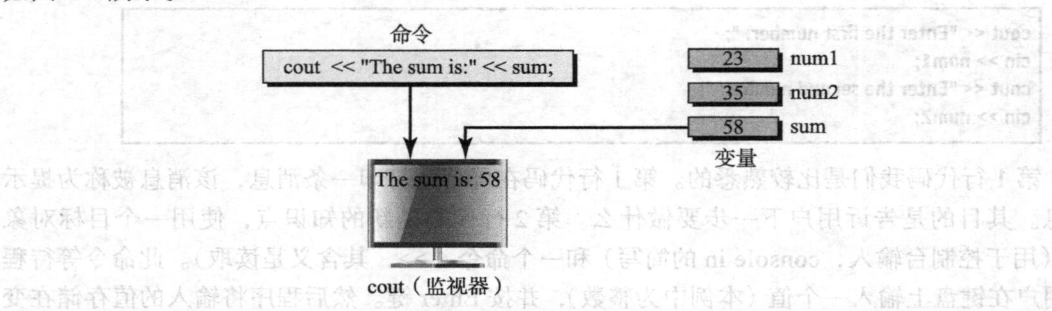

图2-8　第22行语句执行后变量和监视器的状况

cin 和 cout 对象

变量和值的概念与上一个程序中的 cin 和 cout 对象的概念有紧密联系。cin 对象是数据源；cout 对象是数据的目标。cin 对象使用输入设备（例如键盘）作为数据源。cout 对象使用输出设备（例如监视器）作为数据的目标。cin 对象使用运算符（>>）获取数据；cout 对象使用运算符（<<）传递数据。如果把这些对象与它们的内存联系在一起，就可以更好地理解这些对象的作用，示意图如图 2-9 所示。

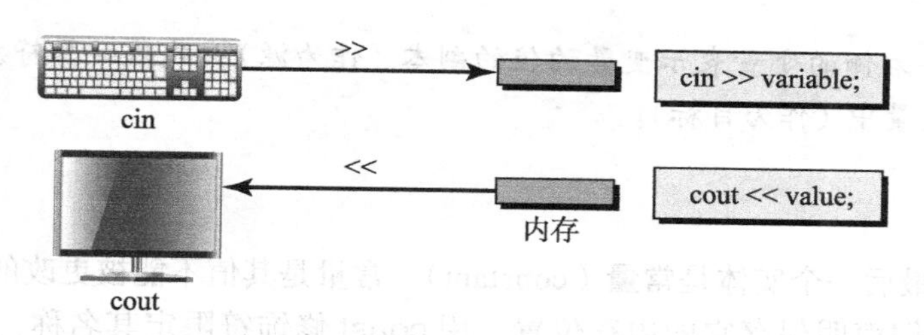

图 2-9　键盘和监视器分别作为数据的输入源和输出目的地

为了记住我们需要使用哪个运算符，我们可以将（>>）和（<<）看作数据移动方向上的双箭头。在输入的情况下，数据向右移动；在输出的情况下，数据向左移动。

关于 cin 对象，我们需要记住的最重要的一点是我们需要一个变量名（作为数据的目的地）。另外，我们需要一个值作为 cout 对象的数据源。我们已经看到我们使用了诸如“cout << sum”的命令，其中 sum 是一个变量的名称。需要知道的是，在这种情况下，我们指的是 sum 的值，而不是 sum 的名称。这是我们需要牢记的一个重要问题。变量的名称有时表示作为目的地的物理变量，有时表示存储在其中的值的副本。

与 cout 对象一起使用的值不一定是变量的值，它可以是独立的值，例如“cout << 12”，其含义是将值 12 发送到监视器。

> cin 对象需要一个变量名；cout 对象需要一个值。

下面的例子阐明了这些重要概念。

```
cin >> x;                // 从键盘获取一个值并存储到变量 x 中
cout << x;               // 获取变量 x 的值并显示，x 的值保持不变
cout << 4;               // 在屏幕上显示值 4
cin >> 4                 // 错误，cin 对象需要一个变量，而不是一个值
```

赋值运算符

与值和变量概念相关的另一个实体是赋值运算符。赋值运算符（=）左侧需要一个变量，右侧需要一个值，如图 2-10 所示。

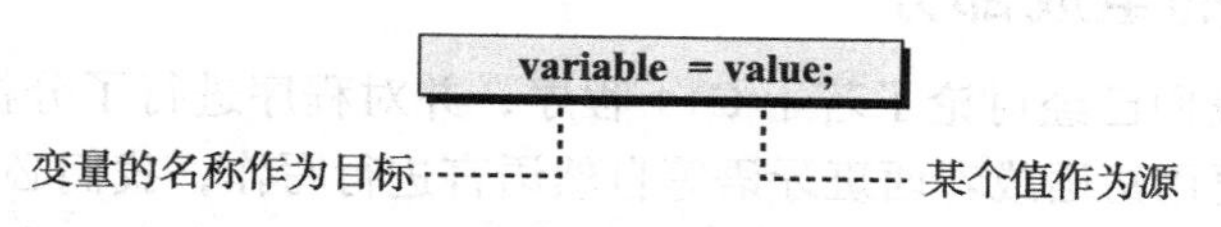

图 2-10　赋值语句的源和目标

当我们讨论一个与赋值运算符相关的值时，我们指的是一个给定名称的变量的值，既可以是单独的值，也可以是一组值的组合形成的单一的值。示例如下：

```
x = y;          // 把变量 y 的值存储到变量 x 中
x = 5;          // 把 5 存储到变量 x 中
x = y + 6;      // 获取 y 的值，加上 6，并把结果存储到变量 x 中
x = x + 3;      // 获取 x 的值，加上 3，并把结果存储到变量 x 中
x = x + y;      // 获取 x 的值，加上 y 的值，并把结果存储到 x 中
```

注意，赋值运算符左侧和右侧的 x 和 y 表示不同的内容。

赋值运算符右侧的变量表示变量的值的副本（作为源）；赋值运算符左侧的变量表示将值存储在该变量中（作为目标）。

2.2.3 常量

本节讨论的最后一个实体是**常量**（constant）。常量是其值不能被更改的存储实体。它的值是固定的。我们声明保存它的内存位置，用 const 修饰符限定其名称，并使用赋值运算符来存储我们希望要保存的值，不管我们要引用它多少次。下面代码显示了如何定义表示数学中常量 π 的语法格式。请注意，术语 double 表示带小数部分的数值，例如 3.14159 或者 7.2。

```
const double PI = 3.14159;        // 常量的定义
```

上面的定义表明，我们希望把 π 的值存储在名为 PI 的内存位置，但是我们希望它的值固定不变，并且在程序运行期间不会被改变。换言之，我们可以访问该值，但不能更改该值。这意味着我们不能在赋值运算符的左侧使用名称 PI（定义时除外），也不能将它与 cin 对象一起使用，如下所示：

```
PI = x;         // 错误：不能改变 PI 的值
cin >> PI;      // 错误：PI 不能接收值
cout << PI;     // PI 的值可以显示在监视器上
x = PI;         // PI 的值可以存储到一个变量中
```

在程序中使用常量值的另一种方法是将其用作**字面量**（literal）。可以在程序中使用一个字面量，而无须将其存储在内存位置。然而，计算机专家不建议在程序中使用字面量，除非每个查阅代码的人都完全明了其用途。下面代码显示了如果给定圆的半径，如何使用存储常量和字面量常量计算圆的周长（本例中的星号 * 表示乘法）。

```
perimeter = 2 * PI * radius;        // 2 是字面量，PI 是存储常量
```

2.3 C++ 程序的组成部分

到目前为止，我们已经讨论了若干 C++ 程序，并对程序进行了分析，介绍了 C++ 程序的组成部分。如果要用法语或者西班牙语等自然语言进行写作，我们必须了解该语言的组成部分。我们必须知道我们所使用的一组词的含义，以及如何将它们组合成句子。我们还必须熟悉语言中使用的标点符号。学习用一种新的计算机语言（如 C++）编写一个程序要简单得多，因为计算机语言中需要掌握的单词和符号数量有限。

C++ 程序（通常称为源代码）的内容通常是由两种相互交织的东西组成的。第一种是编

译器用来创建可运行程序（采用机器语言的方式）的代码。第二种是由程序员添加的注释，用于解释整个程序或者一部分程序的作用。第一种使用 C++ 语言的标记符；第二种使用自然语言（例如英语）中的单词和少量预定义的符号。

2.3.1 标记符

一个不包含注释的 C++ 程序是一个标记符序列。标记符（token）包含标识符、字面量或者符号。接下来将展开阐述。

标识符

标识符（identifier）是 C++ 语言中的实体的名称。有效的标识符名称必须以字母或者下划线开头，包含零个或者多个数字、字母或者下划线。标识符中的字符数没有限制。

> 标识符必须以字母或者下划线开头，包含零个或者多个数字、字母或者下划线。

如前所述，C++ 语言是区分大小写的。许多程序员使用由小写字母组成的标识符。但是使用两个或者三个单词的标识符越来越常见：第一个单词都是小写的，后续单词首字母大写（例如 lowerValue、upperValue、firstTest、secondTest、timeOfWithdraw 等）。C++ 的语言库还使用由下划线分隔的若干单词组成的标识符，但是我们在程序中应避免使用。

【例 2-3】 程序 2-1 中包含的标识符列表如下所示：

```
include, iostream, int, main, std, cout, endl, return
```

这些标识符可以归属于关键字、预定义标识符和用户自定义标识符的类别。接下来我们讨论每个类别。

关键字。**关键字**（有时又称为保留字）是由 C++ 语言保留的标识符，程序员或者语言库不能重新定义。表 2-1 显示了 C++ 中 84 个关键字的列表。这些关键字是根据我们前面讨论过的标识符规则生成的。

表 2-1 关键字

alignas	alignof	and	and_eq	asm
auto	bitand	bitor	bool	break
case	catch	char	char16_t	char32_t
class	compl	const	const_cast	constexpr
continue	decltype	default	delete	do
double	dynamic_cast	else	enum	explicit
export	extern	false	float	for
friend	goto	if	inline	int
long	mutable	namespace	new	noexcept
not	not_eq	nullptr	operator	or
or_eq	private	protected	public	register
reinterpret_cast	return	short	signed	sizeof
static	static_assert	static_cast	struct	switch
template	this	thread_local	throw	true
try	typedef	typeid	typename	union
unsigned	using	virtual	void	volatile
wchar_t	while	xor	xor_eq	

【例 2-4】 请找出程序 2-1 中有哪些标识符是关键字。我们发现其中只有两个是关键字。

```
int, return
```

预定义标识符。我们发现在 C++ 语言中包含一些标识符，它们不是关键字，但它们在语言中属于预定义的。尽管我们可以重新定义它们并在程序中使用它们，但最好不要这样做以避免混淆。

【例 2-5】 请找出程序 2-1 中有哪些标识符是预定义标识符。其中包含六个预定义标识符，如下所示：

```
include, iostream, main, std, cout, endl
```

用户自定义标识符。语言的用户（程序员）可以定义一组在程序中使用的标识符，前提是要遵循我们前面所讨论的标识符规则。以下是一些有效的程序员标识符的示例：

```
z        z12        _3        sum        average        result2        counter
```

但是，我们不建议使用前三个标识符。前两个不是描述性的（名称没有说明任何有关实体的内容）；第三个以下划线开头，通常只用于预定义的标识符。

以下都是无效（非法）的标识符，将会导致编译错误。前三个标识符以数字开头。第三、第四和第五个标识符除了数字或者字母之外还包含其他符号。最后两个是关键字，不能由程序员重新定义。

```
3z        13        3?2        sum@        count-2        delete        double
```

【例 2-6】 程序 2-1 和程序 2-2 中没有定义任何标识符。但是，在程序 2-3 中定义了三个标识符，如下所示：

```
num1, num2, sum
```

字面量

我们在程序中遇到的另一组标记符是字面量。字面量是不同类型的常量值。在本章讨论 C++ 语言中的数据类型之后，我们将讨论一些字面量。

【例 2-7】 在程序 2-1 中，我们使用了五个字面量：一个整数字面量和四个字符串字面量。

```
0                                              //数值字面量
"This is a simple program in C++ "             //字符串字面量
"to show the main structure."                  //字符串字面量
"We learn more about the language ";           //字符串字面量
"in this chapter and the rest of the book.";   //字符串字面量
```

符号

C++ 使用非字母符号作为运算符和标点符号。我们将在本章后面讨论若干运算符。这里我们讨论一些示例程序中的标点符号。表 2-2 显示了 C++ 语言中使用的符号。

表 2-2 C++ 中的符号

{	}	[	]	#	##	(	)	<:	:>
<%	%>	%:	%:%:	;	:	...	,	.	+

（续）

<table>
<tr><td>-</td><td>*</td><td>/</td><td>%</td><td>^</td><td>&</td><td>|</td><td>?</td><td>::</td><td>.*</td></tr>
<tr><td>-></td><td>->*</td><td>~</td><td>!</td><td>=</td><td>==</td><td><</td><td>></td><td><=</td><td>>=</td></tr>
<tr><td>|=</td><td><<</td><td>>></td><td><<=</td><td>>>=</td><td>-=</td><td>+=</td><td>*=</td><td>/=</td><td>%=</td></tr>
<tr><td>&=</td><td>!=</td><td>^=</td><td>++</td><td>- -</td><td></td><td></td><td></td><td></td><td></td></tr>
</table>

【例 2-8】 找出程序 2-1 中使用的符号。结果如下所示：

```
#  (  ) {  ::  <<  ;  }
```

2.3.2　注释

在介绍第二个程序时，我们简要讨论了注释。注释是添加到程序中的解释，以帮助阅读源代码的人理解程序。预处理器将删除注释文本。预处理器是为编译准备源代码的程序。

单行注释

如果我们需要一个简短的注释，我们可以使用单行注释。单行注释以 // 开头，并在行尾终止。它可以从一行中的任何位置开始，并包含该行的剩余部分。编译器将忽略 // 后面的该行剩余内容。

```
cout << "Hello World";          // 输出第一行内容
```

多行注释

多行注释可以跨若干行。它有两个标记，一个用于定义注释的开始，另一个用于定义注释的结束。注释的开始标记为 /*，而注释的结束标记为 */。当预处理器遇到开始标记（/*）时，文本的其余部分将被忽略，直到预处理器遇到结束标记（*/）。下面显示多行注释：

```
/* This is a program to show how we can print a message of two lines
using the output object cout defined in the iostream file. */
```

嵌套注释

C++ 不支持嵌套注释，尽管初学者经常尝试嵌套注释。稍微理解一下编译器处理注释的原理，将很容易澄清这个问题。

1）在行注释中不能嵌套行注释，因为当编译器发现行注释标记（//）时，它会忽略该行后面的其他所有内容。注释行会被直接输出。

2）在多行注释中嵌套一个行注释或者另一个多行注释也不起作用，因为编译器同样会忽略从注释开始标记到注释结束标记之间的所有内容（/* … */）。

3）导致错误的一种情况是在单行注释中嵌套多行注释。让我们仔细观察这种情况。

```
// Single line comment. /* Start of multiple-line comment.
Line two of comments
*/ End of multiple-line comments
```

上述代码包含如下几处错误：

①编译器看不到多行注释的开始标记，因为它忽略了行注释标记（//）后的所有内容。

②第二行和后续行的注释内容很可能是英文或者数学描述。它们将生成一条或者多条错

误消息。

③当发现注释结束标记（*/）时，编译器会生成一条错误消息，因为没有注释开始标记。还记得吗？注释开始标记在第一行就被忽略了。

2.4 数据类型

在面向过程或者面向对象的程序设计语言范式中，用C++语言编写的程序都需要处理数据。换言之，C++程序就像一台机器，它接收一些输入数据，处理数据，并产生输出数据。为了有效地处理数据，C++区分不同的**数据类型**。在C++语言中定义的数据类型的种类非常多，但是我们可以将这些类型分为两大类：内置数据类型和用户自定义数据类型，每个类别又分为两组，如图2-11所示。

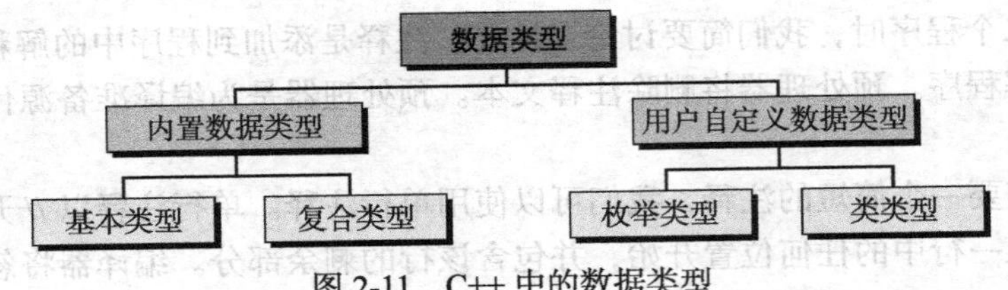

图2-11 C++中的数据类型

内置数据类型（built-in data type）是由语言定义的类型。内置数据类型分为两类：基本（fundamental，或者原始primitive）类型和复合（compound）类型。基本类型是非常基本的数据类型，我们可以直接使用；复合类型是从基本类型派生出来的（将在后续章节中讨论）。用户自定义的数据类型也分为两类：**枚举类型**和**类类型**。

根据其定义的性质，基本数据类型本身分为五类，如图2-12所示。

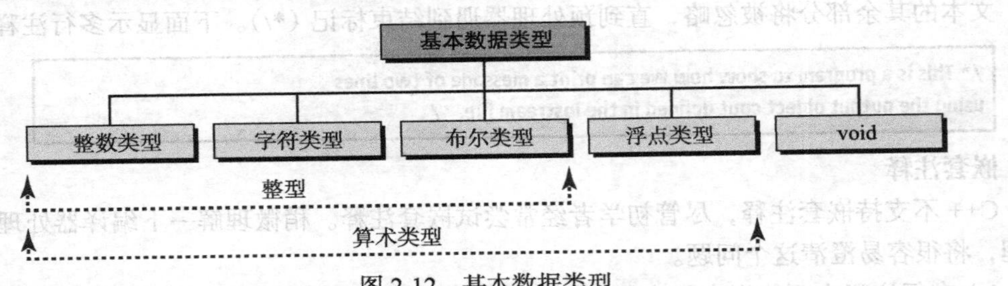

图2-12 基本数据类型

在本节中，我们只讨论基本数据类型和其中一个类类型（字符串类）。我们将在后续章节中讨论复合数据类型。

2.4.1 整数类型

没有小数部分的数值称为**整数**。例如，数值1376是一个整数，但数值1376.25则不是。C++允许三种不同大小的整数：short int、int、long int（在附录P中，我们将讨论C++11中定义的新的整数类型）。在没有歧义的情况下，可以省略术语int。这些类型的大小取决于计算机平台，这意味着long int在一台机器上可以是4个字节，而在另一台机器上可以是8个字节。这三种不同类型的整数都可以是有符号的或者无符号的（缺省是有符号的）。我们使用前缀来加以区分表示。换言之，C++实际上定义了六种不同的**整数类型**。

表2-3显示了典型机器上使用的数值范围。注意，我们假设long int的大小为4个字节

（32 位），然而有些平台则使用 8 个字节。

表 2-3 典型机器上整数类型的范围

类型	符号	范围	
short int	signed	−32 768	+32 767
	unsigned	0	65 536
int	signed	−2 147 483 648	+2 147 483 647
	unsigned	0	4 294 967 295
long int	signed	−2 147 483 648	+2 147 483 647
	unsigned	0	4 294 967 295

虽然 C++ 语言没有定义不同整数类型的精确大小和范围，但强调了它们的相对大小。它定义了以下关系，其中符号≤表示“小于或等于”。

```
short int ≤ int ≤ long int
```

为了更好地可视化显示这些不同的整数类型及其大小，我们可以参考图 2-13。

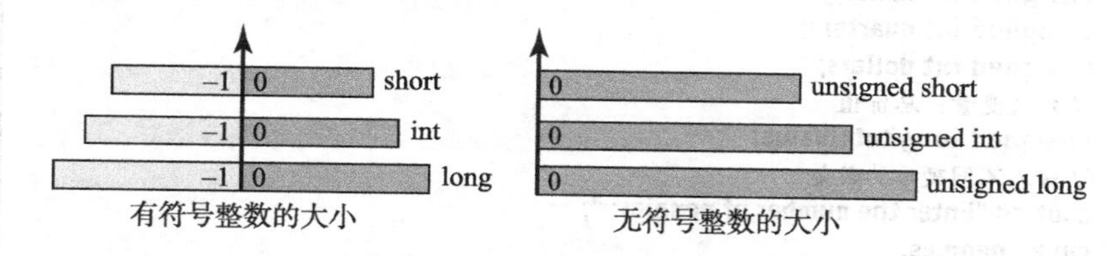

图 2-13 不同整数类型的相对大小

有符号的数据类型可以是正的或者负的；无符号的数据类型只能是正的。请注意，在有符号版本和无符号版本中，数据类型的总范围是相同的，但在有符号版本中，该范围被分为两个等分的部分（0 被视为正数的一部分）。

如果没有显式定义整数的符号，那它就是有符号的。

整数变量

如果显式地定义了一个整数类型的变量：short int(短整数)、int(整数) 和 long int(长整数)（short int 和 long int 可以缩写为 short 和 long），则可以在程序中使用该整数变量。同时，我们可以定义整数的符号：signed（有符号）或者 unsigned（无符号）(默认为 signed)。

【例 2-9】

```
// 全助记符格式
unsigned short int first;
signed short int second;
unsigned int third;
signed int fourth;
unsigned long int fifth;
signed long int six;

// 短助记符格式
unsigned short first;
short second;
unsigned int third;
int fourth;
unsigned long fifth;
long int six;
```

【例 2-10】 假设我们需要计算一套硬币（1 美元、25 美分、10 美分、5 美分、1 美分）的总价值。由于每个硬币的价值都是预先定义的并且总是正数，所以我们可以使用无符号整数常量，如程序 2-4 所示。

程序 2-4　使用无符号整数类型

```
1  /******************************************************************
2   * 计算一套硬币的总价值                                                    *
3   ******************************************************************/
4  #include <iostream>
5  using namespace std;
6  
7  int main ()
8  {
9    // 定义常量
10   const unsigned int pennyValue = 1;
11   const unsigned int nickelValue = 5;
12   const unsigned int dimeValue = 10;
13   const unsigned int quarterValue = 25;
14   const unsigned int dollarValue = 100;
15   // 定义变量（每种硬币的数量）
16   unsigned int pennies;
17   unsigned int nickels;
18   unsigned int dimes;
19   unsigned int quarters;
20   unsigned int dollars;
21   // 定义变量：总价值
22   unsigned long totalValue;
23   // 输入不同硬币的数量
24   cout << "Enter the number of pennies: ";
25   cin >> pennies;
26   cout << "Enter the number of nickels: ";
27   cin >> nickels;
28   cout << "Enter the number of dimes: ";
29   cin >> dimes;
30   cout << "Enter the number of quarters: ";
31   cin >> quarters;
32   cout << "Enter the number of dollars: ";
33   cin >> dollars;
34   // 计算总价值
35   totalValue = pennies * pennyValue + nickels * nickelValue +
36        dimes * dimeValue + quarters * quarterValue + dollars * dollarValue;
37   // 显示结果
38   cout << "The total value is: " << totalValue << " pennies." ;
39   return 0;
40 }
```

运行结果：

```
Enter the number of pennies: 20
Enter the number of nickels: 5
Enter the number of dimes: 10
Enter the number of quarters: 4
Enter the number of dollars: 6
The total value is: 845 pennies.
```

【例 2-11】 在处理银行交易时，我们需要使用有符号整数。存款是正价值交易；取款是负价值交易。程序 2-5 显示了当我们以零美元的余额开立账户时，三次交易后的账户余额。注意，因为我们同时使用正整数和负整数，所以即使余额为负，我们也允许客户取款。

程序 2-5 使用有符号整数类型

```
1  /*****************************************************************
2   * 计算三次交易后的账户余额                                      *
3   *****************************************************************/
4  #include <iostream>
5  using namespace std;
6  
7  int main ()
8  {
9    // 定义变量
10   int balance = 0;
11   int transaction;
12   // 输入第一次交易并调整账户余额
13   cout << "Enter the value of the first transaction: ";
14   cin >> transaction;
15   balance = balance + transaction;
16   // 输入第二次交易并调整账户余额
17   cout << "Enter the value of the second transaction: ";
18   cin >> transaction;
19   balance = balance + transaction;
20   // 输入第三次交易并调整账户余额
21   cout << "Enter the value of the third transaction: ";
22   cin >> transaction;
23   balance = balance + transaction;
24   // 输出最终账户余额
25   cout << "The total balance is: " << balance << " dollars. ";
26   return 0;
27 }
```

运行结果：

```
Enter the value of the first transaction: 70
Enter the value of the second transaction: -50
Enter the value of the third transaction: 35
The total balance in your account is: 55 dollars.
```

【例 2-12】 在本书后面章节，我们将详细讨论每种数据类型的大小、最大值和最小值。在这里我们编写一个小程序（程序 2-6），并使用 sizeof 运算符来检查当前工作的计算机平台中整数数据类型的大小。这可以帮助我们避免使用超出数据类型范围的数据值。

请注意，int 和 long int 的大小在我们用来运行程序的计算机中是相同的。（译者注：实际运行结果根据读者使用的计算机的不同而可能略有不同。）

程序 2-6 检查整数数据类型的大小

```
1  /*****************************************************************
2   * 检查所有三种整数数据类型的大小的程序                          *
3   *****************************************************************/
4  #include <iostream>
5  using namespace std;
6  
7  int main ()
8  {
9    cout << "Size of short int is " << sizeof (short int) << " bytes." << endl;
```

```
10    cout << "Size of int is " << sizeof (int) << " bytes." << endl;
11    cout << "Size of long int is " << sizeof (long int) << " bytes." << endl;
12    return 0;
13  }
```

运行结果：
Size of short int: 2 bytes.
Size of int: 4 bytes.
Size of long int: 4 bytes.

整数字面量

当我们在程序中显式地使用任何数据类型的值时，该值被称为字面量。在程序中使用的**整数字面量**（integer literal）具有一个常量值，其值在程序执行时不变。如果程序员没有定义整数字面量的大小和符号，系统将使用适合该值的最小的字节大小的整数类型（int或者long）。如果整数为正，则字面量为无符号整数（unsigned）；如果为负，则字面量为有符号整数（signed）。程序员可以使用表2-4中定义的后缀显式定义字面量的大小和符号。

表2-4 用于确定整数字面量的大小和符号的后缀

大小	后缀	符号	后缀
int	无	signed int 或者 signed long	无
long	l或者L	unsigned int 或者 unsigned long	u或者U

请注意，short int不适用于字面量。整数字面量的默认数据类型为int（无后缀）。为了告诉编译器我们需要数据类型long，可以使用后缀l或者L（小写字母或者大写字母）。数值字面量在默认情况下是有符号的：如果要明确地指定无符号整数，则需要使用后缀u或者U（小写字母或者大写字母）。但是请注意，我们不建议使用小写字母格式（l或者u）。小写字母l很容易与数字1混淆。

为了显式定义整数字面量的大小或者符号，我们可以使用后缀。

【例2-13】 下面示例显示了使用整数字面量的四种情况：

```
1234        // 系统使用有符号整数 int
1234U       // 系统使用无符号整数 unsigned int
1234L       // 系统使用有符号长整数 long
1234UL      // 系统使用无符号长整数 unsigned long
```

整数字面量可以用于两个目的：用作计算中的独立值，也可以用于初始化变量。

整数字面量用于初始化。当我们声明一个变量时，我们可以将其初始化为一个值。这并不意味着我们以后不能更改变量的值。这仅仅表明如果没有把一个新的值赋给变量，那么这个变量就保留着初始值。在本书的后面章节，我们将讨论变量的默认初始化值，但在此之前，我们应该记住使用适当的值初始化变量。

为了说明在使用整数字面量初始化变量时需要小心谨慎的原因，让我们运行程序2-7。

程序2-7 使用整数字面量进行初始化

```
1  /***************************************************************
2   * 使用若干整数字面量作为变量的初始值                                *
3   ***************************************************************/
4  #include <iostream>
```

```
5   using namespace std;
6
7   int main ()
8   {
9     // 定义和初始化
10     int x = -1245;
11     unsigned int y = 1245;
12     unsigned int z = -2367;
13     unsigned int t = 14.56;
14    // 输出初始化值
15     cout << x << endl;
16     cout << y << endl;
17     cout << z << endl;
18     cout << t;
19     return 0;
20  }
```

```
运行结果:
Value of x: -1245          // 正确
Value of y: 1245           // 正确
Value of z: 4294964929     // 逻辑错误：负数值被转换为正数值
Value of t: 14             // 结果值被截断
```

结果表明，第 12 行存在逻辑错误。我们将变量 z 定义为无符号整数，并将值 -2367 存储在其中。该数值被解释为 4294964929，并存储在 z 中。我们将在下一章中讨论这个逻辑错误，但这不是我们期望的结果。第 18 行的输出是另一个意外结果。由于我们将变量 t 定义为无符号整数类型，并在其中存储了一个非整数值，因此系统将丢弃小数点后面的部分，从而把字面值更改为整数值。大多数纯粹主义者认为这是一个逻辑错误。

整数字面量用作独立值。我们也可以在程序中使用整数字面量作为独立值。在这种情况下，程序根据字面量及其内容来确定其值。

【例 2-14】 程序 2-8 演示了一些测试上述情况的示例。

程序 2-8 使用整数字面量作为独立值

```
1   /****************************************************************
2    * 使用一些独立字面量值                                          *
3    ****************************************************************/
4   #include <iostream>
5   using namespace std;
6
7   int main ()
8   {
9     // 定义变量
10     int x;
11     unsigned long int y;
12    // 赋值
13     x = 1456;
14     y = -14567;
15    // 输出结果
16     cout << x << endl;
17     cout << y << endl;
```

```
18    cout << 1234 << endl;
19    cout << 143267L << endl;
20    return 0;
21  }
```

```
运行结果：
1456           // 正确
4294952729     // 错误。变量为无符号整数，值为有符号整数
1234           // 正确
143267         // 正确
```

注意，在第14行中，必须把无符号字面量存储到变量y中，但我们却为其赋了一个负值。在下一章中，我们将解释这种行为的原因。

2.4.2 字符类型

我们讨论的第二种整型是**字符类型**，被称为char。我们可以把char看作一个比unsigned short int小的整数。在C++中最初定义的char的大小是一个字节，默认为无符号数。然而，今天我们有时会看到大小为一个字节、两个字节甚至四个字节的字符。我们甚至还听说过有符号和无符号的字符。这些新类型的设计支持C++语言的国际化。我们将在附录A中讨论其中的一些类型。在本章中，我们假设char数据类型是一个字节的整数类型，它定义了ASCII编码系统中的字符（参见附录A）。ASCII编码系统使用从0到127的整数值定义128个字符。

字符变量

在程序中我们可以使用字符变量，就像我们使用整数变量一样。但是，我们在本章中避免使用有符号和无符号限定符。下面是声明char类型的两个变量的示例。

```
char first;
char second;
```

字符字面量

我们可以使用两种类型的**字符字面量**（character literal）。我们可以在两个**单引号**内使用ASCII表（附录A）中定义的字母字符。我们还可以使用ASCII表中定义的字符所对应的整数值。只要我们指定的字符字面量没有产生歧义，编译器就会将相应的字符存储在内存中。程序2-9给出了一个非常简单的例子，使用了四个字符变量，并使用四个字符字面量进行了初始化。

程序2-9 字符变量和字面量

```
1   /**********************************************************************
2    * 使用若干字符变量并进行初始化                                        *
3    **********************************************************************/
4   #include <iostream>
5   using namespace std;
6
7   int main ()
8   {
9     // 定义若干字符类型变量并进行初始化
10    char first = 'A';
```

```
11    char second = 65;
12    char third = 'B';
13    char fourth = 66;
14    //输出结果值
15    cout << "Value of first: " << first << endl;
16    cout << "Value of second: " << second << endl;
17    cout << "Value of third: " << third << endl;
18    cout << "Value of fourth: " << fourth;
19    return 0;
20  }
```

```
运行结果：
Value of first: A
Value of second: A
Value of third: B
Value of fourth: B
```

注意，第一个和第二个变量的输出值是相同的。我们在第一个变量中使用了字符字面量，在第二个变量中使用了整数字面量。第三个和第四个变量是相类似的情况。

字符字面量总是使用一对单引号括起来。

我们可以使用转义序列（反斜杠后跟符号）定义一些特殊字符，如表 2-5 所示。

在前五个字符前面使用了反斜杠，因为它们不是可打印字符。单引号用于分隔字符，因此必须使用反斜杠来表示单引号本身，而不是分隔符。双引号与之相同，但用于分隔字符串。因此，我们需要使用反斜杠来告诉程序我们确实需要这个字符，而不是分隔符。如果我们真的想用反斜杠字符，则必须用另一个反斜杠进行转义。

表 2-5　若干特殊字符

转义序列	说明	转义序列	说明
\n	新行（换行）符	\f	换页符
\t	制表符	\'	单引号
\b	退格符	\"	双引号
\r	回车符	\\	反斜杠

【例 2-15】 尽管可以定义变量并把这些转义序列赋值给变量，但展示其效果的更好方法是在字符串中使用它们。程序 2-10 演示了如何在七个字符串中使用七个转义序列。换页符无法用于监视器，它是为向打印机进纸而设计的。

程序 2-10　若干特殊字符的效果

```
1   /************************************************************
2    * 在字符串中使用若干特殊字符                                *
3    ************************************************************/
4   #include <iostream>
5   using namespace std;
6
7   int main ()
8   {
9     cout << "Hello\n";
10    cout << "Hi\t friends." << endl;
11    cout << "Buenos dias  \bamigos." << endl; // dias 后面有两个空格
12    cout << "Hello\rBonjour mes amis." << endl;
13    cout << "This is a single quote\'." << endl;
```

```
14      cout << "This is a double quote\"." <<endl;
15      cout << "This is how to print a backslash \\.";
16      return 0;
17  }
```

运行结果：

```
Hello                                 // \n 的作用与 endl 相同
Hi friends.                           // 制表符的效果
Buenos dias amigos.                   // \b 删除前面的一个字符（前面两个空格中的一个）
Bonjour mes amis.                     // \r 把控制移动到行首（Hello 将不存在）
This is a single quote'.              // 显示一个单引号
This is a double quote".              // 显示一个双引号
This is how to a print backslash \.  // 显示一个反斜杠
```

2.4.3 布尔类型

C++ 语言定义了一种被称为**布尔**（Boolean）的类型（以法国数学家 / 哲学家 George Bool 命名），表示一个值要么是真（true），要么是假（false）。该类型被称为布尔类型（Boolean），但程序中使用的类型名实际上是 bool，它是一个关键字。

布尔类型主要用于表示两个值的比较结果。例如，如果比较整数 23 和 24 的相等性，则结果为 false。但是，如果比较二者是否不相等，则结果为 true。

布尔变量

在几乎所有的实现中，bool 类型的值都存储在一个字节的内存块中。换言之，其大小是一个字节或者说是八个二进制位。如果我们将 char 类型的变量与 bool 类型的变量进行比较，我们会发现两者都使用一个字节的内存，并且我们可以在两者中存储整数。区别在于 char 类型中存储的整数被解释为字符，而 bool 类型中存储的整数被解释为逻辑值（0/1 或者 false/true）。

bool 数据类型的大小为一个字节。

布尔数据类型用于决策过程，我们将在以后的章节中讨论。

布尔字面量

因为布尔类型实际上是一个单字节的整数，所以可以使用一个小整数来表示**布尔字面量**（Boolean literal）。传统上，任何零值都被解释为 false，任何非零值都被解释为 true。当输出布尔类型的值时，结果是 0 或者 1。在后续章节中，我们将介绍如何打印字面量“false”或者“true”。

【例 2-16】 程序 2-11 是一个测试布尔值的简单程序。

程序 2-11 布尔类型

```
1   /***********************************************************
2    * 布尔变量和布尔值的使用                                    *
3    ***********************************************************/
4   #include <iostream>
5   using namespace std;
6
7   int main ()
8   {
```

```
9   //变量定义
10    bool x = 123;
11    bool y = -8;
12    bool z = 0;
13    bool t = -0;
14    bool u = true;
15    bool v = false;
16   //输出值
17    cout << "Value of x: " << x << endl;
18    cout << "Value of y: " << y << endl;
19    cout << "Value of z: " << z << endl;
20    cout << "Value of t: " << t << endl;
21    cout << "Value of u: " << u << endl;
22    cout << "Value of v: " << v << endl;
23    return 0;
24 }
```

```
运行结果:
Value of x: 1      // 123 被解释为 1 (true)
Value of y: 1      // -8 被解释为 1 (true)
Value of z: 0      // 0 被解释为 0 (false)
Value of t: 0      // -0 被解释为 0 (false)
Value of u: 1      // true 的输出为 1
Value of v: 0      // false 的输出为 0
```

请注意，任何非零值（正或者负）都被解释为 true，任何零值都被解释为 false。布尔值显示为 1 或者 0，除非显式地将其更改为 true 或者 false（将在后续章节中讨论如何实现）。

2.4.4 浮点类型

在 C++ 语言中，带小数部分的数值被称为**浮点类型**（floating-point）。为了提高计算效率，C++ 定义了三种不同的浮点大小：float、double 和 long double。所有浮点数都是有符号数。

所有浮点数都是有符号数。

在计算机中，浮点数使用 IEEE 标准进行存储。虽然语言没有定义不同浮点类型的精确大小和范围，但它强调了它们的相对大小。它定义了以下关系，其中符号 <= 表示“小于或者等于”。

```
float ≤ double ≤ long double
```

图 2-14 以图形方式显示了这种关系。注意，所有浮点类型都是有符号数，因此其关系比整数数据类型更简单。

浮点变量

定义浮点变量的方法与定义整数变量的方法相同。可以在声明浮点变量时进行初始化，或者稍后为它们赋值。

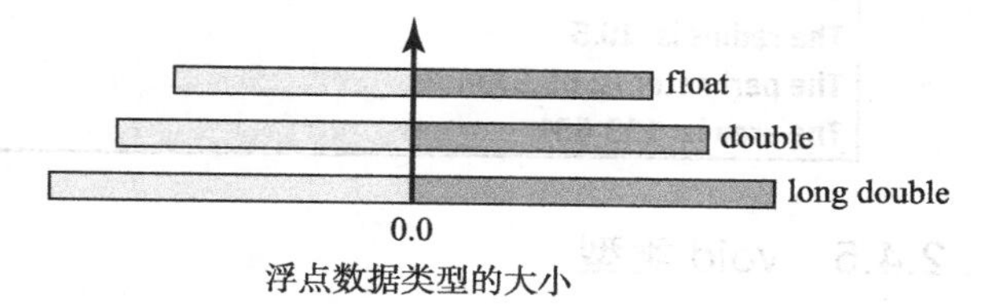

图 2-14 浮点数据类型的相对大小

浮点字面量

浮点字面量（floating-point literal）

是带小数的数值，如 32.78、141.123、123456.0 和 -2472.657。程序员可以使用如表 2-6 所示的后缀显式定义其大小。

浮点字面量的缺省大小为 double。

表 2-6　定义浮点字面量大小的后缀

浮点类型	后缀	示例
float	f 或者 F	12.23F，12345.45F，-1436F
double	无	1425.36，1234.34，123454
long double	l 或者 L	2456.23L，143679.00004L，-0.02345L

【例 2-17】 前面章节讨论过如何计算圆的面积和周长。程序 2-12 计算给定半径的圆的面积和周长。

程序 2-12　计算圆的面积和周长

```
1  /***************************************************************
2   *本程序计算圆的面积和周长                                      *
3   ***************************************************************/
4  #include <iostream>
5  using namespace std;
6
7  int main ().
8  {
9    // 定义一个常量
10   const double PI = 3.14159;
11   // 定义三个变量
12   double radius;
13   double perimeter;
14   double area;
15   // 输入半径的值
16   cout << "Enter the radius of the circle: ";
17   cin >> radius;
18   // 计算周长和面积并把它们存储到变量中
19   perimeter = 2 *  PI * radius;     // 2 是一个常数字面量
20   area = PI * PI * radius;
21   // 输出半径、周长和面积
22   cout << "The radius is: " << radius << endl;
23   cout << "The perimeter is: " << perimeter << endl;
24   cout << "The area is: " << area;
25   return 0;
26 }
```

运行结果：

```
Enter the radius of the circle: 10.5
The radius is: 10.5
The perimeter is: 65.9734
The area is: 103.631
```

2.4.5　void 类型

void（空）类型是没有值的特殊类型。但是，它很有用，因为它可以显示出缺少值的情况。例如，我们可以使用 void 来表示一个函数不返回任何值。缺少值与我们在现实生活中使用的概念相同，例如当我们说空头支票时，空头支票是一种用于其他目的但没有价值的支票。例如，可以使用 void 类型来显示函数什么也不返回。

2.4.6 字符串类

到目前为止，我们在本章中使用的字符串都是从 C 语言继承而来的。它是以空字符结尾的字符集合。例如，字符串 "John" 可以看作五个字符 'J' 'o' 'h' 'n' 和 '\0' 的集合，最后一个字符定义了一个空字符（终止符）。在后续章节中讨论数组（元素的集合）时，我们将详细阐述这种类型的字符串。

C++ 定义了一种新的**字符串类**，它是用户自定义类型（参见图 2-11）。string 类在 C++ 库中定义，虽然比较复杂，但更易于使用，因为它定义了许多操作。后续章节中，随着我们全面讨论这个类，我们将逐步使用 **C++ 字符串**代替 **C 字符串**。

要使用 C++ 字符串，我们必须在程序中包含头文件 <string>，如下所示：

```
# include <string>
```

我们使用以下定义来声明 string 类型的变量：

```
string name;
```

其中 name 是变量的名称。字符串类的对象也可以使用双引号创建，例如："John"。C 字符串的字面量和 C++ 字符串对象之间的其中一个差异是：前者由五个字符组成，而后者只有四个字符（C++ 字符串中不需要空字符）。当我们逐渐学会使用字符串类的对象上的操作时，C++ 字符串的主要区别和优势就会凸显出来。

可以使用 C++ 字符串做的有趣的事情之一是字符串拼接。可以简单地使用加法运算符（+）将两个或者多个字符串连接在一起，如下例所示。

【例 2-18】 在这个例子中，我们编写了一个使用 C++ 字符串的简单程序（程序 2-13）。首先声明一个人的名字、姓氏和中间名（中间名的首字母），然后使用字符串拼接来打印整个姓名。这只是一个简单的字符串类的示例，更强大的示例将在后续章节中讨论。

程序 2-13 使用字符串类

```
1   /***************************************************************
2    * 这个程序根据某人的名字、中间名和姓氏，                          *
3    * 打印这个人的全名                                              *
4    ***************************************************************/
5   #include <iostream>
6   #include <string>            // 需要使用字符串类
7   using namespace std;
8
9   int main ()
10  {
11    // 定义变量
12    string first;
13    string initial;
14    string last;
15    string space = " ";
```

```
16    string dot = ".";
17    string fullName;
18    //输入名字、中间名和姓氏
19    cout << "Enter the first name: ";
20    cin >> first;
21    cout << "Enter the initial: ";
22    cin >> middle;
23    cout << "Enter the last name: ";
24    cin >> last;
25    //使用字符串拼接运算符拼接全名
26    fullName = first + space + initial + dot + space + last;
27    //输出全名
28    cout << "The full name is: " << fullName;
29    return 0;
30  }
```

运行结果：

```
Enter the first name: John
Enter the initial: A
Enter the last name: Brown
The full name is: John A. Brown
```

本章小结

C++ 语言是区分大小写的。我们通常把程序分成多行，以提高可读性。程序中的头文件位于程序顶部，包含预定义的代码。每个 C++ 程序至少需要一个名为 main 的函数。

C++ 程序使用变量，每个变量都有一个名称和数据类型。

程序中使用的 cin 和 cout 对象与变量和值的概念密切相关。cin 对象从键盘获取数据并将其存储在变量中；cout 对象向监视器发送一个值。

C++ 程序中的源代码通常由两个相互交织的部分组成：代码和注释。代码使用 C++ 语言的标记符，注释是英文句子。标记符可以分为标识符、字面量或者符号。

用 C++ 语言编写的程序用于处理数据。C++ 可以识别不同类型的数据：内置的（基本的和复合的）数据类型和用户自定义的数据类型（枚举和类）。基本数据类型分为五类：整数类型、字符类型、浮点类型、布尔类型和 void 类型。整数是没有小数部分的数值。字符类型定义字母。布尔类型表示 true 或者 false。浮点类型表示带小数部分的数值。

C++ 语言使用两种字符串类型：第一种是 C 字符串，第二种是 C++ 字符串。

思考题

1. 请指出以下程序中的错误（如果存在）。

```
1  #include <iostream>
2  using namespace std
3
4  int main ()
5  {
6      cout << 25;
7      cout << first;
8      return 0;
9  }
```

2. 请指出以下程序中的错误（如果存在）。

```
1  #include <iostream>
2
3  int main ()
4  {
5      cout << 35;
6      cout << 45;
7      return 0;
8  }
```

3. 请指出以下程序中的错误（如果存在）。

```
1  #include <iostream>
2  using namespace std
3
4  int main ();
5  {
6      std :: cout << 35 << endl;
7      std :: cout << 45;
8  }
```

4. 请指出以下程序中的错误（如果存在）。

```
1  #include <iostream>
2  using namespace std
3
4  int main ()
5  {
6      int a = 25;
7      a + 30 = a ;
8      cout << a;
9      return 0;
10 }
```

5. 请指出以下程序中的错误（如果存在）。

```
1  #include <iostream>
2  using namespace std;
3
4  int main ()
5  {
6      string name;
7      cout << name;
8      return 1;
9  }
```

6. 请指出以下程序中的错误（如果存在）。

```
1  using namespace std;
2
3  int main ()
4  {
5      double = 24;
6      cout << double ;
7      return 0
8  }
```

7. 请指出以下程序中的错误（如果存在）。

```
1  #include <iostream>
2  using namespace std;
3
4  int main ()
5  {
6       int y = 32.34;
7       cout << y ;
8       return 0;
9  }
```

8. 请指出以下程序中的错误（如果存在）。

```
1  #include <iostream>
2  using namespace std;
3
4  int main ()
5  {
6       float y = 32.34;
7       cout << y ;
8       return 0;
9  }
```

9. 请指出以下程序中的错误（如果存在）。

```
1  #include <iostream>
2  using namespace std;
3
4  int main ()
5  {
6       double y = 32.34;
7       cout << ;
8       return 0;
9  }
```

10. 请指出以下程序中的错误（如果存在）。

```
1  #include <iostream>
2  using namespace std;
3
4  int main ()
5  {
6       char y = 32;
7       char t = 78;
8       char z = y + t;
9       cout << z;
10      return 0;
11 }
```

11. 请问以下程序的输出结果是什么。

```
1  #include <iostream>
2  using namespace std;
3
4  int main ()
5  {
6       int x = 12;
```

```
7       int y = 14;
8       int z = x + y;
9       cout << z;
10      return 0;
11  }
```

12. 请问以下程序的输出结果是什么。

```
1   #include <iostream>
2   using namespace std;
3
4   int main ()
5   {
6       double x = 12.24;
7       double y = 14.32;
8       cout << x  << " + " << y;
9       return 0;
10  }
```

13. 请问以下程序的输出结果是什么。

```
1   #include <iostream>
2   using namespace std;
3
4   int main ()
5   {
6       char x = 'A';
7       char y = 'B';
8       cout << x  << y;
9       return 0;
10  }
```

14. 请问以下程序的输出结果是什么。

```
1   #include <iostream>
2   using namespace std;
3
4   int main ()
5   {
6       bool truth = true;
7       bool lie = false;
8       bool result = truth + lie;
9       cout << result;
10      return 0;
11  }
```

15. 请列出以下程序中的所有关键字。

```
1   #include <iostream>
2   using namespace std;
3
4   int main ()
5   {
6       int x = 0;
7       int y = 1;
8       cout  << x << y;
9       return 0;
10  }
```

16. 请找出以下程序中的所有非关键字标识符。

```
1  #include <iostream>
2  using namespace std;
3  
4  int main ()
5  {
6      int x = 4;
7      int y = 22;
8      cout  << x << y;
9      return 0;
10 }
```

17. 请找出以下程序中的所有变量。

```
1  #include <iostream>
2  using namespace std;
3  
4  int main ()
5  {
6      int x = 4;
7      int y = 22;
8      cout << x << y;
9      return 0;
10 }
```

18. 请指出以下程序中包含行注释的行号。

```
1  #include <iostream>
2  using namespace std;
3  
4  int main ()
5  {
6      // Declaration and initialization of two variables
7      int x = 4;
8      int y = 22;
9      // Print the value of x and y
10     cout << x << " " << y;
11     return 0;
12 }
```

19. 请指出以下程序中包含块注释的行号。

```
1  /***********************************************************
2   * A small program to print two values.                   *
3   ***********************************************************/
4  #include <iostream>
5  using namespace std;
6  
7  int main ()
8  {
9      int x = 4;
10     int y = 22;
11     cout << x << " " << y;
12     return 0;
13 }
```

编程题

1. 请编写一个程序，打印以下星号组成的三角形。

```
*
**
***
```

2. 请编写一个程序，打印以下星号组成的图形。

```
*
**
***
**
*
```

3. 请编写一个程序，打印如下所示的大写字母 H。

```
H  H
H  H
HHHH
H  H
H  H
```

4. 请编写一个程序，根据给定任务的持续时间（小时、分钟和秒），计算以秒为单位的持续时间。
5. 请编写一个程序，输入四个整数值，计算并打印它们的累加和。
6. 请编写一个程序，给定一个正方形的边长，计算并打印正方形的面积和周长。
7. 请编写一个程序，使用 C++ 字符串，在提示输入你的名字和姓氏之后，按下面的格式打印你的姓名。注意，姓氏应该在名字之前，如下所示。

   Your full name is: *last, first*

8. 请编写一个程序，提示用户输入两个整数，然后打印它们的累加和。运行程序若干次，每次使用不同的变量值。
9. 请编写一个程序，根据销售金额计算交易的销售税，假设税率为 9%。使用一个常量来定义税率。运行程序若干次，每次使用不同的销售金额值。输出时使用以下格式。

   Sale amount: ...
   Tax amount: ...
   Total amount due: ...

10. 请编写一个程序，给定街道号、街道名、城市名、州名和邮政编码，按以下格式打印地址。

    street-number, street city, state zip-code

第 3 章
C++ Programming: An Object-Oriented Approach

表达式和语句

本章将介绍程序的两个主要组件：表达式和语句。C++ 程序中的表达式看起来像自然语言（例如英语）中的短语，C++ 程序中的语句看起来像英语中的句子。如果想正确使用自然语言，我们需要理解短语和句子以及它们是如何形成的。为了理解 C++ 语言以及如何用这种语言创建程序，我们需要理解表达式和语句。

C++ 语言包括若干类型的表达式和若干类型的语句。本章不会覆盖全部内容，但是我们讨论的内容足够编写简单的程序。在接下来的几章中，我们将继续讨论表达式和语句，直到读者掌握所有这些内容。

学习目标

阅读并学习本章后，读者应该能够

- 讨论表达式、表达式的构成、表达式的求值，并讨论基本表达式、一元表达式、乘法类表达式、加法类表达式和赋值表达式。
- 讨论表达式中的类型转换，包括隐式转换和显式转换。
- 利用子表达式的优先级和结合性，讨论复杂表达式中简单表达式的计算顺序。
- 讨论整数和浮点数据类型中的上溢问题和下溢问题。
- 演示如何使用不带参数和带一个参数的操作符格式化输入和输出数据。
- 讨论语句并阐述一些简单的语句类型，包括声明语句、表达式语句、空语句、复合语句和返回语句。
- 讨论程序设计的三个步骤：理解问题、开发算法和编写代码。

3.1 表达式

我们已经看到，程序是一个值处理器。程序获取值，处理值，并创建新的值。为此，程序定义了几个称为表达式的小实体。**表达式**（expression）是一个具有可以更改内存状态（称为副作用）的值的实体。

表达式是一个具有值并可能具有副作用的实体。

表达式可以是简单值，还可以使用运算符组合值以创建新值。在 C++ 中，一个表达式可以不包含运算符，也可以包含一个运算符、两个运算符或者三个运算符。我们还可以组合表达式来创建另一个表达式。

要处理表达式，我们需要一个表达式列表，其中包含用于创建表达式的相应运算符。附录 C 中包含了 C++ 表达式列表，但实现本章的目标只需要 C++ 表达式列表的一部分，如表 3-1 中所示。注意，在该列表中使用了一些缩写，例如 expr（expression，表达式）和 op（operator，运算符）。

表 3-1 部分 C++ 表达式列表

组别	名称	运算符	表达式	优先级	结合性
基本表达式	字面量 名称 括号表达式		literal name (expr)	19	→
一元表达式	正数 负数 大小	+ – sizeof	+expr –expr sizeof expr	17	←
乘法类表达式	乘法 除法 余数	* / %	expr * expr expr / expr expr % expr	14	→
加法类表达式	加法 减法	+ –	expr + expr expr – expr	13	→
赋值表达式	简单赋值 复合赋值	= op=	variable = expr variable op= expr	3	←

在讨论各类别中的表达式之前，让我们先给出一些关于表 3-1 的信息。

1）表 3-1 仅列出了 5 组表达式，但目前 C++ 包括 19 组表达式，如附录 C 所示。

2）第 1 列定义了表达式组别信息。第 2 列定义了通常用于每个子组的名称。第 3 列定义了用作运算符的符号或者术语。第 4 列显示了运算符如何组合表达式以创建新的表达式。

3）第 5 列定义了运算符的优先级。我们将在本章后面讨论优先级。当我们有一组具有不同优先级的表达式时，优先级会告诉我们计算的顺序：首先计算具有最高优先级的表达式。请注意，我们有 19 个优先级别，但在表 3-1 中只显示了其中的 5 个。

4）第 6 列定义了运算符的结合性。我们在本章后面还会讨论结合性。当我们有多个具有相同优先级的表达式时，结合性告诉我们计算的顺序。结合性可以是从左到右（→）或从右到左（←）。

接下来讨论表 3-1 中定义的五组表达式。

3.1.1 基本表达式

基本表达式（primary expression）是没有运算符的简单表达式。它是构造更复杂表达式的基本构造块。基本表达式具有优先级 19，这意味着它在所有表达式中具有最高优先级。C++ 包含若干基本表达式，如附录 C 所示，但本章仅讨论表 3-1 所示的表达式。

字面量

前文已经讨论过**字面量**（literal，程序中使用的值）。程序中的字面量是基本表达式，它有一个值，但没有副作用。一些作为基本表达式的字面量如下所示。请注意，没有短整数字面量。

字面量	说明	字面量	说明
false	// 布尔字面量	12897234L	// 长整数字面量
'A'	// 字符字面量	245.78F	// 浮点数字面量
"Hello"	// 字符串字面量	114.7892	// 双精度数字面量
234	// 整数字面量	245.784321L	// 长双精度数字面量

【例 3-1】 如前所述，一个字面量可以用于使用其值，如程序 3-1 所示。

程序 3-1　字面量表达式

```
1   /****************************************************************
2    * 本程序演示若干字面量表达式的使用方法                              *
3    ****************************************************************/
4   #include <iostream>
5   using namespace std;
6   
7   int main ()
8   {
9       cout << false << " " << 'A' << " " << "Hello" << endl;
10      cout << 23412 << " " << 12897234L << endl;
11      cout << 245.78F << " " << 114.782 << " " << 2.051L;
12      return 0;
13  }
```

运行结果：

```
0 A Hello                 // 一个布尔值、一个字符、一个字符串
23412 12897234            // 一个整数、一个长整数（没有短整数字面量）
24.78 114.782 2.051       // 一个浮点数、一个双精度数、一个长双精度数
```

名称

在程序中使用**名称**时，我们使用的是基本表达式。作为名称的标识符可以是变量、对象名、函数名等。在程序中使用的最简单的名称是变量名。它是一个具有值的基本表达式，值存储在相应的变量中。如前述章节所示，也可以使用限定名称（由双冒号分隔的组名称和个体名称所组成的名称）。

```
表达式          说明
x               // 可以是一个变量的名称
cout            // 一个对象的名称
std :: cout     // 使用其名称空间限定的名称
```

括号表达式

如果希望把一个优先级较低的表达式更改为基本表达式，可以将其包含在括号中。这样可以在需要基本表达式的地方使用复杂表达式。有时候还需要在计算表达式的其余部分之前对表达式的这部分进行求值。基本表达式具有最高的优先级，因此我们可以添加括号以强制优先计算括号内的表达式。我们将在本章后面使用一些括号表达式，下面给出了两个示例。

```
表达式          说明
(x + 3) * 5     // 先计算 x 和 3 的和，然后将结果乘以 5
12 / ( x + 2)   // 先计算 x 和 2 的和，然后用结果去除 12
```

在这两种情况下，如果移除括号，则表达式的值将改变，而且不符合预期，请参考下一个程序的运行结果。

【例 3-2】　程序 3-2 演示了使用圆括号时结果会产生巨大差异。我们使用上面描述的两个带括号的表达式和不带括号的表达式来比较其差异。

程序 3-2　括号表达式

```
1   /****************************************************************
2    * 本程序演示括号表达式的使用方法                                    *
```

```
3    *********************************************************************/
4   #include <iostream>
5   using namespace std;
6
7   int main ()
8   {
9     // 变量声明
10    int x = 4;
11    // 打印第一个表达式的带括号和不带括号版本
12    cout << "Value with parentheses: " << (x + 3) *  5 << endl;
13    cout << "Value without parentheses: " << x + 3  *  5 << endl << endl;
14    // 打印第二个表达式的带括号和不带括号版本
15    cout << "Value with parentheses: " << 12 / ( x + 2) << endl;
16    cout << "Value without parentheses: " << 12 /  x + 2;
17    return 0;
18  }
```

```
运行结果:
Value with parentheses: 35
Value without parentheses: 19

Value with parentheses: 2
Value without parentheses: 5
```

3.1.2 一元表达式

一元表达式（unary expression）是将运算符作用于单个值（称为操作数）构成的表达式，该值必须是基本表达式（如果不是，则必须首先将其转换为基本表达式，如本章后面所述）。一元表达式的结果是基本表达式。在一元表达式中，运算符位于操作数之前。

正数表达式和负数表达式

C++ 包含两种表达式：**正数表达式**（plus expression）和**负数表达式**（minus expression），但我们一起加以讨论。在这两种表达式类型中，运算符都在操作数之前，如图 3-1 所示。

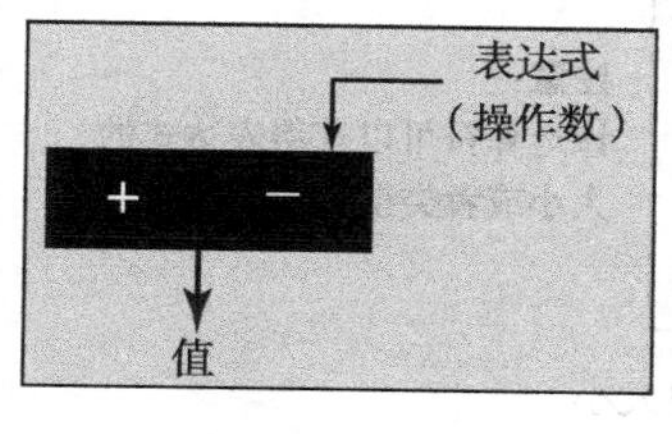

值 0 操作数

操作数 0 值

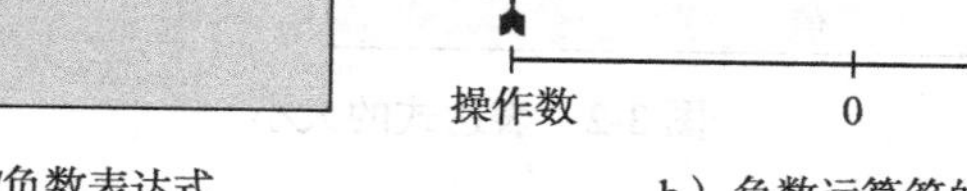

a）正数/负数表达式　　b）负数运算符的效果

图 3-1　正数表达式和负数表达式

正数运算符不更改其操作数的值（仅用于强调），负数运算符更改其操作数的值（翻转值）。如果该值最初为正值，则负数运算符将其更改为负值，反之亦然。

【例 3-3】 程序 3-3 演示了如何将正数 / 负数运算符应用于基本表达式。

程序 3-3　测试正数 / 负数表达式

```
1   /*********************************************************************
2    * 本程序演示正数 / 负数表达式的使用方法                                *
3    *********************************************************************/
```

```
4   #include <iostream>
5   using namespace std;
6
7   int main ()
8   {
9     // 变量声明和初始化
10    int x = 4;
11    int y = -10;
12    // 将正数 / 负数运算符应用于变量 x
13    cout << "Using plus operator on x: " << +x << endl;
14    cout << "Using minus operator on x: " << -x << endl;
15    // 将正数 / 负数运算符应用于变量 y
16    cout << "Using plus operator on y: " << +y << endl;
17    cout << "Using minus operator on y: " << -y;
18    return 0;
19  }
```

运行结果：

```
Using plus operator on x: 4
Using minus operator on x: -4
Using plus operator on y:-10
Using minus operator on y: 10
```

sizeof 表达式

另一个一元表达式是使用 sizeof 运算符的表达式，如图 3-2 所示。此运算符有两个版本：一个用于获得表达式的大小，另一个用于获得类型的大小。第一个版本先计算表达式（获得其值），然后再获得值的大小。第二个版本用于获得由 C++ 实现定义的类型的大小。以下是这两个版本的示例：

```
sizeof expression    // 获得表达式的大小
sizeof (type)        // 获得类型的大小
```

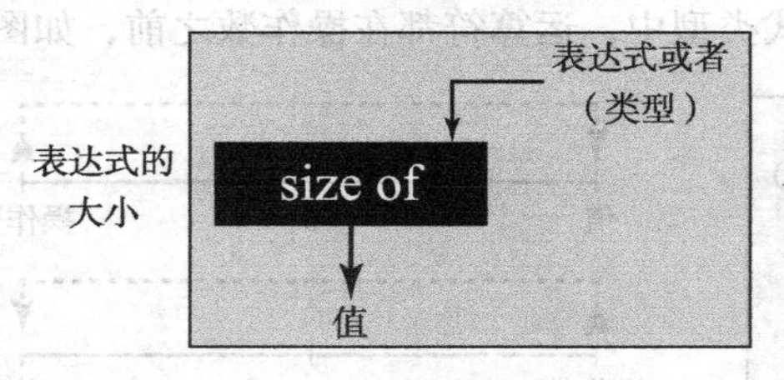

注意：
该运算符可以获得表达式的大小或者类型的大小。

图 3-2　表达式的大小

但是，请注意第一个版本中的表达式要求是基本表达式。所以，如果我们有一个包含运算符的表达式，则需要将它包含在括号中。当然，第二个版本的结果依赖于系统，因为数据类型的大小依赖于系统。我们将在后续章节中讨论其他一元表达式。

3.1.3　乘法类表达式

乘法类表达式（multiplicative expression）是一种二元表达式，其中有两个操作数：左操作数和右操作数。创建乘法类表达式可以使用三种运算符：乘法运算符、除法运算符和余数运算符，如图 3-3 所示。

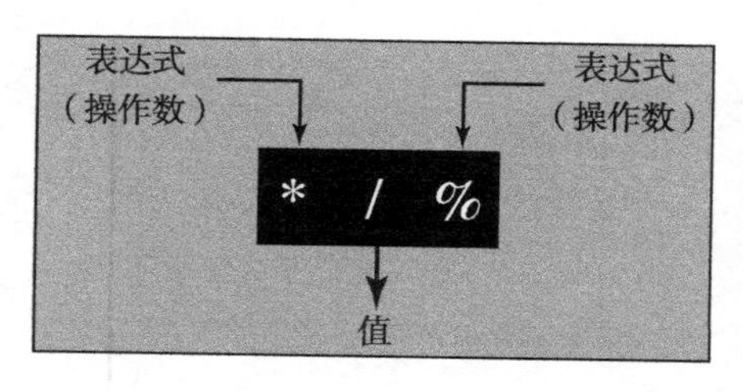

注意：对于乘法运算符和除法运算符，如果两个操作数都是正数或者都是负数，则运算结果是正数，否则运算结果是负数。对于余数运算符，结果的正负取决于系统实现。

图 3-3 乘法类表达式

乘法运算符

为了将两个值相乘，我们使用符号为 *（星号）的乘法运算符。如果两个操作数不是同一类型，则可能会进行类型转换，我们将在稍后讨论。

除法运算符

为了将一个值除以另一个值，我们使用除法运算符，其符号为 /（正斜杠）。如果两个操作数都是整数，则结果为整数值。如果其中一个操作数是浮点数，则结果是浮点数值。

余数运算符

为了获取一个整数值除以另一个整数值的余数，我们使用余数运算符，其符号是百分号（%）。这两个操作数必须是正整数类型。如果其中一个操作数是负整数类型，则结果取决于系统实现。

【例 3-4】 程序 3-4 测试了一些乘法类表达式。

程序 3-4 乘法类表达式

```
1  /***************************************************************
2   * 展示乘法类表达式的效果                                        *
3   ***************************************************************/
4  #include <iostream>
5  using namespace std;
6
7  int main ()
8  {
9    // 乘法
10   cout << "Testing multiplication operator" << endl;
11   cout << "Value of 3 * 4 = " << 3 * 4 << endl;
12   cout << "Value of 2.4 * 4.1 = " << 2.4 * 4.1 << endl;
13   cout << "Value of -3 * 4 = " << -3 * 4  << endl;
14   // 除法
15   cout << "Testing division operator" << endl;
16   cout << "Value of 30 / 5 = " << 30 / 5  << endl;
17   cout << "Value of 4 / 7 =  " << 4 / 7 << endl;
18   // 余数
19   cout << "Testing remainder operator" << endl;
20   cout << "Value of 30 % 5 = " << 30 % 5 << endl;
21   cout << "Value of 30 % 4 = " << 30 % 4 << endl;
22   cout << "Value of 3 % 7 = " << 3 % 7 << endl;
23   return 0;
24 }
```

运行结果：

```
Testing multiplication operator
```

```
Value of 3 * 4 = 12
Value of 2.4 * 4.1 = 9.84
Value of -3 * 4 = -12
Testing division operator
Value of 30 / 5 = 6
Value of 4 / 7 = 0
Testing remainder operator
Value of 30 % 5 = 0
Value of 30 % 4 = 2
Value of 3 % 7 = 3
```

3.1.4　加法类表达式

加法类表达式是一种二元表达式，其中有两个操作数：左操作数和右操作数。C++ 包括两种加法类表达式：**加法**（addition）和**减法**（subtraction）（都被称为加法类表达式），如图 3-4 所示。

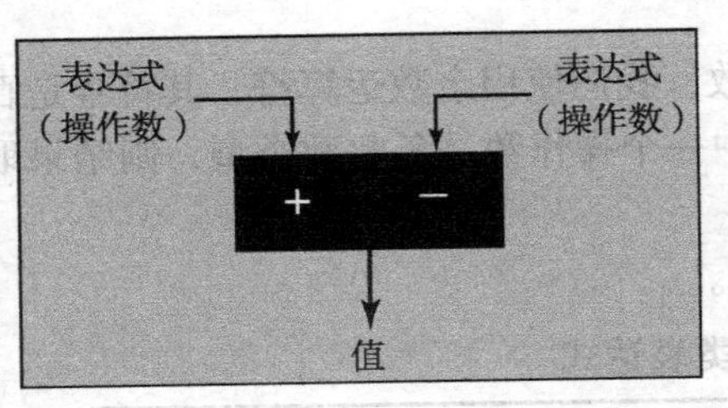

图 3-4　加法类表达式

注意： 当一个操作数减去另一个操作数时，运算结果的正负由较大操作数的正负所决定。

【例 3-5】 程序 3-5 演示了如何测试一些加法类表达式。

程序 3-5　加法和减法

```
 1 /**************************************************************
 2  * 展示加法和减法运算符的效果                                   *
 3  **************************************************************/
 4 #include <iostream>
 5 using namespace std;
 6
 7 int main ()
 8 {
 9   // 测试一些加法运算
10   cout << "Some addition operations" << endl;
11   cout << "Value of 30 + 5 = " << 30 + 5 << endl;
12   cout << "Value of 20.5 + 6.2 = " << 20.5 + 6.2 << endl;
13   // 测试一些减法运算
14   cout << "Some subtraction operations" << endl;
15   cout << "Value of 5 - 30 = " << 5 - 30 << endl;
16   cout << "Value of 51.2 - 30.4 = " << 51.2 - 30.4 << endl;
17   return 0;
18 }
```

运行结果：

```
Some addition operations
Value of 30 + 5 = 35
Value of 20.5 + 6.2 = 26.7
```

```
Some subtraction operations
Value of 5 - 30 = -25
Value of 51.2 - 30.4 = 20.8
```

3.1.5 赋值表达式

赋值表达式创建一个值并具有副作用。它改变计算机的内存状态。存在两种赋值表达式：简单赋值表达式和复合赋值表达式。我们将在这里讨论简单赋值表达式和一些复合赋值表达式，其余的复合赋值表达式将在后续章节中讨论。

简单赋值表达式

简单赋值运算符使用“=”符号。虽然该符号看上去与数学中的等式类似，但在C++中它被称为赋值运算符。它是一个有两个操作数的二元运算符：左操作数是变量名，右操作数是要由运算符计算的表达式。其示意图如图3-5所示。

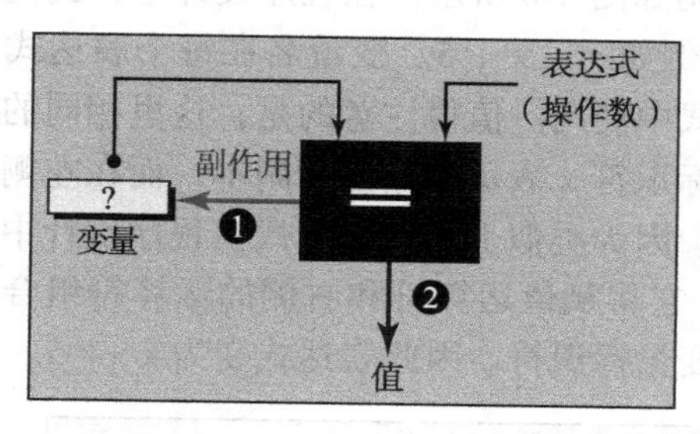

注意：

左操作数必须是一个变量。变量的原始值丢失，右边表达式的值存储到该变量中，这就是副作用。首先实施副作用，然后返回值。

图3-5 简单赋值表达式

一个简单赋值表达式的操作过程可以总结为两个步骤：

1）赋值运算符将表达式的值存储到变量中。这被称为副作用，因为变量的前一个值丢失，并在其中存储了一个新值。

2）赋值运算符返回在步骤1中获得的值，该值可以在更复杂的表达式中使用。

在赋值表达式中，先将右侧表达式的值存储到变量中（副作用），然后返回表达式的值。

【例3-6】 程序3-6演示了如何测试返回值和副作用。首先执行赋值，该赋值返回用于打印的值（我们需要使用括号来使赋值表达式首先被完成）。然后打印变量的值以检查副作用。

程序3-6 简单赋值运算符

```
1  /*************************************************************
2   * 测试一些简单赋值表达式                                        *
3   *************************************************************/
4  #include <iostream>
5  using namespace std;
6
7  int main ()
8  {
9    // 变量声明
10   int x;
11   int y;
12   // 第一个赋值
```

```
13    cout << "Return value of assignment expression: " << (x = 14) << endl;
14    cout << "Value of variable x: " << x << endl;
15    // 第二个赋值
16    cout << "Return value of assignment expression: " << (y = 87) << endl;
17    cout << "Value of variable y: " << y;
18    return 0;
19  }
```

```
运行结果：
Return value of assignment expression: 14
Value of variable x: 14
Return value of assignment expression: 87
Value of variable y: 87
```

复合赋值表达式

复合赋值（compound assignment）及其说明如图 3-6 所示。在程序设计中，我们经常需要更改变量的内容后再将结果存储回变量中，例如 x = x + 5。变量名在每个表达式中使用两次：一次在左表达式中使用，一次在右表达式中使用。值得注意的是，这里相同的标识符代表两个不同的东西。在赋值运算符的右侧，标识符 x 表示原始值的副本，而在左侧则表示计算右表达式的值，并将结果存储到变量 x 中。因为类似上面的表达式在程序设计中非常常见，所以 C++ 提供了这种表达式的简单形式。它将赋值运算符和右侧的运算符组合成一个称为复合赋值的单个运算符，并且只使用一次变量标识符。因此表达式变为 x += 5。

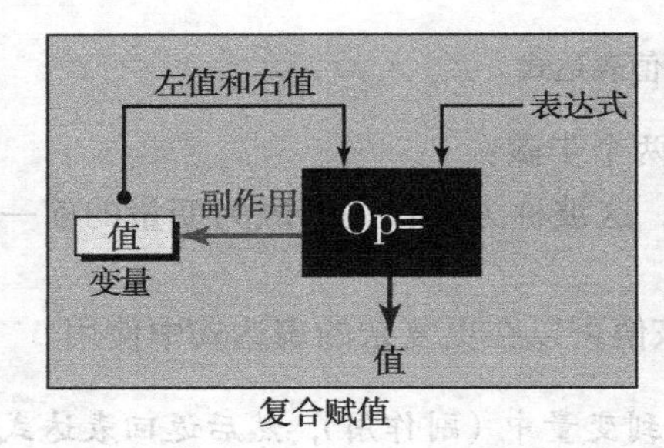

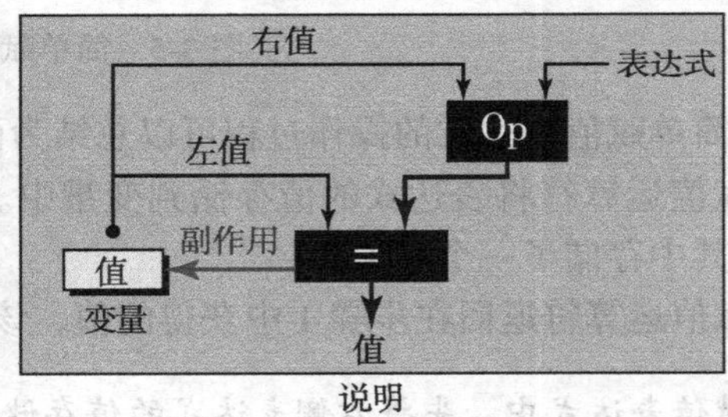

图 3-6　复合赋值及其说明

若干复合赋值表达式及其编译器的解释如下所示：

```
x += 5        // 编译器解释为：x = x + 5
y -= 3        // 编译器解释为：y = y - 3
z *= 10       // 编译器解释为：z = z * 10
t /= 8        // 编译器解释为：t = t / 8
u %= 7        // 编译器解释为：u = u % 7
```

请注意，每个复合赋值运算符由两个运算符组成，我们不能将它们彼此分隔。换言之，如果使用 z * = 10 而不是 z *= 10，则将导致错误。

【例 3-7】 程序 3-7 测试了某些复合赋值表达式的副作用。

程序 3-7　测试复合赋值表达式

```
1   /***************************************************************
2    * 测试某些复合赋值表达式                                      *
3    ***************************************************************/
4   #include <iostream>
```

```
5   using namespace std;
6
7   int main ()
8   {
9     //声明五个变量
10    int x = 20;
11    int y = 30;
12    int z = 40;
13    int t = 50;
14    int u = 60;
15    //使用复合赋值表达式
16    x += 5;
17    y -= 3;
18    z *= 10;
19    t /= 8;
20    u %= 7;
21    //输出结果
22    cout << "Value of x: " << x << endl;
23    cout << "Value of y: " << y << endl;
24    cout << "Value of z: " << z << endl;
25    cout << "Value of t: " << t << endl;
26    cout << "Value of u: " << u;
27    return 0;
28  }
```

运行结果：

```
Value of x: 25
Value of y: 27
Value of z: 400
Value of t: 6
Value of u: 4
```

第一个结果可以被认为是 x = x + 5 或者 x = 20 + 5，实际上是 x = 25。最后一个结果可以被认为是 u = u % 7 或者 u = 60 % 7，实际上是 u = 4。

3.1.6 左值和右值的概念

在 C++ 中，任何可以放在赋值运算符左侧的实体都被称为**左值**（lvalue），任何可以放在赋值运算符右侧的实体都被称为**右值**（rvalue）。这两个名字有时会引起误解。更清楚地说，可以作为值目标的实体称为左值，可以作为值源的实体称为右值。变量位于赋值的左侧并充当目标时，它是左值。当同一个变量位于赋值的右侧并充当源时，它就是右值。在复合赋值中，变量既是左值又是右值。当我们将类似 x += 3 的表达式展开为 x = x + 3 的表达式时，可以看到这一点。

左值（lvalue）可以被认为是值的目的地，右值（rvalue）可以被认为是值的来源。

3.2 类型转换

我们已经讨论了若干算术数据类型、若干运算符和若干表达式，但是需要记住，运算符对数据进行操作。在获取算术运算的返回值和副作用（如果有的话）之前，我们需要考虑如下两个问题。

1）当对诸如布尔或者字符等非算术数据类型应用算术运算符时会发生什么？例如，如果把两个布尔数据项相加，或者将两个字符相乘，结果会如何？

2）使用二元运算符时，如果操作数的类型不同，会发生什么情况？返回值的类型是什么？

为了回答这两个问题，我们需要讨论两个不同的过程：隐式类型转换（implicit type conversion）和显式类型转换（explicit type conversion）。在第一个过程中，数据类型被隐式地转换以满足要求；在第二个过程中，我们强制进行**类型转换**。

C++ 语言提供了一种测试任何表达式的类型的方法，如下所示：

```
typeid (expression).name()
```

其中，expression 是需要知道其类型的表达式。该方法返回类型的缩写（I 表示 int，D 表示 double，依此类推）。要使用此方法，我们需要包含 <typeinfo> 头文件。

3.2.1　隐式类型转换

每当在没有定义某种运算操作的数据类型上使用该运算操作时，C++ 编译器在给出结果之前会执行**隐式类型转换**。隐式类型转换意味着将操作数的类型更改为可以应用该运算进行操作的那种类型。这可以通过如下两个步骤完成：隐式类型提升（将操作数的类型提升到更大的大小）和隐式类型更改（在二元运算操作中将一个操作数的类型更改为另一个操作数的类型，使两者成为同一种类型）。我们分别讨论这两个步骤。

隐式类型转换由编译器自动完成。

隐式类型提升

隐式类型提升（implicit type promotion）被自动应用于任何操作数，以使其适合算术运算。这样做有两个原因。首先，操作数的类型（例如，布尔和字符）不适用于算术运算。其次，没有为类型（例如，short 和 float）定义算术运算符，因为如果我们对 short 或者 float 值应用算术运算符，结果可能超出 short 或者 float 的取值范围（考虑将两个数据项相乘）。为了避免这些问题，编译器对隐式类型提升应用了五条规则，如表 3-2 所示。

表 3-2　隐式类型提升

规则	原始类型	提升后的类型	规则	原始类型	提升后的类型
1	bool	int	4	unsigned short	unsigned int
2	char	int	5	float	double
3	short	int			

【例 3-8】 程序 3-8 演示了发生隐式类型提升的情况：把数据类型 bool、char、short 提升到数据类型 int，以及把 float 数据类型提升到 double 数据类型。请注意针对 bool、char、short 和 float，没有定义任何算术运算操作。因此，如果针对这些数据类型执行算术运算操作（如本例中的加法运算），则意味着会执行隐式类型提升。

程序 3-8　数据类型的隐式提升

```
1  /***************************************************************
2   * 测试没有定义任何算术运算操作的类型的隐式类型转换              *
```

```
3   * 包括：bool、char、short 和 float                              *
4   ****************************************************************/
5   #include <iostream>
6   #include <typeinfo>
7   using namespace std;
8
9   int main ( )
10  {
11    // 声明
12    bool x = true;
13    char y = 'A';
14    short z = 14;
15    float t = 24.5;
16    // 从 bool 到 int 的类型转换
17    cout << "Type of x + 100: " << typeid (x + 100).name() << endl;
18    cout << "Value of x + 100: " << x + 100 << endl;
19    // 从 char 到 int 的类型转换
20    cout << "Type of y + 1000: " << typeid (y + 1000).name() << endl;
21    cout << "Value of y + 1000: " << y + 1000 << endl;
22    // 从 short 到 int 的类型转换
23    cout << "Type of z * 100: " << typeid (z * 10).name() << endl;
24    cout << "Value of z * 100: " << z * 100 << endl;
25    // 从 float 到 double 的类型转换
26    cout << "Type of t + 15000.2: " << typeid (t + 15000.2).name() << endl;
27    cout << "Value of t + 15000.2: " << t + 15000.2;
28    return 0;
29  }
```

```
运行结果：
Type of x + 100: i                    // 类型是整数 int
Value of x + 100: 101
Type of y + 1000: i                   // 类型是整数 int
Value of y + 1000: 1065
Type of z * 100: i                    // 类型是整数 int
Value of z * 100: 1400
Type of t + 15000.2: d                // 类型是双精度数 double
Value of t + 15000.2: 15024.7
```

隐式类型更改

根据前面讨论的规则，在隐式提升每个操作数的类型后，编译器可能执行隐式类型更改，也可能不执行隐式类型更改。**隐式类型更改**（implicit type change）发生在两个操作数具有不同数据类型的时候。数据类型更改后，两个操作数具有相同的数据类型。如果运算符是一元运算符，则不需要隐式转换；如果运算符是二元运算符，则可能需要隐式类型更改，以使两个操作数的类型相同。我们先讨论两种情况。

没有副作用的表达式。这种情况下，需要根据图 3-7 中定义的类型层次结构，将层次结构中较低级别（较小）的操作数转换为层次结构中较高级别的类型。请注意，前面已提升的类型不包含在此列表中，例如 bool、char、short、unsigned short 和 float 类型的数据项不在此列表中，因为它们已经如前所述进行了隐式类型提升。

接着编译器采用的规则如下：先查找具有更高级别的操作数，并将另一个操作数转换为该级别。

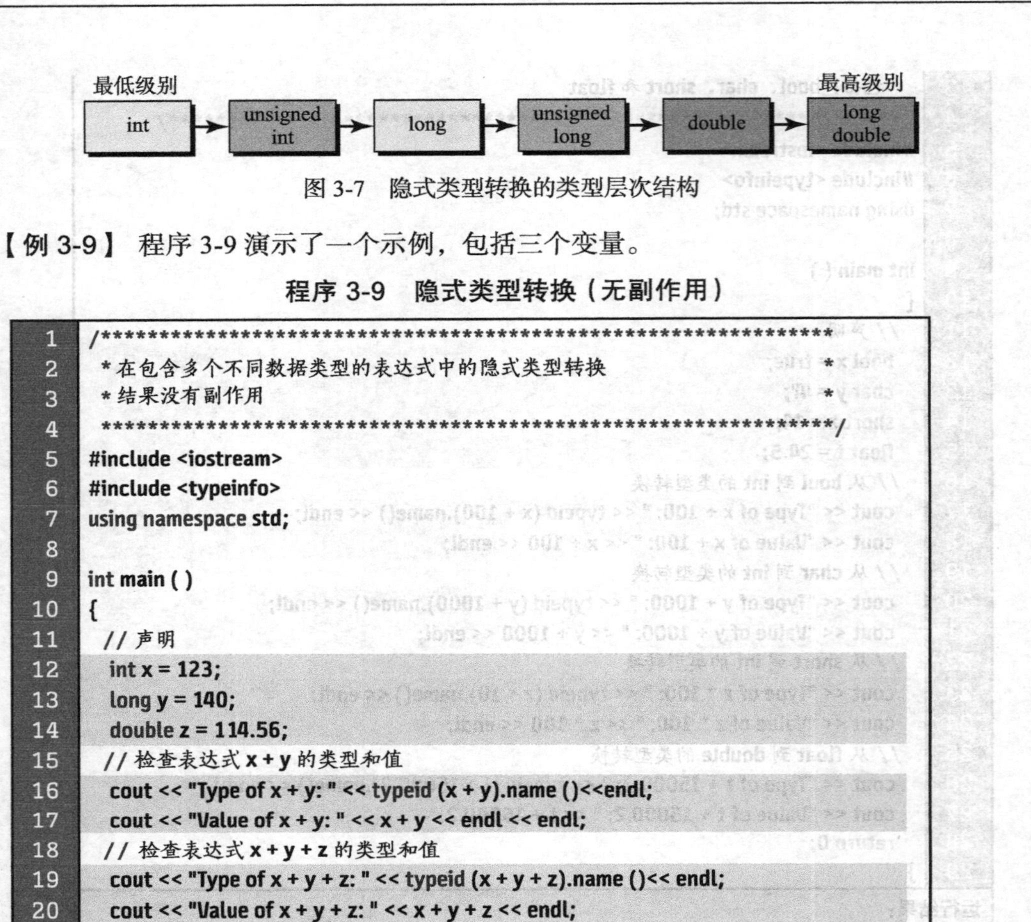

图 3-7　隐式类型转换的类型层次结构

【例 3-9】 程序 3-9 演示了一个示例，包括三个变量。

程序 3-9　隐式类型转换（无副作用）

```
1   /***************************************************************
2    * 在包含多个不同数据类型的表达式中的隐式类型转换              *
3    * 结果没有副作用                                              *
4    ***************************************************************/
5   #include <iostream>
6   #include <typeinfo>
7   using namespace std;
8
9   int main ( )
10  {
11    // 声明
12    int x = 123;
13    long y = 140;
14    double z = 114.56;
15    // 检查表达式 x + y 的类型和值
16    cout << "Type of x + y: " << typeid (x + y).name ()<<endl;
17    cout << "Value of x + y: " << x + y << endl << endl;
18    // 检查表达式 x + y + z 的类型和值
19    cout << "Type of x + y + z: " << typeid (x + y + z).name ()<< endl;
20    cout << "Value of x + y + z: " << x + y + z << endl;
21    return 0;
22  }
```

```
运行结果:
Type of x + y: l                  // 类型是长整数 long
Value of x + y: 263

Type of x + y + z: d              // 类型是双精度数 double
Value of x + y + z: 377.56
```

带副作用的表达式。在某些操作或者行为中，如果将某个值存储到某个预定义类型的目标，我们无法更改该目标的类型，因为它事先已被定义。例如，当我们将一个值赋给不同类型的变量时，就会发生这种情况。在这些情况下，编译器执行隐式类型更改。编译器更改源类型以适应目标类型。例如，如果我们将浮点值赋值给一个整数变量，编译器将截断源值并将整数部分赋给目标变量。另外，如果我们试图将整数值赋值给浮点类型的变量，那么编译器会增加小数部分（0 值），使其成为浮点值。

【例 3-10】 程序 3-10 演示了如何将浮点值截断为整数，并存储到 int 类型变量中，以及如何将整数值更改为浮点类型，并存储到 double 类型变量中。

程序 3-10　隐式类型转换（带副作用）

```
1   /***************************************************************
2    * 检查在包含多个不同数据类型的表达式中的隐式类型转换          *
```

```
3  *******************************************************************/
4  #include <iostream>
5  #include <typeinfo>
6  using namespace std;
7
8  int main ( )
9  {
10   // 声明
11   int x;
12   double y;
13   // 赋值
14   x = 23.67;
15   y = 130;
16   // 检查 x 的类型和值
17   cout << "Type of x = 23.67: " << typeid (x = 23.67).name ()<<endl;
18   cout << "Value of x after assignment: " << x << endl << endl;
19   // 检查 y 的类型和值
20   cout << "Type of y = 130: " << typeid (y = 130).name ()<<endl;
21   cout << "Value of y after assignment: " << y << endl;
22   return 0;
23 }
```

```
运行结果:
Type of x = 23.67: i                    // 类型是整数 int
Value of x after assignment: 23

Type of y = 130: d                      // 类型是双精度数 double
Value of y after assignment: 130
```

注意，y 的值打印输出为 130，但实际上是 130.0。我们需要一个操作符来显示小数部分（将在本章后面讨论）。

3.2.2 显式类型转换（强制转换）

有时我们需要或者希望显式地更改操作数的类型。这可以使用**显式类型转换**（explicit type conversion)，一种被称为**强制转换**（casting）的过程来完成。在 C++ 中，强制转换可以通过多种方式来实现，但是我们只阐述其中一种方式：static_cast。

```
static_cast <type> (expression)
```

【例 3-11】 程序 3-11 演示了隐式类型转换和显式类型转换之间的区别。程序中包含两个不同类型的变量。我们可以把这两个变量的值相加，从而允许隐式类型转换，或者使用显式类型转换。在第一种情况下，编译器将 y 的值更改为双精度 double；在第二种情况下，我们强制将 x 的值更改为整数 int。注意，在隐式类型转换中，y 的值被改为 30.0，加上 23.56，结果是 53.56。在显式类型转换中，23.56 的值被改为 23，加上 30，结果为 53。

程序 3-11 显式类型转换

```
1  /*******************************************************************
2   * 比较表达式中的显式类型转换和隐式类型转换                          *
3   *******************************************************************/
4  #include <iostream>
```

```
5   using namespace std;
6
7   int main ( )
8   {
9     // 声明
10    double x = 23.56;
11    int y = 30;
12    // 允许隐式类型转换
13    cout << "Without casting: " << x + y <<endl;
14    // 强制显式类型转换
15    cout << "With casting: " << static_cast <int> (x) + y;
16    return 0;
17  }
```

运行结果:

```
Without casting: 53.56
With casting: 53
```

3.3 表达式的求值顺序

在上述两个小节中，我们讨论了一些表达式类型。读者可能会提出疑问，如果我们有一个包含多个运算符的复杂表达式，结果会如何呢？计算机如何计算表达式？例如，下面的表达式由四个运算符组成，其结果值是多少？

```
3 + 4 * 7 / 22 - 8
```

要回答这个问题，我们需要考虑运算符的两个属性：优先级和结合性。附录C给出了C++中使用的所有运算符和表达式的列表，但是考虑到本章的目的，我们只需要部分运算符和表达式的列表，如表3-1所示。

3.3.1 优先级

当我们有一个由具有不同优先级的几个简单表达式组成的复杂表达式时，需要使用以下步骤来计算复杂表达式的结果值：

1）首先对最高优先级的简单表达式求值，并用结果值替换它。结果生成一个新的表达式。

2）重复步骤1，直到整个表达式被求值。

【例3-12】 程序3-12演示了通过C++来获得表达式5 + 7 * 4的求值结果。程序将同时显示结果和求值步骤。注意，乘法的优先级是14，而加法的优先级是13。

程序3-12 对一个简单的表达式求值

```
1   /*********************************************************************
2    * 对一个包含两个不同级别优先级的简单表达式求值                              *
3    *********************************************************************/
4   #include <iostream>
5   using namespace std;
6
7   int main ( )
8   {
```

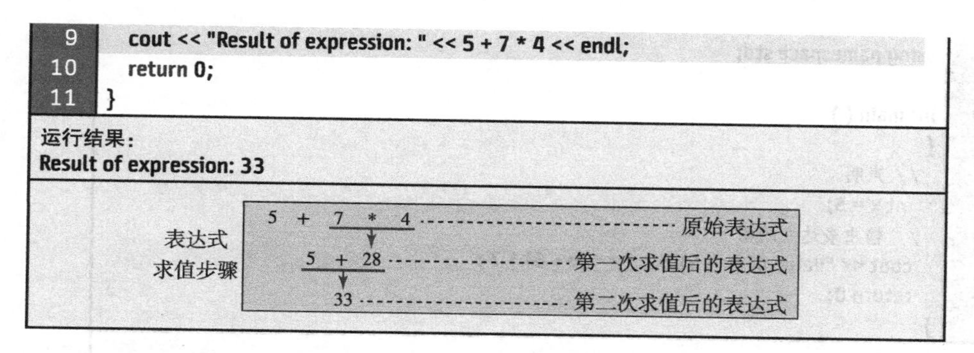

```
9   cout << "Result of expression: " << 5 + 7 * 4 << endl;
10  return 0;
11 }
```

运行结果：
Result of expression: 33

【例 3-13】 程序 3-13 演示了我们如何对复杂表达式 result = 5 − 15 % 4 求值，它由三个运算符组成：赋值运算符、减法运算符和余数运算符，每个运算符都具有不同的优先级。

程序 3-13　对一个复杂的表达式求值

```
1  /***************************************************************
2   * 对一个包含三个不同级别优先级的表达式求值                      *
3   ***************************************************************/
4  #include <iostream>
5  using namespace std;
6
7  int main ( )
8  {
9    // 声明一个变量
10   int result;
11   // 对表达式求值，并把结果存储到变量中
12   result = 5 - 15 % 4;
13   // 输出存储在变量中的值
14   cout << "The value stored in result: " << result;
15   return 0;
16 }
```

运行结果：
The value stored in result: 2

表达式求值步骤：

```
result = 5 - 15 % 4 -------------------- 原始表达式
result = 5 - 3 ------------------------- 第一次求值后的表达式
结果 2 <-- 副作用 -- result = 2 --------- 第二次求值后的表达式
2 -------------------------------------- 返回值
```

【例 3-14】 现在假设我们需要有目的地改变一个简单表达式的优先级。例如，我们需要将字面量 6 加到变量 x 的值中，然后将结果乘以 7。这里乘法的优先级高于加法，但我们需要先做加法运算。解决方法是使用括号。如前所述，一对括号创建了一个优先级最高的基本表达式。程序 3-14 演示了如何编写和求值表达式。

程序 3-14　带括号的简单表达式

```
1  /***************************************************************
2   * 对一个带括号的简单表达式求值                                  *
3   ***************************************************************/
4  #include <iostream>
```

```
5   using namespace std;
6
7   int main ( )
8   {
9     // 声明
10    int x = 5;
11    // 输出表达式的值
12    cout << "Value of (x + 6) * 7: " << (x + 6) * 7 ;
13    return 0;
14  }
```

运行结果:

Value of (x + 6) * 7: 77

表达式求值步骤

```
(x + 6) * 7 ---------------- 原始表达式
   11 * 7 ----------------- 第一次求值后的表达式
     77 ------------------- 第二次求值后的表达式
```

【例 3-15】 作为另一个例子，假设 x 和 y 的原始值分别为 8 和 10。程序 3-15 演示了如何对以下表达式求值后获得 y 的值：y *= x + 5。注意，当我们将复合赋值扩展为简单赋值时，复合赋值运算符右侧的表达式被视为基本表达式。换句话说，复合赋值实际上等价于 y *= (x+5)，它被扩展为 y = y * (x + 5)。

程序 3-15　带副作用的表达式

```
1   /***************************************************************
2    * 对一个带副作用的简单表达式求值
3    ***************************************************************/
4   #include <iostream>
5   using namespace std;
6
7   int main ( )
8   {
9     // 变量的声明和初始化
10    int x = 8;
11    int y = 10;
12    // 赋值
13    y *= x + 5;
14    // 输出变量 y 的值
15    cout << "Value of y: " << y ;
16    return 0;
17  }
```

运行结果:

Value of y: 130

表达式求值步骤

```
        y *= x + 5 ---------------- 原始表达式
        y = y * (x + 5) ----------- 解释后的表达式
        y = y * 13 ---------------- 对括号求值后的表达式
      副作用
y [130] <---- y = 130 ------------- 对乘法求值后的表达式
```

3.3.2 结合性

在前面的示例中，我们使用的复杂表达式只包含某个表达式类型（乘法类表达式、加法类表达式或者赋值表达式）的一个实例。如果一个复杂表达式包含多个具有相同优先级的表达式（例如，两个乘法类表达式），结果会如何呢？在这种情况下，我们需要使用简单表达式的结合性。运算符的结合性可以是从左到右或者从右到左（参见表 3-1）。当需要使用某个表达式类型的多个实例来对特定表达式求值时，我们需要根据表达式的结合性来对表达式求值。

【例 3-16】 让我们计算一个包含两个乘法类运算符和两个加法类运算符的表达式：5 − 30 / 4 * 8 + 10。程序 3-16 演示了 C++ 如何对该表达式求值。

程序 3-16 优先级和结合性

```
1  /**************************************************************
2   * 对涉及优先级和结合性的表达式求值                              *
3   **************************************************************/
4  #include <iostream>
5  using namespace std;
6
7  int main ( )
8  {
9    cout << "Value of expression: " << 5 - 30 / 4 * 8 + 10 ;
10   return 0;
11 }
```

运行结果：
Value of expression: −41

表达式求值步骤

```
5 - 30 / 4 * 8 + 10 ---------- 原始表达式
5 - 7 * 8 + 10 --------------- 对除法求值后的表达式
5 - 56 + 10 ------------------ 对乘法求值后的表达式
-51 + 10 --------------------- 对减法求值后的表达式
-41 -------------------------- 对加法求值后的表达式
```

在这个复杂的表达式中，因为乘法类表达式的优先级高，所以首先对它们进行计算。此外，由于它们的结合性（从左到右），所以从左侧开始求值。

【例 3-17】 程序 3-17 演示了系统如何对包含两个相同优先级的运算符（复合赋值运算符）的表达式求值，其结合性是从右到左。

程序 3-17 具有相同优先级的表达式

```
1  /**************************************************************
2   * 对具有从右到左的结合性的表达式求值                            *
3   **************************************************************/
4  #include <iostream>
5  using namespace std;
6
7  int main ( )
8  {
9    // 变量的声明和初始化
10   int x = 10;
```

```
11    int y = 20;
12    // 赋值
13    y += x *= 40;
14    // 输出变量 x 和 y 的值
15    cout << "Value of x: " << x << endl;
16    cout << "Value of y: " << y;
17    return 0;
18  }
```

运行结果：
Value of x: 400
Value of y: 420

表达式求值步骤

y += x *= 40 ---------- 原始表达式

y += x *= 40 —副作用→ x 400 ---------- 第一次求值后的表达式

y 420 ←副作用— y += 400 ---------- 第二次求值后的表达式

3.4 上溢和下溢

在编写更多程序之前，我们先讨论所有程序员都需要理解的一个问题：上溢和下溢。在现实生活中，有可能在纸上写下一个很大的数值。但是，每个计算机系统中数据类型所占的内存分配是有限的。如果我们试图存储一个大于某种类型的最大值或者小于其最小值的数值，会发生什么情况？其结果就是**上溢**或者**下溢**。C++ 在编译过程中不会产生错误消息，尽管可能会给出警告。所得结果并不是我们所期望的。对于整数和浮点数，上溢和下溢行为是不同的（如前所述，其他类型被提升为整数或者浮点数）。

3.4.1 整数的上溢和下溢

如前所述，整数数据类型包括无符号整数和有符号整数。无符号整数和有符号整数的上溢和下溢错误是不同的。下面我们分别讨论这两种情况。

无符号整数的上溢和下溢

无符号整数的范围是从零到正的最大值。我们想知道当尝试使用大于最大值或者小于零的值时会发生什么。在第一种情况下，会出现向上溢出；在第二种情况下，会出现向下溢出。图 3-8 显示了这两种情况。在这两种情况下，值都被回绕到有效范围。循环标记更好地显示出上溢和下溢的效果。加法按顺时针方向，减法按逆时针方向。

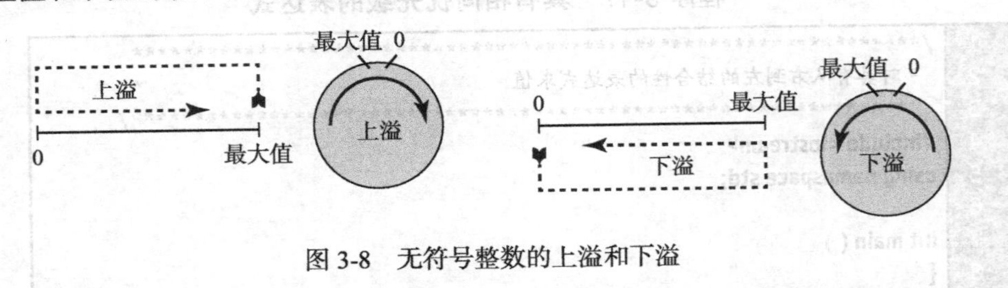

图 3-8　无符号整数的上溢和下溢

程序 3-18 演示了在无符号整数类型中上溢和下溢的运算结果。我们将变量（num1）初始化为最大值，然后尝试把它加上 1。我们将另一个变量（num2）初始化为零，然后尝试把它减去 1。输出结果显示发生了回绕现象。由于最大值和最小值取决于系统，所以我们使用第 11 行和第 12 行中的库函数来设置这些值。

程序 3-18 测试无符号整数的上溢和下溢

```
1  /***************************************************************
2   * 本程序用于测试无符号整数的上溢和下溢                          *
3   ***************************************************************/
4  #include <iostream>
5  #include <limits>
6  using namespace std;
7
8  int main ( )
9  {
10   // 创建两个分别为最大值和最小值的无符号整数
11   unsigned int num1 = numeric_limits <unsigned int> :: max();
12   unsigned int num2 = numeric_limits <unsigned int> :: min();
13   // 打印最大值和最小值
14   cout << "The value of maximum unsigned int: " << num1 << endl;
15   cout << "The value of minimum unsigned int: " << num2 << endl;
16   // 强制使得整数溢出
17   num1 += 1;
18   num2 -= 1;
19   // 输出溢出后的值
20   cout << "The value of num1 + 1 after overflow: " << num1 << endl;
21   cout << "The value of num2 - 1 after underflow: " << num2 << endl;
22   return 0;
23 }
```

```
运行结果:
The value of maximum unsigned int: 4294967295
The value of minimum unsigned int: 0
The value of num1 + 1 after overflow: 0
The value of num2 - 1 after underflow: 4294967295
```

有符号整数的上溢和下溢

有符号整数的范围是从负最小值到正最大值。我们想知道当尝试使用大于最大值或者小于最小值的值时会发生什么。在第一种情况下，会出现向上溢出；在第二种情况下，会出现向下溢出。图 3-9 显示了这两种情况。换言之，值被回绕到有效范围。大于正最大值的正值变为负值，小于负最小值的负值变为正值。

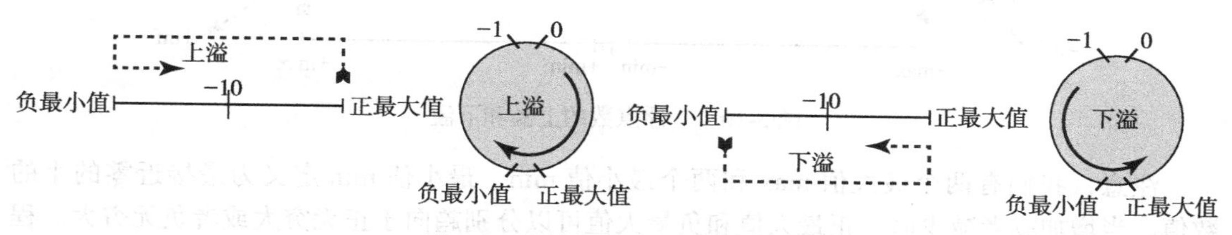

图 3-9 有符号整数的上溢和下溢

程序 3-19 演示了在有符号整数类型中上溢和下溢的运算结果。我们将变量（num1）初始化为最大值，然后尝试把它加上 1。我们将另一个变量（num2）初始化为最小值，然后尝试把它减去 1。输出结果显示发生了回绕现象。由于最大值和最小值取决于系统，所以我们使用第11行和第12行中的库函数来设置这些值。

程序 3-19 测试有符号整数的上溢和下溢

```
1  /*********************************************************************
2   *本程序用于测试有符号整数的上溢和下溢                              *
3   *********************************************************************/
4  #include <iostream>
5  #include <limits>
6  using namespace std;
7
8  int main ( )
9  {
10   // 创建两个分别为最大值和最小值的有符号整数
11   int num1 = numeric_limits <int> :: max();
12   int num2 = numeric_limits <int> :: min();
13   // 打印最大值和最小值
14   cout << "Value of maximum signed int: " << num1 << endl;
15   cout << "Value of minimum signed int: " << num2 << endl;
16   // 强制使得 num1 和 num2 溢出
17   num1 += 1;
18   num2 -= 1;
19   // 输出溢出后的值
20   cout << "The value of num1 + 1 after overflow: "  << num1 << endl;
21   cout << "The value of num2 - 1 after underflow: " << num2 << endl;
22   return 0;
23 }
```

运行结果：

```
The value of maximum signed int: 2147483647
The value of minimum signed int: -2147483648
The value of num1 + 1 after overflow: -2147483648
The value of num2 - 1 after underflow: 2147483647
```

3.4.2 浮点数的上溢和下溢

浮点数值中也存在上溢和下溢，但是这里有两点区别：首先，所有浮点值都是有符号的；第二，当发生上溢和下溢时，没有回绕现象，而是下沉现象。上溢导致下沉到正无穷大(+infinity)；下溢导致下沉到负无穷大（−infinity）。其示意图如图 3-10 所示。

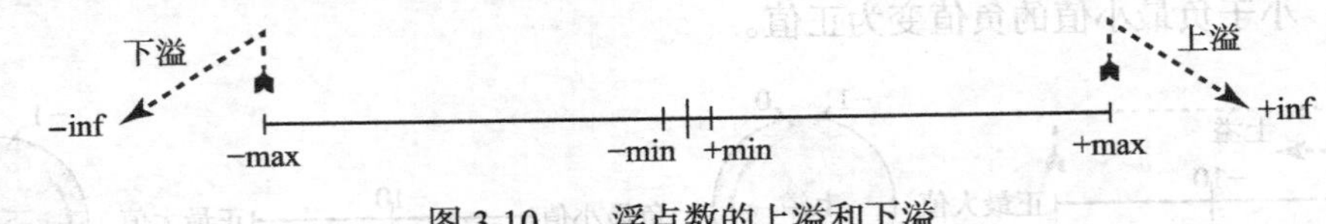

图 3-10 浮点数的上溢和下溢

注意，我们有两个最大值 max 和两个最小值 min。最小值 min 定义为最接近零的小的数值。当增加或者减少时，正最大值和负最大值可以分别趋向于正无穷大或者负无穷大。程序 3-20 将 num1 和 num2 初始化为正最大值和负最大值，然后将每个数值乘以 1000.00，使得其值溢出到正无穷大 +inf 和负无穷大 −inf。由于正最大值和负最大值取决于系统，因此我们使用第 11 行和第 12 行中的库函数来设置这些值。

程序 3-20 双精度数的上溢和下溢

```
/*********************************************************************
 * 本程序用于测试双精度数的上溢和下溢                                 *
 *********************************************************************/
#include <iostream>
#include <limits>
using namespace std;

int main ( )
{
  // 获取双精度数的正最大值和负最大值
  double num1 = +numeric_limits <double> :: max ();
  double num2 = -numeric_limits <double> :: max ();
  // 打印双精度数的正最大值和负最大值
  cout << "The value of maximum double: " << num1 << endl;
  cout << "The value of minimum double: " << num2 << endl;
  // 把两个变量分别乘以 1000.00
  num1 *= 1000.00;
  num2 *= 1000.00;
  // 输出溢出后的值
  cout << "The value of num1 * 1000 after overflow: " << num1 << endl;
  cout << "The value of num2 * 1000 after underflow: " << num2 << endl;
  return 0;
}
```

运行结果：

```
The value of maximum double: 1.79769e+308
The value of minimum double: -1.79769e+308
The value of num1 * 1000 after overflow: INF
The value of num2 * 1000 after underflow: -INF
```

3.5 格式化数据

在前面章节中，我们编写的程序使用标准（默认）格式输入或者输出数据。例如，布尔数据的输入和输出为 0 或者 1。整数值以十进制格式输入和输出（请参见附录 B）。浮点数值以标准格式输入（整数部分后接小数点和小数部分），并根据应用程序的要求输出。有时我们需要改变这种行为，需要根据所喜好的方式格式化数据。解决方法是借助于被称为**操作符**（manipulator）的预定义对象。C++ 包含用于输出数据的操作符和用于输入数据的操作符。下面先讨论用于输出数据的操作符，因为输出操作符更常见，其使用更广泛，然后讨论用于输入数据的操作符。

3.5.1 用于输出的操作符

如前所述，输出操作是使用输出对象完成的。我们使用预定义的输出对象（cout）。到目前为止，我们仅仅把数据项传递给插入运算符（<<），还没有定义在显示或者打印数据时如何格式化数据。输出操作符是可以传递给插入运算符以更改输出行为的对象。本书介绍两种不同类型的输出操作符：不带参数的输出操作符和带一个参数的输出操作符。接下来分别进行讨论。

不带参数的输出操作符

C++ 包含若干不带参数的输出操作符。无参数操作符是 <iostream> 的一部分，这意味着不需要额外的头文件。常见的无参数操作符如表 3-3 所示。

表 3-3　用于输出流的不带参数的操作符

操作符	布尔	字符	整数	浮点
endl	√	√	√	√
noboolalpha, boolalpha	√			
dec, oct, hex			√	
noshowbase, showbase			√	
(none), fixed, scientific				√
noshowpoint, showpoint				√
noshowpos, showpos			√	√
nouppercase, uppercase			√	√
left, right, internal	√	√	√	√

每个类别（endl 除外）的操作符分为两组或者三组。每组中都有一个默认的操纵符。如果没有使用该组中的操作符，则被设置为默认值。在表 3-3 中，我们以**粗体**显示默认值。注意，术语（none）意味着没有默认值：系统将使用基于浮点数值的固定格式或者科学计数法格式，除非我们明确定义了其中的一个格式。除 endl 之外的所有操作符都会更改输出流的状态。当我们选择组中的一个操作符时，输出流使用该操作符，直到使用不同的操作符更改其状态。这意味着我们可以将其中一个操作符应用于所有输出数据，直到需要更改该操作符为止。

除 endl 之外的所有操作符都会更改输出流的状态。

endl 操作符。我们从本书的开头就使用了这个操作符。其设计目的是在数据结尾添加一个 '\n' 字符，以强制下一个输出移到下一行。它是唯一不更改流状态的无参数操作符。我们每次需要 endl 操作符时都必须重复指定。

用于显示布尔字面量的操作符（noboolalpha, boolalpha）。表 3-3 中的第一组操作符只能用于布尔数据类型。noboolalpha 操作符将布尔值作为整数（0 或者 1）输出，boolalpha 操作符将布尔值作为字面量（false 或者 true）输出。程序 3-21 演示了其测试代码。注意，第 17 行并不需要操作符，因为第 16 行中已经改变了输出状态。

程序 3-21　测试 boolalpha 操作符

```
1   /***************************************************************
2     * 本程序用于测试逻辑值的 boolalpha 操作符                          *
3     ***************************************************************/
4   #include <iostream>
5   using namespace std;
6
7   int main ( )
8   {
9     // 声明变量
10      bool x = true;
11      bool y = false;
```

```
12   // 测试不使用操作符的值
13   cout << "Value of x using default: " << x << endl;
14   cout << "Value of y using default: " << y << endl;
15   // 测试使用操作符的值
16   cout << "Value of x using manipulator: " << boolalpha << x << endl;
17   cout << "Value of y: " << y;
18   return 0;
19 }
```

运行结果：

```
Value of x using default: 1
Value of y using default: 0
Value of x using manipulator: true
Value of y: false
```

用于显示不同基数的操作符（dec, oct, hex）。尽管所有整数都以二进制基数（请参见附录B）存储在计算机中，但是在程序中也有可能需要将其输出为三种不同格式之一：dec（基数为10）、oct（基数为8）或者hex（基数为16）。默认值是dec（十进制）。显示的结果不会影响计算机中的大小、符号或者值，这只是为了程序员的方便而已。程序员可以选择使用最方便的数制格式。例如，我们可以将整数1237打印为1237（十进制）、2325（八进制）或者4D5（十六进制）。

用于显示基数前缀的操作符（noshowbase, showbase）。当我们以不同的基数输出一个整数时，可以选择不显示数值的基数（默认值），也可以显示数值的基数。基数显示为前缀（十进制没有前缀，八进制前缀为0，十六进制前缀为0x）。

【例3-18】 程序3-22演示了如何使用操作符把一个整数转换为不同基数的整数，结果包括显示基数前缀的输出以及不显示基数前缀的输出。

程序3-22 测试用于整数基数的操作符

```
1  /***************************************************************
2   * 本程序用于测试以不同的基数输出数据                              *
3   *(十进制、八进制、十六进制)                                     *
4   ***************************************************************/
5  #include <iostream>
6  using namespace std;
7
8  int main ( )
9  {
10   // 声明变量x
11   int x = 1237;
12   // 把x以三种不同基数输出，并且不显示基数前缀
13   cout << "x in decimal: " << x << endl;
14   cout << "x in octal: " << oct << x << endl;
15   cout << "x in hexadecimal: " << hex << x << endl << endl;
16   // 把x以三种不同基数输出，并且显示基数前缀
17   cout << "x in decimal: " << x << endl;
18   cout << "x in octal: " << showbase << oct << x << endl;
19   cout << "x in hexadecimal: " << showbase << hex << x;
20   return 0;
21 }
```

```
运行结果:
x in decimal: 1237
x in octal: 2325
x in hexadecimal: 4d5

x in decimal: 1237
x in octal: 02325              // 前缀 0 表示该数值为八进制
x in hexadecimal: 0x4d5        // 前缀 0x 表示该数值为十六进制
```

用于显示固定格式或者科学计数法格式的操作符。C++ 用于显示浮点类型的值的方式有两种：固定格式或者科学计数法格式。如图 3-11 所示。

符号	整数部分	•	小数部分

固定格式

符号	整数部分固定为一位数	e 或 E	指数符号	指数

科学计数法格式

图 3-11　用于输出固定格式或者科学计数法格式的操作符

在**固定格式**（fixed format）中，浮点数值显示为整数部分和小数部分，两者由小数点分隔，例如“dddd.ddd”，其中每个 d 都表示一个数字。如果是负数，则前面加一个负号。在**科学计数法格式**（scientific format）中，如果数值特别大，则以固定格式的数值乘以指数（10^n），如果数值特别小，则乘以指数（10^{-n}）。

用于显示小数点的操作符。如果浮点数的小数部分为零，则 C++ 默认不打印十进制小数点。我们可以使用 showpoint 操作符强制输出小数部分为 0 的十进制小数点。

【例 3-19】 程序 3-23 演示了用于浮点类型的操作符的使用方法。

程序 3-23　测试浮点数的操作符

```
1  /***************************************************************
2   * 本程序用于测试一些用于浮点类型的操作符                      *
3   ***************************************************************/
4  #include <iostream>
5  using namespace std;
6
7  int main ( )
8  {
9    // 声明变量
10   double x = 1237;
11   double y = 12376745.5623;
12   //使用固定格式（默认选项）操作符和 showpoint 操作符
13   cout << "x in fixed_point format: " << x << endl;
14   cout << "x in fixed_point format: " << showpoint << x << endl;
15   //使用科学计数法格式的操作符
16   cout << "y in scientific format: " << y << scientific;
17   return 0;
18 }
```

```
运行结果:
x in fixed_point format: 1237            // 没有使用 showpoint 操作符
x in fixed_point format: 1237.00         // 使用了 showpoint 操作符
y in scientific format: 1.23767e+007
```

用于显示正号的操作符。如果数值为正，则 C++ 默认不显示正号（+）。负号则总是显示。要强制使得正数打印正号，可以使用 showpos 操作符。

用于显示大写字母的操作符。如前所述，整数和浮点值有时会包括十六进制表示法中的字母字符（a、b、c、d、e、f）以及表示十六进制前缀的 x 和科学计数法中的 e。默认情况下，这些字符以小写形式打印。如果我们希望它们以大写形式打印，则可以使用 uppercase 操作符。

用于对齐字段中的数值的操作符。稍后我们将讨论如何定义字段的大小（占用的字符数），以便使用带参数的操作符打印值。一旦大小确定后，我们需要决定如何调整字段中的值和符号（如果有的话）。C++ 使用了三种格式，如图 3-12 所示。

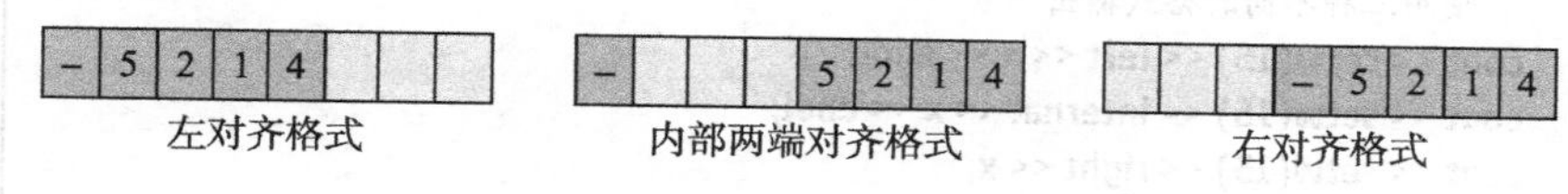

图 3-12　调整字段中的值和符号

在左对齐格式（left，默认值）中，数值和符号位于字段的左侧，字段其余部分用填充符填充（稍后我们将对此进行解释）。在内部两端对齐格式（internal）中，符号占据字段的最左边部分，数值占据最右边部分，其余部分用填充符填充。在右对齐格式（right）中，符号和数值占据最右边的部分，而左边的部分用填充符填充。请注意，此操作符不会更改流的状态。

带参数的操作符

我们只讨论三个采用一个参数（整数或者字符）的操作符，如表 3-4 所示。要使用这些操作符，需要在程序中包含 <iomanip> 头。它们并没有在 <iostream> 中定义。

表 3-4　用于输出流的带一个参数的操作符

操作符	布尔	字符	整数	浮点
setprecision(n)				√
setw(n)	√	√	√	√
setfill(ch)	√	√	√	√

操作符 setprecision(n)。此操作符仅用于固定格式（非科学计数法）的浮点值。括号内的整数（*n*）定义小数点后的位数。

操作符 setw(n)。此操作符用于定义希望值所占据的字段的大小。注意，对于浮点数据类型，我们需要考虑整数部分、小数点和小数部分所占据的空间。请注意，此操作符不会更改流的状态。它需要为每个值单独设置。

操作符 setfill(ch)。当要显示的值的实际大小小于 setw(n) 定义的大小时，此操作符决定如何使用填充符填充字段。括号内的参数是用作填充的字符字面量。前面讨论过的对齐操作符（left、internal 和 right）可以确定填充的位置。

> 除了 setw 之外，所有带参数的操作符都会更改输出流的状态。我们必须在需要的地方重复使用 setw。

【例 3-20】 程序 3-24 演示了如何测试带参数的操作符。

程序 3-24　使用带参数的操作符

```
1  /*****************************************************************
2   * 本程序用于测试用于浮点类型的其他操作符                           *
```

```
3   ***********************************************************/
4   #include <iostream>
5   #include <iomanip>
6   using namespace std;
7
8   int main ( )
9   {
10    // 声明变量
11    double x = 1237234.1235;
12    // 应用常用的格式
13    cout << fixed << setprecision(2) << showpos << setfill('*');
14    // 使用三种不同的格式输出
15    cout << setw(15) << left << x << endl;
16    cout << setw(15) << internal << x << endl;
17    cout << setw(15) << right << x;
18    return 0;
19  }
```

运行结果：

```
+1237234.12****
+****1237234.12
****+1237234.12
```

3.5.2 用于输入的操作符

有一些操作符可以与输入一起使用。本节仅讨论其中的两个，剩余的将在第16章中讨论。本节讨论的用于输入的操作符如表3-5所示。请注意，用于输入的操作符与用于输出流的前两组操作符相同，但它们用于输入流。在第一种情况下，我们可以将布尔值输入为false或者true，而不是0或者1。在第二种情况下，我们可以用八进制或者十六进制输入整数值。

表3-5　用于输入流的操作符

操作符	布尔	字符	整数	浮点
noboolalpha, boolalpha	√			
dec, oct, hex			√	

【例3-21】 程序3-25演示了如何将布尔值输入为false或者true，然后输出为0或者1。

程序3-25　测试布尔值的输入

```
1   /***********************************************************
2    * 本程序用于测试将布尔值输入为 false 或者 true              *
3    ***********************************************************/
4   #include <iostream>
5   using namespace std;
6
7   int main ( )
8   {
9     // 声明变量
10    bool flag;
11    // 使用操作符输入值
12    cout << "Enter true or false for flag: ";
13    cin >> boolalpha >> flag;
14    // 输出值
15    cout << flag ;
```

```
16    return 0;
17  }
```

运行结果：

```
Enter true or false for flag: false
0
```

运行结果：

```
Enter true or false for flag: true
1
```

【例 3-22】 程序 3-26 演示了如何使用十进制、八进制和十六进制输入三个变量的值，然后把这三个值输出为十进制值。

程序 3-26　使用不同基数输入整数

```
1   /***************************************************************
2    * 本程序使用不同基数输入整数                                     *
3    ***************************************************************/
4   #include <iostream>
5   using namespace std;
6
7   int main ( )
8   {
9     // 声明变量
10    int num1, num2, num3;
11    // 使用十进制（无操作符）输入第一个数值
12    cout << "Enter the first number in decimal: ";
13    cin >> num1;
14    // 使用八进制输入第二个数值
15    cout << "Enter the second number in octal: ";
16    cin >> oct >> num2;
17    //使用十六进制输入第三个数值
18    cout << "Enter the third number in hexadecimal: ";
19    cin >> hex >> num3;
20    //输出值
21    cout << num1 << endl;
22    cout << num2 << endl;
23    cout << num3;
24    return 0;
25  }
```

运行结果：

```
Enter the first number in decimal: 124
Enter the second number in octal: 76
Enter the third number in hexadecimal: 2ab
124
62
683
```

3.6 语句

我们已经讨论了表达式，但是在 C++ 语言中创建一个程序需要使用语句。一个 C++ 程序是由一系列**语句**（statement）组成的。计算机按照程序规定的顺序执行每一条语句，以实现程序的目标。

如果将C++语言与自然语言（如英语）进行比较，我们会发现C++中的语句与自然语言的句子的作用相同。例如，当阅读英文文章时，句子是一个信息单元，在阅读下一个句子之前，我们必须解释并理解当前的句子。C++程序中的语句是C++运行环境必须在执行下一条语句之前执行的命令。然而，程序设计语言（例如C++）包括一些语句，它们会改变执行顺序，从程序中的一个点移动到另一个点。

C++提供了许多不同类型的语句，每条语句都有一个预定义的任务。本节只讨论其中的一些基本语句，后续章节中将逐步讨论其余语句。

有些语句需要使用英文分号来终止，而有些语句则不需要分号，因为它们在末尾有另一个用作终结符的标记。

有些语句在结尾处需要英文分号作为终结符，但有些语句已经有了内置的终结符。

3.6.1　声明语句

C++中有两个术语必须解释：声明和定义。**声明**（declaration）通过指定实体的类型并给它一个名称（标识符）来引入一个实体。**定义**（definition）则表示为实体分配内存。对于一个复杂的程序，有许多实体需要声明和定义，但目前为止我们只限于变量和常量的声明。

变量的声明

我们在前面章节中讨论了变量的概念。要在程序中使用变量，必须先声明变量。**变量声明**（variable declaration）实际上是声明和定义，除非我们添加一个额外的修饰符（称为extern）来将定义推迟到程序的其他部分。变量声明语句需要英文分号来指示语句终止。

变量声明语句需要英文分号来终止。

单个变量的声明（single declaration）。单个变量的声明为一个变量命名并定义其类型。它还在内存中保留一个物理位置，适合保存所声明类型的数据项。下面的代码片段中声明和定义了三个变量：test、sum和average。注意，每个声明都需要一个英文分号来终止语句。

```
short test;
int sum;
double average;
```

多个变量的声明（multiple declaration）。如果需要声明同一类型的多个变量，则可以进行组合并且只使用一个声明语句。例如，下面的代码片段显示了如何声明三个int类型的变量、两个double类型的变量和一个char类型的变量。注意，需要用英文逗号分隔变量名，但是只需要一个英文分号来终止语句。还请注意，虽然不需要在逗号后添加空格，但我们强烈建议每个逗号后至少有一个空格，以提高程序的可读性。

```
int first, second, third;
double average, mean;
char ch;
```

初始化（initialization）。声明一个变量时，我们为变量指定一个名称，并要求计算机分配相应类型的内存位置。然而，声明并没有规定最初应该存储在内存位置中的内容，但是我们可以在声明变量时初始化该变量，如下所示。

```
int first = 0;
double average = 0.0, mean = 0.0;
char ch = 'a';
```

读者也许会提出疑问：如果没有初始化，那么结果是什么？这需要分两种情况讨论。

全局变量（global variable）。如果变量是全局变量（在任何函数外部声明），则将其初始化为默认值（整数初始化为 0，浮点数初始化为 0.0）。请注意，字符和布尔数据类型是特殊的整数（小整数），因此被初始化为 0。

局部变量（local variable）。如果变量是局部变量（在函数内部声明），则不会初始化该变量，但该变量保留了以前使用时留下的一些垃圾数据。因此，在使用局部变量之前，我们需要将其初始化或者通过其他方法更改存储在其中的垃圾数据。

> 在使用局部变量之前，我们需要在其中存储一个值。

常量的声明

在前面的章节中，我们讨论了常量是一个内存位置，它的值通过初始化设定，并且不能被更改。要使用常量，我们需要声明常量（在这种情况下，声明还意味着定义）。**常量声明**（constant declaration）与变量声明类似，但有如下四个差异。第一，需要在类型前面使用关键字 const。第二，必须在声明常量时对其进行初始化。第三，习惯上使用大写字母命名常量，以将其与程序中的变量区分开来（如果名称涉及多个单词，则单词之间使用下划线分隔）。第四，常量通常在程序的全局区域中（在任何函数之前，包括 main）声明，这使得常量在所有函数中都可见。下面的代码片段是常量声明的一些示例。

```
const int FOOT_TO_INCH = 12;
const double TAX_RATE = 8.5;
const double PI = 3.1415926536;
```

> 常量在声明时必须初始化。

3.6.2 表达式语句

表达式语句（expression statement）是以英文分号结尾（作为结束符）的表达式。我们在前面章节中了解到表达式有一个值，而且可能还有副作用。当在程序中出现表达式语句时，计算机确定其值并执行其副作用。然后该值被丢弃，但副作用会改变计算机的内存状态。这意味着没有副作用的表达式作为语句是无用的，我们需要避免这种情况，尽管编译器不会创建错误信息。

> 计算机对表达式进行求值并执行其副作用，然后该值被丢弃。

【例 3-23】 下面的代码片段是表达式语句。其中一些表达式语句有作用，另外一些则毫无用处。只有最后一条语句会导致编译错误。

```
num = 24;               // 表达式语句
```

```
num *= 10;              // 表达式语句
num = data + 6;         // 表达式语句
num1 + num2;            // 没有作用。因为没有副作用
num1 * 6;               // 没有作用。因为没有副作用
num;                    // 没有作用。因为没有副作用
6;                      // 没有作用。因为没有副作用
cout << "Hello!";       // 输出
cin >> data             // 输入
```

在一些文献中，表达式语句被称为赋值语句。C++ 语言的语法中没有赋值语句，只有表达式语句。我们遵循正式的语言规范。

虽然我们需要将数据输入到程序中，并将结果从程序中发送出来，但是 C++ 中没有输入 / 输出语句。在 C++ 中，输入和输出操作是使用流对象（cin 和 cout）来完成的，如图 3-13 所示。

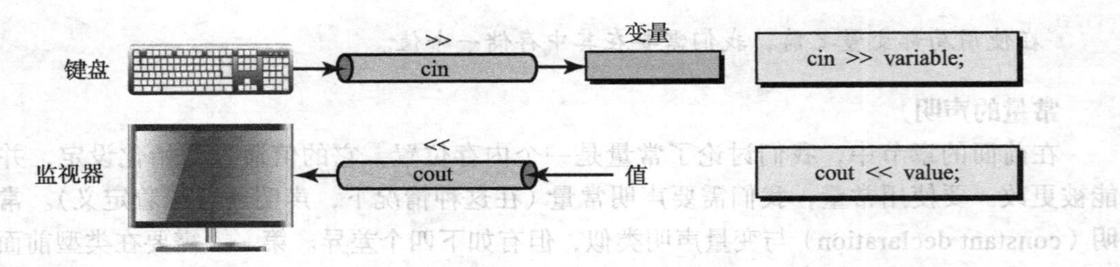

图 3-13　输入输出对象和运算符

流对象可以被认为是将程序连接到物理输入设备（例如键盘）或者物理输出设备（例如监视器）的管道。连接到键盘的流被称为 cin，连接到监视器的流被称为 cout。符号 >>（称为提取运算符）实际上是为 cin 对象定义的运算符，符号 <<（称为插入运算符）实际上是为 cout 对象定义的运算符。

从表面上看，当我们使用输入 / 输出命令时，它看起来像是一种新的语句类型，但事实上并非如此。输入 / 输出命令实际上是一条表达式语句，它的结尾需要一个英文分号。

> 在程序中使用的输入 / 输出命令实际上是表达式语句，需要使用英文分号作为结束符。

3.6.3　空语句

空语句（null statement）是不执行任何操作的语句。我们将在后面的两章中看到 C++ 语法需要语句但不需要副作用的场合。我们将在后续章节中讨论空语句的一些应用，但目前假定我们错误地在表达式的末尾添加了两个英文分号。编译器不会产生错误，因为它认为我们使用了两条语句：一条表达式语句和一条空语句。示例代码片段如下所示。

```
num = 24; ;             // 包括一条表达式语句和一条空语句
```

3.6.4　复合语句

有时我们需要将几条语句视为单条语句。在这种情况下，我们将多条语句括在一对花括号中。可以在一对花括号内组合任意数量的语句（零个或者多个），结果是创建一条称为复

合语句（compound statement）的单条语句，也称为语句块（block）。每当 C++ 语法需要单条语句但实际需要多条语句时，则必须使用复合语句。后续章节中将讨论，有时我们甚至将一条语句放在一对花括号中，以提高程序的可读性并避免混淆。注意，在复合语句的末尾不需要英文分号，右花括号则用作结束符。下面代码片段显示了一条包含两条语句的复合语句。

```
{
  int num = 8;
  cout << num << endl;
}
```

如果在复合语句末尾添加英文分号，会发生什么情况？什么也不会发生。编译器会认为一条复合语句后面跟着一条空语句。

【例 3-24】 在函数头之后，函数体需要一个复合语句，如下所示。

```
int main ()
{
  ...
  return 0;
}
```

3.6.5 返回语句

在程序中使用的另一种语句是**返回语句**（return statement），它可以向调用该函数的实体返回一个值，例如 main() 函数向系统返回 0。返回语句属于将在第 4 章中详细讨论的控制语句类别。

【例 3-25】 在这个例子中，我们用 C++ 编写了一个程序。程序将输入三个整数，计算它们的和，并打印出结果（程序 3-27）。程序中使用了各种不同的语句。

程序 3-27 输入三个整数并输出它们的和

```
1   /***************************************************************
2    * 本程序用于输入三个整数并输出这三个整数之和                      *
3    ***************************************************************/
4   #include <iostream>
5   #include <iomanip>
6   using namespace std;
7
8   int main ( )
9   {
10    // 声明变量
11    int first, second, third, sum;
12    // 提示和输入三个整数
13    cout << "Enter the first integer: ";
14    cin >> first;
15    cout << "Enter the second integer: ";
16    cin >> second;
17    cout << "Enter the third integer: ";
18    cin >> third;
19    // 计算
```

```
20    sum = first + second + third;
21    // 输出
22    cout << "The sum of the three integers is: " << sum;
23    return 0;
24  }
```

运行结果：
```
Enter the first integer: 20
Enter the second integer: 30
Enter the third integer: 10
The sum of the three integers is: 60
```

运行结果：
```
Enter the first integer: 10
Enter the second integer: 8
Enter the third integer: 3
The sum of the three integers is: 21
```

让我们分析一下在这个程序中使用的语句。程序从主函数（main）的定义开始。主函数的主体（第10行到第23行）是一个复合语句。第11行是声明语句。第13行到第18行是表达式语句（输入/输出命令）。第20行是表达式语句（加法运算）。第22行是表达式语句（输出命令）。第23行是返回语句。从第10行到第23行的复合语句由两个分隔符（左花括号和右花括号）之间的多条语句（注释除外）组成。

3.7 程序设计

在本节中，我们将设计并实现四个程序。我们将使用由以下三个步骤组成的软件开发过程。

- **理解问题**。在第一步中，我们讨论是否理解所要解决的问题。
- **开发算法**。在第二步中，我们为要解决的问题开发一个算法。开发算法可以使用若干不同的工具来实现，但在本章中只使用非正式的描述，因为我们编写的程序非常简单。
- **编写代码**。在第三步中，我们使用所开发的算法来编写C++代码。

3.7.1 提取浮点数的整数部分和小数部分

让我们设计并实现一个程序：在给定一个浮点数值的情况下，提取并打印其整数部分和小数部分。

理解问题

给定了一个浮点数，我们要分离其整数部分和小数部分。例如，给定123.78，我们希望输出123和0.78。

开发算法

这种情况下的算法很简单。我们输入一个浮点数，提取它的整数部分和小数部分，然后输出原始数值的整数部分和小数部分。算法如下：

1）输入

　a）输入一个浮点数。

2）处理

　a）提取整数部分。

b）提取小数部分。

3）输出

a）输出原始数值。

b）输出整数部分。

c）输出小数部分。

编写代码

我们已经将程序输入部分、处理部分和输出部分的所有元素进行了形式化，接下来可以编写代码（参见程序 3-28）。

程序 3-28 提取一个浮点数的整数部分和小数部分

```
1  /***************************************************************
2   * 本程序演示了如何提取                                        *
3   * 一个浮点数值的整数部分和小数部分                            *
4   ***************************************************************/
5  #include <iostream>
6  #include <iomanip>
7  using namespace std;
8
9  int main ( )
10 {
11   // 声明变量
12   double number;
13   int intPart;
14   double fractPart;
15   // 输入
16   cout << "Enter a floating-point number: ";
17   cin >> number;
18   // 处理
19   intPart = static_cast <int> (number);
20   fractPart = number - intPart;
21   // 输出
22   cout << fixed << showpoint << setprecision (2);
23   cout << "The original number: " << number << endl;
24   cout << "The integral part: " << intPart << endl;
25   cout << "The fractional part: " << fractPart;
26   return 0;
27 } // main 函数结束
```

运行结果：
```
Enter the floating-point number: 145.72
The original number: 145.72
The integral part: 145
The fractional part: 0.72
```

运行结果：
```
Enter the floating-point number: -546
The original number: -546.00
The integral part: -546
The fractional part: 0.00
```

运行结果：
```
Enter the floating-point number: -0.14
```

```
The original number: -0.14
The integral part: 0
The fractional part: -0.14
```

程序中使用了操作符（fixed、showpoint 和 setprecision）来格式化输出。注意，在每次运行中，我们只输入一个浮点数值。测试我们的程序是否可以正常工作的一种方法是在每次运行中，把整数部分和小数部分相加，然后判断结果是否为原始数值。在第一次运行中，结果为 145 + 0.72 = 145.72。

3.7.2 提取整数的个位数

让我们设计并实现一个程序：给定一个整数，提取并打印该整数的个位数（最右边的那个数字）。

理解问题

给定一个整数，我们要提取该整数的个位数。例如，给定 6759，我们希望输出 9。

开发算法

这种情况下的算法很简单。我们输入一个整数，并提取它的个位数字，然后打印原始整数和个位数。算法如下：

1）输入

a）输入一个整数。

2）处理

a）提取个位数（最右边那位数字）。

3）输出

a）输出原始数值。

b）输出个位数。

编写代码

我们使用定义的输入、处理和输出部分编写程序（程序 3-29）。

程序 3-29　提取给定整数的个位数

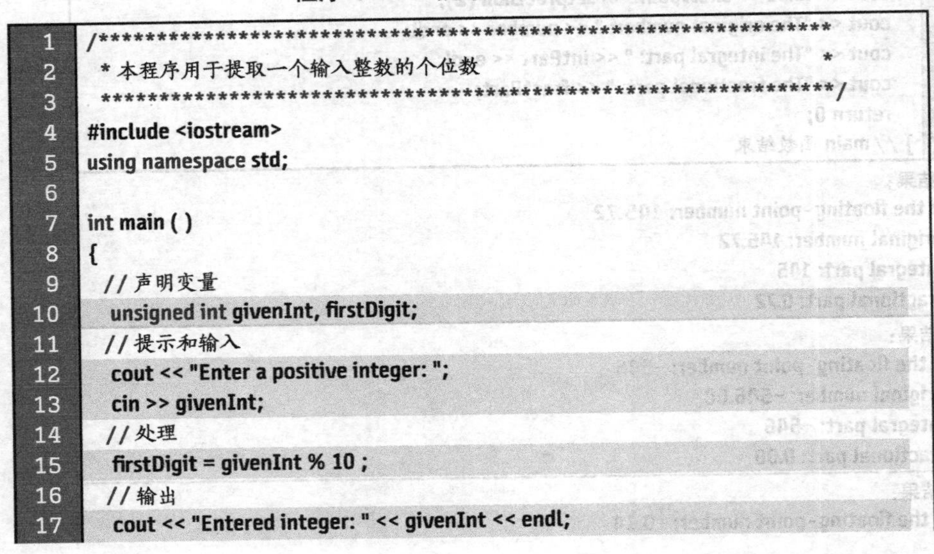

```
1   /***************************************************************
2    * 本程序用于提取一个输入整数的个位数                              *
3    ***************************************************************/
4   #include <iostream>
5   using namespace std;
6
7   int main ( )
8   {
9     // 声明变量
10    unsigned int givenInt, firstDigit;
11    // 提示和输入
12    cout << "Enter a positive integer: ";
13    cin >> givenInt;
14    // 处理
15    firstDigit = givenInt % 10 ;
16    // 输出
17    cout << "Entered integer: " << givenInt << endl;
```

```
18    cout << "Extracted first digit: " << firstDigit << endl;
19    return 0;
20  }
```

运行结果：
Enter a positive integer: 253
Entered integer: 253
Extracted first digit: 3

运行结果：
Enter a positive integer: 45672
Entered integer: 45672
Extracted first digit: 2

3.7.3　把时间分解为时分秒

让我们设计和实现另一个问题：找出在以秒为单位的时间段内有多少小时、多少分钟和多少秒。

理解问题

给定一个表示执行任务耗费的以秒为单位的时间的较大的整数，例如 234 572，我们要计算执行任务耗费了多少小时、多少分钟和多少秒。

开发算法

这种情况下的算法很简单。我们输入一个表示秒的时间数值，并提取它的小时数、分钟数和秒数，然后打印小时数、分钟数和秒数。算法如下：

1）输入

　a）输入一个表示以秒为单位的时间间隔的整数。

2）处理

　a）提取秒数中包含的小时数。

　b）提取剩余秒数中包含的分钟数。

　c）获取剩余的秒数。

3）输出

　a）输出以秒为单位的时间间隔。

　b）输出提取的小时数。

　c）输出提取的分钟数。

　d）输出剩余的秒数。

编写代码

接下来基于上述算法编写程序代码（程序 3-30）。

程序 3-30　把时间分解为时分秒

```
1  /***************************************************************
2   * 本程序用于把以秒为单位的时间间隔                              *
3   * 转换为小时数、分钟数和秒数                                    *
4   ***************************************************************/
5  #include <iostream>
6  using namespace std;
7
8  int main ( )
```

```
9	{
10	  // 声明变量
11	  unsigned long duration, hours, minutes, seconds;
12	  // 提示和输入
13	  cout << "Enter a positive integer for the number of seconds: ";
14	  cin >> duration;
15	  // 处理
16	  hours = duration / 3600L;
17	  minutes = (duration - (hours * 3600L)) / 60L;
18	  seconds = duration - (hours * 3600L) - (minutes * 60);
19	  // 输出
20	  cout << "Given Duration in seconds: " << duration << endl;
21	  cout << "Result: ";
22	  cout << hours << " hours, ";
23	  cout << minutes << " minutes, and ";
24	  cout << seconds << " seconds.";
25	  return 0;
26	} // main 函数结束
```

```
运行结果:
Enter a positive integer for the number of seconds: 4000
Given duration in seconds: 4000
Result: 1 hours, 6 minutes, and 40 seconds.
```

```
运行结果:
Enter a positive integer for the number of seconds: 39250
Given duration in seconds: 39250
Result: 10 hours, 54 minutes, and 10 seconds.
```

我们不能只看结果就判断答案的正确性。我们应该测试一些结果，以确保所设计程序的正确性。例如，我们可以从输出结果反向计算其输入值，如下所示：

```
运行结果1: (1 * 3600) + (6 * 60) + 40 = 3600 + 360 + 40 = 4000（秒）
运行结果2: (10 * 3600) + (54 * 60) + 10 = 36000 + 3240 + 10 = 39250（秒）
```

因为得到的结果值与给定的时间值相同，因此程序工作正常。

3.7.4　计算平均值和偏差

在这个问题中，我们要读取三个整数，计算它们的平均值，并确定每个整数与平均值的偏差。

理解问题

一个数值列表的平均值的计算方法如下：将列表中的每个数值累加在一起，并将结果除以列表的大小。每个数值的偏差表示该数值与平均值的距离，可能为正或者为负。例如，假设有三个数值：10、14 和 15。这些数值之和是 39。平均值是 39/3，即 13。第一个数的偏差为 10 – 13 = – 3，第二个数的偏差为 14 – 13 = 1，第三个数的偏差为 15 – 13 = 2。

开发算法

这种情况下的算法稍微有点复杂。首先输入三个整数，然后计算总和、平均值，接着计算每个数值与平均值的偏差，最后输出总和、平均值、每个数值的偏差。算法如下：

1）输入

　a）输入三个整数。

2）处理

a）把三个数值累加在一起计算总和。

b）把总和除以 3，计算平均值。

c）计算每个数值与平均值的偏差。

3）输出

a）输出总和。

b）输出平均值。

c）输出每个数值的偏差。

编写代码

接下来编写程序（程序 3-31）。

程序 3-31　计算总和、平均值和偏差

```
/***************************************************************
 * 本程序用于输入三个整数，                                      *
 * 计算它们的总和、平均值、每个数值与平均值的偏差。              *
 ***************************************************************/
#include <iostream>
#include <iomanip>
using namespace std;

int main ( )
{
  // 声明变量
  int num1, num2, num3;
  int sum;
  double average;
  double dev1, dev2, dev3;
  // 提示和输入
  cout << "Enter the first integer: ";
  cin >> num1;
  cout << "Enter the second integer: ";
  cin >> num2;
  cout << "Enter the third integer: ";
  cin >> num3;
  // 处理
  sum = num1 + num2 + num3;
  average = static_cast <double> (sum) / 3;
  dev1 = num1 - average;
  dev2 = num2 - average;
  dev3 = num3 - average;
  // 输出
  cout << fixed << setprecision (2) << showpos;
  cout << "Sum of three numbers: " << sum << endl;
  cout << "Average: " << setw(9) << average << endl;
  cout << "Deviation of number 1: " << setw(9) << dev1 << endl;
  cout << "Deviation of number 2: " << setw(9) << dev2 << endl;
  cout << "Deviation of number 3: " << setw(9) << dev3 << endl;
  return 0;
}
```

```
运行结果：
Enter the first integer: 100
Enter the second integer: 101
Enter the third integer: 103
Sum of three numbers: 304
Average:    101.33
Deviation of number 1:    -1.33
Deviation of number 2:    -0.33
Deviation of number 3:    +1.67
```

我们需要确保结果的正确性。如果把每一个偏差加在平均值上，则结果应该是原始的数值。然而，这种计算可能会得出近似值。计算结果如下：

a）101.33 + (−1.33) = 100（第一个数值）。

b）101.33 + (−0.33) = 101（第二个数值）。

c）101.33 + (+1.67) = 103（第三个数值）。

本章小结

表达式是一个具有值的实体，可能还有副作用。表达式可以由一个简单值或者多个值与运算符组合而成。基本表达式仅仅包含一个值，不包含运算符。一元表达式包含一个值和一个运算符。二元表达式包含一个运算符和两个值。三元表达式由两个运算符和三个值组成。表达式求值可能涉及类型转换，包括隐式类型转换和显式类型转换。

为了确定复杂表达式中简单表达式的计算顺序，我们需要考虑运算符的两个属性：优先级和结合性。当表达式具有不同优先级时使用优先级，当表达式的优先级相同时使用结合性。

向上溢出和向下溢出定义了当数值超过变量的范围（太大或者太小）时使用的规则。

有时我们需要格式化数据。C++ 借助于被称为操作符的预定义对象来实现数据格式化，包括用于输出数据的操作符和用于输入数据的操作符。

在 C++ 中，编写一个程序需要语句。C++ 提供了许多不同类型的语句，每条语句都有一个预先定义的任务。本章讨论了五种类型的语句：声明语句、表达式语句、空语句、复合语句和返回语句。

程序设计是一个过程，至少包含以下三个必须严格遵循的步骤：理解程序、开发算法和编写代码。

思考题

1. 计算以下整数表达式的十进制值。指出存在的错误。

```
a. 0723                b. 0491
c. -0241               d. -0412
```

2. 计算以下整数表达式的十进制值。指出存在的错误。

```
a. -0x1A12             b. -0xH21B
c. 0xA2EF              d. 0x4
```

3. 计算以下浮点数表达式的值。指出存在的错误。

```
a. 0.0712F                          b. 6.0712
c. 1.234E2                          d. 1123E-2
```

4. 计算以下表达式的值。

```
3 + 7 * 2              4 * 2 + 7              7 / 4 + 3 * 4
```

5. 计算以下表达式的值。

```
4 / 3 * 6 / 2 + 8 % 3                24 % 5 + 16 % 5 * 4 / 2
```

6. 假设 x = 3、y = 5、z = 6，计算以下表达式的值（假设表达式彼此独立）。

```
x - 3 + y * 2 - z / 2                (x + 2) % (y * y)
```

7. 假设 x = 3、y = 5、z = 6，计算以下表达式的值（假设表达式彼此独立）。

```
(y / 3 + x) * z % x                  y - 2 - x * y
```

8. 假设 x = 4、y = 2，计算以下各表达式求值后 x 的值。

```
x -= y - 2;                          x += y + 2;
```

9. 假设 x = 4、y = 2、z = 5，计算以下各表达式求值后 z 的值。

```
z /= x + y + 4;                      z %= y * 2 + 3;
```

10. 以下各表达式的求值结果是什么？

```
2 + 4            3 % 5;            3.2 * 2;            5 + 3.2;
```

11. 以下各表达式的求值结果是什么？

```
'A' - 2          'B' + 3.2;        4.5 + 'D';          'C' * 2;
```

12. 如果 x 是整数，y 是浮点数，z 是双精度数，则以下各表达式求值后 x、y、z 的值分别是什么？

```
x = 'A'          y = 3             z = 2               z = 2.5 + 2
```

13. 把以下简单赋值语句修改为复合赋值语句。

```
x = x + 4;       y = y - 2;        z = z + 2 - t       t = t  / 5;
```

14. 把以下复合赋值语句修改为简单赋值语句。

```
x += 8           y -= 8 + x        z *= t + 2          t %= x + 5
```

15. 假定 x 是字符、y 是短整数、z 是整数、t 是长整数，指出以下语句中的错误。

```
x = 105.2;       y = 4.6;          z = 27;             t = 'A';
```

16. 以下哪些变量声明是正确的？

```
a. char letter = 'A';
b. int first, second;
c. double average, tax = 8.5;
d. long double 3x;
```

17. 以下哪些变量声明是正确的？

```
a. char letter, grade;
b. int sum = 5;
c. double average * 3;
d. long double 34.123;
```

18. 假设以下语句在程序中依次执行。当全部语句执行完成后，x、y 和 z 的值分别是什么？

```
int x = 2, y = 4, z = 5;
x = y - 2;
y += x;
z /= x + y + 4;
z %= y * 2 + 3;
```

19. 假设以下语句在程序中依次执行。当全部语句执行完成后，x、y 和 z 的值分别是什么？

```
double x = 2.5, y = 3.2, z = 5.1;
x = y + 2 ;
y += x - 1.5;
z += x + y * 4;
```

20. 声明两个 int 类型的变量、两个 float 类型的变量、三个 double 类型的变量、一个 char 类型的变量。
21. 声明 x、y、z 和 t 分别为 char、short、int 和 float 类型的常量。为每个常量指定一个适当的值。
22. 以下语句哪些会导致编译错误？请说明原因。

```
a. 2 + x = 4;                    b. z = z + 2 = 6;
c. y = x + 2 % 5;                d. t = z = 4 + 5;
```

23. 假设以下语句在程序中依次执行。当全部语句执行完成后，x、y 和 z 的值分别是什么？

```
int x = 2, y, z;
x = y = x + 2 ;
y += y - 6;
z = y + 5;
```

24. 假设以下语句在程序中依次执行。当全部语句执行完成后，x 和 y 的值分别是什么？ x 和 y 的值是否被交换？请解释。

```
int x = 2, y = 10;
y = x ;
x = y;
```

25. 假设以下语句在程序中依次执行。当全部语句执行完成后，x 和 y 的值分别是什么？ x 和 y 的值是否被交换？请解释。

```
int x = 2, y = 10;
int temp;
temp = x ;
x = y;
y = temp ;
```

26. 以下代码行的输出结果是什么？

```
cout << fixed << setprecision (2) << 124.78560 << endl;
cout << fixed << setprecision (2) << 0.14 << endl;
cout << fixed << setprecision (2) << 20 << endl;
cout << fixed << setprecision (2) << 14767.0 << endl;
```

27. 以下代码行的输出结果是什么？

```
cout << fixed << setprecision (4) << 124.79 << endl;
cout << fixed << setprecision (4) << 0.14 << endl;
cout << fixed << setprecision (4) << 20.0 << endl;
cout << fixed << setprecision (4) << 14767.00 << endl;
```

28. 在 C++ 中编写下面的表达式语句。每条语句都是独立的。

```
递增 x 的值。
递减 y 的值。
将 t 的值设置为 5.14。
将变量 first 的当前值复制到变量 second 中。
```

29. 在 C++ 中编写下面的表达式语句。每条语句都是独立的。

```
将变量 x 的内容加上 2。
将变量 y 的值乘以 7。
将 x 与 y 的值相加，并将结果存储到 z 中。
将 flag（布尔类型）的值设置为 false。
```

30. 指出以下语句中的错误（x、y、z 和 t 是数值类型）。

```
x += 7;
y *= y;
z + 6 = z;
3t = 12;
```

31. 指出以下语句中的错误（假设所有的变量都是数值类型）。

```
x = 7 = 4;
y *= y = 6;
z = z + 5 = 8;
hello = 12;
```

32. 假设 x 是 bool 类型、y 是 char 类型、z 是 short 类型、t 是 float 类型。以下各表达式中，发生了哪些隐式类型转换？各表达式的结果值分别是多少？

```
x + 4
y * 2
z – 24;
t + 23.4;
```

33. 假设 x 是 int 类型、y 是 long 类型、z 是 long long 类型、t 是 long double 类型。以下各表达式中，发生了哪些隐式类型转换？

```
x + y
y * z
z – t;
```

34. 以下程序代码片段的输出结果是什么？

```
double x = –909.245;
cout << fixed << setprecision (2) << x << endl;
cout << fixed << setw (15) << setprecision (3) << setfill ('*') << x;
```

35. 以下程序代码片段的输出结果是什么？

```
int x = 50;
cout << setw(10) << setfill ('*') << x;
```

36. 以下程序代码片段的输出结果是什么？

```
int n1 = 10, n2 = 20;
char ch = '$';
cout << setw(n1) << setfill (ch) << 241 << endl;
cout << setw(n2) << setfill (ch) << 14672;
```

37. 以下程序代码片段的输出结果是什么？

```
int n1 = 8, n2 = 12;
char ch = '#';
cout << setw(n1) << setprecision (3) << setfill (ch) << 475.12 << endl;
cout << setw(n2) << setprecision (4) << setfill (ch) << 0.151;
```

38. 执行以下程序代码片段，输入布尔值 false，则输出结果是什么？

```
bool x;
cin >> boolalpha >> x;
cout << x ;
```

39. 执行以下程序代码片段，输入 72，则输出结果是什么？

```
int x;
cin >> oct >> x;
cout << x;
```

40. 执行以下程序代码片段，输入 72，则输出结果是什么？

```
int x;
cin >> hex >> x;
cout << x;
```

编程题

1. 编写一个程序，分别获取你的计算机系统的 short 和 unsigned int 类型的最大值和最小值。
2. 编写一个程序，分别获取你的计算机系统的 long 和 long long 类型的最大值和最小值。
3. 编写一个程序，分别获取你的计算机系统的 float 和 double 类型的最大值和最小值。
4. 编写一个程序，提取一个 int 类型的输入值的十位数。
5. 编写一个程序，分别提取出一个 int 类型的输入值的右侧三位数。
6. 编写一个程序，给定一个三位数的整数，构造并打印另一个整数，其位数与给定整数的位数相反。例如，给定 372，程序打印 273。
7. 编写一个程序，给定以小时为单位的时间间隔，计算其包含多少星期、多少天、多少小时。
8. 编写一个程序，给定一个任务持续的时间（时、分、秒），计算以秒为单位的持续时间。
9. 编写一个程序，输入以秒为单位的时间（一个长整数，long 值），计算其包含多少天、多少小时、多少分钟、多少秒。
10. 编写一个程序，输入四个成绩（int 值），计算并输出它们的平均值（double 值）。
11. 编写一个程序，帮助商店的出纳员计算找零：以美元和美分表示顾客付款的金额。要求找零单位为美元、25 美分、10 美分、5 美分和 1 美分。
12. 编写一个程序，在给定摄氏温度的情况下，使用公式 F = (9/5)C + 32，计算并打印对应的华氏温度。
13. 编写一个程序，在给定华氏温度的情况下，使用公式 C = (F − 32) * 5/9，计算并打印对应的摄氏温度。

14. 假设有三对夫妻组成一个团队在一家餐厅共同用餐。第一个家庭有两个孩子。其他每个家庭各有一个孩子。编写一个程序，计算每个家庭分担的账单费用：假设一个孩子收取成人份额3/4的费用。总费用（税前）作为输入，税率是9.5%，另加20%的服务费。
15. 时薪制雇员通常每周工作40小时，每小时拿固定薪酬。如果工作超过40小时，就会每小时多得到60%的薪酬。编写一个程序，要求员工输入上周加班的小时数和每周固定工资，然后计算并打印总薪酬。
16. 编写一个程序，为公司创建客户账单。该公司只销售三种产品：电视机、DVD播放器和遥控器，销售单价分别为1400.00美元、220.00美元和35.20美元。要求程序从键盘上读取每种设备购买的数量。然后，计算每种设备的金额、总金额以及计算8.25%的增值税之后的总金额。

选择结构

我们在前几章中编写的程序是基于语句的顺序执行。程序主要由输入语句、计算语句和输出语句组成。每次运行程序时，程序中的所有语句都会依次执行。按顺序执行语句可以解决简单的问题，但我们也需要有控制语句。在本章中，我们将讨论第一组控制语句：选择语句。选择语句允许程序员根据一个或者多个条件在不同的操作之间进行选择执行。

学习目标

阅读并学习本章后，读者应该能够

- 讨论实现简单选择时对关系表达式和等性表达式的需求。
- 理解如何使用 if 语句实现单分支选择结构。
- 理解如何使用 if-else 语句实现双分支选择结构。
- 理解如何使用嵌套的 if-else 语句实现多分支选择结构。
- 演示如何通过将关系 / 等性表达式和逻辑表达式进行组合，以实现复杂的判断逻辑。
- 理解如何使用 switch 语句进行基于离散值的多路选择。
- 演示如何使用条件表达式来代替简单的 if-else 语句。

4.1 简单选择结构

为了解决一些问题，我们必须在检验一个真假条件的基础上做出决定。如果条件为真，则需要执行一组语句；如果条件为假，则需要执行另一组语句（或者不执行任何语句）。这个过程被称为选择。问题是，我们如何创建要测试的条件呢？答案是使用求值结果为真或者为假的表达式。

4.1.1 关系和等性表达式

为了做出简单的决策，我们需要关系表达式或者等性表达式。这些表达式如附录 C 所示，我们将其显示在表 4-1 中，以供快速参考。我们把这些表达式区分为两组，因为它们拥有不同的优先级（11 和 10）。

表 4-1 关系和等性表达式

组别	名称	运算符	表达式	优先级	结合性
关系表达式	小于	<	expr < expr	11	→
	小于或等于	<=	expr <= expr		
	大于	>	expr > expr		
	大于或等于	>=	expr >= expr		
等性表达式	等于	==	expr == expr	10	→
	不等于	!=	expr != expr		

关系表达式

如表 4-1 所示，**关系表达式**使用四个关系运算符来比较两个值。请注意，如果 a 等于或者小于 b，则 (a <= b) 的结果为 true；如果 a 大于 b，则结果为 false。同样，如果 a 大于或者等于 b，则 (a >= b) 的结果为 true；如果 a 小于 b，则结果为 false。对比较关系表达式求值时，左右表达式的类型必须相同。如果这两个表达式不是同一类型，则应用第 3 章中讨论的类型转换方法。下面是一些关系运算符的示例。

```
3 < 4                // 结果是 true，因为 3 小于 4
12.78 < 7.36         // 结果是 false，因为 12.78 不小于 7.36
5 <= 3.24            // 左边的操作数首先被提升为 5.0。结果是 false
5.0 <= 5             // 右边的操作数首先被提升为 5.0。结果是 true
3 > false            // 右边的操作数被提升为 0。结果是 true
5 >= 2.2             // 左边的操作数被提升为 5.0。结果是 true
```

等性表达式

表 4-1 还包括了用于确定两个实体是否相等的两个**等性表达式**。请注意，如果 a 等于 b，则 (a == b) 的结果为 true；如果 a 不等于 b，则结果为 false。同样地，如果 a 的值与 b 的值不同，则 (a != b) 的结果为 true；如果两者相等，则结果为 false。对比较关系表达式求值时，左右表达式的类型必须相同。如果这两个表达式不是同一类型，则应用第 3 章中讨论的类型转换方法。下面是一些等性运算符的示例。

```
3 == 4               // 结果为 false
true == false        // 表达式转换为 (1 == 0)，结果为 false
true != false        // 表达式转换为 (1 != 0)，结果为 true
3 != false           // 右侧的操作数被转换为 0，结果为 true
65 != 'A'            // 结果为 false，因为 'A' 被转换为 65
```

优先级和结合性

当尝试对一个包含关系表达式和等性表达式的复杂表达式求值时，我们需要注意这两种分别属于不同组别的表达式的优先级和结合性。在复杂表达式中，括号有助于明确计算顺序。以下是一些示例：

```
3 < 4 == 1           // 使用优先级，等同于 (3 < 4) == 1。结果为 true
2 < 3 < 0            // 使用结合性，等同于 (2 < 3) < 0。结果为 false
3 <= 6 < 5           // 使用结合性，等同于 (3 <= 6) < 5。结果为 true
4 == 4 < 2           // 使用优先级，等同于 4 == (4 < 2)，即 4 == 0。结果为 false
```

陷阱

我们应该避免针对浮点值使用等性运算符，因为无法判断浮点值存储在内存中的精度。换言之，浮点值的比较结果取决于系统。

```
x == 2.78123         // 即使 x 的值是 2.78123，结果也可能是 false
x != 14.67823        // 即使 x 的值约等于 14.67823，结果也可能是 true
```

4.1.2 单分支选择结构：if 语句

最常用的选择结构是单分支选择结构，C++ 中使用 **if 语句**来实现。图 4-1 显示了 C++ 中单分支选择结构的逻辑和 if 语句的语法。

在这种类型的语句中，我们使用表达式来测试条件。如果测试结果为真，则程序执行语句（或者语句集）；如果结果为假，则跳过语句（或者语句集）。注意，如果要执行多条语句，则需要使用复合语句。

这种类型的选择结构通常被称为**单分支选择结构**（one-way selection）：执行任务还是不执行任务。注意，只有在每次运行程序时，结果是不可预测的（基于测试条件），选择表达式才有意义。换言之，关系表达式或者等性表达式的结果应该基于对程序输入的判断，否则，决策就没有意义。

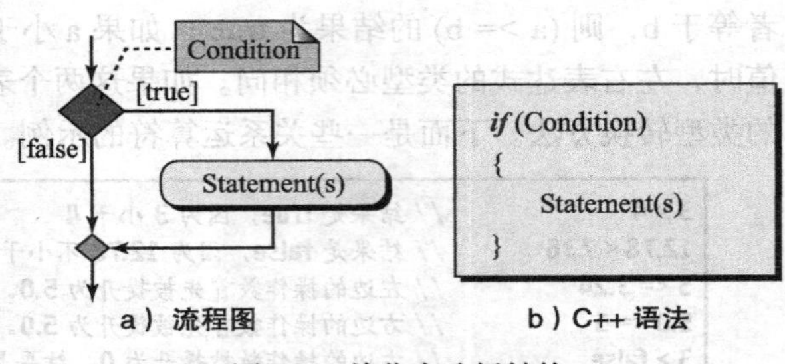

图 4-1 单分支选择结构

注意，当测试结果为真时要执行的语句是一条语句，但是这条语句可以是一条复合语句（两个花括号内的一组语句）。当只有一条语句要执行时，一些程序员会省略花括号，我们强烈反对这种做法。花括号可以清楚地显示程序的流程，即使只有一条语句要执行，也不建议省略。

【例 4-1】 程序 4-1 演示了一个非常简单的 if 语句的使用方法：程序根据用户所输入的整数，确定其绝对值。如果用户输入一个负整数，程序会将其更改为相同大小的正整数，否则，程序将不执行任何操作。程序退出决策部分（第 15 行到第 18 行）后，将打印所输入整数的绝对值。

注意，在这个简单的示例中，我们不需要花括号（第 16 行和第 18 行），但是使用花括号是一种良好的编程实践。

程序 4-1　打印绝对值

```
1   /***************************************************************
2    * 使用 if 语句打印一个数值的绝对值                                *
3    ***************************************************************/
4   #include <iostream>
5   using namespace std;
6
7   int main ()
8   {
9     // 声明变量
10    int number;
11    // 获取输入
12    cout << "Enter an integer: ";
13    cin >> number;
14    // 计算绝对值
15    if (number < 0)
16    {
17        number = −number;
18    }
19    // 输出绝对值
20    cout << "Absolute value of the number you entered is: " << number;
21    return 0;
22  }
```

运行结果：

```
Enter an integer: 25
Absolute value of the number you entered is: 25
```

```
运行结果:
Enter an integer: -17
Absolute value of the number you entered is: 17
```

【例 4-2】 程序 4-2 同样演示了 if 语句的使用方法。该程序计算员工的每周总收入。如果员工在一周内工作超过 40 小时，则加班小时工资为基本小时工资的 130%。

程序 4-2 员工的总薪酬

```
1  /***************************************************************
2   * 使用 if 语句计算员工的总薪酬                                    *
3   ***************************************************************/
4  #include <iostream>
5  #include <iomanip>;
6  using namespace std;
7
8  int main ()
9  {
10   // 声明变量
11    double hours;
12    double rate;
13    double regularPay;
14    double overPay;
15    double totalPay;
16   // 输入
17    cout << "Enter hours worked: ";
18    cin >> hours;
19    cout << "Enter pay rate: ";
20    cin >> rate;
21   // 不依赖于条件的计算
22    regularPay = hours * rate;
23    overPay = 0.0;
24   // 如果工作时间不超过 40 小时，则跳过下面的计算
25    if (hours > 40.0)
26    {
27        overPay = (hours - 40.0) * rate * 0.30;
28    } // if 语句结束
29   // 剩下的计算（计算总薪酬）
30    totalPay = regularPay + overPay;
31   // 打印结果
32    cout << fixed << showpoint;
33    cout << "Regular pay  = " << setprecision (2) << regularPay << endl;
34    cout << "Over time pay = " << setprecision (2) <<  overPay << endl;
35    cout << "Total pay = " << setprecision (2) << totalPay << endl;
36    return 0;
37 }
```

```
运行结果:
Enter hours worked: 30
Enter pay rate: 22.00
Regular pay = 660.00
Over time pay = 0.00
Total pay = 660.00
```

```
运行结果：
Enter hours worked: 45
Enter pay rate: 25.00
Regular pay = 1125.00
Over time pay =  37.50
Total pay = 1162.50
```

在程序第一次运行中，由于员工工作不足40小时，程序忽略第27行中的语句。加班工资保持为初始化时的0.0。在第二次运行中，由于员工工作超过40小时，程序执行第27行的语句，计算加班工资为37.50美元，并累加到常规工资（regularPay）中，结果总工资（totalPay）为1162.50美元。

一些陷阱

当我们使用if语句进行单分支选择时，必须小心谨慎以避免陷阱。这些错误通常不会被编译器捕获，因为它们是逻辑错误，而不是语法错误。程序能够正常编译并运行，但结果不正确。

使用赋值运算符而不是相等运算符。常见的错误是使用赋值运算符（=）而不是相等运算符（==）。以下是一个示例：

```
if (x = 0)              // 该语句的结果始终为 false
{
   statement;           // 该语句永远不会被执行
}
```

注意，在这种情况下，不管x的值是什么，条件永远是false。表达式将x的值更改为0（而这并不是我们的本意）。表达式的结果是x的值，即0。前文已经讨论过，在需要布尔值的表达式中，0值被解释为false。这意味着语句永远不会被执行。

同样，如果我们使用(x = 5)而不是(x == 5)，则结果始终为真，因此语句总是被执行，如下所示：

```
if (x = 5)              // 该条件始终为 true
{
   statement;           // 该语句总是被执行
}
```

省略花括号。另一个常见的错误是当需要执行多条语句时，却省略了花括号：

```
if (x == 0)
   statement1;          // 如果 x 等于 0，则执行该语句
   statement2;          // 该语句总是会被执行
```

在上面的示例中，编译器将if语句的语句体解释为仅包括一条语句（statement1），因此总是会执行statement2。我们建议总是使用花括号来避免这种混淆。

多余的英文分号。if语句是一个由头和一个作为主体的复合语句组成的语句。如果在头的末尾多放一个英文分号，则编译器会认为其主体是空语句。在下面的示例中，无论x的值是什么，都会执行statement1和statement2。换句话说，statement1和statement2被视为独立于if语句的复合语句。

```
if (x == 0);            // 英文分号表示 if 语句的语句体是空语句
```

```
{
  statement1;      // statement1 总是会被执行
  statement2;      // statement2 总是会被执行
}
```

4.1.3 双分支选择结构：if-else 语句

我们讨论的第二个选择结构是使用 if-else 语句的**双分支选择结构**（two-way selection），如图 4-2 所示。当程序在执行期间到达 **if-else 语句**时，它首先计算布尔表达式。如果此表达式的值为 true，则程序执行 statement1 并跳过 statement2，否则，程序执行 statement2 并忽略 statement1。注意，在每次运行中，只执行两条语句中的一条，即 statement1 或者 statement2，但不能同时执行这两条语句。还要注意的是，statement1 和 statement2 可以是一条简单语句，也可以是一条复合语句。但是，我们始终建议使用花括号（一条复合语句），即使每个分支只包含一条语句也建议使用花括号。if-else 语句用于无论决策的结果是什么都必须执行某种操作的情况。

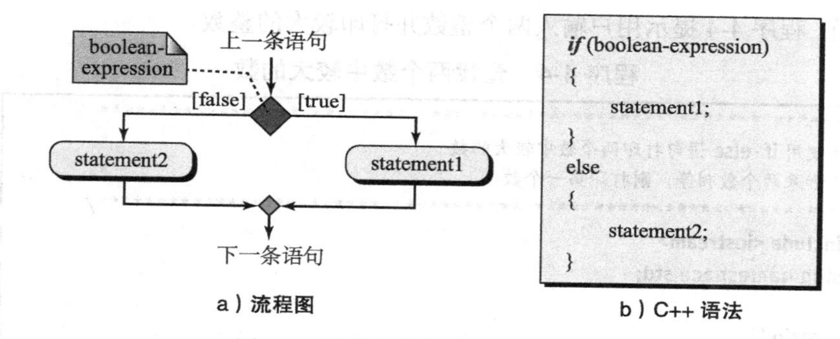

图 4-2 双分支选择结构：if-else 语句

【例 4-3】 程序 4-3 根据考试分数打印学生的及格 / 不及格等级。输入考试分数后，使用 if-else 语句检查考试分数是否大于或等于 70。在这种情况下，if 分支和 else 分支都只有一条语句。程序要么执行第 17 行要么执行第 21 行，但不会同时执行这两行代码。

注意，在这种情况下，我们将测试结果打印为及格 / 不及格等级，不管等级的值是多少。但是，也可以在使用变量保存结果时仅使用 if 语句来实现，这将在后续章节中讨论。

程序 4-3 计算考试分数的及格 / 不及格等级

```
1   /*****************************************************************
2    * 使用 if-else 语句计算考试分数的及格 / 不及格等级              *
3    *****************************************************************/
4   #include <iostream>
5   using namespace std;
6
7   int main ()
8   {
9     // 声明局部变量
10    int score;
11    // 输入
12    cout << "Enter a score between 0 and 100: ";
13    cin >> score;
```

```
14   //判断决策
15   if (score  >= 70)
16   {
17       cout << "Grade is pass" << endl;
18   } // if 结束
19   else
20   {
21       cout << "Grade is nopass" << endl;
22   } // else 结束
23   return 0;
24 }
```

运行结果：
Enter a score between 0 and 100: 65
Grade is nopass.

运行结果：
Enter a score between 0 and 100: 92
Grade is pass.

【例 4-4】 程序 4-4 提示用户输入两个整数并打印较大的整数。

程序 4-4　查找两个数中较大的数

```
1  /***************************************************************
2   * 使用 if-else 语句打印两个数中较大的数                        *
3   * 如果两个数相等，则打印第一个数                               *
4   ***************************************************************/
5  #include <iostream>
6  using namespace std;
7
8  int main ()
9  {
10   // 声明变量
11   int num1, num2;
12   int larger;
13   // 输入语句
14   cout << "Enter the first number: ";
15   cin >> num1;
16   cout << "Enter the second number: ";
17   cin >> num2;
18   // 判断决策
19   if (num1 >=  num2)
20   {
21       larger = num1;
22   } // if 结束
23   else
24   {
25       larger = num2;
26   } // else 结束
27   // 输出结果
28   cout << "The larger number is: " << larger;
29   return 0;
30 }
```

```
运行结果:
Enter the first number: 40
Enter the second number: 25
The larger number is: 40
```

```
运行结果:
Enter the first number: 22
Enter the second number: 67
The larger number is: 67
```

注意，在上面的程序中，我们将输出与决策过程分开。我们不在决策部分打印较大的值。我们将较大的值存储在一个变量中（称为 larger），然后在决策过程完成后打印出来。这个策略比上一个程序中使用的策略更好，因为我们将任务分开。

另一个陷阱：多余的英文分号。头部中的额外英文分号会与前文所述的单分支选择结构一样产生问题。但是，这里的错误是不同的，它是一个编译错误。程序编译不成功，因为编译器认为我们使用了没有对应的 if 子句的 else 子句。这个错误并不危险，因为它强制我们修改程序并重新编译。编译时错误没有运行时错误那么严重。

// 编码	// 编译器解释结果

if (b); // 多余的英文分号
{
  statement1;
}
else
{
  statement2;
}
	if (b) ; // 空语句
	{
	  statement1
	}
	else // 此处编译错误：else 没有对应的 if
	{
	  statement2
	}

嵌套 if-else 语句

我们在前面讨论了 if-else 语句中的每个分支都可以是任何类型的语句。这意味着每个分支都可以是另一个 if-else 语句。这种情况称为嵌套的 if-else 语句。尽管我们可以有任意多个层次的嵌套，但是程序在嵌套几个层次之后会变得非常复杂。图 4-3 显示了具有两个级别的嵌套 if-else 语句。请注意，嵌套可以发生在 if 分支或者 else 分支或者两者中。图 4-3 显示了两个分支中都包含嵌套的情况。

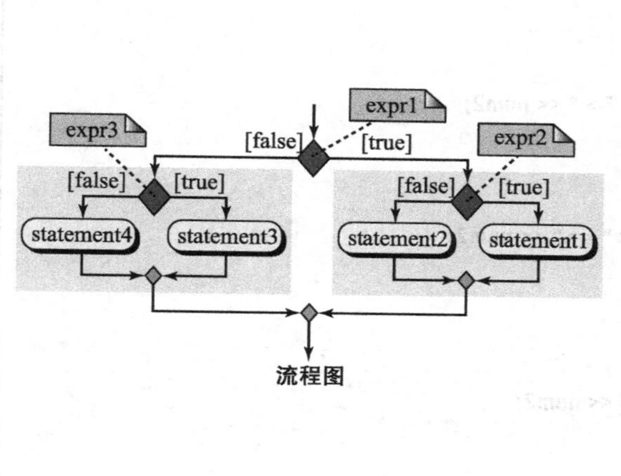

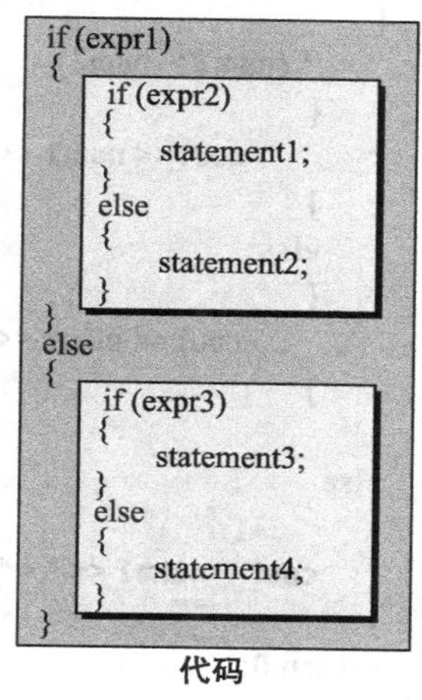

```
if (expr1)
{
    if (expr2)
    {
        statement1;
    }
    else
    {
        statement2;
    }
}
else
{
    if (expr3)
    {
        statement3;
    }
    else
    {
        statement4;
    }
}
```

代码

图 4-3　嵌套 if-else 语句

【例 4-5】 让我们编写一个程序来确定一个数值是大于、等于还是小于第二个数值。流程图如图 4-4 所示。注意，本例中的 else 分支只有一个任务。

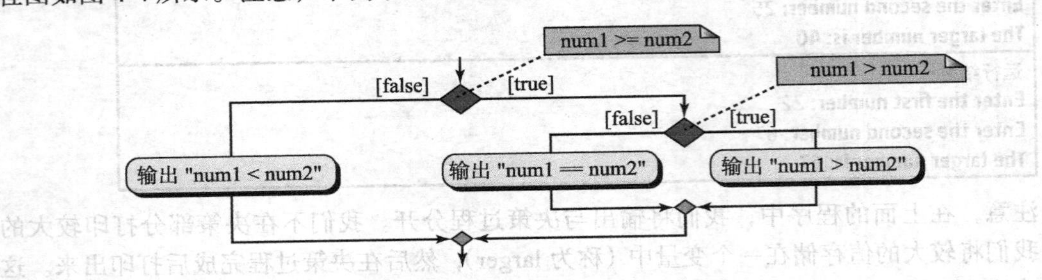

图 4-4　确定两个数值大小关系的流程图

解决方案首先确定第一个数值是否大于或者等于第二个数值。如果此条件为假，则程序已确定 num1 小于 num2。如果 num1 大于或等于 num2，程序将进一步测试以找到确切的关系。这是一个嵌套 if-else 语句的示例，其中左分支只是一个简单语句。代码如程序 4-5 所示。

程序 4-5　确定两个数值的大小关系

```
1  /*************************************************************
2   * 确定一个数值是大于、等于还是小于第二个数值                    *
3   *************************************************************/
4  #include <iostream>
5  using namespace std;
6  
7  int main ()
8  {
9    // 声明变量
10   int num1, num2;
11   // 获取输入
12   cout << "Enter the first number: ";
13   cin >> num1;
14   cout << "Enter the second number: ";
15   cin >> num2;
16   // 使用嵌套 if-else 语句进行判断决策
17   if (num1 >= num2)
18   {
19       if (num1 > num2)
20       {
21           cout << num1 << " > " << num2;
22       }
23       else
24       {
25           cout << num1 << " == " << num2;
26       }
27   }
28   else
29   {
30       cout << num1 << " < " << num2;
31   }
32   return 0;
33 }
```

```
运行结果:
Enter the first number: 42
Enter the second number: 32
42 > 32
```

```
运行结果:
Enter the first number: 12
Enter the second number: 12
12 == 12
```

```
运行结果:
Enter the first number: 12
Enter the second number: 28
12 < 28
```

一个严重的陷阱：悬空的 else 问题。嵌套的 if-else 语句可能会产生一个称为**悬空的 else** 的经典问题。如果在嵌套的 if-else 语句中，存在的 if 分支比 else 分支多，则需要判断哪个 if 分支应该与哪个 else 分支配对。答案是编译器将 else 与一个最近未配对的 if 匹配。这个匹配原则与我们书写代码的意图无关。

> else 总是与一个最近未配对的 if 匹配。

图 4-5 显示了一个流程图和两个编码解决方案。

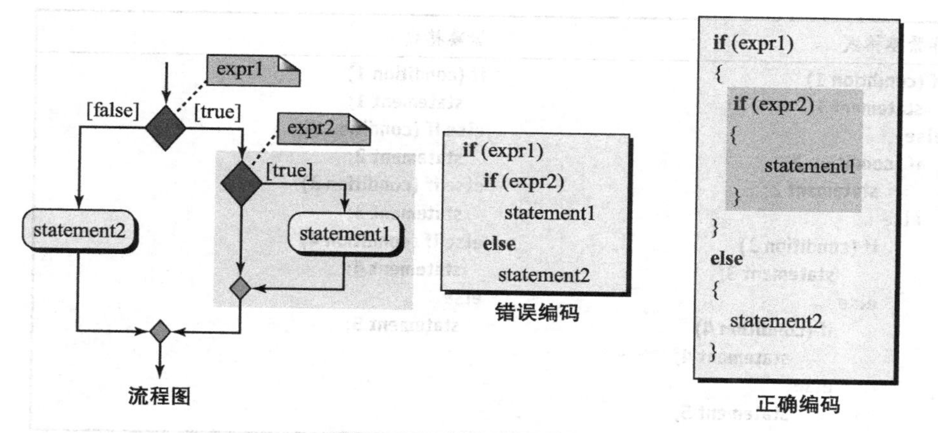

图 4-5 悬空的 else 问题

第一个代码片段与流程图不对应，因为 if 语句和 else 语句被错误配对。在这种情况下，我们必须使用如第二个代码片段所示的复合语句，以确保 else 语句与第一个 if 语句配对，而不是与第二个 if 语句配对。稍后我们将看到这个问题还有另一个解决方案：使用逻辑表达式。

> 始终使用复合语句以避免悬空的 else 问题。

4.1.4 多分支选择结构

有时我们需要解决多个问题，如下所示。请注意，每条语句都可以是复合语句。

```
如果 condition 1为真，执行 statement 1
```

如果 **condition 2**为真，执行 **statement 2**
...
如果上述条件皆不为真，执行 **statement n**

这就是所谓的**多分支选择结构**（multiway selection）。我们可以使用嵌套的 if-else 语句来实现多路选择。图 4-6 显示了具有五种不同语句的多路条件的流程图。

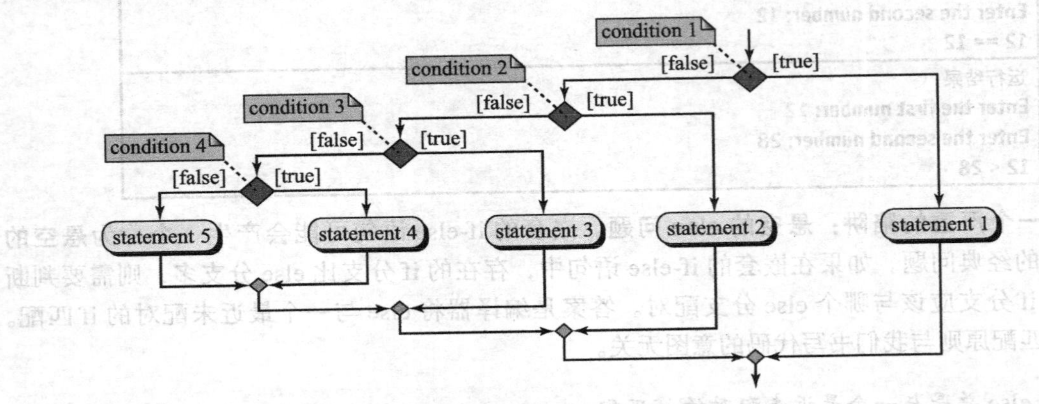

图 4-6　多路 if-else 分支选择结构

我们可以用两种不同的方式格式化图 4-6 中的多分支选择结构的实现代码，如下所示：

非紧凑格式	紧凑格式

<pre>if (condition 1)
  statement 1;
else
  if (condition 2)
      statement 2;
  else
      if (condition 3)
          statement 3;
      else
          if (condition 4)
              statement 4;
          else
              statement 5;</pre>
	<pre>if (condition 1)
	   statement 1;
	else if (condition 2)
	   statement 2;
	else if (condition 3)
	   statement 3;
	else if (condition 4)
	   statement 4;
	else
	   statement 5;</pre>

注意，左边的实现在语法上与右边的实现相同，除了 else 和 if 的位置。如果在两个连续代码行之间没有其他任何语句，则 else 和 if 将组合在一起，使用一个空白字符分隔。就 C++ 而言，一个换行符和一个空白字符是相同的。紧凑格式在行数上较短，并且需要较少的缩进。同时，紧凑格式更容易阅读理解。为了使代码简短，我们没有使用花括号。程序中应该始终使用花括号，如下一个示例所示。

【例 4-6】 假设一位教授想要编写一个程序，以确定 0 到 100 之间的测试分数所对应的字母等级（A、B、C、D 和 F）成绩。使用多分支选择结构的代码如程序 4-6 所示。

程序 4-6　确定给定的百分制分数的成绩等级

```
1  /***************************************************************
2   * 使用多分支选择结构确定一个百分制分数的成绩等级                  *
3   ***************************************************************/
```

```
4   #include <iostream>
5   using namespace std;
6
7   int main ()
8   {
9     //声明变量
10    int score;
11    char grade;
12    //获取输入
13    cout << "Enter a score between 0 and 100: ";
14    cin >> score;
15    //使用if-else的多分支选择结构
16    if (score >= 90)
17    {
18        grade = 'A';
19    }
20    else if (score >= 80)
21    {
22        grade = 'B';
23    }
24    else if (score >= 70)
25    {
26        grade = 'C';
27    }
28    else if (score >= 60)
29    {
30        grade = 'D';
31    }
32    else
33    {
34        grade = 'F';
35    }
36    //输出
37    cout << "The grade is: " << grade;
38    return 0;
39  } // main 结束
```

运行结果：

```
Enter a score between 0 and 100: 83
The grade is B
```

运行结果：

```
Enter a score between 0 and 100: 65
The grade is D
```

运行结果：

```
Enter a score between 0 and 100: 95
The grade is A
```

4.2 复杂条件决策

在前一节中，我们使用关系表达式和等性表达式解决了一些决策问题。有时决策非常复杂，因此需要更多的表达式类型来定义。例如，不能基于特定范围内的单一值做出决定，如

x 应该在 5 到 10 之间。我们需要使用更多的表达式。在本节中，我们将介绍三种逻辑表达式，它们有助于更复杂的决策。这些逻辑表达式可以用于计算机程序设计的其他领域（例如循环结构），我们将在后续章节中讨论。

4.2.1 逻辑表达式

逻辑表达式（logical expression）使用逻辑运算符，这些逻辑运算符作用于一个或者两个布尔类型的操作数，并创建布尔类型值。C++ 包含三个逻辑运算符：逻辑非（!）、逻辑与（&&）和逻辑或（||）。附录 C 列举了所有 C++ 表达式；表 4-2 中仅包含本节需要的表达式。表 4-2 中还包含了关系表达式和等性表达式，这样方便读者理解创建选择语句所必需的所有表达式之间的关系。请注意，**逻辑非表达式**的优先级较高（甚至高于关系表达式和等性表达式），但是**逻辑与**和**逻辑或表达式**的优先级则较低。值得注意的是，逻辑非的结合性是从右到左。

表 4-2　一些表达式的优先级和结合性

组别	名称	运算符	表达式	优先级	结合性
一元表达式	逻辑非	!	! expr	17	←
关系表达式	小于 小于或等于 大于 大于或等于	< <= > >=	expr < expr expr <= expr expr > expr expr >= expr	11	→
等性表达式	等于 不等于	== !=	expr == expr expr != expr	10	→
逻辑与	逻辑与	&&	expr && expr	6	→
逻辑或	逻辑或	||	expr || expr	5	→

显示逻辑表达式结果的最佳方法是使用真值表。图 4-7 显示了三个逻辑运算符及其解释。

真值表

操作数	结果
false	true
true	false

逻辑非

真值表

左操作数	右操作数	结果
false	false	false
false	true	false
true	false	false
true	true	true

逻辑与

真值表

左操作数	右操作数	结果
false	false	false
false	true	true
true	false	true
true	true	true

逻辑或

图 4-7　逻辑表达式及其真值表

对于逻辑非表达式，只有一个输入。请注意，如果逻辑非运算符应用于非 true 非 false 的操作数，则会自动进行类型转换。下面是一些使用逻辑非表达式的示例。

```
!true          // 结果为 false
!false         // 结果为 true
!5             // 首先把 5 转换为布尔类型值 true。结果为 false
!0             // 首先把 0 转换为布尔类型值 false。结果为 true
```

逻辑与表达式需要两个布尔值作为其操作数。如果两个操作数都为 true，则结果为布尔值 true；如果其中一个操作数为 false，则结果为 false。同样请注意，如果逻辑与运算符应用于非 true 非 false 的操作数，则会自动进行类型转换。下面是一些使用逻辑与表达式的示例。

```
true && true              //结果为 true
false && true             //结果为 false
(3 < 5) && (4 > 2)        //即 true && true，结果为 true
(4 < 6) && (2 == 3)       //即 true && false，结果为 false
6 && false                //首先把 6 转换为 true，结果为 false
4 && 0                    //首先把 4 转换为 true，把 0 转换为 false，结果为 false
```

逻辑或表达式同样需要两个布尔值作为其操作数。仅当两个操作数都为 false 时，结果为布尔值 false；如果其中一个操作数为 true，则结果为 true。同样请注意，如果逻辑或运算符应用于非 true 非 false 的操作数，则会自动进行类型转换。下面是一些使用逻辑或表达式的示例和结果。

```
true || true              //结果为 true
false || false            //结果为 false
(1 < 4) || (2 ==3)        //即 (true || false)，结果为 true
(4 < 3) || (4 <= 3)       //即 (false || false)，结果为 false
3 || false                //首先把 3 转换为 true，结果为 true
0 || false                //首先把 0 转换为 false，结果为 false
```

下面显示优先级和结合性如何影响复杂表达式的求值。在表达式中添加括号有助于我们确定计算结果。

```
!false && true            // 等同于 (!false) && true。结果为 true
false || true && false    // 等同于 false || (true && false)。结果为 false
2 == 2 || 5 < 6 && false  // 等同于 true || (false && false)。结果为 true
```

4.2.2 逻辑表达式的应用

可以使用关系 / 等性表达式和逻辑表达式的组合来编写复杂的选择标准。接下来我们讨论一些应用案例。

逻辑与表达式的使用

因为当两个操作数都为真时，逻辑与表达式的结果为 true，所以当需要两个条件都必须为真时执行某些操作的决策，可以使用逻辑与表达式。换言之，以下表达式只有当 condition1 和 condition2 都为 true 时，结果才为 true。

```
condition1 && condition2
```

例如，要判断一个数值是否位于某个范围内，则需要两个条件，如图 4-8 所示。

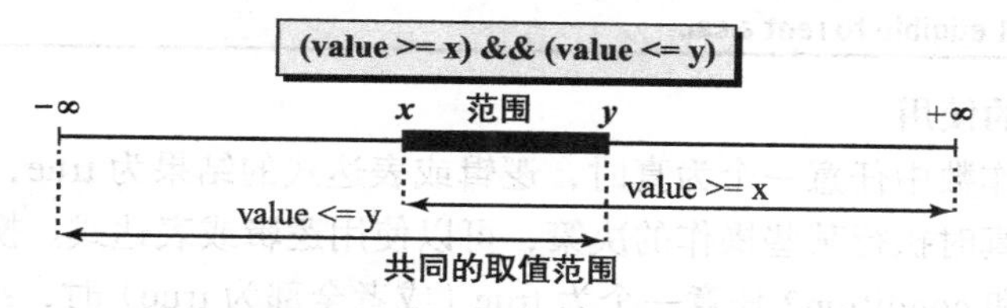

图 4-8 两个关系的组合作为共同的取值范围

【例 4-7】 假设某租车公司将租车的最小和最大年龄分别定义为 25 岁和 100 岁。程序 4-7 演示了如何使用一段简单的决策代码来测试一个人的年龄。

程序 4-7 确定年龄资格

```
1  /*************************************************************
2   * 确定租车的年龄资格                                           *
3   *************************************************************/
4  #include <iostream>
5  using namespace std;
6
7  int main ()
8  {
9    // 声明变量
10   int age;
11   bool eligible;
12   // 获取输入
13   cout << "Enter your age: ";
14   cin >>  age;
15   // 设置条件
16   eligible = (age >=25) && (age <= 100);
17   // 测试条件并输出
18   if (eligible)
19   {
20        cout << "You are eligible to rent a car.";
21   }
22   else
23   {
24        cout << "Sorry! You are not eligible to rent a car.";
25   }
26   return 0;
27 }
```

```
运行结果：
Enter your age: 27
You are eligible to rent a car.
```

```
运行结果：
Enter your age: 54
You are eligible to rent a car.
```

```
运行结果：
Enter your age: 103
Sorry! You are not eligible to rent a car.
```

```
运行结果：
Enter your age: 21
Sorry! You are not eligible to rent a car.
```

逻辑或表达式的使用

因为当两个操作数中任意一个为真时，逻辑或表达式的结果为 true，所以当需要两个条件中至少有一个为真时执行某些操作的决策，可以使用逻辑或表达式。换言之，以下表达式只有当 condition1 和 condition2 任意一个为 true（或者全部为 true）时，结果为 true。

```
condition1 || condition2
```

例如，要判断一个数值是否位于某个范围之外，则需要测试两个条件，如图 4-9 所示。请注意，要使测试正常工作，x 和 y 必须是不同的，并且 y 必须大于 x。否则，会有重叠，并且范围不明显。

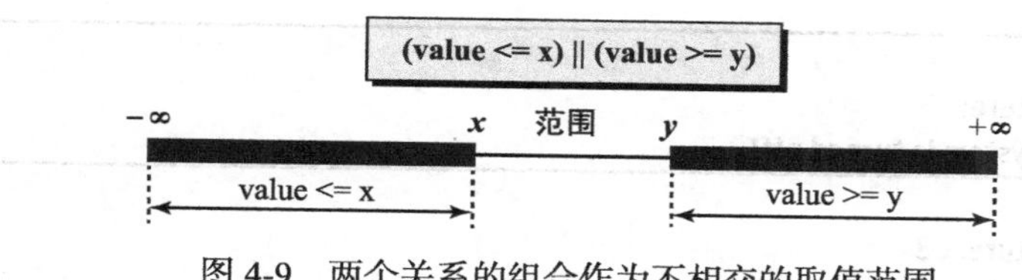

图 4-9 两个关系的组合作为不相交的取值范围

【例 4-8】 假设室内温度高于 75 华氏度或者低于 60 华氏度时，我们希望开启室内的空调系统。可以通过输入温度和打开空调系统来模拟这一过程，如程序 4-8 所示。当空调系统打开后，程序可以决定打开加热器还是冷却器。注意，要打开加热器或者冷却器，必须先打开空调系统。

程序 4-8 打开和关闭空调系统

```
1  /*****************************************************************
2   * 当温度高于某个值或者低于某个值时，                              *
3   * 打开空调系统                                                    *
4   *****************************************************************/
5  #include <iostream>
6  using namespace std;
7
8  int main ()
9  {
10     //声明变量
11     int temperature;
12     bool hot;
13     bool cold;
14     //输入温度
15     cout << "Enter the temperature: ";
16     cin >>  temperature;
17     //设置两个条件
18     hot = temperature >= 75;
19     cold = temperature <= 65;
20     //做出决策
21     if (hot || cold)
22     {
23         cout << "The air condition system is turned on!" << endl;
24         if (hot)
25         {
26             cout << "The cooler is working!" << endl;
27         }
28         else
29         {
30             cout << "The heater is working!" << endl;
31         }
32     }
33     else
34     {
35         cout << "The air condition system is turned off!" << endl;
36     }
```

```
37    return 0;
38 }
```

运行结果：
Enter the temperature: 73
The air condition system is turned off!

运行结果：
Enter the temperature: 63
The air condition system is turned on!
The heater is working!

运行结果：
Enter the temperature: 82
The air condition system is turned on!
The cooler is working!

逻辑非表达式的使用

逻辑非运算符翻转逻辑表达式的值。如果逻辑表达式最初为true，则结果为false；如果逻辑表达式最初为false，则结果为true。换言之，下面表达式的结果是给定条件的相反值。

```
!condition
```

【例4-9】 在我们目前的日历中，二月有28天而不是29天。为了判断是否为闰年，需要检查三个条件：如果某年份可以被400整除，则确定是闰年，否则，如果该年份可以被4整除，但不能被100整除，那么该年份也是一个闰年。可以使用如下所示的逻辑表达式来编写条件：

```
leapYear = (divisibleBy400) || (divisibleBy4 && !(divisibleBy100))
```

程序4-9显示了我们如何使用上述条件来测试给定的年份是否为闰年。

程序4-9　确定指定的年份是否为闰年

```
1  /***************************************************************
2   * 使用三个测试条件确定某年份是否为闰年                              *
3   ***************************************************************/
4  #include <iostream>
5  using namespace std;
6
7  int main ()
8  {
9    //声明变量
10   int year;
11   bool divBy400, divBy4, divBy100;
12   bool leapYear;
13   //输入年份
14   cout << "Enter the year: ";
15   cin >>  year;
16   //设置条件
17   divBy400 = ((year % 400) == 0);
18   divBy4 = ((year % 4) == 0);
19   divBy100 = ((year % 100) == 0);
20   leapYear = (divBy400) || (divBy4 && !(divBy100));
21   //测试条件并输出
```

```
22  if (leapYear)
23  {
24      cout << "Year " << year  << " is a leap year." << endl;
25  }
26  else
27  {
28      cout << "Year " << year  << " is not a leap year." << endl;
29  }
30  return 0;
31 }
```

```
运行结果：
Enter the year: 2000
Year 2000 is a leap year.
```

```
运行结果：
Enter the year: 1900
Year 1900 is not a leap year.
```

```
运行结果：
Enter the year: 2012
Year 2012 is a leap year.
```

```
运行结果：
Enter the year: 2014
Year 2014 is not a leap year.
```

消除逻辑非运算符

当我们编写一个条件时，结果在表达式前面出现了一个逻辑非运算符，这有时会很难理解或者难以解释。

涉及关系或者等性运算符的表达式。如果逻辑非运算符后面的表达式是关系或者等性表达式，则很容易消除逻辑非运算符并更改表达式，因为我们很容易找到另一个表达式，该表达式与现有的表达式相反。下面演示如何消除关系表达式或者等性表达式前面的逻辑非运算符。

```
!(x < 7)    →    (x >= 7)        !(x >= 7)    →    (x < 7)
!(x > 7)    →    (x <= 7)        !(x <= 7)    →    (x > 7)
!(x == 7)   →    (x != 7)        !(x != 7)    →    (x == 7)
```

涉及逻辑与或者逻辑或运算符的表达式。当表达式涉及逻辑与或者逻辑或运算符时，情况有些复杂。我们需要使用被称为**德摩根定律**（De Morgan's law）的两种形式中的一种，德摩根定律是以逻辑学家奥古斯都·德·摩根（August De Morgan，1806～1871）命名的。德摩根定律的两种形式如下：

```
!(x && y)    →    (!x || !y)        !(x || y)    →    (!x && !y)
```

换言之，如果将逻辑与运算符更改为逻辑或运算符，并在每个表达式前面插入逻辑非运算符，则可以消除逻辑与表达式前面的逻辑非运算符。相反，如果我们将逻辑或运算符更改为逻辑与运算符，并在每个表达式前面插入逻辑非运算符，则可以消除逻辑或表达式前面的逻辑非运算符。

乍一看，我们似乎是消除了一个逻辑非运算符，但是添加了两个逻辑非运算符，结果使得表达式变长，但事实上是简化了表达式并使其更易于理解。

【例 4-10】 假设当温度低于 35 华氏度（太冷）或者高于 90 华氏度（太热）时停止慢跑，则停止慢跑的条件如下：

```
Stop Condition → (temp < 35) || (temp > 90)
```

如果想确定继续慢跑的条件，则可以很容易地将上述表达式取反，结果为：

```
Continue Condition → !((temp < 35) || (temp > 90))
```

根据德摩根定律的第二种形式，可以把表达式转换为：

```
Continue Condition → (!(temp < 35) && !(temp > 90))
```

尽管该表达式中有两个逻辑非运算符，但是可以简化该表达式（使用我们在本节中学习的规则），以获得更好的效果。

```
Continue Condition → (temp >= 35) && (temp <= 90)
```

最后一个表达式表明：只要温度在 35 到 90 之间，我们就可以继续慢跑。

交换 if 和 else 块

有时我们希望交换 if-else 语句中的 if 和 else 块。我们可以通过翻转 if 子句中的条件来实现。当 if 块为空时，这相当实用，如下所示。通过交换 if 和 else 块，我们可以使 else 块为空，并且可以将空语句块删除，从而简化决策。注意，我们不能消除空 if 语句块，但是我们可以消除空 else 语句块。以下代码片段演示了交换 if 和 else 块有助于简化代码：

```
//原始代码              //交换后的代码            //简化后的代码
if (x)                  if (!x)                   if (!x)
{                       {                         {
                          statement;                statement;
}                       }                         }
else                    else
{                       {
  statement;
}                       }
```

短路行为：需要注意的问题

C++ 力图高效，因此会使用所谓的**短路行为**（short-circuit behavior）。短路行为可以发生在逻辑与和逻辑或表达式中。

逻辑与运算符的短路行为。在逻辑与表达式中，C++ 先计算左操作数。如果结果为 false，则不会计算右操作数，因为无论右操作数的值如何，结果都为 false。如果左操作数为 true，则会计算右操作数。虽然这种行为在很多情况下不会影响我们的程序，但当右操作数有副作用时，它可能会影响我们的程序。这意味着我们应该避免使用带有副作用的表达式作为右操作数。以下是两个示例：

```
(3 < 2) && (x = 2)        // 忽略了第二个操作数。x 不会改变
(2 < 6) && (x = 2)        // 计算了第二个操作数。把整数 2 存储到 x 中
```

逻辑或运算符的短路行为。在逻辑或表达式中，C++ 先计算左操作数。如果为 true，则不会计算右操作数，因为无论右操作数的值如何，结果都为 true。如果左操作数为 false，则

会计算右操作数。虽然这种行为在很多情况下不会影响我们的程序，但当右操作数有副作用时，它可能会影响我们的程序。这意味着我们应该避免使用带有副作用的表达式作为右操作数。以下是两个示例。

```
(2 < 5) || (x = 3)          // 忽略了第二个操作数。x 不会改变
(7 < 6) || (x = 5)          // 计算了第二个操作数。把整数 5 存储到 x 中
```

我们建议在逻辑表达式中避免副作用。

4.3 基于特定值的选择结构

在前面两节中，我们讨论了依赖于布尔条件的决策过程。我们测试了一个条件和多个条件。在每一种情况下，如果条件是 true，则执行某些操作，否则，执行另一些操作或者不执行任何操作。有时我们需要做出多个决策，但是在每种情况下要做出的决策不是基于布尔条件，而是基于一些特定的整数值。例如，我们可能必须在一周中的每一天执行不同的任务。这是一个基于与某天对应的七个值（星期几）如 1、2、3、4、5、6、7 的决策过程。虽然我们仍然可以使用七个条件——(day == 1), (day == 2) ，…，(day == 7)，并使用多个 if-else 语句来实现，但 C++ 为此提供了更好的解决方案：switch 语句。接下来我们展开讨论。

4.3.1 switch 语句

C++ 语言中，另一种多分支选择结构是 **switch 语句**，其中决策是基于特定值。图 4-10 显示了一个示例，即测试特定值（1 到 n），并根据测试结果执行 n 个任务中的一个。

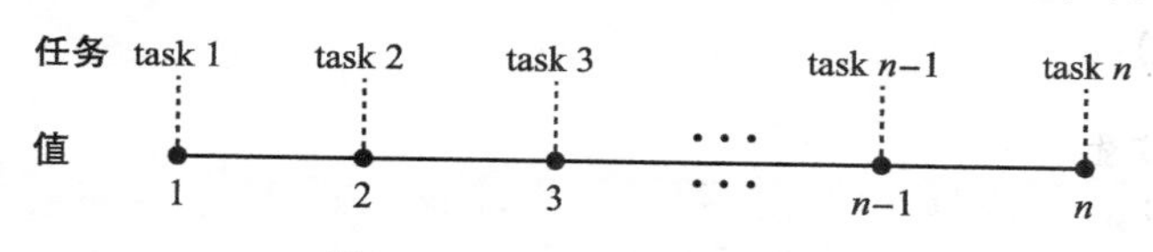

图 4-10 基于整数值的决策

switch 语句的结构

在讨论 switch 语句如何用于多值决策之前，让我们先讨论该语句的结构。基本的 switch 语句是为所谓的贯穿（fall through）流而设计的。图 4-11 显示了包含四种情况的 switch 语句的流程图和语法（我们可以根据需要拥有任意多种情况）。

当表达式的值与其中一种情况中定义的值匹配时，switch 语句做出决策进入贯穿流。换言之，基于一个值进入贯穿流，但是当进入贯穿流后，它会一直走到最后并执行所有剩余的贯穿语句。

在图 4-11 中，首先将表达式（expression）的值与 value1 进行比较。如果匹配成功，则执行 Statement1 到 Statement4。如果表达式（expression）的值与 value1 不匹配，但与 value2 匹配，则执行 Statement2 到 Statement4（将跳过 Statement1）。如果所有值都不匹配，则不执行任何语句。case 子句中的值应该是唯一的。

【例 4-11】 为了演示 switch 语句的工作原理，我们使用程序 4-10。在该程序中，我们将整数值 0 到 6 分配给一周中对应的第几天（星期日到星期六），并查看 switch 语句如何在特定日期（由输入值指定）进入该星期，并打印该日期的星期名称和剩余的星期名称。

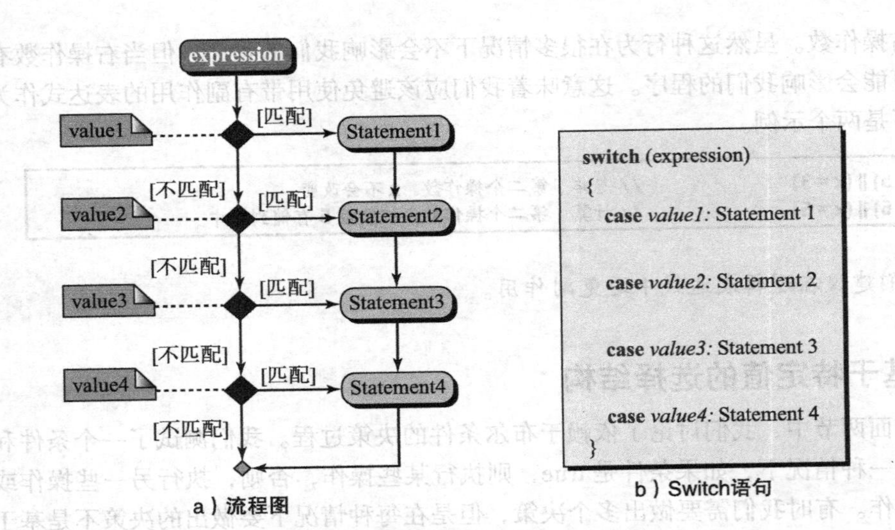

图 4-11　switch 语句的瀑布式流程图

程序 4-10　打印星期名称

```
 1 /*************************************************************
 2  * 使用 switch 语句，                                          *
 3  * 打印从指定日开始直到一周结束的所有星期名称                    *
 4  *************************************************************/
 5 #include <iostream>
 6 using namespace std;
 7
 8 int main ()
 9 {
10   // 声明变量
11   int day;
12   // 输入
13   cout << "Enter a number between 0 and 6: ";
14   cin >> day;
15   // switch 语句（决策和输出）
16   switch (day)
17   {
18       case 0: cout << "Sunday" << endl;
19       case 1: cout << "Monday" << endl;
20       case 2: cout << "Tuesday" << endl;
21       case 3: cout << "Wednesday" << endl;
22       case 4:  cout << "Thursday" << endl;
23       case 5: cout << "Friday" << endl;
24       case 6: cout <<"Saturday" << endl;
25   } // switch 结束
26   return 0;
27 }// main 结束
```

```
运行结果：
Enter a number between 0 and 6: 2
Tuesday
Wednesday
Thursday
```

```
Friday
Saturday
```

运行结果：

```
Enter a number between 1 and 6: 4
Thursday
Friday
Saturday
```

运行结果：

```
Enter a number between 1 and 6: 6
Saturday
```

运行结果：

```
Enter a number between 1 and 6: 8
```

我们运行程序四次。在第一次运行中，我们输入 2，switch 语句在 case 2 进入，并打印一周的其余日期（星期二到星期六）。在第二次运行中，我们输入 4，switch 语句在 case 4 进入，并打印一周的其余日期（星期四到星期六）。在第三次运行中，我们输入 6，程序只打印一周的最后一天（星期六）。在第四次运行中，我们输入 8，它不在整数值列表中，不打印任何内容。

带 break 的 switch 语句

要将 switch 语句用于多值选择，必须添加 break 语句。在遇到 break 语句时将中断程序流，并跳到 switch 语句的末尾。图 4-12 显示的 switch 语句，在每个分支处添加了 break 语句。我们可以看到，每次进入 switch 语句时，只有一条路径，即与表达式匹配的路径。比较图 4-11 和图 4-12 的流程图，以查看其差异。图 4-11 使用贯穿流；图 4-12 使用选择流。

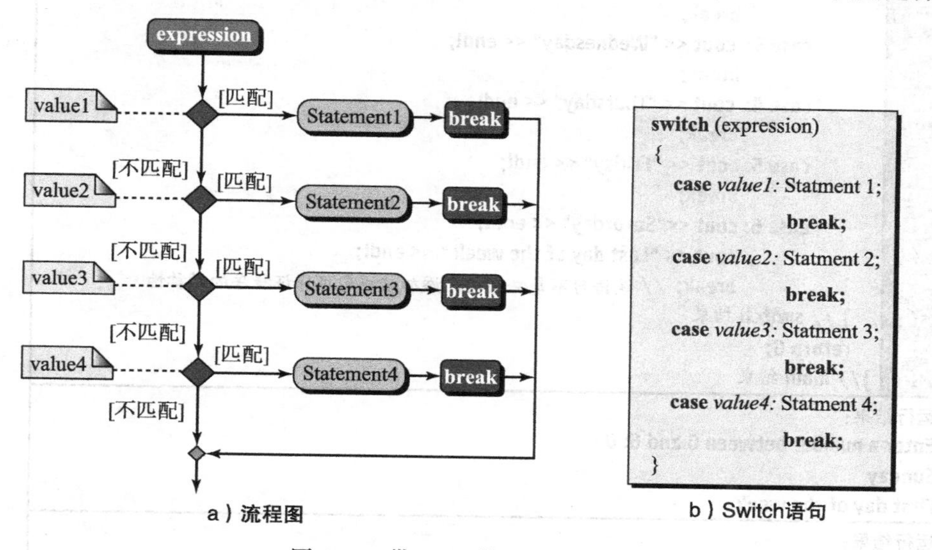

图 4-12　带 break 的 switch 语句

【例 4-12】 让我们将程序 4-10 改为带 break 的 switch 语句。程序 4-11 显示了其结果。在程序的每次运行中，只打印一个星期名称。和前面一样，我们在第一次运行中输入 2，结果是星期二。在第二次运行中，我们输入 4，结果是星期四。当我们输入 8 时，没有匹配的情况，也没有打印任何内容。请注意，最后一个 break 并不是真正必需的，但是优秀的程序员会添加 break 来实现对称性。还要注意，我们在第一种和最后一种情况中添加了第二个输出，以证明在任何 case 分支中也可以执行多条语句。

程序 4-11　打印给定日的星期名称

```
1  /*****************************************************************
2   * 使用带 break 的 switch 语句，                                  *
3   * 打印指定日的星期名称                                            *
4   *****************************************************************/
5  #include <iostream>
6  using namespace std;
7
8  int main ()
9  {
10   // 声明变量
11   int day;
12   // 输入
13   cout << "Enter a number between 0 and 6: ";
14   cin >> day;
15   // switch 语句（决策和输出）
16   switch (day)
17   {
18       case 0: cout << "Sunday" << endl;
19               cout << "First day of the week " << endl;
20               break;
21       case 1: cout << "Monday" << endl;
22               break;
23       case 2: cout << "Tuesday" << endl;
24               break;
25       case 3:  cout << "Wednesday" << endl;
26               break;
27       case 4:  cout << "Thursday" << endl;
28               break;
29       case 5: cout << "Friday" << endl;
30               break;
31       case 6: cout <<"Saturday" << endl;
32               cout << "Last day of the week " << endl;
33               break; // 该语句不是必须的。添加该语句可保证程序的对称性
34    } // switch 结束
35    return 0;
36  }// main 结束
```

```
运行结果:
Enter a number between 0 and 6: 0
Sunday
First day of the week
```

```
运行结果:
Enter a number between 0 and 6: 4
Thursday
```

```
运行结果:
Enter a number between 0 and 6: 6
Saturday
Last day of the week
```

```
运行结果:
Enter a number between 0 and 6: 8
```

添加默认情况

有时存在多种情况下执行相同的操作。在这种情况下，如果所有的情况都在 switch 语句的末尾，那么我们可以通过添加一个**默认情况**（default case）来缩短语句的长度。当所有情况都不匹配时，将进入默认情况。

默认情况也可以用于错误检测。当所有情况都与给定值不匹配时，它可以用来检测错误。但是，注意默认情况只能使用一次，并且是作为最后一种情况。

> 默认情况必须是 switch 语句中的最后一种情况。

这意味着，如果我们使用默认情况捕获多个 case 值，则不能将其用于错误检测。我们需要使用其他工具。下一个示例显示了默认情况的使用。

【例 4-13】 让我们演示如何使用 switch 语句将测试分数更改为字母等级成绩。我们需要解决一个问题：成绩等级基于一定范围的测试分数，而不是一个单一的值。例如，如果分数在 90 到 100 之间，则成绩等级应分配为字母“A”。然而，如果使用除法运算符将范围更改为单个值，则可以使用 switch 语句来解决此问题。如果我们将 0 到 100 之间的分数除以 10，则得到 0 到 10 之间的值。程序 4-12 演示了如何使用此技术解决问题。

程序 4-12　打印测试成绩的等级

```
1  /***************************************************************
2   * 使用 switch 语句打印给定测试分数的等级                          *
3   ***************************************************************/
4  #include <iostream>
5  using namespace std;
6  
7  int main ()
8  {
9    // 声明变量
10   int score;
11   char grade;
12   // 获取输入
13   cout << "Enter a score between 0 and 100: ";
14   cin >> score;
15   // 使用 switch 语句进行决策
16   switch (score / 10)
17   {
18       case 10: grade = 'A';
19                break;
20       case 9 : grade = 'A';
21                break;
22       case 8 : grade = 'B';
23                break;
24       case 7 : grade = 'C';
25                break;
26       case 6 : grade = 'D';
27                break;
28       default: grade = 'F';
29       } // switch 结束
30   // 输出
31   cout << "Score: " << score << endl;
```

```
32      cout << "Grade: " << grade << endl;
33      return 0;
34  }
```

运行结果：
Enter a score between 0 and 100: 71
Score: 71
Grade: C

运行结果：
Enter a score between 0 and 100: 93
Score: 93
Grade: A

运行结果：
Enter a score between 0 and 100: 24
Score: 24
Grade: F

在程序4-12中，我们使用了一个默认情况（default）来处理几种情况（5、4、3、2、1、0），即所有分数小于60的情况。但是，仍然存在一个问题。如果用户输入的分数小于0或者大于100怎么办？程序将给出等级F。为了解决这个问题，我们需要首先测试输入的有效性。一种解决方案如下所示：

```
cout << "Enter a score between 0 and 100: ";
cin >> score;
if (score > 100)
{
   score = 100;
}
if (score < 0)
{
   score = 0;
}
```

在下一章中讨论循环结构时，我们将讨论一种更好的输入验证策略。

合并多种情况

如果多种情况执行的任务相同，我们可以合并这些情况。由于switch语句的贯穿特性允许程序在遇到break语句之前遍历这些情况，故可以合并多种情况。我们将在下一个示例中演示该方法。

【例4-14】 在程序4-13中，我们读取成绩的字母等级并打印通过/不通过的结果。default语句用作错误检测器。

程序4-13　判断所输入的成绩等级是否通过

```
1   /**************************************************************
2    * 使用 switch 语句判断输入的成绩是否通过                          *
3    **************************************************************/
4   #include <iostream>
5   using namespace std;
6
7   int main ()
8   {
```

```
9   //声明变量
10  char grade;
11  //输入
12  cout << "Enter a grade (A, B, C, D, F): ";
13  cin >> grade;
14  //使用 switch 语句的决策部分
15  switch (grade)
16  {
17      case 'A':
18      case 'B':
19      case 'C': cout << "Grade is pass";
20              break;
21      case 'D':
22      case 'F': cout << "Grade is nopass";
23              break;
24      default: cout <<"Error in the input. Try again.";
25  } // switch 结束
26  return 0;
27 }// main 结束
```

```
运行结果:
Enter a grade (A, B, C, D, F): A
Grade is pass
```

```
运行结果:
Enter a grade (A, B, C, D, F): D
Grade is nopass
```

```
运行结果:
Enter a grade (A, B, C, D, F): F
Grade is nopass
```

```
运行结果:
Enter a grade (A, B, C, D, F): G
Error in the input. Try again.
```

```
运行结果:
Enter a grade (A, B, C, D, F): B
Grade is pass
```

```
运行结果:
Enter a grade (A, B, C, D, F): C
Grade is pass
```

4.4 条件表达式

另一种可以用于决策的构造是**条件表达式**（conditional expression）。它是 C++ 中唯一的三元表达式。它使用两个操作符和三个操作数，接下来展开讨论。

4.4.1 条件表达式的结构

条件表达式的语法如下所示：

```
condition ? expression1 : expression2
```

表 4-3 显示了条件表达式在附录 C 表达式列表中的内容。

表 4-3　条件表达式

组别	名称	运算符	表达式	优先级	结合性
条件表达式	条件表达式	? :	expr1 ? expr2 : expr3	4	←

expr1 的类型必须是一个布尔表达式，其值必须为 true 或者 false。expr2 和 expr3 类型必须相同，但通用类型可以是前面章节中讨论的任何类型。整个构造也是一个带有值的表达式。如果 expr1 为 true，则整个表达式的值为 expr2 的值，否则，表达式的值为 expr3 的值。

条件表达式可以在任何可以使用表达式的地方使用。

4.4.2　比较

让我们将条件表达式与双分支 if-else 语句进行比较，如图 4-13 所示。

- 双分支 if-else 语句由一个布尔表达式构成，该语句有一个真分支和一个假分支。条件表达式则由一个布尔表达式和两个表达式组成，这两个表达式的其中一个用于 true 的情况，另一个用于 false 的情况。
- 双分支 if-else 语句是一条语句，它可以在任何可以使用语句的地方使用。条件表达式是一个表达式，它也可以在任何可以使用表达式的地方使用。我们可以通过以下方式之一使用条件表达式：
  a）可以在语句中作为表达式使用。
  b）可以在末尾添加分号，将其更改为独立语句。当第二个和第三个表达式都有副作用时，此方法非常有用。

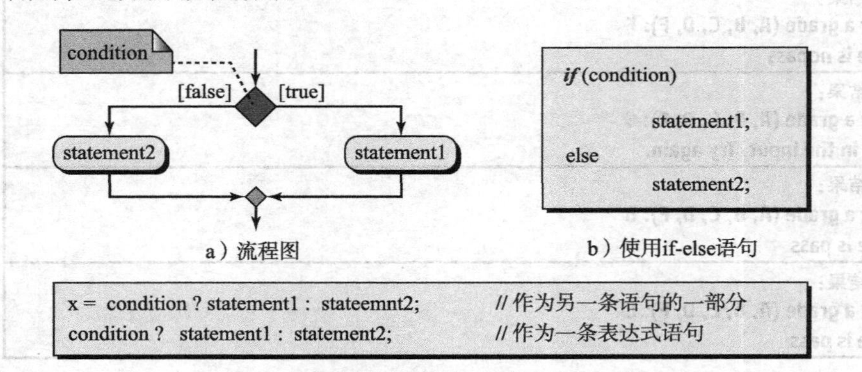

图 4-13　条件表达式与双分支 if-else 语句的比较

【例 4-15】 在程序 4-14 中，我们使用条件表达式打印两个整数中较大的整数。

程序 4-14　条件表达式

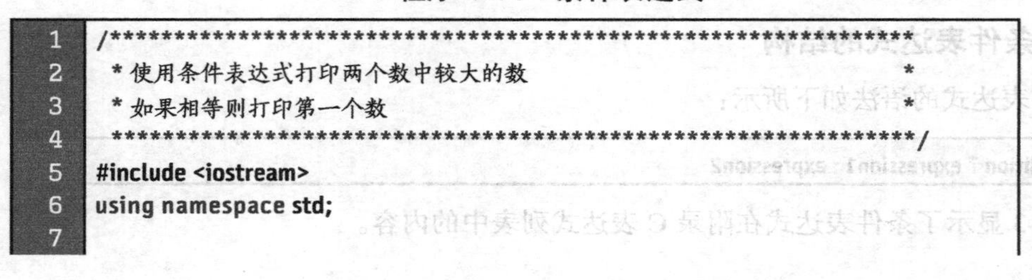

```
1  /*************************************************************
2   * 使用条件表达式打印两个数中较大的数                          *
3   * 如果相等则打印第一个数                                      *
4   *************************************************************/
5  #include <iostream>
6  using namespace std;
7
```

```
8   int main ()
9   {
10    // 声明变量
11    int num1, num2;
12    int larger;
13    // 输入
14    cout << "Enter the first number: ";
15    cin >> num1;
16    cout << "Enter the second number: ";
17    cin >> num2;
18    // 进行决策
19    larger = num1 >= num2 ? num1 : num2;
20    // 输出
21    cout << "The larger is: " << larger;
22    return 0;
23  }// main 结束
```

```
运行结果:
Enter the first number: 40
Enter the second number: 25
The larger is: 40
```

```
运行结果:
Enter the first number: 22
Enter the second number: 67
The larger is: 67
```

4.5　程序设计

在本节中，我们将设计并实现三个程序。我们将使用第 3 章中讨论的软件开发过程：理解问题、开发算法、编写代码。

4.5.1　学生成绩

让我们创建一个程序，帮助教授根据班级学生的三次测试的结果计算其成绩。教授认为学生的分数应该是三个分数中最高分和最低分的平均分。

理解问题

我们需要输入、处理和输出。首先需要输入三次测试成绩。接下来，我们需要找到三个输入的测试成绩中的最大值和最小值。最后，我们需要计算这两个值的平均值，并将学生的成绩设置为平均值。虽然在以后的章节中会讨论更有效的方法，但是我们在这里还是使用嵌套 if-else 结构找到最高成绩，然后找到最低成绩。学生的成绩是最低成绩和最高成绩的平均值。

开发算法

根据上述的理解，我们可以设计以下算法：

1）输入

　a）输入成绩 1（score1）

　b）输入成绩 2（score2）

　c）输入成绩 3（score3）

2）处理

a）查找最高成绩（maxScore）

①如果 score1 大于其他两个成绩，则 maxScore 等于 score1

②否则，如果 score2 大于其他两个成绩，则 maxScore 等于 score2

③否则，maxScore 等于 score3

b）查找最低成绩（minScore）

①如果 score1 小于其他两个成绩，则 minScore 等于 score1

②否则，如果 score2 小于其他两个成绩，则 minScore 等于 score2

③否则，minScore 等于 score3

c）计算学生成绩

①计算 maxScore 和 minScore 之和

②把结果除以 2，得到学生的成绩

3）输出

a）输出三个成绩

b）输出 minScore 和 maxScore

c）输出学生的成绩

编写代码

接下来可以基于开发的算法编写代码（程序 4-15）。程序中唯一需要解释的部分是第 47 行到第 52 行。最高成绩和最低成绩为整数类型，我们希望学生成绩为整数类型。如果最高成绩和最低成绩之和是奇数，则加 1 使其和为偶数。加 1 有效地将除法的结果四舍五入到较高的分数，否则，分数将被截断到较低的分数。

程序 4-15　根据三次测试成绩计算学生的成绩

```
1   /***************************************************************
2    * 根据三次测试成绩计算学生的成绩                                  *
3    * 程序中计算学生的成绩方法如下：                                  *
4    * 学生成绩等于三次测试成绩中最高分和最低分的平均值                  *
5    ***************************************************************/
6   #include <iostream>
7   using namespace std;
8
9   int main ( )
10  {
11    // 声明变量
12    int score1, score2, score3, maxScore, minScore, score;
13    // 输入
14    cout << "Enter the first score: ";
15    cin >> score1;
16    cout << "Enter the second score: ";
17    cin >> score2;
18    cout << "Enter the third score: ";
19    cin >> score3;
20    // 查找最高成绩
21    if (score1 > score2 && score1 > score3)
22    {
```

```
23          maxScore = score1;
24      }
25      else if (score2 > score1 && score2 > score3)
26      {
27          maxScore = score2;
28      }
29      else
30      {
31          maxScore = score3;
32      }
33      // 查找最低成绩
34      if (score1 < score2 && score1 < score3)
35      {
36          minScore = score1;
37      }
38      else if (score2 < score1 && score2 <= score3)
39      {
40          minScore = score2;
41      }
42      else
43      {
44          minScore = score3;
45      }
46      // 计算学生成绩并四舍五入取整
47      int temp = maxScore + minScore;
48      if (temp % 2 == 1)
49      {
50          temp += 1;
51      }
52      score = temp / 2;
53      // 输出结果
54      cout << "Scores: " << score1 << " " << score2 << " " << score3 << endl;
55      cout << "minimum and maximum scores: ";
56      cout << minScore << " " << maxScore << endl;
57      cout << "Student score: " << score;
58      return 0;
59  }
```

运行结果:

```
Enter the first score: 78
Enter the second score: 92
Enter the third score: 79
Scores: 78 92 79
minimum and maximum scores: 78 92
Student score: 85
```

运行结果:

```
Enter the first score: 65
Enter the second score: 93
Enter the third score: 60
Scores: 65 93 60
minimum and maximum scores: 60 93
Student score: 77
```

4.5.2　计算给定收入的税款

让我们设计一个程序来计算和打印给定收入的税款。

理解问题

许多国家的所得税都采用分段计算的方法，即不同收入段的税率各不相同，如图 4-14 所示。

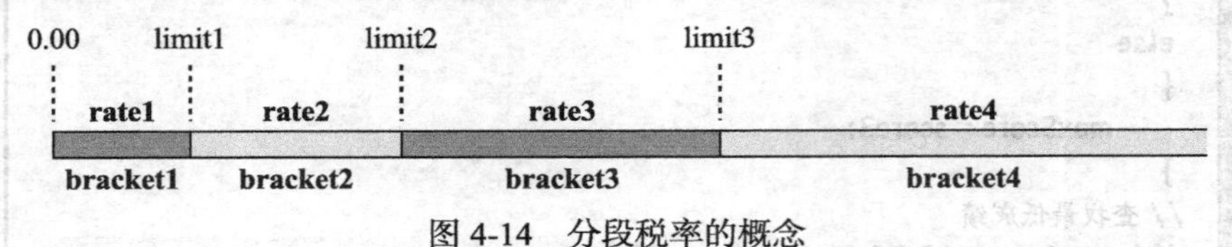

图 4-14　分段税率的概念

总税款是各纳税区段的税款的总和（表 4-4）。位于纳税区段 1 的纳税人的纳税额为 tax1。位于纳税区段 2 的纳税人的纳税额为 tax1 和 tax2 之和。位于纳税区段 3 的纳税额人的纳税额为 tax1、tax2 和 tax3 之和。位于纳税区段 4 的纳税人的纳税额为 tax1、tax2、tax3 和 tax4 之和。表中的 diff 表示收入和前一个 limit 的差。

表 4-4　纳税区段计算表

纳税区段	tax1	tax2	tax3	tax4
bracket1	diff × rate1			
bracket2	limit1 × rate1	diff × rate2		
bracket3	limit1 × rate1	limit2 × rate2	diff × rate3	
bracket4	limit1 × rate1	limit2 × rate2	limit3 × rate3	diff × rate4

开发算法

这种情况下的算法比较简单。首先输入收入，然后使用嵌套的 if-else 结构计算税额，最后打印税额，如图 4-15 所示。

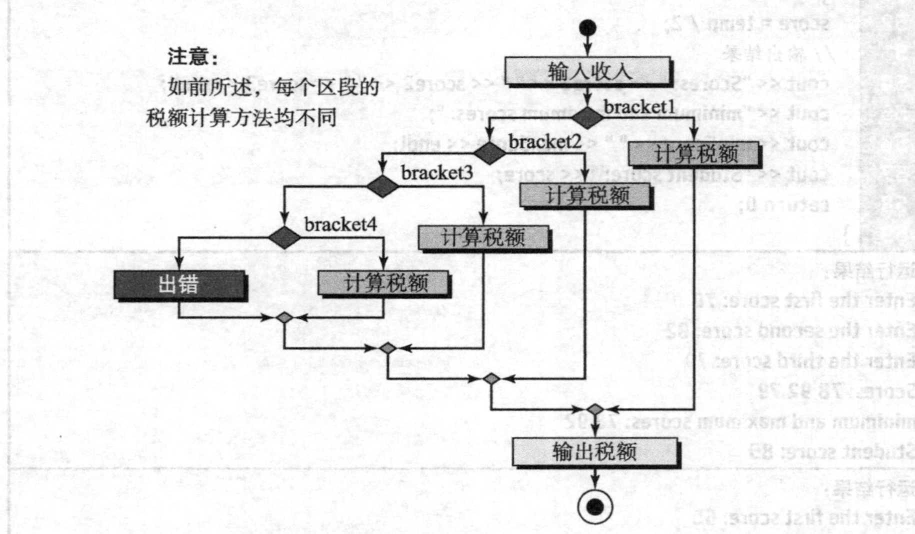

图 4-15　计算所得税的算法设计流程图

编写代码

我们基于前面的两部分内容编写了一个程序。因为税率和纳税区段将来可能发生变化，我们设计了可重用的程序，而且，我们在程序中声明税率和区段上限为变量，而不是使用字面量固定其值。程序中还使用了四个布尔变量，分别用于定义收入属于哪个纳税区段（bracket1 到 bracket4）。如果 bracket1 为 true，则收入在第一个纳税区段，依此类推。为了防止错误（例如输入的收入为负值），我们在 if-else 语句中创建了另一个分支来处理错误条件。如果用户输入了负收入，程序将在第 44 行打印一条消息，并立即在第 45 行返回。第 45 行的返回语句对于打印第 48 行和第 49 行的结果是必需的。参见程序 4-16。

程序 4-16 个人税赋计算

```
1   /***************************************************************
2    * 基于四个纳税区段                                             *
3    * 计算个人税赋                                                 *
4    ***************************************************************/
5   #include <iostream>
6   using namespace std;
7
8   int main ( )
9   {
10    // 变量的声明和初始化
11    double income, tax ;
12    bool bracket1, bracket2, bracket3, bracket4;
13    double limit1 = 10000.00, limit2 = 50000.00, limit3 = 100000.00;
14    double rate1 = 0.05, rate2 = 0.10, rate3 = 0.15, rate4 = 0.20;
15    // 输入
16    cout << "Enter income in dollars: " ;
17    cin >> income;
18    // 定义纳税区段
19    bracket1 = (income <= limit1) && (income >=0) ;
20    bracket2 = (income > limit1) && (income <= limit2);
21    bracket3 = (income > limit2) && (income <= limit3);
22    bracket4 = (income > limit3);
23    // 计算税额
24    if (bracket1)
25    {
26        tax = income * rate1;
27    }
28    else if (bracket2)
29    {
30        tax = limit1 * rate1 + (income - limit1) * rate2 ;
31    }
32    else if (bracket3)
33    {
34        tax = limit1 * rate1 + (limit2 - limit1) * rate2 +
35                                (income - limit2) * rate3 ;
36    }
37    else if (bracket4)
38    {
39        tax = limit1 * rate1 + (limit2 - limit1) * rate2 +
40              (limit3 - limit2) * rate3 + (income - limit3) * rate4 ;
41    }
42    else
43    {
44        cout << "Error! Invalid income!";
```

```
45        return 0;
46      }
47      // 打印收入和税赋
48      cout << "Income: " << income << endl;
49      cout << "Tax due: " << tax;
50      return 0;
51  }
```

```
运行结果：
Enter income in dollars: 8500
Income: 8500
Tax due: 425
```

```
运行结果：
Enter income in dollars: 14500
Income: 14500
Tax due: 950
```

```
运行结果：
Enter income in dollars: -5
Error! Invalid income!
```

```
运行结果：
Enter income in dollars: 123000
Income: 123000
Tax due: 16600
```

第三次运行中，收入（−5）无效。除此以外，其他每次运行中的收入均位于其中一个纳税区段。在第一次运行中，纳税人位于纳税区段1中，将其收入的5%支付给税收部门（425美元）。在第二次运行中，纳税人位于纳税区段2中，为纳税区段1（前10000美元）支付500美元，为纳税区段2（剩余的4500美元）支付450美元，依此类推。请读者验证每个结果的正确性。

4.5.3 日期编号

假设我们给一年中的所有日期进行编号。例如，1月1日是第1天，12月31日是第365天。我们想找出给定月和日的日期在一年中的日期编号。

理解问题

为了解决这个问题，我们需要知道月和日。我们需要知道月，因为必须计算过去几个月的总天数。

开发算法

在这个算法中，我们需要执行两个任务：累计过去所有月份的天数，并累加当前月份的天数。第一个任务是通过switch语句完成的，每个月是一种情况。第二个任务是在switch语句之后执行。使用贯穿方法的switch语句（不带break语句）非常适合这个问题，因为我们需要累计前几个月经过的天数。为了使switch语句正确工作，我们从一年中的最后一个月开始直到第一个月。例如，如果我们在七月，我们进入第七个条件（case 7），但是我们将六月到一月的天数相加。当我们从switch语句中出来时，我们再累加当前月份的天数。

编写代码

程序4-17实现了上述算法。请注意，我们忽略了闰年的情况，闰年时2月份包含29天。我们把闰年的情况留作练习。我们也忽略了输入错误检查，这将在以后的章节中讨论。

程序 4-17 日期编号

```
1  /**************************************************************
2   * 计算当前日期的日期编号                                        *
3   * 从年初开始所经过的天数（包括当前日期）                        *
4   **************************************************************/
5  #include <iostream>
6  using namespace std;
7
8  int main ( )
9  {
10   // 变量的声明
11   int month;
12   int day;
13   int totalDays = 0;
14   // 输入当前日期的月和日
15   cout << "Enter month: ";
16   cin >> month;
17   cout << "Enter day of month: ";
18   cin >> day;
19   // 各月份的天数
20   int m01 = 31;
21   int m02 = 28;
22   int m03 = 31;
23   int m04 = 30;
24   int m05 = 31;
25   int m06 = 30;
26   int m07 = 31;
27   int m08 = 31;
28   int m09 = 30;
29   int m10 = 31;
30   int m11 = 30;
31   // 使用贯穿的 switch 语句计算总天数
32   switch (month)
33   {
34       case 12 : totalDays += m11;
35       case 11 : totalDays += m10;
36       case 10 : totalDays += m09;
37       case 9  : totalDays += m08;
38       case 8  : totalDays += m07;
39       case 7  : totalDays += m06;
40       case 6  : totalDays += m05;
41       case 5  : totalDays += m04;
42       case 4  : totalDays += m03;
43       case 3  : totalDays += m02;
44       case 2  : totalDays += m01;
45       case 1  : totalDays += 0;
46   }
47   // 把当月的天数累加到总天数中
48   totalDays += day;
49   // 打印结果
50   cout << "Day number: " << totalDays;
51   return 0;
```

```
52 }
```

运行结果：

```
Enter month: 1
Enter day of month: 23
Day number: 23
```

运行结果：

```
Enter month: 4
Enter day of month: 12
Day number: 102
```

运行结果：

```
Enter month: 11
Enter day of month: 24
Day number: 328
```

运行结果：

```
Enter month: 12
Enter day of month: 31
Day number: 365
```

本章小结

为了解决一些问题，我们需要在检测结果为 true/false 的条件的基础上做出决策。这被称为选择。最常用的选择结构是单分支选择结构，在 C++ 中通过 if 语句来实现。我们讨论的第二种选择结构是使用 if-else 语句的双分支选择结构。

有时决策问题过于复杂，以至于无法用关系表达式和等性表达式来解决。我们可以通过逻辑表达式将关系表达式和等性表达式结合在一起以实现选择。我们讨论了三个逻辑运算符：逻辑非、逻辑与和逻辑或。

C++ 中的另一种多分支选择结构是基于特定值进行决策的 switch 语句。

另一个可以用于决策的构造称为条件表达式。它是 C++ 中唯一的三元表达式。条件表达式使用两个运算符和三个操作数。

思考题

1. 如果最初 x = 4 并且 y = 0，那么在执行以下代码后，x 和 y 的值是什么？

```
if (x != 0)
{
    y = 3;
}
```

2. 如果最初 x = 4、y = 0 并且 z = 2，那么在执行以下代码后，x、y 和 z 的值是什么？

```
if ( z == 2 )
{
    y = 1;
}
else
{
    x = 3;
}
```

3. 如果最初 x = 4、y = 0 并且 z = 2，那么在执行以下代码后，x、y 和 z 的值是什么？

```
if ( x > y || y < z )
```

```
{
    x = 10;
}
```

4. 如果最初 x = true、y = false 并且 z = true，那么在执行以下代码后，x、y 和 z 的值是什么？

```
if ( x )
{
    if ( y )
    {
        z = false;
    }
    else
    {
        y = true;
    }
}
```

5. 如果最初 x = 4、y = 0 并且 z = 2，那么在执行以下代码后，x、y 和 z 的值是什么？

```
if ( z == 0 || y != 0 )
{
    if ( z <= 2 )
    {
        z = 4;
    }
}
else
{
    y = 5;
    z = y + x;
}
```

6. 如果最初 x = 0、y = 0 并且 z = 1，那么在执行以下代码后，x、y 和 z 的值是什么？

```
switch ( x )
{
    case 0 :  x = 2;
              y = 3;
    case 1 :  x = 4;
    default : y = 5;
              x = 1;
}
```

7. 如果最初 x = 2、y = 1 并且 z = 1，那么在执行以下代码后，x、y 和 z 的值是什么？

```
switch ( x )
{
    case 0 :  x = 3;
              y = 2;
    case 1 :  x = 2;
    default : y = 3;
              x = 4;
}
```

8. 用 switch 语句重写以下多分支选择语句。

```
if ( x == 2 )
{
    x++;
```

```
}
else if (x == 3)
{
    x--;
}
else
{
    cout << "End!";
}
```

9. 用 if-else 多分支选择语句重写以下 switch 语句。

```
switch (x)
{
    case 1: cout << "One" << endl;
        break;
    case 2: cout << "Two" << endl;
        break;
    default: cout << "Any" << endl;
        break;
}
```

10. 用 switch 语句重写以下代码片段。

```
if (ch == 'A' || ch == 'a')
    countA++;
else if (ch == 'E' || ch == 'e')
    countE++;
else if (ch == 'I' || ch == 'i')
    countI++;
else
    cout << "Error--Not A, E, or I " << endl;
```

11. 编写一个代码片段，如果整数变量 score 大于或者等于 90，则将值 1 赋给变量 best。
12. 编写一个代码片段，如果浮点数变量 amount 大于 5.4，则将值 4 累加到整数变量 num。
13. 编写一个代码片段，如果变量 flag 为 true，则打印整数变量 num 的值。
14. 编写一个代码片段，执行如下操作：如果变量 divisor 不等于 0，则将变量 dividend 除以 divisor，结果存储到 quotient；如果 divisor 等于 0，则把它赋值给 quotient。然后输出所有三个变量。假设变量 dividend 和 divisor 是整数，而 quotient 是双精度数。
15. 编写一个代码片段，检查变量 flag 是否为 true。如果为 true，则读取整数变量 num1 和 num2 的值。然后计算和打印这两个输入数的和、平均值。如果变量 flag 为 false，则不执行任何操作。
16. 编写一个代码片段，测试一个整数变量 num1 的值。如果值为 10，则计算其平方。如果值为 9，则读入一个新值到 num1。如果值为 2 或者 3，则将 num1 乘以 99，并打印结果。使用嵌套 if 语句（而不是 switch 语句）实现代码。

编程题

1. 编写一个程序，接收用户输入的一个两位无符号整数，然后反转这些数字并打印出结果。如果输入数字超过两位，使用 if 语句终止程序。
2. 编写一个程序，给定三个整数，打印其中最小数。
3. 编写一个程序，给定一个 1 到 12（包括 1 和 12）之间的整数，打印出对应的月份。
4. 编写一个程序，给定车辆类型（“c”表示小汽车 car，“b”表示公共汽车 bus，“T”表示卡车 truck）和车辆在停车场的时间，根据下面显示的费率返回停车费。

```
car: $2 per hour          bus:$3 per hour          truck: $4 per hour
```

5. 编写一个程序，确定学生的成绩等级。程序读取三个测试成绩（0 到 100 之间），根据以下规则计算成绩等级：
   a. 如果平均分大于或者等于 90，则成绩等级为‘A’。
   b. 如果平均分在 80 和 90 之间，则检查第三个成绩，如果第三个成绩大于 90 分，则成绩等级为‘A’，否则成绩等级为‘B’。
   c. 如果平均分在 70 和 80 之间，则检查第三个成绩，如果第三个成绩大于 80 分，则成绩等级为‘B’，否则成绩等级为‘C’。
   d. 如果平均分在 60 和 70 之间，则检查第三个成绩，如果第三个成绩大于 70 分，则成绩等级为‘C’，否则成绩等级为‘D’。
   e. 如果平均分小于 60，则检查第三个成绩，如果第三个成绩大于 60 分，则成绩等级为‘D’，否则成绩等级为‘F’。
6. 编写一个程序，计算并打印学生在大学的总学费。学生每学分支付 10 美元的费用，最多 12 个学分。一旦他们支付了 12 个学分，他们就没有额外的每学分费用。注册费是每名学生 10 美元。
7. 批发商店对购买的商品按购买数量给予折扣，如下所示：

数量	折扣	数量	折扣
1～9	0%	50～99	5%
10～49	3%	≥100	10%

   编写一个程序，给定一个商品的购买数量和单价，计算折扣后的总金额。
8. 编写一个程序，给定一个点的 x 和 y 值坐标，打印其位于笛卡尔（直角）坐标系中的象限（1、2、3 和 4）。例如，如果 x 和 y 都是正数，则该点位于第一个象限中。如果 x 和 y 都为负，则该点位于第三象限，依此类推。
9. 修改程序 4-17 以考虑闰年的额外天数（闰年的 2 月份是 29 天而不是 28 天）。程序必须从用户获取年份，并使用以下公式来确定该年份是否为闰年。

```
leapYear = (year % 400) || (year % 4 && ! (year % 100))
```

10. 编写一个程序，使用泽勒同余法（Zeller's congruence）找出任意给定日期是星期几。Zeller 发现了以下公式，使用年、月和日计算该日期是星期几。

```
weekday = (day + 26 * (month + 1) / 10 +
                   year + year / 4 - year /100 + year /400 ) % 7
```

    该公式基于以下内容：
    a. 一周有七天，所以计算必须取 7 的模数。
    b. 第一项 day，表示当月的每一天将星期数向前移动 1。
    c. 第二项 26*（month+1）/10 是泽勒同余。泽勒没有考虑每个月的天数，而是设计了这个解决方案。泽勒的公式适用于太阳系，即一年从三月份开始，而不是从一月份开始。因此，我们必须将 1 月和 2 月视为上一年的 13 月和 14 月。换句话说，2017 年 1 月必须被视为 2016 年的第 13 个月。
    d. 下一项 year，将各年的星期数向前移动 1，因为**非闰年**包括 365 天，当除以 7 时，余数为 1。
    e. 下一项 year/4，是考虑到闰年的情况。我们知道，能被 4 整除的年份**可能**是闰年，因此星期数加 1。
    f. 下一项 year/100，用于从前一年中减去能被 100 整除的年份。
    g. 下一项 year/400，定义能被 400 整除的年份为闰年，星期数加 1。
11. 编写一个程序，给定一个以美元为单位的金额，打印出面额为 100、50、20、10、5 和 1 的最少钞票数量。使用条件表达式（:?）打印非零的钞票数量。

循环结构

我们在前几章中编写的程序是基于顺序语句结构或者选择语句结构。在大多数程序中，我们需要重复代码行。循环结构允许程序多次迭代执行一段代码。C++ 提供了三种用于重复执行一组语句的结构：while 语句、for 语句和 do-while 语句，使用循环结构无须在代码中多次重复书写代码。为了使用这三种循环结构，我们需要学习更多的表达式类型。

学习目标

阅读并学习本章后，读者应该能够

- 复习在循环语句中可以用作计数器的表达式。
- 理解 while 语句的语法，及其在创建计数器控制或者事件控制的循环结构中的使用。
- 理解如何在计数器控制的循环语句中初始化和更新计数器。
- 理解如何在事件控制的循环语句中使用哨兵（sentinel）、文件尾标记（EOF）或者标志（flag）。
- 理解如何使用 for 语句的语法以替换计数器控制的 while 语句。
- 理解如何在 for 语句的头部中实现初始化、测试和计数器更新。
- 理解 do-while 语句的语法，以及如何将其用于数据验证。
- 分析和比较三种循环结构。
- 讨论与循环结构有关的其他语句，以及如何通过改变循环语句的结构来避免使用这些语句。

5.1 概述

虽然仅仅使用顺序结构和选择结构就可以解决计算机科学中的许多问题，但这些方法并不能解决所有问题。假设我们要将 100 个测试分数更改为相应的成绩等级。我们有三个选择：

1）仅仅使用一条选择语句编写一个程序，并运行 100 次。这种方法非常耗时。

2）使用 100 条选择语句编写一个程序。这种方法程序代码会很长，并且如果学生人数发生变化，则需要修改程序代码。

3）使用循环语句（也称为循环结构），它允许我们重复一个活动或者一组活动，需要重复多少次就可以重复多少次。

5.1.1 前缀表达式和后缀表达式

当使用循环语句时，我们通常需要使用计数器（counter）来检查循环的次数，并在完成任务后终止循环。C++ 设计了两组表达式来模拟计数器：后缀递增 / 递减（postfix increment/decrement）和前缀递增 / 递减（prefix increment/decrement）。这两种表达式是附录 C 中讨论的 C++ 表达式的一部分。为了快速引用，表 5-1 中重复列举了这些表达式。请注意，后缀递增 / 递减表达式属于后缀表达式组，前缀递增 / 递减表达式属于一元表达式组。

表 5-1 前缀表达式和后缀表达式

组别	名称	运算符	表达式	优先级	结合性
后缀表达式	后缀递增 后缀递减	++ --	lvalue++ lvalue--	18	→
一元表达式	前缀递增 前缀递减	++ --	++lvalue --lvalue	17	←

后缀递增表达式和后缀递减表达式

后缀递增表达式和后缀递减表达式由单个操作数和紧跟在操作数后的一个运算符（++ 或者 --）组成，如图 5-1 所示。

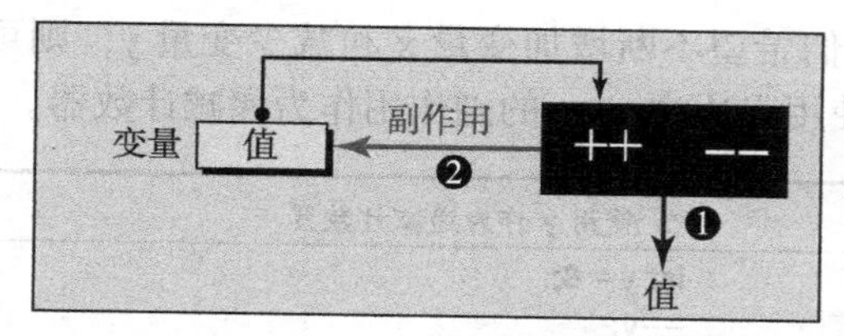

图 5-1 后缀递增表达式和后缀递减表达式

在这两个表达式中，操作数是一个变量的名称（前文讨论过的左值）。表达式的返回值是变量的前一个值，但变量的值被递增或者递减（副作用）。换句话说，如果变量 x 的值为 3，则表达式 x++ 返回 3，并将 x 的内容递增到 4。同样，如果 y 的值为 8，则表达式 y-- 返回 8，并将 y 递减为 7，如图 5-2 所示。

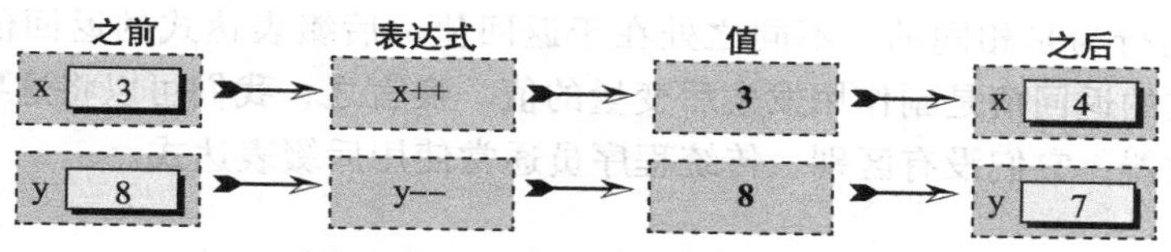

图 5-2 后缀表达式的值和副作用

【例 5-1】 图 5-2 表明，如果我们希望不断增加变量 x 和减少变量 y，则可以使用表达式 x++ 的副作用作为递增计数器，使用表达式 y-- 的副作用作为递减计数器。

```
//使用 x 作为递增计数器
int x = 3;
x++;
x++;
x++;
cout << x;  // 打印 6
```

```
// 使用 y 作为递减计数器
int y = 8;
y--;
y--;
y--;
cout << y;  // 打印 5
```

前缀递增表达式和前缀递减表达式

前缀递增表达式和前缀递减表达式由单个操作数和在操作数前面的一个运算符（++ 或 --）组成，如图 5-3 所示。

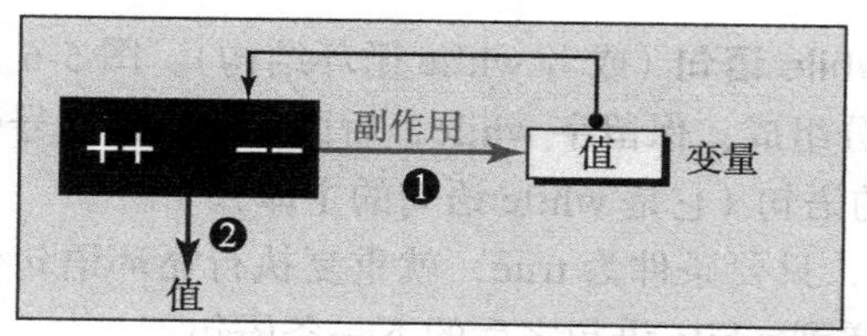

图 5-3 前缀递增表达式和前缀递减表达式

在这两个表达式中，副作用（变量被递增或者被递减）在返回值之前产生。图 5-4 显示了使用这些表达式递增或者递减变量的结果。

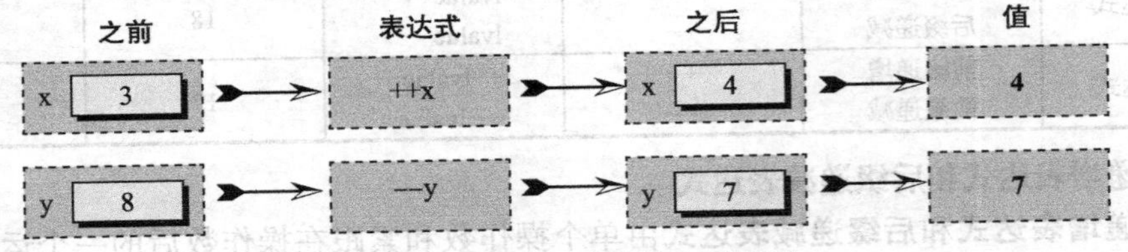

图 5-4　前缀表达式的值和副作用

【例 5-2】 图 5-4 表明，如果我们希望不断增加变量 x 和减少变量 y，则可以使用表达式 ++x 的副作用作为递增计数器，使用表达式 --y 的副作用作为递减计数器。

// 使用 x 作为递增计数器	// 使用 y 作为递减计数器

int x = 3;
x++;
x++;
x++;
cout << x;　// 打印 6
	int y = 8;
	--y;
	--y;
	--y;
	cout << y;　// 打印 5

前缀表达式和后缀表达式的比较

当比较后缀表达式和前缀表达式时，我们发现如果只对它们的副作用（变量的内容）感兴趣时，它们的行为是相同的。不同之处在于返回值。后缀表达式的返回值是变量的原始值；前缀表达式的返回值是副作用发生后变量的值。换言之，我们可以将这些表达式中的任何一个用作计数器，它们没有区别。传统程序员通常使用后缀表达式。

5.1.2　循环语句

C++ 中使用了三种循环语句：while 语句、for 语句和 do-while 语句，如图 5-5 所示。我们将在本章中讨论这三种循环语句。

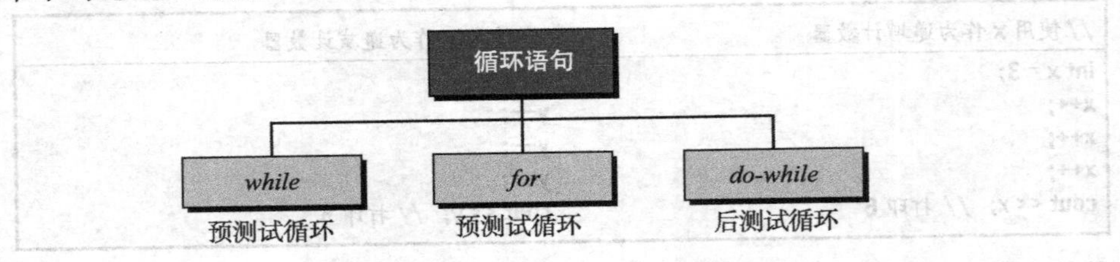

图 5-5　C++ 中的三种循环语句

5.2　while 语句

我们讨论的第一种循环语句是 **while 语句**（或者 while 循环结构）。图 5-6 显示了该语句的流程图和语法。while 语句由三部分组成：保留字 while，后面是包含在括号中的布尔表达式（称为条件），再后面是一个单独的语句（它是 while 语句的主体）。

while 循环结构重复对条件求值，只要条件为 true，就重复执行循环语句体。当条件变为 false 时，循环停止。接着控制传递到 while 语句之后的下一条语句。

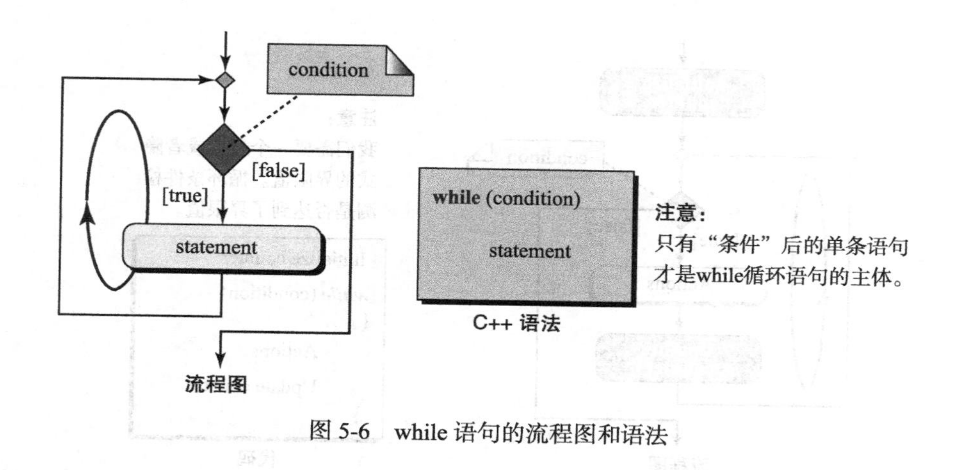

图 5-6 while 语句的流程图和语法

while 语句末尾不需要英文分号。

循环语句体必须是单条语句。如果我们需要在循环体中执行多条语句，则将这些语句包含在一条复合语句中，以使它们成为 while 循环的单条语句。然而，无论需要多少条语句来执行主体，最佳实践是始终使用复合语句作为循环语句体。常用的 while 语句的代码片段如下所示。

```
while (condition)
{
  statement-1;
  ...
  statement-n;
}
```

while 语句的主体必须是单条语句。

图 5-6 中的结构是 while 语句背后的一般思想。它没有定义条件检查的内容，也没有定义当我们想要重复语句时如何将条件设计为真，或者当我们需要停止重复语句时如何将条件设计为假。C++ 提供了两种类型的 while 语句：计数器控制循环和事件控制循环。每种类型用于解决不同的问题。

5.2.1 计数器控制 while 语句

通常，我们事先知道循环体必须重复多少次。在这种情况下，我们可以使用计数器。我们可以在循环之前将计数器设置为初始值，在每次迭代中递增或者递减计数器，完成任务后停止循环。例如，如果我们决定每天早上做 20 个俯卧撑，我们使用的是一个**计数器控制循环结构**。我们事先知道要重复多少次。我们数一数做俯卧撑的次数，做完后停下来。

为了将如图 5-6 所示的一般 while 循环结构更改为计数器控制的循环，必须在循环之前添加一条语句来初始化计数器，并在循环体中增加语句更新计数器。现在，循环体必须是一条复合语句，其中包括要执行的操作以及计数器的更新。图 5-7 显示了流程图和 C++ 代码。除了计数器之外，我们还需要一个界限值来检查计数器的值。界限值可以是特定的，也可以是隐式的。

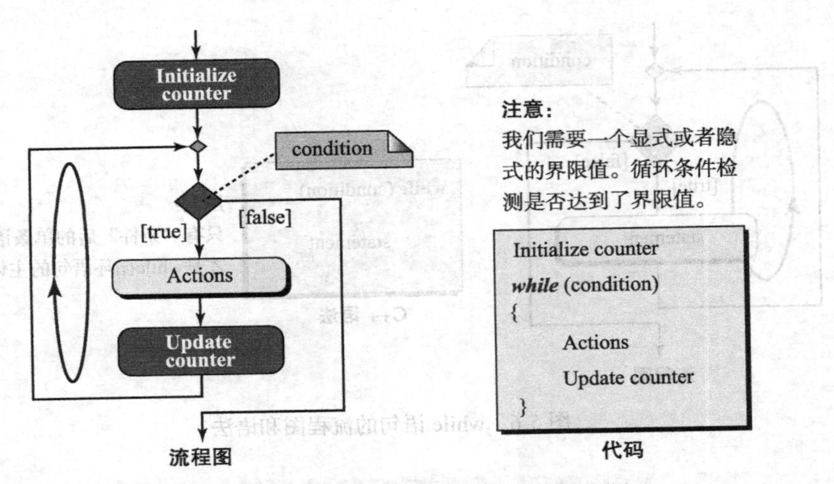

图 5-7　计数器控制 while 语句的流程图和语法

在计数器控制的 while 循环中，计数器在进入循环之前被初始化，在每次迭代中被检查是否到达界限值，并在循环中被更新。

【例 5-3】 假设我们需要打印信息“Hello world!”10 次。我们可以使用 while 循环，而不是重复打印信息 10 次。在这里，界限值（10）可以在循环条件中直接指定，如程序 5-1 所示。

程序 5-1　使用 while 语句输出信息 10 次

```
1  /*************************************************************
2   * 使用 while 循环打印某信息 10 次                              *
3   *************************************************************/
4  #include <iostream>
5  using namespace std;
6
7  int main ( )
8  {
9     // 声明和初始化计数器
10    int counter = 0;
11    // while 语句
12    while (counter < 10)  // 重复次数固定为 10
13    {
14        cout << "Hello world!" << endl;
15        counter++;
16    }
17    return 0;
18 }
```

```
运行结果:
Hello world!
Hello world!
Hello world!
Hello world!
Hello world!
Hello world!
```

```
Hello world!
Hello world!
Hello world!
Hello world!
```

我们使用了一个计数器，它在第10行被初始化，在第15行被更新。我们想要使用的界限值（10）在第12行中被字面量编码为循环条件。当计数器为0、1、2、3、4、5、6、7、8和9时，循环体将重复。当计数器变为10时，我们退出循环，如表5-2所示。

表5-2 每次循环中计数器和布尔表达式的值

计数器	0	1	2	3	4	5	6	7	8	9	10
条件	true	true	true	true	true	true	true	true	true	true	false
是否执行语句体？	yes	yes	yes	yes	yes	yes	yes	yes	yes	yes	no

无论我们运行程序多少次，消息都会被打印10次，因为重复次数是固定的。

【例5-4】 假设教授需要编写一个程序来确定每个学生在学期内的平均分数。教授每学期给每个学生考核四次。在这种情况下，迭代界限值也是固定的，但是教授可以使用不同学生的分数运行程序多次。程序5-2演示了该情况。

程序5-2 计算一系列分数的平均值

```
1  /***************************************************************
2   * 使用计数器控制的 while 循环结构，                             *
3   * 计算每个学生的平均成绩                                        *
4   ***************************************************************/
5  #include <iostream>
6  #include <iomanip>
7  using namespace std;
8
9  int main ()
10 {
11   // 声明变量
12   int score;
13   int sum = 0;
14   double average;
15   // 循环
16   int counter = 0; // 初始化计数器
17   while (counter < 4)  // 测试计数器
18   {
19        // 处理（读取成绩并累加到 sum）
20        cout << "Enter the next score (between 0 and 100): ";
21        cin >> score;
22        sum = sum + score;
23        counter++; // 增加计数器的值
24   }
25   // 结果
26   average = static_cast <double> (sum) / 4;
27   cout << fixed << setprecision (2) << showpoint;
28   cout << "The average of scores is: " << average;
29   return 0;
30 }
```

```
运行结果：
Enter the next score (between 0 and 100): 78
Enter the next score (between 0 and 100): 68
Enter the next score (between 0 and 100): 92
Enter the next score (between 0 and 100): 88
The average of scores is: 81.50
```

```
运行结果：
Enter the next score (between 0 and 100): 80
Enter the next score (between 0 and 100): 90
Enter the next score (between 0 and 100): 76
Enter the next score (between 0 and 100): 74
The average of scores is: 80.00
```

【例 5-5】 在上一个程序中，控制循环迭代的界限值被设置为固定值。我们也可以设计一个计数器控制的循环，在这个循环中，每次运行的界限值都可以改变。例如，程序 5-3 显示了我们如何打印从 0 到界限值的整数值，每次运行中，界限值由用户指定。注意，在这个程序中，我们实际上打印了计数器变量 count 的值。

程序 5-3　打印 *n* 个整数

```
1   /***********************************************************************
2    *基于 while 语句循环打印从 0 到 n 的整数                                *
3    ***********************************************************************/
4   #include <iostream>
5   using namespace std;
6
7   int main ()
8   {
9     // 声明界限值和计数器
10    int n, count;
11    // 输入界限值 n
12    cout << "Enter the number of integers to print: ";
13    cin >> n;
14    // 打印整数
15    count = 0;
16    while (count < n)
17    {
18        cout << count << endl;
19        count++;
20    }
21    return 0;
22  }
```

```
运行结果：
Enter the number of integers to print: 5
0
1
2
3
4
```

```
运行结果：
Enter the number of integers to print: 0
```

```
运行结果:
Enter the number of integers to print: -3
```

注意，在最后两次运行中，第 16 行中的条件（count < n）为 false，程序永远不会进入循环体。

【例 5-6】 我们从数学上知道，级数求和的计算公式如下所示：

$$1 + 2 + 3 + 4 + \cdots + n = n\,(n+1) / 2$$

$$1 + 4 + 9 + 16 + \cdots + n^2 = n\,(n+1)\,(2n + 1) / 6$$

$$1 + 8 + 27 + 64 + \cdots + n^3 = n^2\,(n+1)^2 / 4$$

我们可以使用一个简单的计数器控制 while 语句来检查这些公式的结果，如程序 5-4 所示。注意，在每种情况下，计数器都是上述公式中 n 的值。还要注意，在本例中，计数器需要从 1（而不是 0）开始，而且循环条件应该是（counter <= n）。

程序 5-4　使用 while 循环计算级数的和

```
1   /*****************************************************************
2    * 使用 while 循环计算三个级数的和                                   *
3    *****************************************************************/
4   #include <iostream>
5   using namespace std;
6
7   int main ()
8   {
9     // 声明和初始化
10    int sum1 = 0, sum2 = 0, sum3 = 0;
11    int n;
12    // 输入界限值 n
13    cout << "Enter the value of n: ";
14    cin >> n;
15    // while 语句
16    int counter = 1;   // 初始化计数器
17    while (counter <= n)
18    {
19          sum1 += counter;
20          sum2 += counter * counter;
21          sum3 += counter * counter * counter;
22          counter++; // 更新计数器
23    }
24    // 打印结果
25    cout << "Value of n: " << n << endl;
26    cout << "Value of sum1: " << sum1 << endl;
27    cout << "Value of sum2: " << sum2 << endl;
28    cout << "Value of sum3: " << sum3 << endl;
29    return 0;
30  }
```

```
运行结果:
Enter the value of n: 5
Value of n: 5
Value of sum1: 15
```

```
Value of sum2: 55
Value of sum3: 225
```

```
运行结果：
Enter the value of n: 15
Value of n: 15
Value of sum1: 120
Value of sum2: 1240
Value of sum3: 14400
```

5.2.2　事件控制 while 语句

有时我们想重复一个行为，但我们并不知道需要执行的次数。我们知道当一个事件发生时我们应该停止。这也可能发生在现实生活中。例如，我们想在早上做俯卧撑，但是我们不知道我们能持续做多久。我们觉得累的时候应该停下来。换言之，疲劳是我们想要停止循环的事件。图 5-8 显示了一个**事件控制循环**。

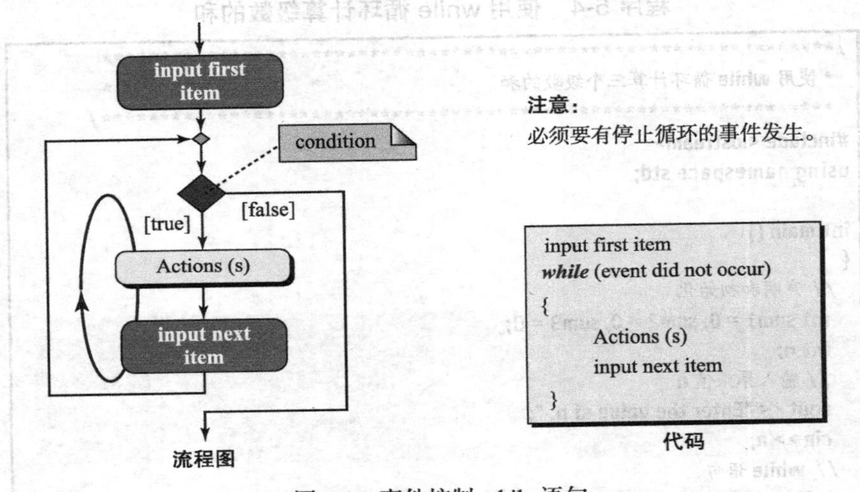

图 5-8　事件控制 while 语句

在计算机程序设计中发生的事件包括多个种类，但最常见的事件包括：哨兵（sentinel）的出现、文件结束（end-of-file）标记的出现、某种条件（condition）的发生。

哨兵控制的 while 循环结构

哨兵（sentinel）是防止未经授权的人通过某位置的警卫。在数据处理中，哨兵是一个添加到数据列表中的值，用于显示何时需要停止处理。哨兵的类型与其余数据相同，但其值必须与之前的所有数据项不同。注意，当我们使用 while 循环作为哨兵控制的循环时，不会处理哨兵（while 循环是一个预测试循环）。

> 哨兵控制的 while 循环结构中，不会处理哨兵。

【例 5-7】 假设我们有一长串需要累计的正数。我们不想在把数值输入计算机之前先统计其个数。由于我们知道列表中没有负数，我们可以在列表末尾添加一个负数作为哨兵，如下所示：

14 23 71 87 … 66 12 −1

在这种情况下，哨兵是整数 -1。使用哨兵可以对若干数值求和，如程序 5-5 所示。注意，我们在循环之前和循环当中分别使用了一个提示语句，以提示用户如何操作。程序 5-5 演示了如何使用哨兵来控制迭代。

程序 5-5　哨兵控制的循环

```
 1 /***************************************************************
 2  *使用哨兵控制的 while 循环结构                                  *
 3  *计算若干数值之和                                              *
 4  ***************************************************************/
 5 #include <iostream>
 6 using namespace std;
 7 
 8 int main ()
 9 {
10   //变量的声明
11   int sum = 0;
12   int num;
13   //循环，包括第一次输入
14   cout << "Enter an integer (-1 to stop): ";
15   cin >> num;
16   while (num != -1)
17   {
18        sum = sum + num;
19        cout << "Enter an integer (-1 to stop): ";
20        cin >> num ; //更新哨兵
21   }
22   //输出结果
23   cout << "The sum is: " << sum;
24   return 0;
25 }
```

```
运行结果：
Enter an integer (-1 to stop): 25
Enter an integer (-1 to stop): 22
Enter an integer (-1 to stop): 12
Enter an integer (-1 to stop): 67
Enter an integer (-1 to stop): -1
The sum is: 126
```

文件结束标记控制的 while 循环结构

我们可以从键盘或者文件中读取数据。在这两种情况下，我们都可以在一个**文件结束标记（EOF）控制的循环结构**中使用 EOF 标记作为哨兵。键盘可以输入 EOF，文件中始终存在 EOF。EOF 是一个标记，表示不再通过键盘输入数据，或者已经到达了文件的末尾。EOF 标记允许我们读取数据项并控制循环的重复。如果我们使用键盘作为数据源，则 EOF 在 UNIX 环境中为 ctrl+d（按住 ctrl 键的同时按 d 键），在 Windows 环境中为 ctrl+z（按住 ctrl 键的同时按 z 键）。另外，如果我们使用文件作为数据源，则在创建文件时 EOF 标记被添加。图 5-9 显示了这两种情况。

我们该如何使用 EOF 呢？幸运的是，>> 运算符有副作用，并返回一个值。它读取流中的下一项。如果下一个项目是 EOF，则丢弃它并返回 false（循环终止）。如果下一个项目不是 EOF，它将下一个项目存储在运算符 >> 之后的变量中，并返回 true（循环继续）。

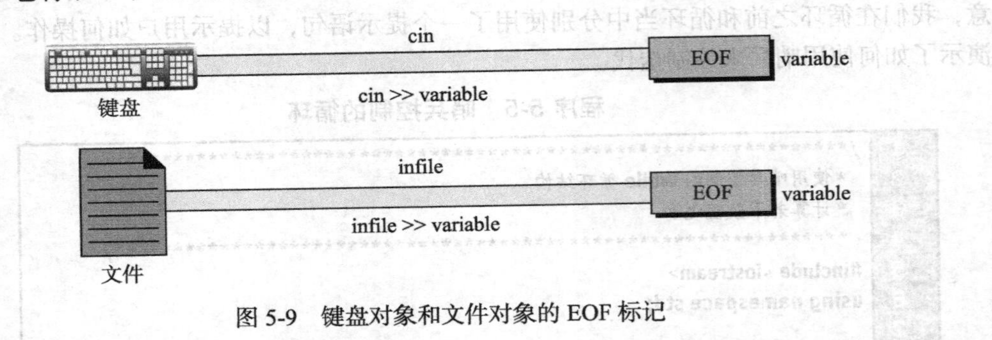

图 5-9 键盘对象和文件对象的 EOF 标记

//从键盘读取数据	//从文件读取数据

while (cin >> num)
{
process;
}
	while (infile >> num)
	{
	process;
	}

【例 5-8】 程序 5-6 使用 EOF 控制循环重新实现了上一个示例。

程序 5-6 使用 EOF

```
1   /*****************************************************************
2    * 使用文件结束标记（EOF）控制的 while 循环结构                     *
3    * 计算从键盘输入的若干数值之和                                     *
4    *****************************************************************/
5   #include <iostream>
6   using namespace std;
7   
8   int main ()
9   {
10    // 变量的声明
11    int sum = 0;
12    int num;
13    // 循环，包括初始化
14    cout << "Enter the first integer (EOF to stop): ";
15    while (cin >> num)
16    {
17        sum = sum + num;
18        cout << "Enter the next integer: ";
19    }
20    // 输出结果
21    cout << "The sum is: " << sum;
22    return 0;
23  }
```

运行结果：

```
Enter the first integer: 24
Enter the next integer: 12
Enter the next integer: 123
Enter the next integer: 14
Enter the next integer: ^Z
The sum is: 173
```

【例 5-9】 我们可以读取名为 numbers.dat 的文件并执行同样的操作，该文件中有五个数值：100、200、300、400 和 500，如程序 5-7 所示。

程序 5-7 使用文件的 EOF

```
1  /***************************************************************
2   * 使用文件结束标记（EOF）控制的 while 循环结构                    *
3   * 计算包含在文件中的若干数值之和                                  *
4   ***************************************************************/
5  #include <iostream>
6  #include <fstream>
7  using namespace std;
8
9  int main ()
10 {
11   // 变量的声明
12   int sum = 0;
13   int num;
14   ifstream infile;
15   // 打开文件
16   infile.open ("numbers.dat");
17   // while 循环
18   while (infile >> num)
19   {
20       sum = sum + num;
21   }
22   // 输出结果
23   cout << "The sum is: " << sum;
24   infile.close ();
25   return 0;
26 }
```

```
运行结果：
The sum is: 1500
```

标志控制的 while 循环结构

有时，我们寻找的事件并不像哨兵或者 EOF 一样，一定发生在一个项目列表的末尾。事件可以是项目列表中任何位置都可能发生的条件。例如，假设我们正在搜索整数列表中大于或者等于 100 的第一个整数。我们不能使用哨兵，因为我们不是在寻找一个特定的停止循环的值（比如 100）。停止循环的数值可以是 100、132、150 等。我们不能使用 EOF，因为我们不想处理所有整数。我们正在寻找一个可以定义为（number >= 100）的条件。解决方案是一个**标志控制的循环结构**。我们创建一个称为标志（flag）的布尔变量，在进入循环之前将标志设置为 false，当条件发生时将其设置为 true。但是，如果条件不发生，我们需要将标志与其他方法组合以避免死循环。换言之，我们将标志控制条件与哨兵控制条件或者 EOF 控制条件结合起来。

> 标志控制的循环机制通常与另一种循环机制相结合，以防止死循环。

【例 5-10】 假设我们要打印整数列表中大于或者等于 150 的第一个整数。我们可以使用一个布尔标志，在进入循环之前将其设置为 false。如果找到这个数值，我们将标志设置为 true 并跳出循环。否则，我们让 EOF 标记终止循环。程序 5-8 使用了这种设计方法。

程序 5-8　使用标志

```
 1  /*****************************************************************
 2   * 使用 EOF 和标志终止循环                                        *
 3   *****************************************************************/
 4  #include <iostream>
 5  #include <fstream>
 6  using namespace std;
 7
 8  int main ()
 9  {
10    // 声明变量
11    ifstream infile;
12    int num;
13    bool flag;
14    // 打开文件
15    infile.open ("numbers.dat");
16    // 循环查找大于或等于 150 的数值
17    flag = false;
18    while (infile >> num && !flag )
19    {
20          if (num >= 150)
21          {
22                cout << "The number is: " << num;
23                flag = true;
24          }
25    }
26    // 检查标志
27    if (!flag)
28    {
29          cout << "The number was not found!";
30    }
31    infile.close ();
32    return 0;
33  }
```

```
运行结果：
The number is: 170
```

```
运行结果：
The number was not found!
```

注意，我们使用相同的文件（每次包含不同的内容）测试程序。在第一次运行时，文件的内容如下：

```
90 110 120 135 170 200 230 EOF
```

当 170 被读取时，程序退出循环，因为它与我们正在查找的条件相匹配（大于或等于 150），标志（flag）被设置为 true，即 !flag 为 false，尽管其他条件（infile >> num）为 true，依然会终止循环。在第二次运行中，文件的内容是：

```
90 110 120 135 137 140 EOF
```

标志（flag）永远不会变为 true，但是当我们到达 EOF 时，另一个条件（infile >> num）变为 false，因此循环被终止。

如果在循环中找到满足条件的数值，则打印出来。如果循环被终止，标志（flag）仍然是 false，则表明已经到达了文件的结尾。

陷阱

虽然 while 语句有许多要避免的陷阱，但我们在这里重点讨论其中的四个。

零迭代循环（zero-iteration loop）。我们必须避免一个从不迭代的循环，当布尔表达式在第一次测试时即为假时会发生这种情况。在这种情况下，循环的主体永远不会执行，不管我们运行程序多少次。不幸的是，这个问题在编译时或者运行时不会被捕获。我们必须非常小心以避免这种情况。以下是零迭代的两种情况。

```
while (false)                    while (2 == 3)
{                                {
  ...                              ...
}                                }
```

无限迭代循环（infinite-iteration loop，死循环）。while 语句陷阱的另一个例子是一个死循环，它一直重复执行其主体。当每次测试布尔表达式的结果都为 true 时，就会发生这种情况。在这种情况下，循环永远不会停止。通常应该避免这种类型的循环，但是我们也会创建这样的循环来让服务器程序一直运行，我们将在下一节中详细讨论。

```
while (true)                     while (2 != 3)
{                                {
  ...                              ...
}                                }
```

空循环体（empty-body loop）。有时我们不小心编写了空循环体的循环结构。例如，如果在 while 语句的头部后面添加了英文分号，就会发生这种情况。编译器认为我们不希望使用花括号，并且分号前面有一个空语句。实际上的循环体被解释为在循环之后执行一次的独立语句。编译器不会产生错误，但程序不会按我们预期的方式执行（逻辑错误）。

```
while (expression); // 空表达式（忽略了复合语句）
{
  ...
}
```

在布尔表达式中使用浮点值。正如我们在前一章中所讨论的，浮点值取决于计算机系统中数值的精度。例如，3.141516 和 3.1415 不相等，但如果精度仅为小数点后四位，则二者可能看起来相等。因此，在测试浮点数是否相等时必须非常小心。

创建延迟

C++ 提供了一些工具（将在后续章节中讨论）来创建程序中的延迟。我们也可以使用 while 语句来创建延迟。我们可以使用一个计数器控制的 while 语句，它的主体只有更新语句。循环不做任何事情，只进行迭代，结果会造成延迟。下面是一个例子。如果计算机在一秒钟内执行 10 000 次迭代，下面的代码会产生 10 秒的延迟，这意味着这种类型的延迟取决于计算机的速度。

```
int counter = 0;
while (counter < 100000)
{
  counter++;
}
```

5.2.3 while 语句分析

如果我们将 while 语句视为如图 5-10 所示的决策语句的组合，则可以更好地理解 while 语句的行为。在每个循环执行过程中，有一个真路径（true path）和一个假路径（false path）。图 5-10 显示，测试次数总是比重复次数多一次。如果循环体重复 n 次，则需要进行 $n+1$ 次测试。在前 n 次测试中，结果为 true；在最后一次测试中，结果为 false。这意味着即使没有执行主体，我们也需要进行一次测试，其结果是 false。以上分析表明 while 语句是一个**预测试循环**（pre-test loop）。在每次迭代执行主体之前，必须先执行测试。

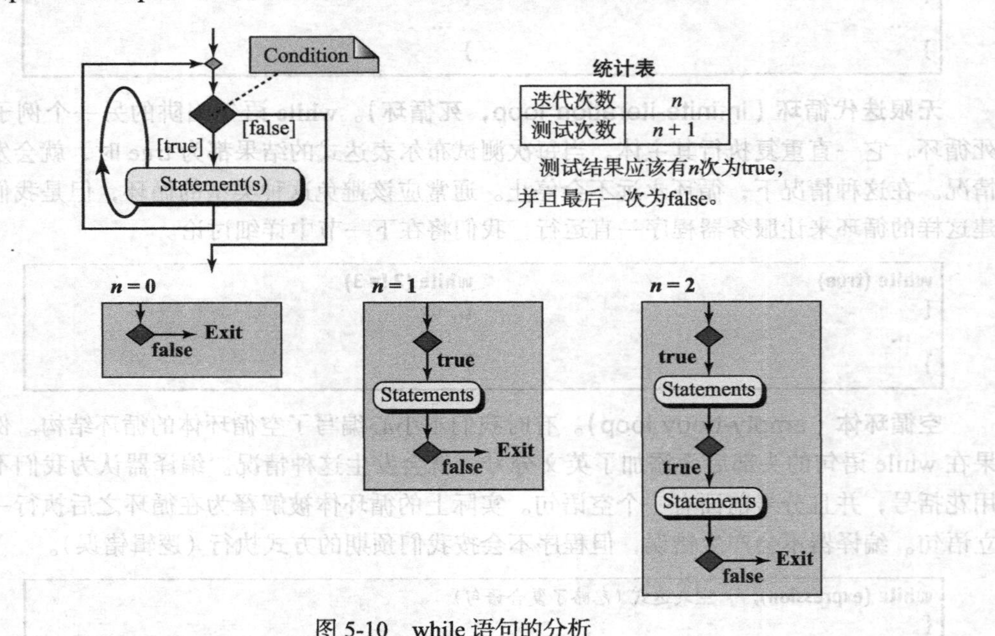

图 5-10 while 语句的分析

while 语句是一个预测试循环。如果执行了 n 次循环体，则条件被测试了（$n+1$）次。

5.3 for 语句

在前一节中我们讨论了可以使用 while 语句作为计数器控制的循环。但是 C++ 提供了另一种循环构造（称为 **for 语句**），特别适用于需要计数器控制迭代的场合。它将循环初始化、条件测试和更新这三个部分组合到循环构造本身中。简单比较这两种循环结构，结果表明，对于计数器控制的循环结构，for 语句比 while 语句更加紧凑，如图 5-11 所示。

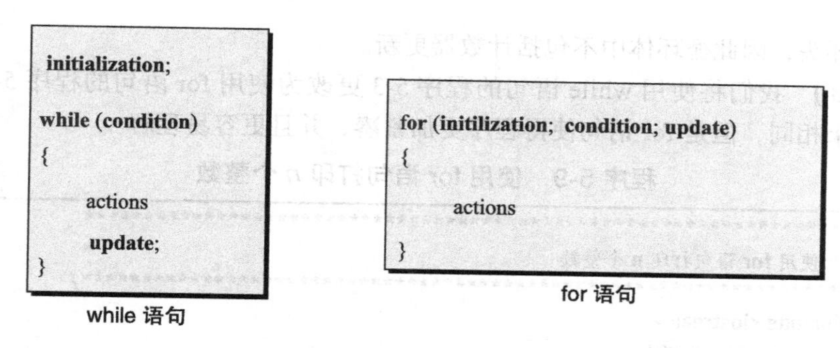

图 5-11 while 循环和 for 循环的比较

5.3.1 循环头

我们现在详细讨论 for 语句循环头中的三个组成部分。

初始化

初始化部分是初始化计数器的表达式。按惯例，计数器通常使用小写字母变量名，例如 i、j、k。我们可以在循环头之前声明计数器，或者在初始化计数器的同时声明计数器。

```
// 在循环头之前声明计数器              // 在循环头中声明计数器
int i;
for (i = 0 ; ... ; ... )              for (int i = 0 ; ... ; ... )
{                                     {
  actions                               actions
}                                     }
```

在第一种方法中，计数器的作用域扩展到循环边界之外；在第二种方法中，计数器的作用域仅在循环内部。

循环条件

循环条件部分是循环头中的第二部分。与 while 语句一样，for 语句中的循环条件也是布尔表达式。如果循环条件为 true，则执行循环体，否则，退出循环，这表示 for 语句是一个预测试循环。如果循环条件为空（在第一个和第二个分号之间），则默认为 true。

如果循环条件为空，则默认为 true。

【例 5-11】 以下 for 循环语句无法通过编译，因为循环头中缺少循环条件。

```
for (int i = 0 ; i++) {...}
```

【例 5-12】 以下 for 循环语句编译成功并且创建死循环，因为缺少循环条件，故默认为 true。

```
for (int i = 0 ; ; i++) {...}
```

更新

更新部分在执行循环体之后执行。更新后，再次测试循环条件，以查看是否应该再次执行循环体。更新部分也可以省略，这意味着必须在循环体中显式或者隐式地进行更新。

5.3.2 循环体

for 语句的循环体与相应的 while 语句的循环体相同，但有一个例外：由于计数器更新已移动到循环头，因此循环体中不包括计数器更新。

【例 5-13】 我们将使用 while 语句的程序 5-3 更改为使用 for 语句的程序 5-9。我们可以看到，结果相同，但是 for 语句使得程序更加紧凑，并且更容易理解。

程序 5-9　使用 for 语句打印 *n* 个整数

```
1   /***************************************************************
2    * 使用 for 语句打印 n 个整数                                    *
3    ***************************************************************/
4   #include <iostream>
5   using namespace std;
6   
7   int main ()
8   {
9     // 声明变量
10    int n;
11    // 获取 n 的值
12    cout << "Enter the number of integers to print: ";
13    cin >> n;
14    // 循环
15    for (int counter = 0; counter < n; counter++)
16    {
17        cout << counter << " ";
18    }
19    return 0;
20  }
```

```
运行结果:
Enter the number of integers to print: 5
0 1 2 3 4
```

```
运行结果:
Enter the number of integers to print: 3
0 1 2
```

【例 5-14】 假设我们要打印 1 到 300 之间可以被 7 整除的整数。但是，输出结果要打印在一张由 10 列组成的表中。程序 5-10 显示了解决方案。

我们知道最后一个数值必须小于或者等于 300，但我们不知道输出结果的行数和列数。我们在一行中打印数据，直到列数变为 10，然后转移到下一行。

程序 5-10　打印可以被 7 整除的整数

```
1   /***************************************************************
2    * 打印 1 到 300 之间的可以被 7 整除的整数,                      *
3    * 结果打印在一张由 10 列组成的表中                              *
4    ***************************************************************/
5   #include <iostream>
6   #include <iomanip>
7   using namespace std;
8   
9   int main ()
10  {
11    // 变量的声明和初始化
12    int lower = 1;
```

```
13     int higher = 300;
14     int divisor = 7;
15     int col = 1;
16    // 处理循环
17     for (int i = lower; i < higher ; i++)
18     {
19         if (i % divisor == 0)
20         {
21              cout << setw(4) << i;
22              col++;
23              if (col > 10 )
24              {
25                  cout << endl;
26                  col = 1;
27              }
28         }
29     }
30     return 0;
31  }
```

运行结果：

```
  7  14  21  28  35  42  49  56  63  70
 77  84  91  98 105 112 119 126 133 140
147 154 161 168 175 182 189 196 203 210
217 224 231 238 245 252 259 266 273 280
287 294
```

【例 5-15】 假设我们要打印某个月的日历，给定该月的天数和该月的第一天是星期几（0 到 6 分别表示星期日到星期六）。程序 5-11 给出了解决方案。

程序 5-11　打印一个月的日历

```
1   /*****************************************************************
2    * 打印某个月的日历，                                              *
3    * 给定该月的天数和该月的第一天是星期几                              *
4    *****************************************************************/
5   #include <iostream>
6   #include <iomanip>
7   using namespace std;
8
9   int main ()
10  {
11    // 变量的声明和初始化
12    int startDay;
13    int daysInMonth;
14    int col = 1;
15    // 验证一个月中的天数
16    do
17    {
18        cout << "Enter the number of days in the month (28, 29, 30, or 31): ";
19        cin >> daysInMonth;
20    } while (daysInMonth < 28 || daysInMonth > 31);
21    // 验证第一天是星期几
```

```
22    do
23    {
24        cout << "Enter start day (0 to 6): ";
25        cin >> startDay;
26    } while (startDay < 0 || startDay > 6);
27    // 打印标题
28    cout << endl;
29    cout << "Sun Mon Tue Wed Thr Fri Sat" << endl;
30    cout << "--- --- --- --- --- --- ---" << endl;
31    // 打印第一天前面的空白符
32    for (int space = 0; space < startDay; space++)
33    {
34        cout << "    ";
35        col++;
36    }
37    // 打印日历
38    for (int day = 1; day <= daysInMonth; day++)
39    {
40        cout << setw(3) << day << " ";
41        col++;
42        if (col > 7)
43        {
44            cout << endl;
45            col = 1;
46        }
47    }
48    return 0;
49 } // main 结束
```

运行结果：

```
Enter the number of days in the month (28, 29, 30, or 31): 31
Enter start day (0 to 6): 3

Sun Mon Tue Wed Thr Fri Sat
--- --- --- --- --- --- ---
              1   2   3   4
  5   6   7   8   9  10  11
 12  13  14  15  16  17  18
 19  20  21  22  23  24  25
 26  27  28  29  30  31
```

5.4 do-while 语句

我们讨论的最后一个重复语句是 **do-while 语句**，在该语句中，逻辑表达式在每次迭代结束时而不是在开始时进行测试。因此，do-while 循环是一个**后测试循环**。由于每次迭代中的逻辑表达式都是在执行循环体之后测试的，所以 do-while 语句中的循环体总是至少执行一次（如图 5-12 所示）。

do-while 语句由五个部分组成：保留字 do、循环体语句（通常是用花括号括起来的复合语句）、保留字 while、括号中的逻辑表达式和分号。注意，在 do-while 语句的末尾需要英文分号。

在 do-while 语句的末尾需要英文分号。

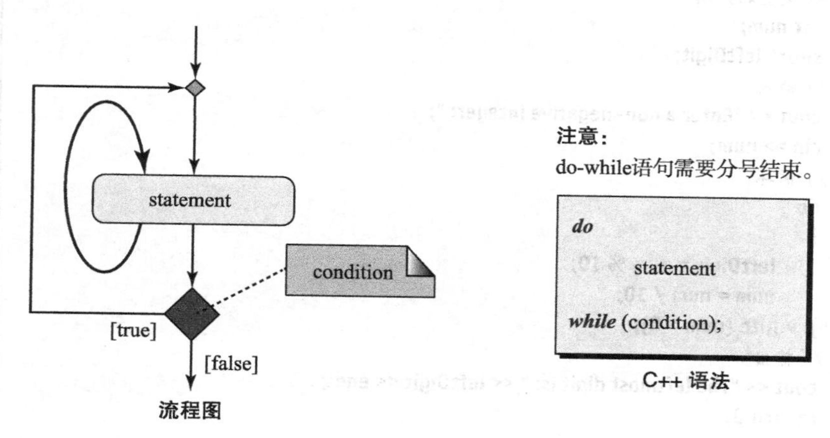

图 5-12　do-while 语句的流程图和语法

5.4.1　事件控制的循环结构

do-while 语句被设计用于事件控制的循环。但是，do-while 循环的后测试特性需要不同的设置来控制循环。

至少执行一次迭代

当问题要求循环体至少执行一次时，建议使用 do-while 循环。

当问题要求循环体至少执行一次时，我们使用 do-while 循环。

【例 5-16】 我们要编写一个程序来提取和打印任何非负整数的最左边的数字。换言之，如果输入整数为 247，则应打印数字 2；如果输入整数为 7562，则应打印数字 7。程序 5-12 包含此问题的解决方案。

在第一次运行中，整数只有一个数字。循环的第一次迭代提取唯一的数字，num 的值变为 0，这意味着循环终止。

在第二次运行中，整数有三个数字。在这里，第一次迭代提取最右边的数字（1），但是第二次迭代丢弃这个数字并用新的最右边的数字（3）替换它，因为当我们进入第二次迭代时，num 的值是 23。一个三位数的数值要经过三次迭代。这个问题表明我们确实需要至少一次迭代，因为输入的整数至少有一个数字（即使整数是 0）。

程序 5-12　提取一个整数的最左边数字

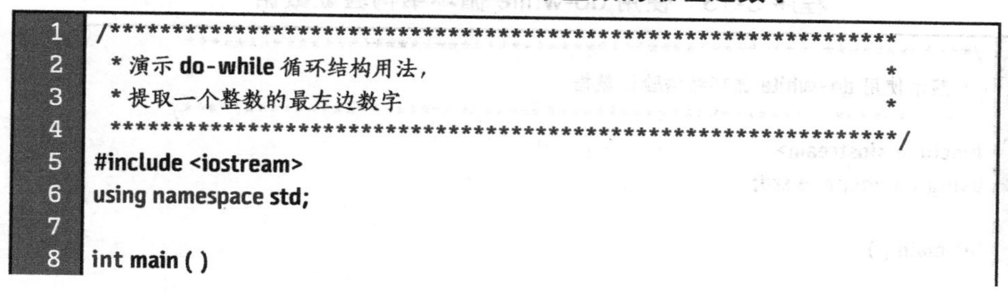

```
1  /**************************************************************
2   * 演示 do-while 循环结构用法，                                   *
3   * 提取一个整数的最左边数字                                         *
4   **************************************************************/
5  #include <iostream>
6  using namespace std;
7
8  int main ( )
```

```
9   {
10      // 变量的声明
11      int num;
12      short leftDigit;
13      // 输入
14      cout << "Enter a non-negative integer: ";
15      cin >> num;
16      // 循环
17      do
18      {
19          leftDigit = num % 10;
20          num = num / 10;
21      } while (num > 0);
22      // 输出
23      cout << "The leftmost digit is: " << leftDigit << endl;
24      return 0;
25  }
```

```
运行结果:
Enter a non-negative integer: 5
The leftmost digit is: 5
```

```
运行结果:
Enter a non-negative integer: 231
The leftmost digit is: 2
```

数据验证

另一个需要至少迭代一次的常见示例是**数据验证**（data validation）。当用户输入的数据必须与一组标准匹配时，数据验证应该包括在程序中。例如，当我们在航空公司预订系统中输入航班信息时，一些数据是必需的，一些数据是可选的。数据验证确保我们以正确的格式输入了所有必需的数据。

【例 5-17】 数据输入通常要求数据在预定义的范围内。例如，当我们将百分制成绩转换为等级成绩时，测试分数通常在 0 和 100 之间。如果用户输入的数值超出范围，程序要求用户重新输入，直到用户输入的数值位于正确的范围内。程序 5-13 演示了该数据验证。注意，使用逻辑或表达式和不相等表达式来定义不可接受的范围。

在第一次运行中，用户输入位于有效范围内的百分制成绩。循环体只执行一次，输入的成绩被接受。在第二次运行中，用户两次都输入了超出范围的成绩。循环体重复三次，前两次迭代中输入的成绩被丢弃，第三次迭代中输入的成绩被接受。程序表明，do-while 循环结构的设计适用于数据验证。

程序 5-13　使用 do-while 循环结构验证数据

```
1   /***************************************************************
2    * 演示使用 do-while 循环结构验证数据                              *
3    ***************************************************************/
4   #include <iostream>
5   using namespace std;
6
7   int main ( )
```

```
8   {
9     // 变量的声明
10    int score;
11    char grade;
12    // 输入验证循环结构
13    do
14    {
15        cout << "Enter a score between 0 and 100: ";
16        cin >> score;
17    } while (score < 0 || score > 100);
18    // 决策
19    switch (score / 10)
20    {
21        case 10: grade = 'A';
22        break;
23        case 9: grade = 'A';
24        break;
25        case 8: grade = 'B';
26        break;
27        case 7: grade = 'C';
28        break;
29        case 6: grade = 'D';
30        break;
31        default: grade = 'F';
32    }
33    // 输出
34    cout << "The grade is " << grade << endl;
35    return 0;
36  }
```

```
运行结果：
Enter a score between 0 and 100: 83
The grade is B
```

```
运行结果：
Enter a score between 0 and 100: 111
Enter a score between 0 and 100: -10
Enter a score between 0 and 100: 97
The grade is A
```

5.4.2 do-while 循环结构的分析

在前面一节中，我们分析了 while 循环，结果发现 while 循环是一个预测试循环，其中测试的次数比迭代的次数多一次。我们还讨论了 for 循环是 while 循环的变体，这意味着它遵循 while 循环的规范。这里有必要分析一下 do-while 循环并将其与 while 循环进行比较。为此，我们像处理 while 循环一样展开 do-while 循环（图 5-13）。

在 do-while 循环中，如果循环体被执行了 n 次，那么循环条件也被测试了 n 次。

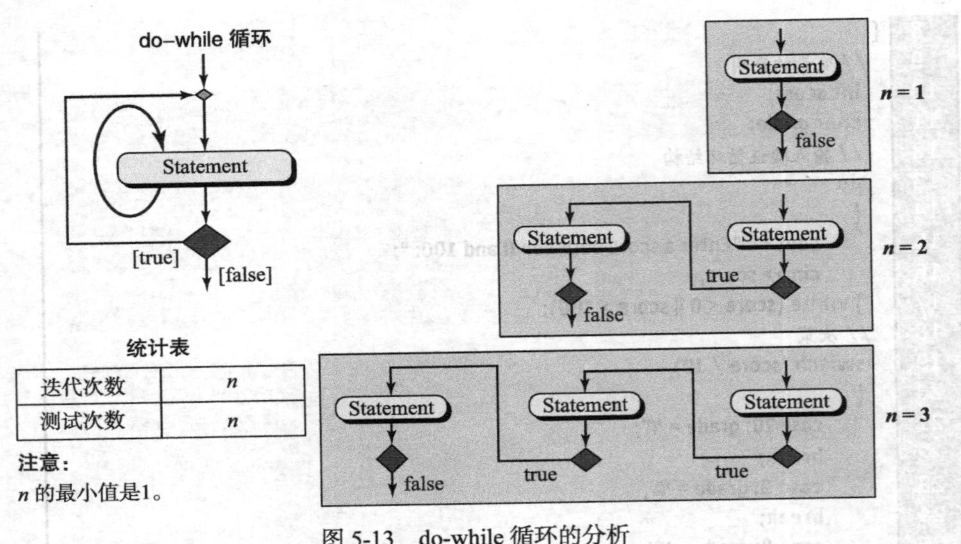

图 5-13　do-while 循环的分析

5.5　有关循环结构的详细信息

在本节中，我们首先比较三种循环结构：while 循环、for 循环和 do-while 循环。然后，我们将看到如何将它们组合在一起以创建嵌套循环，这有助于解决复杂问题。

5.5.1　三种循环结构的比较

前几节讨论的三种循环语句具有一些共同特性和不同特性。表 5-3 显示了 n 次迭代的比较汇总。

- while 循环和 for 循环都是预测试循环；do-while 循环是后测试循环。
- while 和 do-while 循环的设计目的主要是用于事件控制的循环；for 循环的设计目的主要是用于计数器控制的循环。
- while 循环和 for 循环可以迭代零次；do-while 循环至少迭代一次。
- 如果我们希望在一个循环中有 n 次迭代，那么在 while 循环和 for 循环中需要 $n + 1$ 次测试，但是在 do-while 循环中只需要 n 次测试。

表 5-3　while 循环、for 循环和 do-while 循环的比较

特点	while 循环	for 循环	do-while 循环
测试类型	预测试循环	预测试循环	后测试循环
主要设计目的	事件控制	计数器控制	事件控制
最少迭代次数	0	0	1
测试次数	$n+1$	$n+1$	n

5.5.2　嵌套循环

我们可以创建一个**嵌套循环**，即在一个外部循环中包括另一个内部循环。内部循环与外部循环的类型不必相同。

【例 5-18】 我们可以使用嵌套 for 循环打印一组水平和垂直排列的星号，也就是说，打印一个由星号组成的矩形。外部循环控制行；内部循环控制列。在外部循环的每次迭代中，

内部循环对每个列重复一次。要打印每行有 8 个星号（即 8 列）的 4 行矩形，外部循环使用 4 次迭代，内部循环使用 32（4×8）次迭代。程序 5-14 列出了代码和两次运行结果。

程序 5-14　包含在一个 for 循环中的另一个 for 循环

```
1  /**********************************************************
2   * 使用位于一个 for 循环中的另一个 for 循环                  *
3   * 打印水平和垂直排列的星号图案                              *
4   **********************************************************/
5  #include <iostream>
6  using namespace std;
7
8  int main ( )
9  {
10   // 变量声明
11   int rows;   // 行数
12   int cols;   // 列数
13   // 输入
14   cout << "Enter the number of rows: ";
15   cin >> rows;
16   cout << "Enter the number of columns: ";
17   cin >> cols;
18   // 输出
19   for (int count1 = 1; count1 <= rows; count1++)
20   {
21       for (int count2 = 1; count2 <= cols; count2++)
22       {
23           cout << "*";
24       }
25       cout << endl;
26   }
27   return 0;
28 }
```

```
运行结果:
Enter the number of rows: 3
Enter the number of columns: 8
********
********
********
```

```
运行结果:
Enter the number of rows: 2
Enter the number of columns: 6
******
******
```

【例 5-19】 让我们来看另一个例子。这一次，外部循环计数器的值控制内部循环计数器的起始值（程序 5-15）。

程序 5-15　打印一个数字图案

```
1  /**********************************************************
2   * 使用位于一个 for 循环中的另一个 for 循环                  *
3   * 打印水平和垂直排列的数字图案                              *
4   **********************************************************/
```

```
5   #include <iostream>
6   using namespace std;
7
8   int main ( )
9   {
10    // 变量声明
11    int rows;    // 行数
12    int cols;    // 列数
13    // 输入
14    cout << "Enter the number of rows: ";
15    cin >> rows;
16    cout << "Enter the number of columns: ";
17    cin >> cols;
18    // 嵌套循环
19    for (int i = 1; i <= rows; i++)
20    {
21        for (int j = i; j <= i + cols -1; j++)
22        {
23              cout << j << " ";
24        } // 内循环结束
25        cout << endl;
26    }
27    return 0;
28  }
```

```
运行结果:
Enter the number of rows: 3
Enter the number of columns: 6
1 2 3 4 5 6
2 3 4 5 6 7
3 4 5 6 7 8
```

```
运行结果:
Enter the number of rows: 2
Enter the number of columns: 8
1 2 3 4 5 6 7 8
2 3 4 5 6 7 8 9
```

【例 5-20】 程序 5-16 演示了如何创建从 2 到 10 的乘法表。表的大小由用户输入。注意，第 i 行第 j 列中的值实际上是 $(i * j)$ 的值。

程序 5-16　创建一个乘法表

```
1   /*****************************************************************
2    * 使用位于一个 for 循环中的另一个 for 循环                       *
3    * 打印一个从 2 到 10 的乘法表                                    *
4    *****************************************************************/
5   #include <iostream>
6   #include <iomanip>
7   using namespace std;
8
9   int main ()
10  {
11    // 声明变量 size
12    int size;
```

```
13  // 输入和验证
14  do
15  {
16      cout << "Enter table size (2 to 10): " ;
17      cin >> size;
18  } while (size < 2 || size > 10);
19  // 打印乘法表（使用嵌套循环）
20  for (int i = 1 ; i <= size; i++)
21  {
22      for (int j = 1 ; j <= size ; j++)
23      {
24          cout << setw (4) << i * j;
25      }
26      cout << endl;
27  }
28  return 0;
29 }
```

```
运行结果：
Enter table size (2 to 10): 4
   1   2   3   4
   2   4   6   8
   3   6   9  12
   4   8  12  16
```

```
运行结果：
Enter table size (2 to 10): 11
Enter table size (2 to 10): 9
   1   2   3   4   5   6   7   8   9
   2   4   6   8  10  12  14  16  18
   3   6   9  12  15  18  21  24  27
   4   8  12  16  20  24  28  32  36
   5  10  15  20  25  30  35  40  45
   6  12  18  24  30  36  42  48  54
   7  14  21  28  35  42  49  56  63
   8  16  24  32  40  48  56  64  72
   9  18  27  36  45  54  63  72  81
```

5.6 其他相关语句

除了三种循环语句之外，C++ 还定义了另一组语句，这些语句通常用于循环语句：return（返回语句）、break（中断语句）、continue（继续语句）和 goto（跳转语句）。图 5-14 显示了这些语句的分类。

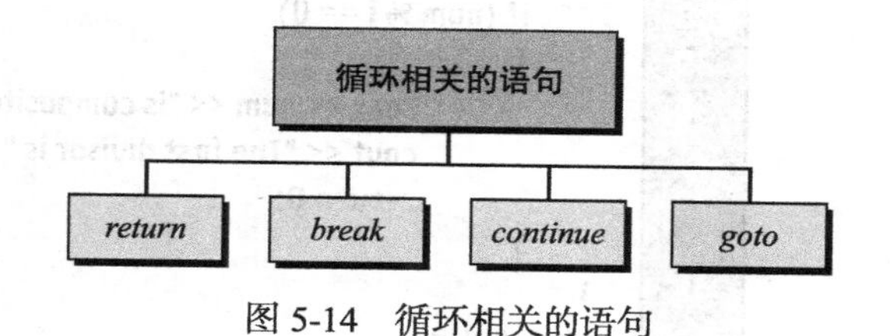

图 5-14 循环相关的语句

5.6.1 return 语句

return 语句立即终止当前函数，并将控制权返回给函数调用方。在循环中使用 return 语句会导致循环终止，同时也会终止包含正在迭代的循环的函数（例如 main）。

【例 5-21】 在数学上，我们可以将正整数分为三组：1、素数（质数）和合数。质数只能被它本身和 1 整除。合数不是素数。素数在计算机科学和安全中起着非常重要的作用。为了确定一个数值是否是素数，我们可以使用以下步骤（尽管存在更有效的方法，但更加复杂）。

1）如果数值是 1，则该数值不是素数。

2）如果数值可以被小于它本身的任何数值（1 除外）整除，则不是素数，该数值是合数。

请注意，我们不必测试从 2 到数值本身的所有数值。只要找到可以整除的大于或者等于 2 的整数，就可以判断该数不是素数，并终止循环。程序 5-17 演示了一个解决方案。

请注意，我们在第 23 行插入了一个 return 语句，在第 32 行插入了一个 return 语句，在第 37 行插入了一个 return 语句。这三条语句都可以终止程序。

程序 5-17　判断一个数是否为素数

```
1  /*****************************************************************
2   * 使用 return 语句判断一个数是否为素数。                           *
3   * 当检查到要判断的数为 1 或者为合数时，                            *
4   * 直接退出 main 函数                                              *
5   *****************************************************************/
6  #include <iostream>
7  using namespace std;
8
9  int main ()
10 {
11   // 声明变量
12   int num;
13   // 输入验证循环
14   do
15   {
16       cout << "Enter a positive integer: " ;
17       cin >> num;
18   } while (num <= 0);
19   // 测试输入的数是否为 1
20   if (num == 1)
21   {
22       cout << "1 is not a composite nor a prime.";
23       return 0;
24   }
25   // 测试输入的数是否为合数
26   for (int i = 2; i < num ; i++)
27   {
28       if (num % i == 0)
29       {
30            cout << num << "is composite." << endl;
31            cout << "The first divisor is " << i << endl;
32            return 0;
33       }
34   }
35   // 输出结果
36   cout << num << " is a prime." << endl;
37   return 0;
38 }
```

运行结果：

```
Enter a positive integer: 1
1 is not a composite nor a prime.
```

```
运行结果：
Enter a positive integer: 12
12 is composite.
The first divisor is 2
```

```
运行结果：
Enter a positive integer: 23
23 is a prime.
```

```
运行结果：
Enter a positive integer: 97
97 is a prime.
```

5.6.2 break 语句

C++ 中的 **break 语句**可以用于循环体或者 switch 语句中。我们在第 4 章的 switch 语句中讨论了 break 语句的用法。我们还可以使用 break 语句在需要时提前从循环中跳转出来。所有三种循环语句都可以通过 break 语句终止。break 语句将控制权转移到循环的末尾。大多数计算机科学专业人士认为，在循环结构中使用 break 语句会破坏程序的结构性。他们认为应该重新设计循环中的布尔表达式，以避免使用 break 语句。下面的代码显示了两个不同的循环，其中一个使用 break 语句，另一个不使用 break 语句，两个代码段执行相同的操作。（Statement1 和 Statement2 是语句符号，可以是任何语句，包括复合语句。）

```
while (expression)
{
   Statement1
   if (condition)
       break;
   Statement2
}
```

```
while (expression && !condition)
{
   Statement1
   if (!condition)
       Statement2
}
```

5.6.3 continue 语句

下一个与循环相关的语句是 **continue 语句**。如前所述，如果条件发生，break 语句将终止循环。但有时我们不想终止循环，我们只想终止一次迭代，然后继续剩下的迭代。在这种情况下，我们可以使用 continue 语句，但是我们需要小心，因为在 while 和 do-while 循环中，控制权被转移到布尔表达式。因此，在 continue 语句之前，必须包含所有必需的更新语句。但是，在 for 循环中，由于更新显式存在，控制权被转移到循环头中的更新语句。图 5-15 显示了各循环结构中的转移位置。

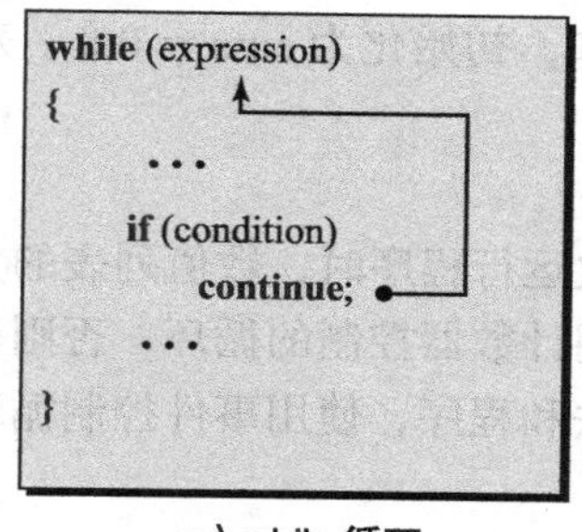

a）while 循环

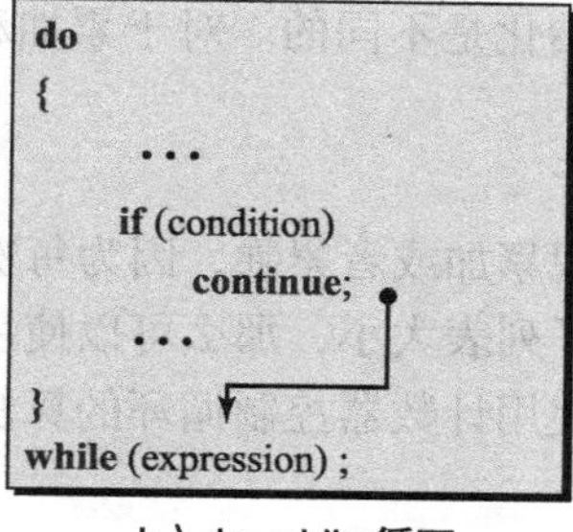

b）do-while 循环

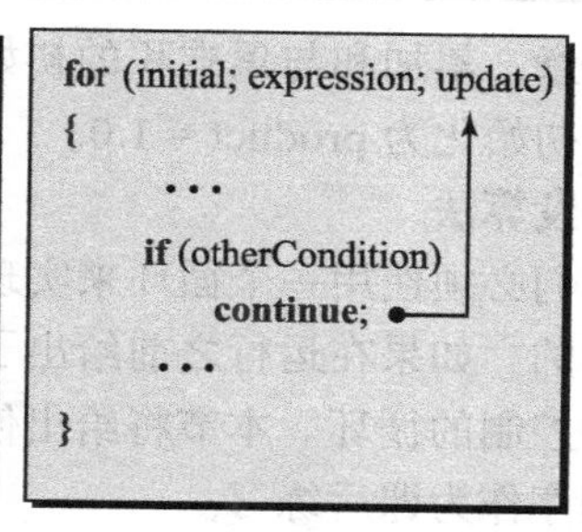

c）for 循环

图 5-15　continue 语句

5.6.4 goto 语句

C++ 语言中，另一个可以用于循环结构的语句是 goto 语句。其语法格式如下：

```
goto label
```

其中 label 是引用同一函数中带标签语句的标识符。虽然 goto 语句仍然是 C++ 语言的一部分，但纯粹主义的程序员认为使用 goto 语句会破坏程序的结构性。使用 goto 语句可能会导致杂乱无章的代码片段，从而使得调试困难并且费时。基于上述原因，我们在这里不深入讨论这条语句，因此也不需要讨论 label 语句。

5.7 程序设计

在本节中，我们演示如何使用选择语句和循环语句来解决计算机科学中的一些经典问题。我们为这些问题提供了一些基本的解决方案，但在后续章节中，我们将提供更结构化和更高效的解决方案。

5.7.1 累加和与累乘积

尽管在前面的章节中我们编写了一些求若干数值之和的程序，但在本小节中，我们将讨论涉及累加和与累乘积的问题的规范化解决方案。累加和是求一个数值列表之和；累乘积是求一个数值列表之积。下面是实现一组实数的累加和与累乘积的算法思想。

sum	=	17.0	+	14.4	+	…	+	71.2
product	=	17.0	×	14.4	×	…	×	71.2

理解问题

我们有一个需要累加（或者累乘）的浮点数列表。我们可以初始化一个变量 sum（或者 product）来保存累加和（或者累乘积）的结果，并使用一个循环将每个数与 sum 相加（或者将每个数与 product 相乘），如下所示：

初始化	sum = 0.0	product = 1.0
迭代 1 迭代 2 … 迭代 *n*	sum += 17.0 sum += 14.4 … sum += 71.2	product *= 17.0 product *= 14.4 … product *= 71.2
打印结果	sum	product

注意，累加和与累乘积的初始化是不同的。对于累加和，初始化为 sum = 0.0；对于累乘积，初始化为 product = 1.0。

开发算法

我们必须使用一个循环来实现累加或者累乘，因为每次运行程序时，数值列表的大小不是固定的。如果在运行之前给出了列表大小，那么可以使用计数器控制的循环，否则可以使用事件控制的循环。本节将给出使用计数器控制循环的算法和程序，使用事件控制循环的算法和程序作为课后练习。

1）**原始输入**。在开始循环之前，我们必须确定输入的内容。在计数器控制的循环中，我们需要输入列表的大小。

2）**初始化**。我们必须初始化 sum 和 product 的值。换句话说，我们必须设置（sum = 0.0）和（product = 1.0）。在使用计数器控制循环时，我们还必须初始化计数器。

3）**每次迭代中的处理**。在每次迭代中，我们必须读取下一个数值，然后把该数与 sum 相加（或者把该数与 product 相乘）。我们还必须递增计数器。迭代次数由计数器控制。

4）**输出结果**。循环结束后，可以输出结果（sum 和 product）。

通过以下步骤，我们可以给出非正式算法：

1）输入

a）输入列表大小（size）

2）初始化

a）把 sum 初始化为 0.0

b）把 product 初始化为 1.0

c）把计数器初始化为 0

3）当计数器小于 size 时重复处理

a）读取下一个数

b）把新的累加和（sum）设置为前一个 sum 值与该数之和

c）把新的累乘积（product）设置为前一个 product 值与该数之乘积

d）更新计数器

4）输出

a）输出 sum

b）输出 product

编写程序

程序 5-18 是基于该算法的代码。sum 和 product 的数据类型是长双精度（long double），所以几乎不会发生上溢或者下溢。但是，如果需要，可以添加代码来测试程序。

程序 5-18　计算若干数值的累加和与累乘积

```
1   /***************************************************************
2    * 本程序演示如何计算一个数值列表的累加和与累乘积                *
3    * 假设给定列表的大小                                          *
4    ***************************************************************/
5   #include <iostream>
6   #include <iomanip>
7   using namespace std;
8
9   int main ( )
10  {
11    //声明变量
12    int size;
13    long double number;
14    long double sum, product;
15    //输入验证列表的大小 size
16    do
17    {
18        cout << "Enter a non-negative integer value for size: " ;
19        cin >> size;
```

```
20      } while (size < 0);
21      //初始化
22      sum = 0;
23      product = 1;
24      //处理
25      for (int i = 1; i <= size; i++)
26      {
27          cout << "Enter the next integer: ";
28          cin >> number;
29          sum += number;
30          product *= number;
31      }
32      //输出
33      cout << fixed << setprecision (2);
34      cout << "sum =  " << sum << endl;
35      cout << "product =  " << product;
36      return 0;
37  }
```

```
运行结果:
Enter a non-negative integer value for size: 6
Enter the next number: 12
Enter the next number: 13.45
Enter the next number: 15
Enter the next number: 22.10
Enter the next number: 11.34
Enter the next number: 14
sum = 87.89
product = 8494310.92
```

```
运行结果:
Enter a non-negative integer value for n: 0
sum = 0.00
product = 1.00
```

第一次运行将六个数相加与相乘。第二次运行结果表明，如果列表为空（size = 0），则结果是变量 sum 和 product 的初始值。这表明列表是空的。

5.7.2 阶乘

我们在编程中遇到的另一个计算是求一个数的阶乘。

理解问题

阶乘是一种特殊的乘法，其中要被乘的数都是正整数（从 1 到阶乘数）。换言之，我们要求解（$1 \times 2 \times 3 \times \cdots\cdots \times n$）的结果，其中 n 是列表的大小。这被称为 n 的阶乘（在数学上表示为 $n!$）。程序要求用户输入的唯一信息是序列中的最后一个数（n）。

开发算法

求解阶乘的算法与我们在上一个问题中讨论的乘法算法相同。但是，因为在每次迭代中，我们必须将计数器乘以当前的阶乘结果 factorial，所以我们的计数器必须从 1（而不是零）开始，否则，阶乘的值在第一次迭代中变为零，因而在算法的其余部分中也将保持为零。因为计数器从 1 开始，所以当计数器是 $n + 1$ 而不是 n 时，我们需要退出循环。

1）输入

    a）输入 *n*
2）初始化
    a）把 factorial 初始化为 1
    b）把计数器初始化为 1
3）当计数器小于 *n* + 1 时重复处理
    a）把新的 factorial 设置为前一个 factorial 与计数器之乘积
    b）更新计数器
4）输出
    a）输出 factorial

编写程序

基于程序组成，编写这个问题的代码非常简单。然而，我们必须关注一个潜在的问题：结果值的溢出。根据系统中无符号长整数（unsigned long）的大小，对于较大的 *n*，结果可能会溢出（程序 5-19）。

程序 5-19 阶乘

```
1  /***************************************************************
2   * 本程序使用列表累乘积的算法思想                                    *
3   * 计算 n! 的值（n 的阶乘）                                         *
4   ***************************************************************/
5  #include <iostream>
6  using namespace std;
7
8  int main ( )
9  {
10   // 变量声明
11   int n;
12   unsigned long long factorial;
13   // 输入
14   do
15   {
16        cout << "Enter the factorial size: ";
17        cin >> n;
18   } while (n < 0);
19   // 初始化
20   factorial = 1;
21   // 处理
22   for (int i = 1; i < n + 1; i++)
23   {
24        factorial *= i;
25   }
26   // 输出
27   cout << n << "! = " << factorial;
28   return 0;
29 }
```

```
运行结果：
Enter the factorial size: 0
0! = 1
```

```
运行结果：
Enter the factorial size: 4
4! = 24
```

```
运行结果：
Enter the factorial size: 12
12! = 479001600
```

```
运行结果：
Enter the factorial size: 22
22! = 17196083355034583040
```

```
运行结果：
Enter the factorial size: 30
30! = 9682165104862298112
```

输出结果分析如下：第一次输出定义了 0! = 1，这是阶乘的定义。第二次输出很明显，当 n 为 4 时，$4! = 1 \times 2 \times 3 \times 4 = 24$。第三次输出表明，当 n 为 12 时，阶乘的计算结果变得非常大。第四次和第五次输出的结果不大正常。30! 的结果居然小于 22! 的结果，这意味着产生了溢出，但并不确定溢出的位置。可能对于 $n < 22$ 的值中就已经产生了溢出现象。

5.7.3　乘幂

乘法的另一个有趣的例子是计算一个数的幂（b^n），其中 b 被称为基数，n 被称为指数。

理解问题

这是一个乘法的例子，其中列表是相同值（称为基）的重复。换言之，我们把基数本身相乘 n 次。

$$power = b^n = b \times b \times b \times \cdots \times b \times b \text{（}n\text{ 次）}$$

C++ 的库提供了一个计算任何基的幂的函数。此函数中使用的数值是浮点值。有时我们需要使用无符号整数，本节计算整数的乘幂。

开发算法

乘幂的非正式算法如下所示：

1）输入
  a）输入 base
  b）输入 exponent
2）初始化
  a）把 power 初始化为 1
  b）把计数器初始化为 0
3）当计数器小于 exponent 时重复处理
  a）把新的 power 设置为前一个 power 与 base 之乘积
  b）更新计数器
4）输出
  a）输出 power

编写程序

因为我们知道如何处理溢出，所以我们在程序 5-20 中给出了该程序的最终版本。

程序 5-20 计算乘幂（b^n）

```
1  /******************************************************************
2   * 本程序使用列表累乘积的算法思想                                      *
3   * 计算 base 的 exponent 方（b^n）                                    *
4   ******************************************************************/
5  #include <iostream>
6  using namespace std;
7
8  int main ( )
9  {
10   // 变量声明
11    int base, exponent;
12    unsigned long int power, temp;
13    bool overflow;
14   // 输入并验证 base
15    do
16    {
17       cout << "Enter a non-negative integer value for b: " ;
18       cin >> base;
19    } while (base < 0);
20   // 输入并验证 exponent
21    do
22    {
23       cout << "Enter a non-negative integer value for n: " ;
24       cin >> exponent;
25    } while (exponent < 0);
26   // 初始化
27    power = 1;
28    temp = power;
29    overflow = false;
30   // 处理
31    for (int i = 1; (i <= exponent) && (!overflow); i++)
32    {
33       power * = base;
34       if (power / base != temp)
35       {
36           overflow = true; // 终止循环
37       }
38       temp = power;
39    }
40   // 输出
41    if (overflow)
42    {
43       cout << "Overflow occurred! Try again with smaller b or n." ;
44    }
45    else
46    {
47       cout << base << "^" << exponent << " = " << power;
48    }
49    return 0;
50 }
```

```
运行结果：
Enter a non-negative integer value for b: 22
Enter a non-negative integer value for n: 5
22^5 = 5153632
```

```
运行结果：
Enter a non-negative integer value for b: 5
Enter a non-negative integer value for n: 25
5^25 = 298023223876953125
```

```
运行结果：
Enter a non-negative integer value for b: 5
Enter a non-negative integer value for n: 28
Overflow occurred! Try again with smaller b or n.
```

关于程序5-20，第一点需要注意的是，我们可能尝试使用负值基数 b（base）。如果基数 b 为负数，则当指数 n（exponent）是奇数时，结果将是负数。因此，我们使用无符号值作为power和temp的数据类型，以保证溢出检测的简单性。

第二点需要注意的是，在最后一次运行测试中，变量power会溢出，这意味着我们的系统不能接受 5^{28}。这个问题和以前的问题的区别在于，溢出取决于 b 和 n 的值。因此，更难确定是什么导致了溢出。

5.7.4　最小值和最大值

计算机科学中另外两个常见的问题是查找一系列数值中的最小值和最大值。

理解问题

假设我们有以下数值列表：17、14、12、18和71。在进入循环之前，我们将变量smallest初始化为+infinity（正无穷大）。在每个步骤中，我们确定上一次迭代的结果和下一个数值之间的较小值，这可以通过使用if语句来实现。下面演示了查找最小值的过程。查找最大值的过程与此相似。在进入循环之前，我们将变量largest设置为−infinity（负无穷大）。在每次迭代中，我们确定前一个最大值和下一个数值之间的较大值。

数值	smallest = + infinity
17	smallest = 17 和前一个最小值（+infinity）之间的较小值 = 17
14	smallest = 14 和前一个最小值（17）之间的较小值 = 14
12	smallest = 12 和前一个最小值（14）之间的较小值 = 12
71	smallest = 71 和前一个最小值（12）之间的较小值 = 12
结果	12

有些文献将smallest（或者largest）初始化为第一个数值。尽管这种做法几乎适用于所有情况，但如果列表为空，则不起作用。换言之，这不是一个通用的解决方案。读者可能会提出疑问，为什么把最小值设为正无穷大（+infinity）。原因是我们希望在第一次迭代之后，smallest是列表中第一个数值的值。这个过程与求最大值（largest）相类似。

开发算法

求最小值和最大值的非正式算法如下所示：

1）输入

　a）输入size

2）初始化

a）把 smallest 初始化为 +infinity

b）把 largest 初始化为 -infinity

c）把计数器初始化为 0

3）当计数器小于 size 时重复处理

a）读取下一个数

b）把新的 smallest 设置为前一个 smallest 与所读取数之间的较小值

c）把新的 largest 设置为前一个 largest 与所读取数之间的较大值

d）更新计数器

4）输出

a）输出 smallest

b）输出 largest

编写程序

程序 5-21 演示了如何结合前面讨论的内容，查找一个整数列表中的最小值和最大值。

程序 5-21　查找整数列表中的最小值和最大值

```
1  /***************************************************************
2   * 本程序查找一个整数列表中的最小值和最大值                        *
3   * 列表的大小未知                                                  *
4   ***************************************************************/
5  #include <iostream>
6  #include <limits>                  // 用于数值大小限制的头文件
7  using namespace std;
8
9  int main ( )
10 {
11   // 变量声明
12    int size;
13    int number, smallest, largest;
14   // 初始化
15    smallest = numeric_limits <int> :: max();
16    largest = numeric_limits <int> :: min() ;
17   // 输入 size
18    do
19    {
20         cout << "Enter the size of the list (non-negative): ";
21         cin >> size;
22    } while (size <= 0);
23   // 处理
24    for (int i = 1; i <= size; i++)
25    {
26         cout << "Enter the next item: ";
27         cin >> number;
28         if (number < smallest)
29         {
30              smallest = number;
31         }
32         if (number > largest)
33         {
```

```
34                largest = number;
35            }
36        } // for 结束
37        // 输出
38        cout << "The smallest item is: " << smallest << endl;
39        cout << "The largest item is: " << largest << endl;
40        return 0;
41    }
```

运行结果：

```
Enter the size of the list (non-negative): 6
Enter the next item: 12
Enter the next item: -3
Enter the next item: 14
Enter the next item: 15
Enter the next item: 27
Enter the next item: -7
The smallest item is: -7
The largest item is: 27
```

运行结果：

```
Enter the size of the list (non-negative): 3
Enter the next item: 1
Enter the next item: 87
Enter the next item: 45
The smallest item is: 1
The largest item is: 87
```

5.7.5 any 或者 all 查询

计算机科学中的另一个常见问题是使用指定的标准搜索项目列表。特别是我们经常需要查看列表中是否存在符合某标准的项目，或者是否所有项目都符合标准。

理解问题

如果我们要查看是否存在符合标准的项目（any 查询），我们会在找到第一个符合标准的项目后立即停止搜索。如果我们想检查是否所有项目都符合标准（all 查询），我们会在找到第一个不符合标准的项目后立即停止搜索。图 5-16 显示了搜索过程。

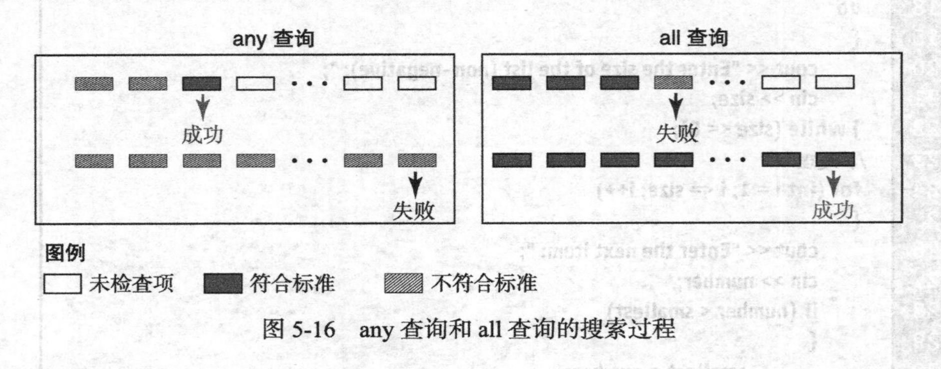

图 5-16　any 查询和 all 查询的搜索过程

开发算法

由于我们需要在找到符合标准的项目（any 查询）或者发现不符合标准的项目（all 查询）时停止搜索，因此我们需要结合使用计数器控制（或哨兵控制或 EOF 控制）循环和标志控制循环。我们称这个标志为 success。对于 any 查询，我们将标志 success 初始化为 false；对于 all 查询，我们将标志 success 初始化为 true。在 any 查询情况下，如果在循环过程中 success 变为 true，则查询结果成功。在 all 查询情况下，如果 success 在整个过程中都是 true，则查询结果成功。对于这两种循环，找到其循环条件的最佳方法是考虑退出循环的条件，并使用德摩根定律（De Morgan's Law）将其更改为继续循环的条件。

问题	终止循环条件	继续循环条件
any:	（没有剩余项目）or (success)	（还有剩余项目）and (not success)
all:	（没有剩余项目）or (not success)	（还有剩余项目）and (success)

有关 any 查询和 all 查询算法的非正式描述如下所示。注意，any 查询和 all 查询的算法分别阐述，因为二者终止循环的条件和输出结果是不同的。

any 查询算法的描述	all 查询算法的描述
1）输入 a）输入 size 2）初始化 a）success = false b）counter = 0 3）重复（counter < size 或者 !success） a）读取下一个数 b）success = criteria c）counter++ 4）如果 success 为 true a）输出查询成功信息 5）否则 a）输出查询不成功信息	1）输入 a）输入 size 2）初始化 a）success = true b）counter = 0 3）重复（counter < size 或者 success） a）读取下一个数 b）success = criteria c）counter++ 4）如果 success 为 true a）输出查询成功信息 5）否则 a）输出查询不成功信息

编写程序

由于这两个程序非常相似，我们给出了 any 查询的代码，并将 all 查询作为课后练习。程序 5-22 列出了代码。程序包括一个验证循环、一个搜索循环、一个后循环决策，以检查是否找到符合给定条件的项目。

程序 5-22 判断在列表中是否存在满足给定条件的项目

```
1  /***************************************************************
2   * 本程序搜索一个项目列表，                                    *
3   * 判断是否存在被 7 整除的项目                                 *
4   ***************************************************************/
5  #include <iostream>
6  using namespace std;
7
8  int main ( )
9  {
10   //变量声明
11   bool success;
```

```
12    int size;
13    int item;
14    //输入验证
15    do
16    {
17        cout << "Enter the number of items in the list: ";
18        cin >> size;
19    } while (size < 0);
20    //处理
21    for (int i = 0; (i < size ) && (!success); i++)
22    {
23        cout << "Enter the next item: ";
24        cin >> item;
25        success = (item %7 == 0);
26    }
27    //检查成功或者失败
28    if (success)
29    {
30        cout << "The number " << item <<  " is divisible by 7." << endl;
31    }
32    else
33    {
34        cout << "None of the numbers is divisible by 7." << endl;
35    }
36    return 0;
37 }
```

```
运行结果:
Enter the number of items in the list: 5
Enter the next item: 12
Enter the next item: 32
Enter the next item: 28
The number 28 is divisible by 7.
```

```
运行结果:
Enter the number of items in the list: 5
Enter the next item: 6
Enter the next item: 12
Enter the next item: 15
Enter the next item: 17
Enter the next item: 22
None of the numbers is divisible by 7.
```

本章小结

循环结构允许程序多次迭代执行一段代码。当我们使用循环语句时，我们需要使用一个计数器来检查重复的次数。C++ 设计了两组表达式来模拟计数器：后缀递增 / 递减表达式和前缀递增 / 递减表达式。C++ 中有三种循环语句：while 循环、for 循环和 do-while 循环。

while 循环包含一个循环条件（一个布尔表达式），后跟一个称为循环体的语句。for 语句由以下三个要素组成：循环初始化、循环条件测试和计数器更新。do-while 循环与 while 循环相似，只是在每次迭代结束时测试逻辑表达式。

还有四个可以与循环结构一起使用的语句：return、break、continue 和 goto。return 语句终止函数。break 语句提前跳转到循环的末尾，并停止循环。循环中的 continue 语句用于立即终止当前迭代，但将继续执行下一次迭代。goto 语句将控制权转移到带标签的语句，它被认为是非结构化的，因此不推荐使用。

思考题

1. 以下循环结构中的循环体执行了多少次？循环终止后 x 的值是多少？

```
int x = 5;
while ( x < 9 )
{
    x++;
}
```

2. 以下循环结构中的循环体执行了多少次？循环终止后 x 的值是多少？

```
int x = 5;
while ( x < 11 )
{
    x += 2;
}
```

3. 以下循环结构中的循环体执行了多少次？循环终止后 x 的值是多少？

```
int x = 7;
while ( x < 3 )
{
    x++;
}
```

4. 以下循环结构中的循环体执行了多少次？循环终止后 x 和 y 的值是多少？

```
int x = 7;
int y = 5;
while ( x < 11 && y > 3 )
{
    x++;
    y--;
}
```

5. 以下循环结构中的循环体执行了多少次？循环终止后 x 的值是多少？

```
int x = 7;
while ( false )
{
    x++;
}
```

6. 以下循环结构中的循环体执行了多少次？循环终止后 x 的值是多少？

```
int x = 10;
while ( true )
{
    x--;
}
```

7. 以下循环结构中的循环体执行了多少次？循环终止后 x 的值是多少？

```
int x = 5;
do
{
    x -= 2;
} while (x != 4);
```

8. 以下循环结构中的循环体执行了多少次？循环终止后 x 的值是多少？

```
int x = 15;
do
{
    x -= 2;
    if (x < 9)
    break;
} while (true);
```

9. 以下程序片段输出的结果是什么？

```
int x = 13;
while (x > 7)
{
    cout << x << " ";
    x--;
}
```

10. 以下程序片段输出的结果是什么？

```
for (int x = 13; x > 7; x--)
{
    cout << x << "  ";
}
```

11. 以下程序片段输出的结果是什么？

```
int x = 13;
do
{
    cout << x << "  ";
    x--;
} while (x > 7);
```

12. 把以下 do-while 循环结构修改为 while 循环结构。

```
int x = 10;
do
{
    cout << x << end;
    x++;
} while (true);
```

do-while 循环结构和 while 循环结构都是无限循环。

13. 把以下 while 循环结构修改为 for 循环结构。

```
int x = 10;
while (x < 20)
{
    cout << x << endl;
```

```
        x++;
    }
```

14. 把以下 for 循环结构修改为 while 循环结构。

```
for (int x = 20; x < 10; x--)
{
    cout << x << endl;
}
```

编程题

1. 编写一个程序，打印以下四种图形模式中的任何一种。模式类型（1 到 4）和模式大小（1 到 9）由用户提供。

```
type 1          type 2          type 3          type 4
*               *****           *****               *
**              ****             ****              **
***             ***               ***             ***
****            **                 **            ****
*****           *                   *           *****
```

2. 编写一个程序，打印以下两种图形模式中的任何一种。模式类型（1 到 2）和模式大小（行数，1 到 6）由用户提供。

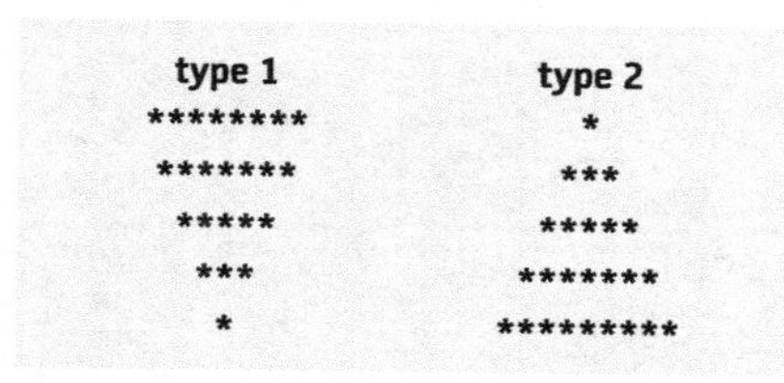

3. 编写一个程序，从键盘读取一个由若干小于 1000 的正整数所组成的列表。程序打印所读取数值的总和与平均值。当用户输入 1000 作为哨兵时，程序停止执行。
4. 编写一个程序，要求用户输入一个由正整数或者负整数所构成的列表，使用整数 0 作为哨兵。然后，程序对所输入的正整数和负整数进行计数。
5. 编写一个程序，要求用户输入两个正整数，然后程序打印位于给定整数之间的偶数列表和奇数列表。
6. 编写一个程序，打印 1 到 100 之间的所有可以被 7 整除的整数。
7. 编写一个程序，打印 1 到 100 之间的所有可以同时被 5 和 7 整除的整数。
8. 编写一个程序，读取一个 1 到 100 之间的正整数，并打印其因子（除数）。因子是可以整除一个数的整数。例如，10 的因子是 1、2、5 和 10。12 的因子是 1、2、3、4、6 和 12。
9. 编写一个程序，读取 1 到 100 之间的两个整数并打印它们的公因数（除数）。
10. 编写一个程序，使用以下算法查找两个整数 m 和 n 的最大公约数：
    a）令 $m = m - n$，当 $m < n$ 时，交换 m 和 n。
    b）重复步骤 a 直到 n 为 0。得到最大公约数为 m。使用以下数据测试所编写的程序：9 和 12、7 和 11、12 和 140。
11. 编写一个程序，读取一个成绩列表（成绩在 0 到 100 之间），查找并打印最低成绩和最高成绩。程序需要用户输入成绩的份数，要求成绩的份数必须小于或等于 10。
12. 编写一个程序，显示 2000 年至 2099 年中的所有闰年。闰年是可以被 4 整除的年份，但如果一个年份可以被 100 整除，那该年份也应该可以被 400 整除才是闰年。

13. 编写一个程序，要求用户输入一个正整数。然后程序输出其每位数的数字之和。
14. 编写一个程序，要求用户输入一个正整数。然后程序按相反的顺序打印整数。例如，如果用户输入 359，程序将打印 953。
15. 编写一个程序，要求用户输入一个正整数。然后程序检查这个数值是否是回文数。回文整数是当数字顺序颠倒时与原数相同的整数。例如，353 和 376673 就是回文整数。提示：使用第 14 题中的程序结果。
16. 编写一个程序，创建一个在 0 到 99 摄氏度之间的摄氏度 / 华氏度的转换表。换算公式如下：

```
Fahrenheit = Celcius * 180.0 / 100.0 + 32
```

函　　数

到目前为止，我们介绍的程序非常简单。它们解决的问题很容易就能理解。然而，当我们讨论规模越来越大、越来越复杂的程序时，我们会发现，要理解大多数程序的所有方面，首先必须将它们简化为更基本的部分。

一种常见的实践方法是把一个复杂的问题分成若干小问题。我们将程序划分为可处理的模块。在本章中，我们将讨论函数。函数是计算机程序设计中的一种抽象，它有助于将一个大程序分成更小的部分。

学习目标

阅读并学习本章后，读者应该能够

- 介绍函数并讨论将程序的任务分成更小的任务（每个小任务都分配给一个函数）的优点。
- 讨论函数的组成元素：定义、声明和调用。
- 将 C++ 中的函数分为两组：库函数和用户自定义函数。
- 讨论库函数。库函数无须定义就可以直接使用，例如数学函数、字符函数、时间函数和随机数生成器。
- 介绍用户自定义函数的基本思想，并将其分为四组：不带参数的 void 函数、带参数的 void 函数、不带参数有返回值的函数、带参数有返回值的函数。
- 讨论主调函数和被调用函数之间的数据交换机制：数据传递和数据返回。
- 讨论默认参数和函数重载，函数重载允许定义两个或者多个具有不同签名的函数。
- 讨论程序中实体的作用域：局部范围和全局范围。
- 讨论程序中实体的生命周期，包括自动变量和静态变量。
- 使用函数设计和编写程序。

6.1 概述

函数是被设计用来执行任务的实体。函数包含函数头和一组用花括号括起来的语句。在前面的章节中，我们使用一个单独的函数 main 来完成程序的整个任务。换言之，我们把解决问题的全部责任都交给 main 函数。这种方法适用于任务非常简单并且程序很短的小型程序。

当任务较复杂时，我们将任务分解成更小的任务，其中每个小任务负责一部分功能。这种方法在许多情况下十分常见。例如，制造一辆汽车可以分为若干任务。车身在一个地方制造，发动机在另一个地方制造，轮胎在第三个地方制造，等等。当所有的零部件都制作完毕后，汽车最后被组装在一起。

同样的方法也可以应用于程序设计。我们可以将程序划分为不同的部分，其中每个部分负责任务的一部分功能。这种情况下的每个部分在 C++ 中被称为函数，但也可以使用其他具有相同含义的术语，例如方法或者过程。示意图如图 6-1 所示。

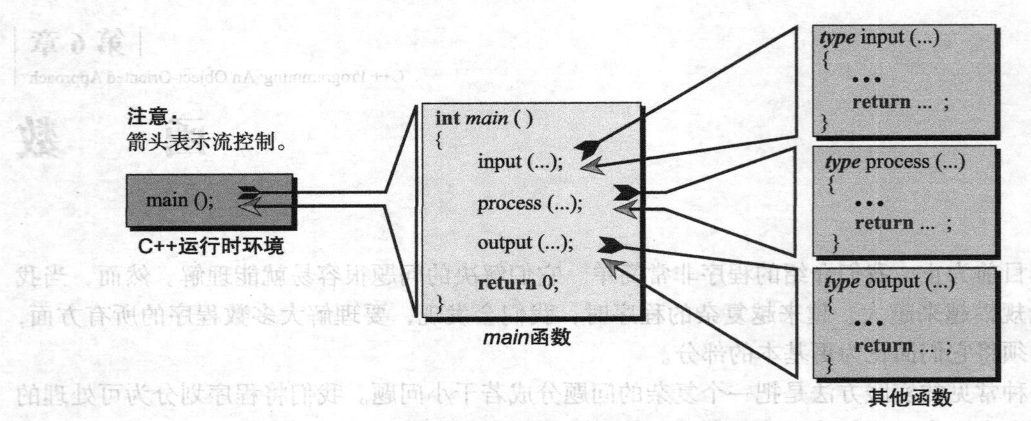

图 6-1　由若干函数组成的程序

在图 6-1 中，程序的职责仍然落在主函数（main）中，但是 main 函数调用其他函数来完成部分工作。语句 input(…)、process(…) 和 output(…) 是函数调用，调用相应函数并让它们完成部分工作。当每个函数终止时，控制权返回到 main 函数。该图还显示了 main 函数由 C++ 运行时环境调用。当我们运行程序时，调用 main 函数并执行其任务。换言之，自我们在本书第 1 章中编写的第一个程序开始，我们就使用了函数的概念。在本章中，我们将进一步展开阐述有关函数的概念，并学习如何编写更多的函数。

6.1.1　函数的优点

当一个任务被划分为几个小任务时，由于每个函数涉及的开销，因此需要编写更多的代码，读者可能想知道这样做的好处。尽管会有额外的工作，但任务划分具有如下优点。

更容易编写更简单的任务

每个人都知道完成简单的任务比完成复杂的任务更容易。集中精力制造汽车轮胎比制造整辆汽车更容易。

错误检查（调试）

程序设计的一大难题是发现错误（称为 bug）。当程序被划分为小函数时，错误检查（或称为调试，debug）就简单得多。每个函数都可以单独调试，然后组装成一个程序。

可重用性

小任务可以在许多大任务中重用。如果我们隔离这些小任务并为每个小任务编写一个函数，则可以通过组装这些小函数来创建许多大程序。我们不需要一遍又一遍地重写这些小任务。

函数库

一些涉及与操作系统和计算机硬件交互的常见任务被预先编写成函数，并提供给用户使用。我们可以直接使用这些函数而不需要为这些任务编写代码。

6.1.2　函数的定义、声明和调用

要使用函数，必须考虑三个实体：函数定义、函数声明和函数调用。我们在这里简要讨论这些问题。

函数定义

函数定义用于创建函数。就像 C++ 中的其他实体一样，函数的定义必须遵循语法规则。

图 6-2 显示了函数定义的基本语法。

如图 6-2 所示，函数定义由两部分组成：**函数头**（function header）和**函数体**（function body）。函数头定义函数的名称、从函数返回的数据类型以及**参数列表**（parameter list）。参数列表包含传递给函数的数据项（及其数据类型）。函数体定义由函数执行的操作。

```
return-type function-name (parameter list)
{
        Body
}
```

图 6-2　函数定义的语法

【例 6-1】 以下代码片段显示了一个函数的定义，该函数查找并返回两个给定整数中较大的整数。

```
int larger (int first, int second)    // 函数头
{
   int temp;
   if (first > second)
   {
        temp = first;
   }
   else
   {
        temp = second;
   }
   return temp;
}
```

函数的返回类型为整数。函数带两个整数类型的参数。函数体是一条复合语句，这与在前几章中的 main 函数相同。

函数声明

函数声明（function declaration，也称为函数原型 function prototype），仅包括函数头，后跟英文分号。参数名称是可选的；类型是必需的。函数声明用于显示如何调用该函数。函数声明告诉我们传递哪些参数给函数，函数返回什么值。函数声明不定义函数执行的操作。

【例 6-2】 以下代码片段显示了例 6-1 中定义的函数的声明。我们给出了两个版本：有参数名和无参数名。

```
int larger (int first, int second);              // 有参数名
int larger (int, int );                          // 无参数名
```

函数调用

函数调用是一个后缀表达式，用于调用一个函数来完成工作。正如我们在前几章中学习的表达式一样，我们需要知道这个表达式的语法才能调用它。函数调用属于后缀组（类似于后缀递增表达式和后缀递减表达式）。表 6-1 显示了函数调用表达式与附录 C 中讨论的其他表达式之间的关系。

表 6-1　函数调用

组别	名称	运算符	表达式	优先级	结合性
后缀表达式	函数调用	(…)	name (expr, …)	18	→

图 6-3 显示了有关函数调用表达式的详细信息。

如果我们将函数调用的语法与其他后缀表达式（例如后缀递增表达式或者后缀递减表达式）进行比较，我们会发现在函数调用的情况下，操作数是一个名称（函数名），运算符是一

对带零个或者多个值的列表的圆括号。函数调用可以有副作用或者返回值，或者既有副作用又有返回值。如果函数只有副作用，则需要将其用作表达式语句。如果函数有返回值（既可以有副作用，也可以没有副作用），则可以在任何需要值的情况下使用该函数。调用前文求较大值的函数的代码为：larger(value1, value2)，如下一个示例所示。

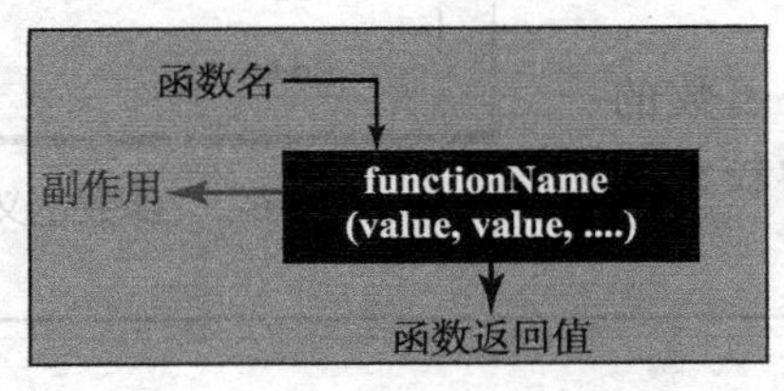

注意：
函数调用可以有副作用或者返回值，或者既有副作用又有返回值。

图 6-3　函数调用操作符

【例 6-3】 以下代码片段显示了在 main 函数中可以根据需要多次调用 larger 函数。

```
int main ()
{
   cout << larger (3, 13);
   cout << larger (10, 12);
   cout << larger (2, 12);
   return 0;
}
```

读者可能已经注意到，我们在前几章中使用的 main 函数本身就是函数的定义。该函数由 C++ 运行时环境调用。main 函数的函数头表明，它返回一个整数，但是它的参数列表为空。

实际参数和形式参数

在结束本节之前，我们将讨论函数中使用的两个术语：**实际参数**（argument）和**形式参数**（parameter）。形式参数是在函数定义的函数头中声明的变量；实际参数是在调用函数时初始化参数的值。换言之，形式参数类似于赋值语句左侧的变量，对应的实际参数类似于赋值语句右侧使用的值。当调用函数时，系统显式地初始化形式参数。

以下代码片段显示了带有赋值语句的函数调用。形式参数 x 使用实际参数 5 进行初始化。

```
int main ()
{
   ...
   fun (5);              // 函数调用，5 是实际参数
   ...
}
void fun (int x)         // 函数定义，x 是形式参数
{
   ...
}
```

稍后我们将讨论，实际参数可以是一个变量，甚至是一个对象，但只有它的值参与了函数调用。

6.1.3　库函数和用户自定义函数

要在程序中使用函数，我们必须有一个函数定义和一个函数调用。但是，我们必须在两

种类型之间进行选择：库函数和用户自定义函数。

库函数

C++ 库包含预定义函数。我们只需要进行声明就可以调用这些函数。我们将在本章后面讨论其中的一些函数。

用户自定义函数

我们需要很多函数，但是没有为它们预先定义的函数。我们必须先定义这些函数，然后才能调用它们。我们将在本章后面学习如何实现自定义函数。

6.2 库函数

如前所述，如果在库中存在某个预定义的函数，我们就不需要担心该函数的定义。我们只需要添加定义该函数的对应头文件。但是，我们必须调用函数，这意味着我们必须知道函数的声明。

6.2.1 数学函数

C++ 定义了一组可以在程序中调用的数学函数。这些函数都是预定义的并且收集在 <cmath> 头文件中。头文件名称中的前缀“c”强调它是从 C 语言继承来的（但是有一些更改）。

> 使用数学函数需要 <cmath> 头文件。

数值函数

数值函数用于数值计算。参数通常是一个或者多个数值，返回结果也是一个数值。表 6-2 显示了该类别中的一些常见函数以及函数声明。

表 6-2　<cmath> 中定义的一些数值函数

函数声明	说明
type abs(type x);	返回 x 的绝对值
type ceil(type x);	返回大于或等于 x 的最小值
type floor(type x);	返回小于或等于 x 的最大值
type log(type x);	返回 x 的自然（以 e 为底）对数
type log10(type x);	返回 x 的常用（以 10 为底）对数
type exp(type x);	返回 e^x
type pow(type x, type y);	返回 x^y，注意在这种情况下 y 也可以为 int 类型
type sqrt(type x);	返回 x 的平方根（$x^{1/2}$）

表中的 type 是 float、double 或者 long double。对于 pow 函数，第二个参数也可以是整数。所有这些函数都可以接收不同类型的参数，并且可以返回不同类型的值。这意味着这些函数是被重载的（我们将在本章后面讨论重载）。

【例 6-4】 程序 6-1 显示了表 6-2 中给出的函数的返回值。

程序 6-1　测试数值函数

```
1  /***************************************************************
2   * 本程序演示在 <cmath> 头文件中定义的一些                        *
3   * 数值函数的应用                                                *
```

```
4      **********************************************************/
5   #include <iostream>
6   #include <cmath>
7   using namespace std;
8
9   int main ( )
10  {
11     // abs 函数，使用两个不同的参数值
12     cout << "abs (8) = " << abs (8) << endl;
13     cout << "abs (-8) = " << abs (-8) << endl;
14     // floor 函数和 ceil 函数，使用相同的参数值
15     cout << "floor (12.78) = " << floor (12.78) << endl;
16     cout << "ceil (12.78) = " << ceil (12.78) << endl;
17     // log 函数和 log10 函数
18     cout << "log (100) =  " << log (100) << endl;
19     cout << "log10 (100) =  " << log10 (100) << endl;
20     // exp 函数和 pow 函数
21     cout << "exp (5) =  " << exp (5) << endl;
22     cout << "pow (2, 3) = " << pow (2,3) << endl;
23     // sqrt 函数
24     cout << "sqrt (100)     " << sqrt (100);
25     return 0;
26  }
```

运行结果：

```
abs(8) = 8
abs(-8) = 8
floor(12.78) = 12
ceil(12.78) = 13
log(100) = 4.60517
log10 (100) = 2
exp(5) = 148.413
pow(2, 3) = 8
sqrt(100) = 10
```

注意，由于这些函数都没有副作用，当这些函数被用作表达式语句时，返回值都将被丢弃。因此，我们必须在需要值的地方使用这些函数，例如作为插入运算符（<<）的参数。

【例 6-5】 作为这些函数中的 pow 函数的应用，让我们讨论求解一元二次方程 $ax^2+bx+c=0$ 的根。我们已经在数学课本上学习过，根据（b^2-4ac）的值（我们称之为项，term），存在以下三种情况：

1）当 term 为负数时，没有实数根（根为复数）。

2）当 term 等于 0 时，存在两个相同根：$-b/2a$。

3）当 term 为正数时，存在两个不同的根，分别为：$(-b+(b^2-4ac)^{1/2})/2a$ 和 $(-b-(b^2-4ac)^{1/2})/2a$。

使用上述三条规则，可以求解任意一元二次方程的根，参见程序 6-2。

程序 6-2 求解一元二次方程的根

```
1   /**********************************************************
2     * 本程序用于求解一元二次方程的根                        *
3     **********************************************************/
4   #include <iostream>
```

```
5   #include <cmath>
6   using namespace std;
7
8   int main ( )
9   {
10      // 变量的声明
11      int a, b, c;
12      double term;
13      // 输入三个系数的值
14      cout << "Enter the value of coefficient a: ";
15      cin >> a;
16      cout << "Enter the value of coefficient b: ";
17      cin >> b;
18      cout << "Enter the value of coefficient c: ";
19      cin >> c;
20      // 计算项（b²−4ac）的值
21      term = pow (b, 2) − 4 * a * c;
22      if (term < 0 )
23      {
24          cout << "There is no root!" << endl;
25      }
26      else if (term == 0)
27      {
28          cout << "The two roots are equal." << endl;
29          cout << "x1 = x2 = " << −b / (2 * a) << endl;
30      }
31      else
32      {
33          cout << "There are two distinct roots: " << endl;
34          cout << "x1 = " << (−b + sqrt (term)) / (2 * a) << endl;
35          cout << "x2 = " << (−b − sqrt (term )) / (2 * a) << endl;
36      }
37      return 0;
38  }
```

运行结果：

```
Enter the value of coefficient a: 3
Enter the value of coefficient b: 5
Enter the value of coefficient c: 4
There is no root!
```

运行结果：

```
Enter the value of coefficient a: 1
Enter the value of coefficient b: 2
Enter the value of coefficient c: 1
The two roots are equal.
x1 = x2 = −1
```

运行结果：

```
Enter the value of coefficient a: 4
Enter the value of coefficient b: -9
Enter the value of coefficient c: 2
There are two distinct roots:
x1 = 2
x2 = 0.25
```

三角函数

表 6-3 给出了常用三角函数的列表。注意，所有这些函数都是被重载的。数据类型 type 表示任意浮点类型（float、double 和 long double）。

注意前三个函数的参数必须是弧度，其中 180 度 =π 弧度（π 近似等于 3.14159265358979323846 2）。

表 6-3　三角函数

函数声明	函数说明	参数单位	返回值范围
type cos(type x);	返回余弦值	弧度	[−1, +1]
type sin(type x);	返回正弦值	弧度	[−1, +1]
type tan(type x);	返回正切值	弧度	任意
type acos(type x);	返回反余弦值	[−1, +1]	[0, π]
type asin(type x);	返回反正弦值	[−1, +1]	[0, π]
type atan(type x);	返回反正切值	任意	[−π/2, +π/2]

【例 6-6】 程序 6-3 演示了一些三角函数的应用。

程序 6-3　三角函数

```
/***********************************************************************
 *本程序用于测试一些三角函数的使用                                        *
 ***********************************************************************/
#include <iostream>
#include <cmath>
using namespace std;

int main ( )
{
   // 定义常量 PI 和 45 度角的弧度
   const double PI = 3.14159265358979323846 2;
   double degree = PI / 4;
   // 计算 45 度角的正弦值、余弦值和正切值
   cout << "sin (45): " << sin (degree) << endl;
   cout << "cos (45): " << cos (degree) << endl;
   cout << "tan (45): " << tan (degree);
   return 0;
}
```

运行结果：

```
sin (45): 0.707107
cos (45): 0.707107
tan (45): 1
```

【例 6-7】 根据三角函数学，给定正多边形的边数和每边的长度，我们可以求解一个正四边形或者正多边形的周长和面积，如图 6-4 所示。

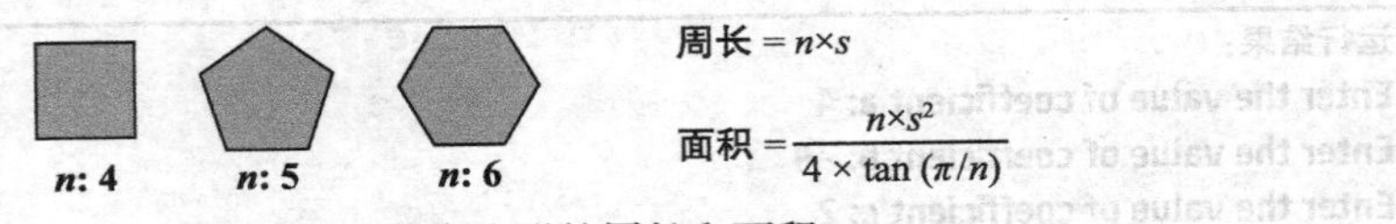

图 6-4　正多边形的周长和面积

程序 6-4 演示了如何计算任意正多边形的周长和面积。

程序 6-4　计算正多边形的周长和面积

```
1  /*****************************************************************
2   * 给定一个正多边形的边数和边长                                    *
3   * 计算正多边形的周长和面积                                        *
4   *****************************************************************/
5  #include <iostream>
6  #include <cmath>
7  using namespace std;
8
9  int main ( )
10 {
11   // 声明
12   const double PI = 3.14159265358979323846;
13   int n;
14   double s, peri, area;
15   // 输入边的数量
16   do
17   {
18       cout << "Enter the number of sides (4 or more): ";
19       cin >> n;
20   } while (n < 4);
21   // 输入边的长度
22   do
23   {
24       cout << "Enter length of each side: ";
25       cin >> s;
26   } while (s <= 0.0);
27   // 计算周长和面积
28   peri = n * s;
29   area = (n * pow (s, 2)) / (n * tan (PI / n));
30   // 打印结果
31   cout << "Perimeter: " << peri << endl;
32   cout << "Area: " << area;
33   return 0;
34 }
```

运行结果：

```
Enter the number of sides (4 or more): 2
Enter the number of sides (4 or more): 3
Enter the number of sides (4 or more): 4
Enter length of each side: 5
Perimeter: 20
Area: 25
```

运行结果：

```
Enter the number of sides (4 or more): 5
Enter length of each side: 5
Perimeter: 25
Area: 34.4095
```

6.2.2　字符函数

我们在第 2 章中讨论了作为基本类型的字符。可以将我们讨论的针对基本类型的大多数操作应用于字符类型。我们还可以对字符使用输入 / 输出操作。

此外，在 C++ 中还包含几个处理字符的库函数。所有字符处理函数都包含在 <cctype> 函数库中（C character type 的缩写），它们继承自 C 语言。

> 使用字符函数，需要包含 <cctype> 头文件。

字符分类函数

所有字符分类函数名都以前缀 is 开头，例如 iscontrol、isupper 等。如果参数属于由函数名定义的类别，则函数返回 1（可以解释为真），否则，函数返回 0（可以解释为假）。表 6-4 显示了这些函数的列表。注意，每个函数中的参数都是 int 类型，但我们总是传递一个字符作为参数（字符是小于短整数的整数）。

表 6-4　字符分类函数

函数声明	说明
int isalnum (int x);	判断参数是否为字母数字字符
int isalpha (int x);	判断参数是否为字母字符
int iscntrl (int x);	判断参数是否为控制字符
int isdigit (int x);	判断参数是否为十进制数字字符（0 到 9）
int isgraph (int x);	判断参数是否为除空格之外的可打印字符
int islower (int x);	判断参数是否为小写字母（a 到 z）
int isprint (int x);	判断参数是否为可打印字符（包含空格）
int ispunct (int x);	判断参数是否为标点符号
int isspace (int x);	判断参数是否为空白字符（空格、回车或者制表符）
int isupper (int x);	判断参数是否为大写字母（A 到 Z）
int isxdigit (int x);	判断参数是否为十六进制数字字符（0 ～ 9、a ～ f 或者 A ～ F）

字符转换函数

C++ 语言包含两个函数，它们用于将字符从一类转换为另一类。这些函数以前缀 to 开头，并返回一个整数（即转换后的字符的值）。表 6-5 列举了这两个函数。

表 6-5　字符转换函数

函数声明	说明
int tolower (int x)	返回参数所对应的小写字母
int toupper (int x)	返回参数所对应的大写字母

注意，这些函数的返回值被定义为一个整数，但是我们总是可以将返回值（隐式或者显式地）转换为字符，参见程序 6-5。

程序 6-5 显示了一个使用预定义函数 isalpha 和 toupper 将文本中的字符更改为大写并计算字母字符个数的示例。

程序 6-5　使用字符函数

```
1  /**********************************************************************
2   * 本程序把小写字母转换为大写字母                                    *
3   * 并计算字母字符的个数                                              *
4   **********************************************************************/
5  #include <iostream>
6  #include <cctype>
7  using namespace std;
8
```

```
9   int main ( )
10  {
11    // 声明
12    char ch;
13    int count = 0;
14    // 输入若干字符并处理
15    while (cin >> noskipws >> ch)
16    {
17        if (isalpha (ch))
18        {
19              count++;
20        }
21        ch = toupper (ch);
22        cout << ch ;
23    }
24    // 输出字母字符个数
25    cout << "The count of alphabetic characters is: " << count;
26    return 0;
27  }
运行结果:
This is a line made of more than 10 characters.
THIS IS A LINE MADE OF MORE THAN 10 CHARACTERS.
^Z
The count of alphabetic characters is: 35
```

6.2.3　处理时间

在 C++ 程序设计中，经常遇到的库函数之一是在 <ctime> 头文件中定义的 time 函数。我们将在附录 I 中更详细地讨论该函数，但本节中读者只需要了解，当传递给该函数的参数为 0 时，该函数返回从 UNIX 纪元（1970 年 1 月 1 日午夜）到现在经过的秒数。换言之，time(0) 返回从 UNIX 纪元到调用函数时经过的秒数，如图 6-5 所示。

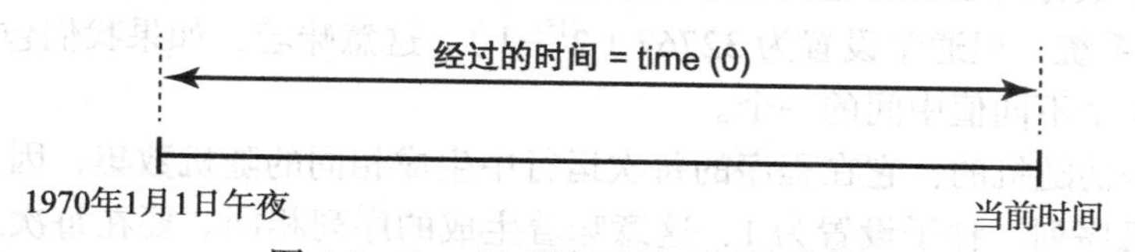

图 6-5　time(0) 所定义的时间的秒数

要找到当前时间的日期、月份和年份，我们需要更复杂的函数，我们将在后续章节中讨论（必须考虑闰年）。但是，找到当前时间的小时、分钟和秒则非常容易，参见下一个示例。

【例 6-8】 程序 6-6 演示了如何使用 time(0) 函数打印当前时间。值得注意的是，计算的时间是格林尼治标准时间（GMT），而不是当地时间。

注意在程序 6-6 中使用的 / 和 % 运算符。第一个运算符（/）用于计算经过的时间单位；第二个运算符（%）用于计算当前时间单位。还要注意的是，如果需要计算你所在位置的当前时间，则必须加上或者减去当地时间和格林威治标准时间之间的差值。

程序 6-6　计算当前时间

```
1   /*****************************************************************
2    * 本程序使用 time(0) 函数计算当前时间                              *
```

```
3    ******************************************************************/
4    #include <iostream>
5    #include <ctime>
6    using namespace std;
7
8    int main ( )
9    {
10       // 计算经过了多少秒和当前的秒数
11       long elapsedSeconds = time (0);
12       int currentSecond = elapsedSeconds % 60;
13       // 计算经过了多少分钟和当前的分钟数
14       long elpasedMinutes = elapsedSeconds / 60;
15       int currentMinute = elpasedMinutes % 60;
16       // 计算经过了多少小时和当前的小时数
17       long elapsedHours = elpasedMinutes / 60;
18       int currentHour = elapsedHours % 24;
19       // 打印当前时间
20       cout << "Current time: ";
21       cout << currentHour << " : " << currentMinute << " : " << currentSecond;
22       return 0;
23   }
```

运行结果：
Current time: 20 : 57 : 59

运行结果：
Current time: 20 : 58 : 22

6.2.4 随机数生成

C++11标准定义了用于创建概率论中定义的任意分布的随机数的类。我们将在后续章节中讨论这些类。在本章中，我们将介绍在<cstdlib>头文件中定义的继承自C语言的随机数生成器。这个函数称为rand，它生成一个介于0和RAND_MAX之间的随机整数，RAND_MAX值依赖于系统，但通常设置为32767（$2^{15}-1$）。这意味着，如果我们包含0，则产生的随机数是32768个不同值中间的一个。

这个函数是伪随机的：它在程序的每次运行中生成相同的随机数集，因为它使用相同的种子来启动随机序列。种子设置为1，这意味着生成的序列相同。要在每次运行中获得不同的序列，我们必须在每次运行程序时为rand函数提供不同的种子。最好的选择是使用函数time(0)，如前所述，它是从纪元开始经过的秒数。每次运行程序时，time(0)是不同的，这意味着我们可以得到一组不同的随机数。为此，我们只需将time(0)的返回值传递给另一个名为srand（seed for random，随机数的种子）的函数，然后再调用rand函数。

我们常常需要创建一个从a到b（包含）的范围内的随机数。为此，我们必须使用rand函数创建一个随机数。然后，我们将其缩放到0到（$b-a$）的范围，最后进行平移。示意图如图6-6所示。

【例6-9】 作为生成随机数的应用，我们创建了一个数值猜测游戏（参见程序6-7）。我们在两个限制数（低和高）之间生成一个随机数，让用户在有限次数中猜测这个随机数。在每次尝试中，如果猜测不正确，我们会提示用户。

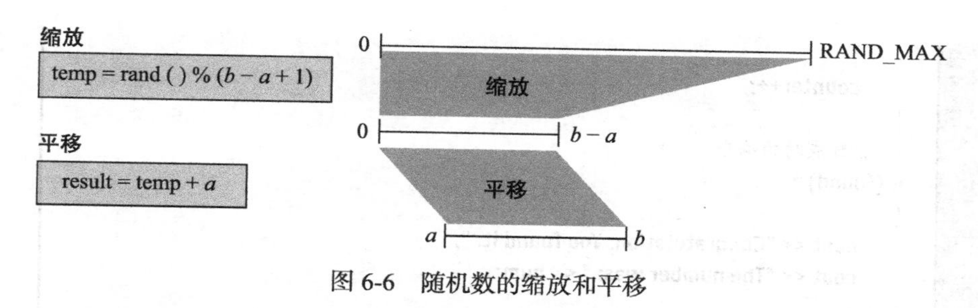

图 6-6　随机数的缩放和平移

程序 6-7　猜数值游戏

```
/*****************************************************************
 * 本程序使用随机数模拟猜数值游戏                                *
 *****************************************************************/
#include <iostream>
#include <cstdlib>
#include <ctime>
using namespace std;

int main ( )
{
  // 声明和初始化
  int low = 5;
  int high = 15;
  int tryLimit = 5;
  int guess;
  // 生成随机数
  srand (time (0));
  int temp = rand();
  int num = temp % (high - low + 1) + low;
  // 循环猜数
  int counter = 1;
  bool found = false;
  while (counter <= tryLimit  &&  !found)
  {
      do
      {
            cout << "Enter your guess between 5 to 15 (inclusive): ";
            cin >> guess;
      } while (guess < 5 || guess > 15);

      if (guess == num)
      {
            found = true;
      }
      else if (guess > num)
      {
            cout << "Your guess was too high!" << endl;
      }
      else
      {
            cout << "Your guess was too low!" << endl;
```

```
42          }
43          counter++;
44      }
45      // 猜数成功的响应
46      if (found)
47      {
48          cout << "Congratulation: You found it. ";
49          cout << "The number was: " << num;
50      }
51      // 猜数失败的响应
52      else
53      {
54          cout << "Sorry, you did not find it! ";
55          cout << "The number was: " << num;
56      }
57      return 0;
58  }
```

```
运行结果:
Enter your guess between 5 to 15 (inclusive): 7
Your guess was too low!
Enter your guess between 5 to 15 (inclusive): 8
Congratulation: You found it. The number was: 8
```

```
运行结果:
Enter your guess between 5 to 15 (inclusive): 15
Your guess was too high!
Enter your guess between 5 to 15 (inclusive): 14
Your guess was too high!
Enter your guess between 5 to 15 (inclusive): 13
Your guess was too high!
Enter your guess between 5 to 15 (inclusive): 12
Your guess was too high!
Enter your guess between 5 to 15 (inclusive): 11
Your guess was too high!
Sorry, you did not find it! The number was: 10
```

6.3 用户自定义函数

我们需要的所有函数并不是都在C++库中定义的。要调用库中未定义的函数，我们必须首先在程序中定义该函数。

6.3.1 函数的四种类型

图6-2显示了函数定义的一般语法，图6-3显示了函数调用运算符的一般语法。在本节中，我们进一步展开阐述有关函数的概念，详细阐述C++中常见的四种函数的语法：不带参数的void函数、带参数的void函数、不带参数有返回值的函数、带参数有返回值的函数，如图6-7所示。我们还将讨论每种类型的应用，这样当我们想要设计一个函数时，就知道需要选择哪种类型。

不带参数的void函数

也许最简单函数类型就是不带参数的void函数。函数不接收任何参数，不返回任何内容。这种类型的函数仅被设计用于其副作用，副作用发生在函数内部，否则，函数就没有任何意义。因为函数不返回值，所以我们不能在需要值的地方使用这些函数。我们只能将此类函数用作使用其副作用的后缀表达式。

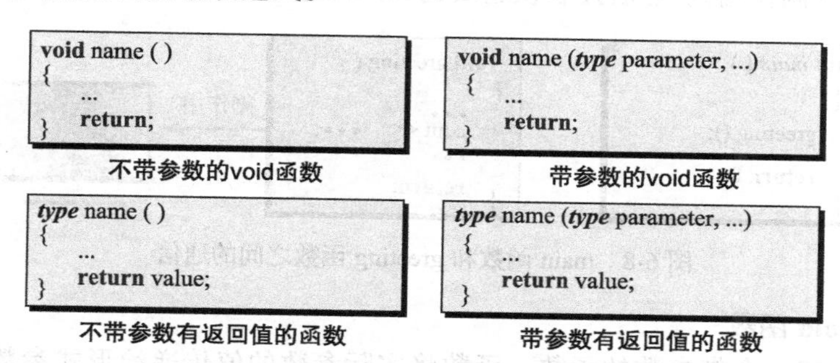

图 6-7　C++ 中的四种函数

【例 6-10】 让我们创建一个不带参数的 void 函数：在屏幕上打印一条消息，例如问候语（副作用），参见程序 6-8。注意，函数定义（第 13 行到第 19 行）位于 main 函数中的函数调用（第 23 行）之前。实际上，我们有两个函数定义，一个是 greeting 函数，一个是 main 函数。greeting 函数在 main 函数内部被调用，而 main 函数由 C++ 运行时环境调用。

程序 6-8　使用不带参数的 void 函数

```
1  /*************************************************************
2   * 本程序使用一个 void 函数打印一个方框中的问候语               *
3   *************************************************************/
4  #include <iostream>
5  using namespace std;
6
7  /*************************************************************
8   * greeting 函数的函数定义。                                  *
9   * 这是一个不带参数的 void 函数。                             *
10  * 函数打印三行消息，不返回任何值。                           *
11  * 函数仅具有副作用（显示三行消息）                           *
12  *************************************************************/
13 void greeting ( )
14 {
15   cout <<"*******************" << endl;
16   cout <<"* Hello Friends     *" << endl;
17   cout <<"*******************";
18   return;
19 }
20
21 int main ( )
22 {
23   greeting(); // 调用 greeting 函数（表达式语句）
24   return 0;
25 }
运行结果：
*****************
* Hello Friends     *
*****************
```

图6-8显示了main函数和greeting函数之间的通信。尽管main函数和greeting函数之间没有数据交换，但存在一个副作用。main函数首先调用greeting函数。greeting函数在屏幕上打印问候语（副作用）。然后控制权返回到main函数（void返回）。

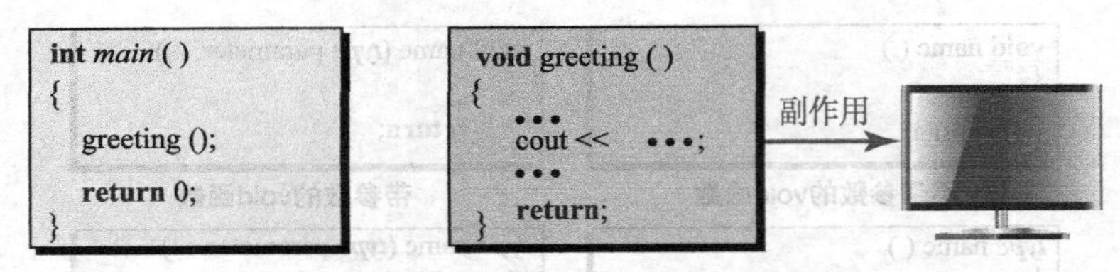

图6-8 main函数和greeting函数之间的通信

带参数的void函数

接着我们创建一个带参数的函数。函数将实际参数的值传递给形式参数。函数有副作用，但不向调用方返回任何数据。当我们只需要函数的副作用，同时还需要传递一些信息（例如每次调用函数时要输出什么内容）的时候，就使用这种类型的函数。稍后我们将看到，在“输入-处理-输出”设计模式中，带参数的void函数是输出模块的最佳选择。

【例6-11】程序6-9显示了一个示例，其中参数指示函数在每次执行中如何格式化图案模式。第14行到第25行是函数定义，第38行是函数调用。

程序6-9 带参数的void函数

```
1   /***************************************************************
2    * 本程序演示如何使用带参数的 void 函数                          *
3    * 创建不同的图案模式                                            *
4    ***************************************************************/
5   #include <iostream>
6   using namespace std;
7
8   /***************************************************************
9    * pattern 函数的函数定义。                                      *
10   * 这是一个带参数的 void 函数。                                  *
11   * 每次运行时，函数接收用户输入的 size，                         *
12   * 参数 size 用于创建不同大小的图案模式                          *
13   ***************************************************************/
14  void pattern (int size)
15  {
16    for (int i = 0; i < size; i++)
17    {
18        for (int j = 0; j < size; j++)
19        {
20              cout << "*" ;
21        }
22        cout << endl;
23    }
24    return;
25  }
26
27  int main()
28  {
29    // 变量声明
```

```
30    int patternSize;  //传递给 pattern 函数的实际参数
31    //输入验证
32    do
33    {
34        cout << "Enter the size of the pattern: ";
35        cin >> patternSize;
36    } while (patternSize <=0);
37    //函数调用
38    pattern (patternSize); // patternSize 是实际参数
39    return 0;
40 }
```

```
运行结果：
Enter the size of the pattern: 3
***
***
***
```

```
运行结果：
Enter the size of the pattern: 4
****
****
****
****
```

图 6-9 显示了 main 函数和 pattern 函数之间的通信。每次运行程序时，我们都会获取 patternSize 的值，并将该值传递给函数调用以控制打印的图案的大小。换言之，每次调用函数时，patternSize 的值都存储在 size 参数中。

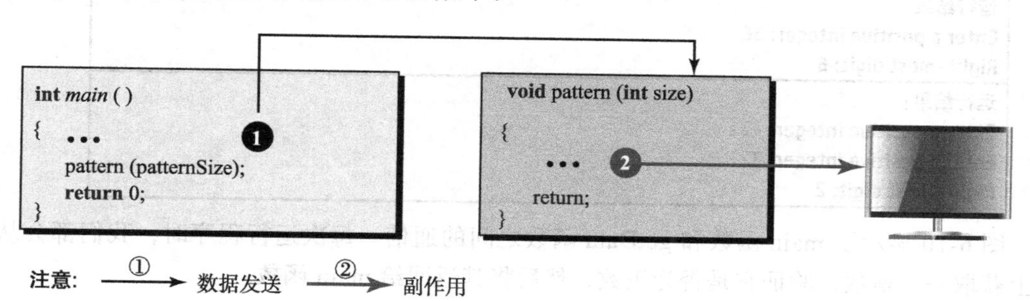

图 6-9　main 函数和 pattern 函数之间的通信

不带参数有返回值的函数

这种类型的函数仅被设计用于其返回值。函数通常具有输入副作用，并将输入值返回给主调函数。稍后我们将看到，在“输入 – 处理 – 输出”设计模式中，不带参数有返回值的函数是输入模块的最佳选择。程序 6-10 展示了这种设计理念，它调用一个输入函数 getdata 从键盘读取一个整数，然后返回给 main 函数进行处理。

程序 6-10　不带参数有返回值的函数

```
1  /*****************************************************************
2   *本程序演示如何定义一个函数从键盘读取一个正整数，                *
3   *并打印其最右侧的数字                                           *
4   *****************************************************************/
```

```
5   #include <iostream>
6   using namespace std;
7
8   /**************************************************************
9    * getData 函数是一个不带参数有返回值的函数。                    *
10   * 它接收用户的键盘输入。                                      *
11   * 验证输入值是正整数后，                                      *
12   * 把结果返回给 main 函数                                      *
13   **************************************************************/
14  int getData()
15  {
16    int data;
17    do
18    {
19        cout << "Enter a positive integer: ";
20        cin >> data;
21    } while (data <= 0);
22    return data;
23  }
24
25  int main()
26  {
27    int number = getData(); // 不带参数的函数调用
28    cout << "Right-most digit: " << number % 10;
29    return 0;
30  }
```

运行结果：

```
Enter a positive integer: 56
Right-most digit: 6
```

运行结果：

```
Enter a positive integer: -72
Enter a positive integer: 72
Right-most digit: 2
```

图 6-10 显示了 main 函数和 getData 函数之间的通信。每次运行程序时，我们都会从键盘上获取一个整数，验证它是否为正数，然后将其返回给 main 函数。

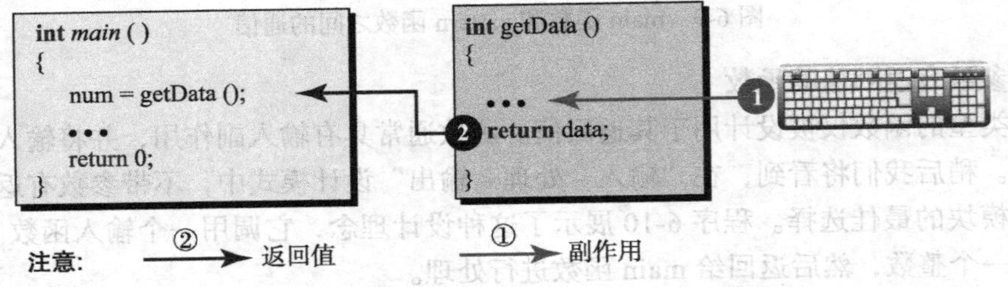

图 6-10 main 函数和 getData 函数之间的通信

带参数有返回值的函数

程序 6-11 演示了如何创建并使用带参数有返回值的函数。larger 函数接收来自 main 函数的两个值，查找其中较大的值，并返回结果。

程序 6-11　带参数有返回值的函数

```
1  /****************************************************************
2   * 本程序演示如何定义一个函数                                  *
3   * 查找由用户给定的两个正整数中的较大值                        *
4   ****************************************************************/
5  #include <iostream>
6  using namespace std;
7
8  /****************************************************************
9   * larger 函数是一个带两个参数并且有返回值的函数。             *
10  * 它接收主调函数传递的两个值，返回其中的较大值。              *
11  * 函数没有副作用                                              *
12  ****************************************************************/
13 int larger (int fst, int snd)
14 {
15   int max;
16   if (fst > snd)
17   {
18       max = fst;
19   }
20   else
21   {
22       max = snd;
23   }
24   return (max);
25 } // larger 结束
26
27 int main()
28 {
29   // 声明
30   int first, second;
31   // 获取输入
32   cout << "Enter the first number: ";
33   cin >> first;
34   cout << "Enter the second number: ";
35   cin >> second;
36   // 函数调用
37   cout << "Larger: " << larger (first, second); // 函数调用
38   return 0;
39 }
```

运行结果：

```
Enter the first number: 56
Enter the second number: 71
Larger: 71
```

运行结果：

```
Enter the first number: -10
Enter the second number: 8
Larger: 8
```

图 6-11 显示了 main 函数和 larger 函数之间的通信。main 函数将两个值（实际参数）传递给 larger 函数。larger 函数查找两者中较大的值，并将较大值返回给 main 函数。

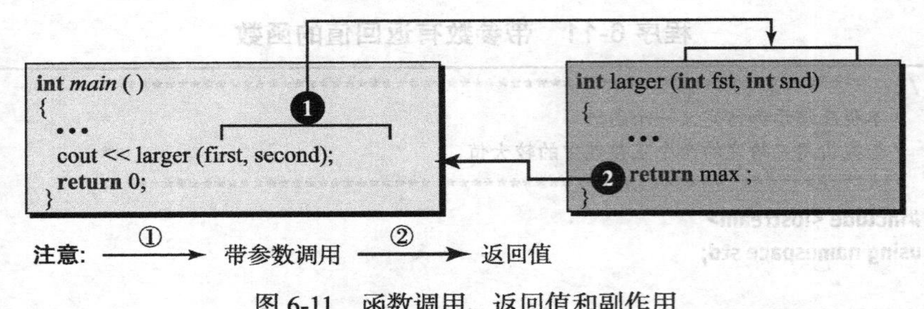

图 6-11　函数调用、返回值和副作用

6.3.2　使用声明

在前面的所有示例中，我们都将函数定义放在函数调用之前。另一种方法是在函数调用之前放置函数的声明（原型），在函数调用之后放置函数定义，如图 6-12 所示。注意函数声明的结尾必须有英文分号。

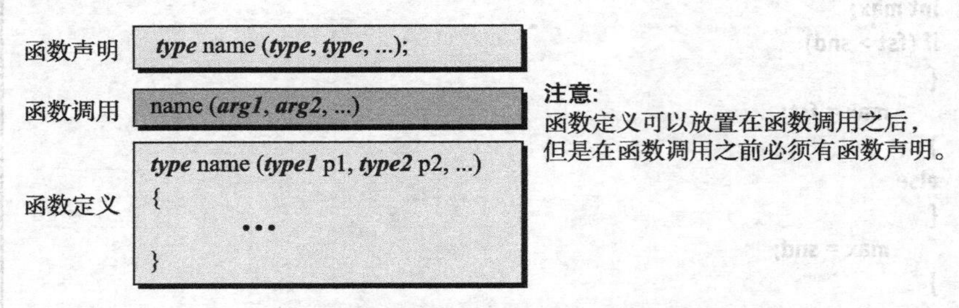

图 6-12　在函数调用之后设置函数定义

函数声明末尾需要加上英文分号。

【例 6-12】 程序 6-12 遵循模块程序设计的原则：输入、处理、输出。main 函数只负责三个函数调用：调用输入函数、调用处理函数和调用输出函数。其余分别由三个函数完成：input 函数、process 函数、output 函数。程序 6-12 旨在从用户处获取年份（输入），判断给定年份是否为闰年（处理），并打印结果（输出）。我们之前讨论过，给定年份需要满足两个标准才能称为闰年。首先，该年份必须能被 4 整除。其次，如果该年份能被 100 整除，则必须也能被 400 整除。

程序 6-12　判断给定年份是否为闰年

```
1   /***************************************************************
2    * 本程序演示如何使用三个函数:                                  *
3    * 输入一个代表年份的整数、判断给定年份是否为闰年、              *
4    * 打印结果                                                    *
5    ***************************************************************/
6   #include <iostream>
7   using namespace std;
8
9   // 函数声明（函数原型）
10  int input ();
11  bool process (int year);
12  void output (int year, bool result);
13
```

```
14  int main ()
15  {
16    // 输入、处理、输出
17    int year = input();
18    bool result = process (year);
19    output (year, result);
20  }
21  /*******************************************************************
22   * 输入函数的定义，被 main 函数调用。                                 *
23   * 它接收用户输入的年份值（副作用），                                 *
24   * 验证输入的年份值是否大于 1582，                                    *
25   * 并把结果返回给 main 函数。                                         *
26   * 该函数是不带参数的有返回值的函数，具有副作用                       *
27   *******************************************************************/
28  int input ()
29  {
30    int year;
31    do
32    {
33        cout << "Enter a year after 1582: " ;
34        cin >> year;
35    } while (year <= 1582);
36    return year;
37  }
38  /*******************************************************************
39   * 处理函数的定义。                                                   *
40   * 它接收 main 函数传递的年份值作为参数，                             *
41   * 判断给定年份是否为闰年，                                           *
42   * 并把一个布尔值返回给 main 函数。                                   *
43   * 该函数是带参数的有返回值的函数，没有副作用                         *
44   *******************************************************************/
45  bool process (int year)
46  {
47    bool criteria1 = (year % 4 == 0);
48    bool criteria2 = (year % 100 != 0) || (year % 400 == 0);
49    return (criteria1) && (criteria2);
50  }
51  /*******************************************************************
52   * 输出函数接收年份值和处理函数返回的布尔值，                         *
53   * 在监视器上打印出结果。                                             *
54   * 该函数是带两个参数的 void 函数                                     *
55   *******************************************************************/
56  void output (int year, bool result)
57  {
58    if (result)
59    {
60        cout << "Year " << year << " is a leap year.";
61    }
62    else
63    {
64        cout << "Year " << year << " is not a leap year.";
65    }
```

```
66      return;
67  }
```

运行结果：
Enter a year after 1582: 1900
Year 1900 is not a leap year.

运行结果：
Enter a year after 1582: 1500
Enter a year after 1582: 1600
Year 1600 is a leap year.

图 6-13 显示了 main 函数与其他三个函数之间的通信。其中数字序号定义了通信的顺序。main 函数首先调用 input 函数，该函数从键盘获取用户输入并返回一个值（在验证之后）给 main 函数。然后，main 函数调用 process 函数并将年份值传递给 process 函数，以确定给定年份是否为闰年并返回结果。然后，main 函数调用 output 函数来打印年份值，并指示该年份是否为闰年。

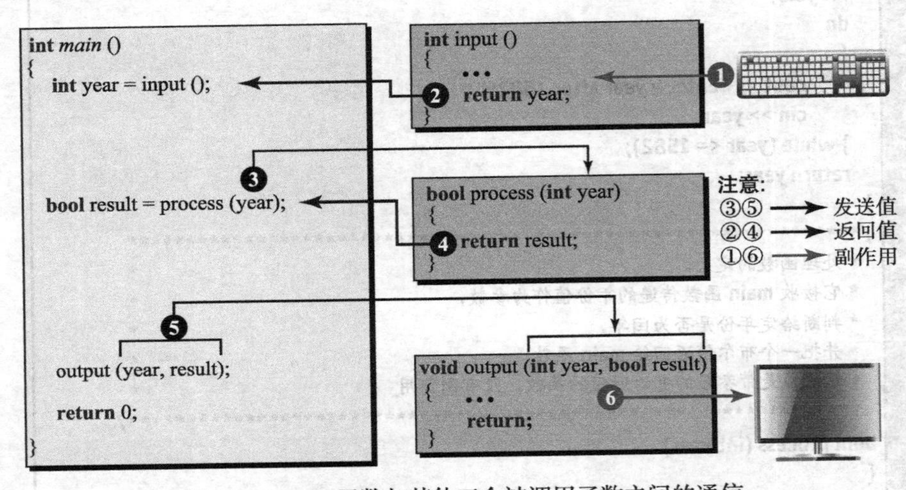

图 6-13　main 函数与其他三个被调用函数之间的通信

6.4　数据交换

在上一节中，我们展示了主调函数可以与被调用函数之间进行通信。在通信过程中，两个函数可以双向交换数据：正向交换数据和反向交换数据，如图 6-14 所示。接下来分别讨论这两种情况。

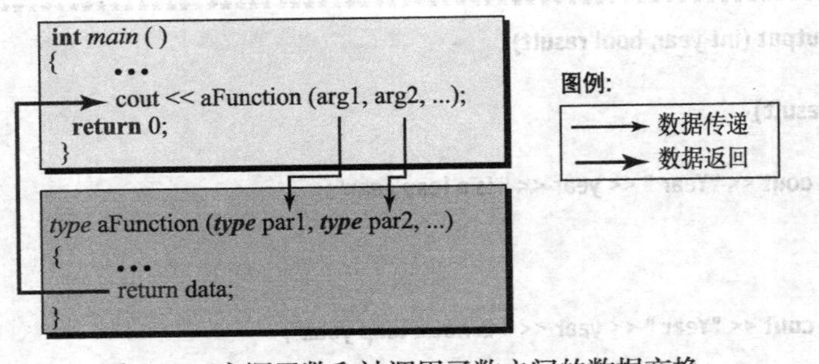

图 6-14　主调函数和被调用函数之间的数据交换

6.4.1 传递数据

如果被调用函数的参数列表不为空，则数据将从每个实际参数传递到对应的形式参数。根据应用程序的不同，我们可以使用三种机制将数据从实际参数传递到形式参数：按值传递、按引用传递和按指针传递。

按值传递

在**按值传递**（pass-by-value）机制中，实际参数将数据项的副本传递给对应的形式参数。形式参数接收并存储该值。由于交换是通过一个值进行的，所以实际参数可以是一个字面量值或者一个变量的值，如图 6-15 所示。

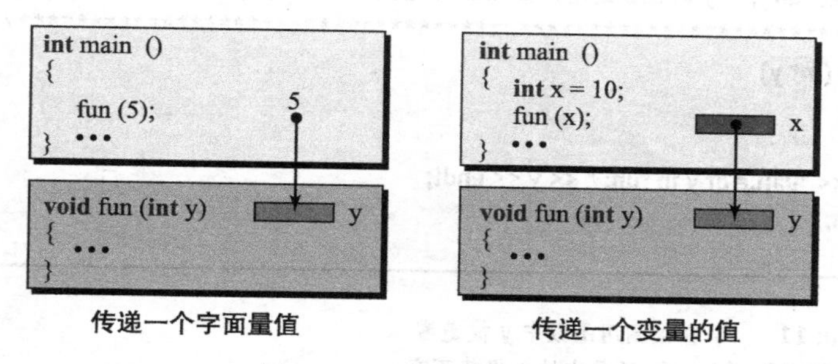

图 6-15　按值传递的概念

当不希望被调用函数更改传递给它的实际参数的值时，我们使用按值传递。换言之，被调用的函数只能读取实际参数的值，但不能修改实际参数的值。在计算机术语中，这被称为只读访问。

按值传递方法使用实际参数的副本初始化对应的形式参数。

【例 6-13】 程序 6-13 给出了一个简单的例子，用来演示按值传递背后的思想。我们在被调用函数（fun）中递增了形式参数 y 的值，但对应的实际参数 x 的值不会增加。

在现实生活中也存在按值传递的例子。当一个朋友想向你借一份珍贵的文件时，你可以给他一份该文件的副本。你的朋友可以随心所欲地处理文件副本，但你的原始文件仍然保持原样。

程序 6-13　测试按值传递

```
1   /***************************************************************
2    * 本程序演示按值传递时                                        *
3    * 修改形式参数无法影响对应的实际参数                          *
4    ***************************************************************/
5   #include <iostream>
6   using namespace std;
7
8   // 函数声明（函数原型）
9   void fun (int y);
10
11  int main ()
12  {
13    // 声明和初始化实际参数
14    int x = 10;
```

```
15      // 调用函数 fun，把 x 作为实际参数传递
16      fun (x);
17      // 打印 x 的值，结果无变化
18      cout << "Value of x in main: " << x << endl;
19      return 0;
20  }
21  /*********************************************************************
22   * fun 是一个函数，                                                  *
23   * 它接收 x 的值并储存到形式参数 y 中。                              *
24   * x 和 y 位于两个独立的内存位置，                                   *
25   * 在函数 fun 中，y 的值被递增，但 main 函数中 x 的值保持不变        *
26   *********************************************************************/
27  void fun (int y)
28  {
29      y++;
30      cout << "Value of y in fun: " << y << endl;
31      return;
32  }
```

运行结果：

```
Value of y in fun: 11      // 在被调用函数中 y 被递增
Value of x in main: 10  // main 函数中的 x 保持不变
```

按引用传递

在按引用传递（pass-by-reference）机制中，实际参数和对应的形式参数之间共享内存位置。主调函数和被调用函数可以用相同的名称或者不同的名称（建议使用不同的名称）引用该内存位置。形式参数名是实际参数名的别名。示意图如图 6-16 所示。

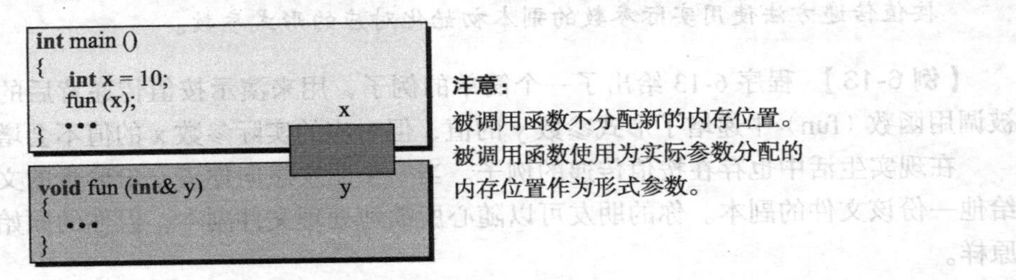
int main ()
{
int x = 10;
fun (x);
• • •
}
x
void fun (int& y)
{
• • •
}
y
注意：
被调用函数不分配新的内存位置。
被调用函数使用为实际参数分配的
内存位置作为形式参数。

图 6-16　按引用传递的概念

在图 6-16 中，x 和 y 这两个名称实际上是指同一个变量位置，由主调函数创建并由被调用函数使用。但是，为了告诉编译器 y 不是新的内存位置，C++ 要求我们将 y 声明为整数位置的别名（int& y），而不是新的整数位置（int y）。

【例 6-14】 程序 6-14 重复上一个示例，但我们使用了按引用传递而不是按值传递。

程序 6-14　测试按引用传递

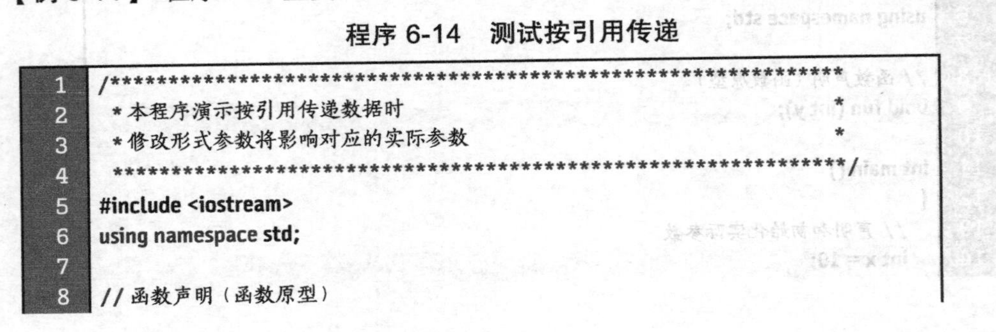

```
1   /*********************************************************************
2    * 本程序演示按引用传递数据时                                        *
3    * 修改形式参数将影响对应的实际参数                                  *
4    *********************************************************************/
5   #include <iostream>
6   using namespace std;
7
8   // 函数声明（函数原型）
```

```
 9  void fun (int& y);   // & 符号表示 y 是一个联盟（按引用传递的参数）
10
11  int main ()
12  {
13      // 声明和初始化实际参数
14      int x = 10;
15      // 调用函数 fun，把 x 作为实际参数传递
16      fun (x);
17      // 打印 x 的值以查看 x 值的变化
18      cout << "Value of x in main: " << x << endl;
19      return 0;
20  }
21  /***************************************************************
22   * fun 是一个函数，按引用接收 y 的值。                              *
23   * 这意味着在函数调用中，形式参数 y 是实际参数 x 的别名。               *
24   * 在该函数中递增了形式参数，                                       *
25   * 结果 main 函数中的实际参数也随之被递增                             *
26   ***************************************************************/
27  void fun (int& y)
28  {
29      y++;
30      cout << "Value of y in fun: " << y << endl;
31      return;
32  }
运行结果：
Value of y in fun: 11
Value of x in main: 11        // 在函数 fun 中更改 y，main 函数中的 x 随之改变
```

注意，我们用值为 10 的参数 x 来调用函数 fun。fun 函数增加 y 的值并将其打印为 11。由于 main 函数和 fun 函数共享相同的内存位置，因此 main 函数中的 x 值也会增加，具体请参见运行结果。

在现实生活中也存在按引用传递的例子。当我们将一个文档传递给一个朋友来编辑时，我们将与他共享原始文档。我们的朋友对文档所做的任何更改，都会在返回的文档中得到体现。

在计算机术语中，我们有时将按引用传递称为读写通信（read-write communication），这意味着被调用函数可以读取主调函数中参数的值，并且也可以更改这些值。

【例 6-15】 作为一个经典的例子，我们讨论交换两个变量（first 和 second）的过程。交换后，变量 first 保存最初存储在 second 中的内容，变量 second 保存最初存储在 first 中的内容。例如，如果 first 和 second 变量的值分别为 10 和 20，则交换变量后，first 应该为 20，second 应该为 10。我们可以证明，使用简单的赋值语句 first = second 和 second = first，会将两个变量都设置为 10 或者 20。我们必须使用一个临时变量和三个赋值语句，如图 6-17 所示。请注意，每个变量的名称必须在赋值运算符的左侧和右侧各出现一次。

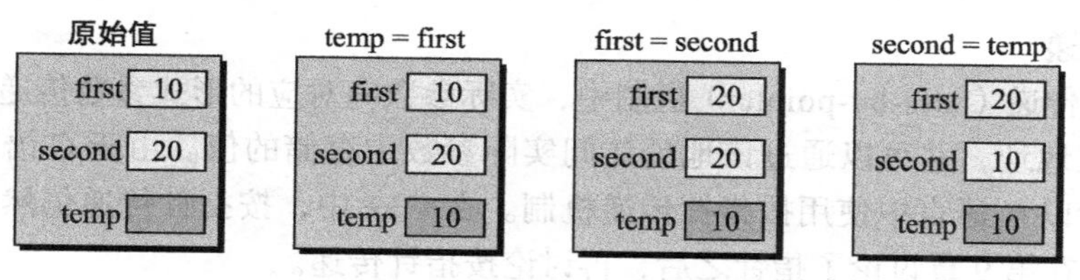

图 6-17　使用一个临时变量来交换两个值

两个变量的值交换被用于排序和其他算法，并且在每个算法中被重复使用多次，因此有必要创建一个函数并在需要时调用该函数。但是，如果我们使用按值传递，被调用函数不能交换主调函数中的变量，因为数据是在被调用函数中交换的，而不是在主调函数中交换的。在这种情况下，我们需要使用按引用传递，如程序6-15所示。

程序 6-15　使用交换函数

```
1  /******************************************************************
2   * 本程序演示如何使用按引用传递                                    *
3   * 交换主调函数中的两个变量值                                      *
4   ******************************************************************/
5  #include <iostream>
6  using namespace std;
7
8  void swap (int& first, int& second);  // 函数声明
9
10 int main ( )
11 {
12   // 声明
13   int first = 10;
14   int second = 20;
15   swap (first, second); // 函数调用
16   // 打印以测试交换结果
17   cout << "Value of first in main: " << first << endl;
18   cout << "Value of second in main: " << second;
19   return 0;
20 }
21 /******************************************************************
22  * swap 函数带两个按引用传递的参数，                                *
23  * 然后使用一个临时变量 temp 来交换 fst 和 snd 的值。               *
24  * 由于数据是按引用传递，                                          *
25  * 交换形式参数                                                    *
26  * 意味着交换实际参数（main 函数中的 first 和 second）              *
27  ******************************************************************/
28 void swap (int& fst, int& snd)
29 {
30   int temp = fst;
31   fst = snd;
32   snd = temp;
33   return;
34 }
运行结果：
Value of first in main: 20
Value of second in main: 10
```

按指针传递

在**按指针传递**（pass-by-pointer）机制中，实际参数向对应的形式参数传递内存地址。形式参数存储该地址，并可以通过该地址访问实际参数中存储的值。由于C语言中不存在按引用传递，所以C语言中使用按指针传递机制。在C++中，按指针传递仍然有一定程度的使用。我们将在第9章讨论了指针之后，再讨论按指针传递。

优点和缺点

在结束本节之前，我们简要总结每种方法的优缺点。

1）按值传递非常简单，可以防止被调用函数更改实际参数。但是，按值传递存在一个缺点：需要将实际参数对象从主调函数复制并传递到被调用函数。如果对象很小，就像我们到目前为止使用的内置数据类型一样，复制产生的消耗不大，我们应该使用这种方法。然而，我们将在第 9 章中看到，当对象很大时，我们不会在面向对象的程序设计中使用按值传递。我们会使用其他传递方法（按引用传递和按指针传递）。

2）按引用传递时，如果被调用函数更改形式参数，主调函数中的实际参数也会被更改。如果这是函数的目的，例如交换变量的情况，那么按引用传递是最佳选择。此方法的优点是不需要复制，主调函数和被调用函数中的对象相同，名称不同。这意味着，对于大型对象，当复制成本很高时，我们应该使用此方法。

3）按指针传递与按引用传递具有相同的优势。通常在 C++ 中应避免使用按指针传递，但是如果要传递的数据的本质涉及指针（例如 C 类型字符串和数组，这两个都将在后续章节中讨论），则可以使用按指针传递。

6.4.2　返回值

我们已经看到，void 函数的设计目的在于其副作用，它不会向主调函数返回任何内容。非空函数则需要向其调用方返回一个值。返回值可以通过三种方式完成：按值返回、按引用返回和按指针返回。

按值返回

最常见的返回方法是**按值返回**（return-by-value）。被调用函数在其函数体中创建表达式并将其返回给主调函数。在这种情况下，必须在需要值的情况下使用该函数调用表达式。图 6-18 显示了把表达式值返回给主调函数的两个示例。

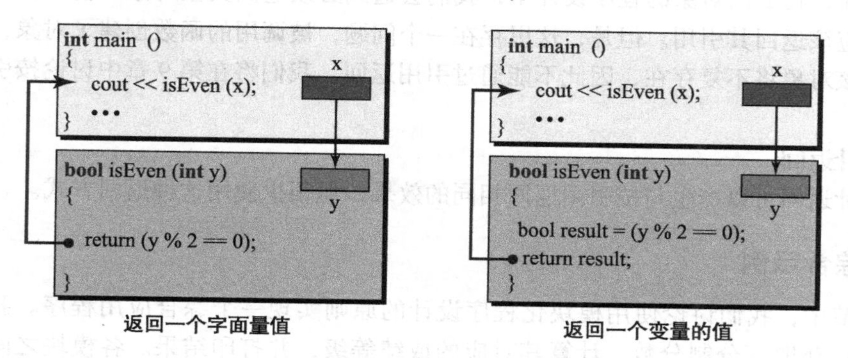

图 6-18　按值返回

【例 6-16】 程序 6-16 演示了按值返回的示例。

程序 6-16　从函数返回字面量值

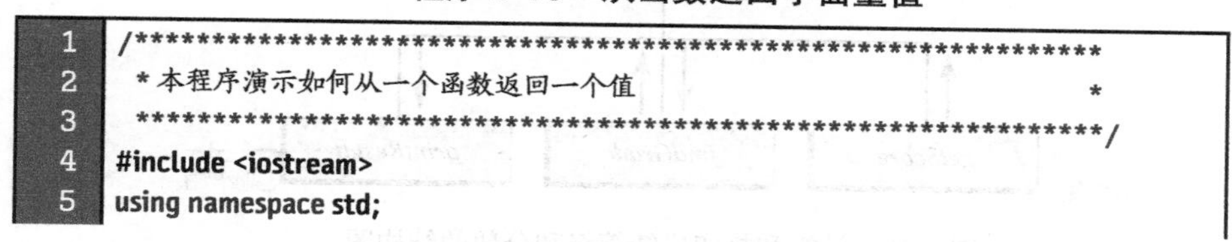

```
1  /***************************************************************
2   * 本程序演示如何从一个函数返回一个值                          *
3   ***************************************************************/
4  #include <iostream>
5  using namespace std;
```

```
6
7   // 函数声明
8   bool isEven (int y);
9
10  int main ( )
11  {
12     // 函数调用
13     cout << boolalpha << isEven (5) << endl;
14     cout << boolalpha << isEven (10);
15     return 0;
16  }
17  /*****************************************************************
18   * isEven 函数带一个参数 y，                                      *
19   * 通过取 y 的余数判断 y 是否为偶数。                             *
20   * 最后函数返回判断结果值（字面量值）                             *
21   *****************************************************************/
22  bool isEven (int y)
23  {
24     return ((y % 2) == 0);
25  }
运行结果：
false
true
```

按引用返回

我们在 C++（当用作面向过程的程序设计语言时）中遇到的大多数函数是按值返回数据。使用按值返回的方式既简单又直接。但是，按值返回有一个缺点。编译器必须复制被调用函数返回的对象，然后将其副本返回给主调函数。对于内置数据类型，这通常是不存在问题的。然而，在面向对象的程序设计中，我们会遇到必须返回大型对象的情况。为了提高效率，我们应该返回其引用。但是，这里存在一个问题，被调用的函数创建了对象，但当函数终止后，该对象将不复存在，因此不能通过引用返回。我们将在第 9 章中讨论按引用返回的问题。

按指针返回

按指针返回可以产生与按引用返回相同的效果，但很少使用这种返回方式。

6.4.3　综合示例

在本节中，我们讨论使用模块化程序设计的原则实现一个综合应用程序。我们将编写一个程序：获取百分制分数，计算其对应的成绩等级，并打印结果。各模块之间的关系如图 6-19 所示。

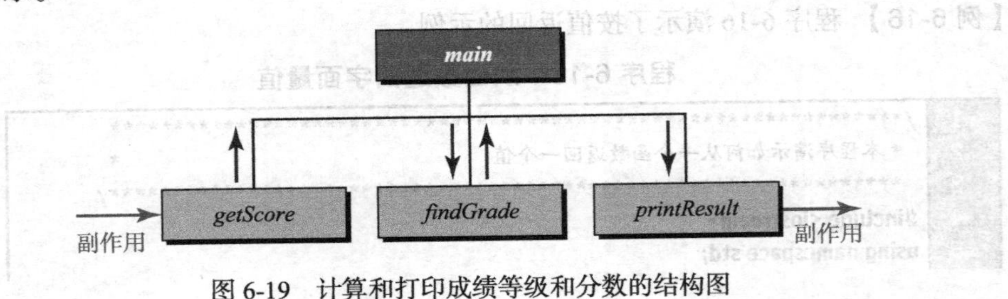

图 6-19　计算和打印成绩等级和分数的结构图

【例 6-17】 程序 6-17 展示了我们如何为每个任务创建函数。前两个函数是有返回值的函数，第三个函数是 void 函数。

程序 6-17　根据百分制分数计算成绩等级

```
1   /*****************************************************************
2    * 本程序演示如何使用三个函数，                                     *
3    * 输入、计算、并打印一个学生的百分制分数和成绩等级                 *
4    *****************************************************************/
5   #include <iostream>
6   using namespace std;
7   
8   // 函数声明
9   int getScore ();
10  char findGrade (int score);
11  void printResult (int score, char grade);
12  
13  int main ()
14  {
15    // 声明
16    int score;
17    char grade;
18    // 函数调用
19    score = getScore ();
20    grade = findGrade (score);
21    printResult (score, grade);
22    return 0;
23  }
24  /*****************************************************************
25   * getScore 函数是一个具有副作用的输入函数。                        *
26   * 它获取用户的输入，                                               *
27   * 按值返回给 main 函数。                                           *
28   * 它使用一个局部变量（score），将其值返回给 main 函数。            *
29   * 该函数还验证输入的分数是否位于 0 到 100 之间                     *
30   *****************************************************************/
31  int getScore ()
32  {
33    int score; // 声明局部变量
34    do
35    {
36        cout << "Enter a score between 0 and 100: ";
37        cin >> score;
38    } while (score < 0 || score > 100);
39    return score;
40  }
41  /*****************************************************************
42   * findGrade 函数计算传递给它的百分制分数的成绩等级，               *
43   * 并返回一个字母等级（A、B、C、D、F）。                            *
44   * 它使用按值传递方式获取百分制分数（score）。                      *
45   * 它使用按值返回方式把成绩等级返回给 main 函数。                   *
46   * 我们使用了嵌套 if-else 结构，                                    *
47   * 当然也可以使用 switch 语句来实现                                 *
48   *****************************************************************/
```

```
49  char findGrade (int score)
50  {
51    char grade; // 声明局部变量
52    if (score >= 90)
53    {
54        grade = 'A';
55    }
56    else if (score >= 80)
57    {
58        grade = 'B';
59    }
60    else if (score >= 70)
61    {
62        grade = 'C';
63    }
64    else if (score >= 60)
65    {
66        grade = 'D';
67    }
68    else
69    {
70        grade = 'F';
71    }
72    return grade;
73  }
74  /*************************************************************
75   * 最后一个函数打印百分制分数和对应的成绩等级                      *
76   *************************************************************/
77  void printResult (int score, char grade)
78  {
79    cout << endl << "Result of the test." << endl;
80    cout << "Score: " << score << " out of 100"<< endl;
81    cout << "Grade: ";
82  }
```

```
运行结果:
Enter a score between 0 and 100: 87
Result of the test.
Score: 87 out of 100
Grade: B
```

```
运行结果:
Enter a score between 0 and 100: 93
Result of the test.
Score: 93 out of 100
Grade: A
```

图 6-20 显示了 main 函数和三个被调用函数之间的通信。

在分析图 6-20 中的小程序时，我们发现 main 函数负责协调其他三个函数的工作。第一个函数 getScore 从外部获取数据（score），并将 score 返回给 main 函数。第二个函数 findGrade 接收来自 main 函数的数据（score），对其进行处理，并将结果（grade）返回给 main 函数。第三个函数 printResult 从 main 函数接收两个数据项（score 和 grade），并将它们发送到监视器。

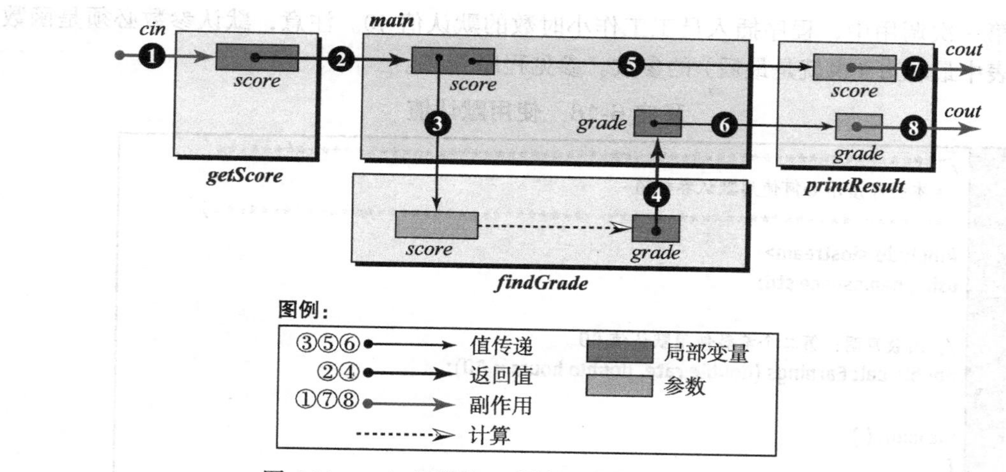

图 6-20　main 函数和三个被调用函数之间的数据传递

注意，我们对处理百分制分数的形式参数和局部变量使用了相同的名称。它们都使用了名称 score，在程序中不会造成任何混淆，其原因是它们位于不同的领域（或者称为作用域，稍后我们将展开阐述）。我们有四个称为 score 的存储位置，每个实体中有一个。getScore 函数和 main 函数中的 score 是局部变量；findGrade 和 printResult 中的 score 是形式参数。我们还有三个名为 grade 的存储位置，其中包括两个局部变量和一个形式参数。

6.5　有关参数的进一步讨论

还有另外两个与参数相关的必须讨论的问题：默认参数和重载函数。

6.5.1　默认参数

程序可以多次调用一个函数。如果函数被设计为按值传递参数，那么在很多时候参数的值可能是相同的。在这种情况下，我们可以对一个或者多个参数（包括所有参数）使用**默认参数值**。如果只有某些参数具有默认参数值，则它们必须是最右边的参数。

【例 6-18】 假设我们设计了一个名为 calcEarnings 的函数，用于计算一组员工的每周薪酬。该函数使用两个参数：rate（小时工资）和 hours（工作小时数）。大多数员工工作 40 小时，但有些员工工作时间较短。我们假设员工工作时间不能超过 40 小时（公司规定）。我们可以通过简单地将默认值分配给参数列表中的参数，从而使用 40 作为参数 hours 的默认值。以下声明语句声明了一个参数 hours 默认为 40 的 calcEarnings 函数。

```
double calcEarnings (double rate, double hours = 40.0);
```

注意，如前所述，如果没有函数声明（当函数定义出现在函数调用之前），那么函数定义同时充当函数声明。这意味着默认参数必须在函数定义头中定义。

当员工工作 40 小时时，我们可以只传递 rate 参数对应的实际参数来调用 calcEarnings 函数。当员工工作时间少于 40 小时时，我们会同时传递参数 rate 和 hours。以下示例演示了默认参数的概念。

```
calcEarnings (payRate);
calcEarnings (payRate, hourWorked);
```

在第一次调用中，程序插入员工工作小时数的默认值 40。注意，默认参数必须是函数参数列表中最右边（也就是最后）的参数。参见程序 6-18。

程序 6-18　使用默认值

```
1  /*****************************************************************
2   * 本程序演示如何使用默认参数值                                   *
3   *****************************************************************/
4  #include <iostream>
5  using namespace std;
6  
7  // 函数声明：第二个参数使用默认值 40
8  double calcEarnings (double rate, double hours = 40);
9  
10 int main ( )
11 {
12    // 第一次函数调用使用默认值
13    cout << "Emplyee 1 pay: " << calcEarnings (22.0) << endl;
14    cout << "Emplyee 2 pay: " << calcEarnings (12.50, 18);
15    return 0;
16 }
17 /*****************************************************************
18  * 本函数包含两个参数，                                            *
19  * 在函数定义中无须指定默认参数值，                                  *
20  * 因为在函数声明中包含了默认参数值                                  *
21  *****************************************************************/
22 double calcEarnings (double rate, double hours)
23 {
24    double pay;
25    pay = hours * rate;
26    return pay;
27 } // calcEarnings 结束
```

运行结果：

```
Emplyee 1 pay: 880
Emplyee 2 pay: 225
```

6.5.2　函数重载

C++ 是否允许两个同名的函数呢？如果参数列表不同（参数类型不同、参数个数不同或者参数顺序不同），则答案是肯定的。在 C++ 中，这被称为**函数重载**。编译器允许程序中包含两个同名函数的条件被称为**函数签名**。函数的签名是函数名称和参数列表中的类型的组合。如果两个函数定义有不同的签名，则编译器可以区分它们。请注意，函数的返回类型不包含在签名中，因为 C++ 在函数调用时必须在重载函数之间进行选择，返回类型不包含在函数调用的语法中。

> 函数的返回值不包含在函数签名中。

【例 6-19】 以下代码片段演示了如何定义不同的函数来查找两个整数或者两个浮点值之间的最大值。

```
int max (int a, int b)                    double max (double a, double b)
```

```
{                                     {
  ...                                   ...
}                                     }
```

以上两个函数被认为是两个不同的（重载的）函数，因为它们的函数签名不同。第一个函数（左边的函数）的函数签名为：max(int, int)；第二个函数的函数签名为：max(double, double)。

【例 6-20】 不能认为以下两个函数是两个重载函数，因为它们的签名相同。

```
int get ()                            double get ()
{                                     {
  ...                                   ...
}                                     }
```

第一个函数的函数签名为：get( )；第二个函数的函数签名同样为：get( )。如果我们使用上述两个定义编写一个程序，则不能编译通过。

编译器不能编译包含两个具有相同签名的函数定义的程序。

【例 6-21】 重载可以帮助我们从一个函数的内部调用另一个函数。例如，我们可以编写三个重载函数分别用于：查找两个整数的最大值、查找三个整数的最大值、查找四个整数的最大值。第二个函数在其定义中调用第一个函数；第三个函数在其定义中调用第二个函数。参见程序 6-19。

程序 6-19　创建三个重载函数

```
1   /***************************************************************
2    * 本程序演示如何使用不同的函数签名                              *
3    * 创建三个重载函数                                              *
4    ***************************************************************/
5   #include <iostream>
6   using namespace std;
7
8   // 函数声明
9   int max (int num1, int num2);
10  int max (int num1, int num2, int num3);
11  int max (int num1, int num2, int num3, int num4);
12
13  int main ()
14  {
15    //分别调用三个重载的 max 函数
16    cout << "maximum (5, 7): " << max (5, 7) << endl;
17    cout << "maximum (7, 9, 8): " << max (7, 9, 8) << endl;
18    cout << "maximum (14, 3, 12, 11): " << max (14, 3, 12, 11);
19    return 0;
20  }// main 结束
21  /***************************************************************
22   * 以下是带两个参数 num1 和 num2 的                              *
23   * 函数定义,                                                     *
24   * 它返回两个参数中的最大值                                      *
25   ***************************************************************/
```

```
26  int max (int num1, int num2)
27  {
28    int larger; //局部变量
29    if (num1 >= num2)
30    {
31       larger = num1;
32    }
33    else
34    {
35       larger = num2;
36    }
37    return larger;
38  }
39  /*********************************************************
40   * 以下是带三个参数 num1、num2 和 num3 的                    *
41   * 函数定义，                                              *
42   * 它调用带两个参数的 max 函数先查找前两个值的最大值，         *
43   * 然后再次调用带两个参数的 max 函数，                       *
44   * 查找前两个参数最大值与 num3 之间的最大值                  *
45   *********************************************************/
46  int max (int num1, int num2, int num3)
47  {
48    return max(max(num1, num2), num3);
49  }
50  /*********************************************************
51   * 以下是带四个参数 num1、num2、num3 和 num4 的              *
52   * 函数定义，                                              *
53   * 它调用带三个参数的 max 函数先查找前三个值的最大值，         *
54   * 然后再调用带两个参数的 max 函数，                         *
55   * 查找前三个参数最大值与 num4 之间的最大值。                *
56   * 最后得到四个值中的最大值                                 *
57   *********************************************************/
58  int max (int num1, int num2, int num3, int num4)
59  {
60    return max(max(num1, num2, num3), num4);
61  } // 带四个参数的 max 函数的结束
```

运行结果：

```
maximum (5, 7): 7
maximum (7, 9, 8): 9
maximum (14, 3, 12, 11): 14
```

6.6 作用域和生命周期

在本节中，我们将讨论影响函数设计和使用的两个概念：作用域和生命周期。

6.6.1 作用域

作用域（scope）定义源代码中命名实体（常量、变量、对象、函数等）的可见范围。要定义实体的作用域，必须首先定义作用域。作用域是源代码中具有一个或者多个声明的区域。换言之，当我们声明一个实体时，我们为该实体创建了一个作用域。

作用域是一个声明的区域。

局部作用域

具有**局部作用域**（local scope）的实体的可见范围从其声明位置开始到语句块的结尾（由结束的右花括号定义）。局部作用域可以从语句块前面的头部开始（例如函数头部或者循环头部），也可以直接从语句块开始处（开始的左花括号）之后开始，有时也可以从语句块的中间开始。但是，局部作用域的结尾始终是在语句块结束的右花括号处。

图 6-21 显示了局部作用域的三种情况：第一种情况显示了在函数头部中声明的参数的作用域；第二种情况显示了在 for 循环中声明的计数器的作用域；第三种情况显示了在 main 函数体中声明的局部变量的作用域。

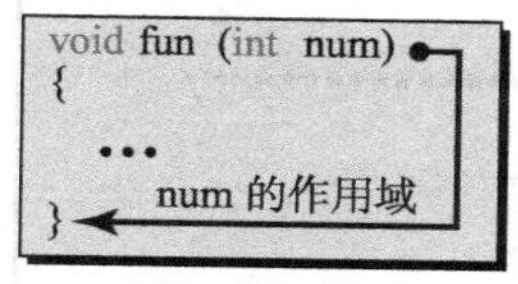

形式参数的作用域

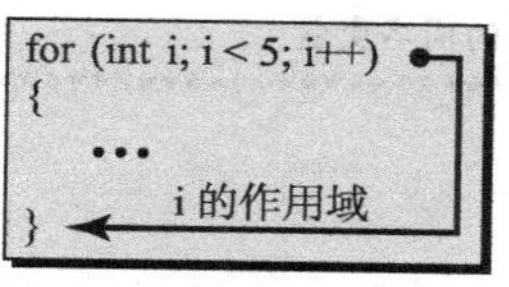

for循环计数器的作用域

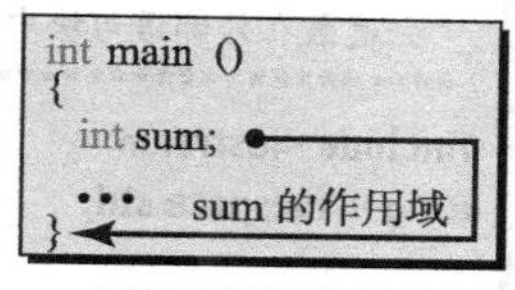

局部变量的作用域

图 6-21　局部作用域的三种情况

重叠作用域。在同一个语句块中不能有两个同名的实体，这样做会导致编译错误。图 6-22 显示了我们声明两个同名变量的情况：一个在函数头部，另一个在函数体中。这两个变量的作用域都是函数体，尽管一个作用域在另一个作用域之前声明，但这两个作用域有重叠，因此在编译时会产生错误。

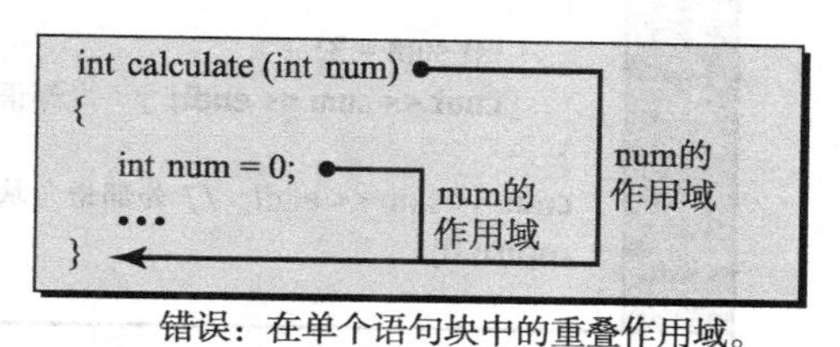

错误：在单个语句块中的重叠作用域。

图 6-22　重叠作用域（本例错误）

嵌套语句块。当我们编写程序时，我们经常看到**嵌套语句块**：一个语句块包含在另一个语句块内。外部语句块和内部语句块中声明的实体的作用域是什么？外部语句块中的实体具有包含内部块的较大的作用域，而内部语句块中声明的实体仅具有包含内部块的较小作用域。图 6-23 显示了两者的差异。

变量 sum 在外部语句块中声明，它在外部语句块和内部语句块中均可见。变量 i（循环计数器）在内部语句块的头部中声明，它仅在内部语句块中可见。如果我们尝试在内部语句块之外使用变量 i，就会产生一个编译错误。

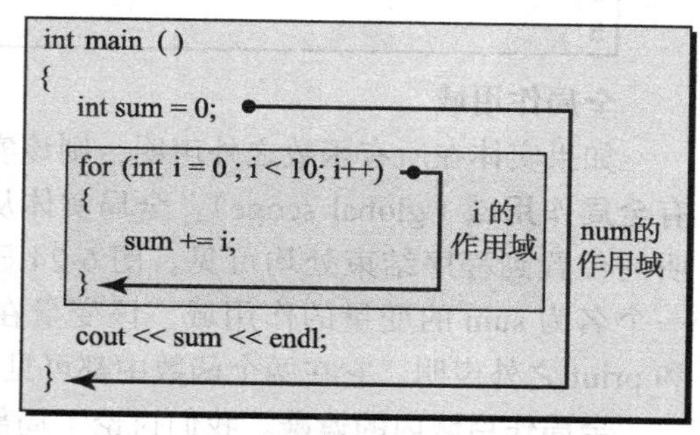

嵌套语句块

图 6-23　嵌套语句块中的作用域

局部作用域内的遮蔽。在前面的嵌套块中，实体（变量）的名称是不同的。如果我们声明两个同名的实体，一个在外部语句块中，一个在内部语句块中，结果会发生什么情况？其可见性如何？虽然这看起来像重叠的作用域范围，但情况并不相同，因为有两个不同的作用域（内部和外部）。在这种情况下，称为**函数遮蔽**（function shadowing），内部语句块中实体的可见性会遮蔽外部语句块中实体的可见性。换言之，一旦在内部语句块中声明了相同的名称，则只有相应的实体才可见，直到超出语句块末尾的范围。外部语句块实体在定义内部语句块的实体之前以及内部语句块终止之后可见。

程序6-20演示了一个简单的可见性示例。我们有两个名为sum的变量。外部语句块中变量sum被初始化为5，内部语句块中变量sum被初始化为3。当我们在内部语句块中打印sum时，外部语句块中声明的变量sum将被遮蔽。系统只看到初始化为3的变量sum，并打印其内容（3）。当我们离开内部语句块时，外部语句块中的变量sum再次可见，其内容（5）被打印出来。

程序6-20　局部作用域内的遮蔽

```
1  /***************************************************************
2   *在内部语句块中声明的变量，                                      *
3   *会遮蔽在外部语句块中声明的同名变量                               *
4   ***************************************************************/
5  #include <iostream>
6  using namespace std;
7
8  int main ( )
9  {
10    int sum = 5;
11    cout << sum << endl;
12    {
13        int sum = 3;
14        cout << sum << endl; // 内部语句块中的变量 sum 可见
15    }
16    cout << sum << endl; // 外部语句块中的变量 sum 可见
17    return 0;
18 }
运行结果：
5
3
5
```

全局作用域

如果实体在所有函数之外声明，则该实体具有**全局作用域**（global scope）。全局实体从其声明的位置到程序结束处均可见。图6-24显示了一个名为sum的变量的作用域，该变量在main和print之外声明。它在两个函数中都可见。

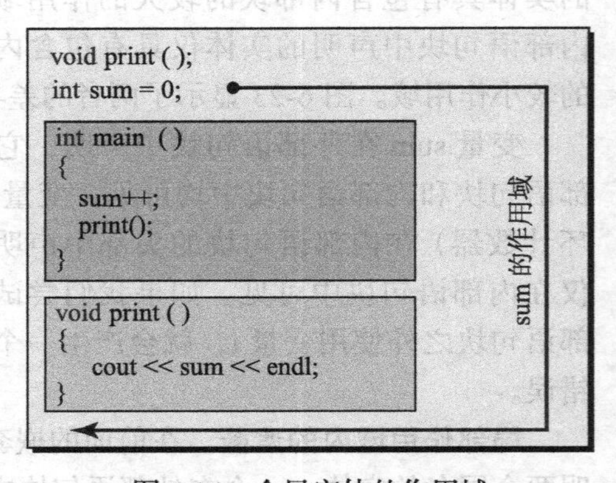

图6-24　全局实体的作用域

全局作用域内的遮蔽。我们讨论了局部作用域内的遮蔽。如果我们将全局作用域视为程序中从程序开始到程序结束的最大语句块，我们可以看到遮蔽也适用于全局作用域。换言之，局部作用域遮蔽全局作用域。程序6-21显示了这一点。我们有两个名为num的变量：全局变量和局部变量。局部变量遮蔽全局变量。

程序 6-21　全局作用域内的遮蔽

```
1  /*********************************************************************
2    * 本程序用于测试全局作用域内的遮蔽                                     *
3    *********************************************************************/
4  #include <iostream>
5  using namespace std;
6
7  int num = 5;  // 全局变量
8
9  int main ( )
10 {
11    cout << num << endl;  // 全局变量 num
12    int num = 25;  // 局部变量 num
13    cout << num;  // 局部变量 num 遮蔽全局变量 num
14    return 0;
15 } // End of main
运行结果:
5
25
```

作用域解析运算符。有时我们可能需要忽略遮蔽，并在局部语句块内访问全局实体。C++ 提供了一个运算符（：：），它可以显式地或者隐式地定义实体的作用域。图 6-25 显示了该运算符的两个版本。第一个版本使用一个右操作数；第二个版本使用两个操作数。

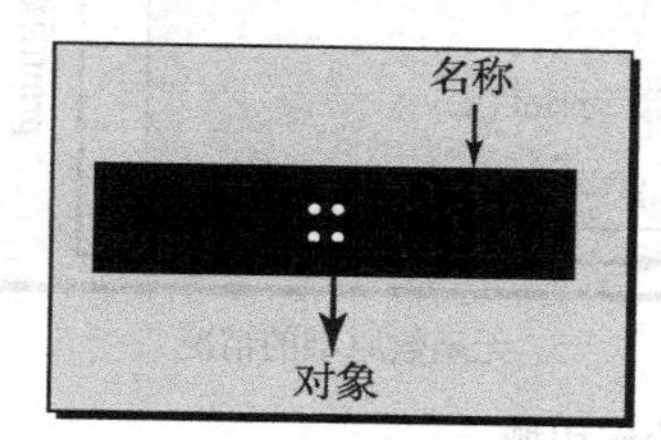

只有一个操作数的版本

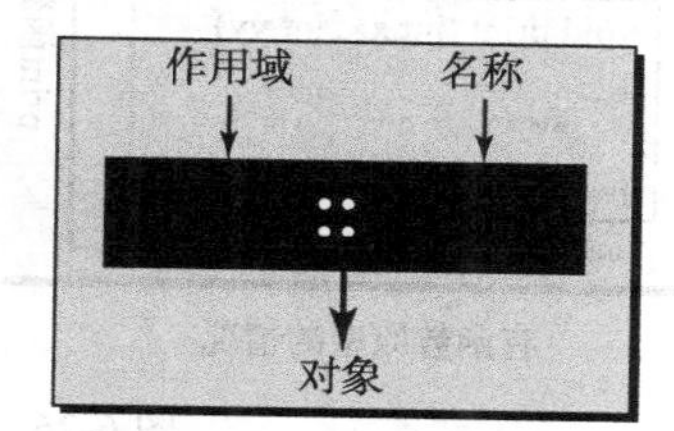

有两个操作数的版本

图 6-25　作用域解析运算符的两个版本

在第一个版本中，作用域的名称是隐式的（它表示程序的全局作用域）。在第二个版本中，我们必须给出作用域的名称。程序 6-22 演示了第一个版本的作用域运算符的使用方法。

程序 6-22　使用作用域解析运算符

```
1  /*********************************************************************
2    * 本程序用于测试作用域解析运算符的使用方法                             *
3    *********************************************************************/
4  #include <iostream>
5  using namespace std;
6
7  int num = 5; // 全局变量
8
9  int main ( )
10 {
11    int num = 25; // 局部变量
12    cout <<" Value of Global num: " << ::num << endl;
```

```
13    cout <<" Value of Local num: " << num << endl;
14    return 0;
15  } // main 函数结束
```

运行结果：
Value of Global num: 5
Value of Local num: 25

函数的作用域

理解函数的作用域具有一定的挑战性。函数有名称，函数的参数也有名称。我们必须将这两者区分开来。

函数名的作用域。函数是具有名称的实体，因而它具有作用域。**函数的作用域**是从其声明位置到程序结束为止。我们曾经讨论过，在 main 函数之前插入函数定义，可以省略函数声明（函数原型）。在这种情况下，函数定义既是函数声明又是函数定义。图 6-26 显示了两种情况下函数名的作用域。

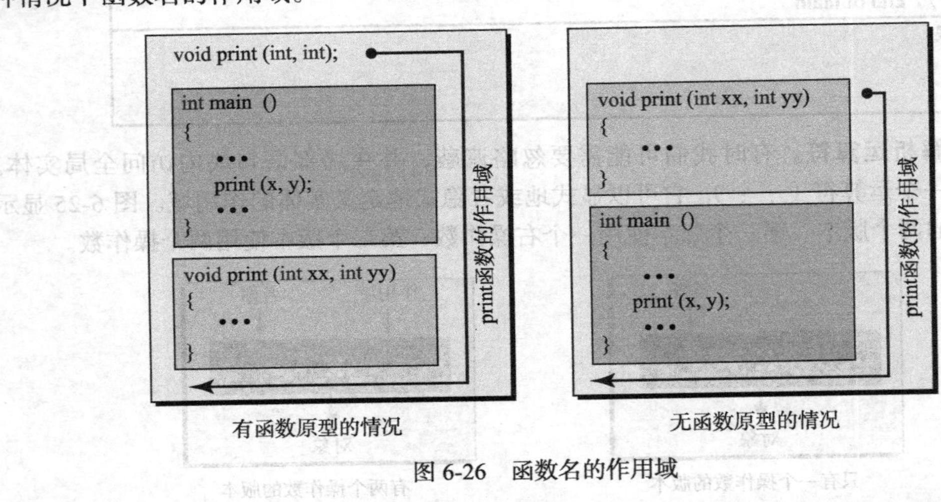

图 6-26　函数名的作用域

在第一种情况下，函数名（print）的作用域从其声明位置开始，到程序结束为止。在第二种情况下，函数定义既是函数声明也是函数定义，这意味着函数名（print）的作用域从函数头部开始，到程序结束为止。

函数参数的作用域。除了函数名之外，函数参数也是具有名称的实体，这意味着它们也具有作用域。它们的作用域从其声明的位置开始，直到函数块结束为止。但是，我们必须注意，参数没有在函数声明（函数原型）中声明，它们在函数定义的头部中声明。函数声明只是声明函数的名称并给出其结构（参数的类型）。这就是我们可以有一个函数声明而不需要提到参数名的原因。图 6-27 显示了图 6-26 中 print 函数的参数的作用域。我们在声明中省略了参数的名称，以强调编译器不需要这些名称，并且其作用域不是从该位置开始的。但是，我们强烈建议将其包含在声明中，以增强文档的可读性。

注意，在这两种情况下，参数范围都是函数定义的局部范围。这就是我们需要按值传递、按引用传递和按指针传递的原因，即使形式参数与相应的实际参数使用相同的名称也是如此。形式参数的作用域是被调用函数定义的局部作用域；实际参数的作用域是主调函数的局部作用域。

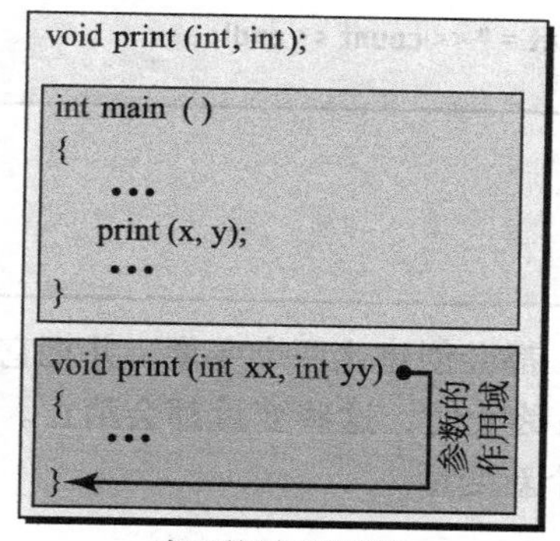

有函数原型的情况

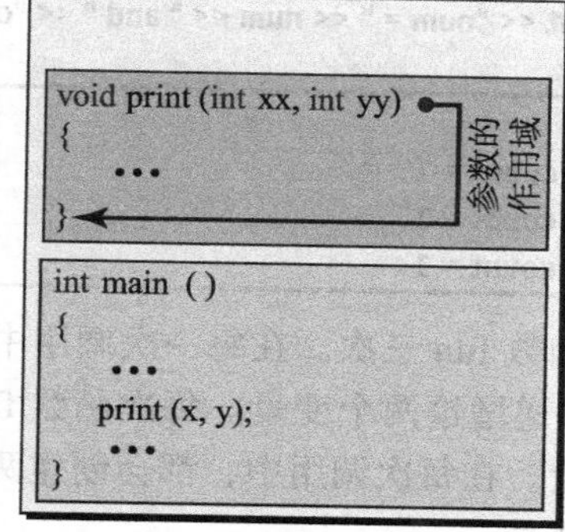

无函数原型的情况

图 6-27　函数参数的作用域

6.6.2　生命周期

程序中的每一个实体都有一个**生命周期**：它被创建，也会消亡。然而，在本节中，我们只讨论函数中局部变量的生命周期。函数中变量的生命周期很重要，因为我们可以多次调用函数。函数中有两种类型的局部变量：自动变量和静态变量。

自动局部变量

自动局部变量（automatic local variable）在函数被调用时产生，在函数终止时消亡。默认情况下，函数中的所有局部变量都有自动生命周期，但是我们也可以在变量声明前面显式地使用修饰符 auto，以强调每次调用函数时局部变量都会重新被创建。参见程序 6-23。

程序 6-23　测试自动局部变量

```
1   /***************************************************************
2    * 本程序用于测试自动局部变量的使用方法                          *
3    ***************************************************************/
4   #include <iostream>
5   using namespace std;
6
7   void fun ( );
8
9   int main ( )
10  {
11    fun ( );
12    fun ( );
13    fun ( );
14    return 0;
15  }
16  /***************************************************************
17   * 本函数包含两个自动局部变量：num 和 count。                    *
18   * 每次函数调用，这些变量都会被初始化                            *
19   ***************************************************************/
20  void fun ( )
21  {
22    int num = 3;                    // 隐式自动变量
23    auto int count = 0;             // 显式自动变量
24    num++;
25    count++;
```

```
26      cout << "num = " << num << " and " << "count = " << count << endl;
27  }
```

运行结果：
num = 4 and count = 1
num = 4 and count = 1
num = 4 and count = 1

程序调用函数 fun 三次。在每一次调用中，都会创建这两个变量，并把它们分别初始化为 3 和 0，然后递增这两个变量。每次函数 fun 终止时，这些变量都会消亡。换言之，系统不跟踪这些变量。在每次调用中，都会创建两个新变量。

静态局部变量

在 C++ 中，修饰符 static 有三种应用，但是这里我们只讨论与局部变量相关的 static 修饰符。**静态局部变量**（static local variable）是使用 static 修饰符限定的局部变量。静态变量的生命周期是定义该变量的程序的生命周期。静态变量只初始化一次，但只要程序运行，系统就会跟踪其内容。这意味着静态变量是在对函数的第一次调用中初始化的，但可以在随后的函数调用中更改。我们可以将前一个程序中的变量 count 更改为静态变量，并查看系统如何在每次调用中保持 count 的值。程序 6-24 演示了如何使用静态变量来判断总共调用了多少次函数 fun。

程序 6-24　测试静态变量

```
1   /***************************************************************
2    * 本程序用于测试静态变量的使用方法                              *
3    ***************************************************************/
4   #include <iostream>
5   using namespace std;
6
7   void fun ( );
8
9   int main ( )
10  {
11    fun ( );
12    fun ( );
13    fun ( );
14    return 0;
15  }
16  /***************************************************************
17   * 本函数包含一个静态变量：count。                               *
18   * 在第一次调用函数时初始化该变量，但会保持其值，以用于下一次调用。 *
19   * 这意味着每次调用都会递增其值                                  *
20   ***************************************************************/
21  void fun ( )
22  {
23    static int count = 0;  // 显式静态变量
24    count++;
25    cout << "count = " << count << endl;
26  }
```

运行结果：
count = 1
count = 2
count = 3

注意，程序忽略了第二次和第三次调用中的初始化，但程序在每次调用后都保留 count 的值。第一次调用结束后，count 的值为 1；第二次调用结束后，count 的值为 2；第三次调用结束后，count 的值为 3。

在前一个程序中，我们使用静态变量来统计调用 fun 函数的次数。静态变量还可以用于其他方式。例如，当函数被多次调用时，我们可以使用静态变量来查找实际参数的最大值或者最小值。

初始化

有必要比较一下全局变量、自动局部变量和静态局部变量的初始化。如果一个自动局部变量没有被显式初始化，它将保留以前使用时遗留的垃圾数据。另外，如果未显式初始化全局变量和静态局部变量，则它们被初始化为默认值（对于整型为 0，对于浮点类型为 0.0，对于布尔类型为 false）。换言之，全局变量和静态局部变量在初始化方面的行为相同。

> 如果未显式初始化，则全局变量和静态局部变量将被初始化为默认值，但自动局部变量则包含以前使用时留下的垃圾数据。

程序 6-25 演示了如何隐式初始化三个变量（一个全局变量、一个静态局部变量和一个自动局部变量）。全局变量和静态局部变量被初始化为默认值；自动局部变量保留以前使用时留下的垃圾值。

程序 6-25　初始化不同的变量类型

```
1   /***************************************************************
2    * 本程序用于测试变量的初始化                                   *
3    ***************************************************************/
4   #include <iostream>
5   using namespace std;
6   
7   int global;  // 全局变量
8   
9   int main ( )
10  {
11     static int sLocal;  // 静态局部变量
12     auto int aLocal;    // 自动局部变量
13     // 打印值
14     cout << "Global = " << global << endl;
15     cout << "Static Local = " << sLocal << endl;
16     cout << "Automatic Local = " << aLocal << endl;
17     return 0;
18  }
```

```
运行结果:
Global = 0
Static Local = 0
Automatic Local = 4202830  // 垃圾值
```

6.7 程序设计

在本节中，我们将讨论如何在两个程序中使用函数。这两个程序都与金融有关。它们是经典程序，在计算机科学文献中有相关讨论。

6.7.1 固定投资的未来价值

我们要编写一个使用函数的程序，并引入因子（factor）和乘数（multiplier）的概念。在接下来的两个程序中，我们将使用这些概念来计算定期投资的未来价值，以及计算贷款的每月还款额。

理解问题

固定投资的未来价值可以按如下公式进行计算：

$$\text{future value} = \text{investment} \times (1 + \text{rate})^{\text{term}}$$

其中，rate 是周期性利率，term 是期数（年或者月）。我们可以把（1 + rate）作为一个因子（factor），表示下一期结束时一美元投资的价值。我们也可以把（$\text{factor}^{\text{term}}$）作为一个乘数（multiplier），表示一美元投资在期末的价值。基于上述假设，则计算公式如下：

$$\text{factor} = (1 + \text{rate})$$
$$\text{multiplier} = \text{factor}^{\text{term}}$$
$$\text{future value} = \text{investment} \times \text{multiplier}$$

换言之，要找到一项投资的未来价值，我们必须找到一美元的未来价值（使用 multiplier），然后将其乘以投资金额。

开发算法

结构图。在图 6-28 所示的结构图中，我们将任务划分为三个函数：输入函数（input）、处理函数（process）和输出函数（output）。这些函数又会调用其他函数。input 函数调用 getInput 函数三次以获取投资额（investment）、利率（rate）和期数（周期数 term）。process 函数调用另一个名为 findMultiplier 的函数来获取乘数（multiplier），并将其乘以投资额（investment），以计算投资的未来价值（futureValue）。输出函数首先调用 printData 函数来打印给定的数据，然后调用 printResult 打印结果（乘数和未来价值）。我们将输出分为两类，以显示给定的数据（由输入函数收集的数据）和计算的数据（在处理函数中计算得到的数据）。

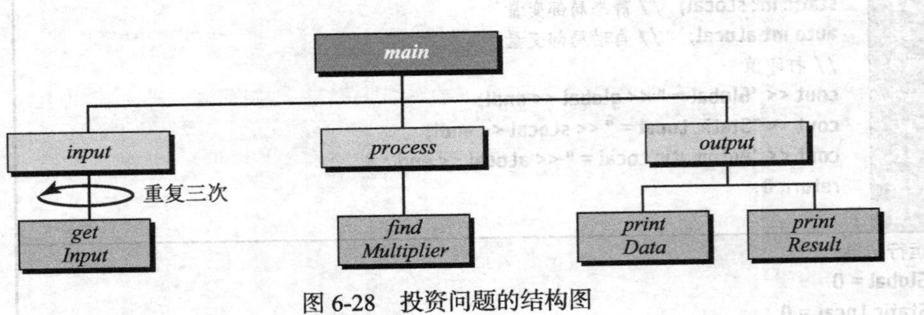

图 6-28　投资问题的结构图

main 函数。main 函数调用三个函数：input、process 和 output。

input 函数。input 函数是一个 void 函数，用于获取用户输入的三个值，并使用按引用传递方式传递给 main 函数。为了获取输入，input 函数调用另一个函数 getInput，每次调用时显示不同的消息，然后获取投资额、利率和期数（年数）。

process 函数。process 函数用于计算未来价值。前三个参数使用按值传递方式，最后两个参数使用按引用传递方式，因此需要更改 main 函数中的 multiplier 和 futureValue 的值。process 函数调用另一个函数 findMultiplier，以计算 multiplier。multiplier 是一个数值，该数值乘以最初的投资额可以计算期末的未来价值。我们将其定义为（multiplier = $factor^{term}$），其中 factor 为 (1 + rate / 100)。

output 函数。output 调用两个函数：printData 和 printResult，以显示用户输入的数据和计算的结果数据。

编写程序

程序 6-26 显示了 main 函数和 main 函数调用的三个函数：input、process 和 output。它还给出了 getInput（由 input 调用）、findMultiplier（由 process 调用）、printData 和 printResult（由 output 调用）的定义。

程序 6-26　计算固定投资的未来价值

```
1  /***************************************************************
2   * 本程序演示如何通过函数                                        *
3   * 计算基于复合利率的投资的未来价值                              *
4   ***************************************************************/
5  #include <iostream>
6  #include <iomanip>
7  #include <cmath>
8  using namespace std;
9  
10 // 声明顶层函数
11 void input (double& invest, double& rate, double& term);
12 void process (double invest, double rate, double term,
13                          double& multiplier, double& futureValue);
14 void output (double invest, double rate, double term,
15                          double multiplier, double futureValue);
16 // 声明底层函数
17 double getInput (string message);
18 double findMultiplier (double rate, double period);
19 void printData (double invest, double rate, double term);
20 void printResult (double multiplier, double value);
21 
22 int main ( )
23 {
24   // 变量声明
25   double invest, rate, term;          // 用于输入
26   double multiplier, futureValue;     //用于结果
27   // 调用顶层函数
28   input (invest, rate, term);
29   process (invest, rate, term, multiplier, futureValue);
30   output (invest, rate, term, multiplier, futureValue);
31   return 0;
32 }
33 /***************************************************************
34  * input 函数通过调用 getInput 函数三次来获取三个值。            *
35  * 它使用按引用传递方式把获取的值传递给 main 函数。              *
36  * 函数结束后，                                                  *
37  * 数据值分别存储在变量 invest、rate 和 term 中                  *
```

```
38  *****************************************************************/
39  void input (double& invest, double& rate, double& term)
40  {
41    invest = getInput ("Enter the value of investment: ");
42    rate = getInput ("Enter the interest rate per year: ");
43    term = getInput("Enter the term (number of years): ");
44  }
45  /****************************************************************
46   * process 函数通过调用 findMultiplier 函数来计算 multiplier，   *
47   * 然后把返回的值与投资额相乘，                                   *
48   * 以计算未来价值                                                 *
49   *****************************************************************/
50  void process (double invest, double rate, double term,
51                       double& multiplier, double& futureValue)
52  {
53    multiplier = findMultiplier (rate, term);
54    futureValue  = multiplier * invest;
55  }
56  /****************************************************************
57   * output 函数通过调用 printData 函数打印三个给定的值，           *
58   * 然后调用 printResult 函数打印两个计算结果值                    *
59   *****************************************************************/
60  void output (double invest, double rate, double term,
61                        double multiplier, double futureValue)
62  {
63    printData (invest, rate, term);
64    printResult (multiplier, futureValue);
65  }
66  /****************************************************************
67   * getInput 函数用于获取用户输入。                                *
68   * 其唯一参数是一个 string 类型的对象，                           *
69   * 用于在每次调用时提示用户输入对应数据。                         *
70   * 它验证输入的数据，                                             *
71   * 然后把结果返回给 input 函数中的调用表达式                      *
72   *****************************************************************/
73  double getInput (string message )
74  {
75    double input;
76    do
77    {
78        cout << message;
79        cin >> input;
80    } while (input < 0.0);
81    return input;
82  }
83  /****************************************************************
84   * findMultiplier 函数非常简单。                                  *
85   * 两个值 rate 和 term 被按值传递给该函数。                       *
86   * 先计算 factor，然后使用 pow 函数返回 multiplier                *
87   *****************************************************************/
88  double findMultiplier (double rate, double term)
89  {
```

```
90      double factor = 1 + rate/100;
91      return pow (factor , term);
92  }
93  /*************************************************************
94   * printData 函数打印用户输入的三个值,                          *
95   * 同时打印相关的解释信息。                                     *
96   * 它是一个 void 函数，仅具有副作用                              *
97   *************************************************************/
98  void printData (double invest, double rate, double term)
99  {
100     cout << endl << "Information about investment" << endl;
101     cout << "Investment: " << fixed << setprecision (2) << invest << endl;
102     cout << "Interest rate: " << rate << fixed << setprecision (2);
103     cout << " percent per year" << endl;
104     cout << "Term: " << term << " years" << endl << endl;
105 }
106 /*************************************************************
107  * printResult 函数打印程序的两个结果值。                        *
108  * 它打印投资的 multiplier 和 futureValue。                     *
109  * 它是一个 void 函数，仅具有副作用                              *
110  *************************************************************/
111 void printResult (double multiplier, double futureValue)
112 {
113     cout << "Investment is multiplied by: " << fixed << setprecision (8);
114     cout << multiplier << endl;
115     cout << "Future value: " << fixed << setprecision(2);
116     cout << futureValue << endl;
117 }
```

```
运行结果：
Enter the value of investment: 360000
Enter the interest rate per year: 5
Enter the term (number of years): 30
Information about investment
Investment: 360000.00
Interest rate: 5.00 percent per year
Term: 30.00 years

Investment is multiplied by: 4.32194238
Future value: 1555899.26
```

结果表明，multiplier 为 4.32194238。年利率为 5% 的情况下，1 美元的投资 30 年会产生 4.32194238 美元。我们可以使用不同的投资额但使用相同的 rate 和 term 来测试程序，结果表明 multiplier 是相同的，只有未来价值会改变。

6.7.2 周期性投资的未来价值

除了一次性投资外，我们还可以周期性投资按复利回报的银行账户。例如，我们每个月投资相同的金额。

理解问题

1）我们假设每个周期的投资额相同。

2）我们假设投资间隔和支付利息的间隔是相同的。换言之，如果我们按月投资，则利息也会按月计算并累加到前一个月的金额上。

开发算法

基于前两个假设，我们只需要改变前一个程序中乘数（multiplier）的计算方法，就可以使其适用于周期性投资。在这种情况下，我们实际上有 n 个乘数（multiplier），每个周期一个，其中 n = term × period。第二次投资的乘数应计算为（n−1）个期间。第三次投资的乘数应计算为（n−2）个期间。最后一次投资的乘数只能计算一个期间。换言之，总乘数的计算公式为：

$$\text{multiplier} = (1 + \text{factor})^{n} + (1 + \text{factor})^{n-1} + \cdots + (1 + \text{factor})^{1}$$

这是一个幂级数，在数学中存在一个求解公式，但是我们可以很容易地用一个循环来模拟它。因此，可以修改前一个程序中的 findMultiplier 函数，在 for 循环中模拟该序列。

```
double findMultiplier (double rate, double term)
{
   double multiplier = 0;
   double factor = 1 + rate/100;
   for (int i = term; i >=0; i--)
   {
       multiplier += pow (factor , i);
   }
   return multiplier;
}
```

编写程序

根据前面的解释，我们可以重写固定投资的程序，以求解周期性投资的未来价值（程序 6-27）。

程序 6-27　计算周期性投资的未来价值

```
1  /***********************************************************************
2   * 本程序演示如何通过函数                                                *
3   * 计算周期性投资（每期投资额相同）的未来价值                              *
4   ***********************************************************************/
5  #include <iostream>
6  #include <iomanip>
7  #include <cmath>
8  #include <string>
9  using namespace std;
10
11 // 声明顶层函数
12 void input(double& invest, double& rate, double& term);
13 void process (double invest, double rate, double term,
14                      double& multiplier, double& futureValue);
15 void output (double invest, double rate, double term,
16                     double multiplier, double futureValue);
17 // 声明底层函数
18 double getInput (string message);
19 double findMultiplier (double rate, double period);
20 void printData (double invest, double rate, double term);
```

```
21  void printResult (double multiplier, double value);
22
23  int main ( )
24  {
25    // 变量声明
26    double invest, rate, term;              // 用于输入
27    double multiplier, futureValue;         //用于结果
28    // 调用顶层函数
29    input (invest, rate, term);
30    process (invest, rate, term, multiplier, futureValue);
31    output (invest, rate, term, multiplier, futureValue);
32    return 0;
33  }
34  /***********************************************************
35   * input 函数通过调用 getInput 函数三次来获取三个值。          *
36   * 它使用按引用传递方式把获取的值传递给 main 函数。             *
37   * 函数结束后,                                              *
38   * 数据值分别存储在变量 invest、rate 和 term 中               *
39   ***********************************************************/
40  void input (double& invest, double& rate, double& term)
41  {
42    invest = getInput ("Enter the value of periodic investment: ");
43    rate = getInput ("Enter the interest rate per year: ");
44    term = getInput("Enter the term (number of years): ");
45  }
46  /***********************************************************
47   * process 函数通过调用 findMultiplier 函数来计算 multiplier, *
48   * 然后把返回的值与投资额相乘,                                *
49   * 以计算未来价值                                            *
50   ***********************************************************/
51  void process (double invest, double rate, double term,
52                double& multiplier, double& futureValue)
53  {
54    multiplier = findMultiplier (rate, term);
55    futureValue  = multiplier * invest;
56  }
57  /***********************************************************
58   * output 函数通过调用 printData 函数打印三个给定的值,         *
59   * 然后调用 printResult 函数打印两个计算结果值                 *
60   ***********************************************************/
61  void output (double invest, double rate, double term,
62               double multiplier, double futureValue)
63  {
64    printData (invest, rate, term);
65    printResult (multiplier, futureValue);
66  }
67  /***********************************************************
68   * getInput 函数用于获取用户输入。                            *
69   * 其唯一参数是一个 string 类型的对象,                         *
70   * 用于在每次调用时提示用户输入对应数据。                       *
71   * 它验证输入的数据,                                         *
72   * 然后把结果返回给 input 函数中的调用表达式                   *
```

```
73   ****************************************************************/
74   double getInput(string message )
75   {
76     double input;
77     do
78     {
79          cout << message;
80          cin >> input;
81     } while (input < 0.0);
82     return input;
83   }
84   /****************************************************************
85    * 本例中的 findMultiplier 函数需要计算各年的 multiplier 并累加在一起，     *
86    * 各 multiplier 的指数各不相同。                                          *
87    * 函数最后把结果返回给 process 函数                                        *
88    ****************************************************************/
89   double findMultiplier (double rate, double term)
90   {
91     double multiplier = 0;
92     double factor = 1 + rate/100;
93     for (int i =  term; i > 0 ; i– –)
94     {
95          multiplier += pow (factor , i);
96     }
97     return multiplier;
98   }
99   /****************************************************************
100   * printData 函数打印用户输入的三个值，                                    *
101   * 同时打印相关的解释信息                                                  *
102   ****************************************************************/
103  void printData (double invest, double rate, double term)
104  {
105    cout << endl << "Information about period invesment" << endl;
106    cout << "Periodic Investment: " << fixed << setprecision (2)
107                                           << invest << endl;
108    cout << "Interest rate: " << rate << fixed << setprecision (2);
109    cout << " percent per year" << endl;
110    cout << "Term: " << term << " years" << endl << endl;
111  }
112  /****************************************************************
113   * printResult 函数打印程序的两个结果值。                                  *
114   * 它打印投资的 multiplier 和 futureValue                                  *
115   ****************************************************************/
116  void printResult (double multiplier, double futureValue)
117  {
118    cout << "Result of investment" << endl;
119    cout << "Investment is multiplied by :" << fixed << setprecision (8);
120    cout << multiplier << endl;
121    cout << "Future value: " << fixed << setprecision(2);
122    cout << futureValue << endl;
123  }
```

```
运行结果：
Enter the value of periodic investment: 12000
Enter the interest rate per year: 5
Enter the term (number of years): 30

Information about period invesment
Periodic Investment: 12000.00
Interest rate: 5.00 percent per year
Term: 30.00 years

Result of investment
Investment is multiplied by: 69.76078988
Future value: 837129.48
```

请注意，程序 6-26 和程序 6-27 中的总投资额均为 36 万美元，但第二个程序的未来价值较少。这是很自然的，因为当我们周期性投资时，投资额并不是在整个期限内全部投资。

本章小结

函数是一个实体，由两部分组成：函数头部和旨在执行任务的一系列语句的函数体。随着程序变得越来越复杂，我们将任务划分为更小的任务，每个任务负责一部分工作。把一个任务分成几个小任务具有如下四种优点：更容易编码、更容易调试、可以重复使用并且可以保存供将来使用。与函数相关的三个实体为：函数定义、函数声明和函数调用。

对于库函数，其定义已经在 C++ 库中创建，我们只需要在程序中包含相应的头文件，就可以调用该函数。对于用户自定义的函数，我们需要声明、定义和调用该函数。

程序中的函数可能需要交换数据。我们将此活动称为数据传递和数据返回。每种活动都可以使用以下三种方法中的一种：按值传递、按引用传递或者按指针传递。

我们可以在函数声明中使用一个或者多个具有预定义值的默认参数。

函数重载为具有相同名称但不同签名的函数提供了多个定义，签名不同意味着参数的数量不同，或者参数的类型不同，或者两者都不同。函数的返回值不是其签名的一部分。

作用域和生命周期是影响函数设计和使用的两个概念。作用域定义源代码中命名实体的可见范围。程序中的每个实体都有生命周期：它被创建，也会消亡。

思考题

1. 请指出以下各函数声明的函数签名。

```
int firstFunction (int x, float y, int z);
void secondFunction (int x, boolean y);
void thirdFunction (double x, double y);
void fourthFunction ();
```

2. 给定以下两个函数声明，确定它们是否互为重载函数。

```
int fun (int x, int y);
void fun (int a, int b);
```

3. 给定以下两个函数声明，确定它们是否互为重载函数。

```
int fun (int x, int y);
void fun (float a, float b);
```

4. 给定以下两个函数声明，确定它们是否互为重载函数。

```
int fun (int x, int y, int z);
float fun (int a, int b);
```

5. 给定以下两个函数声明，确定它们是否互为重载函数。

```
int  functionOne (int x, int y);
int  functionTwo (int a, int b);
```

6. 给定以下两个函数声明，确定它们是否互为重载函数。

```
int fun ();
float fun ();
```

7. 找出以下函数声明中正确的函数声明。

```
a. float one (int a, int b);
b. boolean two (int a, b);
c. float (int a, int b);
d. void three (void);
e. int ();
```

8. 指出以下函数定义的错误之处。

```
void one (int a)
{
    return a;
}
```

9. 指出以下函数定义的错误之处。

```
int two (int a)
{
    int b = a * a;
}
```

10. 指出以下函数定义的错误之处。

```
int three (int a, int b)
{
    c = a * b;
    return c;
}
```

11. 指出以下函数定义的错误之处。

```
void one ()
{
    cout << "In One" << endl;
    void two ()
    {
        cout << "In Two" << endl;
        return;
    }
    return;
}
```

12. 指出以下函数定义的错误之处。

```
int wrong (int x)
```

```
{
    double x = 2.7;
    return x;
}
```

13. 当调用以下函数时，其返回值是什么？

```
int test ()
{
    return 3.25;
}
```

14. 当调用以下函数时，其返回值是什么？

```
char test ()
{
    return 67;
}
```

15. 当调用以下函数时，其返回值是什么？

```
double test ()
{
    return 9;
}
```

16. 以下程序的输出结果是什么？

```
include <iostream>
using namespace std;

int main ()
{

        int x;
        cout << x;
        return 0;
}
```

17. 以下程序的输出结果是什么？

```
# include <iostream>
using namespace std;

int x;

int main ()
{
    cout << x;
    return 0;
}
```

18. 以下程序的输出结果是什么？

```
# include <iostream>
using namespace std;

int x;
```

```
int main ()
{

    int x;
    cout << x;
}
```

19. 以下程序的输出结果是什么？

```
#include <iostream>
#include <cmath>
using namespace std;

int main ()
{
    double x = 23.671;
    cout << floor (x * 10 + 0.5) / 10 << endl;
    return 0;
}
```

20. 以下程序的输出结果是什么？

```
#include <iostream>
#include <cmath>
using namespace std;

int main ()
{
    double x = −23.671;
    cout << floor (x * 10 − 0.5) / 10  << endl;
    return 0;
}
```

编程题

1. 编写一个简单的程序，确定以下函数调用的结果。

```
a. abs (25) and abs (−23)
b. floor (44.56) and floor (−23.78)
c. ceil (25.23) and ceil (−2.89)
```

2. 编写一个简单的程序，确定以下函数调用的结果。

```
a. pow (5.0, 3) and  pow (5, −3)
b. sqrt (44.56)
c. exp (−6.2) and exp (44.26)
d. log (16.2) and log10 (14.24)
```

3. 编写一个简单的程序，确定以下函数调用的结果。其中 PI 的值参见本书中的定义。

```
a. sin (0) and sin (PI)
b. cos (0) and cos (PI)
c. tan (0) and tan (1)
d. asin (0) and asin (1)
e. acos (0) and acos (1)
f. atan (0) and atan (1)
```

4. 编写一个程序，演示如何使用 round 函数对一个值进行四舍五入。例如，数值 23.2 四舍五入为 23，数值 23.8 四舍五入为 24，数值 −23.2 四舍五入为 −23，数值 −23.8 四舍五入为−24。

5. 编写一个程序，演示随机数生成器的行为。请生成 5 组随机数，其中每组包含 10 到 99 之间（两位数的值）的 10 个随机数。然后计算每一组数值的总和，看看它们的变化。
6. 编写一个程序，演示如何调用 rand 函数来创建一个为 0 或者为 1 的随机数，以模拟抛硬币。
7. 编写一个程序，演示如何调用 rand 函数来创建一个介于 0.1 和 0.9 之间的随机数，小数点后只有一个数字。提示：先找到 1 到 9 之间的随机数，然后将结果除以 10 得到一个双精度值。在程序中用 10 个数值来测试你的答案。
8. 编写一个程序，演示如何调用 rand 函数来创建一个属于以下集合的随机数：2、4、6、8、10。
9. 编写一个程序，使用 <cmath> 头文件中的相应函数打印前 10 个整数的平方根。用适当的标题将结果制成表格形式。
10. 编写一个程序，使用 <cmath> 头文件中的 pow 函数打印前 10 个整数的立方根。用适当的标题将结果制成表格形式。
11. 编写一个函数，使用 <cmath> 头文件中的 log 函数计算 $\log_2 x$。注意，可以使用公式：$\log_a x = \log_b x / \log_b a$。再编写一个程序测试你编写的函数，其中 x 取值为 1 到 10。使用适当的标题将结果制成表格形式。
12. 编写一个函数，在给定摄氏温度的情况下，返回华氏温度。再编写一个程序测试你编写的函数，使用摄氏温度值 0、37、40 和 100 进行测试。从摄氏温度到华氏温度的转换公式如下：

    Fahrenheit = Celcius * 180.0 / 100.0 + 32

13. 编写一个函数，在给定华氏温度的情况下，返回摄氏温度。再编写一个程序测试你编写的函数，使用华氏温度值 32、98.6、104 和 212 进行测试。从华氏温度到摄氏温度的转换公式如下：

    Celcius = (Fahrenheit − 32) * (100.0 / 180.0)

14. 编写一个函数来求解一个正数的阶乘，计算公式如下所示。再编写一个程序测试你编写的函数，从用户处获取 n 的值并打印其阶乘值。使用介于 1 和 20 之间（为避免溢出）的 n 值测试程序。

    $$\text{factorial}(n) = n * (n-1) * (n-2) * \cdots * 3 * 2 * 1$$

15. 每次对 n 个对象中的 k 个进行排列，可以定义为如下公式。这个方程告诉我们 n 个对象一次可以形成多少个 k 个元素的排列。

    $$P(n, k) = \text{factorial}(n) / \text{factorial}(n-k)$$

16. 每次对 n 个对象中的 k 个进行组合，可以定义为如下公式。这个方程告诉我们 n 个对象一次可以形成多少个 k 个元素的组合。

    $$C(n, k) = \text{factorial}(n) / (\text{factorial}(n-k) * \text{factorial}(k))$$

17. 帕斯卡三角形（Pascal triangle）定义了二项式展开式中各项的系数（C_n）。

    $$(x+y)^n = C_0 x^0 y^n + C_1 x^1 y^{n-1} + \cdots + C_{n-1} x^{n-1} y^1 + C_n x^n y^0$$

    以下内容显示了当 n 为 0 到 5 时的系数。请注意，行和列都从 0 开始。

```
n = 0       1
n = 1       1  1
n = 2       1  2  1
n = 3       1  3  3  1
n = 4       1  4  6  4  1
n = 5       1  5  10 10 5  1
```

请注意，行和列中的每个系数的值是前一行前一列中的系数与前一行同一列中的系数之和，如下所示。

$$Pascal (row, col) = pascal (row - 1, col - 1) + pascal (row - 1 , col)$$

编写一个函数来计算任意 n 的系数。测试你所编写的函数，按行和列打印当 n 从 0 到 10 时的系数。

18. 三角学中 PI（π）的计算公式如下：

$$\pi = 4 \times [1 - 1/3 + 1/5 + \cdots + (-1)^{i+1} / (2i - 1)]$$

其中 i 的取值从 1 到 n。随着 n 越来越大，π 的值越来越接近实际值。编写一个函数，返回给定 n 值时的 π 值。然后编写一个程序，以表格形式打印 i 从 1 到 2001（以 200 为增量）时 i 和 π 的值。

19. 编写一个函数，创建位于 1111 到 9999 范围内的四位随机整数。消除其中包含零的任何数值。然后创建 100 个这样的数值，并将它们打印到一个表中，每行 10 个数值。
20. 编写两个函数。第一个函数用于计算任意 3 个整数的平均值。第二函数用于求任意 3 个整数的中位数。中位数是当集合排序后位于中间的数值（例如，4、9、6 的中位数为 6，因为如果对数值进行排序，则集合为（4，6，9））。然后在程序中调用这两个函数，找出任意 3 个整数的平均值和中位数。至少使用 5 个不同的集合测试你所编写的程序，并将结果制成表格。
21. 编写一个函数，获取任意正整数（大于 0）的位数。从键盘读取整数。例如，367 的位数是 3。提示：将给定整数连续除以 10，然后递增计数，直到整数为 0。请编写程序测试该函数，并以表格形式列出至少 10 个整数的位数。
22. 编写一个函数，获取任意给定整数中各个位数上的数字之和。例如，367 的各个位数上的数字之和为 16。编写一个程序，以表格形式列出至少 5 个整数，并打印其各个位数上的数字之和。
23. 编写一个函数，反转其参数中的各位数字。例如，给定整数 378，函数返回 873。使用几个不同位数的整数测试程序。
24. 如果一个整数正向读和反向读结果一样，则它是一个回文数。例如，5、121、12321、1347431 都是回文数。编写一个函数来确定一个整数是否是回文数。使用第 23 题中开发的函数，在测试给定的整数是否为回文数之前反转该整数。
25. 如果一个整数只能被它本身和 1 整除，那么这个整数就是一个素数。请注意，整数 1 不是素数。编写一个函数来测试给定的整数是否为素数。然后编写一个程序，在 10 列的表格中打印小于 100 的素数。
26. 判断一个整数是否是素数的另一种方法是，判断该整数是否只能被自身整除，而不能被 1 到其平方根范围内的任何其他数整除。使用此方法编写一个函数来测试给定整数是否为素数。然后编写一个程序，在 10 列的表格中打印小于 500 的素数。
27. 如果我们反转一个整数后，得到的仍然是一个素数，则该整数是反素数 emirp（prime 的反向拼写）。编写一个函数，在一行中打印从 1 到 1000 的 emirp 整数。
28. 编写一个程序，获取给定整数的所有因子。因子是小于给定整数的数，可以整除该数。程序应该使用一个函数来测试一个整数是否可以被另一个整数整除。使用 1 到 100 之间的整数测试你的函数。
29. 编写一个函数，找出给定整数的所有素因子。素因子就是既是素数也是因子的数。使用在第 25 题中判断素数的算法。使用 1 到 100 之间的整数测试你的函数。
30. 编写一个程序，给定期望的未来价值、年数和利率，求解要投资的固定金额。提示：首先计算乘数。
31. 编写一个程序，给定按月还款额、年数和利率，求解可以贷款的金额。提示：首先计算乘数。

用户自定义类型：类

本章是 C++ 作为一种面向对象语言的第一章。在前面的章节中，我们讨论了 C++ 的基础知识，我们将利用在这些章节中学习到的知识来理解本章和后续章节中讨论的内容。在本章中，我们将展示如何创建新类型以及如何使用新类型。创建新类型的机制是使用类（class）或者枚举类型。我们将在本章中讨论类，而忽略对枚举类型的讨论，因为它们并不常用。

学习目标

阅读并学习本章后，读者应该能够

- 介绍面向对象的概念，包括类型和实例、属性和行为、数据成员和成员函数。
- 讨论声明数据成员和成员函数的类定义。
- 讨论三种类型的构造函数：参数构造函数、默认构造函数和拷贝构造函数。
- 讨论在回收对象之前用于清理对象的析构函数。
- 讨论实例数据成员和实例成员函数。
- 讨论静态数据成员和静态成员函数。
- 讨论如何将程序的三个部分划分为三个独立的文件：接口文件、实现文件和应用程序文件，并实现面向对象程序设计的其中一个目标：封装。

7.1 概述

在第 1 章中，我们提到 C++ 是面向过程和面向对象语言的结合体。在前几章中，我们主要将该语言用作过程语言；在本章中，我们开始将其用作面向对象的语言。

7.1.1 现实生活中的类型和实例

类型（type）是可以从中创建实例（instance）的概念。换言之，类型是一种抽象，该类型的**实例**则是具体的实体。单词“时钟”是一种类型，它定义了我们熟悉的一般概念。约翰办公室的时钟、苏办公室的时钟或者火车站的时钟就是这种类型的实例。单词“圆”是一种类型。我们可以画不同半径的圆作为该类型的实例。另外，我们可以说单词“人”是一种类型，而约翰、苏和米歇尔就是“人”这种类型的实例。

类型与其实例之间的关系是一对多关系。一个类型可以有多个实例。图 7-1 显示了类型 Circle 和四个实例（名为 circle1、circle2、circle3 和 circle4）。

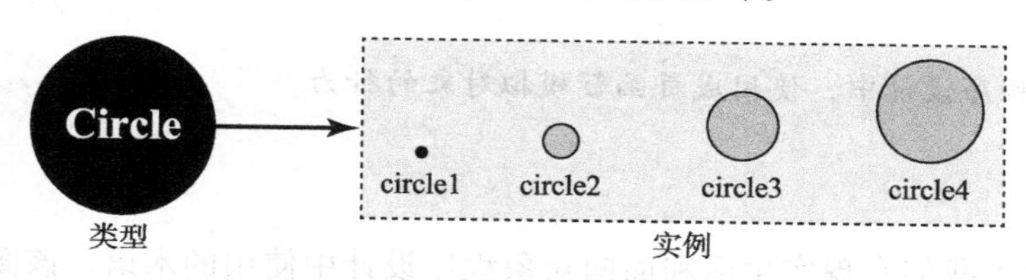

图 7-1　类型及其实例

我们在这个世界上遇到的任何实例都有一组属性和一组行为。

属性

在现实生活中，属性是一个实例的特征。在计算机科学中，**属性**（attribute）是我们在某个实例中感兴趣的任何特征。例如，如果实例是某个雇员，我们可能只对该雇员的姓名、地址、职位和薪水感兴趣。如果实例是某个大学的学生，我们可能只对该学生的姓名、年份、所学课程和成绩感兴趣。如果实例是一个圆，我们可能只关心它的半径、周长和面积。

> 属性是我们感兴趣的实例的特征。

行为

当我们考虑实例时，我们也会考虑实例的行为。从这个意义上讲，**行为**（behavior）是一个实例可以对其自身执行的操作。例如，如果实例是某个雇员，我们假设他可以提供他的姓名、地址和薪水。如果实例是一个圆，我们假设该实例可以提供它的半径、周长和面积。注意，在面向对象程序设计中，我们假设一个实例能够对其本身执行操作。

> 行为是我们假定一个实例可以对其自身执行的操作。

7.1.2　程序中的类和对象

在面向对象程序设计中，我们仍然使用术语类型（type）和实例（instance）。在C++中，可以使用一个名为类（class）的构造来创建用户自定义的类型。类的实例称为**对象**（object）。这意味着我们使用类型和类，我们还使用实例和对象。在面向对象程序设计中，对象的属性和行为被表示为数据成员（data member）和成员函数（member function）。

数据成员

对象的**数据成员**是变量，其值表示属性。例如，一个圆对象的半径可以用double类型的变量值表示。换言之，圆类型的对象的属性可以被有效地表示为double类型的变量，该变量保存该实例的半径值。但是，实例的某些属性并不独立，它们可能依赖于其他属性的值。例如，对于一个数学家来说，一个圆的周长和面积可能是两个属性，但我们并不把它们表示为数据成员，因为圆的周长和面积都取决于半径的值。

> 在面向对象程序设计中，使用数据成员模拟对象的属性。

成员函数

在程序设计中，函数是执行某种操作的实体。在面向对象程序设计中，**成员函数**是模拟对象行为的函数。例如，我们可以编写一个函数，允许一个圆提供其半径、面积和周长。我们也可以编写一个函数，允许一个圆设置其半径。然而设计成员函数时，需要了解依赖属性。

> 面向对象程序设计中，使用成员函数模拟对象的行为。

7.1.3　比较

图7-2显示了我们在现实生活和面向对象程序设计中使用的术语。该图可用于整章内容，可以使我们更好地理解本章所讨论的概念。

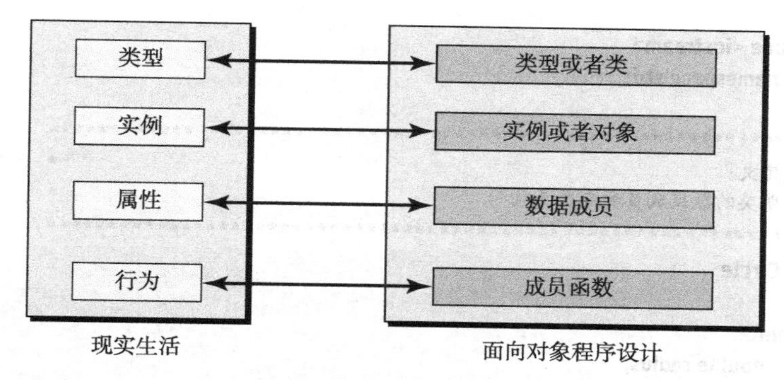

图 7-2　现实生活和面向对象程序设计中的术语比较

7.2　类

尽管 C++ 有许多预定义的（内置的）数据类型，例如 int、double、char 和 bool，C++ 还允许程序员创建新的类型。在 C++ 中，新类型主要是使用**类**创建的。要编写面向对象的程序，我们需要创建一个类作为一个类型，然后将对象实例化为该类型的实例。在面向过程的程序设计中，我们只需要编写一个应用程序（main 函数和其他一些函数），并使用内置类型的对象。在面向对象程序设计中，我们需要三个部分：类定义、成员函数定义和应用程序（该程序使用从类创建的对象）。换言之，如果我们想编写面向对象的程序而不是编写过程式的程序，则应该考虑这三个部分。我们将看到，有时每个部分都是由不同的实体设计和保存的。在某些情况下，前两个部分（类定义和成员函数定义）被创建并存储在 C++ 库中（例如字符串类和文件类），但原则是相同的：我们需要所有三个部分，如图 7-3 所示。

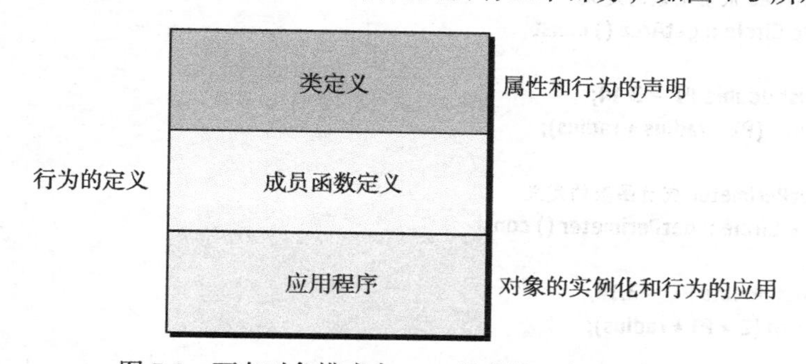

图 7-3　面向对象模式中 C++ 程序的三个组成部分

在类定义部分，我们声明数据成员和成员函数。在成员函数定义部分，我们定义所有成员函数。在应用程序部分，我们实例化对象并将成员函数应用于这些对象。

7.2.1　一个示例

在正式讨论这三个部分之前，我们先在程序 7-1 中使用它们。我们知道有许多问题还没有讨论，但是我们将使用这个程序来逐步展开讨论这些知识点。

程序 7-1　创建并处理两个圆对象

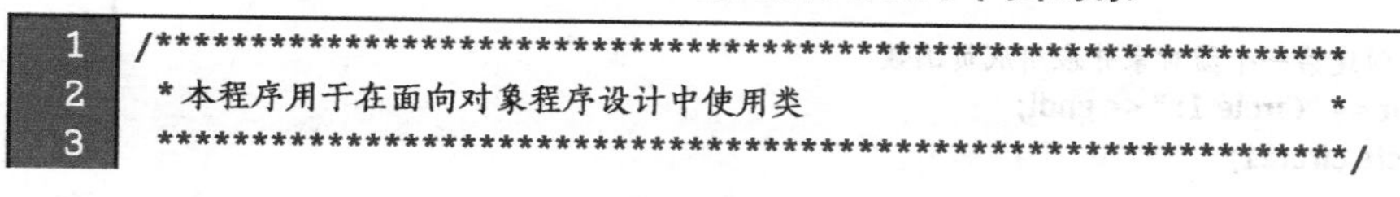

```
1  /***************************************************************
2   * 本程序用于在面向对象程序设计中使用类                          *
3   ***************************************************************/
```

```
4   #include <iostream>
5   using namespace std;
6
7   /**********************************************************************
8    * 类定义:                                                              *
9    * 声明类的数据成员和成员函数                                            *
10  **********************************************************************/
11  class Circle
12  {
13    private:
14          double radius;
15    public:
16          double getRadius () const;
17          double getArea () const;
18          double getPerimeter () const;
19          void setRadius (double value);
20  };
21  /**********************************************************************
22   * 成员函数定义。                                                        *
23   * 类定义中的每个函数都将在这部分中定义                                   *
24  **********************************************************************/
25  // getRadius 成员函数的定义
26  double Circle :: getRadius () const
27  {
28    return radius;
29  }
30  // getArea 成员函数的定义
31  double Circle :: getArea () const
32  {
33    const double PI  = 3.14;
34    return (PI * radius * radius);
35  }
36  // getPerimeter 成员函数的定义
37  double Circle :: getPerimeter () const
38  {
39    const double PI  = 3.14;
40    return (2 * PI * radius);
41  }
42  // setRadius 成员函数的定义
43  void Circle :: setRadius (double value)
44  {
45    radius = value;
46  }
47  /**********************************************************************
48   * 应用程序部分: 在这部分中, 实例化对象。                                 *
49   * 对象使用成员函数来获取或者设置它们的属性                               *
50  **********************************************************************/
51  int main ( )
52  {
53    // 创建第一个圆对象并应用成员函数
54    cout << "Circle 1: " << endl;
55    Circle circle1;
```

```
56    circle1.setRadius (10.0);
57    cout << "Radius: " << circle1.getRadius() << endl;
58    cout << "Area: " << circle1.getArea() << endl;
59    cout << "Perimeter: " << circle1.getPerimeter() << endl << endl;
60    // 创建第二个圆对象并应用成员函数
61    cout << "Circle 2: " << endl;
62    Circle circle2;
63    circle2.setRadius (20.0);
64    cout << "Radius: " << circle2.getRadius() << endl;
65    cout << "Area: " <<  circle2.getArea() << endl;
66    cout << "Perimeter: " << circle2.getPerimeter();
67    return 0;
68  }
```

```
运行结果：
Circle 1:
Radius: 10
Area: 314
Perimeter: 62.8
Circle 2:
Radius: 20
Area: 1256
Perimeter: 125.6
```

第 11 ～ 20 行是类定义，第 25 ～ 46 行是成员函数定义，第 51 ～ 68 行是应用程序。接下来我们将根据该程序详细讨论每个部分。

7.2.2 类定义

为了创建新类型，必须首先编写**类定义**。类定义由三部分组成：类头部、类体和分号。**类头部**由保留字 class 和设计者指定的名称组成。虽然在定义用户自定义的类型时可以自由使用小写或者大写字母，但是我们遵循一个惯例，即类名以大写字母开头，以将它们与以小写字母开头的库中的类区分开来。**类体**是一个语句块（从左花括号开始，到右花括号结束），它包含数据成员和成员函数的声明。类定义的第三个元素是终止类定义的英文分号。以下是程序 7-1 中类定义的代码：

```
class Circle  // 类头部
{
  private:
      double radius;                      // 数据成员声明
  public:
      double getRadius () const;          // 成员函数声明
      double getArea () const;            // 成员函数声明
      double getPerimeter () const;       // 成员函数声明
      void setRadius (double value);      // 成员函数声明
}; // 在类定义尾部需要一个英文分号
```

声明数据成员

类定义的第一部分声明类的数据成员：内置类型或者其他先前定义的类类型的变量或者常量。类的数据成员模拟类的实例化对象的属性。但是，在任何对象中，我们都可能有若干属性，其中一些属性依赖于其他属性，并且可以根据其他属性进行计算。在依赖属性中，我们需要选择最简单和最基本的属性。例如，在圆的类中，我们有三个属性：半径、面积和周长。给定其中一个属性，另外两个属性可以通过给定的属性来计算。因为半径是最原始和最基本的，因此选择半径作为数据成员。在依赖属性中选择多个属性可能会导致程序出错：我们可能会更改其中一个属性，而忘记更改其他属性。例如，如果我们同时选择了半径和面积作为数据成员，则使用成员函数更改半径而不同时更改面积，将创建一个面积计算错误的圆，反之亦然。

类中的数据成员不能相互依赖。

声明成员函数

类定义的第二部分声明类的成员函数，也就是说，它声明用于模拟类行为的所有函数。这部分与我们在前几章中讨论全局函数时使用的函数原型声明类似。但是，不同之处在于，一些函数的末尾有const限定符，而有些函数则没有。那些需要更改对象中某些内容的函数不能使用此限定符；那些不允许更改任何内容的函数需要使用此限定符。我们将在本章的后面部分讨论这个问题。

访问修饰符

访问修饰符（access modifier）用于确定类的可访问性。类中数据成员和成员函数的声明在默认情况下是私有的，无法访问私有数据成员和私有成员函数以进行检索或者更改。

当成员没有访问修饰符时，默认情况下它是私有的。

为了规避这个限制，C++定义了三种访问修饰符。类的设计者可以将访问修饰符应用于成员的声明，以控制对该成员的访问。C++为此提供了三种修饰符：私有（private）、受保护（protected）和公共（public），如图7-4所示。表7-1列举了这些修饰符的一般概念。

图7-4　访问修饰符

表7-1　各访问修饰符的成员访问权限

访问修饰符	是否允许从同一个类访问	是否允许从子类访问	是否允许从任意位置访问
private	是	否	否
protected	是	是	否
public	是	是	是

当成员是私有的时，只能在类内部（通过成员函数）访问它。当成员是公共的时，可以从任何地方（在同一个类、子类和应用程序中）访问它。我们将在第11章讨论子类时讨论受保护成员。

数据成员的访问修饰符。数据成员的访问修饰符通常设置为private，以强调其私有性（尽管没有修饰符也表示private）。这意味着数据成员不能直接访问，必须通过成员函数访问数据成员。

然而，私有性并不意味着不可见，而是意味着访问的私有性。私有数据成员是可见的，但只有通过成员函数才能访问私有数据成员。

类的数据成员通常设置为private。

成员函数的访问修饰符。要对数据成员进行操作，应用程序必须使用成员函数，这意味着成员函数的声明通常必须设置为 public。

类的实例成员函数通常设置为 public。

但是，有时成员函数的修饰符必须设置为 private，例如某个成员函数是用于辅助其他成员函数，但不允许类以外的函数调用。我们将在本章后面讨论这个问题。

分组访问修饰符。读者可能已经注意到，我们在整个类定义中只使用了一个 private 关键字和一个 public 关键字。这被称为分组访问修饰符。我们把所有数据成员分成一组，并使用 private 关键字后跟冒号来表示所有数据成员都是私有的。我们还对所有成员函数进行了分组，并使用一个 public 关键字后跟冒号来表示所有这些函数都是公共的。换言之，当前访问修饰符在我们遇到新的访问修饰符之前是一直有效的。我们可以为每个数据成员或者成员函数定义一个访问修饰符并后跟冒号，但这不是必需的。访问修饰符后面的缩进是为了使代码更清晰。

7.2.3　成员函数定义

成员函数的声明给出了函数原型，每个成员函数也需要一个定义。函数的定义必须单独完成，除非在某些特殊情况下，我们将在本章后面讨论。在程序 7-1 中，我们根据类定义中函数声明，为每个成员函数创建了一个定义。

```
double Circle :: getRadius () const
{
  return radius;
}
double Circle :: getArea () const
{
  const double PI  = 3.14;
  return (PI * radius * radius);
}
double Circle :: getPerimeter () const
{
  const double PI  = 3.14;
  return (2 * PI * radius);
}
void Circle :: setRadius (double value)
{
  radius = value;
}
```

每个成员函数的定义类似于我们在前几章中使用的函数定义，但有两个不同之处。第一个是应用于某个成员函数的限定符（const）。第二个是必须用类名称限定的函数名称。这类似于我们认识两个叫苏的人，一个来自布朗家，另一个来自怀特家。在 C++ 中，我们需要先提到类名（姓氏），接着是类作用域符号（::）来实现这个目标。类作用域是一个基本表达式，如附录 C 所示，但为了方便起见，在表 7-2 中将类作用域重复列出。

表 7-2　类作用域

组别	名称	运算符	表达式	优先级	结合性
基本表达式	类作用域	::	class :: name	19	→

读者可能会提出疑问，我们为什么没有在类定义中包含数据成员或者成员函数的姓氏。原因是这些成员包含在类定义中，它们属于该类。这类似于当一个人在家里时，我们不需要提到他的姓氏，当他不在家庭圈中时，则需要使用姓氏。图 7-5 显示了其差异。

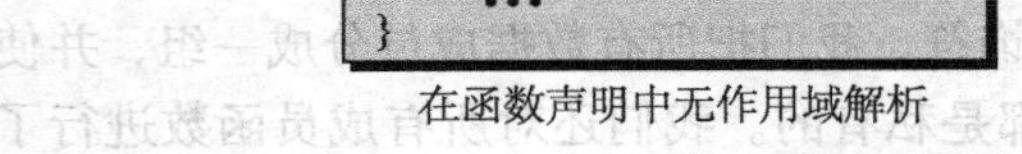

在函数声明中无作用域解析

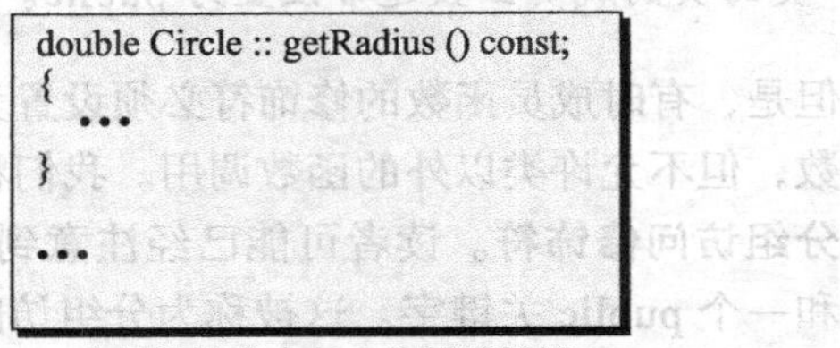

在函数定义中包含作用域解析

图 7-5　在类定义和成员函数中使用名称

注意，函数的返回类型总是在整个名称之前。在类定义中，没有显式的姓氏，因此函数的返回类型出现在函数名之前。在函数定义中，函数的返回类型必须在整个名称之前。

7.2.4　内联函数

当函数主体比较短小时，函数调用（存储参数、转移控制、获取参数和存储返回值）所涉及的执行时间可能大于在函数内执行代码的时间。为了提高程序性能，我们可以将函数声明为**内联函数**（inline function），以指示编译器用函数中的实际代码替换函数调用。但是，编译器可能会忽略此请求。我们在前面的程序中没有使用内联函数，后续的程序也不会这样做，但是你可以自由使用这个选项。

隐式内联函数

当我们将函数的声明（在类定义中）替换为其定义时，该函数被定义为**隐式内联函数**。不建议这样做，原因有二：首先，它使定义更难理解；第二，它违反了封装的原则，我们稍后将对此进行讨论。下面代码片段显示了具有隐式内联函数的 Circle 类的部分定义。

```
class Circle
{
  // 数据成员
  private:
      double radius;
  // 成员函数
  public:
      double getRadius ()  const  { return radius }
      ...
};
```

我们在类定义中包含了 getRadius 函数的定义，这意味着不需要再单独对这个函数进行定义。

显式内联函数

通过在函数定义前面添加关键字 inline，可以将函数定义为**显式内联函数**。在这种情况下，除了添加 inline 关键字外，定义保持不变。

```
inline double Circle :: getRadius()      const
{
  return radius;
}
```

7.2.5 应用程序

类定义和成员函数定义的目的在于应用。我们需要一个应用程序部分（main 函数），以实例化类的对象，并将成员函数应用到这些对象上，如程序 7-1 所示。我们首先使用以下格式实例化一个名为 circle1 的对象（我们将在本章后面讨论这种格式和类似的格式）。

对象实例化

在使用任何成员函数之前，我们必须实例化类的对象，如下所示：

```
Circle circle1;
```

执行此行之后，我们得到一个名为 circle1 的对象，它封装了一个名为 radius 的 double 类型的数据成员（变量），其中包含上一个操作留下的垃圾。在使用这个变量之前，我们必须更改它的值。

操作对象

在实例化之后，我们可以让对象对自己应用一个或者多个在成员函数中定义的操作。

```
circle1.setRadius (10.0);
cout << "Radius: " << circle1.getRadius() << endl;
cout << "Area: " << circle1.getArea() << endl;
cout << "Perimeter: " << circle1.getPerimeter() << endl << endl;
```

第一行代码设置 circle1 的半径。第二行代码获取该圆的半径值。接下来的两行代码计算并打印名为 circle1 的对象的面积和周长。

成员选择

读者可能想知道为什么我们在对象名和成员函数之间使用了一个英文句点，并且这个函数应该在该对象上进行操作。这被称为成员选择运算符，我们将在本章后面加以讨论。换言之，我们可以使用该运算符对不同的对象应用相同的函数，如下所示：

```
circle1.getRadius();                // 将获取 circle1 的半径
circle2.getRadius();                // 将获取 circle2 的半径
```

7.2.6 结构

我们有时会在 C++ 中遇到一种语言构造：struct（结构），它实际上是 C 语言遗留下来的。C++ 语言中的结构（struct）实际上是一个类，但结构与类的唯一区别在于：在结构中，默认情况下所有成员都是公共的，而在类中，默认情况下所有成员都是私有的。我们总是可以创建一个类来模拟一个结构，如下所示：

```
struct                        class
{                             {
    string first;                 public:
    char middle;                      string first;
    string last;                      char middle;
};                                    string last;
                              };
```

一些程序员仍然使用结构而不是类，因为这样做允许他们直接访问结构中的成员（因为结构中的成员是公共的），而无须使用成员函数。这常见于一些将多个数据项聚合为一项的程序，例如链表结构中的节点，我们将在第18章中讨论这些内容。我们的建议是使用类而不是结构。

7.3 构造函数和析构函数

在面向对象程序设计中，实例是对类定义中定义的数据成员进行封装的对象。如果我们希望一个对象对其本身执行一些操作，我们应该首先创建该对象并初始化其数据成员。创建对象是在应用程序中调用被称为**构造函数**（constructor）的特殊成员函数时完成的；初始化是在执行构造函数主体时完成的。

当我们不再需要一个对象时，应该清理该对象，并回收该对象占用的内存。当调用另一个被称为**析构函数**（destructor）的特殊成员函数时，清理将自动完成。对象超出作用域时，析构函数主体将被执行；当程序终止时，回收将完成。

> 构造函数是用于创建和初始化对象的特殊成员函数。析构函数是用于清理和销毁对象的特殊成员函数。

换言之，对象经历了如图7-6所示的五个步骤。它由一个被称为构造函数的特殊成员函数创建和初始化。它执行应用程序请求的一些操作。它被另一个被称为析构函数的特殊成员函数清理和回收。

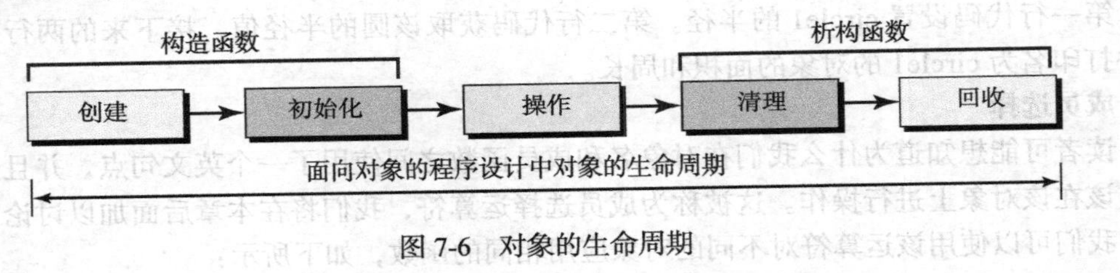

图7-6 对象的生命周期

7.3.1 构造函数

构造函数是一个成员函数，它在被调用时创建对象，并在被执行时初始化对象的数据成员。类定义中数据成员的声明不会初始化数据成员，声明只给出数据成员的名称和类型。

构造函数有两个特征：它没有返回值；其名称与类的名称相同。构造函数不能有返回值（甚至不能是void），因为它不是为返回任何内容而设计的，其用途不同。构造函数用于创建一个对象并初始化该对象的数据成员。尽管我们会看到一个构造函数也可以执行一些其他任务，例如值的验证，但这些任务也被认为是初始化的一部分。

在一个类中，可以包含三种类型的构造函数：参数构造函数（parameter constructor）、默认构造函数（default constructor）和拷贝构造函数（copy constructor）。

构造函数的声明

构造函数是类的成员函数，这意味着它必须在类定义中声明。构造函数没有返回值，其名称与类的名称相同，并且不能具有常量限定符const，因为构造函数要初始化数据成员的值（稍后将详细讨论）。下面显示了如何将三个构造函数的声明添加到类Circle中。

```
class Circle
{
  ...
```

```
public:
    Circle (double radius);                    // 参数构造函数
    Circle ();                                 // 默认构造函数
    Circle (const Circle& circle);             // 拷贝构造函数
    ...
}
```

注意，类的所有构造函数通常都是公共的，以便应用程序可以调用任何构造函数来初始化类的对象。

参数构造函数。通常类定义中会包含一个参数构造函数，它用指定的值初始化每个实例的数据成员。参数构造函数可以重载，这意味着我们可以有多个参数构造函数，每个参数构造函数具有不同的签名。参数构造函数的优点是我们可以用特定的值初始化每个对象的数据成员。换言之，如果我们使用参数构造函数，一个圆的半径可以初始化为 3.1，另一个圆的半径可以初始化为 4.6，依此类推。

类的参数构造函数可以被重载。

默认构造函数。默认构造函数是不带参数的构造函数。它用于创建对象，把对象的所有数据成员设置为字面量值或者默认值。注意，我们不能重载默认构造函数。因为默认构造函数没有参数列表，所以不适合重载。

类的默认构造函数不能被重载。

拷贝构造函数。有时候，我们希望将对象的每个数据成员初始化为与先前创建的对象的相应数据成员相同的值。在这种情况下，我们可以使用拷贝构造函数。拷贝构造函数将给定对象的数据成员值复制到刚刚创建的新对象中。在调用拷贝构造函数之后，源对象和目标对象的每个数据成员具有完全相同的值，尽管它们是不同的对象。拷贝构造函数只有一个参数，按引用接收源对象。参数类型前面的 const 修饰符保证按引用传递时不能更改源对象。请记住，按引用传递有两个特征：第一，不需要物理复制对象；第二，改变目标对象意味着同时会改变源对象。使用 const 修饰符后，我们保留第一个特征，但禁止第二个特征。还要注意，我们不能重载拷贝构造函数，因为参数列表是固定的，并且不能有替代形式。

类的拷贝构造函数不能被重载。

构造函数的定义

如前所述，构造函数是一个成员函数，但却是一个特殊的成员函数。构造函数不能有返回值，并且其名称与类的名称相同。下面代码片段显示了 Circle 类的三个构造函数的定义。

```
// 参数构造函数的定义
Circle :: Circle (double rds)
: radius (rds)         // 初始化列表
{
    // 任何其他语句
}
// 默认构造函数的定义
Circle :: Circle ()
: radius (1.0) // 初始化列表。如果省略，则 radius 被设置为垃圾值
{
```

```
    // 任何其他语句
}
// 拷贝构造函数的定义
Circle :: Circle (const Circle& cr)
: radius (cr.radius)     // 初始化列表
{
    // 任何其他语句
}
```

构造函数定义和其他成员函数定义的主要区别在于，构造函数可以在函数头后面有一个初始化列表来初始化数据成员。初始化列表放在构造函数头之后和构造函数主体之前，并以冒号开头。在 Circle 类中，我们只有一个数据成员要初始化。如果需要初始化多个数据成员，则每个数据成员的初始化必须用逗号分隔开。通常，初始化列表的格式如下：

```
: dataMemember (parameter), ... , dataMember (parameter)
```

我们可以将每个初始化看作一个赋值语句，用于将参数赋值给数据成员，例如 data Member = parameter。初始化列表没有终止符。下一行是构造函数的主体。数据成员的名称必须与数据成员声明中定义的名称相同，但每个参数的名称由程序员确定。

另一个要点是，创建对象时必须初始化对象的常量数据成员。如前面章节所述，常量实体被声明之后，就不能被修改，但是 C++ 允许我们在构造函数的初始化部分初始化对象的常量数据成员。

然而，有时我们必须使用构造函数的主体来初始化复杂的数据成员（通过赋值），这些成员不能在初始化列表中简单地被初始化。构造函数的主体还可以用于其他处理，例如验证参数，根据需要打开文件，甚至打印消息以验证是否调用了该构造函数。

7.3.2　析构函数

与构造函数类似，析构函数也有两个特殊的特性：首先，析构函数的名称是以波浪线符号（~）开头的类的名称，但是波浪线被添加到名字中，而不是姓氏中（所有成员函数的姓氏都相同）；其次，与构造函数一样，析构函数不能有返回值（甚至不能是 void），因为它什么也不返回。当类的实例化对象超出其作用域时，系统保证自动调用和执行析构函数。换言之，如果我们实例化了类的 5 个对象，则会自动调用析构函数 5 次，以确保所有对象都被清理。如果对象构造时调用了诸如文件之类的资源，则清理工作十分重要。程序终止后，分配的内存将被回收。析构函数不能接收任何参数，这意味着析构函数不能被重载。

> 析构函数是没有参数的特殊成员函数，用于清理和回收对象。

析构函数的声明

以下代码片段显示了类定义中析构函数的声明。

```
class Circle
{
    ...
    public:
        ...
        ~Circle ();                              // 析构函数
}
```

析构函数的定义

析构函数的定义与其他三个成员函数的定义类似，但析构函数的名称前面必须有一个波浪号（~）。和所有构造函数一样，析构函数应该是公共的。

```
// 析构函数的定义
Circle :: ~Circle ()
{
  // 任何需要的语句
}
```

7.3.3 创建和销毁对象

我们现在准备好讨论如何在需要对象时实例化对象，以及如何在不需要对象时销毁对象。注意，调用构造函数会创建一个对象。当构造函数执行时，它将初始化数据成员。在执行析构函数时，数据成员会被清除。当程序终止时，内存位置会被释放。

图 7-7 显示了调用参数构造函数、默认构造函数和拷贝构造函数的语法。没有调用析构函数的语法，因为析构函数是由系统调用的。我们必须注意默认构造函数的调用方法。在 C++ 中，默认构造函数不需要空括号。如果我们在调用默认构造函数时使用了空括号，系统会认为我们想要调用一个没有参数的重载的参数构造函数，如果没有定义这样的构造函数，我们会得到一个错误。

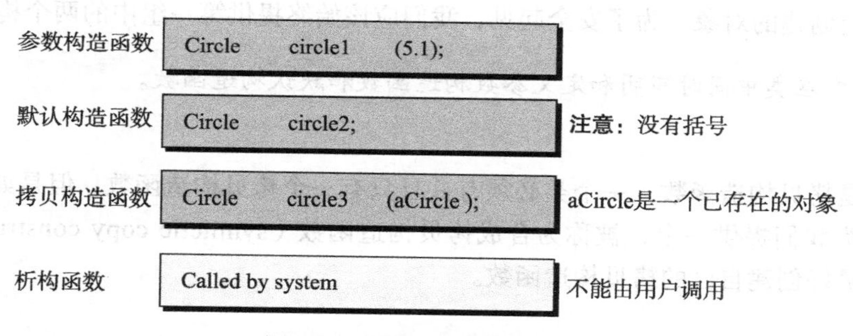

图 7-7 类对象的构造和析构

请注意，文献中有时会将赋值运算符作为语法的一部分，但赋值运算符只是将一个对象的数据成员复制到另一个对象。这两个对象必须已经存在。我们将在第 13 章中讨论赋值运算符，并在讨论时解释赋值运算符与上述四个成员函数的关系。

表 7-3 比较了处理内置类型时变量的创建方法与处理类的对象时对象的创建方法。

表 7-3 变量的创建与类对象的创建的比较

成员	类对象	内置类型
参数构造函数	Circle circle1 (10.0);	double x1 = 10.0;
默认构造函数	Circle circle2;	double x2;
拷贝构造函数	Circle circle3 (circle1);	无
析构函数	不直接调用	不直接调用

7.3.4 必需的成员函数

读者可能会提出疑问：这四个成员函数中哪一个是一个类所必需的？如果我们忘记声明

和定义其中的一个或者多个，就像我们在 Circle 类中所做的那样，结果会如何呢？为了更好地回答这个问题，我们将这些成员函数分为三组，如图 7-8 所示。每组在需求上都是彼此独立的。

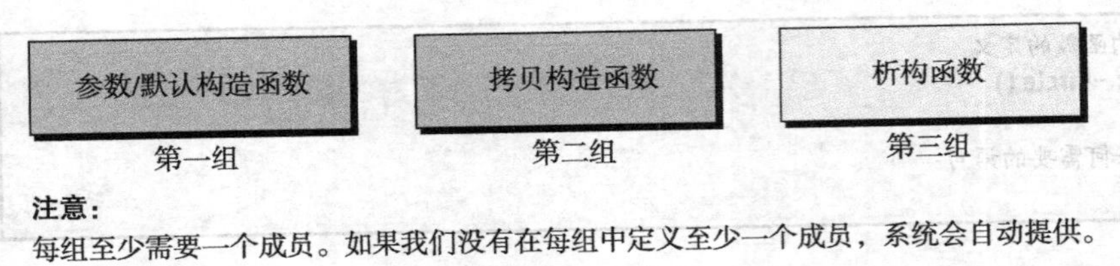
参数/默认构造函数
拷贝构造函数
析构函数
第一组
第二组
第三组
注意：
每组至少需要一个成员。如果我们没有在每组中定义至少一个成员，系统会自动提供。

图 7-8　对构造函数和析构函数分组

第一组

第一组由参数构造函数和默认构造函数组成。我们必须至少有一个这样的构造函数，有时我们可能同时需要这两个构造函数。如果我们提供了其中任何一个，则系统不会为我们提供任何构造函数。如果我们两者都不提供，则系统将提供一个默认构造函数，被称为合成默认构造函数（synthetic default constructor），它将每个数据成员初始化为系统中残留的垃圾值。这就是前面对 Circle 的第一个定义所发生的情况。我们需要第一组中的两个构造函数中的一个，但是我们却没有提供任何一个，所以系统就为我们提供了一个默认构造函数，并在 radius 数据成员中存储了垃圾值。在使用 setRadius 成员函数重新初始化 radius 之前，我们没有使用之前创建的对象。为了安全起见，我们应该始终提供第一组中的两个构造函数。

> 我们通常在类中同时声明和定义参数构造函数和默认构造函数。

第二组

第二组是拷贝构造函数。一个类必须有并且只有一个拷贝构造函数，但是如果我们不提供，系统会为我们提供一个，被称为合成拷贝构造函数（synthetic copy constructor）。大多数情况下，最好创建自己的拷贝构造函数。

第三组

第三组是析构函数。一个类必须有并且只能有一个析构函数，但是如果我们不提供，系统会为我们提供一个析构函数，被称为合成析构函数（synthetic destructor）。大多数时候，合成析构函数并不能满足我们的需求。最好创建自己的析构函数。

【例 7-1】 程序 7-2 使用本节讨论的构造函数和析构函数重构了 Circle 类。我们在每个构造函数和析构函数的函数体中包含一条消息，以显示何时调用了它们。

程序 7-2　一个完整的 Circle 类

```
1   /***************************************************************
2    * 本程序用于在面向对象程序设计中使用类                          *
3    ***************************************************************/
4   #include <iostream>
5   using namespace std;
6
7   /***************************************************************
8    * 类定义：                                                      *
9    * 声明类的参数构造函数、默认构造函数、                          *
10   * 拷贝构造函数、析构函数和其他成员函数                          *
```

```
11    *****************************************************************/
12  class Circle
13  {
14     private:
15          double radius;
16     public:
17          Circle (double radius);       //参数构造函数
18          Circle ();                    //默认构造函数
19          ~Circle ();                   //析构函数
20          Circle (const Circle& circle);      //拷贝构造函数
21          void setRadius (double radius);     //更改器
22          double getRadius () const;          //访问器
23          double getArea () const;            //访问器
24          double getPerimeter () const;       //访问器
25  };
26  /*****************************************************************
27   *成员函数的定义：                                                *
28   *参数构造函数、默认构造函数、                                    *
29   *拷贝构造函数、析构函数和其他成员函数的定义                      *
30   *****************************************************************/
31  //参数构造函数的定义
32  Circle :: Circle (double rds)
33  : radius (rds)
34  {
35     cout << "The parameter constructor was called. " << endl;
36  }
37  //默认构造函数的定义
38  Circle :: Circle ()
39  : radius (0.0)
40  {
41     cout << "The default constructor was called. " << endl;
42  }
43  //拷贝构造函数的定义
44  Circle :: Circle (const Circle& circle)
45  : radius (circle.radius)
46  {
47     cout << "The copy constructor was called. " << endl;
48  }
49  //析构函数的定义
50  Circle :: ~Circle ()
51  {
52     cout << "The destructor was called for circle with radius " ;
53     cout << endl;
54  }
55  //setRadius成员函数的定义
56  void Circle :: setRadius (double value)
57  {
58     radius = value;
59  }
60  //getRadius成员函数的定义
61  double Circle :: getRadius () const
62  {
```

```
63    return radius;
64  }
65  // getArea 成员函数的定义
66  double Circle :: getArea () const
67  {
68    const double PI  = 3.14;
69    return (PI * radius * radius);
70  }
71  // getPerimeter 成员函数的定义
72  double Circle :: getPerimeter () const
73  {
74    const double PI  = 3.14;
75    return (2 * PI * radius);
76  }
77  /******************************************************************
78   * 应用程序部分:                                                    *
79   * 创建 Circle 类的三个对象（circle1、circle2 和 circle3），         *
80   * 并操作这些对象                                                   *
81   ******************************************************************/
82  int main ( )
83  {
84    // 实例化 circle1 并应用成员函数进行操作
85    Circle circle1 (5.2);
86    cout << "Radius: " << circle1.getRadius() << endl;
87    cout << "Area: " << circle1.getArea() << endl;
88    cout << "Perimeter: " << circle1.getPerimeter() << endl << endl;
89    // 实例化 circle2 并应用成员函数进行操作
90    Circle circle2 (circle1);
91    cout << "Radius: " << circle2.getRadius() << endl;
92    cout << "Area: " <<  circle2.getArea() << endl;
93    cout << "Perimeter: " << circle2.getPerimeter() << endl << endl;
94    // 实例化 circle3 并应用成员函数进行操作
95    Circle circle3;
96    cout << "Radius: " << circle3.getRadius() << endl;
97    cout << "Area: " <<  circle3.getArea() << endl;
98    cout << "Perimeter: " << circle3.getPerimeter() << endl << endl;
99    // 此处会调用析构函数
100   return 0;
101 }
```

运行结果：

```
The parameter constructor was called.
Radius: 5.2
Area: 84.9056
Perimeter: 32.656

The copy constructor was called.
Radius: 5.2
Area: 84.9056
Perimeter: 32.656

The default constructor was called.
Radius: 0
```

```
Area: 0
Perimeter: 0

The destructor was called for circle with radius: 0
The destructor was called for circle with radius: 5.2
The destructor was called for circle with radius: 5.2
```

请注意，应用程序使用参数构造函数、拷贝构造函数和默认构造函数创建了三个对象：circle1、circle2 和 circle3。注意，应用程序不调用析构函数，但当对象超出其作用域时，系统会调用析构函数。有趣的是，这些对象的销毁顺序与它们的构造顺序相反。最后创建的对象首先被销毁；第一个创建的对象最后被销毁。在第 14 章中，当我们讨论在栈内存中创建对象时，我们将解释其原因。在堆栈中，最后插入的项是第一个可以被删除的项（如一叠盘子）。

【例 7-2】 有时我们只需要参数构造函数，因为默认构造函数和拷贝构造函数没有实际意义。我们在第 6 章中讨论了随机数生成器的使用。我们知道，对于每个随机数，我们需要使用适当的参数（例如 time(0)）来调用 seed 函数。然后我们调用 random 函数。最后，我们需要对所生成的随机数进行缩放和平移，以获得所需范围内的随机数。所有这些操作都可以在类中完成。我们实例化类的一个对象来创建一个随机数，如程序 7-3 所示。

程序 7-3 定义和创建三个随机数

```
1   /***************************************************************
2    * 本程序用于声明、定义和使用                                     *
3    * 一个用于生成给定区间内的随机数的类，                           *
4    * 在类的构造函数中指定区间范围                                   *
5    ***************************************************************/
6   #include <iostream>
7   #include <cstdlib>
8   #include <ctime>
9   using namespace std;
10
11  /***************************************************************
12   * 一个随机数生成器的类定义：                                     *
13   * 声明类的数据成员和成员函数                                     *
14   ***************************************************************/
15  class RandomInteger
16  {
17    private:
18        int low;    // 数据成员
19        int high;   // 数据成员
20        int value;  // 数据成员
21    public:
22        RandomInteger (int low, int high); // 构造函数
23        ~RandomInteger (); // 析构函数
24        // 阻止使用合成拷贝构造函数
25        RandomInteger (const RandomInteger& random) = delete;
26        void print () const; // 访问器成员函数
27  };
28  /***************************************************************
29   * 随机数生成器类的成员函数定义：                                 *
```

```
30	 * 构造函数、析构函数、访问器成员函数                                    *
31	 ***********************************************************************/
32	// 构造函数
33	RandomInteger :: RandomInteger (int lw, int hh)
34	:low (lw), high (hh)
35	{
36	   srand (time (0));
37	   int temp = rand ();
38	   value = temp % (high - low + 1) + low;
39	}
40	// 析构函数
41	RandomInteger :: ~RandomInteger ()
42	{
43	}
44	// 访问器成员函数
45	void RandomInteger :: print () const
46	{
47	   cout << value << endl;
48	}
49	/***********************************************************************
50	 * 应用程序:                                                            *
51	 * 实例化随机数对象，并打印随机数的值                                    *
52	 ***********************************************************************/
53	int main ( )
54	{
55	   // 生成一个 100 到 200 之间的随机整数
56	   RandomInteger r1 (100, 200);
57	   cout << "Random number between 100 and 200: ";
58	   r1.print ();
59	   // 生成一个 400 到 600 之间的随机整数
60	   RandomInteger r2 (400, 600);
61	   cout << "Random number between 400 and 600: ";
62	   r2.print ();
63	   // 生成一个 1500 到 2000 之间的随机整数
64	   RandomInteger r3 (1500, 2000);
65	   cout << "Random number between 1500 and 2000: ";
66	   r3.print ();
67	   return 0;
68	}
```

```
运行结果:
Random number between 100 and 200: 130
Random number between 400 and 600: 570
Random number between 1500 and 2000: 1720
```

关于程序 7-3，我们需要注意以下四点：

- 这是一个不需要默认构造函数的示例，因为使用字面量创建一个随机数没有实际意义。由于我们已经定义了一个参数构造函数，所以在这种情况下，系统不会创建一个默认构造函数，这正是我们想要的。
- 这个类也不需要拷贝构造函数（创建相同的随机数也没有实际意义），但是我们不能阻止系统定义一个合成拷贝构造函数。在这种情况下，新的 C++11 标准提供了解决方案：它允许我们声明一个拷贝构造函数并将其赋值为关键字 delete，这会阻止系统提供一个合成的拷贝构造函数（参见第 25 行的类定义）。

- 虽然我们有三个数据成员，但是只有两个（low 和 high）在构造函数的初始化列表中被初始化。第三个数据成员（value）是在构造函数的主体中计算的。因为计算逻辑太复杂，无法在初始化列表中完成。唯一的访问器成员函数是 print 函数，它打印所生成的随机数的值。
- 最后一点，本例中没有更改器成员函数，因为我们不希望修改所创建的随机数。

7.4　实例成员

在前三节中，我们学习了如何定义类，如何从类中实例化对象，以及如何在对象的数据成员上应用成员函数。在本节和下一节中，我们将进一步讨论这些成员以及后台成员之间的交互，这样做将帮助我们设计更好的类。当我们设计一个类时，我们可以有两组成员：实例成员和类成员。

7.4.1　实例数据成员

实例数据成员定义对象实例的属性，这意味着每个对象必须封装在类中定义的数据成员集。这些数据成员仅属于相应的实例，其他实例无法访问。这里的术语“封装”意味着为每个对象分配单独的内存区域，并且每个区域为每个数据成员存储不同的值。图 7-9 显示了封装的概念。

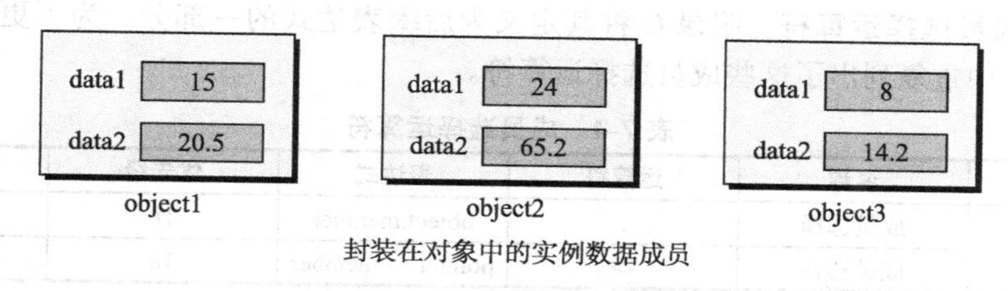

图 7-9　对象中数据成员的封装

实例数据成员的访问修饰符

尽管实例数据成员可以具有私有或者公共访问修饰符，但实例数据成员更适合私有。如果我们将实例数据成员公开，那么应用程序可以直接访问这些成员，而不需要调用实例成员函数。这并不是面向对象程序设计的设计目标。在面向对象程序设计中，我们希望对象在其属性上应用其行为。换言之，我们必须将实例数据成员设置为私有的，以便只能通过实例成员函数来访问它们。

> 类的实例数据成员通常是私有的，只能通过实例成员函数访问。

7.4.2　实例成员函数

实例成员函数定义可以应用于对象的实例数据成员的行为。尽管每个对象都有自己的实例数据成员，但内存中每个实例成员函数只有一个副本，必须由所有实例共享。图 7-10 显示了具有两个实例数据成员和四个实例成员函数的类的情况。

因为我们从这个类创建了三个对象，所以我们有三对实例数据成员和四个实例成员函数，每个成员函数在对象之间共享（一次一个对象）。

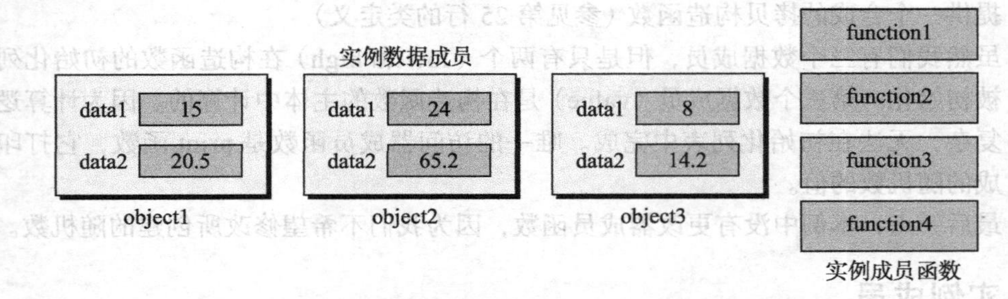

图 7-10　具有两个数据成员和四个成员函数的类

实例成员函数的访问修饰符

与实例数据成员不同，实例成员函数的访问修饰符通常是公共的，并且允许从类外部（应用程序）进行访问，除非实例成员函数仅由类中的其他实例成员函数使用。

> 类的实例成员函数必须是公共的，以便可以从类外部访问它。

实例成员函数的选择运算符

应用程序（例如，main 函数）可以调用实例成员函数对实例进行操作。在面向对象程序设计中，此调用必须通过实例完成。应用程序必须首先创建一个实例，然后让该实例调用实例成员函数。换言之，这有点类似于实例是在对自身进行操作。C++ 语言为此定义了两个运算符，称为**成员选择运算符**。附录 C 将其定义为后缀表达式的一部分。为了更方便地快速参考，表 7-4 中重复列出了这些成员选择运算符。

表 7-4　成员选择运算符

组别	名称	运算符	表达式	优先级	结合性
后缀表达式	成员选择	.	object.member	18	→
	成员选择	->	pointer -> member	18	→

图 7-11 显示了如何使用第一种成员选择运算符选择给定名称的对象的成员函数。稍后在讨论锁定和解锁对象实例时，我们将讨论第二种成员选择运算符的使用方法。

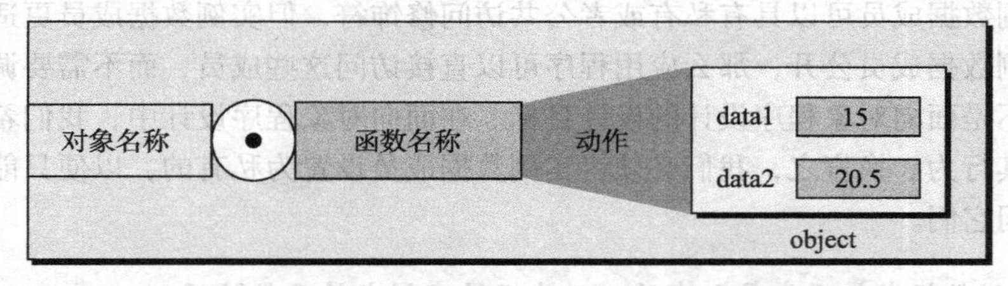

图 7-11　使用成员选择运算符以进行实例操作

我们在程序 7-2 和程序 7-3 中使用了点成员选择运算符，让对象调用它们的成员函数。

锁定和解锁

如果一个成员函数只有一个副本，那么该函数如何在一个时间被一个对象使用，而在另一个时间被另一个对象使用？更重要的问题是，当一个对象正在使用一个函数时，我们如何阻止其他对象使用该函数？换言之，在函数被使用时我们如何将其锁定，在函数终止（返回）时又如何将其解锁，以便以后其他对象可以使用该函数？

答案是：在 C++ 的后台完成锁定和解锁。C++ 将一个指针（一个保存对象地址的变量）添加到每个成员函数中。因此，当我们使用图 7-11 中描述的点成员选择运算符时，编译器将其转换为图 7-12 中描述的指针成员选择运算符，图 7-12 中每个成员函数都有一个名为 **this 指针**的隐藏指针。函数由 this 指针指向的对象使用。换言之，函数代码被应用于 this 指针指向的对象的数据成员。

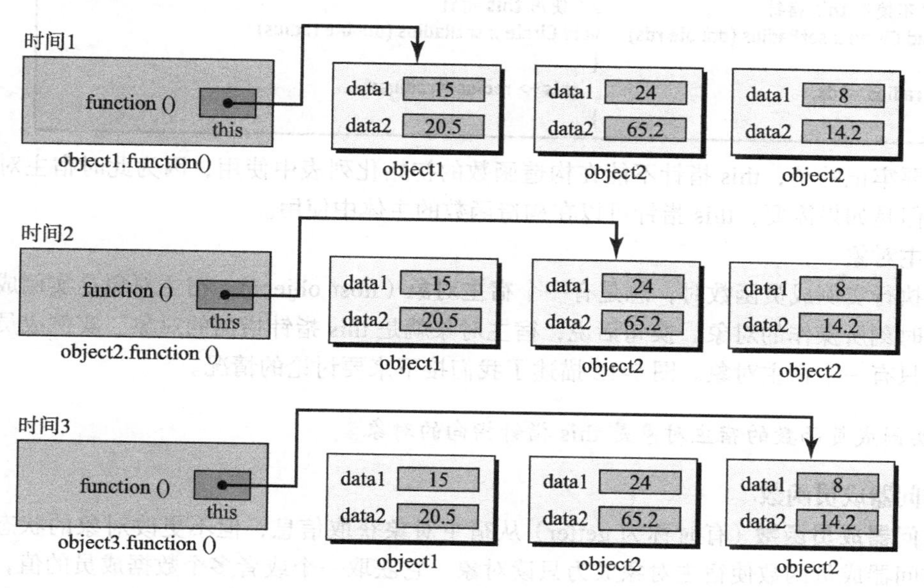

图 7-12 针对特定对象锁定和解锁某一函数

图 7-12 显示了在不同时间被不同对象使用的一个成员函数。虽然函数相同，但在不同的时间操作的对象不同。因为每个函数只有一个 this 指针，我们一次只能调用一个函数，所以我们可以确保锁定和解锁能够正确完成。

隐藏参数。实例成员函数如何获取 this 指针呢？编译器会将 this 指针作为参数添加到实例成员函数中，如下所示：

```
// 用户编写的代码                    // 编译器转换后的代码
double getRadius () const           double getRadius (Circle* this) const
{                                   {
    return radius;                      return (this -> radius);
}                                   }
```

运算符（->）是一个特殊的运算符，它是间接运算符（*）和成员选择运算符（.）的组合。换言之，我们有以下关系：

```
this -> radius                等同于                (*this).radius
```

当我们使用成员选择运算符调用实例成员函数时，编译器将调用语句转换为两条语句，如下所示：

```
// 用户编写的代码                    // 系统转换后的代码
circle1.getRadius();                this = &circle1;
                                    getRadius (this);
```

显式使用 this 指针

我们可以在程序中使用 this 指针来引用数据成员，而不是直接使用数据成员本身，并且可以使用数据成员的名称作为参数。这样，我们就不必像过去那样使用缩写名称了。下面显示了在不使用 this 指针和使用 this 指针的情况下对 Circle 类的 setRadius 所编写的代码。

```
// 不使用 this 指针                          // 使用 this 指针
void Circle :: setRadius (double rds)      void Circle :: setRadius (double radius)
{                                          {
   radius = rds;                              this -> radius = radius;
}                                          }
```

需要牢记的是，this 指针不能在构造函数的初始化列表中使用，因为此时宿主对象尚未构造。但是如果需要，this 指针可以在构造函数的主体中使用。

宿主对象

当执行实例成员函数时，总是有一个**宿主对象**（host object）。宿主对象是实例成员函数在给定时刻所操作的对象。换句话说，宿主对象就是 this 指针指向的对象。实例成员函数执行期间只有一个宿主对象。图 7-12 描述了我们接下来要讨论的情况。

> 实例成员函数的宿主对象是 this 指针指向的对象。

访问器成员函数

访问器成员函数（有时称为 getter）从宿主对象获取信息，但不更改对象的状态。换言之，访问器成员函数使宿主对象成为只读对象。它获取一个或者多个数据成员的值，但不更改其在宿主对象中的值。例如，返回 Circle 对象中 radius（半径）值的实例成员函数是访问器实例成员函数。为了保证访问器实例成员函数不会更改对象的状态，我们必须在声明和定义中的函数头的末尾添加 const 限定符，如下所示：

```
double getRadius () const;           // const 限定符确保宿主对象只读
double getPerimeter () const;        // const 限定符确保宿主对象只读
double getArea() const;              // const 限定符确保宿主对象只读
```

访问器成员函数的参数列表通常为空。函数头末尾的 const 修饰符将宿主对象定义为不能修改的常量对象。

> 访问器实例函数不能更改宿主对象的状态，它需要 const 修饰符。

访问器成员函数可以不必返回数据成员的值。也可以使用访问器成员函数来创建副作用（例如，输出值），只要对象的状态保持不变。例如，我们可以有一个输出函数来打印一个 Circle 类的对象的半径、周长和面积，该函数不带返回值，如下所示。

```
void Circle :: print () const
{
   cout << "Radius: " << radius << endl;
   cout << "Perimeter: " << 2 * radius * 3.14 << endl;
   cout << "Area: " << radius * radius * 3.14 << endl;
}
```

在第 16 章中，我们将讨论插入运算符（<<）实际上是应用于 cout 对象的访问器函数。

更改器成员函数

程序中类类型的对象通常由参数构造函数初始化。这意味着对象的状态是在构造时设置的。然而，有时我们必须改变原始状态。例如，如果我们创建一个表示银行账户的类，则表示余额的数据成员将随时间变化（每次存款和取款）。这意味着我们可能需要实例成员函数来更改其宿主对象的状态。这样的函数称为**更改器成员函数**（有时称为 setter）。此函数不能有常量限定符 const，因为它需要更改宿主对象的状态。在 Circle 类中，只包含一个实例更改器成员函数，如下所示：

```
void setRadius (double rds);      // 更改器函数没有 const 限定符
```

更改器实例成员函数不一定需要通过参数来更改数据成员的值，它可以是一个不带参数但具有副作用（例如，通过输入一个值）的函数。

更改器实例函数更改宿主对象的状态，它不能有 const 修饰符。

例如，我们可以编写一个输入函数，输入一个 Circle 类的对象的半径值，该函数没有返回值，如下所示。

```
void Circle :: input()
{
   cout << "Enter the radius of the circle object: ";
   cin >> radius;
}
```

在第 16 章中，我们将看到提取运算符（>>）实际上是应用于 cin 对象的一个更改器函数。

注意，构造函数和析构函数可以被认为是更改器成员函数，因为它们初始化对象或者清除对象（从而更改了状态）。

7.4.3 类不变式

类设计中的一个重要问题是类不变式。**类不变式**（class invariant）是一个或者多个条件，我们必须对类的某些或者所有实例数据成员施加这些条件，并且我们应该通过实例成员函数来施加这些条件。换言之，它是一个与实例数据成员和实例成员函数相关的问题。当我们讨论 Circle 类时，我们看到一个圆的半径（radius）必须是一个正值，一个带有负的半径值的圆是没有意义的。编译器无法捕捉到这个问题，因为我们将 radius 定义为 double 类型，double 值可以是负数。

不变式是一个或者多个必须施加在某些或者所有类数据成员上的条件。

我们通过创建对象的实例数据成员函数（参数构造函数）或者修改数据成员值的更改器成员函数来强制类的不变式。例如，我们可以修改 Circle 类的参数构造函数来保证类的不变式。下面代码展示了如何更改构造函数。

```
Circle :: Circle (double rds)
: radius (rds)
{
   if (radius <= 0.0))
```

```
        {
            cout << "No circle can be made!" << endl;
            cout << "The program is aborted" << endl;
            assert (false);
        }
    }
```

在这种情况下，我们必须中止程序，因为无法创建相应的对象，而缺少对象可能会影响程序的其余部分。assert 函数是一个库函数。当其参数为 true 时，它不起作用；当其参数为 false 时，它会中止程序。我们将它与 false 参数一起使用，以在不满足不变式的情况下中止程序。要使用 assert 函数，我们必须在程序中包含 <cassert> 头文件。

【例 7-3】在本例中，我们创建了一个新的类 Rectangle，它表示一个有两个数据成员的矩形类型：长度（length）和高度（height），如程序 7-4 所示。请注意，我们没有使用任何更改器成员函数，因为我们不希望在矩形被创建后调整其大小，但可以轻松地添加矩形。另外请注意，我们没有使用两个访问器函数分别获取长度和高度，而是使用了一个访问器函数 print 来打印长度和高度。

程序 7-4　使用 Rectangle 类

```
1   /***************************************************************
2    * 本程序用于声明、定义和使用 Rectangle 类                          *
3    ***************************************************************/
4   #include <iostream>
5   #include <cassert>
6   using namespace std;
7   /***************************************************************
8    * Rectangle 类的类定义                                          *
9    *（类的数据成员和成员函数的声明）                                  *
10   ***************************************************************/
11  class Rectangle
12  {
13      private:
14          double length;                          // 数据成员
15          double height;                          // 数据成员
16      public:
17          Rectangle (double length, double height); // 构造函数
18          Rectangle (const Rectangle& rect);      // 拷贝构造函数
19          ~Rectangle ();                          // 析构函数
20          void print () const;                    // 访问器成员函数
21          double getArea () const;                // 访问器成员函数
22          double getPerimeter () const;           // 访问器成员函数
23  };
24  /***************************************************************
25   * Rectangle 类的成员函数定义：                                    *
26   * 构造函数、析构函数、访问器实例成员函数                            *
27   ***************************************************************/
28  // 参数构造函数
29  Rectangle :: Rectangle (double len, double hgt)
30  : length (len), height (hgt)
31  {
32      if ((length <= 0.0) || (height <= 0.0 ))
```

```
33      {
34          cout << "No rectangle can be made!" << endl;
35          assert (false);
36      }
37  }
38  // 拷贝构造函数
39  Rectangle :: Rectangle (const Rectangle& rect)
40  : length (rect.length), height (rect.height)
41  {
42  }
43  // 析构函数
44  Rectangle :: ~Rectangle ()
45  {
46  }
47  // 访问器成员函数：打印长度和高度
48  void Rectangle :: print() const
49  {
50      cout << "A rectangle of " << length << " by " << height << endl;
51  }
52  // 访问器成员函数：获取面积
53  double Rectangle :: getArea () const
54  {
55      return (length * height);
56  }
57  // 访问器成员函数：获取周长
58  double Rectangle :: getPerimeter () const
59  {
60      return (2 * (length + height));
61  }
62  /**************************************************************
63   * 用于实例化三个对象并使用这三个对象的应用程序。                    *
64   **************************************************************/
65  int main ( )
66  {
67      // 实例化三个对象
68      Rectangle rect1 (3.0, 4.2);         // 使用参数构造函数
69      Rectangle rect2 (5.1, 10.2);        // 使用参数构造函数
70      Rectangle rect3 (rect2);            // 使用拷贝构造函数
71      // 第一个对象的操作
72      cout << "Rectangle 1: ";
73      rect1.print();
74      cout << "Area: " << rect1.getArea() << endl;
75      cout << "Perimeter: " << rect1.getPerimeter() << endl << endl;
76      // 第二个对象的操作
77      cout << "Rectangle 2: ";
78      rect2.print();
79      cout << "Area: " << rect2.getArea() << endl;
80      cout << "Perimeter: " << rect2.getPerimeter() << endl << endl;
81      // 第三个对象的操作
82      cout << "Rectangle 3: ";
83      rect3.print();
84      cout << "Area: " << rect3.getArea() << endl;
```

```
85  cout << "Perimeter: " << rect3.getPerimeter() << endl << endl;
86  return 0;
87  }
```

运行结果：

```
Rectangle 1: A rectangle of 3 by 4.2
Area: 12.6
Perimeter: 14.4

Rectangle 2: A rectangle of 5.1 by 10.2
Area: 52.02
Perimeter: 30.6

Rectangle 3: A rectangle of 5.1 by 10.2
Area: 52.02
Perimeter: 30.6
```

7.5 静态成员

如前所述，类可以有两种类型的成员：实例成员和静态成员。我们在前一节讨论了实例成员，我们在这一节讨论静态成员。与实例成员一样，静态成员可以分为静态数据成员和静态成员函数。我们将分别展开讨论。

7.5.1 静态数据成员

静态数据成员（static data member）是属于所有实例的数据成员，它也属于类本身。作为静态数据成员的一个例子，假设我们希望跟踪程序中当前实例的数量。我们可以创建一个名为 count 的静态数据成员，并将该成员初始化为 0。我们可以修改所有构造函数的定义，以便在每次创建实例时递增这个静态成员。我们可以修改析构函数的定义，以便在对象被销毁（超出作用域）时将该静态成员递减。静态数据成员还有其他应用，我们将在本章稍后和其他章节中讨论。

声明静态数据成员

数据成员属于类，它们的声明必须包含在类定义中。静态成员必须用关键字 static 限定。下面代码片段显示了如何在类定义中声明一个名为 count 的静态数据成员。

```
class Rectangle
{
  private:
    ...
    static int count;                          // 静态数据成员
  public:
    ...
}
```

初始化静态数据成员

我们知道实例数据成员通常是在构造函数中初始化的，但是静态数据成员不属于任何实例，这意味着静态数据成员不能在构造函数中初始化。静态数据成员必须在类定义之后初始化。这意味着静态数据成员必须在程序的全局区域中初始化。我们在类定义的后面添加其初始化编码。我们必须通过向定义中添加类名和类作用域运算符（::）来显示其属于哪个类，但不应添加静态限定符 static。在类的定义部分已经将其限定为静态数据成员。以下代码片段显示了如何初始化静态成员 count，静态成员 count 已在类中声明。

```
int Rectangle :: count = 0;          //静态数据成员的初始化
```

7.5.2 静态成员函数

在声明和初始化静态数据成员之后，我们必须寻求访问静态数据成员的方法（例如，打印它的值）。因为静态数据成员通常是私有的，所以我们需要一个公共成员函数来实现这一点。尽管这可以由实例成员函数完成，但通常我们为此使用**静态成员函数**（static member function）。静态成员函数可以通过对象访问静态数据成员，也可以在不存在对象时通过类的名称访问静态数据成员。换言之，使用静态成员函数，我们可以在实例想要访问相应的静态数据成员时访问它，或者在应用程序需要访问相应的静态数据成员时访问它。注意，静态成员函数没有宿主对象，因为静态成员函数不与任何实例关联。

静态成员函数没有宿主对象。

声明静态成员函数

静态成员函数和静态数据成员一样属于类。静态成员函数应该在类中声明，但必须用关键字 static 限定。以下代码片段显示了如何添加静态成员函数 getCount 来获取 Rectangle 类中静态数据成员 count 的值。

```
class Rectangle
{
  private:
     ...
     static int count;                  //静态数据成员
  public:
     static int getCount();             //静态成员函数
     ...
}
```

定义静态成员函数

与实例成员函数相类似，静态成员函数必须在类外部定义。静态成员函数和实例成员函数的定义没有区别。如果我们想查看函数定义是实例函数还是静态函数，我们需要参考其声明。

```
int Rectangle :: getCount()
{
  return count;
}
```

注意，我们不能使用 const 限定符，因为没有宿主对象。

调用静态成员函数

静态成员函数可以通过实例或者类（例如，Rectangle）调用。要通过实例调用静态成员函数，我们使用与调用实例成员函数相同的语法；要通过类调用静态成员函数，我们使用类的名称和类解析运算符（：：）。下面代码片段显示了这两种调用方法。

```
rect.getCount ();              // 通过实例调用
Rectangle :: getCount();       // 通过类调用
```

警告：不能使用静态成员函数访问实例数据成员，因为静态成员函数没有隐藏的 this 指针，该指针定义了需要引用的实例。

静态成员函数不能用于访问实例数据成员，因为它没有 this 指针参数。

在另一方面，实例成员函数则可以访问静态数据成员（不使用 this 指针），但我们通常避免这样做。最佳编程实践方法是使用实例成员函数访问实例数据成员，使用静态成员函数访问静态数据成员。我们建议程序中实例成员的区域与静态成员的区域以符号方式分离，如图 7-13 所示。

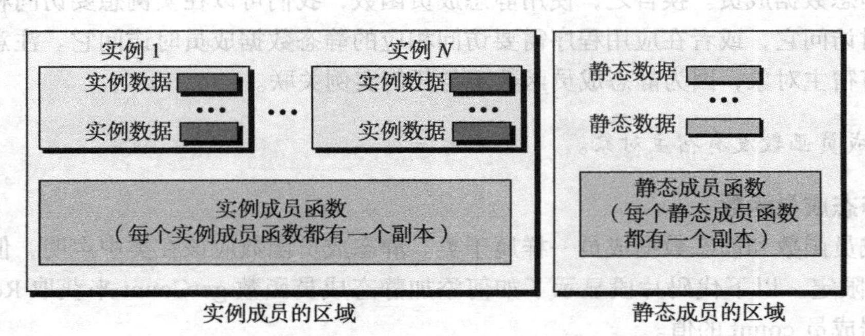

图 7-13 分离实例成员的区域与静态成员的区域

【例 7-4】 在本示例中，我们编写了一个程序，使用 Rectangle 类测试静态数据成员和相应的静态成员函数。程序 7-5 还显示了如何统计对象实例的个数。

程序 7-5 测试静态成员

```
1   /***************************************************************
2    * 本程序用于创建对象并统计对象个数                                  *
3    ***************************************************************/
4   #include <iostream>
5   using namespace std;
6   /***************************************************************
7    * Rectangle 类的定义                                             *
8    ***************************************************************/
9   class Rectangle
10  {
11    private:
12        double length;
13        double height;
14        static int count;   //静态数据成员
15    public:
16        Rectangle (double length, double height);
17        Rectangle ();
18        ~Rectangle ();
19        Rectangle (const Rectangle& rect);
20        static int getCount (); // 静态成员函数
21  };
```

```
22 // 初始化静态数据成员
23 int Rectangle :: count = 0;
24 /*****************************************************************
25  * 实例成员函数的定义                                              *
26  *****************************************************************/
27 // 参数构造函数的定义
28 Rectangle :: Rectangle (double len, double hgt)
29 : length (len), height (hgt)
30 {
31   count++;
32 }
33 // 默认构造函数的定义
34 Rectangle :: Rectangle ()
35 : length (0.0), height (0.0)
36 {
37   count++;
38 }
39 // 拷贝构造函数的定义
40 Rectangle :: Rectangle (const Rectangle& rect)
41 :length (rect.length), height (rect.height)
42 {
43   count++;
44 }
45 // 析构函数的定义
46 Rectangle :: ~Rectangle ()
47 {
48   count--;
49 }
50 /*****************************************************************
51  * 静态成员函数的定义                                              *
52  *****************************************************************/
53 int Rectangle :: getCount ()
54 {
55   return count;
56 }
57
58 /*****************************************************************
59  * 用于创建和统计 Rectangle 对象的应用程序                         *
60  *****************************************************************/
61 int main ( )
62 {
63   {
64       Rectangle rect1 (3.2, 1.2);
65       Rectangle rect2 (1.5, 2.1);
66       Rectangle rect3;
67       Rectangle rect4 (rect1);
68       Rectangle rect5 (rect2);
69       cout << "Count of objects: " << rect5.getCount() << endl;
70   }
71   cout << "Count of objects: " << Rectangle :: getCount();
72   return 0;
73 }
```

```
运行结果:
Count of objects: 5
Count of objects: 0
```

静态数据成员的声明在第 14 行中使用关键字 static 完成。该数据成员的初始化在类外部的第 23 行中完成。静态成员函数的定义在第 53 ～ 56 行中完成。我们在 main 函数中添加了一个嵌套块（第 63 ～ 70 行），以创建一个用于创建对象的作用域。这意味着在这个块之外没有对象实例，但是我们可以使用类名检查静态数据成员的值。在离开块之前，我们创建了五个对象实例，我们通过最后一个对象检查静态数据成员的值，发现其值是 5。在我们离开这个块之后，所有的实例都会被销毁，我们可以通过类名来检查静态数据成员的值，发现其值为 0。

【例 7-5】 在本例中，我们设计了一个表示银行账户的类。我们为这个类设计的两个实例数据成员分别为账号和账户余额。尽管在创建实例时，用户可以初始化账户余额，但用户不能输入账号，因为账号必须是唯一的。为了防止账号重复，我们使用了一个名为 base 的静态数据成员，将它初始化为 0，该成员随着新账户的开设而递增。我们把这个静态数据成员加上 100000，使其成为一个符合惯例的大数值。我们不需要静态成员函数，因为我们不需要检查静态数据成员的值。程序 7-6 显示了这个类及其应用。

程序 7-6 银行账户的示例

```
1  /***************************************************************
2   * 本程序用于声明、定义和使用一个银行账户类                      *
3   ***************************************************************/
4  #include <iostream>
5  #include <cassert>
6  using namespace std;
7
8  /***************************************************************
9   * 类定义（所有成员的声明）                                      *
10  ***************************************************************/
11 class Account
12 {
13    private:
14        long accNumber;
15        double balance;
16        static int base; // 静态数据成员
17    public:
18        Account (double bal); // 构造函数
19        ~Account (); // 析构函数
20        void checkBalance () const;                  // 访问器
21        void deposit (double amount);                // 更改器
22        void withdraw (double amount);               // 更改器
23 };
24 // 初始化静态数据成员
25 int Account :: base = 0;
26 /***************************************************************
27  * 所有成员函数的定义                                            *
28  ***************************************************************/
29 // 参数构造函数
```

```
30 Account :: Account (double bal)
31 :balance (bal)
32 {
33     if (bal < 0.0)
34     {
35         cout << "Balance is negative; program terminates";
36         assert (false);
37     }
38     base++;
39     accNumber = 100000 + base;
40 
41     cout << "Account " << accNumber << " is opened. " << endl;
42     cout << "Balance $" << balance << endl << endl;
43 }
44 // 析构函数
45 Account :: ~Account ()
46 {
47     cout << "Account #: " << accNumber << " is closed." << endl;
48     cout << "$" << balance << " was sent to the customer." << endl <<endl;
49 }
50 // 访问器成员函数
51 void Account :: checkBalance () const
52 {
53     cout << "Account #: " << accNumber << endl;
54     cout << "Transaction: balance check" << endl;
55     cout << "Balance: $" << balance << endl<< endl;
56 }
57 // 更改器成员函数
58 void Account :: deposit (double amount)
59 {
60     if (amount > 0.0)
61     {
62         balance += amount;
63         cout << "Account #: " << accNumber << endl;
64         cout << "Transaction: deposit of $" << amount << endl;
65         cout << "New balance: $" << balance << endl << endl;
66     }
67     else
68     {
69         cout << "Transaction aborted." << endl;
70     }
71 }
72 // 更改器成员函数
73 void Account :: withdraw (double amount)
74 {
75     if (amount > balance)
76     {
77         amount = balance;
78     }
79     balance -= amount;
80     cout << "Account #: " << accNumber << endl;
81     cout << "Transaction: withdraw of $" << amount << endl;
```

```
82    cout << "New balance: $" << balance << endl << endl;
83  }
84  /***************************************************************
85    * 使用 Account 类的应用程序（main 函数）                        *
86    ***************************************************************/
87  int main ( )
88  {
89    // 创建三个账户
90    Account acc1 (2000);
91    Account acc2 (5000);
92    Account acc3 (1000);
93    // 进行账户交易
94    acc1.deposit (150);
95    acc2.checkBalance ();
96    acc1.checkBalance ();
97    acc3.withdraw (800);
98    acc1.withdraw (1000);
99    acc2.deposit (120);
100   return 0;
101 }
```

运行结果：

```
Account 100001 is opened.
Balance $2000

Account 100002 is opened.
Balance $5000

Account 100003 is opened.
Balance $1000

Account #: 100001
Transaction: deposit of $150
New balance: $2150

Account #: 100002
Transaction: balance check
Balance: $5000

Account #: 100001
Transaction: balance check
Balance: $2150

Account #: 100003
Transaction: withdraw of $800
New balance: $200

Account #: 100001
Transaction: withdraw of $1000
New balance: $1150
```

```
Account #: 100002
Transaction: deposit of $120
New balance: $5120

Account #: 100003 is closed.
$200 was sent to the customer.

Account #: 100002 is closed.
$5120 was sent to the customer.

Account #: 100001 is closed.
$1150 was sent to the customer.
```

关于程序 7-6，我们需要解释几处要点。

- 虽然我们可以使用一个成员函数让用户关闭每个账户，但是当程序终止时，我们允许账户被自动关闭。在我们的设计中，当析构函数被调用时，账户被关闭，余额被发送给银行客户。
- 我们定义了一个参数构造函数，这意味着系统不会创建一个合成构造函数（由于账号的初始化，在这个类中不需要合成构造函数）。
- 由于我们没有定义拷贝构造函数，因此系统创建了一个合成的拷贝构造函数，应用程序可以调用它。但我们必须避免这种情况，因为我们不能有两个账号相同的账户。防止这种情况发生的最好方法是声明一个拷贝构造函数并将其赋值为关键字 delete，这样会自动阻止系统创建一个合成的拷贝构造函数，如下所示：

```
Account (const Account& acc) = delete;
```

然而，这个特性在 C++11 标准中才被引入，这意味着程序必须用 C++11 编译器编译。

- 我们在构造函数中使用了 assert 库函数（如前所述），以确保不会开设余额为负数的账户。

7.6　面向对象的程序设计

我们现在从面向过程的程序设计转向面向对象的程序设计。要做到这一点，我们需要更改前面几章中使用的一些过程。我们需要在编译和运行程序的方式中进行一些更改。我们将在以后的章节中使用本节中介绍的方法。

7.6.1　独立文件

如前所述，当 C++ 作为一种面向对象的程序设计语言时，通常有三个代码段：类定义、成员函数定义和应用程序。到目前为止，我们使用了一个包含这些部分的单独文件。稍后我们将讨论，C++ 允许我们创建三个单独的文件，分别包含这三个部分，如图 7-14 所示。

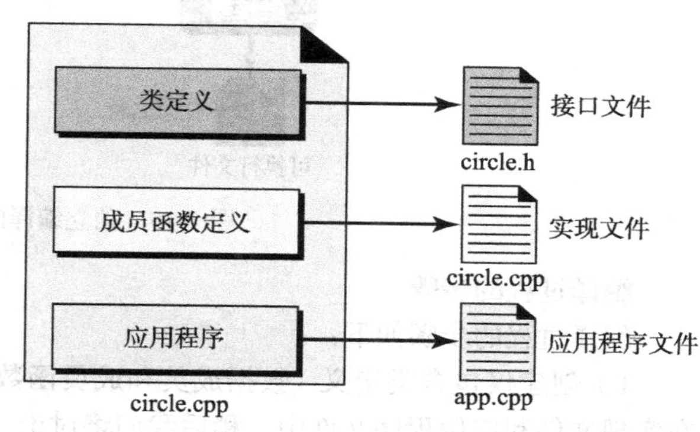

图 7-14　C++ 中为一个类创建三个独立的文件

接口文件

接口文件（interface file）是包含类定义（数据成员声明和成员函数声明）的文件。它给出了类的整体描述，以供其他文件使用，也就是说，接口文件定义了其他两个文件所使用的类型。此文件的名称通常是用类的名称加上扩展名 .h，例如 circle.h。字母 h 将其指定为头文件。

实现文件

实现文件（implementation file）包含成员函数的定义。它为接口文件中给出的所有成员函数声明提供实现代码。该文件的名称通常是用类的名称加上扩展名 .cpp，例如 circle.cpp，尽管在不同的 C++ 环境中其扩展名可能有所不同。

应用程序文件

应用程序文件（application file）包括用于实例化对象并让每个对象执行各自操作的 main 函数。同样，应用程序文件的扩展名也必须为 .cpp，但文件名通常由用户指定。我们使用的名称是 app.cpp，尽管在不同的 C++ 环境中其扩展名可能有所不同。

7.6.2　独立编译

在创建了三个独立的文件之后，我们必须编译这三个文件以创建一个可执行文件。在 C++ 中，这个过程被称为**独立编译**（separate compilation）(图 7-15)。

请注意，尽管每个操作系统的编译过程都相同，但编译命令的名称和所创建文件的名称在不同的环境中可能不同。

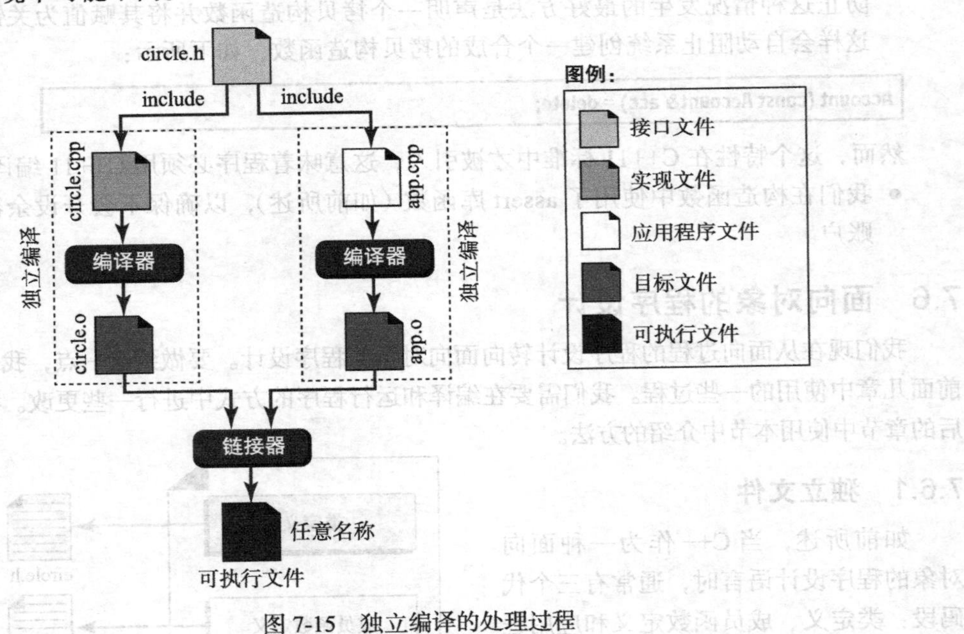

图 7-15　独立编译的处理过程

编译过程的步骤

编译过程的步骤如下：

1）创建仅包含类定义（数据成员和成员函数的声明）的接口文件。这个文件必须包含在实现文件和应用程序文件中，稍后我们将讨论。在我们的 Circle 类中，我们把这个文件命名为 circle.h。

2）创建实现文件（成员函数定义），通过 include 指令包含接口文件。选项- c 指示只进行编译。

```
c++ -c circle.cpp
```

如果编译成功，则会产生一个扩展名为 .o 的目标文件。在我们的示例中，生成的目标文件为 circle.o。

3）创建应用程序文件（main 函数），同样在文件的开始位置使用 include 指定包含接口文件。然后使用以下命令编译该文件。选项- c 指示只进行编译。

```
c++ -c app.cpp
```

如果编译成功，则会产生一个扩展名为 .o 的目标文件。在我们的示例中，生成的目标文件为 app.o。

4）然后使用选项 -o，链接两个目标文件以创建一个可执行文件。如下所示：

```
c++ -o application circle.o  app.o
```

5）结果是一个可执行文件，可以通过如下命令运行：

```
c++ application
```

示例

接下来我们将独立编译过程应用于我们的 Circle 类。从现在开始，我们将为所有设计的类执行相同的过程。

创建接口文件。首先创建包含类定义（数据成员和成员函数的声明）的接口文件，如程序 7-7 所示。此文件中只有三行新内容，它们分别是第 7、8 和 28 行。这三行代码可以防止编译中因其他文件也包含该接口文件而导致的重复包含该文件。我们将很快讨论这个问题。

程序 7-7　接口文件

```
1   /*****************************************************************
2    *这是一个接口文件，用于定义 Circle 类。                         *
3    *文件中包含了数据成员和成员函数的声明。                         *
4    *该文件将被包含在实现文件和                                     *
5    *应用程序文件的顶部                                             *
6    *****************************************************************/
7   #ifndef CIRCLE_H
8   #define CIRCLE_H
9   #include <iostream>
10  #include <cassert>
11  #include "circle.h"
12  using namespace std;
13  // 类定义
14  class Circle
15  {
16    private:
```

```
17          double radius;
18      public:
19          Circle (double radius);               // 参数构造函数
20          Circle ();                            // 默认构造函数
21          Circle (const Circle& circle);        // 拷贝构造函数
22          ~Circle ();                           // 析构函数
23          void setRadius (double radius);       // 更改器函数
24          double getRadius () const;            // 访问器函数
25          double getArea () const;              // 访问器函数
26          double getPerimeter () const;         // 访问器函数
27  };
28  #endif
```

创建实现文件。程序 7-8 为包含所有成员函数定义的实现文件。要独立编译这个文件，编译器必须看到数据成员和成员函数的声明。我们在第 6 行包含接口文件的副本。请注意，include 语句实际上会逐行复制所有声明代码，并在第 6 行之后插入这些声明代码。

程序 7-8 实现文件

```
1   /***************************************************************
2    * 这是实现文件。                                              *
3    * 用于定义所有的成员函数。                                    *
4    * 在文件的顶部包含了一个接口文件的副本，以允许编译该文件      *
5    ***************************************************************/
6   # include "circle.h"
7   /***************************************************************
8    * 带一个参数的参数构造函数。                                  *
9    * 使用给定值初始化一个 circle 对象实例。                      *
10   * 使用 assert 函数来验证半径 radius 是否为正的双精度值。      *
11   * 如果不是，则程序终止                                        *
12   ***************************************************************/
13  Circle :: Circle (double rds)
14  : radius (rds)
15  {
16      if (radius < 0.0)
17      {
18          assert (false);
19      }
20  }
21  /***************************************************************
22   * 默认构造函数，把一个 circle 对象初始化为 0.0。              *
23   * 不需要断言 assert                                           *
24   ***************************************************************/
25  Circle :: Circle ()
26  : radius (0.0)
27  {
28  }
29  /***************************************************************
30   * 拷贝构造函数。                                              *
31   * 把一个 circle 实例的半径拷贝到一个新的 circle 实例。        *
32   * 由于源 circle 实例已经被验证，故不需要再次验证              *
33   ***************************************************************/
34  Circle :: Circle (const Circle& circle)
```

```
35  : radius (circle.radius)
36  {
37  }
38  /**************************************************************
39   * 析构函数。                                                 *
40   * 当应用程序终止时，清理对象                                 *
41   **************************************************************/
42  Circle :: ~Circle ()
43  {
44  }
45  /**************************************************************
46   * setRadius 函数的定义。                                     *
47   * 用于通过减少或者增加半径大小来改变 circle 对象。           *
48   * 需要进行验证，因为新的半径大小必须为正数值                 *
49   **************************************************************/
50  void Circle :: setRadius (double value)
51  {
52    radius = value;
53    if (radius < 0.0)
54    {
55        assert (false);
56    }
57  }
58  /**************************************************************
59   * getRadius 函数的定义。                                     *
60   * 用于返回宿主对象的半径。                                   *
61   * 需要使用 const 修饰符，以防止意外改变宿主对象              *
62   **************************************************************/
63  double Circle :: getRadius () const
64  {
65    return radius;
66  }
67  /**************************************************************
68   * getArea 访问器函数的定义。                                 *
69   * 用于返回宿主对象的面积。                                   *
70   * 需要使用 const 修饰符，以防止意外改变宿主对象              *
71   **************************************************************/
72  double Circle :: getArea () const
73  {
74    const double PI = 3.14;
75    return (PI * radius * radius);
76  }
77  /**************************************************************
78   * getPerimeter 访问器函数的定义。                            *
79   * 用于返回宿主对象的周长。                                   *
80   * 需要使用 const 修饰符，以防止意外改变宿主对象              *
81   **************************************************************/
82  double Circle :: getPerimeter () const
83  {
84    const double PI = 3.14;
85    return (2 * PI * radius);
86  }
```

创建应用程序文件。程序7-9与我们以前看到的程序相同，但有一处例外：在开头使用include宏添加接口文件的内容。

程序7-9　应用程序文件

```
1   /****************************************************************
2    * 这是应用程序文件。                                          *
3    * 用于实例化对象，允许对象使用成员函数操作自己。              *
4    * 为了能够编译，需要包含一个接口文件的副本                    *
5    ****************************************************************/
6   # include "circle.h"
7   int main ( )
8   {
9     // 实例化第一个对象并执行操作
10    Circle circle1 (5.2);
11    cout << "Radius: " << circle1.getRadius() << endl;
12    cout << "Area: " << circle1.getArea() << endl;
13    cout << "Perimeter: " << circle1.getPerimeter() << endl;
14    cout << endl;
15    // 实例化第二个对象并执行操作
16    Circle circle2 (circle1);
17    cout << "Radius: " << circle2.getRadius() << endl;
18    cout << "Area: " <<  circle2.getArea() << endl;
19    cout << "Perimeter: " << circle2.getPerimeter() << endl;
20    cout << endl;
21    // 实例化第三个对象并执行操作
22    Circle circle3;
23    cout << "Radius: " << circle3.getRadius() << endl;
24    cout << "Area: " <<  circle3.getArea() << endl;
25    cout << "Perimeter: " << circle3.getPerimeter() << endl;
26    cout << endl;
27    return 0;
28  }
```

编译、链接和运行。下面的代码显示了编译、链接和运行过程。请注意，这些过程独立于创建其他文件的过程。

```
c++ -c circle.cpp                              // 编译实现文件
c++ -c app.cpp                                 // 编译应用程序文件
c++ -o application circle.o app.o              // 链接两个编译后的文件
application                                    // 运行可执行文件
```

```
Radius: 5.2
Area: 84.9056
Perimeter: 32.656

Radius: 5.2
Area: 84.9056
Perimeter: 32.656

Radius: 0
Area: 0
Perimeter: 0
```

7.6.3　防止多重包含

如果在编译文件中多次包含同一个头文件的内容，编译器将发生错误，编译将中止。为了防止这种情况的发生，我们使用以下预处理器指令：define、ifndef（如果未定义）和endif。在本节中，我们将学习如何使用这三个预处理器指令来忽略头文件的重复包含。在将被包含在另一个文件中的文件（如接口文件）中，我们添加了这三条指令，如图 7-16 所示。

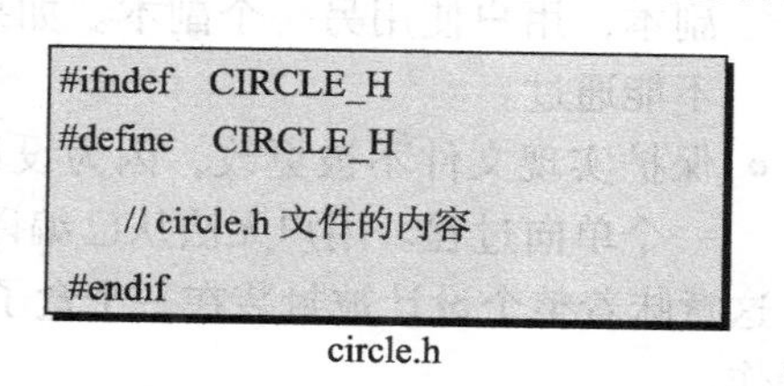

图 7-16　添加了三条指令的头文件的内容

这三条指令作用于一个标志（常量）。我们在图中使用的标志是文件名，然后是下划线再跟字母 H，全部为大写。这是一种约定，只要保持一致，也可以使用任何其他名称。指令 ifndef 遵循 if 语句的语法规则：如果没有定义该指令常量，则包含 ifndef 的主体。如果已经定义了该标志，则预处理将忽略其余代码并跳转到 endif 指令。

现在让我们讨论当程序中多次包含 circle.h 文件或者包含另一个使用 include 指令的文件时的情况。我们在文件中得到了 circle.h 内容的两个副本，如图 7-17 所示。

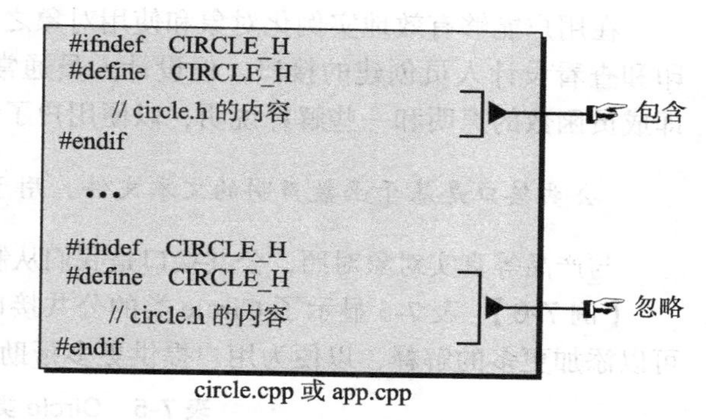

图 7-17　条件指令如何忽略重复包含

当预处理器遇到第一个 ifndef 指令时，由于尚未定义该标志，预处理器将定义该标志（下一行代码），并添加其余代码，直到 endif 指令，这意味着头文件的内容将被添加到源文件中。当预处理器遇到第二个 ifndef 指令时，因为已经定义了该标志，所以预处理器会立即跳到 endif 指令，不再包含头文件的内容。这就是我们希望的结果。你可能想知道一个文件是如何被包含两次的。通常情况下，将两个具有相同 include 文件的不同文件添加到第三个文件时会发生这种情况。

7.6.4　封装

读者可能会提出疑问：为什么需要独立编译？独立编译允许我们实现面向对象程序设计的目标之一：封装。面向对象程序设计中的**封装**（encapsulation）允许我们把类的设计和类的使用区分开来。

类的设计

设计人员创建接口文件和实现文件。设计人员公开接口文件。实现文件被编译，但只有已编译的版本是公开的，源代码仍然是私有的。设计人员可以随时更改实现文件，重新编译，并重新进行发布。

类的使用

用户收到接口文件和已编译版本的实现文件的副本。用户将接口文件添加到应用程序文件中并进行编译。然后，把自己的编译文件和从设计人员处收到的编译文件链接在一起，以创建一个可执行文件。

封装的效果

封装的效果是设计人员可以同时保护接口文件和实现文件不受用户更改的影响，具体如下所示：

- 保护接口文件不被更改，因为在此过程中使用了它的两个副本：设计人员使用一个副本，用户使用另一个副本。如果用户更改了收到的公开副本，则独立编译过程将不能通过。
- 保护实现文件不被更改，因为设计人员将实现文件的编译版本发送给用户。编译是一个单向过程。用户无法从已编译的文件中获取原始文件以进行更改。

这意味着整个设计被封装在一个盒子中，用户不能更改。用户只能创建实例，让实例自己操作。

公共接口

在用户能够有效地实例化对象和使用对象之前，还需要执行一步操作。虽然用户可以打印和查看设计人员创建的接口，但设计人员通常会创建所谓的**公共接口**（public interface），即成员函数的声明和一些解释说明，以便用户了解如何在其应用程序中调用这些成员函数。

> 公共接口是基于函数声明的文本文件，用于告知用户如何使用成员函数。

与产品等真实对象对照，公共接口是我们从制造商处收到的手册，以说明如何使用产品。

【例 7-6】 表 7-5 显示了 Circle 类的公共接口示例（迄今为止开发的）。如果需要，我们可以添加更多的解释，以便为用户提供更多帮助。

表 7-5　Circle 类的公共接口

构造函数和析构函数
Circle :: Circle () 默认构造函数，使用 length = 0.0 和 height = 0.0 创建一个 Circle 对象
Circle :: Circle (double radius) 参数构造函数，使用给定半径 radius 创建一个 Circle 对象
Circle :: Circle (const Circle& circle) 拷贝构造函数，使用一个现有的 Circle 对象创建一个同样的 Circle 对象
Circle :: ~Circle () 析构函数，用于当对象超出作用域时进行清理工作
访问器函数
Circle :: double getRadius () const 访问器函数，用于返回宿主对象的半径
Circle :: double getArea () const 访问器函数，用于返回宿主对象的面积
Circle:: double getPerimeter () const 访问器函数，用于返回宿主对象的周长
更改器函数
Circle:: void setRadius (double radius) 更改器函数，用于修改宿主对象的半径

7.7　设计类

在本节中，我们使用类创建两种新类型：一种表示分数，另一种表示时间。

7.7.1 表示分数的类

分数（fraction），也称为有理数（rational number），是两个整数的比值，如3/4、1/2、7/5等。C++中没有可以表示分数的内置类型，我们需要使用两个整数类型的数据成员为分数创建一个新的类型。我们称第一个为分子numer（numerator的缩写），第二个为分母denom（denominator的缩写）。

我们创建了三个构造函数、一个析构函数、若干访问器函数和若干更改器函数。我们添加了一个名为normalize的私有成员函数，用于处理类不变式。我们还添加了一个名为gcd的私有成员函数，用于查找最大公约数（具体请参见第5章中的讨论）。此函数由normalize函数调用。

注意，我们同时扮演设计师和用户的角色，但在现实生活中，这两个角色通常是由不同的人员担当。

类不变式

在定义一个新类时，我们首先要考虑的是类不变式，正如我们在本章前面讨论的那样。分数对象中与我们考虑的类不变式是如下三个条件：

- 分母不能为0。类似于2/0的分数没有定义。
- 分数的符号由分子符号和分母符号共同决定，并且应设置为分子的符号。
- 分子和分母不应该具有公因数。例如，6/9应该化简为2/3。

我们创建两个私有成员函数来处理类不变式。gcd函数用于查找分子和分母之间的最大公约数。normalize函数使用gcd函数处理三个类不变式。

接口文件

接下来创建接口文件fraction.h。接口文件是类的定义。尽管私有数据成员和私有成员函数（如果有的话）的声明不是公共接口的一部分，但是它们必须包含在这个文件中，因为当这个文件被包含在其他两个编译文件中时，我们需要这些私有数据成员和私有成员函数。

数据成员。我们的分数对象只有两个数据成员：分子，我们称之为numer；分母，我们称之为denom。这两个数据成员都可以是int类型，因为我们通常不使用分子或者分母非常大的分数。

成员函数。我们有许多可以应用于数据成员的成员函数。我们还可以有对从类中创建的多个对象进行操作的成员函数，例如比较两个分数或者计算两个分数之和。我们把第二类成员函数推迟到下一章讨论。

代码。程序7-10显示了接口文件。注意，实例成员函数都是公共的，可以通过类实例访问，但是我们还有两个用来处理类不变式的辅助函数（normalize和gcd）。这两个辅助函数被声明为私有的，因此只能通过类的参数构造函数访问它们。

程序7-10　fraction.h文件

```
1  /***************************************************************
2   * 定义 Fraction 类的接口文件 fraction.h                          *
3   ***************************************************************/
4  #include <iostream>
5  using namespace std;
6
7  #ifndef FRACTION_H
8  #define FRACTION_H
```

```
9
10  class Fraction
11  {
12      // 数据成员
13      private:
14          int numer;
15          int denom;
16
17      // 公共成员函数
18      public:
19          // 构造函数
20          Fraction (int num, int den);
21          Fraction ();
22          Fraction (const Fraction& fract);
23          ~Fraction ();
24          // 访问器函数
25          int getNumer () const;
26          int getDenom () const;
27          void print () const;
28          // 更改器函数
29          void setNumer (int num);
30          void setDenom (int den);
31
32      // 辅助私有成员函数
33      private:
34          void normalize ();
35          int gcd (int n, int m);
36  };
37  #endif
```

实现文件

接下来我们为接口文件中定义的所有成员函数（公共函数和私有函数）编写函数定义。注意，我们在这个程序中添加了接口文件作为头文件（编译器在编译定义之前需要查看声明）。程序7-11是实现文件的源代码清单。

- 我们在第7行中添加了 <cassert> 头文件，以允许在第96行中使用 assert 宏，如果分母为0，则需要终止程序。
- 第99～103行处理类不变式的第二个条件。如果分母为负数，我们同时更改分子和分母的符号，以将负号移动给分子。
- 第105～107行处理类不变式的第三个条件。我们使用最大公约数函数（参见前面章节的讨论）查找分子和分母之间的最大除数，然后用这个除数整除分子和分母。注意，计算最大公约数时，我们使用分子和分母的绝对值。

程序7-11　fraction.cpp文件

```
1  /***************************************************************
2   * 实现文件 fraction.cpp。                                      *
3   * 定义 Fraction 类的实例成员函数和辅助函数                      *
4   ***************************************************************/
5  #include <iostream>
6  #include <cmath>
```

```
7   #include <cassert>
8   #include "fraction.h"
9   using namespace std;
10
11  /*************************************************************
12   * 参数构造函数。                                                *
13   * 获取分子和分母的值，初始化对象，                                  *
14   * 根据类不变式中定义的条件，                                      *
15   * 规范化分子和分母的值                                           *
16   *************************************************************/
17  Fraction :: Fraction (int num, int den = 1)
18  : numer (num), denom (den)
19  {
20    normalize ();
21  }
22  /*************************************************************
23   * 默认构造函数。                                                *
24   * 创建一个分数实例：0/1，不需要验证                                *
25   *************************************************************/
26  Fraction :: Fraction ( )
27  : numer (0), denom (1)
28  {
29  }
30  /*************************************************************
31   * 拷贝构造函数。                                                *
32   * 根据一个现有的分数对象创建一个新的分数对象。                        *
33   * 不需要规范化，因为源对象已经规范化                                *
34   *************************************************************/
35  Fraction :: Fraction (const Fraction& fract )
36  : numer (fract.numer), denom (fract.denom)
37  {
38  }
39  /*************************************************************
40   * 析构函数，仅用于清理分数对象进行回收                              *
41   *************************************************************/
42  Fraction :: ~Fraction ()
43  {
44  }
45  /*************************************************************
46   * getNumer 函数是一个访问器函数，                                 *
47   * 返回宿主对象的分子。需要 const 修饰符                             *
48   *************************************************************/
49  int Fraction :: getNumer () const
50  {
51    return numer;
52  }
53  /*************************************************************
54   * getDenom 函数是一个访问器函数，                                 *
55   * 返回宿主对象的分母。需要 const 修饰符                             *
56   *************************************************************/
57  int Fraction :: getDenom () const
58  {
```

```
59    return denom;
60  }
61  /*********************************************************************
62   * print 函数是一个具有副作用的访问器函数，                                *
63   * 用于以 x/y 格式显示分数对象                                            *
64   *********************************************************************/
65  void Fraction :: print () const
66  {
67    cout << numer << "/" << denom << endl;
68  }
69  /*********************************************************************
70   * setNumer 函数是一个更改器函数，                                         *
71   * 用于修改一个已有的分数对象的分子。分数对象需要规范化                     *
72   *********************************************************************/
73  void Fraction :: setNumer (int num)
74  {
75    numer = num;
76    normalize();
77  }
78  /*********************************************************************
79   * setDenom 函数是一个更改器函数，                                         *
80   * 用于修改一个已有的分数对象的分母。分数对象需要规范化                     *
81   *********************************************************************/
82  void Fraction :: setDenom (int den)
83  {
84    denom = den;
85    normalize();
86  }
87  /*********************************************************************
88   *规范化函数，实现三个分数类不变式                                        *
89   *********************************************************************/
90  void Fraction :: normalize ()
91  {
92    // 处理分母为 0 的情况
93    if (denom == 0)
94    {
95         cout << "Invalid denomination. Need to quit." << endl;
96         assert (false);
97    }
98    //改变分子的正负号
99    if (denom < 0)
100   {
101        denom = - denom;
102        numer = - numer;
103   }
104   //使用最大公约数整除分子和分母
105   int divisor = gcd (abs(numer), abs (denom));
106   numer = numer / divisor;
107   denom = denom / divisor;
108 }
109 /*********************************************************************
110  * 最大公约数函数 gcd，                                                    *
```

```
111    * 查找分子和分母之间的最大公约数                                    *
112    ***************************************************************/
113 int Fraction :: gcd (int n, int m)
114 {
115     int gcd = 1;
116     for (int k = 1; k <= n && k <= m; k++)
117     {
118         if (n % k == 0 && m % k == 0)
119         {
120             gcd = k;
121         }
122     }
123     return gcd;
124 }
```

应用程序文件

应用程序文件由用户创建。程序 7-12 是一个演示示例。用户只需要包含接口文件的副本（第 4 行）。我们运行了一次程序，为分母输入了一个零，程序被中止（此处未显示）。

程序 7-12　应用程序文件（app.cpp）

```
1  /***************************************************************
2   * 应用程序文件（app.cpp），使用 Fraction 对象                        *
3   ***************************************************************/
4  #include "fraction.h"
5  #include <iostream>
6  using namespace std;
7
8  int main ( )
9  {
10     // 实例化一些对象
11     Fraction fract1 ;
12     Fraction fract2 (14, 21);
13     Fraction fract3 (11, -8);
14     Fraction fract4 (fract3);
15     // 打印对象
16     cout << "Printing four fractions after constructed: " << endl;
17     cout << "fract1: ";
18     fract1. print();
19     cout << "fract2: ";
20     fract2. print();
21     cout << "fract3: ";
22     fract3. print();
23     cout << "fract4: ";
24     fract4. print();
25     // 使用更改器函数
26     cout << "Changing the first two fractions and printing them:" << endl;
27     fract1.setNumer(4);
28     cout << "fract1: ";
29     fract1.print();
30     fract2.setDenom(-5);
31     cout << "fract2: ";
```

```
32  fract2.print();
33  //使用访问器函数
34  cout << "Testing the changes in two fractions:" << endl;
35  cout << "fract1 numerator: " << fract1.getNumer() << endl;
36  cout << "fract2 numerator: " << fract2.getDenom() << endl;
37  return 0;
38 }
```

编译、链接和运行

以下是编译、链接和运行过程。请注意，此过程独立于创建其他文件的过程。

```
c++ -c fraction.cpp                          // 编译实现文件
c++ -c app.cpp                               // 编译应用程序文件
c++ -o application fraction.o app.o          // 链接两个编译后的文件
application                                  // 运行可执行文件
```

```
运行结果：
Printing four fractions after constructed:
fract1: 0/1
fract2: 2/3
fract3: -11/8
fract4: -11/8
Changing the first two fractions and printing them:
fract1: 4/1
fract2:-2/5
Testing the changes in two fractions:
Numerator of fract1: 4
Denominator of fract2: 5
```

第二个对象被规范化：14/21 被化简为 2/3。第三个对象也被规范化：11/-8 被更改为 -11/8。第一个对象的分子被更改并重新打印为 4/1。第二个对象的分母被更改为 -5，但它被规范化并打印为 -2/5。最后，我们使用访问器函数来获取第一个对象的分子和第二个对象的分母。

7.7.2　表示时间的类

接下来，我们设计一个表示时间的类。如同实际生活中使用的时间，该类有三个数据成员：小时（hours）、分钟（minutes）和秒（seconds）。我们有一个参数构造函数和一个析构函数，但不需要拷贝构造函数（显示同一个时间的两个对象没有实际意义）。我们有三个访问器函数，分别用于获取小时、分钟和秒。我们只使用一个名为 tick 的更改器函数，每次调用该函数时，时间对象向前移动 1 秒钟。

类不变式

在为这个类编写代码之前，我们必须考虑类的不变式。我们必须关注以下两个条件：

- 所有三个数据成员都必须是非负的，否则，程序将中止。因为时间不能为负数。
- 小时应该在 0 到 23 之间（我们假设为 24 小时制时间），分钟应该在 0 到 59 之间，秒也应该在 0 到 59 之间。我们使用模运算来确保这些值在范围内。如果秒的值大于 59，则必须从中提取相应的分钟数，并将其添加到分钟数中。针对分钟也执行同样的操作。针对小时的情况，则将提取的值丢弃。

我们将使用名为 normalize 的私有成员函数实现类不变式检查。

接口文件

接下来创建接口文件：time.h。接口文件是类的定义。

数据成员。我们需要三个私有数据成员：hours、minutes 和 seconds。它们的数据类型均为 int。

成员函数。有许多成员函数不仅可以应用于数据成员，还可以应用于从类创建的对象。例如，我们可以比较两个时间对象，确定两个时间之间经过的时间差，等等。我们将在后续章节中定义这些操作。在本章中，我们使用成员函数作为构造函数、析构函数、数据成员的访问器函数和数据成员的更改器函数。

代码。程序 7-13 为接口文件的代码清单。这只是一个头文件，将被同时包含在应用程序文件和实现文件中。

程序 7-13　接口文件（time.h）

```
1   /***************************************************************
2    * 定义 Time 类的接口文件 time.h                                *
3    ***************************************************************/
4   #include <iostream>
5   using namespace std;
6   
7   #ifndef TIME_H
8   #define TIME_H
9   
10  class Time
11  {
12    private:
13          int hours;
14          int minutes;
15          int seconds;
16    public:
17          Time (int hours, int minutes, int seconds);
18          Time ();
19          ~Time ();
20          void print() const;
21          void tick();
22  
23    private:
24    void normalize ();                    // 辅助函数
25  };
26  #endif
```

实现文件

接下来我们为接口文件中定义的所有成员函数（公共函数和私有函数）编写函数定义。注意，我们在这个程序中添加了接口文件作为头文件。程序 7-14 是实现文件的源代码清单。

程序 7-14　实现文件（time.cpp）

```
1   /***************************************************************
2    * Time 类的函数的实现文件 time.cpp                             *
3    ***************************************************************/
4   #include <cmath>
5   #include <cassert>
```

```
6   #include "time.h"
7
8   /*******************************************************************
9    * 参数构造函数。                                                  *
10   * 从用户接收对应于三个数据成员的值，                              *
11   * 然后初始化对象。                                                *
12   * 使用规范化函数确保小时、分钟和秒位于指定范围                    *
13   *******************************************************************/
14  Time :: Time (int hr, int mi, int se )
15  : hours (hr), minutes (mi), seconds (se)
16  {
17    normalize ();
18  }
19  /*******************************************************************
20   * 默认构造函数，创建一个时间对象                                  *
21   *******************************************************************/
22  Time :: Time ( )
23  : hours (0), minutes (0), seconds (0)
24  {
25  }
26  /*******************************************************************
27   * 析构函数，仅用于清理时间对象进行回收                            *
28   *******************************************************************/
29  Time :: ~Time ()
30  {
31  }
32  /*******************************************************************
33   * print 函数是一个访问器函数，                                    *
34   * 具有副作用：显示时间                                            *
35   *******************************************************************/
36  void Time :: print () const
37  {
38    cout << hours << ":" << minutes << ":" << seconds << endl;
39  }
40  /*******************************************************************
41   * tick 函数是一个更改器函数，                                     *
42   * 用于递增秒数                                                    *
43   *******************************************************************/
44  void Time :: tick ()
45  {
46    seconds++;
47    normalize();
48  }
49  /*******************************************************************
50   * 规范化函数，检查类的不变式。                                    *
51   * 它要么终止程序，                                                *
52   * 要么规范化时、分、秒                                            *
53   *******************************************************************/
54  void Time :: normalize ()
55  {
56    // 处理负数的数据成员
57    if ((hours < 0) || (minutes < 0) || (seconds < 0))
```

```
58      {
59          cout << "Data are not valid. Need to quit!" << endl;
60          assert (false);
61      }
62      // 处理越界的值
63      if (seconds > 59)
64      {
65          int temp = seconds / 60;
66          seconds = seconds % 60;
67          minutes = minutes + temp;
68      }
69      if (minutes > 59)
70      {
71          int temp = minutes / 60;
72          minutes = minutes % 60;
73          hours = hours + temp;
74      }
75      if (hours > 23)
76      {
77          hours = hours % 24;
78      }
79  }
```

请注意关于程序的以下要点：

- 程序中包含了 <cassert> 头文件，以便我们可以使用 assert 宏，从而防止时、分、秒为负值。
- 在规范化函数（normalize）中，我们使用了三条 if 语句，以处理值越界的情况。

应用程序文件

应用程序文件由用户创建。程序 7-15 是一个演示示例。用户只需要包含接口文件的副本。

程序 7-15　应用程序文件（app.cpp），用于测试 Time 类

```
1   /***************************************************************
2    * 应用程序文件（app.cpp），使用 Time 对象                        *
3    ***************************************************************/
4   #include "time.h"
5
6   int main ( )
7   {
8       // 实例化一个时间对象
9       Time time (4, 5, 27);
10      // 打印初始时间
11      cout << "Original time: " ;
12      time.print();
13      // 向初始时间上增加 143500 秒
14      for (int i = 0; i < 143500; i++)
15      {
16          time.tick ();
17      }
18      // 打印增加了 143500 秒后的时间
19      cout << "Time after 143500 ticks " ;
20      time.print();
```

```
21    return 0;
22 }
```

编译、链接和运行

以下是编译、链接和运行过程。

```
c++ -c time.cpp                                      // 编译实现文件
c++ -c app.cpp                                       // 编译应用程序文件
c++ -o application time.o app.o                      // 链接两个编译后的文件
application                                          // 运行可执行文件
```

```
运行结果：
Original time: 4:5:27
Time after 143500 ticks: 19:57:7
```

注意，在滴答了143500秒之后，一天过去了，我们移动到19点57分7秒。我们可以很容易地验证这一点。

本章小结

面向对象的概念包括类型和实例、属性和行为、数据成员和成员函数。类型是从中创建实例（具体实体）的概念（抽象）。属性是我们感兴趣的对象的特征，例如名称和地址。行为是一个对象的操作。对象的数据成员是表示属性的变量。成员函数则表示对象的一种行为。

类定义由保留字class、类的名称和类的主体组成，类的主体是一个包含数据成员和成员函数声明的语句块。确定数据成员或者成员函数的可见性的修饰符有三种：private、protected和public。只能通过成员函数访问私有数据成员。另外，成员函数通常是公共的，可以通过实例访问。构造函数和析构函数是构造或者销毁类实例的特殊成员函数。对象的状态是存储在其数据成员中的值的组合。

在面向对象的程序设计中，实例化一个对象是通过一个被称为构造函数的特殊成员函数来完成的，而对象清理工作则是由一个被称为析构函数的特殊成员函数来完成的。

实例数据成员定义每个实例的属性，这意味着每个实例都需要封装所有实例数据成员。实例成员函数定义实例的一种行为。

静态数据成员属于所有实例和类。它不能由构造函数初始化，它需要在程序的全局区域中初始化。

在面向对象程序设计中，我们将程序的三个部分拆分成三个独立的文件：接口文件、实现文件和应用程序文件。接口文件包含在实现文件和应用程序文件的开头部分。实现文件和应用程序文件分别编译然后链接在一起。独立编译允许我们向用户公开类的接口，但对用户隐藏实现。这个原理叫作封装。

思考题

1. 以下各项哪个是参数构造函数的声明？哪个是默认构造函数的声明？哪个是拷贝构造函数的声明？

```
Fun ();
Fun (int x);
Fun (const Fun& fun);
```

2. 以下关于Rectangle类的构造函数的声明有什么错误？

```
int Rectangle (int length, int height);
```

3. 以下关于 Rectangle 类的析构函数的声明有什么错误？

```
int ~Rectangle ();
```

4. 修改以下构造函数的定义，使用初始化列表来替代赋值语句。

```
Rectangle :: Rectangle (int len, int wid)
{
    length = len;
    height = wid;
}
```

5. 给定以下类定义：

```
class Sample
{
    private:
        int x;
    public:
        int getX () const;
};
```

指出以下构造函数调用中的错误（如果有的话）。

```
Sample first (4);
Sample second ( );
Sample third;
```

6. 给定以下类定义：

```
class Sample
{
    private:
        int x;
    public:
        Sample (int x);
        int getX () const;
};
```

指出以下实例化中的错误。

```
Sample first (4);
Sample second (4, 8);
Sample third;
```

7. 指出以下类定义中的错误之处。

```
class First
    private:
        double x;
        double y;
    public:
        double getX () const;
        double getY () const;
```

8. 指出以下类定义中的错误之处。

```
class Second
```

```
{
    private:
        double x;
        double y;
    public:
        bool Second (int x, int y);
        double getX () const;
        double getY () const;
};
```

9. 指出以下类定义中的错误之处。

```
class Third
{
    private:
        int x;
        int y;
    public:
        ~Third (int z);
        int getX () const;
        int getY () const;
};
```

10. 如果 x 是类的数据成员，指出以下类的成员函数定义的错误之处。

```
Fun :: int getX ( ) const
{
    return x;
}
```

11. 以下代码片段中，哪个是正确的类 Sample 的访问器函数定义？

```
type Sample :: getValue ( ) const
{
    return . . . ;
}
```

```
type Sample :: getValue ( )
{
    return . . . ;
}
```

12. 以下代码片段中，哪个是正确的类 Sample 的更改器函数定义？

```
void Sample :: setValue ( ) const
{
    . . . ;
}
```

```
void Sample :: setValue ( )
{
    . . . ;
}
```

13. 假设我们有以下类定义。请编写两个成员函数的定义。

```
class Fun
{
    private:
        int x;
    pubic:
        Fun (int);
        Fun (const Fun&);
};
```

14. 演示应用程序如何使用在第 13 题中定义的参数构造函数和拷贝构造函数。

编程题

1. 创建一个名为 One 的类，该类包含两个整数类型的数据成员 x 和 y，以及两个成员函数 getX 和 getY。

定义接口文件、实现文件和应用程序文件。应用程序文件从类中实例化一个对象并打印 x 和 y 的值。

2. 创建一个名为 Two 的类，该类包括一个名为 x 的整数类型的数据成员和一个名为 a 的字符类型的数据成员。定义四个成员函数 getX、getA、setX 和 setA。定义接口文件、实现文件和应用程序文件。应用程序文件从类中实例化一个对象并打印数据成员的值。然后通过更改器函数设置数据成员的值，并再次打印它们的值。
3. 平面笛卡尔坐标中的点通常用两个整数值（x 和 y）定义。用两个数据成员定义一个名为 Point 的类。定义一个打印函数，返回点对象的坐标。定义函数来确定一个点在另一个点的左侧还是右侧、上方还是下方。定义一个函数来计算两点之间的距离，计算公式如下所示：

$$distance = sqrt ((x_2-x_1)^2+(y_2-y_1)^2)$$

4. 为具有以下成员的名为 Person 的类编写接口文件、实现文件和应用程序文件：
   1）数据成员为 name 和 age。
   2）访问器成员函数为 getName 和 getAge。
   3）更改器成员函数为 setName 和 setAge。
   4）包含一个参数构造函数和一个析构函数。
5. 定义一个名为 Triangle 的类，满足以下要求：
   1）数据成员为 firstSide、secondSide 和 thirdSide。
   2）使用构造函数确保任意两边之和大于第三边。
   3）访问器成员函数为 getSides、getPerimeter 和 getArea。计算三角形的周长和面积的公式如下：

$$perimeter = a + b + c$$
$$area = sqrt ((p) * (p-a)*(p-b)*(p-c)) \quad // \ p = perimeter / 2$$

   4）为类定义一个构造函数。
6. 定义一个名为 Address 的类，满足以下要求：
   1）数据成员为 houseNo、streetName、cityName、stateName 和 zipcode。
   2）定义一个参数构造函数和一个析构函数。
   3）定义一个访问器成员函数，用于打印地址。
7. 定义一个名为 Employee 的类，满足以下要求：
   1）数据成员为 name、age、serviceYear 和 salary。
   2）定义一个参数构造函数和一个析构函数。
   3）访问器成员函数包括：getName、getAge、getServiceYear 和 getSalary。
8. 在面向对象程序设计中，我们可以创建一个用于求解数学方程的类。我们在代数中经常需要求解的方程是一元二次方程，如下所示：

$$ax^2 + bx + c = 0$$

该方程的根为：

$$x_1=-b+sqrt(b^2-4*a*c) \qquad x_2=-b-sqrt(b^2-4*a*c)$$

括号内的部分被称为判别式。如果判别式的值为正，则该方程有两个根。如果为零，则该方程有一个根。如果为负，则该方程没有根。创建一个名为 Quadratic 的类，在给定系数 a、b 和 c 时，求解一元二次方程的根。

9. 我们在第 4 章中编写了一个程序，使用 Zeller 同余算法来判断任何给定日期是星期几。

```
weekday=(day+26*(month+1)/10+
        year+year/4-year/100+year/400)%7
```

为了证明用面向过程的程序设计范式编写的程序都可以在面向对象的程序设计范式中实现，设计一个类 Zeller，该类包含三个数据成员，分别为 day、month 和 year，判断给定日期是星期几（Monday 到 Sunday，星期一到星期日）。

10. 设计一个表示复数的 Complex 类。数学中的复数定义为 $x + iy$，其中 x 定义为复数的实部，y 定义为复数的虚部。字母 i 代表 -1 的平方根（也就是说 i^2 是 -1）。Complex 类包括以下函数：将一个复数与宿主对象相加、从宿主对象中减去一个复数、将一个复数乘以宿主对象以及将一个复数除以宿主对象。

$$(x_1+iy_1)+(x_2+iy_2)=(x_1+x_2)+i(y_1+y_2)$$

$$(x_1+iy_1)-(x_2+iy_2)=(x_1-x_2)+i(y_1-y_1)$$

$$(x_1+iy_1)*(x_2+iy_2)=(x_1x_2-y_1y_2)+i(x_1y_2+x_2y_1)$$

$$(x_1+iy_1)/(x_2+iy_2)=((x_1x_2+y_1y_2)+i-x_1y_2+x_2y_1))/\ \text{denominator}$$

其中 denominator = $x_2^2 + y_2^2$

数　　组

在讨论了大多数基本的内置数据类型和用户自定义的类型之后，接下来讨论第一个复合数据类型：数组。我们将在下一章中讨论其他复合数据类型。

学习目标

阅读并学习本章后，读者应该能够

- 介绍一维数组，一维数组是元素序列，其中所有元素都必须是同一类型（基本类型或者用户自定义类型）。
- 讨论一维数组的三个属性：类型、容量和大小。
- 演示如何声明和初始化一个一维数组。
- 讨论如何使用下标表达式访问一维数组的元素。
- 讨论一维数组的非修改性操作和修改性操作。
- 讨论如何使用函数对一维数组进行非修改性操作和修改性操作。
- 讨论并行一维数组及其应用。
- 讨论二维数组以及如何声明和初始化二维数组。
- 讨论如何使用二维数组创建另一个二维数组。
- 演示如何使用线性操作或者之字形操作将二维数组更改为一维数组。

8.1　一维数组

一维数组（one-dimensional array）是具有相同内置或者用户自定义类型的数据项序列。复合类型是包含两种类型的结构：数组是第一种类型，数组中的数据类型是第二种类型。

> 数组是一种复合类型，它定义同一类型的数据项序列。

8.1.1　数组属性

数组有三个属性：类型、容量和大小，如图 8-1 所示。图中显示了三个数组，每个数组具有不同的属性。接下来我们讨论每个属性。

类型

数组的类型是数组中数据项（元素）的类型。例如，我们可以有一个整数（int）数组、一个双精度（double）数组、一个字符（char）数组和一个圆（Circle）数组。

> 数组中所有数据项的类型必须相同，数组类型是数组中元素的类型。

容量

数组的容量是它可以容纳的最大元素个数。此属性是一个字面量值或者常量值，在数组声明后就不能更改容量。我们通常使用大写字母作为容量的名称。

声明数组后，其容量不能更改。

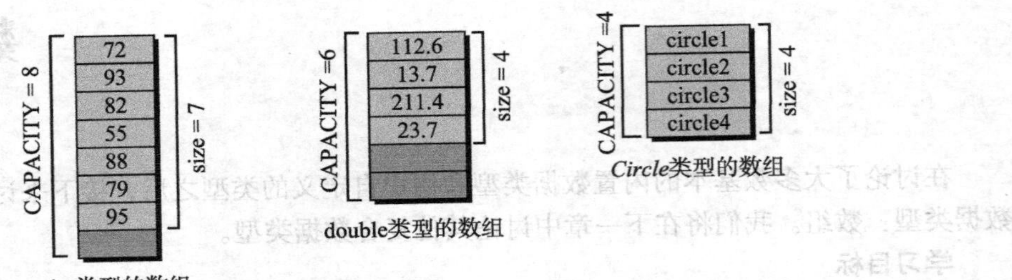

注意：
我们使用大写的CAPACITY表示容量，因为容量是常量或者字面量。
我们用小写的size表示大小，因为大小是一个变量。
深灰色区域是数组的一部分，表示此时还未被占用。

图 8-1　数组的属性

大小

数组的大小定义了每一时刻有多少元素是有效的，即有多少元素包含有效的数据。我们可以创建一个容量为 10 的数组，但在某一时刻我们可能只有 3 个有效元素，在另一时刻，我们可能有 8 个有效元素。换言之，大小是数组的控制属性。

数组大小定义了每一时刻有效元素的数目。

8.1.2　声明、分配和初始化

数组在使用前必须先声明。编译器在数组声明时为其分配内存位置。数组的元素也可以在数组声明时进行初始化。

数组声明

数组声明为数组指定名称、设置元素类型、设置数组的容量以及为数组分配内存位置。换言之，数组的声明也是一个定义。图 8-2 显示了数组声明的语法，并将其与变量的声明进行了比较。

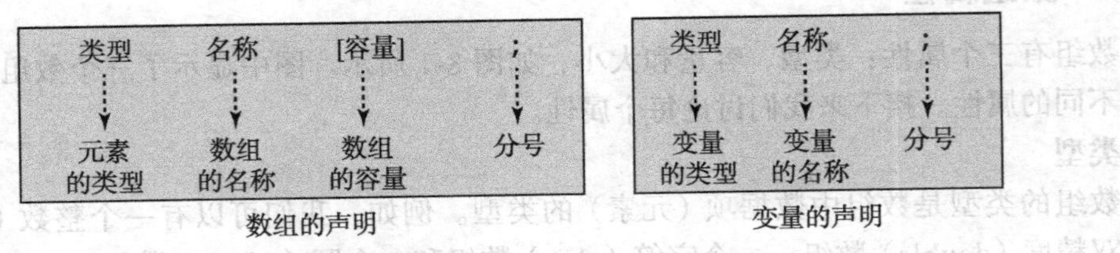

图 8-2　数组声明与变量声明的比较

关于如图 8-2 所示的语法有三个要点：

1）数组声明由四个部分组成：类型、名称、容量和终止分号。

2）如果将数组声明与变量声明进行比较，我们会发现唯一的区别是数组声明中增加了两个方括号括起来的容量。

3）数组的容量必须是常量或者字面量。

【例 8-1】 以下代码片段显示了如何使用容量的字面量值声明图 8-1 中所示的数组。

```
int scores [8];          double nums [6];          Circle crls [4];
```

【例 8-2】 以下代码片段显示了相同的声明，但使用了常量 CAPACITY。

```
const int CAPACITY = 8;      const int CAPACITY = 6;       const int CAPACITY = 4;
int scores [CAPACITY];       double nums [CAPACITY];       Circle crls [CAPACITY];
```

这种做法的优点在于，我们可以使用常量标识符（CAPACITY）来引用数组的容量。如前所述，按惯例通常使用大写字母定义常量。

内存分配

数组声明后，编译器将为该数组分配内存位置。内存分配取决于数组的类型和容量。数组的每个元素都可以通过包含在两个方括号中的索引来访问。但是，索引从 [0] 开始，直到 [CAPACITY−1]。图 8-3 显示了三个数组（scores、nums 和 crls）的声明和内存分配。

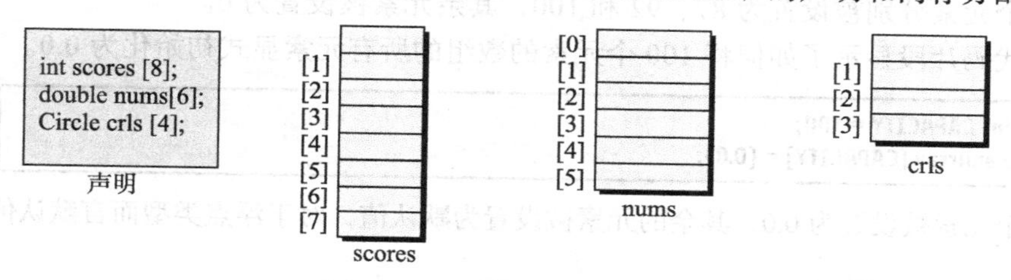

图 8-3　不同类型的三个数组

注意，数组中的元素使用零索引，这意味着索引从 [0] 开始，直到 [CAPACITY−1]。

数组元素使用零索引访问。

初始化

数组的每个元素都像一个单独的变量。当我们声明一个数组时，编译器根据数组类型为每个元素分配内存位置。与单个变量一样，存在两种情况：

1）如果数组在程序的全局区域中声明，则编译器根据数组类型为每个元素提供默认值。布尔类型的默认值为 false，字符类型为空字符，整数类型为 0，浮点类型为 0.0，对象类型为默认构造函数创建的对象。

2）如果数组在函数（包括 main）内声明，则元素会被填充垃圾值（以前使用该内存位置时残留的值）。

显式初始化。为了更好地控制数组元素中存储的初始值，我们可以显式初始化数组元素。但是，初始值必须用花括号括起来，并用逗号分隔。下面显示了如何声明包含 8 个元素的数组 scores，并同时初始化其元素。

```
const int CAPACITY = 8;
int scores [CAPACITY] = {87, 92, 100, 65, 70, 10, 96, 77};
```

当每个元素的初始化都是对构造函数的调用时，我们也可以用这种方式初始化一个类对象数组，如下所示：

```
const int CAPACITY = 4;
Circle circles [CAPACITY] = {Circle (4.0), Circle (5.0), Circle (6.0), Circle (7.0)};
```

该声明将 Circle 对象数组初始化为半径分别为 4.0、5.0、6.0 和 7.0 的四个 Circle 对象。

隐式容量。当初始化元素的数量正好是数组的容量时，我们不需要定义数组容量，如下所示。编译器对初始值进行计数并相应地设置数组的容量。但是，这种方法缺少了引用数组容量的常量的优势。

```
int scores [ ] = {87, 92, 100, 65, 70, 10, 96, 77};
```

部分默认填充。初始化值的数目不能大于数组的容量（否则会产生编译错误），但可以小于数组的容量。在这种情况下，不管数组是在全局区域中声明的还是在函数内部声明的，其余的元素都用默认值填充。

```
const int CAPACITY = 10;
int scores [CAPACITY] = {87, 92, 100};
```

前三个元素分别被设置为 87、92 和 100，其余元素被设置为 0。

以下代码片段显示了如何将 100 个元素的数组的所有元素显式初始化为 0.0。

```
const int CAPACITY = 100;
double anArray [CAPACITY] = {0.0};
```

第一个元素被设置为 0.0，其余的元素被设置为默认值，对于浮点类型而言默认值也是 0.0。

8.1.3　访问数组元素

我们已经了解到，如果要访问变量的内容，必须使用表达式。对于单个变量，变量的名称是一个基本表达式，我们可以使用该基本表达式来获取存储在变量中的值。对于数组，必须使用名为**下标**（subscript）的后缀表达式。下标表达式使用一个操作数的下标运算符（[…]）。操作数是数组的名称。在第 9 章中，我们将讨论数组的名称实际上是指向数组第一个元素的指针的名称。表 8-1 提供了 C++ 表达式的一个例子，如附录 C 所示。

表 8-1　访问数组元素

组别	名称	运算符	表达式	优先级	结合性
后缀表达式	下标	[…]	Name[expr]	18	→

图 8-4 显示了如何使用下标表达式访问数组的元素。

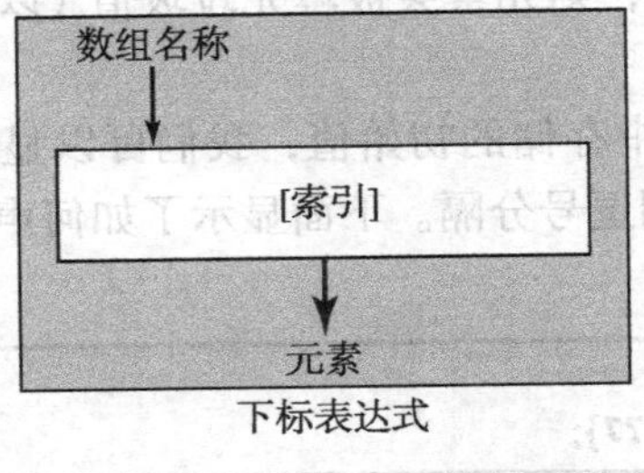

注意：
下标运算符将数组的名称作为其操作数。下标运算符返回数组的一个元素。下标运算符可以用于访问或者更改数组中元素的值。

图 8-4　使用下标表达式访问数组元素

以下代码片段显示了如何访问图 8-3 中声明的三个数组中的元素。

```
scores [4]                              // 访问位置 4 的元素
nums [5]                                // 访问位置 5 的元素
```

```
crls [3]                                    // 访问位置 3 的元素
```

注意，下标表达式返回元素。我们可以在任何可以使用变量的地方使用元素。例如，我们可以将元素用作左值或者右值。图 8-5 显示了我们如何存储（更改或者设置）和检索（获取）元素的值。

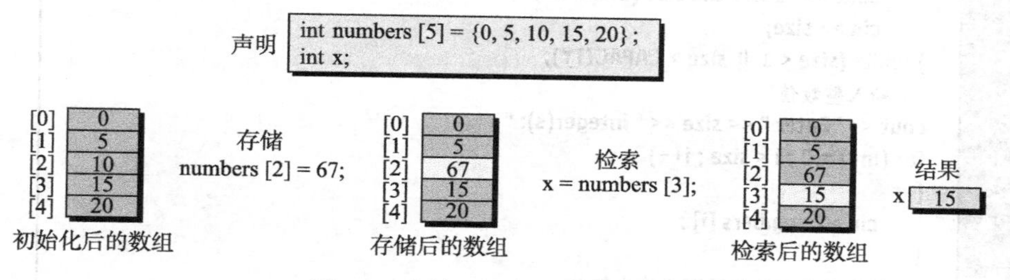

图 8-5　存储和检索数组中的元素

数组越界错误

编译时或者运行时无法捕获的隐藏错误之一是访问超越数组容量限制的数组元素。下标表达式 arrayName[index] 中使用的索引必须在 0 和 CAPACITY−1 之间的范围内。如果使用下标表达式检索超出范围的元素的值，我们将得到一个垃圾值，因为该位置不是数组的一部分。更危险的情况是试图在超出范围的元素中存储值，在这种情况下，我们可能会无意中破坏数据或者程序代码。当程序运行时，结果将不可预测：程序可能会失败或者产生无效的结果。当发生这些错误时，C++ 不会给出编译时或者运行时警告。在下一节中讨论指针时，我们将讨论这种奇怪行为的原因。

> 数组越界错误是一个必须避免的严重问题。

方括号的两种用法

我们讨论了数组中方括号的两种用法，如图 8-6 所示。

```
type array [capacity];        array [index]
```

数组声明　　数组访问

图 8-6　方括号的两种用法

【例 8-3】 当项目列表中的所有项目都需要在内存中进行处理时，数组对于保存相同类型的项目列表非常有用。假设我们要读取一个项目列表并按相反的顺序打印它们，我们可以使用一个容量足以容纳列表中所有元素的数组。程序 8-1 展示了如何使用验证循环来确保数组的预计大小不大于数组的容量。

程序 8-1　按相反顺序打印列表

```
1   /***************************************************************
2    * 使用数组读取一个整数列表，                                    *
3    * 然后按读取顺序的相反顺序输出                                  *
4    ***************************************************************/
5   #include <iostream>
6   using namespace std;
7
8   int main ( )
9   {
10    // 声明
11    const int CAPACITY = 10;
```

```
12    int numbers [CAPACITY];
13    int size;
14    // 从用户处获取大小，并进行验证
15    do
16    {
17        cout << "Enter the size (1 and 10) ";
18        cin >> size;
19    } while (size < 1 || size > CAPACITY);
20    // 输入整数值
21    cout << "Enter " << size << " integer(s): " ;
22    for (int i = 0 ; i < size ; i++)
23    {
24        cin >> numbers [i] ;
25    }
26    // 按输入顺序的相反顺序输出整数
27    cout << "Integer(s) in reversed order: ";
28    for (int i = size - 1 ; i >= 0 ; i--)
29    {
30        cout << numbers[i] << " " ;
31    }
32    return 0;
33 }
```

```
运行结果：
Enter the size (1 to 10): 10
Enter 10 integer(s): 2 3 4 5 6 7 8 9 10 11
Integer(s) in reversed order: 11 10 9 8 7 6 5 4 3 2
```

```
运行结果：
Enter the size (1 to 10): 0
Enter the size (1 to 10): 11
Enter the size (1 to 10): 7
Enter 7 integer(s): 4 11 7 8 2 5 3 8 9          //最后一个整数被忽略
Integer(s) in reversed order: 8 3 5 2 7 8 11 4
```

关于程序 8-1 有如下两个要点：

1）我们在这个数组中没有数组越界的问题，因为我们在验证循环中强制 size 变量的值在 1 和容量（CAPACITY）之间。例如，当我们将容量设置为 10 时，我们可以有 1 到 10 个整数，用索引 0 到 9 将这些整数填充到元素中，没有整数超出数组边界（参见第二次运行）。

2）键盘被当作一个文件，其中的数值被依次输入，这些数值之间至少有一个空格分隔。在读取循环终止之前，逐个读取这些数值。即使用户在输入由 size 变量定义的数目的整数之前按下回车键，程序也会等待用户输入其余的整数（回车键被视为一个空格）。另外，如果用户输入的数目多于由 size 变量定义的数目，则只读取预先定义的数目的整数，其余的则被忽略（如在第二次运行中，最后输入的整数 9 被忽略）。

【例 8-4】 我们可以使用输入文件，而不是从键盘读取数组元素。键盘和输入文件的行为类似（都是充当顺序项的源）。此外，我们可以将数组元素存储在输出文件中，而不是在处理后将它们写入监视器。监视器和输出文件的行为类似（都是充当一系列数据项的目标）。程序 8-2 显示了如何修改程序 8-1，从文件中读取原始数值，并将处理过的数值写入另一个文件。

程序 8-2 使用文件按相反顺序打印数值列表

```
1  /***************************************************************
2    * 使用数组，                                                  *
3    * 从一个文件中读取一个整数列表，                              *
4    * 然后按读取顺序的相反顺序将元素写入另一个文件                *
5    ***************************************************************/
6  #include <iostream>
7  #include <fstream>
8  using namespace std;
9
10 int main ( )
11 {
12   // 声明
13   const int CAPACITY = 50;
14   int numbers [CAPACITY];
15   int size = 0;
16   ifstream inputFile;
17   ofstream outputFile;
18   // 打开输入文件
19   inputFile.open ("inFile.dat");
20   if (!inputFile)
21   {
22       cout << "Error. Input file cannot be opened." << endl;
23       cout << "The program is terminated";
24       return 0;
25   }
26   // 从输入文件中读取数值列表到数组中
27   while (inputFile >> numbers [size] && size <= 50)
28   {
29       size++;
30   }
31   // 关闭输入文件
32   inputFile.close();
33   // 打开输出文件
34   outputFile.open ("outFile.dat");
35   if (!outputFile)
36   {
37       cout << "Error. Output file cannot be opened." << endl;
38       cout << "The program is terminated.";
39       return 0;
40   }
41   // 把倒序的数组元素写入输出文件
42   for (int i = size - 1 ; i >= 0 ; i--)
43   {
44       outputFile << numbers[i] << " " ;
45   }
46   // 关闭输出文件
47   outputFile.close();
48   return 0;
49 }
```

当我们在文本编辑器中打开输入文件和输出文件时，我们得到以下内容，其内容列表是相互反向的。

输入文件	输出文件
12 56 72 89 11 71 61 92 34 13	13 34 92 61 71 11 89 72 56 12

关于程序 8-2，有以下几个要点：

1）程序中包含了 <fstream> 头文件，以便我们可以对文件进行操作。

2）由于不知道输入文件中数值的个数，因此必须谨慎地选择一个较大的数值（在本例中为 50）作为容量。

3）如果输入文件没有被成功打开，我们会显示一条消息并终止程序（第 19 ～ 25 行）。同样，如果输出文件没有被成功打开，我们也会显示一条消息并终止程序（第 34 ～ 40 行）。

4）当到达输入文件的末尾时，数组的大小将被自动设置（第 29 行）。

5）当我们运行程序时，除非打开输入或者输出文件时有问题，否则将没有显示结果。

6）程序从不读取超过 50 个整数（见第 27 行）。

【例 8-5】 虽然大多数时候我们在数组中使用零索引，但有时创建一个额外元素的数组并忽略索引 0 处的元素会比较方便。在本例中，我们的索引从 1 开始。例如，我们可以创建一个包含 12 个元素的数组来保存每个月的天数（非闰年）。但是，为了方便，可以创建一个包含 13 个元素的数组，而不使用索引 0 处的元素，如图 8-7 所示。（请注意，为了节省空间，我们在水平方向显示数组。）

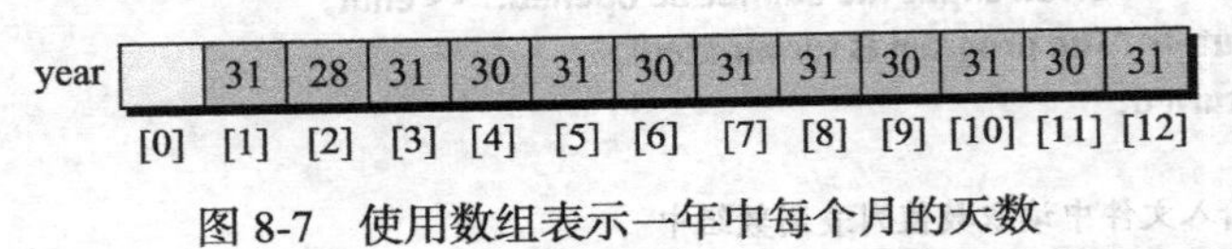

图 8-7 使用数组表示一年中每个月的天数

程序 8-3 展示了如何编写一个简单的程序来使用图 8-7 所示的数组。其作用等同于 switch 语句。

程序 8-3 返回一年中给定月份所包含的天数

```
1   /***************************************************************
2    * 使用数组获取各月包含的天数                                   *
3    ***************************************************************/
4   #include <iostream>
5   using namespace std;
6
7   int main ( )
8   {
9       // 声明
10      int numberOfDays [13] = {0, 31, 28, 31, 30, 31, 30,
11                               31, 31, 30, 31, 30, 31};
12      int month;
13      // 获取输入，并进行验证
14      do
15      {
16          cout << "Enter the month number (1 to 12): ";
17          cin >> month;
18      } while (month < 1 || month > 12);
19      // 输出
```

```
20      cout << "There are " << numberOfDays[month];
21      cout << " days in this month.";
22      return 0;
23  }
```

```
运行结果:
Enter the month number (1 to 12): 1
There are 31 days in this month.
```

```
运行结果:
Enter the month number (1 to 12): 6
There are 30 days in this month.
```

```
运行结果:
Enter the month number (1 to 12): 0
Enter the month number (1 to 12): 13
Enter the month number (1 to 12): 3
There are 31 days in this month.
```

请注意，第三次运行拒绝了小于 1 以及大于 12 的月份。

【例 8-6】 在第 7 章中，我们创建了一个 Circle 类。使用 Circle 实现文件，我们可以创建一个应用程序，它使用 Circle 类的编译版本来创建一个包含 3 个 circle 对象的数组（程序 8-4）。

下面介绍如何编译和链接实现文件和应用程序文件，以查看每个圆的半径、面积和周长。

程序 8-4 创建包含 3 个 circle 对象的数组

```
1   /*************************************************************
2    * 本程序使用 Circle 类的编译版本，                             *
3    * 创建一个包含 3 个 circle 对象的数组                          *
4    *************************************************************/
5   #include <iostream>
6   #include "circle.h"
7   using namespace std;
8
9   int main ( )
10  {
11      // 声明数组
12      Circle circles [3];
13      // 初始化对象
14      circles [0] = Circle (3.0);
15      circles [1] = Circle (4.0);
16      circles [2] = Circle (5.0);
17      // 打印信息
18      for (int i = 0; i < 3 ; i++)
19      {
20          cout << "Information about circle [" << i << "]" << endl;
21          cout << "Radius: " << circles[i].getRadius() << "  ";
22          cout << "Area: " << circles[i].getArea() << "  ";
23          cout << "perimeter: " << circles[i].getPerimeter() << "  ";
24          cout << endl;
25      }
26      return 0;
27  }
```

```
c++ -c circle.cpp
c++ -c app.cpp
c++ -o application circle.o app.o
application
```

```
运行结果:
Information about circle [0]
Radius: 3 Area: 28.26  perimeter: 18.84

Information about circle [1]
Radius: 4 Area: 50.24  perimeter: 25.12

Information about circle [2]
Radius: 5 Area: 78.5  perimeter: 31.4
```

8.2 有关数组的进一步讨论

在本节中，我们将讨论一些可以应用于一维数组的操作。我们将这些操作分为两类：访问数组元素但不更改其值也不修改其顺序的操作；更改其值或者更改数组结构的操作。我们还将解释如何使用带数组的函数来模拟操作。最后，我们将讨论并行数组及其应用。

8.2.1 访问操作

访问操作是既不更改元素值也不更改数组结构（元素顺序）的操作。我们将讨论其中的一些操作。

计算元素的和以及平均值

假设我们希望计算一个名为 numbers 的数组的元素的总和以及平均值。下面代码片段显示了如何使用 for 循环进行此操作。我们假设数组的容量大于或者等于 size。

```
double average;
int sum = 0;
for (int i = 0; i < size; i++)
{
   sum += numbers [i];
}
average = static_cast <double> (sum) / size;
```

查找最小值和最大值

在前面的章节中，我们学习了如何查找数值列表中的最小值和最大值。如果数值被封装在一个数组中，我们可以使用相同的策略来查找数组中元素的最小值和最大值。

```
int smallest = + 1000000;
int largest = - 1000000;
for (int i = 0; i < size; i++)
{
   if (numbers [i] < smallest)
   {
       smallest = numbers [i];
   }
   if (numbers [i] > largest)
   {
```

```
        largest = numbers [i];
    }
}
```

【例 8-7】 在本例中，我们编写一个程序从输入文件（numFile.dat）中读取整数列表，其内容如下所示：

```
14 76 80 33 21 95 22 88 16 39
```

程序 8-5 获取并打印数组元素的总和、平均值、最小值和最大值，并将其显示在监视器上。

程序 8-5　获取数值序列的总和、平均值、最小值和最大值

```
1   /***************************************************************
2    * 使用数组从一个文件中读取一个整数列表，                        *
3    * 然后打印文件中数值的                                        *
4    * 总和、平均值、最小值和最大值                                *
5    ***************************************************************/
6   #include <iostream>
7   #include <fstream>
8   using namespace std;
9
10  int main ( )
11  {
12      // 声明文件
13      ifstream inputFile;
14      // 数组和变量的声明
15      const int CAPACITY = 50;
16      int numbers [CAPACITY];
17      int size = 0;
18      // 初始化
19      int sum = 0;
20      double average;
21      int smallest = 1000000;
22      int largest = -1000000;
23      // 打开输入文件，并进行打开验证
24      inputFile.open ("numFile.dat");
25      if (!inputFile)
26      {
27          cout << "Error. Input file cannot be opened." << endl;
28          cout << "The program is terminated.";
29          return 0;
30      }
31      // 从文件中读取（拷贝）数值
32      while (inputFile >> numbers [size])
33      {
34          size++;
35      }
36      // 关闭输入文件
37      inputFile.close();
38      // 获取总和、平均值、最小值和最大值
39      for (int i = 0; i < size; i++)
```

```
40  {
41          sum += numbers[i];
42          if (numbers[i] < smallest)
43          {
44                  smallest = numbers[i];
45          }
46          if (numbers[i] > largest)
47          {
48                  largest = numbers[i];
49          }
50  }
51  average = static_cast <double> (sum) / size;
52  // 打印结果
53  cout << "There are " << size << " numbers in the list" << endl;
54  cout << "The sum of them is: " << sum << endl;
55  cout << "The average of them is: " << average << endl;
56  cout << "The smallest number is: " << smallest << endl;
57  cout << "The largest number is: " << largest << endl;
58  return 0;
59  }
```

```
运行结果：
There are 10 numbers in the list.
The sum of them is: 484
The average of them is: 48.4
The smallest number is: 14
The largest number is: 95
```

查找某个值

在数组处理中，我们经常需要查找某个值。虽然有先进和高效的算法来搜索数组，但它们要求数组是有序的。我们在这里展示了一个基本搜索，即使数组没有排序，也可以使用该搜索算法。我们有时称之为线性搜索（linear search）。将需要查找的值称为查找参数（search argument）。

【例 8-8】 假设我们要在大小为 100 的数组中查找等于运行时某个输入值的元素。我们可以使用带有两个终止条件的 for 循环。第一个保证我们不会超出数组的大小；第二个测试我们是否找到了要查找的元素。如果搜索成功，就找到了元素的索引下标，这就是我们要查找的内容。

```
bool found = false;
for (int i = 0; (i < size) && (!found); i++)
{
   if (numbers [i] == value)
   {
        index = i;
        found = true;
   }
}
if (found)
{
   cout >> "The value was found at index: " << index;
}
else
```

```
{
  cout >> "The value was not found".;
}
```

8.2.2 修改操作

有些操作可能会更改元素的值、元素的顺序，或者两者都更改。

交换

我们讨论的第一个操作是交换两个元素的内容。这种交换通常用于其他操作（如排序操作）。图 8-8 显示了交换元素 1 和元素 3 时的错误方法和正确方法。

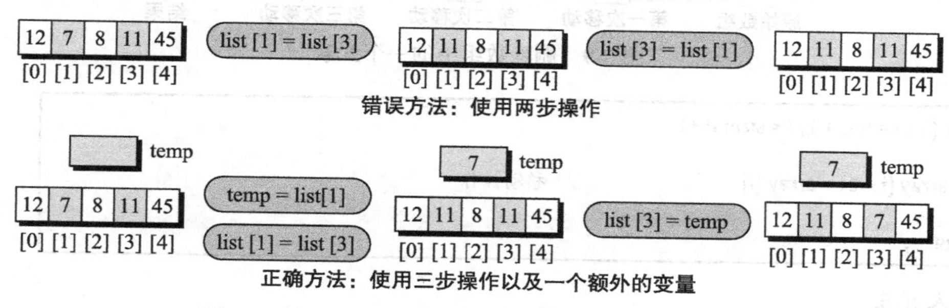

图 8-8 交换数组中两个元素的错误方法和正确方法

在错误的方法中，我们只使用两步操作。如图 8-8 所示，在两步操作之后，list[1] 和 list[3] 都包含整数 11，因为我们在第一步操作中丢失了 list[1] 的内容。

```
list [1] = list [3];              // 把 list[3] 的值复制到 list[1]
list [3] = list [1];              // 把 list[1] 的值复制到 list[3]
```

正确的解决方案是在丢失 list[1] 之前，使用临时变量 temp 来存储它的值。图 8-8 显示，这必须在其他两步操作之前完成。在第二步操作之后，我们会丢失 list[1] 的值，但我们已经在 temp 中保存了一份该元素的副本。我们会在第三步操作中获取这个值。

```
temp = list [1];                  // 保持 list[1] 的原始值
list [1] = list [3];              // 把 list[3] 的值复制到 list[1]
list [3] = temp;                  // 把 temp 的值复制到 list[3]
```

注意，三个变量 temp、list[1]、list[3] 在三步操作的左侧和右侧均只使用了一次。

数组排序

只改变元素顺序的操作称为排序。排序是一个非常复杂的操作，它使用元素交换，直到数组中的元素按顺序排列。我们将在第 17 章和第 19 章中讨论排序技术。

删除元素

数组中效率低下的操作之一是删除元素。假设我们要删除索引 3 处的元素。删除数组中一个元素的方法是将目标索引后的所有元素向前复制（移动）一个位置，如图 8-9 所示。

图中显示我们需要三次移动操作。我们不能删除索引 6 处的元素，但是我们可以减小数组的大小以使该元素不可用。

【例 8-9】 假设我们要删除索引 pos 处的元素。我们可以使用 for 循环向上移动元素，然后减小数组的大小。

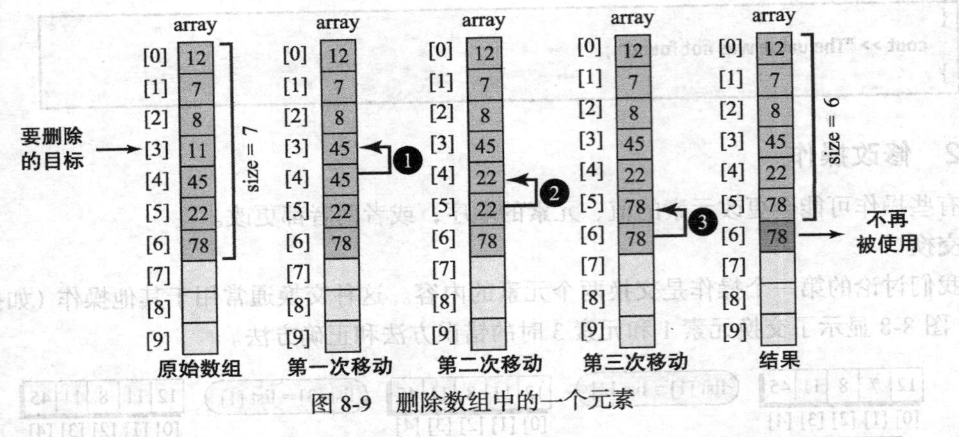

图 8-9　删除数组中的一个元素

```
for (int i = pos + 1; i < size; i++)
{
  array [i – 1] = array [i]                    // 移动操作
}
size––;
```

插入元素

另一个效率低下的操作是将元素插入数组的中间。只有当数组被部分填充时，才能执行此操作，否则，我们将超越数组的边界。图 8-10 显示了如何在索引 5 处插入值为 80 的元素。所有元素都必须朝末端移动一个位置，以便为插入操作留出空间。然而，这次我们需要从最后一个元素开始移动。

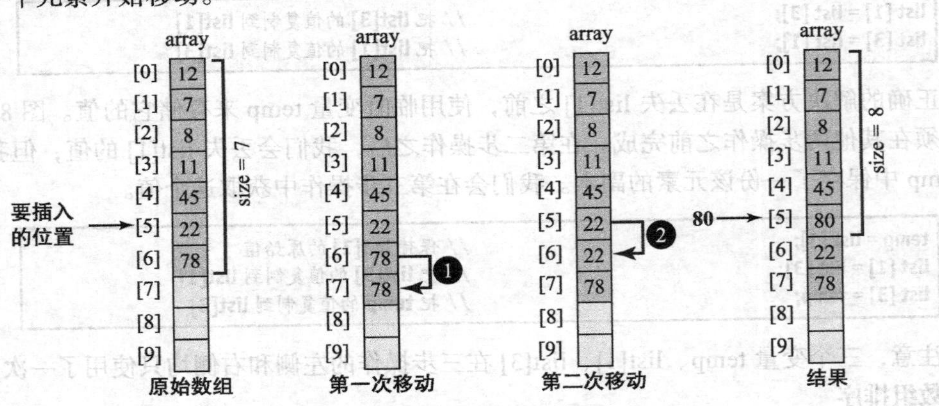

图 8-10　将一个元素插入数组中

【例 8-10】 假设我们要在数组中的特定索引（pos）处插入一项（value），该数组的当前大小由变量 size 定义。如果数组的大小小于其容量，我们将所有从索引 size–1 到索引 pos 的元素向数组末尾移动一个位置，从而为要插入的新元素留出空间。

```
for (int i = size; i > pos; i––)
{
  numbers[i] = number [i –1];                  // 移动操作
}
numbers [pos] = value;
size++;
```

8.2.3　使用带数组的函数

我们已经看到函数是对基本类型和类类型执行操作的代码块。接下来我们讨论函数如何在数组上操作。为此，我们需要知道是否可以将数组作为参数传递给函数，以及是否可以从函数返回数组。

把数组传递给函数

数组的名称是指向数组第一个元素的常量指针（地址），我们将在本章后面讨论这一点。换言之，数组名称定义了内存中的一个固定点。当我们将数组的名称传递给函数时，我们将把这个地址传递给函数。换言之，数组仍然存储在属于主调函数的区域中，但被调用函数可以访问或者修改数组。图 8-11 显示了这种情况。该图显示内存分配只能通过主调函数完成，但被调用函数可以访问或者修改数组的元素。为了访问属于主调函数的数组，被调用函数必须知道数组的起始地址和数组的大小。数组的起始地址是使用包含类型、数组名称和空括号（int array[]）的参数来定义的。数组的大小被定义为单独的参数。

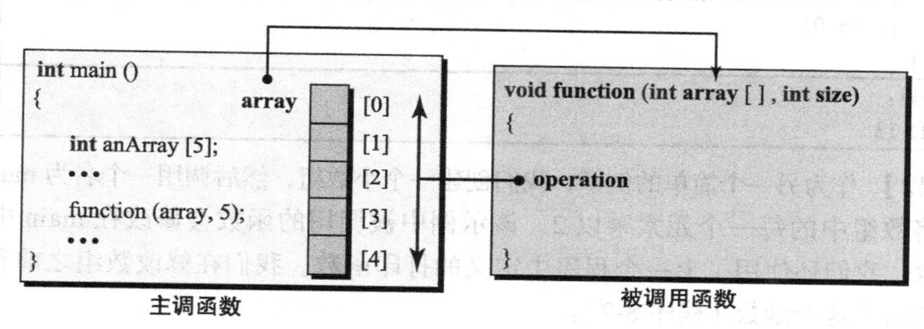

图 8-11　把数组作为参数传递给函数

存在以下两种情况。如果我们希望被调用函数只访问数组的元素而不能修改数组，那么必须在数组名称前面使用 const 修饰符。如果我们希望被调用函数能够修改数组，则不使用 const 修饰符。下面显示了这两种情况下被调用函数的原型：

```
void function (const type array [], int size);     // 仅用于访问操作
void function ( type array [], int size);          // 用于修改操作
```

换言之，我们将 const 修饰符添加到数组的名称前，以防止被调用函数更改数组。

【例 8-11】 作为一个简单的例子，我们创建一个小数组，然后调用一个名为 print 的函数来打印元素的值。该示例中被调用的函数不应该更改传递给它的数组（程序 8-6）。

程序 8-6　使用函数来访问数组

```
1   /***************************************************************
2    * 把数组名称和大小传递给函数，                                   *
3    * 使得函数可以打印数组的元素                                     *
4    ***************************************************************/
5   #include <iostream>
6   using namespace std;
7   /***************************************************************
8    * print 函数接收一个数组的名称和大小，                           *
9    * 然后打印在 main 函数中创建和初始化的该数组的元素，             *
10   * 它不会修改数组                                                 *
```

```
11   ******************************************************************/
12   void print (const int numbers [], int size)
13   {
14     for (int i = 0; i < size; i++)
15     {
16         cout << numbers [i] <<" ";
17     }
18     return;
19   }
20
21   int main( )
22   {
23     // 数组的声明和初始化
24     int numbers [15] = {5, 7, 9, 11, 13};
25     // 调用 print 函数
26     print (numbers, 5);
27     return 0;
28   }
```

运行结果：

```
5 7 9 11 13
```

【例 8-12】 作为另一个简单的例子，我们创建一个小数组，然后调用一个名为 multiplyByTwo 的函数，将数组中的每一个元素乘以 2。该示例中被调用的函数会修改在 main 中创建的数组。请注意，我们还使用了上一个程序中定义的打印函数。我们在修改数组之前和修改数组之后两次调用了这个函数（程序 8-7）。

程序 8-7　使用函数来修改数组

```
1    /******************************************************************
2     * 把数组名称和大小传递给函数，                                    *
3     * 使得函数可以修改数组的元素                                      *
4     ******************************************************************/
5    #include <iostream>
6    using namespace std;
7
8    /******************************************************************
9     * multiplyByTwo 是一个修改函数，用于修改 main 中的数组。          *
10    * 这里没有使用 const 修饰符。                                     *
11    * 该函数访问 main 中的数组并修改其内容                            *
12    ******************************************************************/
13   void multiplyByTwo (int numbers [ ], int size)
14   {
15     for (int i = 0; i < size; i++)
16     {
17         numbers [i]  *= 2;
18     }
19     return;
20   }
21
22   /******************************************************************
23    * print 函数接收一个数组的名称和大小，                            *
24    * 然后打印在 main 函数中创建和初始化的该数组的元素，              *
```

```
25    * 它不会修改数组                                                          *
26    ********************************************************************/
27  void print (const int numbers [ ], int size)
28  {
29      for (int i = 0; i < size; i++)
30      {
31          cout << numbers [i] << " ";
32      }
33      cout << endl;
34      return;
35  }
36
37  int main( )
38  {
39      // 数组的声明和初始化
40      int numbers [5] = {150, 170, 190, 110, 130};
41      // 打印修改前的数组内容
42      print (numbers, 5);
43      // 使用 multiplyByTwo 函数修改数组
44      multiplyByTwo (numbers , 5);
45      // 打印修改后的数组内容
46      print (numbers, 5);
47      return 0;
48  }
```

```
运行结果：
150 170 190 110 130
300 340 380 220 260
```

运行结果的第一行显示了原始数组，第二行显示了修改后的数组。

不能从函数返回数组

C++ 不允许从函数返回数组。换言之，如下的函数原型是错误的：

```
type [ ] function (const type array [], int size);    // 在 C++ 语言中不允许
```

将数组传递给函数时，我们有三个选择，如下所示：

```
// array 不会被改变
void (const type array [ ], int size);
// array 可能会被改变
void (type array [ ], int size);
// array1 不会被改变，但 array2 可以是 array1 的一个可修改版本
void (const type array1 [ ], type array2 [ ], int size);
```

为了模拟从函数返回数组，我们可以使用两个数组（一个常量数组和一个非常量数组）。

【例 8-13】 假设我们要编写一个函数来反转数组的内容，但不更改原始数组。一种可实现的方法是，创建两个相似的数组，原始数组和修改后的数组（最初可以为空），如程序 8-8 所示。请注意，我们同时打印原始数组和修改后的数组。

程序 8-8 通过传递两个数组来模拟返回数组

```
/***************************************************************
 * 通过传递两个数组给函数来模拟返回数组                              *
 ***************************************************************/
#include <iostream>
using namespace std;
/***************************************************************
 * reverse 函数是一个带两个数组参数的函数。                          *
 * 它反转第一个数组的内容并存储到第二个数组中                          *
 ***************************************************************/
void reverse (const int array1[], int array2[], int size)
{
  for (int i = 0, j = size - 1; i < size; i++, j--)
  {
      array2 [j] = array1 [i];
  }
  return;
}
/***************************************************************
 * print 函数接收一个数组的名称和大小,                              *
 * 然后打印数组的元素，而不会修改数组                                *
 ***************************************************************/
void print (const int array [], int size)
{
  for (int i = 0; i < size; i++)
  {
      cout << array [i] << " ";
  }
  cout << endl;
  return;
}

int main ( )
{
  // 声明两个数组
  int array1 [5] = {150, 170, 190, 110, 130};
  int array2 [5];
  // 调用 reverse 函数，将 array2 修改为 array1 的反转
  reverse (array1, array2 , 5);
  // 打印两个数组
  print (array1, 5);
  print (array2, 5);
  return 0;
}
```

运行结果:

```
150 170 190 110 130
130 110 190 170 150
```

8.2.4 并行数组

有时我们需要一个列表，其中每一行由多个数据项组成，各数据项可能是不同类型的。

例如，我们可能需要保存课程中每个学生的姓名、百分制分数和成绩等级的信息。这可以使用三个**并行数组**来完成，如图 8-12 所示。

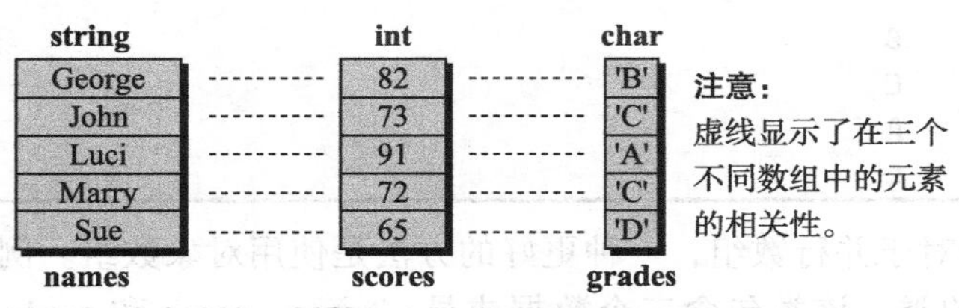

图 8-12 三个不同类型的并行数组

【例 8-14】 程序 8-9 在 main 函数中创建并初始化两个并行数组：names 和 scores。然后，程序调用一个函数来确定学生的成绩等级，并将其存储在第三个称为 grades 的并行数组中。最后打印数组内容。

程序 8-9 使用三个并行数组

```
1  /*********************************************************************
2   * 本程序使用三个并行数组，                                            *
3   * 创建某课程中五名学生的姓名、百分制分数、成绩等级的列表               *
4   *********************************************************************/
5  #include <iostream>
6  #include <iomanip>
7  using namespace std;
8  /*********************************************************************
9   * findGrades 函数接收三个参数：一个常量数组 scores、                    *
10  * 一个非常量数组 grades 以及两个数组的共同大小。                        *
11  * 函数使用第一个数组来创建第二个数组                                    *
12  *********************************************************************/
13 void findGrades (const int scores [], char grades [], int size)
14 {
15   char temp [ ] = {'F', 'F', 'F', 'F', 'F', 'F', 'D', 'C', 'B', 'A', 'A'};
16   for (int i = 0; i < size; i++)
17   {
18           grades [i] = temp [scores [i] /10];
19   }
20   return ;
21 }
22 int main( )
23 {
24   // 声明三个数组并初始化其中两个数组
25   string names [4] = {"George", "John", "Luci", "Mary"};
26   int scores [4] = {82, 73, 91, 72};
27   char grades [4];
28   // 函数调用
29   findGrades (scores, grades, 5);
30   // 打印三个数组的内容
31   for (int i = 0; i < 4; i++)
32   {
33       cout << setw (10) << left << names[i] << "    " << setw (2) ;
34       cout << scores[i] << "     " << setw(2) << grades[i] << endl;
35   }
```

```
36  return 0;
37  }
```

运行结果：
```
George      82      B
John        73      C
Luci        91      A
Mary        72      C
```

【例 8-15】 相对于并行数组，一种更好的方法是使用对象数组。例如，我们可以创建一个名为 Student 的类，该类包含三个数据成员：name、score 和 grade，以解决示例 8-14 中出现的问题。然后，我们创建一个对象数组，其中每个元素都是类 Student 的一个实例。通过这种方式，我们可以创建这个类，并使用独立编译，让每个教授都可以使用这个类。然后，教授可以创建他们的定制应用程序（main 函数），使用该类处理任意数量的学生。程序 8-10 是 Student 类的接口文件。

程序 8-10 Student 类的接口文件

```
1   /*****************************************************************
2    * 本程序是 Student 类的接口文件，                                *
3    * 包含三个私有数据成员和四个公共成员函数                            *
4    *****************************************************************/
5   #ifndef STUDENT_H
6   #define STUDENT_H
7   #include <iostream>
8   #include <string>
9   using namespace std;
10
11  class Student
12  {
13    private:
14        string name;
15        int score;
16        char grade;
17    public:
18        Student ();
19        Student (string name, int score);
20        ~Student ();
21        void print();
22  };
23  #endif
```

程序 8-11 是实现文件的代码清单。

程序 8-11 Student 类的实现文件

```
1   /*****************************************************************
2    * 本程序是 Student 类的实现文件，                                *
3    * 其接口文件由程序 8-10 给定的                                   *
4    *****************************************************************/
5   #include "student.h"
6
7   // 默认构造函数
8   Student :: Student()
```

```
9   {
10  }
11  // 参数构造函数
12  Student :: Student (string nm, int sc)
13  :name (nm), score (sc)
14  {
15     char temp [ ] = {'F', 'F', 'F', 'F', 'F', 'F', 'D', 'C', 'B', 'A', 'A'};
16     grade = temp [score / 10];
17  }
18  // 析构函数
19  Student :: ~Student()
20  {
21  }
22  // 成员函数 print
23  void Student :: print()
24  {
25     cout << setw (12) << left << name;
26     cout << setw (8) << right << score;
27     cout << setw (8) << right << grade << endl;
28  }
```

程序 8-12 是应用程序的代码清单。

程序 8-12 Student 类的应用程序文件

```
1   /***************************************************************
2    * 应用程序文件创建 Student 类的对象实例,                        *
3    * 然后打印每个学生的姓名、百分制分数和成绩等级                  *
4    ***************************************************************/
5   #include "student.h"
6   #include "iomanip"
7
8   int main ( )
9   {
10     // 使用默认构造函数声明 1 个 Student 的数组
11     Student students [5];
12     // 使用参数构造函数初始化 5 个对象
13     students[0] = Student ("George", 82);
14     students[1] = Student ("John", 73);
15     students[2] = Student ("Luci", 91);
16     students[3] = Student ("Mary", 72);
17     students[4] = Student ("Sue", 65);
18     // 打印学生的姓名(name)、百分制分数(score)和成绩等级(grade)
19     for (int i = 0; i < 5; i++)
20     {
21         students[i].print();
22     }
23     return 0;
24  }
```

以下是创建和显示学生信息的应用程序的独立编译和运行过程。

```
c++ – c students.cpp                    // 编译实现文件
c++ – c app.cpp                         // 编译应用程序文件
```

```
c++ – o application student.o app.o      // 链接两个编译后的文件
application                              // 运行可执行文件
运行结果:
George    82    B
John      73    C
Luci      91    A
Mary      72    C
Sue       65    D
```

虽然这种方法看起来更复杂，使用的代码行也更多，但是该方法具有如下优点：当创建了 Student 类后，任何教授都可以通过下载接口和实现来使用该类。每个教授都可以根据自己的需要定制自己的应用程序。

8.3 多维数组

有些应用程序要求一组值按**多维数组**（multidimensional array）的形式来排列。最常见的是二维数组，但我们偶尔也会遇到三维数组。

8.3.1 二维数组

二维数组（two-dimensional array）定义了一种结构化数据类型，其中使用两个索引分别定义行和列中元素的位置。第一个索引定义行；第二个索引定义列。图 8-13 显示了一个名为 scores 的二维数组，该数组包含 5 行 3 列。

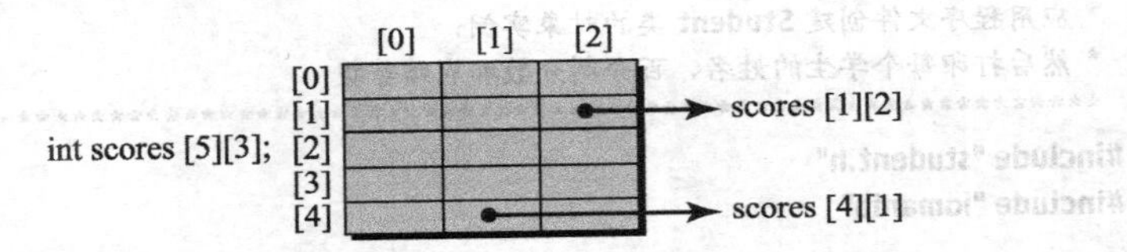

图 8-13　二维数组（5 名学生在 3 次测试中的成绩）

注意，与一维数组一样，第一行是第 0 行，第一列是第 0 列。学生 1 的分数占第 0 行；所有学生的测试 1 的分数占第 0 列。

声明和初始化

声明和定义二维数组类似于声明和定义一维数组，但我们必须定义两个维度：行和列。我们首先定义行数，然后定义列数。下面代码片段显示了如何定义 5 名学生在 3 次测试中的分数数组。

```
int score [5][3];
```

下标运算符。在一维数组中，需要使用一个下标运算符。在二维数组中，则需要使用两个下标运算符。

初始化。数组中的元素是逐行初始化的。但是，如果我们使用花括号来分隔行，那么可以增加列表初始化内容的可读性，如下所示。

```
int scores [5][3] = { {82, 65, 72},
                      {73, 70, 80},
                      {91, 76, 40},
                      {72, 72, 68},
```

```
{65, 90, 80} };
```

要将整个数组初始化为零（当数组被局部声明时），我们只指定第一个值，如下所示：

```
int scores [5][3] = {0};
```

访问元素。我们可以使用由两个索引定义的元素的确切位置来访问数组中的每个元素。访问可以用于在单个元素中存储值或者检索元素的值。

```
scores[1][0] = 5;              // 把 5 存储到第 1 行第 0 列
cin >> scores[2][1];           // 读取输入值到第 2 行第 1 列
x = scores [1][2];             // 拷贝第 1 行第 2 列的内容到变量 x
cout << scores [0][0];         // 打印输出第 0 行第 0 列的内容
```

传递二维数组到函数

将二维数组传递给函数的原理与我们讨论过的一维数组的原理相同。如下所示，第一个参数定义了数组。这里，第一个方括号是空的，但是第二个方括号必须以字面量形式定义第二个维度的大小。第一个维度的大小必须作为单独的参数传递给数组。

```
void function (int array[ ] [3] , int rowSize);
```

【例 8-16】 图 8-14 显示了二维数组和两个一维数组之间的关系。

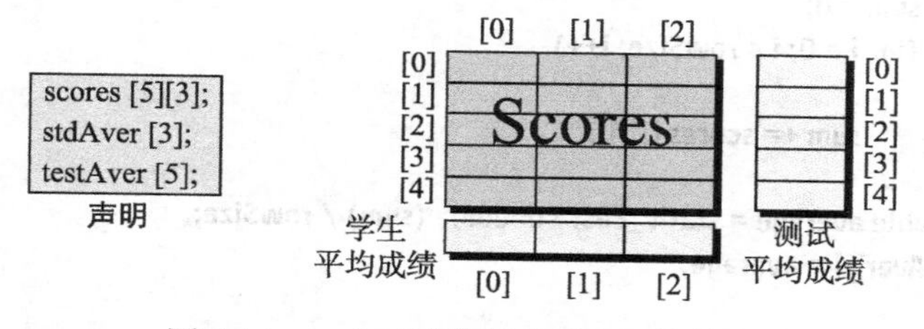

图 8-14　一个二维数组和两个一维数组

程序 8-13 显示了如何使用两个函数获取每名学生的平均成绩和每次测试的平均成绩。每个函数接收两个参数：一个二维数组和一个一维数组。

程序 8-13　使用三个数组

```
1   /*****************************************************************
2    * 本程序基于一个二维测试成绩数组，                                  *
3    * 创建学生平均成绩和测试平均成绩                                    *
4    *****************************************************************/
5   #include <iostream>
6   #include <iomanip>
7   using namespace std;
8
9   /*****************************************************************
10   * 函数接收存储 6 个学生 3 次考试成绩的二维数组作为参数，             *
11   * 然后，                                                          *
12   * 修改 main 函数中表示学生平均成绩的数组                            *
13   *****************************************************************/
14  void findStudentAverage (int const scores [ ][3],
15                           double stdAver [ ], int rowSize, int colSize)
```

```
16  {
17     for (int i = 0; i < rowSize; i++)
18     {
19         int sum = 0;
20         for (int j = 0; j < colSize; j++)
21         {
22             sum += scores[i][j];
23         }
24         double average = static_cast <double> (sum) / colSize;
25         stdAver[i] = average;
26     }
27     return;
28  }
29  /****************************************************************
30   * 函数接收存储 6 个学生 3 次考试成绩的二维数组作为参数,          *
31   * 然后,                                                        *
32   * 修改 main 函数中表示测试平均成绩的数组                        *
33   ****************************************************************/
34  void findTestAverage (int const scores [][3],
35                        double tstAver [], int rowSize , int colSize)
36  {
37     for (int j = 0; j< colSize; j++)
38     {
39         int sum = 0;
40         for (int i = 0; i < rowSize; i++)
41         {
42             sum += scores [i][j];
43         }
44         double average = static_cast <double> (sum) / rowSize;
45         tstAver[j] = average;
46     }
47  }
48
49  int main( )
50  {
51     // 声明三个数组和一些变量
52     const int rowSize = 5;
53     const int colSize = 3;
54     int scores [rowSize][colSize] = {{82, 65, 72},
55                                      {73, 70, 80},
56                                      {91, 67, 40},
57                                      {72, 72, 68},
58                                      {65, 90, 80}};
59     double stdAver [rowSize];
60     double tstAver [colSize];
61     // 调用两个函数来修改两个平均值数组
62     findStudentAverage (scores, stdAver, rowSize, colSize);
63     findTestAverage (scores, tstAver, rowSize, colSize);
64     // 打印标题
65     cout << "              Test Scores              stdAver" << endl;
66     cout << "    ------------------------------   -------  " << endl;
67     // 打印测试成绩和每名学生的平均成绩
```

```
68   for (int i = 0; i < rowSize ; i++)
69   {
70       for (int j = 0 ; j < colSize; j++)
71       {
72           cout << setw (12) << scores[i][j];
73       }
74       cout << fixed << setprecision (2) << "    " << stdAver[i] << endl;
75   }
76   // 打印每次测试的平均成绩
77   cout << "tstAver  ";
78   cout << "----------------------------- ";
79   for (int j = 0 ; j < colSize; j++)
80   {
81       cout << fixed << setprecision (2) << stdAver[j] << "    ";
82   }
83   return 0;
84 }
```

运行结果：

```
              Test Scores                 stdAver
        --------------------------       -------
         82         65         72         73.00
         73         70         80         74.33
         91         67         40         66.00
         72         72         68         70.67
         65         90         80         78.33
tstAver  73.00      74.33      66.00
```

我们在循环中使用索引 i 和 j 来表示行和列。在第 14 行和第 34 行中，我们将二维数组作为常量实体来传递，将第二个索引设置为字面量 3（列数）。在第 15 行和第 35 行中，我们将一维数组作为非常量实体来传递，以允许每个函数存储平均成绩。在第 15 行和第 35 行中，我们传递 rowSize 和 colSize 以供循环使用。

二维数组的操作

前文中为一维数组定义的一些操作同样可以用于二维数组。其他的操作则必须进行修改以适用于二维数组。例如，搜索和排序应该分别应用于每个维度。另外，还有一些操作专门应用于二维数组。

折叠。我们可以围绕水平轴（行折叠）或者垂直轴（列折叠）折叠二维数组，如图 8-15 所示。

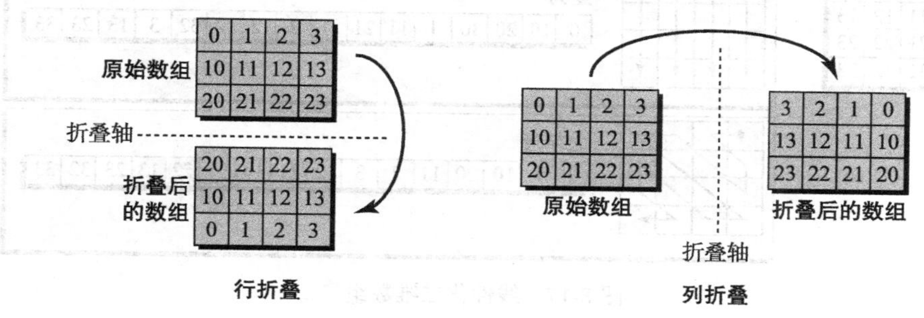

图 8-15　二维数组的折叠

【例 8-17】 以下代码片段显示了如何使用嵌套循环实现行折叠。我们将列折叠的代码留作练习。

```
for (int i = 0 ; i < rowSize ; i++)
{
    for (int j = 0 ; j < colSize ; j++)
    {
        foldedArray [rowSize - 1 - i][j] = originalArray [i][j];
    }
}
```

转置。使用二维数组时，可能需要将其转置。例如，如果我们想用矩阵来求解一组方程，就需要转置操作。转置操作是指改变行和列的角色。一行变成一列；一列变成一行。图 8-16 显示了应用于 3 × 4 数组的转置，其结果是一个 4 × 3 的数组。要转置数组，需要将行索引更改为列索引，反之亦然。

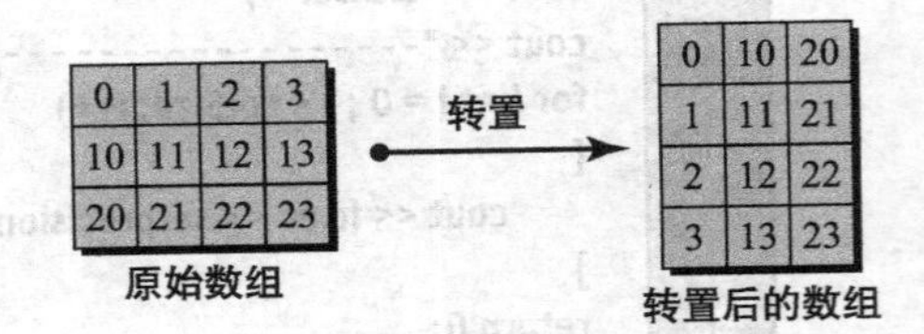

图 8-16　二维数组的转置操作

【例 8-18】 以下代码片段显示了如何使用嵌套循环实现转置操作。

```
for (int i = 0 ; i < orgRowSize ; i++)
{
    for (int j = 0 ; j < orgColSize ; j++)
    {
        trasposedArray [j][i] = originalArray [i][j];
    }
}
```

线性化。我们可能需要通过网络发送二维数组的内容（例如，当我们发送视频时）。在传输之前，必须将二维数组转换为一维数组（通过称为**线性化**的过程）。图 8-17 显示了三种方法：逐行、逐列和之字形。

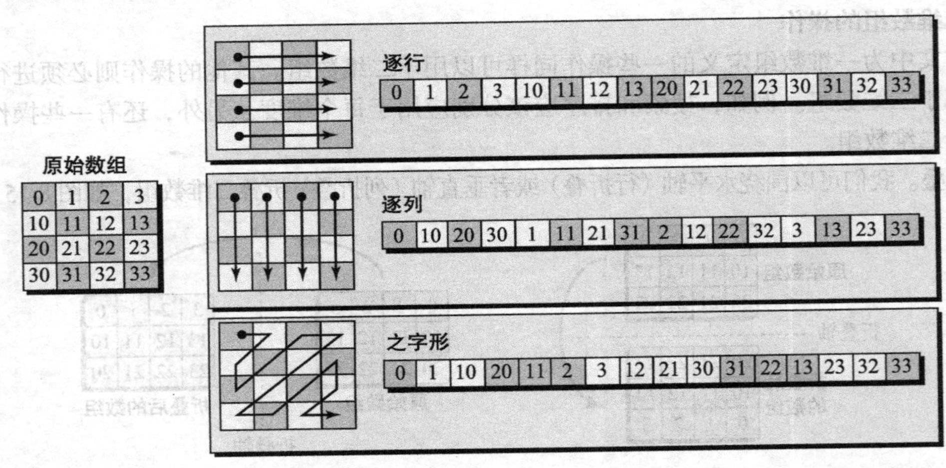

图 8-17　线性化二维数组

8.3.2　三维数组

C++ 中并没有限制数组一定要是两个维度。然而，三维以上的数组并不常见。图 8-18

显示了如何表示**三维数组**（three-dimensional array）。图 8-18 所示的业务在三个州运营，每个州最多包括四个办事处，每个办事处最多有十二名员工。我们将编写一个程序来处理一个三维数组作为练习留给读者。

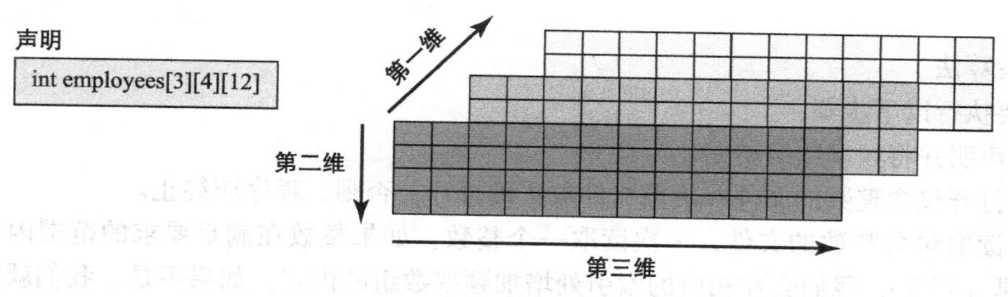

图 8-18　三维数组

8.4　程序设计

在本节中，我们将展示如何使用数组来解决计算机科学中的经典问题。

8.4.1　频率数组和直方图

数组可以用于创建显示元素在整数列表中分布情况的**频率数组**（frequency array）。数组还可用于创建**直方图**（histogram），直方图是频率数组的图形化表示。图 8-19 显示了典型数据和根据数据创建的频率数组。

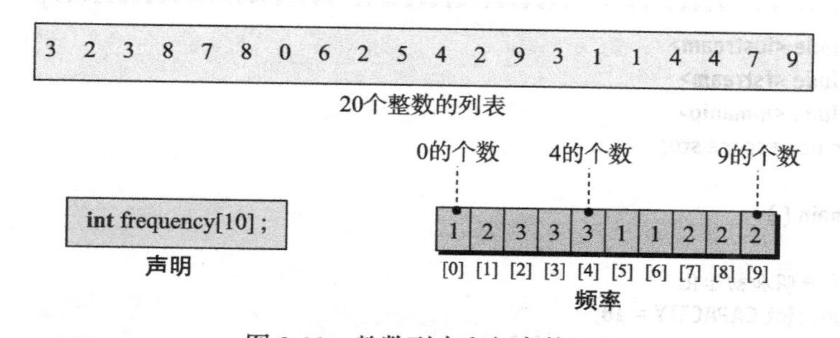

图 8-19　整数列表和频率数组

假设我们收集了一个整数序列，其中每个整数都在 0 到 9 之间（图 8-19 只显示了 20 个整数，但列表可以是任意长度）。我们想知道 0 的个数、1 的个数、2 的个数，等等。我们创建一个包含 10 个元素的数组来保存每个值的个数，称之为频率数组。注意，如果把频率数组中的所有值相加，则结果为列表中整数的个数。整数的个数和整数的范围通常很大，并且整数是从文件中读取的。图中的列表可以被认为是文件的内容。我们从文件中逐个读取整数，并创建一个数组 frequency。例如，每次读取整数 4 时，我们都会把 frequency[4] 的值加 1。注意，我们只能使用整数列表，因为频率数组的索引是整数。

理解问题

我们需要创建一个数组 frequency，其中从列表中读取的每个整数都与数组的索引相关。换言之，如果我们在列表中遇到 6，则需要把 frequency[6] 的值加 1。因此，frequency 数组的大小等于数据列表中的值的范围。数据列表中的值介于 0 和 9 之间，frequency 数组的索引为 [0] 到 [9]。为了创建 frequency 数组，我们需要逐个元素地遍历列表，当我们看到值为

x 的元素时，我们将 1 累加到 frequency[x]。一旦构建了数组，我们就可以创建一个直方图，以图形方式显示频率数组。注意，一些文献为这个问题创建了两个数组：数据数组和频率数组。除非我们出于某种目的希望在程序执行期间将所有数据项保存在内存中，否则不需要数据数组。

开发算法

需要执行以下步骤：

1）声明并将频率数组初始化为全 0。

2）打开包含整数的文件并确保文件被正确打开，否则，程序应终止。

3）读取包含整数的文件，一次读取一个整数。如果整数在满足要求的范围内（在本例中范围是 0 到 9），我们会在相应的索引处增加频率数组的内容。如果不是，我们就忽略它。

4）逐元素打印存储在频率数组中的值。同时，我们必须创建直方图的对应行（在本例中，星号的个数等于该元素的值）。

编写程序

程序 8-14 显示了基于上述四个步骤的算法的实现代码。

程序 8-14　创建频率数组和直方图

```
1   /**************************************************************
2    * 本程序从一个文件中读取一个整数列表，                          *
3    * 然后针对介于 0 到 9（包含）的整数元素，                       *
4    * 创建一个频率数组和一个直方图                                  *
5    **************************************************************/
6   #include <iostream>
7   #include <fstream>
8   #include <iomanip>
9   using namespace std;
10
11  int main ( )
12  {
13      // 声明和初始化
14      const int CAPACITY = 10;
15      int frequencies [CAPACITY] = {0};
16      ifstream integerFile;
17      // 打开包含整数的文件
18      integerFile.open ("integerFile.dat");
19      if (!integerFile)
20      {
21          cout << "Error. Integer file cannot be opened." << endl;
22          cout << "The program is terminated.";
23          return 0;
24      }
25      // 从包含整数的文件中读取数据，并创建频率数组
26      int data;
27      int size = 0;
28      while (integerFile >> data )
29      {
30          if (data >= 0 && data <= 9)
31          {
32              size++;
```

```
33                  frequencies[data]++;
34          }
35      }
36      // 关闭包含整数的文件
37      integerFile.close();
38      // 打印频率数组和直方图
39      cout << "There are " <<  size << " valid data items." << endl;
40
41      for (int i = 0; i < 10 ; i++)
42      {
43          cout << setw (3) << i << " ";
44
45          for (int f = 1; f <= frequencies [i] ; f++)
46          {
47              cout << '*' ;
48          }
49          cout << " " << frequencies [i] << endl;
50      }
51      return 0;
52  }
```

运行结果：

```
There are 202 valid data items.
0 ********************** 22
1 **** 4
2 ******************* 19
3 ************* 13
4 ******************** 20
5 ********************************* 33
6 **************************** 28
7 **************************** 28
8 ************************* 25
9 ********** 10
```

每行中的星号数实际上是对应整数的计数。

以下是我们使用的包含整数的文件的内容。请注意，列表中有 205 个整数，但其中 3 个（以**灰色**显示）不在 0 到 9 的范围内。这意味着程序读取所有 205 个整数，但频率数组仅根据 202 个整数形成。

```
1 3 2 2 5 7 3 2 8 0 6 4 6 7 0 7 8 5 4 2 3 0 6 7 5 8 5 4 8 9 6 5 5 9 2 3 5 2 6 7 8
0 6 4 6 7 0 7 8 5 4 2 13 0 6 7 5 8 5 4 8 9 6 7 0 7 8 5 4 2 3 0 6 7 5 8 5 4 8 9 6 5
2 7 0 7 8 5 4 2 3 0 6 7 5 8 5 4 8 9 6 15 5 9 7 0 7 8 5 4 2 3 0 6 7 5 8 9 6 5 6 6 4
5 9 7 0 7 8 5 4 2 3 0 6 7 5 8 9 6 5 6 6 4 2 7 0 7 8 5 4 2 3 0 6 7 5 8 5 4 8 9 6 5
1 3 2 2 5 7 3 2 8 0 6 4 6 17 0 7 8 5 4 2 3 0 0 6 4 6 7 0 7 8 5 4 2 3 0 6 7 5 8 1 1
```

8.4.2　线性转换

我们前面提到，在二维数组到一维数组的线性转换中，我们有两种选择：行转换和列转换。在第一种情况下，逐行进行转换，在第二种情况下，逐列进行转换（图 8-20）。

理解问题

我们必须找到转换逻辑，使我们能够将 N 行 M 列数组中的数据转换为具有 $N \times M$ 个元素的一维行数组。

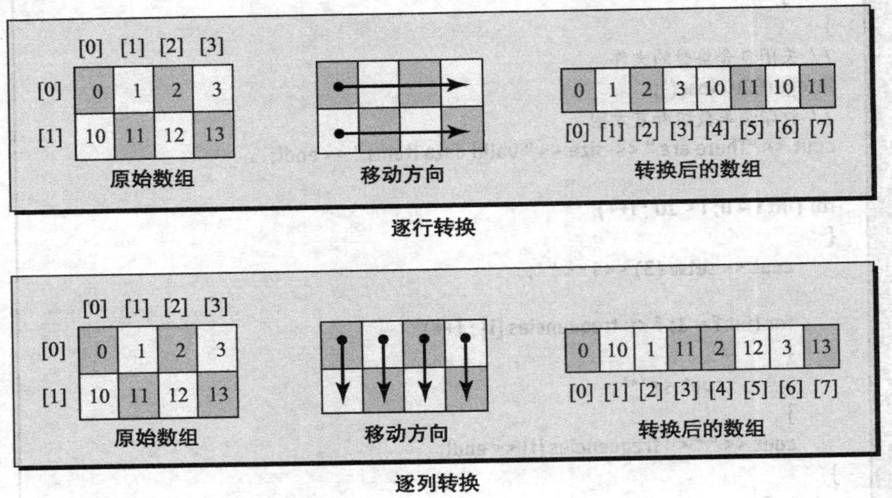

图 8-20　线性转换

开发算法

首先声明并初始化二维数组。对于小数组，我们只需输入值即可；对于大数组，我们将从文件中读取数据。我们定义了四个函数：

1）第一个函数用于逐行转换数组并创建一个名为 rowArray 的数组。

2）第二个函数用于逐列转换数组并创建一个名为 colArray 的数组。

3）第三个函数用于打印二维数组。

4）第四个函数用于打印一维数组。

main 函数的功能十分简单。在声明并初始化二维数组之后，调用第一个函数和第二个函数来创建一维数组。然后调用对应的函数来打印二维数组和转换后的每个一维数组。

编写程序

程序 8-15 列出了基于上述算法的代码清单。

程序 8-15　数组转换

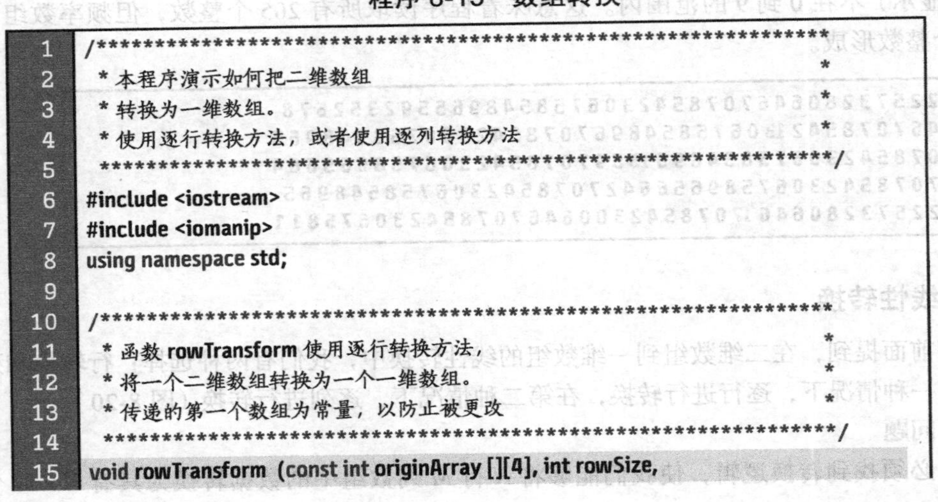

```
1   /***************************************************************
2    * 本程序演示如何把二维数组                                      *
3    * 转换为一维数组。                                              *
4    * 使用逐行转换方法，或者使用逐列转换方法                        *
5    ***************************************************************/
6   #include <iostream>
7   #include <iomanip>
8   using namespace std;
9
10  /***************************************************************
11   * 函数 rowTransform 使用逐行转换方法，                          *
12   * 将一个二维数组转换为一个一维数组。                            *
13   * 传递的第一个数组为常量，以防止被更改                          *
14   ***************************************************************/
15  void rowTransform (const int originArray [][4], int rowSize,
```

```
16                                                      int rowArray[])
17  {
18     int i = 0;
19     int j = 0;
20     for (int k = 0 ; k < 8; k++)
21     {
22         rowArray [k] = originArray [i] [j];
23         j++;
24         if (j > 3)
25         {
26                 i++;
27                 j = 0;
28         }
29     }
30  }
31  /*****************************************************************
32   * 函数 colTransform 使用逐列转换方法,                               *
33   * 将一个二维数组转换为一个一维数组。                                  *
34   * 传递的第一个数组为常量, 以防止被更改                                *
35   *****************************************************************/
36  void colTransform (const int originArray [][4], int rowSize,
37                                                      int colArray[])
38  {
39     int i = 0;
40     int j = 0;
41     for (int k = 0 ; k < 8; k++)
42     {
43         colArray[k] = originArray [i][j];
44         i++;
45         if (i > 1)
46         {
47                 j++;
48                 i = 0;
49         }
50     }
51  }
52  /*****************************************************************
53   * 该函数打印一个二维数组的内容,                                      *
54   * 二维数组被作为一个常量参数传递给该函数                               *
55   *****************************************************************/
56  void printTwoDimensional (const int twoDimensional [][4],
57                                                      int rowSize)
58  {
59     for (int i = 0; i < rowSize; i++)
60     {
61         for (int j = 0; j < 4; j++)
62         {
63                 cout << setw (4) << twoDimensional [i][j];
64         }
65         cout << endl;
66     }
67     cout << endl;
```

```
68  }
69
70  /*********************************************************
71   * 该函数打印一个一维数组的内容，                            *
72   * 一维数组被作为一个常量参数传递给该函数                     *
73   *********************************************************/
74  void printOneDimensional (const int oneDimensional[], int size)
75  {
76    for (int i = 0; i < size; i++)
77    {
78        cout << setw (4) << oneDimensional[i];
79    }
80    cout << endl;
81  }
82
83  int main ( )
84  {
85    // 声明三个数组并初始化第一个数组
86    int originArray [2][4] = {{0, 1, 2, 3}, {10, 11, 12, 13}};
87    int rowArray [8];
88    int colArray [8];
89    // 调用两个函数来转换数组
90    rowTransform (originArray, 2, rowArray);
91    colTransform (originArray, 2, colArray);
92    // 打印二维数组
93    cout << "   Original Array   " << endl;
94    printTwoDimensional (originArray, 2);
95    // 打印逐行转换后的一维数组
96    cout << "Row-Transformed Array:      ";
97    printOneDimensional (rowArray, 8);
98    // 打印逐列转换后的一维数组
99    cout << "Column-Transformed Array:   ";
100   printOneDimensional (colArray, 8);
101   return 0;
102 }
```

运行结果：

```
  Original Array
   0   1   2   3
  10  11  12  13

Row-Transformed Array:        0   1   2   3  10  11  12  13
Column-Transformed Array:     0  10   1  11   2  12   3  13
```

本章小结

数组是具有相同类型的数据项序列。数组是复合类型，我们可以有除 void 之外的任何类型的数组。数组有三个属性：类型、大小和容量。为了声明一个数组，我们必须指定它的类型、名称和用方括号括起来的容量。在程序全局区域中声明的数组被初始化为默认值。在函数内部声明的数组被初始化为上一次操作遗留的垃圾值。

在一维数组中，可以使用单个索引访问每个元素。访问二维数组中的元素需要两个索引，分别表示行和列。在二维数组上有几种常见的操作：折叠、转置和线性化。数组的维度不局限于二维，但是，超过三维的数组非常罕见。

思考题

1. 给定以下声明，编写代码打印索引下标为偶数的数组元素。

```
int arr [10] = {0, 1, 2, 3, 4, 5, 6, 7, 8, 9};
```

2. 给定以下声明，编写代码打印数组的前五个元素。

```
int arr [12] = {0, 10, 20, 39, 40, 50, 60, 70, 80, 90};
```

3. 编写代码，使用以下声明，把数组中偶数索引下标的元素填充为整数 0，奇数索引下标的元素填充为整数 1。

```
int arr [10];
```

4. 编写代码，使用以下声明，把数组的前五个元素填充为整数 5，后五个元素填充为整数 10。

```
int arr [10];
```

5. 以下代码片段的输出结果是什么？

```
int arr [10] = {0, 1, 2, 3, 4, 5, 6, 7, 8, 9};
for (int i = 0; i < 5; i++)
{
    cout << arr [i * 2] << " ";
}
```

6. 以下代码片段的输出结果是什么？

```
int arr [10] = {0, 1, 2, 3, 4, 5, 6, 7, 8, 9};
for (int i = 0; i < 5; i++)
{
    arr [i ] = arr [i * 2 + 1];
    cout << arr [i] << endl;
}
```

7. 以下代码片段的输出结果是什么？

```
int arr1 [5] = {0, 1, 2, 3, 4};
int arr2 [5];
for (int i = 0; i < 5; i++)
{
    arr2 [i] = arr1 [i];
    cout << arr1 [i] + arr2 [i] << endl;
}
```

8. 以下代码片段的输出结果是什么？

```
int arr [5] = {0, 1, 2, 3, 4};
for (int i = 1; i <= 4; i++)
{
    cout << arr [i] << " ";
}
cout << endl;
```

9. 编写代码，创建另一个数组 arr2，其内容是以下名为 arr1 的数组的倒排序。

```
int arr1 [10] = {0, 10, 20, 39, 40, 50, 60, 70, 80, 90};
```

10. 编写一个函数的定义，给定两个整数数组 arr1 和 arr2 作为参数，比较两个数组是否相等并返回 true 或者 false。假设两个数组的容量相同。
11. 编写一个函数的定义，给定一个整数数组 arr1（容量为 *n*），创建另一个数组 arr2，使得 arr2 中的每个元素都是 arr1 中对应元素的两倍。假设 arr2 的容量与 arr1 相同。
12. 编写一个函数的定义，给定一个整数数组 arr（大小为 *n*），检查数组元素是否按升序排列（每个元素的值大于或者等于前一个元素）。要求函数返回 true 或者 false。
13. 编写一个函数的定义，给定一个整数数组 arr（大小为 *n*），在数组中搜索给定的值。要求函数返回 true 或者 false。
14. 编写一个函数的定义，给定一个整数数组 arr，交换两个给定索引下标对应的数组元素。注意，要求函数检查给定索引下标的有效性。
15. 编写一个函数的定义，给定一个整数数组 arr，删除给定索引下标对应的数组元素。注意，要求函数检查给定索引下标的有效性。
16. 编写一个函数的定义，给定一个整数数组 arr，在给定索引下标处插入一个给定的值。注意，要求函数检查给定索引下标的有效性。
17. 编写代码，声明和初始化两个容量为 10 的数组，其中第一个数组保存公司员工的姓名，第二个数组保存公司员工的工资。
18. 编写一个函数的定义，给定一个名为 table 的二维数组（*n* 行 3 列），打印第 *r* 行的元素。注意，要求函数检查作为参数传递的 *r* 的有效性。
19. 编写一个函数的定义，给定一个名为 table 的二维数组（*n* 行 3 列），打印第 *c* 列的元素。注意，要求函数检查作为参数传递的 *c* 的有效性。
20. 显示由以下二维数组通过逐行线性转换得到的一维数组中的元素值。

```
int sample [2][4] = {{1, 2, 3, 4} , {5, 6, 7, 8}};
```

21. 显示由以下二维数组通过逐列线性转换得到的一维数组中的元素值。

```
int sample [2][4] = {{1, 2, 3, 4} , {5, 6, 7, 8}};
```

22. 显示由以下二维数组通过之字形转换得到的一维数组中的元素值。

```
int sample [3][3] = {{1, 2, 3},{4, 5, 6},{7, 8, 9}};
```

编程题

1. 编写一个程序，随机创建一个包含 100 个元素的数组，并用 100 到 200 之间的随机整数填充元素。程序通过调用 print 函数将数组元素打印成 10 行。
2. 编写一个程序，创建一个包含 10 个随机整数（位于 1 到 100 之间）的数组，然后打印数组元素、最小元素和最大元素。
3. 编写一个程序，创建一个包含 10 个随机整数（位于 1 到 100 之间）的数组，然后打印数组元素、元素的平均值、元素的标准方差（使用以下公式，其中 *n* 为数组大小，aver 为元素的平均值）。

```
stdDev = sqrt ( (num[0] - aver)² + ··· + (num [n-1] - aver)² ) / n
```

4. 编写一个程序，创建一个包含 10 个随机整数（位于 1 到 100 之间）的数组。编写一个函数，实现循环移位：将每个元素向数组末尾移动一个位置，这样第一个元素变为第二个元素，第二个元素变为第三个元素，…，最后一个元素变为第一个元素。要求程序打印原始数组和移位后的数组。
5. 编写一个程序，创建一个包含 10 个随机整数（位于 1 到 100 之间）的数组。然后从数组中删除最大值和最小值，最后打印原始数值以及删除两个元素之后的数组。

6. 编写一个程序，创建一个包含 10 个随机整数（位于 1 到 100 之间）的数组。然后创建另一个数组，其元素顺序与前一个数组相反。最后打印两个数组。
7. 编写一个程序，创建一个包含 20 个随机整数（位于 1 到 100 之间）的数组。然后再创建两个数组：一个数组包含奇数值，另一个数组包含偶数值。
8. 编写一个程序，创建一个 6×6 的二维数组，数组包含随机整数（位于 100 到 199 之间）。然后创建两个函数：第一个函数创建一个数组，包含从左到右对角线的元素；第二个函数创建一个数组，包含从右到左对角线的元素。最后打印二维数组和两个一维数组。
9. 编写一个程序，创建两个大小为 5 的一维数组，数组包含随机整数（位于 100 到 199 之间）。然后使用一个函数把两个一维数组合并成一个新数组。合并方法采用依次选取各数组的一个值。要求打印出所有数组的内容。
10. 编写一个程序，创建两个大小为 5 的一维数组，数组包含随机整数（位于 100 到 199 之间）。然后使用一个函数创建一个大小为 2×5 的二维数组，其每行内容为最初的一维数组之一。要求打印三个数组的内容。
11. 编写一个程序，创建两个大小为 5 的一维数组，数组包含随机整数（位于 100 到 199 之间）。然后使用一个函数创建一个大小为 5×2 的二维数组，其每列内容为最初的一维数组之一。要求打印三个数组的内容。

第 9 章

C++ Programming: An Object-Oriented Approach

引用、指针和内存管理

在第 8 章中，我们讨论了一种复合数据类型：数组。在本章中，我们将讨论另外两种复合数据类型：引用和指针。我们还将讨论与堆内存中存储被指对象相关的内存管理。

学习目标

阅读并学习本章后，读者应该能够

- 讨论引用类型和引用变量。
- 讨论如何在引用变量和原始变量之间创建永久绑定。
- 讨论如何通过引用变量检索和更改原始变量中的数据。
- 展示与函数相关的引用的应用：按引用传递和按引用返回。
- 讨论指针类型和指针变量。
- 讨论如何在指针变量和数据变量之间创建永久绑定或者临时绑定。
- 讨论如何通过指针变量检索和更改数据变量中的数据。
- 展示与函数相关的指针的应用：按指针传递和按指针返回。
- 讨论数组和指针之间的关系，讨论指针的算术运算。
- 讨论程序员及其应用程序可用的四个内存区域。
- 讨论对象何时会被存储在栈内存中，以及何时需要被存储在堆内存中。
- 讨论动态内存分配。

9.1 引用

引用（reference）是对象的替代名称。引用已被添加到 C++ 中，以简化实体（例如函数）之间的通信，我们将在本章后面讨论。

9.1.1 概述

声明为 type& 的变量是声明为 type 的变量的替代名称。当我们声明一个**引用变量**时，不会在内存中创建一个新的对象，而只是声明一个现有变量的替代名称。原始变量和引用变量实际上是由不同名称调用的相同内存位置。因此，我们需要在声明引用变量时立即将其绑定到原始变量。绑定是通过使用原始变量的名称初始化引用变量来完成的。以下代码片段显示了如何创建一个名为 score 的 int 类型的原始变量，以及如何将一个 int& 类型的引用变量绑定到该原始变量。

```
int score = 92;                    // 声明并初始化类型为 int 的变量 score
int& rScore = score;               // 声明类型为 int& 的引用变量 rScore，并将其绑定到 score
```

图 9-1 显示了内存中的情况。我们只有一个内存位置，但是有两个名称。其中一个名称定义了原始变量；另一个名称定义了对该变量的引用。

有趣的是，相同的值 92 在通过 score 变量访问时被视为 int 类型，通过 rScore 变量访问时被视为 int& 类型。

我们用于引用变量的名称与原始名称相同，但在开头添加了一个 r，并将原始名称的第一个字母设为大写。这不是强制要求，而是我们的约定，有助于我们记住引用变量绑定到哪个原始变量。

类型：int 类型：int&
原始变量名称 score 92 rScore 引用变量名称
值

图 9-1 原始变量和引用变量

复合类型

尽管原始变量和引用变量在内存中定义了相同的位置，且它们所看到的值相同，但原始变量和引用变量的类型不同。例如，在图 9-1 中，变量 score 的类型是 int，而变量 rScore 的类型是 int&。换言之，引用变量的使用创建了一个新的复合类型：**引用类型**。引用类型是复合类型，因为存在对 int 的引用（int&）、对 double 的引用（double&）、对 bool 的引用（bool&）等。当我们将变量绑定到引用变量时，引用变量的类型是对原始变量类型的引用。以下的代码片段会产生编译错误：不能初始化 double& 的引用变量来引用 int 类型的变量。

```
int num = 100;
double& rNum = num;            // 编译错误。类型不匹配
```

永久绑定

在引用变量被声明并绑定到变量名之后，它们之间就创建了引用关系，并且在变量被销毁（超出作用域范围）之前无法断开引用关系。在 C++ 语言中，我们说变量和相应的引用变量之间的关系是一个恒定关系。图 9-2 显示了在创建引用关系之后，不能更改引用关系。

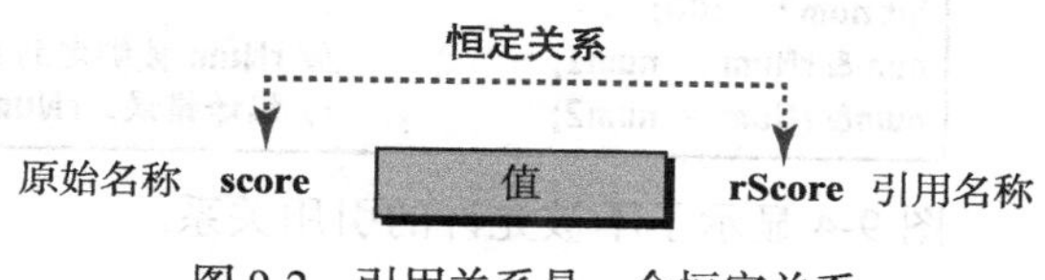

图 9-2 引用关系是一个恒定关系

我们将在本章后面讨论关于引用关系中的恒定性的更多内容。

引用关系一旦建立后，就不能被更改。

【例 9-1】 如果先将 rScore 绑定到 score，然后尝试打破这个关系，将 rScore 绑定到 num，就会出现编译错误，如下所示：

```
int score = 92;
int& rScore = score;
int num = 80;
int& rScore = num;   // 编译错误，破坏了引用关系
```

【例 9-2】 有时我们看到的语句不应与破坏引用关系混淆。例如，考虑以下内容：

```
int score = 92;
int& rScore = score;
int num = 80;
rScore = num;
```

在上述代码片段中，我们并没有破坏引用关系。我们将 num 中的值的副本存储到由引用关系建立的公共内存位置。换言之，最后一条语句并不是绑定，它是对公共变量的赋值。我们也可以使用语句 score = num，结果相同。

多重性

可以将多个引用变量绑定到同一个变量，但反过来则不行：不能将一个引用变量绑定到多个变量。

【例 9-3】 以下代码片段显示了如何把三个引用变量（rNum1、rNum2 和 rNum3）绑定到同一个变量 num。

```
int num = 100;
num& rNum1 = num;          //rNum1 被绑定到 num
num& rNum2 = num;          //rNum2 被绑定到 num
num& rNum3 = num;          //rNum3 被绑定到 num
```

这表明，同一个内存位置可以通过四个名称（num、rNum1、rNum2 和 rNum3）调用。如图 9-3 所示。

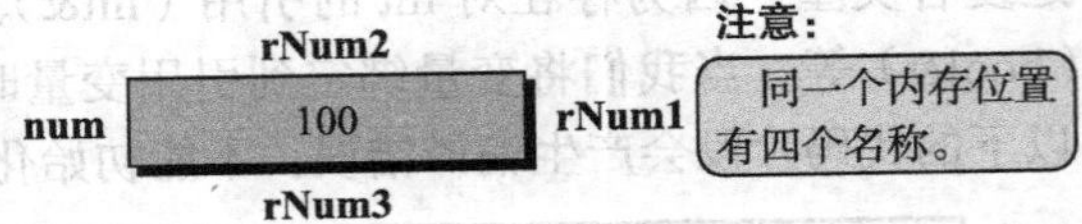

图 9-3　允许将多个引用变量绑定到同一个变量

【例 9-4】 如果试图将一个引用变量绑定到多个变量，则会出现编译错误，因为这样做意味着破坏恒定引用关系并创建一个新的引用关系，这是不允许的。

```
int num1 = 100;
int num2 = 200;
num& rNum  = num1;         // rNum 被绑定到 num1
num& rNum  = num2;         // 编译错误。rNum 不能被绑定到 num2
```

图 9-4 显示了不被允许的引用关系。

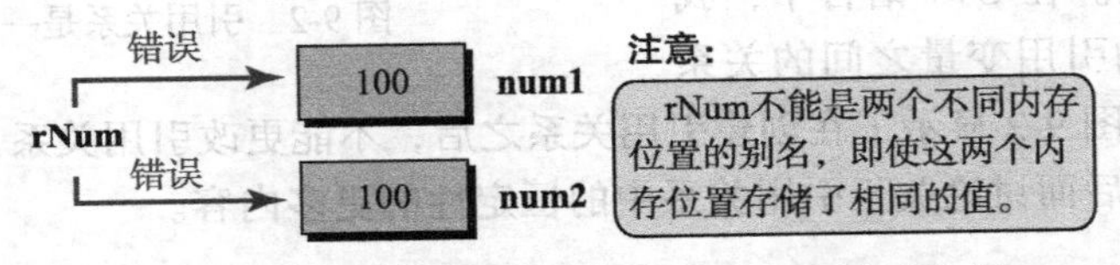

图 9-4　不允许将同一个引用变量绑定到多个变量

不能绑定到值

请注意，引用变量不能绑定到值。例如，以下语句将产生一个编译错误。

```
int& x = 92;               // 编译错误。不能绑定到值
```

9.1.2　检索值

建立引用关系后，可以通过原始变量或者引用变量来检索存储在公共内存位置中的值。

【例 9-5】 程序 9-1 显示了原始变量及其引用的使用。然而，此应用程序仅用于演示。当我们在两个不同的函数中使用变量及其引用时，它们之间的实际关系才会体现其优越性，我们将在本章后面讨论。

程序 9-1　访问值

```
1  /***************************************************************
2   * 本程序演示                                                  *
3   * 如何声明并初始化原始变量和引用变量，                        *
4   * 然后使用原始变量或者引用变量访问公共值                      *
5   ***************************************************************/
```

```
 6  #include <iostream>
 7  using namespace std;
 8
 9  int main ( )
10  {
11    // 创建引用关系
12    int score = 92;
13    int& rScore = score;
14    // 使用数据变量
15    cout << "Accessing value through data variable." << endl;
16    cout << "score: " << score << endl;
17    // 使用引用变量
18    cout << "Accessing value through reference variable." << endl;
19    cout << "rScore: " << rScore;
20    return 0;
21  }
运行结果:
Accessing value through data variable.
score: 92
Accessing value through reference variable.
rScore: 92
```

9.1.3 修改值

一个引用关系中只有一个值，可以通过原始变量或者任何引用变量修改该值，除非我们使用 const 修饰符（图 9-5）。

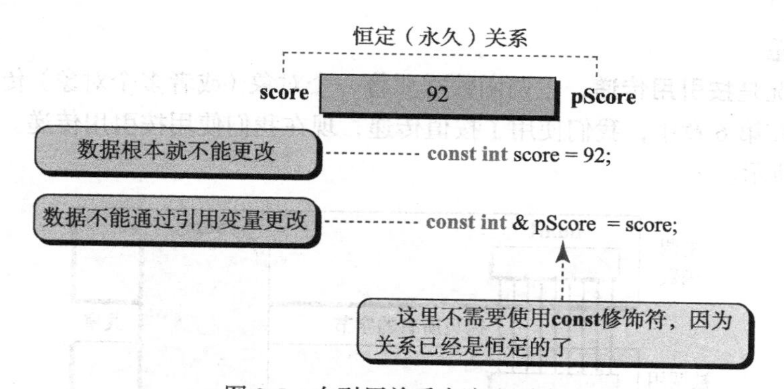

图 9-5 在引用关系中防止更改

在原始变量前面或者引用变量前面可以放置 const 修饰符。关系（绑定）本质上是恒定的，不能被破坏。图 9-5 显示了对原始变量和引用变量使用 const 修饰符的情况。

表 9-1 显示了四种可能的组合，但是第二种组合会产生编译错误，我们将在稍后讨论。

表 9-1 四种可能的组合

组合情况	数据变量	引用变量	状态
1	int name = value;	int& rName = name;	OK
2	const int name = value;	int& rName = name;	编译错误
3	int name = value;	const int& rName = name;	OK
4	const int name = value;	const int& rName = name;	OK

第一种情况

在第一种情况下，通过原始变量或者引用变量更改值没有限制。

第二种情况

第二种情况会产生编译错误，因为我们试图将一个非常量引用变量绑定到一个常量变量。由于原始变量已经是常量，因此无法通过引用变量更改其值。

第三种情况

在第三种情况下，可以通过数据变量更改数据，但我们希望通过引用变量限制数据的更改。

第四种情况

在第四种情况下，我们希望创建一个原始变量和一个引用变量，但原始变量和引用变量都不能更改公共的值。由于数据变量和引用变量只能用于检索数据，而不能用于更改数据，因此本例的应用很少见。

9.1.4　引用的应用

没有必要在同一个名称空间中（例如在同一个函数中）使用引用，因为我们总是可以使用原始变量而不需要使用引用变量。两个变量使用相同的内存位置。当两个变量处于不同的作用域（比如分别在主调函数和被调用函数中）时，引用的理念是有益的。我们可以通过使用一个内存位置来节省内存，并在两个函数中分别使用原始变量和引用变量来访问它。

在本节中，我们将讨论在两个函数之间进行通信时引用的应用方法。我们将讨论在将数据传递给函数和从函数返回数据时使用引用。第一种应用被称为按引用传递；第二种应用被称为按引用返回。

按引用传递

第一种情况是**按引用传递**，主调函数需要将一个对象（或者多个对象）传递给被调用函数进行处理。在第6章中，我们使用了按值传递，现在我们使用按引用传递。两种方法的示意图如图9-6所示。

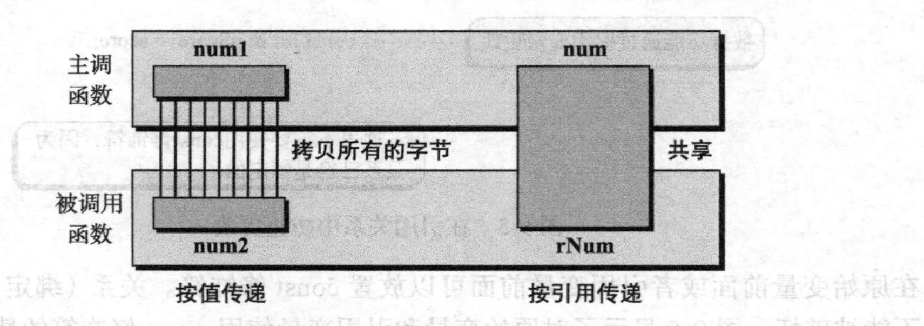

图9-6　按值传递和按引用传递的比较

我们编写两个小程序，以并排的方式比较这两种方法。假设我们需要一个函数来处理传递给它的整数。函数可以设计为按值传递或者按引用传递，如下所示：

```
#include <iostream>                    #include <iostream>
using namespace std;                   using namespace std;
// 函数原型                             // 函数原型
void doIt (int);                       void doIt (int&);
```

```
int main ()
{
   int num = 10;
   doIt (num);
   return 0;
}
// 按值传递
void doIt (int num)
{
   // 代码
}
```

```
int main ()
{
   int num = 10;
   doIt (num);
   return 0;
}
// 按引用传递
void doIt (int& rNum)
{
   // 代码
}
```

按值传递的特点。在按值传递的方法中，主调函数将实际参数的副本发送给被调用函数。副本将成为被调用函数中的形式参数。换言之，在后台将完成以下代码：

```
int num2 = num1;
```

- 在按值传递中，我们有两个独立的对象：实际参数和形式参数。这意味着形式参数的改变（无论是有意的还是意外的）不会影响实际参数。在某些情况下，这可能是一种优势，而在另一些情况下，这可能是一种劣势。
- 使用按值传递的另一个问题是复制成本。如果要复制的对象很大，那么复制实际参数的所有字节可能会产生很大开销。这意味着，如果要传递的对象是基本类型的，则不必担心，因为要复制的字节数很少（通常少于 8 个字节）。但是，如果我们需要复制一个具有数千字节的类的对象，则应该考虑其他方法。

按引用传递的特点。在按引用传递的方法中，形式参数仅仅是实际参数的引用。这两者之间的绑定在后台作为运行环境的一部分进行，如下所示：

```
int& rNum  = num;
```

- 在按引用传递中，实际参数和形式参数完全是同一个对象，因此会节省内存分配的空间。形式参数的任何改变都意味着实际参数会发生相同的改变，除非我们使用一个常量引用（如前所述）。
- 很显然，按引用传递消除了复制成本。实际参数和形式参数是同一个对象。如果要传递大对象（例如类对象）给被调用函数，则应该考虑使用该传递方法。

建议。上面我们讨论了向函数传递参数的各种方法的优缺点，这里我们将给出以下建议，以帮助读者决定使用哪种方法。

1）如果需要防止更改操作，则应该：

a）对于小对象，使用按值传递。

b）对于大对象，使用按常量引用传递。

2）如果需要进行更改操作，则应该使用按引用传递。

警告。不能把引用的形式参数绑定到实际参数值。例如，以下代码片段将导致编译错误。

```
void fun (int& rX) { ... }          // 函数定义
fun (5);                            // 函数调用（编译错误）
```

形式参数 rX 是一个引用参数。函数调用的实际参数要求是一个变量名称，而不是一个值。

【例 9-6】 假设我们想要编写一个函数来打印基本数据类型的值。我们不想更改主调函数中的数据值。这是我们建议的第一种情况，可以使用按值传递或者按常量引用传递。由于对象很小，可以使用按值传递。我们以前已经举过几个这样的例子。

【例 9-7】 假设我们要为类编写一个拷贝构造函数。这也是我们建议的第一种情况，因为我们不希望函数更改原始对象。但是，按值传递的开销较大（对象可能很大），更重要的是，按值传递是不可能的，因为按值传递需要调用拷贝构造函数来复制对象，这意味着我们需要拷贝构造函数来创建拷贝构造函数（恶性循环）。基于 C++ 标准，必须使用按常量引用传递的方法，如下所示：

```
// 在拷贝构造函数中的按引用传递
Circle :: Circle (const Circle& circle)
: radius (circle.radius)
{
}
```

【例 9-8】 假设我们想要编写一个函数来交换两个数据项。这是我们建议的第二种情况。我们需要修改参数，因此可以使用按引用传递的方法，如程序 9-2 所示。

程序 9-2　使用交换函数，采用按引用传递

```
1   /***************************************************************
2     * 本程序演示如何使用按引用传递，                                *
3     * 以允许被调用函数交换主调函数中的两个值                        *
4     ***************************************************************/
5   #include <iostream>
6   using namespace std;
7
8   void swap (int& first, int& second) ; // 函数原型
9
10  int main ( )
11  {
12    // 定义两个变量
13    int x = 10;
14    int y = 20;
15    // 打印交换前的 x 和 y 的值
16    cout << "Values of x and y before swapping." << endl;
17    cout << "x: " << x << "  " << "y: " << y << endl;
18    // 调用 swap 函数来交换 x 和 y 的值
19    swap (x , y);
20    // 打印交换后的 x 和 y 的值
21    cout << "Values of x and y after swapping." << endl;
22    cout << "x: " << x << "  " << "y: " << y;
23    return 0;
24  }
25  /***************************************************************
26    * swap 函数交换形式参数的值，                                   *
27    * 按引用传递允许 main 函数中                                    *
28    * 对应的实际参数同样被交换                                      *
29    ***************************************************************/
30  void swap (int& rX, int& rY)
31  {
```

```
32    int temp = rX;
33    rX = rY;
34    rY = temp;
35  }
```

运行结果：

```
Values of x and y before swapping.
x: 10  y: 20
Values of x and y after swapping.
x: 20  y: 10
```

按引用返回

第二种情况是**按引用返回**，被调用函数具有必须返回到主调函数的对象。

按值返回的特点。在**按值返回**的方法中，被调用函数使用以下函数原型返回指定类型的对象。

```
type function (...);
```

按值返回十分简单，可以在任何地方使用。我们可以返回形式参数或者局部变量的值。唯一的缺点是复制成本。如果要返回的对象是一个基本类型，则没有问题；如果要返回的是一个类的对象，由于需要调用拷贝构造函数，因此复制成本可能很高。

按引用返回的特点。在**按引用返回**的方法中，要返回的对象类型是对另一个对象的引用，其函数原型如下所示。

```
type& function (...);
```

此方法消除了复制的成本，但它有一个缺点：如果对象是值参数或者局部变量（静态变量除外），则不能通过引用返回该对象。我们将在本章后面讨论其原因：当函数终止时，所有局部变量和值参数都会被销毁。当一个对象被销毁时，我们不能给它一个别名，但是可以返回一个对引用参数的引用。

【例 9-9】 假设我们要编写一个函数来获取两个整数之间的较大值。我们可以使用按值传递和按值返回的组合，也可以使用按引用传递和按引用返回的组合。以下代码片段以并排的方式比较了这两种组合方法。

```
#include <iostream>
using namespace std;

// 按值返回
int larger (int x, int y )
{
   if (x > y)
   {
        return x;
   }
   return y;
}
int main ()
{
   int x = 10;
   int y = 20;
```

```
#include <iostream>
using namespace std;

// 按引用返回
int& larger (int& x, int& y )
{
   if (x > y)
   {
        return x;
   }
   return y;
}
int main ()
{
   int x = 10;
   int y = 20;
```

```
    int z = larger (x, y);                    int z = larger (x, y);
    cout << z;                                cout << z;
    return 0;                                 return 0;
}                                         }
Run:                                      Run:
20                                        20
```

在这种情况下，我们演示了这两种方法的示例，因为复制的成本在传递值和返回值方面都很小。

【例 9-10】 在本例中，我们将获取两个类的对象之间的较大者。在第 7 章中，我们创建了一个分数类。我们不在这里重复其接口文件和实现文件（独立编译）。我们仅创建一个应用程序文件，它使用按引用传递和按引用返回来获取两个分数中较大的分数。程序 9-3 显示了应用程序文件的代码清单。

程序 9-3　获取两个分数中的较大者

```
1   /*****************************************************************
2    * 本程序创建两组分数对象实例，                                    *
3    * 然后调用一个名为 larger 的函数以获取各组中的较大者               *
4    *****************************************************************/
5   #include "fraction.h"
6
7   Fraction& larger (Fraction&, Fraction&); // 函数原型
8
9   int main ( )
10  {
11    // 创建第一组分数，并获取其中较大者
12    Fraction fract1 (3, 13);
13    Fraction fract2 (5, 17);
14    cout << "Larger of the first pair of fraction: " ;
15    larger (fract1, fract2).print ();
16    // 创建第二组分数，并获取其中较大者
17    Fraction fract3 (4, 9);
18    Fraction fract4 (1, 6);
19    cout << "Larger of the second pair of fractions: " ;
20    larger (fract3, fract4).print ();
21    return 0;
22  }
23  /*****************************************************************
24   * 该函数按引用接收两个分数对象的实例，                            *
25   * 比较两个分数，并返回较大者                                      *
26   *****************************************************************/
27  Fraction& larger (Fraction& fract1, Fraction& fract2)
28  {
29    if (fract1.getNumer() * fract2.getDenom() >
30                              fract2.getNumer() * fract1.getDenom())
31    {
32        return fract1;
33    }
34    return fract2;
35  }
```

```
运行结果：
Larger of first pair of fractions: 5/17
Larger of second pair of fractions: 4/9
```

值得注意的是，在每次调用中，返回的项都是左值，是被传递给函数的对象之一。这就是为什么我们可以将类中定义的打印函数应用于该对象（第 15 行和第 20 行）。

9.2 指针

指针类型是表示内存位置地址的复合类型。指针变量是其内容为指针类型的变量。在本节中，我们将讨论地址、指针类型、指针变量以及一些相关问题。

9.2.1 地址

当我们谈论地址时，必须区分两个独立的概念：内存中的地址和变量的地址。

内存中的地址

我们曾在第 1 章中简要讨论过，计算机内存是一个字节序列。换言之，最小的可访问单元是一个字节（byte）。当我们说计算机的内存（随机存取内存或者 RAM）为 1KB 时，其含义是该内存由 2^{10} 或者 1024 个字节组成。如今的计算机内存在兆字节（MB，2^{20}）、千兆字节（GB，2^{30}）甚至太字节（TB，2^{40}）的范围内。

内存的每个字节都有一个地址。地址以十六进制格式表示。例如，在内存为 1KB 的计算机中，字节的编号从 0x000 到 0x3ff（十进制为 0 到 1023），如图 9-7 所示。

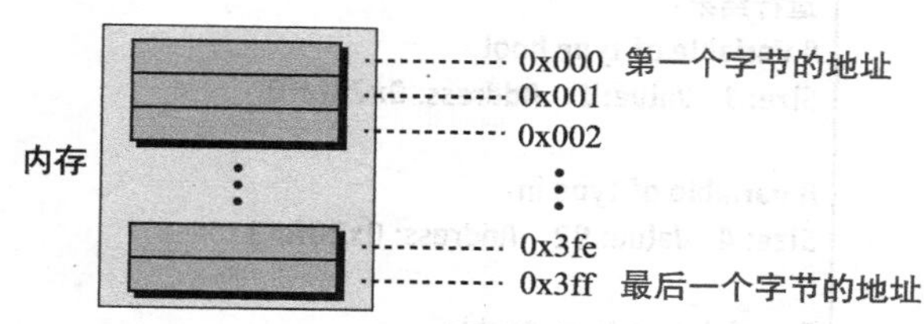

图 9-7 1KB 的随机存取内存的地址

变量的地址

在程序中，可以定义不同类型（布尔、字符、整数、浮点和类）的变量。每个变量占用一个或者多个字节的内存。bool 或者 char 类型的变量通常占用 1 个字节的内存。int 类型的变量可能占用 4 个或者更多字节的内存。double 类型的变量也可能占用 4 个或者更多字节的内存（sizeof 运算符可以用于确定字节数）。变量的地址是其所占用的第一个字节的地址。这是我们在处理地址或者指针时必须牢记的一个要点，稍后将讨论。为了获得变量的地址，我们在变量前面使用地址运算符（&）。

> 变量的地址是该变量所占用的第一个字节的地址。

【例 9-11】 在本例中，我们编写了一个简单的程序来打印三个不同类型（bool、int 和 double）变量的地址（程序 9-4）。

程序 9-4 打印变量的大小、值和地址

```
1  /***************************************************************
2   * 本程序先定义三个变量，                                       *
3   * 然后打印它们的值以及它们在内存中的地址                       *
4   ***************************************************************/
5  #include <iostream>
6  using namespace std;
7  
```

```
 8  int main ( )
 9  {
10     // 声明三个数据变量,
11     bool flag = true;
12     int score = 92;
13     double average = 82.56;
14     // 打印变量 flag 的大小、值、地址
15     cout << "A variable of type bool" << endl;
16     cout << "Size: " << sizeof (flag) << "   ";
17     cout << "Value: " << flag << "   ";
18     cout << "Address: "<< &flag << endl << endl;
19     // 打印变量 score 的大小、值、地址
20     cout << "A variable of type int" << endl;
21     cout << "Size: " << sizeof (score) << "   ";
22     cout << "Value: " << score << "   ";
23     cout << "Address: "<< &score << endl << endl;
24     // 打印变量 average 的大小、值、地址
25     cout << "A variable of type double" << endl;
26     cout << "Size: " << sizeof (average) << "   ";
27     cout << "Value: " << average << "   ";
28     cout << "Address: "<< &average << endl;
29     return 0;
30  }
运行结果:
A variable of type bool
Size: 1  Value: 1  Address: 0x28fef0

A variable of type int
Size: 4  Value: 92  Address: 0x28fef1

A variable of type double
Size: 8  Value: 82.56  Address: 0x28fef5
```

关于程序 9-4 需要注意以下四点：

- 插入运算符（<<）被重载，因此它可以接收地址。我们可以使用表达式“cout << &score”来打印变量的地址。
- 地址用十六进制表示法来显示。地址不是整数，它们是指针类型，稍后我们将讨论。
- 不同系统中的大小和地址可能不同。地址可以从最小到最大，也可以从最大到最小。地址可能会不连续。

图 9-8 显示了程序 9-5 中输出的地址。

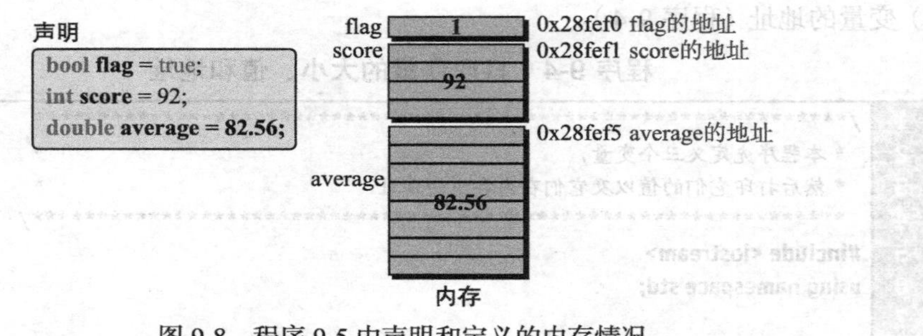

图 9-8　程序 9-5 中声明和定义的内存情况

9.2.2 指针类型和指针变量

我们可以操作地址，因为 C++ 定义了指针类型并允许我们使用指针变量。

指针类型

指针类型是一个复合类型，其字面量值是地址。存在指向 char 的指针（char 变量的地址）、指向 int 的指针（int 变量的地址）、指向 double 的指针（double 变量的地址）等，从这个意义上说，它是一个复合类型。注意，我们只讨论变量的地址，而不是值，因为值总是存储在内存的变量中。为了创建指向某个类型的指针，我们在类型后面添加星号（*）。以下代码片段显示了一些指针类型。

```
bool*                          //指向一个 bool 类型对象的指针
int*                           //指向一个 int 类型对象的指针
double*                        //指向一个 double 类型对象的指针
Circle*                        //指向一个 Circle 类型对象的指针
```

为了创建指向某个类型的指针，我们将星号（*）添加到类型名称后面。

指针变量

可以将指针类型的值存储在指针变量中。虽然可以有不同的指针类型，但是存储在指针变量中的是一个 4 字节的地址。换言之，指针变量的大小在 C++ 中是固定的，这意味着指针变量可以是一个从 0x00000000 到 0xFFFFFFFF 的地址。

声明。要使用指针，需要声明指针变量。要声明指针变量，则需要告诉计算机它是指向特定类型的指针。以下代码片段是声明指针变量的示例。名称由我们自己决定，因为我们稍后会将这些指针变量绑定到相应的数据变量。

```
bool* pFlag;                   // pFlag 是指向 bool 类型的指针变量
int* pScore;                   // pScore 是指向 int 类型的指针变量
double* pAverage;              // pAverage 是指向 double 类型的指针变量
```

我们从右向左读取声明（pFlag 是一个指针，指向 bool 类型）。

初始化。指针变量和数据变量一样，在使用之前必须初始化。指针变量必须用内存中的有效地址初始化，这意味着我们不能用字面量地址初始化指针变量，地址必须是现有变量的地址。为了满足这个标准，可以使用地址运算符（和号，&）。稍后我们将讨论此运算符，以下代码片段说明了如何初始化三个已声明的指针变量。

```
bool* pFlag = &flag;           //使用 flag 的地址初始化 pFlag
int* pScore = &score;          //使用 score 的地址初始化 pScore
double* pAverage = &average;   //使用 average 的地址初始化 pAverage
```

指针初始化有一些限制。由于 pFlag 是指向 bool 类型变量的指针，因此必须使用 bool 类型变量的地址对其进行初始化。同样，由于 pScore 是指向 int 类型变量的指针，因此必须使用 int 类型变量的地址对其进行初始化。图 9-9 显示了指针变量（pScore）与相应的被指变量（score）之间的实际关系和符号关系。

【例 9-12】 以下代码片段将产生编译错误：无法将指针字面量值赋给指针变量。

```
double* pAverage = 0x123467;   // 编译错误。不能使用字面量地址
```

【例 9-13】 以下代码片段将产生编译错误：无法将一个 int 变量的地址赋值给一个

double 类型的指针变量。

```
int num;
double* p = &num;                    // 编译错误。变量 p 的类型是 double*
```

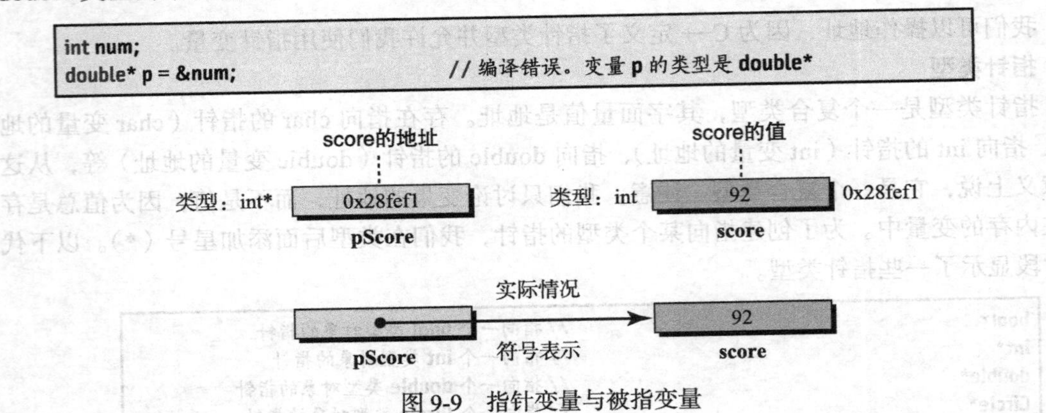

图 9-9 指针变量与被指变量

间接引用（解引用）

指针变量的声明和初始化允许我们使用指针变量中存储的地址来访问被指变量中存储的值。这是通过使用星号的**间接运算符**（也称为解引用运算符）完成的，如下所示：

```
*pFlage;                  // 存储在 flag 中的值
*pScore;                  // 存储在 score 中的值
*pAverage;                // 存储在 average 中的值
```

读者可能会提出疑问，既然可以直接检索或者更改变量 flag 的值，为什么需要使用间接运算符呢？正如我们将在本节后面看到的，答案是需要在不同的场合中使用不同的方法（直接访问或者间接访问）。例如，在主调函数中使用直接方法，在相应的被调用函数中使用间接方法。

两个相关的运算符

我们在本节中使用的两个运算符是地址运算符（&）和间接运算符（*）。这两个运算符是附录 C 中定义的一元运算符，如表 9-2 所示。

表 9-2 地址表达式

组别	名称	运算符	表达式	优先级	结合性
一元表达式	取地址 间接访问	& *	&lvalue *pointer	17	←

图 9-10 显示了如何使用这两个运算符。地址运算符（&）接收一个变量并返回其地址；间接运算符（*）接收一个地址并返回相应的被指变量的值。

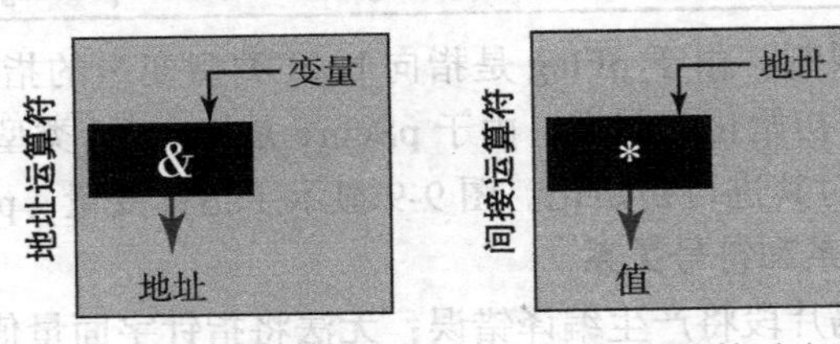

图 9-10 地址运算符（&）和间接运算符（*）

C++ 语言中，符号 & 和 * 有三种用途，如表 9-3 所示。

表 9-3　& 和 * 运算符的用途

符号	类型定义	一元运算符	二元运算符
&	type&	&variable	x & y
*	type*	*pointer	x * y

符号 & 可以用于定义引用类型，还可以用作一元运算符来获取变量的地址，也可以用作二进制按位与运算符（在附录 C 中讨论）。当符号 & 用于定义引用类型时，把它放在类型之后（int&）。当符号 & 用于获取变量的地址时，把它放在变量名之前（&score）。当符号 & 用作二元运算符时，把它放在两个操作数之间。

符号 * 也可以在三种情况下使用：作为指针类型，作为一元运算符以获取被指变量的值，以及作为二元乘法运算符。当符号 * 用于定义指针类型时，把它放在类型之后（int*）。当符号 * 用于间接访问一个值时，把它放在指针之前（*pScore）。当符号 * 用作二元运算符时，把它放在两个操作数之间。

9.2.3　检索值

在建立了指针关系之后，我们可以通过数据变量或者指针变量来检索存储在内存位置中的值。

【例 9-14】 在本例中，我们将编写一个简单的程序，直接和间接地访问数据变量的值（程序 9-5）。

程序 9-5　直接和间接检索数据

```
1   /***************************************************************
2    * 本程序演示如何直接或者间接                                     *
3    * 访问（检索或者修改）数据的值                                   *
4    ***************************************************************/
5   #include <iostream>
6   using namespace std;
7   
8   int main ( )
9   {
10      // 声明和初始化变量
11      int score = 92;
12      int* pScore = &score;
13      // 直接和间接检索数据变量的值
14      cout << "Direct retrieve of score: " << score << endl;
15      cout << "Indirect retrieve of score: " << *pScore;
16      return 0;
17  }
```

```
运行结果：
Direct retrieve of score: 92
Indirect retrieve of score 92
```

9.2.4　使用 const 修饰符

一个容易引起混淆和错误的问题是如何控制对数据变量和指针的更改。可以使用 const 修饰符来控制更改。我们最多可以使用三个 const 修饰符，每个修饰符都有不同的用途，如图 9-11 所示。前两个与更改数据有关，第三个与更改数据变量和指针变量之间的关系有关。

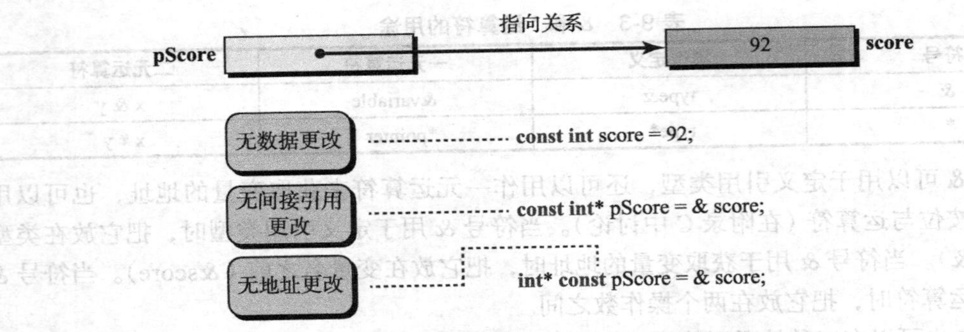

图 9-11　与指针相关的 const 修饰符的三种使用方式

控制数据更改

表 9-4 列举了可以用来控制数据更改的四种可能的组合。

表 9-4　四种可能的组合

情况	数据变量	指针变量
1	int name = value;	int* pName = &name;
2（非法）	const int name = value;	int* pName = &name;
3	int name = value;	const int* pName = &name;
4	const int name = value;	const int* pName= &name;

第一种情况。在第一种情况下，既可以通过数据变量更改值，也可以通过指针变量更改值，两种方法没有任何限制条件。

```
int score = 92;
int* pScore = &score;
score = 80;                       // 通过数据变量更改
*pScore = 70;                     // 通过指针变量更改
```

第二种情况。第二种情况是非法的。当我们尝试将数据变量绑定到指针变量时，会产生一个编译错误。由于数据变量已经是常量，因此无法将其绑定到非常量指针类型。

```
const int score = 92;
int* pScore = &score;             // 编译错误
```

第三种情况。在第三种情况下，可以通过数据变量更改数据，但我们希望防止通过指针变量更改数据。如果我们试图通过指针变量更改数据，结果会产生一个编译错误。

```
int score = 92;
const int* pScore = &score;
score = 80;
*pScore = 80;                     // 编译错误
```

第四种情况。在第四种情况下，我们希望创建一个不能更改数据的数据变量和指针变量。由于数据变量和指针变量只能用于检索数据，而不能更改数据，所以这种情况的应用很少。

```
const int score = 92;
const int* pScore = &score;
score = 80;                       // 编译错误
*pScore = 80;                     // 编译错误
```

更改指针（绑定）

可以使用 const 修饰符在程序的生命周期中将数据变量绑定到指针变量。换言之，前面四种情况中的每一种都可以与非常量指针或者常量指针组合在一起（表 9-5）。

表 9-5 八种可能的组合

更改数据变量和指针变量之间的关系
前四种情况的任何一种 pName = &name; // 可更改
前四种情况的任何一种 const pName = &name; // 不可更改

图 9-12 显示了如果指针关系为常量，则无法中断其关系并使指针指向另一个数据变量。注意，在这种情况下，const 修饰符必须放在指针变量的名称前面。

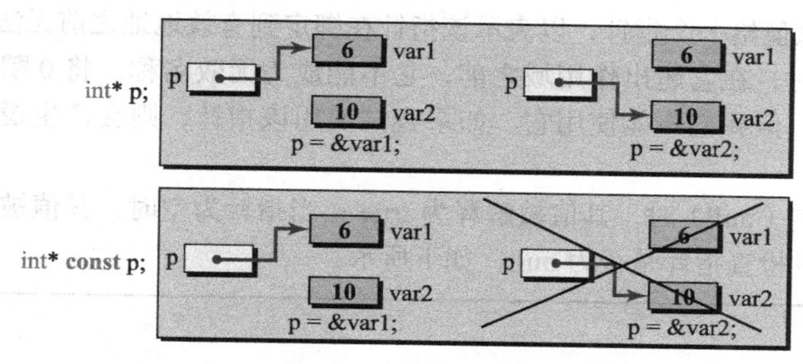

图 9-12 不允许打破常量指向关系

9.2.5 指向指针的指针

指针变量是内存中的一个位置。我们能把这个内存位置的地址存储在另一个内存位置吗？答案是肯定的。存在新的数据类型：指向"指向 int 的指针"的指针，指向"指向 double 的指针"的指针，等等。我们甚至可以更进一步，让一个指针指向"指向'指向 int 的指针'的指针"，依此类推。但在实践中很少看到两个以上的层次。当我们在一个类型后面使用双星号（**）时，其含义是指向"指向类型的指针"的指针，如下所示：

```
int score = 92;                  // 声明和初始化变量 score
const int* pScore = &score;      // 绑定 pScore 到 score
int** ppScore = &pscore;         // 绑定 ppScore 到 pScore
```

指向指针的指针的概念示意图如图 9-13 所示。

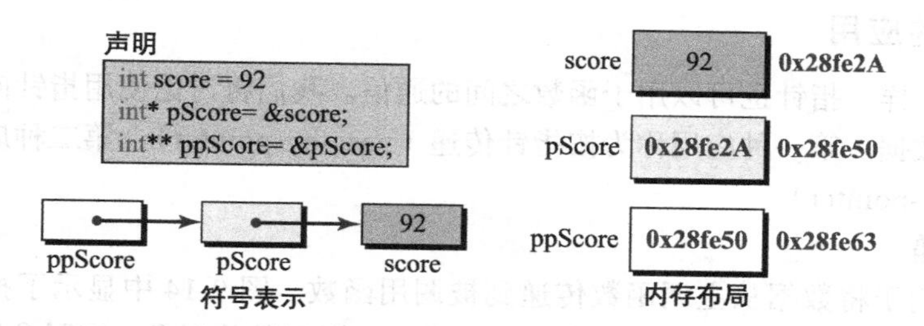

图 9-13 指向 int 指针的指针

9.2.6 两种特殊的指针

有时我们会遇到两种特殊的指针：不指向任何对象的指针（null 指针）和指向 void 的指

针（void 指针）。本节简单地解释这两种指针，在后续章节中将讨论其更多的应用。

不指向任何对象的指针：null 指针

不指向任何对象的指针（有时称为 **null 指针**）是不指向任何内存位置的指针。虽然 C 语言使用单词 NULL 来定义这样的指针，但是 C++ 语言更倾向于使用字面量 0。换言之，如果我们想表示指针在当前时刻没有指向任何内存位置，则可以将它绑定到 0，如下所示。

```
int* p1 = 0;
double* p2 = 0;
```

注意，这两条语句并不表示 p1 或者 p2 指向 0x0 处的内存位置。此地址处的字节用于系统，用户无法使用。字面量 0 仅表示指针此时不指向任何位置。

程序员将 0 赋值给一个指针，以表示该指针在绑定到有效地址之前无法使用。注意，当一个名称被声明时，在它超出作用域之前，它不能成为无效名称。将 0 赋给指针变量意味着声明仍然有效，但我们不能使用它。如果尝试使用该指针，则会产生逻辑错误，程序将中止。

当指针不为空（null）时，其值被解释为 true；当指针为空时，其值被解释为 false。这表明我们总是可以检查指针是否为 null，如下所示：

```
int x = 7;
int* p = &x;
if (p) {...}                              // 此处测试结果为 true（p 非 null）
p = 0;
if (p) {...}                              // 此处测试结果为 false（p 为 null）
```

指向 void 的指针：通用指针

void 指针是一个**通用指针**。它可以指向任何类型的对象。以下代码片段显示了如何使用通用指针指向任何类型。假设我们首先需要将它指向一个 int 对象，然后再指向一个 double 对象。请注意，在将 void 指针转换为适当的类型之前，不能引用相应的对象。

```
void* p;                                  // 声明一个 void 指针
int x = 10;
p = &x;                                   // 让 p 指向 int 类型
double y = 23.4;
p = &y;                                   // 让 p 指向 double 类型
```

9.2.7　指针的应用

就像引用一样，指针也可以用于函数之间的通信。我们将讨论使用指针向函数传递数据和从函数返回数据。第一种应用称为按指针传递（pass-by-pointer）；第二种应用称为按指针返回（return-by-pointer）。

按指针传递

指针可以用于将数据从主调函数传递到被调用函数。图 9-14 中显示了按值传递、按引用传递和按指针传递。第三种情况是**按指针传递**，主调函数将对象（实际参数）的地址传递给被调用函数，被调用函数将其存储在指针（形式参数）中。与按引用传递不同，这里没有共享。运行时系统必须复制主调函数中实际参数的地址，并将其传递给被调用函数。但是，在按指针传递的情况下，复制成本并不高，是复制 4 字节地址的固定成本。

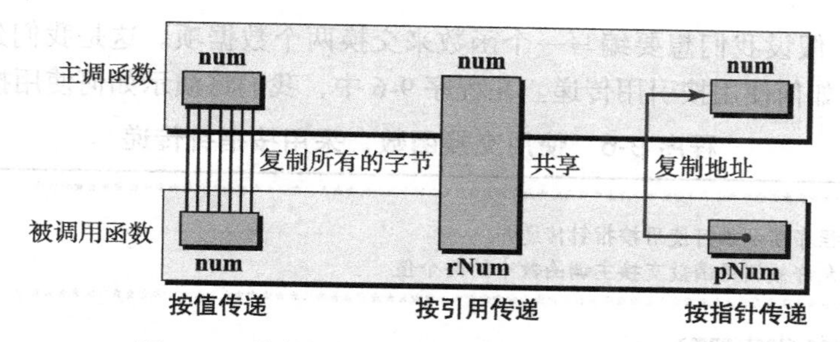

图 9-14 将数据传递给函数的三种方法的比较

我们编写两个小程序，以并排方式比较按值传递和按指针传递这两种方法。假设我们想要一个函数对传递给它的整数进行处理。函数可以设计为按值传递或者按指针传递，如下所示：

```
#include <iostream>
using namespace std;
// 函数原型
void doIt (int);

int main ()
{
   int num1 = 10;
   doIt (num1);
   return 0;
}
// 按值传递
void doIt (int num2)
{
   // 代码
}
```

```
#include <iostream>
using namespace std;
//函数原型
void doIt (int*);

int main ()
{
   int num = 10;
   doIt (&num);
   return 0;
}
//按指针传递
void doIt (int* pNum)
{
   //代码
}
```

按指针传递的特点。我们在前面讨论了按值传递和按引用传递的特点。在按指针传递方法中，形式参数是实际参数的地址。这两者之间的绑定在后台作为运行环境的一部分进行，如下所示：

```
int* pNum  = &num;
```

- 在按指针传递中，实际参数和形式参数绑定在一起。形式参数的任何更改都意味着实际参数的相同更改，除非我们像前面讨论的那样使用常量指针。
- 显然，按指针传递减少了复制成本：只复制一个 4 字节的地址。当我们想将一个大对象（例如类的对象）传递给一个被调用函数时，可以考虑这种方法。

建议。我们将前面的建议拓展为三种实践方法：按值传递、按引用传递和按指针传递。

1）如果需要防止更改操作，则应该：

a）对于小对象，使用按值传递方法。

b）对于大对象，使用按常量引用传递或者按常量指针传递。

2）如果需要进行更改操作，则应该使用按引用传递或者按指针传递。

【例 9-15】 假设我们想要编写一个函数来打印基本数据类型的值。我们不想更改主调函数中数据的值。这是我们建议的第一种情况。由于对象很小，可以使用按值传递。

【例 9-16】 假设我们想要编写一个函数来交换两个数据项。这是我们建议的第二种情况。前文演示了如何使用按引用传递。在程序 9-6 中，我们将演示如何使用按指针传递。

程序 9-6　使用交换函数，采用按指针传递

```
1  /***************************************************************
2   * 本程序演示如何使用按指针传递，                                *
3   * 以允许被调用函数交换主调函数中的两个值                        *
4   ***************************************************************/
5  #include <iostream>
6  using namespace std;
7
8  void swap (int* first, int* second) ; // 函数原型
9
10 int main ( )
11 {
12   // 定义两个变量
13   int x = 10;
14   int y = 20;
15   // 打印交换前的 x 和 y 的值
16   cout << "Values of x and y before swapping." << endl;
17   cout << "x: " << x << "   " << "y: " << y << endl;
18   // 调用 swap 函数来交换 x 和 y 的值
19   swap (&x , &y);
20   // 打印交换后的 x 和 y 的值
21   cout << "Values of x and y after swapping." << endl;
22   cout << "x: " << x << "   " << "y: " << y;
23   return 0;
24 }
25 /***************************************************************
26  * swap 函数交换形式参数的值，                                   *
27  * 使用按指针传递的方法，                                        *
28  * 使得 main 函数中对应的实际参数也相应地被交换                  *
29  ***************************************************************/
30 void swap (int* pX, int* pY)
31 {
32   int temp = *pX;
33   *pX= *pY;
34   *pY = temp;
35 }
```

```
运行结果：
Values of x and y before swapping.
x: 10  y: 20
Values of x and y after swapping.
x: 20  y: 10
```

按指针返回

第二种情况是**按指针返回**，被调用函数具有必须返回到主调函数的对象。我们可以使用按指针返回。

按指针返回的特点。在按指针返回的方法中，要返回的对象类型是一个指向函数形式参数的指针，其本身则是指向主调函数中的一个对象，函数原型如下所示。

```
type* function (type* ...);
```

在讨论按引用返回的情况时我们曾指出，此方法有若干缺点，最值得注意的是：我们不能返回指向值参数或者局部变量的指针。当被调用函数终止时，所有局部对象和值参数都将被销毁，并且不存在可指向的对象。

【例 9-17】 假设我们要编写一个函数来获取两个整数中较大的整数。我们可以使用按值传递和按值返回的组合，也可以使用按指针传递和按指针返回的组合。以下代码片段以并排方式显示两个程序并进行比较。

```
#include <iostream>
using namespace std;

int larger (int x, int y )
{
   if (x > y)
   {
        return x;
   }
   return y;
}
int main ()
{
   int x = 10;
   int y = 20;
   int z = larger (x, y);
   cout << z;
   return 0;
}
```

运行结果：
20

```
#include <iostream>
using namespace std;

int* larger (int* x, int* y )
{
   if (*x > *y)
   {
        return x;
   }
   return y;
}
int main ()
{
   int x = 10;
   int y = 20;
   int z = *larger (&x, &y);
   cout << z;
   return 0;
}
```

运行结果：
20

我们将看到这两种实践的示例，因为复制的成本在传递指针和返回指针方面并不高。

【例 9-18】 在本例中，我们获取两个类的对象中较大的一个。在前文中，我们曾经使用了按引用返回，现在使用按指针返回。我们创建了一个应用程序文件，它使用按指针传递和按指针返回的方法，来获取两个分数中较大的分数。程序 9-7 显示了应用程序文件的代码清单。

注意，成员选择表达式（object->member）实际上是表达式（*object.member）的快捷方式。这两个表达式都在第 7 章中讨论过。还要注意，指针返回的对象是一个左值，这意味着我们可以把成员函数 print 应用于返回的对象，如第 17 行和第 22 行所示。

程序 9-7 获取两个对象中的较大者

```
1  /**************************************************************
2   * 本程序创建两组分数对象实例，                                   *
3   * 然后调用一个名为 larger 的函数以获取各组中的较大者               *
4   **************************************************************/
5  #include "fraction.h"
6  #include <iostream>
7  using namespace std;
8
9  Fraction* larger (Fraction*, Fraction*); // 函数原型
```

```
10
11  int main ( )
12  {
13    //创建第一组分数，并获取其中较大者
14    Fraction fract1 (3, 13);
15    Fraction fract2 (5, 17);
16    cout << "Larger of the first pair of fraction: " ;
17    larger (&fract1, &fract2) -> print ();
18    //创建第二组分数，并获取其中较大者
19    Fraction fract3 (4, 9);
20    Fraction fract4 (1, 6);
21    cout << "Larger of the second pair of fractions: " ;
22    larger (&fract3, &fract4) -> print ();
23    return 0;
24  }
25  /*********************************************************************
26   * 该函数按指针接收两个分数对象的实例，                                *
27   * 比较两个分数，并返回较大者                                          *
28   *********************************************************************/
29  Fraction* larger (Fraction* fract1, Fraction* fract2)
30  {
31    if (fract1 -> getNumer() * fract2 -> getDenom() >
32                        fract2 -> getNumer() * fract1 ->getDenom())
33    {
34        return fract1;
35    }
36    return fract2;
37  }
```

运行结果：
```
Larger of first pair of fractions: 5/17
Larger of second pair of fractions: 4/9
```

9.3 数组和指针

在 C++ 中，数组和指针紧密相关。在本节中，我们将学习如何用指针表示一维数组和二维数组。

9.3.1 一维数组和指针

当我们声明一个名为 arr 的数组（类型为 t，容量为 N）时，系统会创建 N 个类型为 t 的内存位置。然后系统会创建一个指向第一个元素的类型为 t 的常量指针，如图 9-15 所示。

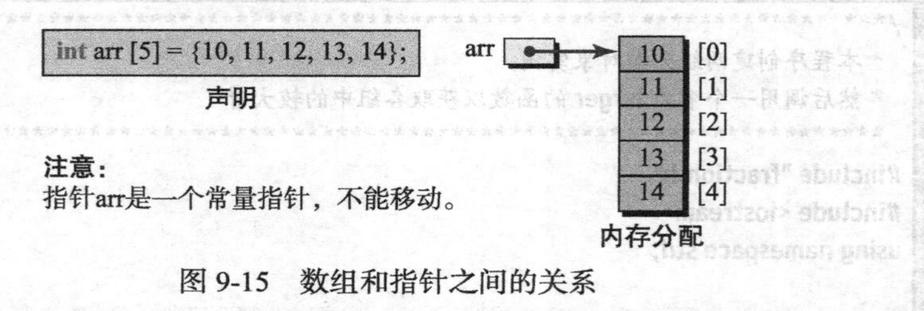

图 9-15　数组和指针之间的关系

指针为常量意味着其内容（地址）不能被更改，它总是指向第一个元素。因为第一个元

素的地址是固定的，所以我们知道所有元素的地址。索引 0 处元素的地址是 arr（或者 arr + 0）；索引 1 处元素的地址是 arr + 1；依此类推。

数组的名称是指向第一个元素的常量指针。

为了演示，我们编写了一个程序打印图 9-15 中定义的数组元素的地址。我们同时使用指针和地址运算符的值来证明它们是相同的（程序 9-8）。程序表明我们可以使用常量指针或者索引访问每个元素的地址。事实上，索引是符号表示，其目的是使元素访问更容易。这意味着 arr[0] 与 *(arr + 0) 等同，但我们必须使用括号，因为星号运算符的优先级高于加法运算符，使用括号将允许我们更改优先级。有趣的是，每个地址都增加了 4 个字节，这表明整数的大小是 4。

当使用指针引用数组元素时，我们需要使用括号来优先计算加法运算符。

程序 9-8　本程序用于检查各数组元素的地址

```
1   /***************************************************************
2    * 本程序说明                                                    *
3    * 系统把数组第一个元素的地址存储在常量指针中                       *
4    ***************************************************************/
5   #include <iostream>
6   using namespace std;
7   
8   int main ()
9   {
10    // 声明一个包含 5 个整数的数组
11    int arr [5];
12    // 分别通过指针和 & 运算符打印地址
13    for (int i = 0; i < 5; i++)
14    {
15        cout << "Address of cell " << i << " Using pointer: ";
16        cout << arr + i << endl;
17        cout << "Address of cell " << i << " Using & operator: ";
18        cout << &arr [i] << endl << endl;
19    }
20    return 0;
21  }
```

运行结果：

```
Address of cell 0 Using pointer: 0x28fee8
Address of cell 0 Using address operator: 0x28fee8

Address of cell 1 Using pointer: 0x28feec
Address of cell 1 Using address operator: 0x28feec

Address of cell 2 Using pointer: 0x28fef0
Address of cell 2 Using address operator: 0x28fef0

Address of cell 3 Using pointer: 0x28fef4
Address of cell 3 Using address operator: 0x28fef4
```

```
Address of cell 4 Using pointer: 0x28fef8
Address of cell 4 Using address operator: 0x28fef8
```

【例 9-19】 程序 9-9 表明使用索引和指针引用数组元素，可以得到相同的结果。

程序 9-9　使用索引和指针访问数组元素

```
 1  /***************************************************************
 2   * 本程序显示如何使用索引或者指向元素的指针                    *
 3   * 访问数组的元素                                              *
 4   ***************************************************************/
 5  #include <iostream>
 6  using namespace std;
 7
 8  int main ( )
 9  {
10     // 声明并初始化一个数组
11     int numbers [5] = {10, 11, 12,13, 14};
12     // 通过索引访问数值元素
13     cout << "Accessing elements through indexes" << endl;
14     for (int i = 0; i < 5; i++)
15     {
16          cout << numbers [i] << "  ";
17     }
18     cout << endl;
19     // 通过指针访问数值元素
20     cout << "Accessing elements through pointers" << endl;
21     for (int i = 0; i < 5; i++)
22     {
23          cout << *(numbers + i) << "  ";
24     }
25     return 0;
26  }
运行结果:
Accessing elements through indexes
10  11  12  13  14
Accessing elements through pointers
10  11  12  13  14
```

指针的算术运算

指针类型不是整数类型。但是，**指针的算术运算**允许对指针类型应用有限的算术运算符。我们必须讨论这些运算符的新定义。当我们使用这些运算符时，必须牢记的是，因为结果是另一个指针值，所以该值必须指向可控范围的内存位置，否则，结果将没有意义。这就是为什么当我们将这些运算符应用于指向数组元素的指针时，使用它们是有意义的：当我们声明数组时，涉及的内存位置在我们的可控范围内。

指针与整数的加法和减法。可以使用加法运算符让指针加上一个整数，或者使用减法运算符从指针中减去一个整数：

```
加法: ptr1 = ptr + n                    减法: ptr2 = ptr − n
```

加法和减法运算符的这种使用方法赋予了它们一种新的定义：创建一个新的指针，指向

$n \times m$ 字节（m 是被指变量的大小，即每个元素所占的字节数）之前或之后的位置。因此该操作将指针向前或者向后移动 n 个元素，如图 9-16 所示。加法创建的指针远离数组的开始位置；减法创建的指针靠近数组的开始位置。虚线箭头表示原始指针。

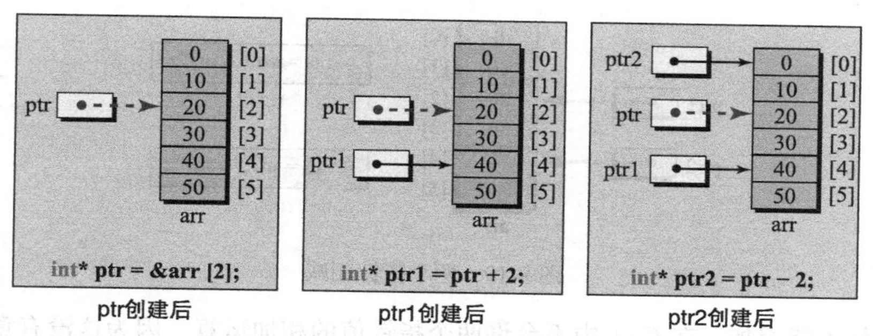

图 9-16　指针与整数的加法和减法

增量运算符和减量运算符。增量运算符和减量运算符被设计用于变量（左值）。它们不能与数组的名称一起使用，因为名称是常量值。ptr++ 的含义与 ptr = ptr + 1 相同，ptr-- 的含义与 ptr = ptr - 1 相同。在这些情况下，原始指针移动方式如图 9-17 所示。增量操作后，原始指针指向 arr[3]。减量操作后，指针重新指向 arr[2]。

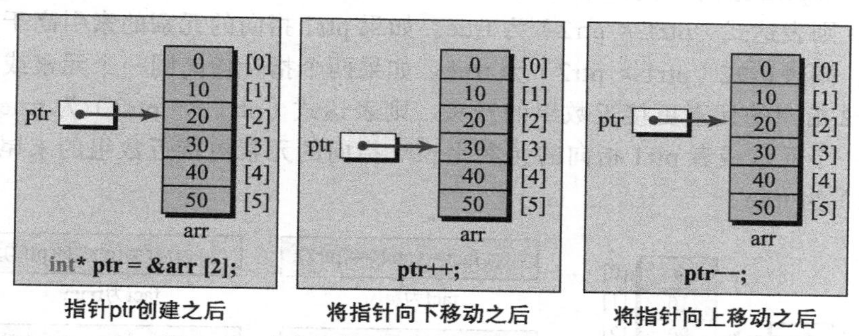

图 9-17　应用于指针的增量运算符和减量运算符

复合加法和减法运算符。我们也可以在指针变量上使用复合加法运算符和复合减法运算符。如果 ptr 是指针变量，则表达式（ptr += 3）等价于（ptr = ptr + 3），表达式（ptr −= 3）等价于（ptr = ptr − 3）。这些操作与增量操作和减量操作一样，将指针移向开头或者结尾，如图 9-18 所示。

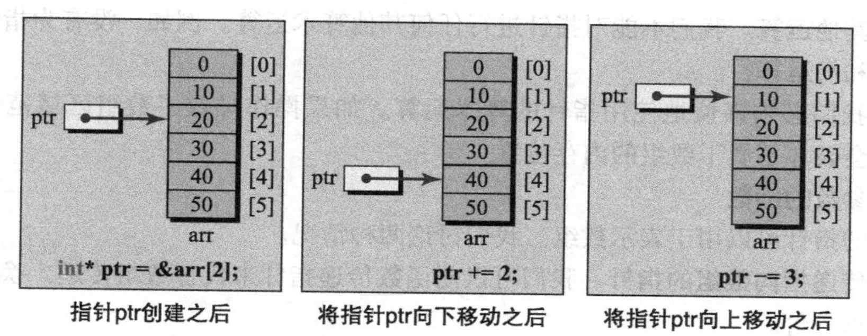

图 9-18　应用于指针的复合加法和复合减法运算符

两个指针相减。我们可以从一个指针中减去另一个指针。如果 ptr1 和 ptr2 都是指针，其中 ptr1 指向较低索引的元素，ptr2 指向较高索引的元素，则表达式（ptr2 – ptr1）返回一个正整数，而（ptr1 – ptr2）返回一个负整数，如图 9-19 所示。换言之，这些操作的结果是整数，而不是指针。

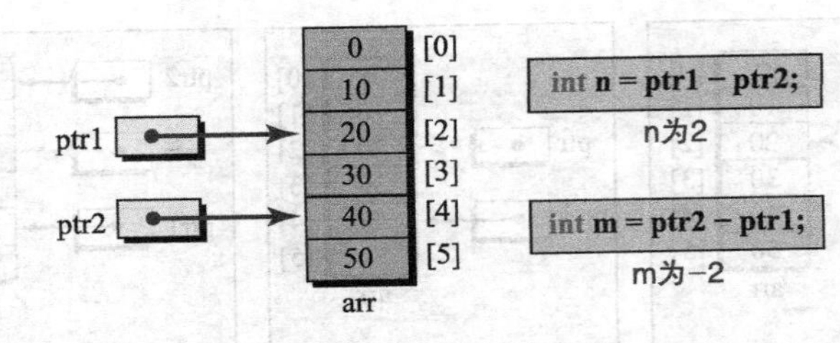

图 9-19 两个指针相减

两个指针不能相加。在 C++ 中不允许两个指针值的相加运算，因为这没有意义。

两个指针不能相加。

比较指针。我们可以比较两个指针，如图 9-20 所示。如果 ptr1 和 ptr2 是指向数组元素的指针，则当 ptr1 和 ptr2 指向同一元素时，表达式（ptr1 == ptr2）为 true。表达式（ptr1 != ptr2）在 ptr1 和 ptr2 指向不同元素时为 true。如果 ptr1 指向的元素的索引低于 ptr2 指向的元素的索引，则表达式（ptr1 < ptr2）为 true。如果 ptr1 指向的元素的索引高于 ptr2 指向的元素的索引，则表达式（ptr1 > ptr2）为 true。如果两个指针指向同一个元素或者 ptr1 指向的元素比 ptr2 指向的元素更接近数组的开头，则表达式（ptr1 <= ptr2）为 true。如果两个指针指向同一个元素或者 ptr1 指向的元素比 ptr2 指向的元素更接近数组的末尾，则表达式（ptr1 >= ptr2）为 true。

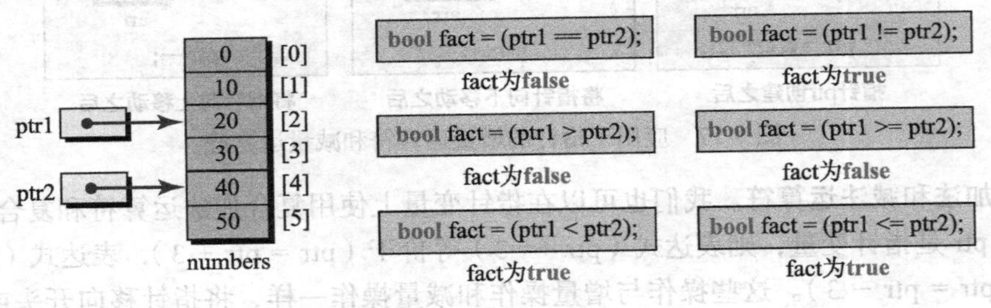

图 9-20 比较指针

不允许其他运算。我们不能对指针进行任何其他算术运算。例如，没有为指针类型定义乘法、除法和模运算符。

警告。我们必须谨慎地使用指针的算术运算。如果操作导致了数组区域范围之外的访问，则可能会破坏不属于数组的内存位置。

带数组参数的函数

函数中的指针可以用于表示数组。我们讨论两种情况。

向函数传递指向数组的指针。我们可以向函数传递指针来代替数组传递。换言之，以下两个函数原型等价。

```
int getSum (const array [ ], int size);      // 使用数组
int getSum (const int* p, int size);         // 使用指针
```

我们使用常量修饰符来防止 getSum 函数更改元素的值。

【例 9-20】 程序 9-10 演示了如何使用上面代码中的第二个版本来获取数组中元素的总和。

在程序 9-10 中，getSum 函数的第一个参数是一个指向常量整数的指针（这样函数就不能更改元素的值）。但是，这并不表示它是一个常量指针，我们可以移动这个指针。如果我们使用原型（const int* const，int）将参数设置为常量指针，那么在第 27 行中就不能使用 p++ 了，我们必须用 *(p + 1) 来代替。

程序 9-10 计算数组中元素的总和

```
1  /*****************************************************************
2   * 本程序演示如何使用指针                                          *
3   * 访问一个数组中的元素                                            *
4   *****************************************************************/
5  #include <iostream>
6  using namespace std;
7
8  int getSum (const int*, int); // 函数原型
9
10 int main ()
11 {
12   // 数组的声明和初始化
13   int arr [5] = {10, 11, 12, 13, 14};
14   // 函数调用
15   cout << "Sum of elements: " << getSum (arr, 5);
16   return 0;
17 }
18 /*****************************************************************
19  * 该函数接收指向一个数组的第一个元素的指针，                      *
20  * 计算并返回该数组中所有元素的总和                                *
21  *****************************************************************/
22 int getSum (const int* p, int size)
23 {
24   int sum = 0;
25   for (int i = 0; i < size ; i++)
26   {
27     sum += *(p++);
28   }
29   return sum;
30 }
```

运行结果：
```
Sum of elements: 60
```

【例 9-21】 假设我们想要编写一个函数，使用指针反转数组中的元素。程序 9-11 显示了一个示例。

程序 9-11 反转数组中的元素

```
1  /*****************************************************************
2   * 本程序演示如何使用指针                                          *
3   * 反转一个数组中的元素                                            *
4   *****************************************************************/
5  #include <iostream>
```

```
6   using namespace std;
7
8   void reverse (int* , int );
9
10  int main ()
11  {
12    // 数组的声明和初始化
13    int arr [5] = {10, 11, 12, 13, 14};
14    // 函数调用
15    reverse (arr, 5);
16    // 打印反转后的数组
17    cout << "Reversed array: ";
18    for (int i = 0; i < 5; i++)
19    {
20          cout << *(arr + i) << "  ";
21    }
22    return 0;
23  }
24  /*****************************************************************
25   * 该函数使用指向一个数组的第一个元素的指针和数组大小,              *
26   * 在原数组中反转数组中的元素                                      *
27   *****************************************************************/
28  void reverse (int* pArr, int size)
29  {
30    int i = 0;
31    int j = size – 1;
32    while (i < size / 2)
33    {
34          int temp = *(pArr + i);
35          *(pArr + i) = *(pArr + j);
36          *(pArr + j) = temp;
37          i++;
38          j– –;
39    }
40  }
运行结果:
Reversed array: 14   13   12   11   10
```

虽然 arr 是一个常量指针，不能移动，但 pArr 不一定是常量指针，它是一个单独的指针变量。还要注意，pArr 不能是指向常量整数的指针，因为它应该在交换时更改所指向的元素。我们只交换了一半的元素。当一半的元素被交换后，交换任务就完成了。

从函数中返回一个数组。尽管我们可能会考虑从函数返回数组，但我们必须记住数组包含两部分信息：指向第一个元素的指针和数组的大小。一个函数只能返回一条信息（除非我们将指针和大小捆绑在一个对象中）。正如我们在第 8 章中提到的，我们不能从函数返回数组。

9.3.2　二维数组和指针

C++ 中的二维数组是数组的数组。如果我们牢记住这一点，那么使用指针和二维数组就非常简单。图 9-21 显示了二维数组的内容以及 C++ 内部的表示。

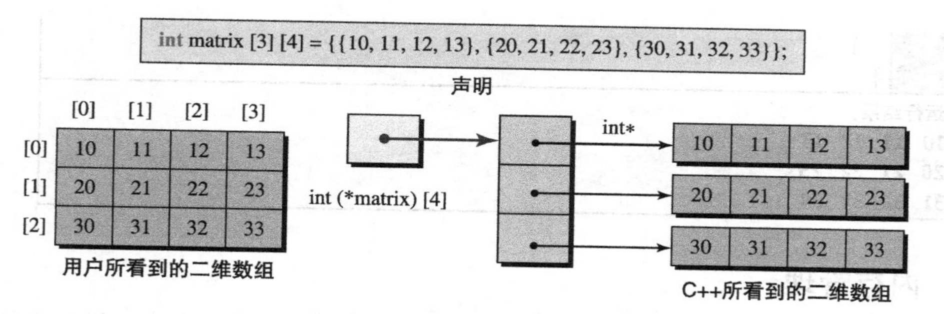

图 9-21 用户和 C++ 所看到的二维数组

与某些观点相反，二维数组的名称不是指向整数指针的指针，它是指向包含 4 个整数的数组或者 int (*matrix)[4] 的指针。数组名的圆括号表示数组被理解为“ matrix 是指向一个包含 4 个整数的数组的指针”。当我们将二维数组传递给一个函数时，我们必须记住这个事实。和往常一样，列的大小必须和指针一起指定，行的大小则作为一个整数单独指定。

【例 9-22】 在本例中，我们编写一个简短的程序，将图 9-21 中定义的数组传递给一个函数，该函数将打印所有的元素，如程序 9-12 所示。

程序 9-12 传递二维数组给函数

```
1  /************************************************************
2   * 本程序演示如何使用指针                                    *
3   * 传递一个二维数组给一个函数                                *
4   ************************************************************/
5  #include <iostream>
6  using namespace std;
7
8  void print (int (*) [4], int);  // 函数原型
9
10 int main ()
11 {
12   int matrix [3][4] = {{10, 11, 12, 13}, {20, 21, 22, 23},
13                                   {31, 32, 33, 34}};
14   // 调用 print 函数
15   print (matrix, 3);
16   return 0;
17 }
18 /************************************************************
19  * 该函数接收一个指针参数和表示行数的 rows 参数，             *
20  * 该指针参数指向 4 个整数的任意数组                          *
21  ************************************************************/
22 void print (int (*m) [4], int rows)
23 {
24   for (int i = 0; i < rows; i++)
25   {
26       for (int j = 0 ; j < 4; j++)
27       {
28             cout << m[i][j] << "  ";
29       }
30       cout << endl;
```

```
31      }
32  }
```

运行结果：

```
10 11 12 13
20 21 22 23
31 32 33 34
```

9.4　内存管理

C++ 编写的程序运行时，会使用内存位置。代码必须存储在内存中，每个基本对象或者用户自定义的对象也必须存储在内存中。然而，C++ 环境将内存划分为不同的区域（如图 9-22 所示），以使内存管理更加高效。请注意，此图不表示计算机中不同内存区域的顺序。

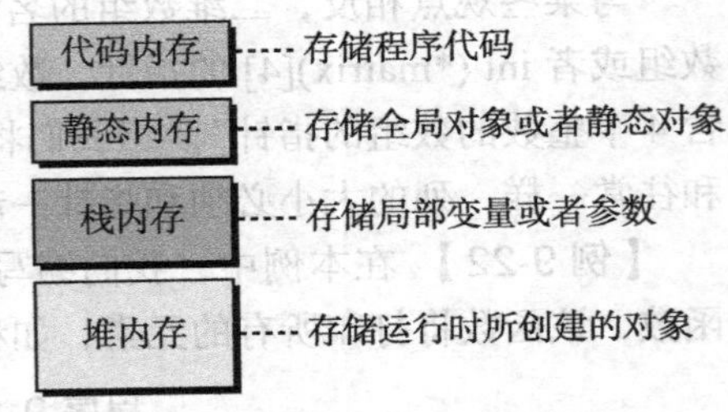

图 9-22　程序所使用的内存区域

在本节中，我们将讨论这些内存区域以及如何使用它们。了解各内存区域的特点并知道如何使用这些内存区域将有助于我们编写更好的程序。

9.4.1　代码内存

代码内存（code memory）存储程序代码。当程序运行时，C++ 的运行环境逐条执行每条语句，或者分支跳转到另一条语句。此内存区域中不存储任何对象。程序终止时释放代码内存。

代码内存存储程序代码，当程序终止时释放其占用的内存。

9.4.2　静态内存

静态内存（static memory）用于保存**全局对象**（不属于任何函数（包括 main 函数）的对象）和在程序中任何位置创建的**静态对象**（在全局区域或者函数内部）。当程序终止时，这些对象会被自动销毁并被释放内存。

静态内存存储全局对象和静态对象，当程序终止时释放其占用的内存。

【例 9-23】 程序 9-13 显示了一个非常简单的程序。三个数据变量（first、second 和 third）的内存位置是在静态内存中创建的。first 是全局变量；second 和 third 是静态变量。

程序 9-13　使用静态内存

```
1   /****************************************************************
2    * 本程序演示全局对象和静态对象                                    *
3    * 在程序的所有位置都可见，                                        *
4    * 它们贯穿于整个程序的生命周期                                    *
5    ****************************************************************/
6   #include <iostream>
7   using namespace std;
8
9   int first = 20;              // 全局变量，位于静态内存
```

```
10  static int second = 30 ;       // 静态变量，在静态内存中创建
11
12  int main ()
13  {
14      static int third = 50;  // 静态变量，位于静态内存
15
16      cout << "Value of Global variable: " << first << endl;
17      cout << "Value of Global static variable: " << second << endl;
18      cout << "Value of local static variable: " << third;
19      return 0;
20  }
运行结果：
Value of Global variable: 20
Value of Global static variable: 30
Value of local static variable: 50
```

9.4.3 栈内存

程序用来保存函数的局部对象或者参数对象的内存部分是**栈内存**（stack memory）。正如我们从日常生活中所知道的，栈是一个后进先出的容器。最后推入的东西最先弹出。栈的这个特性非常适合在函数中存储参数和局部变量。当我们调用函数时，系统将参数和局部变量推入（push）栈内存中。当函数终止时，系统会弹出（pop）这些变量并丢弃它们。图 9-23 显示了一个简单程序的函数调用和栈内存的行为，该程序调用函数 first，函数 first 则调用函数 second。

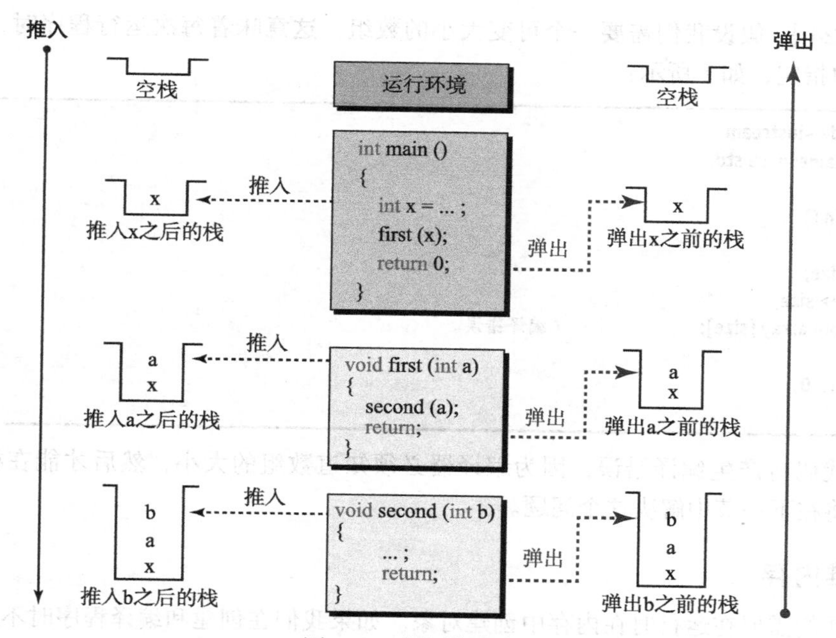

图 9-23 使用栈内存

图 9-23 的右侧显示了栈内存的推入和弹出操作。开始时栈内存为空。当运行环境调用 main 函数时，它将其唯一的局部变量（x）推入栈中（main 函数没有参数）。当 main 调用函数 first 时，系统将其唯一的参数（a）推入栈中。当函数 first 调用函数 second 时，系统将其

唯一的参数（b）推入栈中。

当函数 second 返回时，系统弹出它唯一的参数（b）并将其丢弃（不再需要）。当函数 first 返回时，系统弹出它唯一的参数（a）并将其丢弃（不再需要）。当函数 main 返回时，系统弹出它唯一的局部变量（x）并将其丢弃（不再需要）。当程序终止时，栈内存在结束时再次为空。为了简单起见，我们在函数 first 或者 second 中没有使用局部变量，否则，这些局部变量也将被推入栈中，并从栈中弹出。

优越性

系统在每个程序中都使用栈内存。栈内存的后进先出操作使得它非常有效。只有当对象在函数的作用域中时，它们才被存储并保留在栈内存中。当一个对象超出了函数的作用域时，它会被弹出并丢弃，并且不再是可访问的。

局限性

使用栈内存的效率会对其使用产生两个限制：

- 对象在编译时必须具有名称。堆栈中不能存储未命名的对象。
- 对象的大小必须在编译时定义。除非系统知道对象的确切大小，否则无法为对象分配栈内存。对于单个对象，可以从其类型推断其大小；对于对象列表（例如数组），必须在编译时指定元素的个数。

根据前面的讨论，我们可以将栈内存称为编译时内存。在编译期间，必须明确定义存储在栈内存中的每个对象。

> 在栈内存中创建的对象在编译期间必须具有名称和大小。

【例 9-24】 假设我们需要一个可变大小的数组，这意味着每次运行程序时，数组大小必须由用户指定，如下所示：

```
#include <iostream>
using namespace std;

int main ()
{
    int size;
    cin >> size;
    double array [size];                // 编译错误
    ...
    return 0;
}
```

上面代码将产生编译错误，因为编译器必须知道数组的大小，然后才能在栈中分配内存。我们将在下一节中解决这个问题。

9.4.4　堆内存

有时我们需要在运行时在内存中创建对象。如果我们在创建和编译程序时不知道对象的大小，就会发生这种情况。**堆内存**（heap memory，也称为自由内存或者动态内存）用于存储运行时创建的对象。当一个对象或者一组对象需要大量内存，或者在编译期间无法计算内存量时，就会出现这种情况。在堆内存中创建的对象不能有名称，因此要访问它们，我们需要在栈内存中有一个可以指向它们的指针。换言之，为此我们同时需要栈内存和堆内存。栈内存用于保存指针（一个 4 字节的小对象）；堆内存用于存储指针指向的对象（通常是一个大对象），如图 9-24 所示。

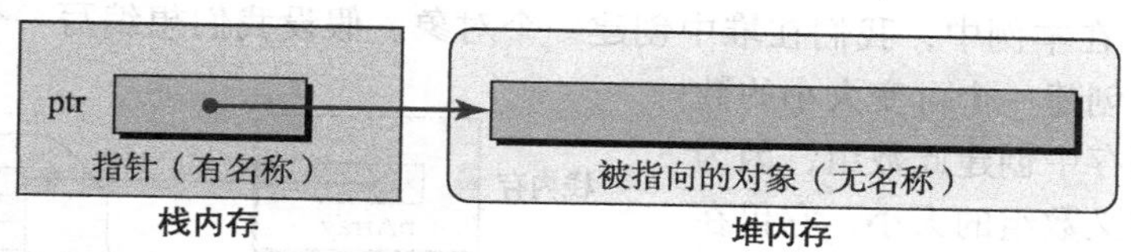

注意：
指针位于栈内存中，并且必须要有一个名称（此处为ptr）。
被指向的对象位于堆内存中，它没有名称。

图 9-24　栈中的指针和堆中的被指向对象

堆内存中的对象不能有名称，它由指向它的指针引用。另外，指针对象在栈中，它必须有一个名称。换言之，指针的名称有助于我们引用它，指针所指向对象的地址使得我们能够访问它。

在堆内存中创建的对象不能有名称，只能通过其地址访问它，该地址由栈内存中的指针访问。

两个运算符：new 和 delete

现在的问题是，在运行时我们如何在堆中创建对象，以及如果对象没有名称，我们如何在不需要时销毁它。这些可以通过四个运算符来完成。这些运算符在附录 C 中定义，并在表 9-6 中列出，以供快速参考。

表 9-6　在堆中分配和释放内存的运算符

组别	名称	运算符	表达式	优先级	结合性
一元表达式	分配对象 分配数组 释放对象 释放数组	new new [ ] delete delete [ ]	new type new type [size] delete ptr delete [ ] ptr	17	←

表 9-6 中的第一个运算符用于在堆中为单个对象分配内存。第二个运算符用于在堆中创建对象数组。第三个运算符用于使用指针删除单个对象。第四个运算符用于删除为堆中的数组分配的内存。图 9-25 显示了使用 new 和 delete 运算符的四种操作。

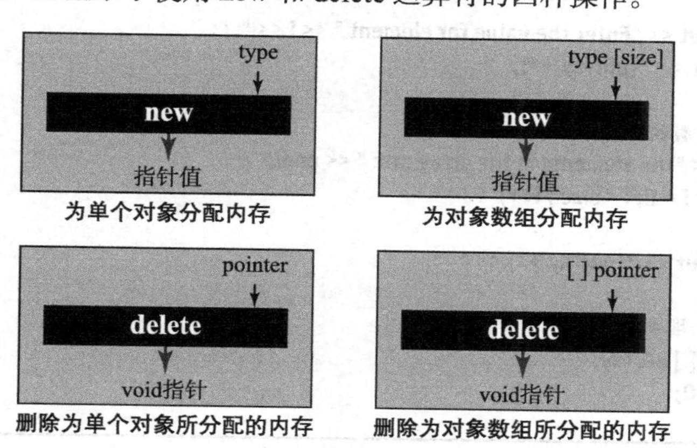

图 9-25　new 和 delete 运算符

注意，delete操作后，指针是悬空指针，我们将在本节稍后讨论。在对其重新应用new运算符之前，不能使用悬空指针。

【例9-25】 在本例中，我们在堆中创建一个对象。假设我们想编写一个程序，在用户每次运行程序时创建一个可变大小的数组。无法在栈内存中创建此数组，因为在编译期间未定义数组的大小，它仅在运行时定义。图9-26显示了这种情况。

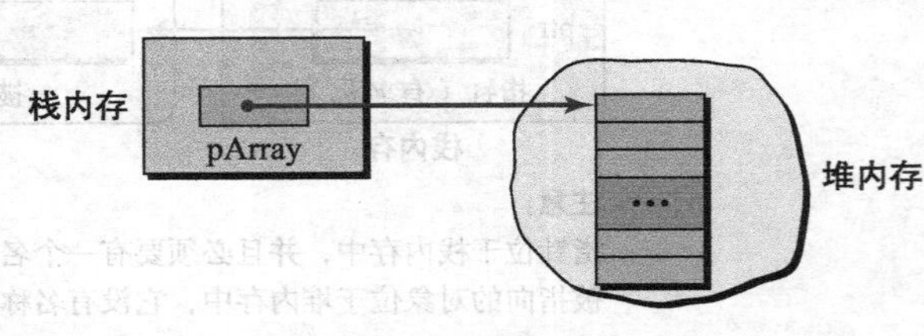

图9-26　栈中的指针和堆中的数组

程序9-14演示了如何实现这种情况。在终止程序之前，我们必须删除堆中创建的数组（第34行）。

程序9-14　使用堆内存存储可变大小的数组

```
1  /*********************************************************************
2   * 本程序演示如何在堆中创建可变大小的数组，                          *
3   * 并使用指针访问其元素                                              *
4   *********************************************************************/
5  #include <iostream>
6  using namespace std;
7  
8  int main ( )
9  {
10   // 在栈中声明数组的大小和指针
11   int size;
12   int* pArray;
13   // 验证 size 是否大于 0
14   do
15   {
16       cout << "Enter the array size (larger than zero): ";
17       cin >> size;
18   } while (size <= 0);
19   // 在堆中创建数组
20   pArray = new int [size];
21   // 输入数组的值
22   for (int i = 0; i < size ; i++)
23   {
24       cout << "Enter the value for element " << i << ": ";
25       cin  >> *(pArray + i);
26   }
27   // 输出数组的值
28   cout << "The elements in the array are: " << endl;
29   for (int i = 0; i < size ; i++)
30   {
31       cout << *(pArray + i) << "   ";
32   }
33   // 删除堆中的数组
34   delete [ ] pArray;
35   return 0;
36 }
```

```
运行结果:
Enter the array size (larger than zero): 3
Enter the value for element 0: 6
Enter the value for element 1: 12
Enter the value for element 2: 5
The elements in the array are:
6 12 5
```

有关堆内存的问题

当我们处理堆内存并使用相关的运算符（new 和 delete）时，必须注意可能发生的一些问题。

释放没有分配的内存。编程中可能发生的错误之一是，我们尝试使用 delete 运算符之前，并没有先使用 new 运算符。这意味着我们试图删除没有在堆中分配的对象。当对象在栈中被分配但我们尝试将其从堆中删除时，通常会发生这种情况。如下所示：

```
double x = 23.4;
double* pX = &x;  // 在栈中分配
delete pX; // 在堆中删除，会导致运行时错误
```

分配了内存但没有释放（内存泄漏）。当我们在堆中分配了一个对象但没有删除该对象时，会出现一个更严重的问题，称为**内存泄漏**（memory leak），如下所示：

```
double* pX = new double;
... // 使用分配的内存
```

这意味着我们已经在堆中创建了一个内存位置，但没有删除它。当指向某内存位置的指针超出作用域时，大多数操作系统都会删除该内存位置。但是，如果该指针重新指向另一个内存位置（在栈或者堆中），则会出现严重问题。在这种情况下，没有超出作用域的指向该对象的指针，因而不会警告操作系统删除已分配的内存。这个严重的问题被称为内存泄漏，应该避免。内存泄漏会使未删除的内存位置不可用，如果内存不足，可能会导致计算机系统崩溃。

悬空指针。另一个问题是**悬空指针**，当我们删除了被指向的对象然后再次尝试使用它时，可能会发生这种情况，如下所示：

```
double* pX = new double;
... // 使用分配的内存
delete pX;
*pX = 35.3;  // 悬空指针
```

最后一行代码会导致一个意外的错误，因为指针为 null，因而无法间接引用。

在处理堆内存时，必须将每个 new 运算符与 delete 运算符配对，并避免悬空指针。

建议

总是显式删除堆中创建的任何内存位置。

9.4.5 二维数组

如前所述，二维数组由行和列组成。创建二维数组时，我们有三种选择。我们将在本节

中讨论这些选择。

仅使用栈内存

如果在编译之前确定了数组的行大小和列大小，那么可以像以前那样完全在栈中创建数组。

同时使用栈内存和堆内存

如果在编译之前确定了行维度，则可以在栈中创建一个指针数组，然后在堆中创建每一行内容，如图 9-27 所示。

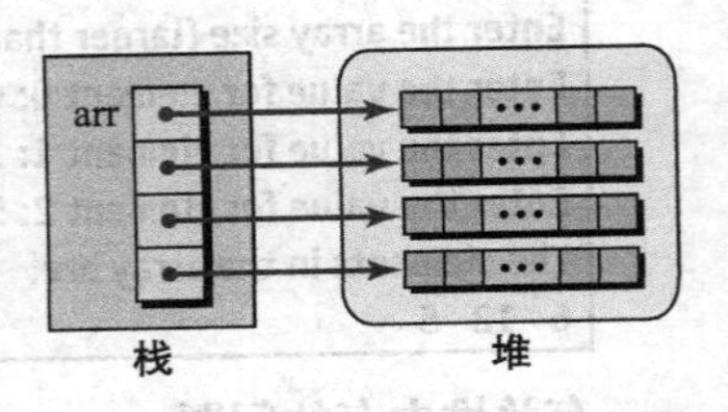

图 9-27　栈中和堆中的二维数组

下面的代码片段显示了创建此类数组的代码。编译之前可以确定行数（4），必须在运行时输入列数。

```
int* arr [4]; // 这是指向整数的 4 个指针的数组，位于栈内存中
cin >> colNums;
for (int i = 0; i < 4 ; i++)
{
  arr[i] = new int [colNums]; // 大小为 colNums 的数组，位于堆内存中
}
```

仅使用堆内存

如果在编译之前行和列两个维度都无法确定，那么我们必须在堆中创建整个二维数组，如图 9-28 所示。

以下代码片段在堆内存中创建一个二维数组。在编译时行和列两个维度都无法确定。

```
cin >> rowNums;
cin >> colNums;
int** arr = new int* [rowNums];
for (int i = 0; i < n + 1 ; i++)
{
  arr[i] = new int [colNums];
}
```

示例：交错数组

我们可以创建一个二维数组，其中每一行的元素个数各不相同，如图 9-29 所示。

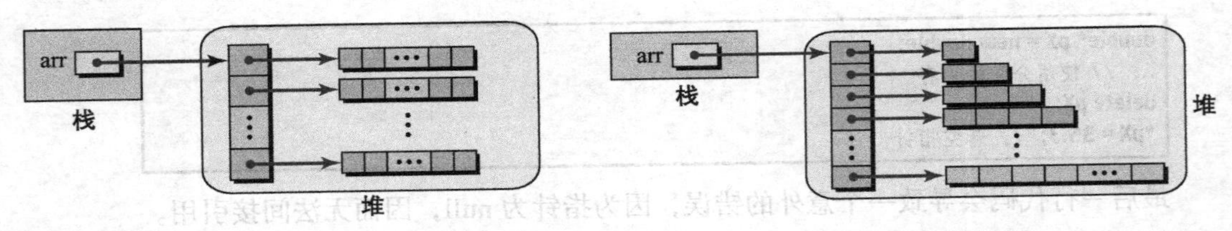

图 9-28　完全在堆中的二维数组

图 9-29　交错数组

换言之，是否允许创建这样一个数组，其中第一行有三个元素，第二行有四个元素？结论是肯定的，这被称为**交错数组**（ragged array，不规则数组）。但我们不能在栈内存中分配元素，它们必须在堆内存中被创建。换言之，我们必须使用一个指针数组，其中每个指针指向一个所需大小的数组，如图 9-29 所示。这种设计比使用二维数组更节省内存，因为二维数组会产生一些空元素。

【例 9-26】 帕斯卡三角形用于确定二项式展开的系数。当二项式（$x+y$）被提升到正整数次幂时，其展开公式如下所示：

$$(x+y)^n = a_0\,x^n + a_1\,x^{n-1}y + a_2\,x^{n-2}y^2 + \cdots + a_{n-1}\,x\,y^{n-1} + a_n y^n$$

每个项的系数（a_0，a_1，a_2，…，a_{n-1}，a_n）可以使用 $n+1$ 行和 $n+1$ 列的三角形来计算。数组中的每个单元格都保存一个项的系数。行数比幂值（n）多一。三角形中的每个元素的计算方法如下：前一行前一列的元素加上前一行同一列的元素，如下所示（$n=4$ 时）。很明显，要计算任意 n 值（0 和 1 以外）的系数，我们必须知道 n 之前的系数值。这表明可以使用二维数组来实现。我们可以使用指针数组动态创建数组并计算系数。

```
n = 0    1
n = 1    1    1
n = 2    1    2    1
n = 3    1    3    3    1
n = 4    1    4    6    4    1
```

【例 9-27】 程序 9-15 用于获取 0 到 9 之间任意 n 值的系数。每行分配的位置数比行数大一。例如，当 $i=0$ 时，我们必须从堆中分配一个位置。当 $i=9$ 时，我们必须从堆中分配十个位置。

程序 9-15　获取帕斯卡系数

```
1   /****************************************************************
2    * 本程序演示如何使用在堆中动态分配交错数组的方法来              *
3    * 创建帕斯卡系数                                               *
4    ****************************************************************/
5   #include <iostream>
6   #include <iomanip>
7   using namespace std;
8   
9   int main ( )
10  {
11      // 声明
12      int maxPower = 10;
13      int n;
14      // 输入验证
15      do
16      {
17          cout << "Enter the power of binomial : ";
18          cin >> n;
19      } while (n < 0 || n > maxPower);
20      // 在堆中分配内存
21      int** pascal = new int* [n + 1];
22      for (int i = 0; i < n + 1 ; i++)
23      {
24          pascal[i] = new int [i];
25      }
26      // 生成系数
27      for (int i = 0; i <= n ; i++)
28      {
29          for (int j = 0; j < i + 1; j++)
```

```
30      {
31          if (j == 0 || i == j)
32          {
33              pascal [i][j] = 1;
34          }
35          else
36          {
37              pascal [i][j] = pascal [i - 1] [j - 1] + pascal [i - 1][j];
38          }
39      }
40  }
41  // 打印系数
42  cout << endl;
43  cout << "Coefficients for (x + y)^" << n << " are:";
44  for (int j = 0; j <= n  ; j++)
45  {
46      cout << setw (5) << pascal [n][j] ;
47  }
48  cout << endl;
49  // 删除分配的内存
50  for (int i = 0; i < n + 1 ; i++)
51  {
52      delete [ ] pascal [i];
53  }
54  delete [ ] pascal;
55  return 0;
56  }
```

运行结果：

```
Enter the power of binomial : 5
Coefficients for (x + y)^5 are: 1    5   10   10    5    1
```

在程序 9-15 的第 21 行中，我们在栈内存中创建了一个名为 pascal 的 int** 类型的变量。在同一行中，我们在堆中创建了一个指针数组，并将返回的指针存储在变量 pascal 中。变量 pascal 现在指向一个指针数组。然后我们使用一个循环来创建（$n+1$）个数组，每个数组在堆中的大小都不同。pascal[i] 中的指针指向各数组。在第 50 ～ 54 行中，我们首先删除了堆中（$n+1$）个整数数组，然后删除了 pascal 变量指向的数组。必须按分配的相反顺序删除堆中分配的内存。

9.5　程序设计

在本节中，我们将创建两个类，其中部分数据成员在堆内存中创建。

9.5.1　课程类

本节我们将创建一个课程类，学校的老师可以使用课程类的对象来创建其所承担的所有课程的统计数据。由于每门课程可能有不同数量的学生，我们必须在每门课程中有一个可变大小的数组来跟踪分数。为了节省每个老师编写程序的时间，课程类由学校管理部门的程序员创建和保存。他们为每个老师提供一个公共接口，该接口可用来创建和运行应用程序。在本节中，我们将讨论管理部门的程序员的开发工作，以及老师应该承担的开发任务。

数据成员

图 9-30 显示了课程类（Course）中数据成员的列表。

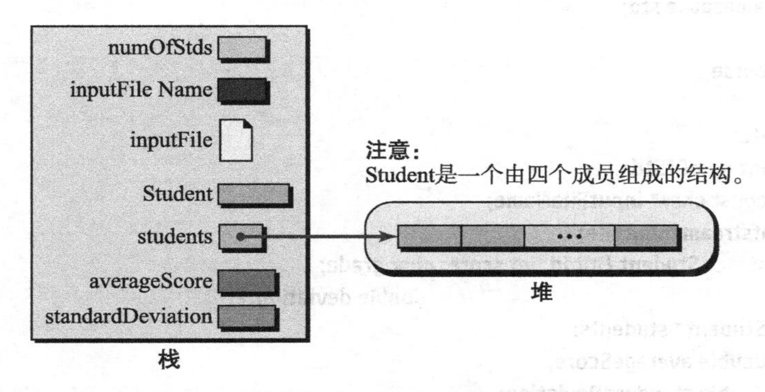

图 9-30　课程类（Course）中数据成员的列表

我们必须为每个学生存储四种信息：身份 ID、百分制分数、成绩等级、相对于平均值的偏离值。我们没有在堆中创建四个并行数组，而是将这四种信息封装成一个结构，并在堆中创建了一个结构数组。

成员函数

当类的对象被实例化时，构造函数会自动调用几个私有成员函数。类的两个公共成员函数是构造函数和析构函数。

输入文件

我们假设输入文件包含以下数据，其中第一列表示学生身份 ID，第二列定义课程中的百分制分数。

```
1000  88
1001  100
1002  92
1003  77
1004  54
1005  82
1006  67
1007  95
1008  93
1009  100
```

接口文件、实现文件和应用程序文件

程序 9-16、程序 9-17 和程序 9-18 分别实现了这三个文件。

程序 9-16　接口文件

```
1  /*****************************************************************
2   * 接口文件包括私有数据成员和公共成员函数。                          *
3   * 私有成员函数是辅助函数，                                        *
4   * 由构造函数调用以完成特定的任务。                                *
5   * 构造函数负责所有的工作。                                        *
6   * 析构函数删除在堆中创建的数组，并关闭输入文件                      *
7   *****************************************************************/
8  #ifndef COURSE_H
9  #define COURSE_H
```

```
10  #include <iostream>
11  #include <fstream>
12  using namespace std;
13
14  class Course
15  {
16     private:
17         int numOfStds;
18         const char* inputFileName;
19         ifstream inputFile;
20         struct Student {int id; int score; char grade;
21                                      double deviation;};
22         Student* students;
23         double averageScore;
24         double standardDeviation;
25         // 私有成员函数
26         void getInput ();
27         void setGrades ();
28         void setAverage ();
29         void setDeviations();
30         void printResult() const;
31     public:
32         Course (int numOfStds, const char* inputFileName);
33         ~Course ();
34  };
35  #endif
```

程序 9-17 实现文件

```
1   /***********************************************************
2    * 实现文件提供了所有私有成员函数                            *
3    * 和公共成员函数的定义                                      *
4    ***********************************************************/
5   #include "course.h"
6   #include <iomanip>
7   #include <cmath>
8
9   /***********************************************************
10   * 构造函数负责初始化学生的人数                              *
11   * 以及包含百分制分数的输入文件的名称。                      *
12   * 随后构造函数打开输入文件，                                *
13   * 并在堆内存中创建一个数组。                                *
14   * 其余的工作由辅助函数完成：                                *
15   * 设置百分制分数、成绩等级、平均值、偏离值，并打印结果      *
16   ***********************************************************/
17  Course :: Course (int num, const char* ifn)
18  :numOfStds (num), inputFileName (ifn)
19  {
20     inputFile.open (inputFileName);
21     students = new Student [numOfStds];
22     getInput ();
23     setGrades ();
```

```
24      setAverage ();
25      setDeviations();
26      printResult();
27  }
28  /*************************************************************
29   * 析构函数                                                  *
30   * 负责使用相应的指针删除在堆内存中创建的数组。              *
31   * 还负责关闭由构造函数打开的输入文件                        *
32   *************************************************************/
33  Course :: ~Course ()
34  {
35      delete [ ] students;
36      inputFile.close ();
37  }
38  /*************************************************************
39   * getInput 函数                                             *
40   * 负责读取包含学生 ID 和百分制分数的文件                    *
41   *************************************************************/
42  void Course :: getInput()
43  {
44      for (int i = 0; i < numOfStds; i++)
45      {
46          inputFile >> students [i].id;
47          inputFile >> students [i].score;
48      }
49  }
50  /*************************************************************
51   * getGrades 函数                                            *
52   * 通过一个字符数组把每个学生的百分制分数转换为成绩等级      *
53   *************************************************************/
54  void Course :: setGrades()
55  {
56      char charGrades [ ] =
57                  {'F', 'F', 'F' , 'F' , 'F' , 'F', 'D' , 'C' , 'B' , 'A' , 'A'};
58      for (int i = 0; i < numOfStds; i++)
59      {
60          int index = students[i].score / 10;
61          students[i].grade = charGrades [index];
62      }
63  }
64  /*************************************************************
65   * setAverage 函数                                           *
66   * 处理数组中的百分制分数，计算班级的平均分                  *
67   *************************************************************/
68  void Course :: setAverage()
69  {
70      int sum = 0;
71      for (int i = 0; i < numOfStds; i++)
72      {
73          sum += students[i].score;
74      }
75      averageScore = static_cast <double> (sum) / numOfStds;
```

```
76  }
77  /****************************************************************
78   *  setDeviations 函数                                            *
79   *  重新处理百分制分数，计算每个学生的分数与平均分的偏离值        *
80   ****************************************************************/
81  void Course :: setDeviations()
82  {
83     standardDeviation = 0.0;
84     for (int i = 0; i < numOfStds; i++)
85     {
86          students[i].deviation = students[i].score - averageScore;
87          standardDeviation += pow(students[i].deviation , 2);
88     }
89     standardDeviation = sqrt (standardDeviation) / numOfStds;
90  }
91  /****************************************************************
92   *  printResult 函数                                              *
93   *  打印有关课程的所有信息                                        *
94   ****************************************************************/
95  void Course :: printResult() const
96  {
97     cout << endl;
98     cout << "Identity    Score    Grade        Deviation" << endl;
99     cout << "--------    -----    -----        ---------" << endl;
100    for (int i = 0; i < numOfStds ; i++)
101    {
102         cout << setw (4) << noshowpoint << noshowpos;
103         cout << right << students[i].id;
104         cout << setw (14) << noshowpoint << noshowpos;
105         cout << right << students[i].score;
106         cout << setw (10) << right << students[i].grade;
107         cout << fixed << setw (20) << right << setprecision (2);
108         cout << showpoint << showpos;
109         cout << students[i].deviation << endl;
110    }
111    cout << "Average score: " << fixed << setw (4);
112    cout << setprecision (2) <<averageScore << endl;
113    cout << "Standard Deviation: " << standardDeviation;
114 }
```

程序 9-18　应用程序文件

```
1   /****************************************************************
2    * 应用程序文件十分简单。                                        *
3    * 用户初始化一个课程类的实例对象，                              *
4    * 传入学生的人数、存储百分制分数的输入文件的名称。              *
5    * 构造函数负责完成所有的工作                                    *
6    ****************************************************************/
7   #include "course.h"
8
9   int main ( )
10  {
```

```
11  // 实例化一个 Course 对象
12  Course course (10, "scores.dat");
13  return 0;
14  }
```

运行结果：

```
Identity    Score    Grade    Deviation
---------   -----    -----    ---------
1000        88       B        +3.20
1001        100      A        +15.20
1002        92       A        +7.20
1003        77       C        -7.80
1004        54       F        -30.80
1005        82       B        -2.80
1006        67       D        -17.80
1007        95       A        +10.20
1008        93       A        +8.20
1009        100      A        +15.20
Average score: +84.80
Standard Deviation: +4.51
```

9.5.2 矩阵类

矩阵是数学中的数值表。在计算机程序设计中，我们可以用二维数组模拟矩阵。在这一节中，我们定义了一个矩阵类（Matrix）来演示如何在矩阵上应用选定的操作。

操作

我们在矩阵上定义了三种运算：加法、减法和乘法。矩阵的除法运算非常复杂，需要对矩阵求逆。

加法运算。如果两个矩阵的行数相同且列数相同，则这两个矩阵可以相加。换言之，两个矩阵必须满足条件 $r1 == r2$ 和 $c1 == c2$。矩阵相加的结果是 $r1$ 行和 $c1$ 列组成的矩阵（图 9-31）。注意，字母是符号值。

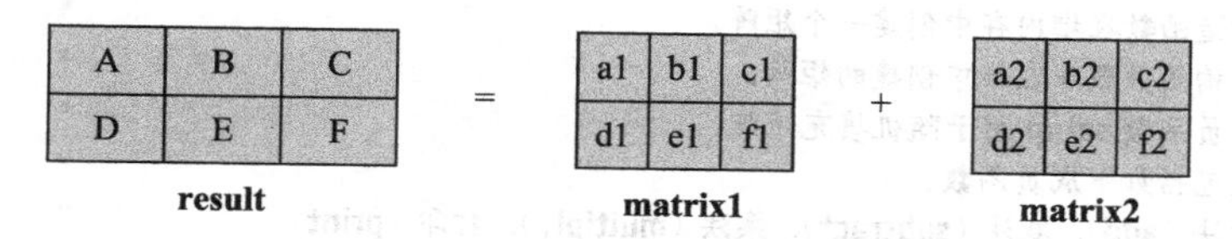

图 9-31　两个矩阵相加

结果矩阵中的各元素是两个矩阵中相应元素之和，如下所示：

```
A = a1 + a2
...
F = f1 + f2
```

减法运算。减法与加法相同，但相应元素值相减（见图 9-32）。

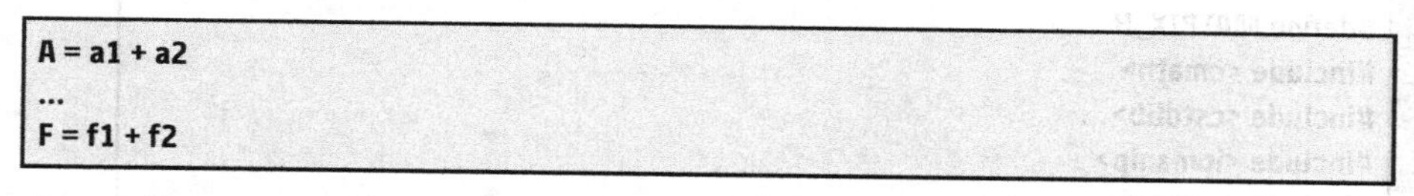

图 9-32　两个矩阵相减

结果矩阵中的各元素是两个矩阵中相应元素之差，如下所示：

```
A = a1 - a2
...
F = f1 - f2
```

乘法运算。如果第一个矩阵中的列数与第二个矩阵中的行数相同（$c1 == r2$），则可以将两个矩阵相乘。结果是一个行数等于 $r1$ 并且列数等于 $c2$ 的矩阵，如图 9-33 所示。

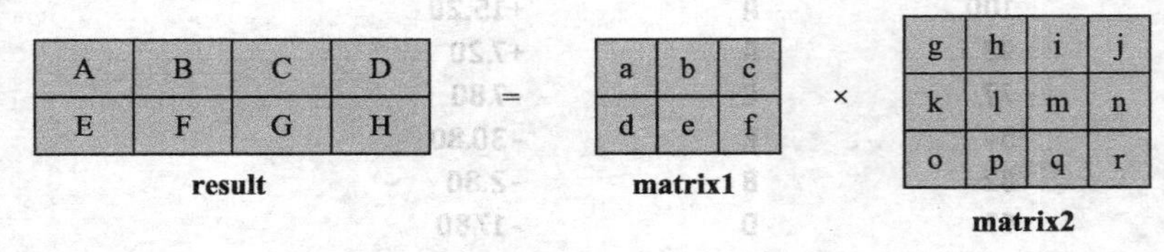

图 9-33　两个矩阵相乘

结果矩阵中的各元素是相应行和列的乘积之和。

```
A = a * g + b * k + c * o
B = a * h + b * l + c * p
...
H = d * j + e * n + f * r
```

实现代码

接下来我们编写接口文件、实现文件和应用程序文件的代码。

接口文件。程序 9-19 是接口文件的代码清单。

程序 9-19　Matrix 类的接口文件

```
1   /***************************************************************
2    * Matrix 类的接口文件。                                        *
3    * 仅有的私有数据成员包括：                                     *
4    * 矩阵的大小和指向堆内存中的矩阵的指针。                       *
5    * 构造函数在堆内存中创建一个矩阵，                             *
6    * 析构函数删除在堆中创建的矩阵。                               *
7    * 成员函数 setup 用于随机填充矩阵。                            *
8    * 还包括如下成员函数：                                         *
9    * 加法（add）、减法（subtract）、乘法（multiply）、打印（print）*
10   ***************************************************************/
11  #include <iostream>
12  #ifndef MATRIX_H
13  #define MATRIX_H
14  #include <cmath>
15  #include <cstdlib>
16  #include <iomanip>
17  #include <cassert>
18  using namespace std;
19
20  // Matrix 类定义
21  class Matrix
22  {
23     private:
```

```
24        int rowSize;
25        int colSize;
26        int** ptr;
27    public:
28    Matrix (int rowSize, int colSize);
29        ~Matrix ();
30        void setup ();
31        void add (const Matrix& second, Matrix& result) const;
32        void subtract (const Matrix& second, Matrix& result) const;
33        void multiply (const Matrix& second, Matrix& result) const;
34        void print () const;
35  };
36  #endif
```

实现文件。程序 9-20 是实现文件的代码清单。请注意，成员函数 setup() 随机填充矩阵，但在实际情况下，我们可以从文件中读取值。

程序 9-20　Matrix 类的实现文件

```
1   /*****************************************************************
2    * 实现文件                                                       *
3    * 实现了接口文件中声明的所有成员函数。                           *
4    * 实现方法遵循前文讨论的操作描述                                 *
5    *                                                                *
6    *****************************************************************/
7   #include "matrix.h"
8
9   // 构造函数：在堆内存中创建一个矩阵
10  Matrix :: Matrix (int r, int c)
11  : rowSize (r), colSize ©
12  {
13    ptr  = new int* [rowSize];
14    for (int i = 0; i < rowSize; i++)
15    {
16        ptr [i] = new int [colSize];
17    }
18  }
19  // 析构函数：删除堆内存中的内存位置
20  Matrix :: ~Matrix ()
21  {
22    for (int i = 0; i < rowSize ; i++)
23    {
24        delete [] ptr [i];
25    }
26    delete [] ptr;
27  }
28  // setup 函数：使用 1 到 5 之间的随机数填充矩阵
29  void Matrix :: setup ()
30  {
31    for (int i = 0; i < rowSize; i++)
32    {
33        for (int j = 0; j < colSize ; j++)
34        {
```

```
35                 ptr [i][j] = rand () % 5 + 1;
36             }
37     }
38 }
39 // add 函数：把第二个矩阵和宿主对象矩阵相加，创建一个结果矩阵
40 void Matrix :: add (const Matrix& second, Matrix& result) const
41 {
42     assert (second.rowSize == rowSize && second.colSize == colSize);
43     assert (result.rowSize == rowSize && result.colSize == colSize);
44 
45     for (int i = 0; i < rowSize ; i++)
46     {
47         for (int j = 0; j < second.colSize; j++)
48         {
49             result.ptr[i][j] = ptr[i][j] + second.ptr[i][j];
50         }
51     }
52 }
53 // subtract 函数：将宿主对象矩阵减去第二个矩阵
54 void Matrix :: subtract (const Matrix& second, Matrix& result) const
55 {
56     assert (second.rowSize == rowSize && second.colSize == colSize);
57     assert (result.rowSize == rowSize && result.colSize == colSize);
58     for (int i = 0; i < rowSize ; i++)
59     {
60         for (int j = 0; j < second.colSize; j++)
61         {
62             result.ptr[i][j] = ptr[i][j] - second.ptr[i][j];
63         }
64     }
65 }
66 // multiply 函数：将宿主对象矩阵乘以第二个矩阵
67 void Matrix :: multiply (const Matrix& second, Matrix& result) const
68 {
69     assert (colSize == second.rowSize);
70     assert (result.rowSize = rowSize);
71     assert (result.colSize = second.colSize);
72     for (int i = 0; i < rowSize ; i++)
73     {
74         for (int j = 0; j < second.colSize; j++)
75         {
76             result.ptr[i][j] = 0;
77             for (int k = 0 ; k < colSize; k++)
78             {
79                 result.ptr[i][j] += ptr[i][k] * second.ptr[k][j];
80             }
81         }
82     }
83 }
84 // print 函数：打印矩阵的元素值
85 void Matrix :: print () const
86 {
```

```
87     for (int i = 0 ; i < rowSize; i++)
88     {
89         for (int j = 0; j < colSize ; j++)
90         {
91             cout << setw (5) << ptr [i][j];
92         }
93         cout << endl;
94     }
95     cout << endl;
96 }
```

应用程序文件。程序 9-21 显示了使用 Matrix 类的应用程序文件的代码清单。

程序 9-21　测试 Matrix 类的应用程序文件

```
1  /***************************************************************
2   * 首先在堆内存中创建几个矩阵对象，                                *
3   * 然后应用一些运算操作                                          *
4   ***************************************************************/
5  #include "matrix.h"
6
7  int main ()
8  {
9      // 矩阵对象 matrix1 的实例化和设置
10     cout << "matrix1" << endl;
11     Matrix matrix1 (3, 4);
12     matrix1.setup ();
13     matrix1.print();
14     // 矩阵对象 matrix2 的实例化和设置
15     cout << "matrix2" << endl;
16     Matrix matrix2 (3, 4);
17     matrix2.setup ();
18     matrix2.print ();
19     // 矩阵对象 matrix3 的实例化和设置
20     cout << "A new matrix3" << endl;
21     Matrix matrix3 (4, 2);
22     matrix3.setup ();
23     matrix3.print ();
24     // 将矩阵 matrix1 和 matrix2 相加，并打印结果矩阵
25     cout << "Result of matrix1 + matrix2" << endl;
26     Matrix addResult (3, 4);
27     matrix1.add (matrix2, addResult);
28     addResult.print ();
29     // 将矩阵 matrix1 减去 matrix2，并打印结果矩阵
30     cout << "  Result of matrix1 - matrix2" << endl;
31     Matrix subResult (3, 4);
32     matrix1.subtract (matrix2, subResult);
33     subResult.print ();
34     // 将矩阵 matrix1 和 matrix3 相乘，并打印结果矩阵
35     cout << "Result of matrix1 * matrix3" << endl;
36     Matrix mulResult (3, 2);
37     matrix1.multiply (matrix3, mulResult);
38     mulResult.print();
```

```
39    return 0;
40  }
```

运行结果：

```
matrix1
    3    2    3    1
    4    4    5    1
    4    1    1    5
matrix2
    4    3    1    3
    3    5    1    5
    4    5    1    4
matrix3
    4    1
    5    5
    3    4
    2    1
Result of matrix1 + matrix2
        7    5    4    4
        7    9    6    6
        8    6    2    9
Result of matrix1 - matrix2
       -1   -1    2   -2
        1   -1    4   -4
        0   -4    0    1
Result of matrix1 * matrix3
       33   26
       53   45
       34   18
```

本章小结

引用类型是一种复合类型，允许使用不同名称引用同一个内存位置。

指针类型是表示内存中地址的复合类型。指针变量是包含地址的变量。

C++ 程序使用的内存由四个不同的区域组成：代码内存（保存程序）、静态内存（保存全局对象和静态对象）、栈内存（保存参数和局部对象）和堆内存（保存运行时创建的对象）。

思考题

1. 给定以下代码片段，给出其输出结果。

```
int x = 10;
int& y = x;
cout << x << " "<< y;
```

2. 给定以下代码片段，给出其输出结果。

```
int x = 100;
int& y = x;
int& z = x;
cout << x << " " << y <<" "<< z;
```

3. 指出以下代码片段的错误之处。

```
int x = 1000;
int& y = 2000;
```

4. 以下代码片段的输出结果是什么？

```
int x = 1;
int y = 2;
int& z = x;
z = y;
cout << x << " " << y << " " << z;
```

5. 指出以下代码片段的错误之处。

```
const int x = 100;
double& y = x;
```

6. 指出以下代码片段的错误之处。

```
const int x = 100;
int& y = x;
```

7. 指出以下代码片段的错误之处。

```
int x = 1000;
const int& y = x;
```

8. 以下程序的输出结果是什么？

```
#include <iostream>
using namespace std;

void fun (int& y);

int main ()
{
    int x = 10;
    fun (x) ;
    cout << x << endl;
    return 0;
}

void fun (int& y)
{
    y++;
}
```

9. 以下程序的输出结果是什么？

```
#include <iostream>
using namespace std;

void fun (int& y);

int main ()
{
    fun (10) ;

    cout << x << endl;
    return 0;
```

```
}

void fun (int& y)
{
    y++;
}
```

10. 以下程序的输出结果是什么？

```
#include <iostream>
using namespace std;

int& fun (int& yy, int& zz);

int main ()
{
    int x = 120;
    int y = 80;
    cout << fun (x , y);
    return 0;
}

int& fun (int& yy, int& zz)
{
    if (yy > zz)
    {
        return yy;
    }
    return zz;
}
```

11. 以下程序的输出结果是什么？

```
#include <iostream>
using namespace std;

int& fun (int yy, int zz);

int main ()
{
    int x = 120;
    int y = 80;
    cout << fun (x , y);
    return 0;
}

int& fun (int yy, int zz)
{
    if (yy > zz)
    {
        return yy;
    }
    return zz;
}
```

12. 以下程序的输出结果是什么？

```
#include <iostream>
```

```
using namespace std;

int& fun (int& yy, int& zz)
{
       if (yy > zz)
       {
              return yy;
       }
       return zz;
}
int main ()
{
       cout << fun (120 , 80);
       return 0;
}
```

13. 给定以下代码片段，给出其输出结果。

```
int x = 10;
int* y = &x;
cout << x << " " << *y;
```

14. 给定以下代码片段，给出其输出结果。

```
int x = 100;
int* y = &x;
int* z = &x;
cout << x << "  " << *y << "  " << *z;
```

15. 指出以下代码片段的错误之处。

```
int* x = 25;
```

16. 指出以下代码片段的错误之处。

```
const int x = 100;
double* y = &x;
```

17. 指出以下代码片段的错误之处。

```
const int x = 100;
int* y = &x;
```

18. 指出以下代码片段的错误之处。

```
const int x = 100;
const int* y = &x;
int* y = &x;
```

19. 指出以下代码片段的错误之处。

```
int x = 1000;
int y = 2000;
int* const z = &x;
z = &y;
```

20. 以下代码片段的输出结果是什么？

```
int sample [5] = {0, 10, 20, 30, 40};
cout << *(sample + 2);
```

21. 以下代码片段的输出结果是什么？

```
int sample [5] = {5, 10, 15, 20, 25};
cout << *sample + 2 << endl;
cout << *(sample + 2);
```

22. 以下代码片段的输出结果是什么？

```
int sample [5] = {0, 10, 20, 30, 40};
cout << *(sample + 7);
```

23. 以下程序的输出结果是什么？

```
#include <iostream>
using namespace std;

void fun (int* x)
{
    cout << *(x + 2);
}

int main ()
{
    int sample [5] = {0, 10, 20, 30, 40};
    fun (sample);
    return 0;
}
```

24. 以下语句在堆内存中创建了什么？绘制示意图，以显示堆内存中的对象。

```
int** arr = new int* [3];
```

25. 以下语句在堆内存中创建了什么？绘制示意图，以显示堆内存中的对象。

```
int** arr = new int* [3];
for (int i = 0; i < 3; i++)
{
    arr[i] = new int [5];
}
```

26. 编写代码，使用键盘上输入的值填充在第25题中创建的二维数组。
27. 如何使用 delete 运算符删除在第25题中创建的数组？

编程题

1. 编写一个函数，使用按引用传递和按引用返回来获取三个整数中最大的一个。可以使用我们前面编写的获取两个整数之间的较大值的函数。在程序中测试所编写的函数。
2. 编写一个函数，使用按引用传递和按引用返回来获取三个分数中最大的一个。可以使用我们前面编写的获取两个分数之间的较大值的函数。在程序中测试所编写的函数。
3. 编写一个函数，使用按引用传递和按引用返回来获取两个分数相乘的结果。在程序中测试所编写的函数。
4. 为了更好地控制数组的使用方式，我们可以创建一个数组类。设计一个名为 Array 的类，包括数据成员：capacity、size 和 arr（指向堆中数组第一个元素的指针）。同时，还可以设计一个名为 insert 的成员函数，该函数将元素添加到数组的末尾。并设计一个名为 print 的函数来打印数组的元素。使用不同的数组元素列表测试该程序。

5. 将第 4 题中的 Array 类重新设计为有序的数组（可能有重复的值）。在有序的数组中，我们可能需要将一些元素向数组末尾的方向移动，以便在正确的位置插入新元素。当我们删除一个元素时，可能还需要将一些元素移到数组的前面。
6. 在面向对象的程序设计中，我们总是可以为要解决的问题创建一个类，并让用户创建类的实例化对象并使用它们。定义一个类，该类可以创建任何大小不超过 10 的乘法表。然后使用应用程序实例化任意大小的乘法表。
7. 将帕斯卡三角形重新设计为一个类 Pascal，在该类中，我们可以创建任意大小（大小由用户传递给构造函数）的二项式的系数列表。例如，Pascal(5) 打印 $(x+y)^5$ 的系数。

第 10 章
C++ Programming: An Object-Oriented Approach

字 符 串

本章将讨论有关字符串的主题，我们几乎在所有的程序中都用到了字符串。C++ 语言继承了 C 语言中的 C 字符串，我们将在本章的第一节中讨论。C++ 还定义了一种更丰富、更安全的字符串，我们将在本章的第二节中讨论。C 字符串是一个以空字符（null，'\0'）终止的字符数组；C++ 字符串则是一个与面向对象的程序设计编程思想相匹配的类。

学习目标

阅读并学习本章后，读者应该能够

- 描述 C 字符串的一般概念和使用方法。
- 提供在 <cstring> 头文件中的 C 字符串库的简要列表。
- 演示 C 字符串在程序中的使用方法。
- 解释库中为 C 字符串定义的操作。
- 提供在 <string> 头文件中的 C++ 字符串库的简要列表。
- 解释库中为 C++ 字符串定义的操作。
- 演示 C++ 字符串在程序中的使用方法。

10.1 C 字符串

虽然 C++ 语言包含 C++ 字符串类的类型（我们将在本章的第二节中讨论），但我们还是先简要地讨论 C 字符串（有时称为 C 样式字符串）。原因有二：首先，有些程序（包括一些库）仍然使用 C 字符串；其次，C++ 字符串类在其定义中使用了一些 C 字符串。因此要了解 C++ 字符串必须先了解 C 字符串的基本知识。要使用 C 字符串，需要 <cstring> 头文件。

> 要使用 C 字符串，需要 <cstring> 头文件。

C 字符串不是类类型。它是一个字符数组，但这并不意味着任何字符数组都是 C 字符串。要成为 C 字符串，数组中的最后一个字符必须是空字符（'\0'）。换言之，C 字符串是以空字符结尾的字符数组。图 10-1 显示了一个 C 字符串。

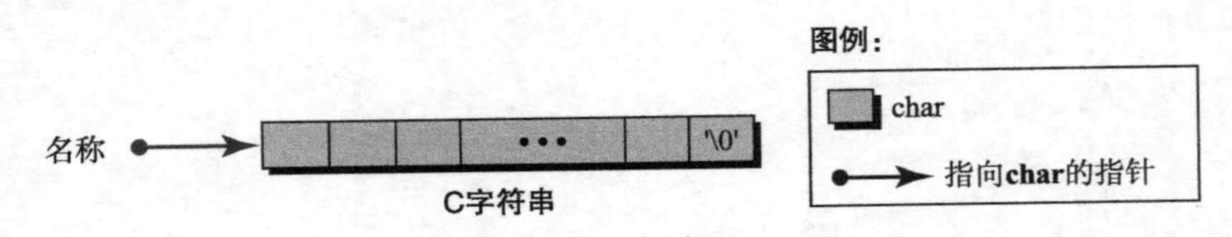

图 10-1　C 字符串的基本知识

因为数组的名称是指向数组中第一个元素的指针，所以 C 字符串的名称是指向字符串中第一个字符的指针。但是，我们必须牢记，C 字符串的名称没有定义变量，它定义了指针值（就像数组的名称）。换言之，C 字符串名称不是左值，而是右值。它是一个常量指针，这意味着该指针不能指向任何其他元素。

C 字符串名称是指向第一个字符的常量指针。

10.1.1 C 字符串库

在讨论如何使用 C 字符串之前，我们在表 10-1 中列出了常见操作及其原型（如果适用）。

表 10-1 <cstring> 头文件中的函数原型

```
// 构造
没有构造函数。用户需要创建一个字符数组。
// 析构
没有析构函数。如果在堆内存中创建数组，则用户需要负责销毁。
// 拷贝构造（设计了两个用于拷贝字符串的成员函数）
char* strcpy (char* str1, const char* str2)
char* strncpy (char* str1, const char* str2, size_t n)
// 获取字符串长度
size_t strlen (const char* str)
// 输入 / 输出
库中没有输入和输出操作，但 >> 和 << 运算符被重载用于输入和输出。
用户还可以使用在 istream 类中定义的 getline( ) 函数来读取一行字符。
// 访问字符
没有访问单个字符的成员函数，但是用户可以使用索引运算符 [...] 来实现字符访问。
// 查找给定字符（正向查找和反向查找）
char* strchr (const char* str, int c)
char* strrchr (const char* str, int c)
// 查找子字符串
char* strstr (const char* str, const char* substr)
// 查找字符集中的任意字符（仅正向查找）
char* strpbrk (const char* str, const char* set)
// 比较两个字符串
int strcmp (char* str1, const char* str2)
int strncmp (char* str1, const char* str2, size_t n)
// 字符串拼接（把一个字符串附加到另一个字符的末尾）
char* strcat (char* str1, const char* str2)
char* strncat (char* str1, const char* str2, size_t n)
// 词法分析
char* strtok (char* str, const char* delimit)
```

10.1.2 C 字符串的操作

本节将讨论如何使用表 10-1 中定义的成员函数。

构造

前面提到 C 字符串不是一个类类型，这意味着在库中没有定义构造函数。要构造一个 C 字符串，必须创建一个字符数组，并把最后一个元素设置为空字符 '\0'。

可以创建两种类型的 C 字符串：非常量字符串和常量字符串。在一个非常量 C 字符串中，可以在创建字符串之后改变其值；在一个常量 C 字符串中，字符串的值不能改变。以下代码片段演示了如何创建和初始化 C 字符串。

```
char str [ ] = {'A', 'B', 'C', 'D', '\0'};              // 非常量字符串
char str [ ] = "ABCD";                                  // 紧凑型非常量字符串
const char str [ ] = {'A', 'B', 'C', 'D', '\0'};        // 常量字符串
const char str [ ] = "ABCD";                            // 紧凑型常量字符串
```

第一种初始化形式是前文讨论过的字符数组的初始化形式。第二种形式有时被称为紧凑形式，使用紧凑型初始化时字符用双引号括起来，不包含空字符。编译器逐个提取字符，并将它们存储在相应的数组单元格中，并自动添加空字符。图 10-2 显示了紧凑形式背后的思想。

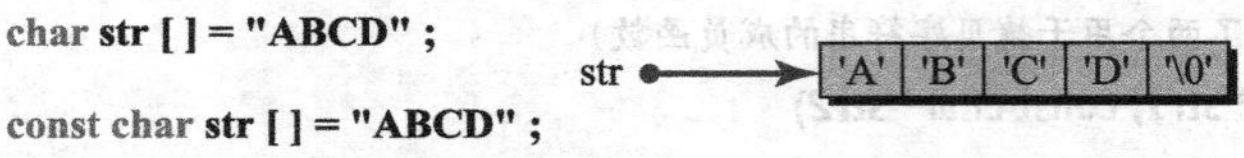

char str [ ] = "ABCD" ;
const char str [ ] = "ABCD" ;
str → 'A' 'B' 'C' 'D' '\0'

图 10-2　使用紧凑型初始化的效果

我们建议使用紧凑形式以避免遗漏空字符。

如前所述，表 10-1 或者图 10-2 中创建的字符串的名称是一个右值指针，而不是一个变量。如果我们想要创建一个变量，我们必须声明一个 char* 或者 const char* 类型的变量，并将字符串的名称赋值给该变量，如图 10-3 所示。

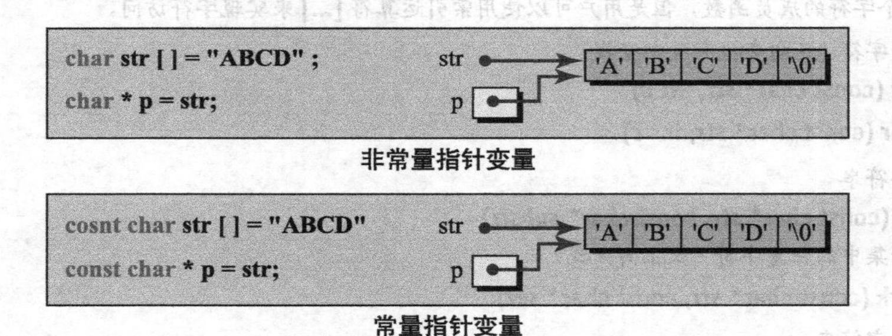

图 10-3　从字符串创建指针变量

在第 8 章中，我们提到过不能从函数返回数组，因为数组需要一个指向第一个元素的指针以及数组的大小。C 字符串的设计消除了第二个需求。C 字符串不需要被告知字符串的大小，因为最后一个字符是空字符，隐式地定义了字符串的大小。这意味着我们可以从函数返回指向 C 字符串的指针变量。

字符串字面量。在前面章节中，我们讨论过数据类型（例如整数、浮点数或者字符）的字面量。例如，3 是一个整数字面量，23.7 是一个浮点数字面量，'A' 是一个字符字面量。我们也可以创建一个字符串字面量。**字符串字面量**是以空字符结尾的字符数组，其名称是由两个引号括起来的数组中的字符序列，如图 10-4 所示。

"Hello dear" → 'H' 'e' 'l' 'l' 'o' ' ' 'd' 'e' 'a' 'r' '\0'

图 10-4　一个字符串字面量

字符串字面量是一个常量字符串，它是 C++ 语言的一部分，C 字符串和 C++ 字符串都使用字符串字面量。一旦创建了一个字符串字面量，就可以在任何可以使用字符串字面量的

地方使用它。我们已经使用了字符串字面量来打印消息。但是，我们必须牢记，字符串字面量是一个常量实体，在创建之后不能被更改。

字符串字面量是常量实体，不能更改。

【例 10-1】 以下代码片段显示，我们可以像使用其他字面量一样使用字符串字面量。例如，我们可以使用 cout 对象和插入运算符来打印整数字面量、浮点数字面量、字符字面量和字符串字面量的值。

```
cout << 5 << endl;                    // 打印整数字面量
cout << 21.3 << endl;                 // 打印浮点数字面量
cout << 'A' << endl;                  // 打印字符字面量
cout << "Hello dear" << endl;         // 打印字符串字面量
```

使用字符串字面量创建字符串。字符串字面量使得创建 C 字符串变得容易。首先创建所需的字符串字面量，然后将其赋值给指向常量字符的指针。图 10-5 显示了其过程。

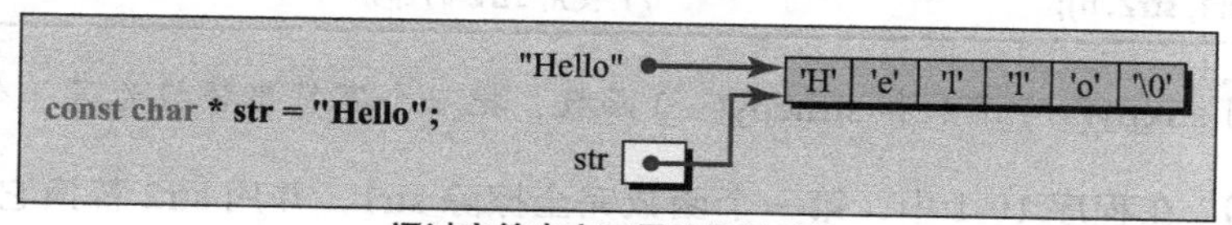

通过字符串字面量给指针赋值

图 10-5　将字符串字面量赋值给指针

C++ 禁止把一个字符串字面量赋值给一个非常量字符指针，如下所示：

```
char* str = "Hello";                  // 编译错误。字符串字面量是一个常量
const char* str = "Hello";            // 正确
```

把一个字符串字面量赋值给一个非常量字符指针将导致编译错误。

紧凑型初始值设定项和字符串字面量。尽管两者看起来相同，但我们必须把紧凑型初始值设定项和字符串字面量区分开来。其区别在于使用的位置。紧凑型初始值设定项是我们讨论的常规初始值设定项的简单形式。使用紧凑型初始值设定项时，编译器会逐个取出字符，并将它们存储在字符数组中。字符串字面量已经是一个常量字符串，可以在任何可以使用字符串的地方使用。字符串字面量是指向在内存中创建的字符串的指针，可以赋值给指向常量字符的指针。以下显示了使用方法的差异：

```
char str1 [ ] = "Hello";              // "Hello" 是紧凑型初始值设定项
const char str2 [ ] = "Hello";        // "Hello" 是紧凑型初始值设定项
const char* str3 = "Hello";           // "Hello" 是字符串字面量
```

在堆内存中构造。由于 C 字符串是一个数组，所以我们可以在堆内存中创建 C 字符串。但是，由于在这种情况下字符串的名称是指向字符的指针，因此不能使用紧凑型初始化。如果是一个非常量字符串，我们必须逐字符进行初始化；如果是常量字符串，我们必须使用字符串字面量来进行初始化。

```
char* str = new char [3];             // 包含 2 个字符的非常量字符串
const char* str = new char [3];       // 包含 2 个字符的常量字符串
```

析构

如表 10-1 所述，C 字符串不是类类型，这意味着库中没有定义析构函数。如果 C 字符串是在栈内存中创建的，那么当 main 函数终止时，它会自动销毁。如果是在堆中创建的，则必须使用 delete 运算符将其删除，这样做可以避免内存泄漏。

```
const char* str = new char [3];                    // 创建
delete [ ] str;                                    // 销毁
```

字符串拷贝

由于 C 字符串没有被定义为类，因此没有拷贝构造函数。然而，库的设计者定义了两个成员函数：strcpy 和 strncpy。函数 strcpy 使用第二个字符串的全部内容替换第一个字符串；函数 strncpy 将第一个字符串中的前 *n* 个字符替换为第二个字符串中的前 *n* 个字符。请注意，拷贝会更改目标字符串，但不会更改源字符串。

```
strcpy (str1, str2);                               // 使用 str2 的全部
strncopy (str1, str2, n);                          // 使用 str2 的部分
```

可以使用 strcpy(…) 函数和 strncpy(…) 函数，把一个字符串替换为另一个字符串。

【例 10-2】 在程序 10-1 中，第一个函数完全擦除 str1，并用 str2 替换它。第二个函数使用 str2 的前 *n* 个字符替换 str1 的前 *n* 个字符，并保持 str1 的其余部分不变。

程序 10-1　使用 C 字符串库拷贝字符串

```
1   /***************************************************************
2    * 本程序演示如何使用 strcpy 和 strncpy                          *
3    * 把一个字符串的全部或者部分替换为                              *
4    * 另一个字符串的全部或者部分                                    *
5    ***************************************************************/
6   #include <cstring>
7   #include <iostream>
8   using namespace std;
9
10  int main ( )
11  {
12    // 把 str2 的全部内容拷贝到 str1，字符串 str1 的内容丢失
13    char str1 [] = "This is the first string.";
14    char str2 [] = "This is the second string.";
15    strcpy (str1, str2);
16    cout << "str1: " << str1 << endl;
17    // 把 str4 的部分内容拷贝到 str3，字符串 str3 的部分内容丢失
18    char str3 [] = "abcdefghijk.";
19    const char* str4 = "ABCDEFGHIJK";
20    strncpy (str3, str4, 4);
21    cout << "str3: " << str3 << endl;
22    return 0;
23  }
```

运行结果：

```
str1: This is the second string.
str3: ABCDefghijk.
```

字符串长度

每个 C 字符串都有一个大小（长度），即字符串中不计算空字符的字符数。C 字符串库中定义了 strlen 函数来获取字符串的长度。strlen 函数接收一个 C 字符串参数，并将其长度作为一个 size_t 类型返回，该类型在库中被定义为无符号整数（unsigned int）。

```
size_t n = strlen (str);                    // 获取 str 的长度
```

strlen(…) 函数返回 C 字符串中不计算空字符的字符数。

【例 10-3】 程序 10-2 演示了如何获取 C 字符串的长度。

程序 10-2　获取两个字符串的长度

```
1  /***************************************************************
2   * 本程序演示如何获取 C 字符串的长度，                          *
3   * C 字符串的长度是空字符前的字符个数                           *
4   ***************************************************************/
5  #include <cstring>
6  #include <iostream>
7  using namespace std;
8
9  int main ( )
10 {
11   // 声明和定义两个字符串
12   const char* str1 = "Hello my friends.";
13   char str2 [] = {'H', 'e', 'l', 'l', 'o', '\0'} ;
14   // 获取并打印每个字符串的长度
15   cout << "Length of str1: " << strlen (str1) << endl;
16   cout << "Length of str2: " << strlen (str2);
17   return 0;
18 }
```

```
运行结果:
Length of str1: 17
Length of str2: 5
```

输入和输出

除了使用紧凑型初始化或者使用字面量字符串赋值之外，我们还可以将字符读入到声明为字符数组的 C 字符串中。当字符串被声明为一个类型（char* 或 const char*）时，则不支持这种操作，因为编译器必须在读取字符之前分配内存。

重载的提取运算符和插入运算符。请回顾一下 <cstring> 库重载了提取运算符（>>）和插入运算符（<<）以用于字符串输入和输出的情况。提取运算符从输入对象（键盘或者文件）中逐个提取字符，并将它们存储在数组中，直到遇到空白字符为止，然后在末尾添加空字符。问题是存储输入的字符数组必须分配足够的内存位置来存储所有输入的字符（在空白之前）以及一个空字符。如果分配的内存不够，则结果无法预测。插入运算符将数组中的字符写入输出设备，直到遇到空字符为止。

```
cin >> str;                                 // 输入
cout << str;                                // 输出
```

【例 10-4】 在程序 10-3 中，在 str4 的字面量字符串中包含一个空字符，以表明字面量字符串以空字符结尾，其余字符不是字面量字符串的一部分（第 15 行）。换言之，第 15 行的字符串只有 8 个字符而不是 19 个字符。我们为 str5 输入字符，但请注意，只有第一个单词存储在字符串中，其余的输入字符均被忽略，因为提取运算符在第一个空白处停止。

程序 10-3　使用 C 字符串

```
1  /***************************************************************
2   * 本程序演示如何创建 C 字符串                                      *
3   * 并应用输入和输出操作                                            *
4   ***************************************************************/
5  #include <iostream>
6  using namespace std;
7
8  int main ( )
9  {
10   // 创建一个常量字符串和一个非常量字符串
11   char str1 [] = {'H', 'e', 'l', 'l', 'o', '\0'} ;
12   const char str2 [] = {'H', 'e', 'l', 'l', 'o', '\0'};
13   // 使用字符串字面量创建两个常量字符串类型
14   const char* str3 = "Goodbye";
15   const char* str4 = "Goodbye\0 my friend";
16   // 打印上面创建的四个字符串
17   cout << "str1: " << str1 << endl;
18   cout << "str2: " << str2 << endl;
19   cout << "str3: " << str3 << endl;
20   cout << "str4: " << str4 << endl << endl;
21   // 创建和输入第五个字符串
22   char str5 [20];
23   cout << "Enter the characters for str5: " ;
24   cin >> str5;
25   cout << "str5: " << str5;
26   return 0;
27 }
```

```
运行结果：
str1: Hello
str2: Hello
str3: Goodbye
str4: Goodbye

Enter the characters for str5: This is the one.
str5: This
```

getline 函数。要读取包含空格的字符行，必须使用为此目的定义的函数：getline 函数。getline 函数是 istream 类的成员，这意味着我们必须使用一个 cin 类型的对象。如果省略 delim 参数，则使用 '\n' 字符。

```
cin.getline (str, n);                        // 使用 '\n' 作为分隔符
cin.get (str, n, 'delimeter');               // 使用特定的分隔符
```

【例 10-5】 程序 10-4 创建一个字符串数组，读取行内容到数组中，并打印这些行。

程序 10-4 使用字符串数组

```
1  /*****************************************************************
2   * 本程序演示如何使用 getline 函数                                  *
3   * 读取一系列的行，并打印这些行                                      *
4   *****************************************************************/
5  #include <iostream>
6  #include <cstring>
7  using namespace std;
8
9  int main ()
10 {
11   // 声明一个字符串数组
12   char lines [3][80];
13   // 输入三行内容
14   for (int i = 0; i < 3; i++)
15   {
16       cout << "Enter a line of characters: ";
17       cin.getline (lines [i], 80);
18   }
19   // 打印三行内容
20   cout << endl;
21   cout << "Output: " << endl;
22   for (int i = 0; i < 3; i++)
23   {
24       cout << lines [i] << endl;
25   }
26   return 0;
27 }
```

```
运行结果：
Enter a line of characters: This is the first line.
Enter a line of characters: This is the second line.
Enter a line of characters: This is the third line.

Output:
This is the first line.
This is the second line.
This is the third line.
```

访问字符

我们讨论的下一个操作是如何访问字符串中的任意字符。如果我们知道字符在字符串中的位置，就可以使用下标运算符来访问字符。当字符串是常量时，访问操作仅意味着检索。当字符串是非常量时，访问操作意味着检索或者更改。

```
char c = str [i];                    // 字符串 str 是一个常量
str[i] = c;                          // 字符串 str 是一个非常量
```

【例 10-6】 程序 10-5 演示了如何访问常量字符串和非常量字符串中的字符。

程序 10-5 访问 C 字符串中的字符

```
1  /*****************************************************************
2   * 本程序演示如何使用下标运算符                                      *
```

```
3     * 访问字符串中的字符                                              *
4     *****************************************************************/
5     #include <cstring>
6     #include <iostream>
7     using namespace std;
8
9     int main ( )
10    {
11      // 创建两个 C 字符串
12      const char* str1 = "Hello my friends.";
13      char str2 [ ] = "This is the second string.";
14      // 检索给定位置的字符
15      cout << "Character at index 6 in str1: " << str1[6] << endl;
16      // 更改给定位置的字符
17      str2 [0]  = 't';
18      cout << "str2 after change: " << str2;
19      return 0;
20    }
运行结果:
Character at index 6 in str1: m
str2 after change: this is the second string.
```

查找字符

我们可以搜索一个字符串来查找某个字符。搜索可以返回指向第一次出现（正向搜索）或者最后一次出现（反向搜索）的字符的指针。正向搜索使用 strchr 函数；反向搜索使用 strrchr 函数。在设置了指向字符的指针之后，如果字符串不是常量，则可以使用该指针来更改字符。如果找不到字符，则返回空指针。

```
char* ptr = strchr (str, 'c');                    // 正向搜索
char* ptr = strrchr (str, 'c');                   // 反向搜索
```

我们可以使用 strchr(…) 和 strrchr(…) 成员函数创建指向字符的指针。

【例 10-7】 程序 10-6 演示了如何搜索一个字符。

程序 10-6　搜索给定字符的开始位置

```
1     /****************************************************************
2      *本程序用于                                                      *
3      *正向搜索给定字符第一次出现的位置                                  *
4      *反向搜索给定字符最后一次出现的位置                                *
5      ****************************************************************/
6     #include <cstring>
7     #include <iostream>
8     using namespace std;
9
10    int main ( )
11    {
12      // 声明一个字符串
13      char str [ ]  = "Hello friends.";
14      // 把第一个出现的字母 e 转换为大写字母 E
15      char* cPtr = strchr (str, 'e');
```

```
16     *cPtr = 'E' ;
17     cout << "str after first change: " << str << endl;
18     // 把最后一个出现的字母 e 转换为大写字母 E
19     cPtr = strrchr (str, 'e');
20     *cPtr = 'E' ;
21     cout << "str after last change: " << str << endl;
22     return 0;
23 }
```

运行结果：

```
str after first change: HEllo friends.
str after last change: HEllo friEnds.
```

查找子字符串

可以使用 strstr 函数搜索字符串以查找子字符串的位置。函数返回指向子字符串中第一个字符的指针。如果在字符串中找不到子字符串，则返回空指针。

```
char* ptr = strstr (str, substr);                    // 搜索子字符串
```

【例 10-8】 程序 10-7 演示了如何搜索子字符串。

程序 10-7　搜索子字符串

```
1  /***************************************************************
2   * 本程序演示如何使用 strstr 函数                                  *
3   * 在一个字符串中查找给定子字符串的出现位置。                        *
4   ***************************************************************/
5  #include <cstring>
6  #include <iostream>
7  using namespace std;
8
9  int main ( )
10 {
11     // 创建一个字符串
12     char str [ ] = "Hello friends of mine.";
13     // 搜索子字符串的位置
14     char* sPtr = strstr (str, "friends");
15     cout << "The substring starts at index:  " << sPtr - str;
16     return 0;
17 }
```

运行结果：

```
The substring starts at index:  6
```

查找字符集中的任意字符

有时我们需要搜索一个字符串来查找在一组字符中定义的任意字符的位置。换言之，我们不是要搜索某个指定字符的位置，我们希望搜索集合中任意字符的第一个位置。在本章后面，我们将了解到该操作可以用于把字符串分成标记符。该集合也定义为一个字符串。函数 strpbrk 就是用于此目的。如果找不到字符，则返回空指针。

```
char* p = strpbrk (str, set);                    // 搜索字符集中的字符
```

成员函数 strpbrk(…) 允许我们搜索一组字符中任意字符第一次出现的位置。

【例 10-9】 程序 10-8 演示了如何使用 strpbrk(…) 函数。

程序 10-8　查找字符集中的任意字符

```
1  /*****************************************************************
2   * 本程序演示如何使用 strpbrk 函数                                  *
3   * 查找字符集中任意字符的位置                                       *
4   *****************************************************************/
5  #include <cstring>
6  #include <iostream>
7  using namespace std;
8
9  int main ( )
10 {
11     // 创建一个字符串
12     char str [ ] = "Hello friends of mine.";
13     // 搜索字符集中任意字符第一次出现的位置
14     char* pPtr = strpbrk (str, "pfmd");
15     cout << "The character " << *pPtr << " was found." << endl ;
16     cout << "It is at index: " << pPtr - str;
17     return 0;
18 }
```

运行结果：

```
The character f was found.
It is at index: 6
```

字符串比较

可以使用 strcmp 和 strncmp 函数来比较两个字符串。strcmp 用于比较两个字符串；strncmp 用于比较两个字符串的前 *n* 个字符。字符串比较逐个字符地进行，直到达到不相同的字符为止（请注意，在比较中也使用空字符）。当发现不相同的字符时，则比较停止。如果第一个字符串中的字符小于第二个字符串中的字符，函数返回一个负数。如果第一个字符串中的字符大于第二个字符串中的字符，则返回一个正数。如果找不到不相同的字符，则函数返回 0。

```
int value = (str1, str2);              //比较 str1 和 str2 的全部
int value = (str1, str2, n);           //比较 str1 和 str2 的前 n 个字符
```

可以使用 strcmp(…) 和 strncmp(…) 成员函数比较两个字符串。

【例 10-10】 程序 10-9 使用了这两个比较函数。注意，比较函数不会更改两个字符串的内容。

程序 10-9　比较 C 字符串

```
1  /*****************************************************************
2   * 本程序演示如何使用 strcmp 和 strncmp 函数                        *
3   * 比较两个字符串                                                   *
4   *****************************************************************/
5  #include <cstring>
6  #include <iostream>
7  using namespace std;
8
```

```
9   int main ( )
10  {
11      // 声明三个 C 字符串
12      const char* str1 = "Hello Alice.";
13      const char* str2 = "Hello John.";
14      const char* str3 = "Hello Betsy.";
15      // 比较字符串的全部
16      cout << "Comparing str1 and str2: ";
17      cout  << strcmp (str1, str2) << endl;
18      cout << "Comparing str2 and str3: ";
19      cout << strcmp (str2, str3) << endl;
20      // 比较字符串的部分
21      cout << "Comparing first 5 characters of str1 and str2: ";
22      cout << strncmp (str1, str2, 5);
23      return 0;
24  }
```

运行结果：

```
Comparing str1 and str2: -1
Comparing str2 and str3: 1
Comparing first 5 characters of str1 and str2: 0
```

拼接（附加）

我们可以在一个字符串的末尾附加另一个字符串。在这种情况下，目标字符串将被更改，但源字符串保持不变。<cstring> 库中为此定义了两个成员函数：strcat 和 strncat。strcat 函数在第一个字符串末尾处拼接第二个字符串中的所有字符。strncat 函数在第一个字符串末尾拼接第二个字符串的前 n 个字符。但是，必须确保 str1 有足够的内存位置来接收所拼接的字符串（程序 10-10）。

```
strcat (str1, str2);            //把 str2 的全部拼接到 str1
strncat (str1, str2, n);        //把 str2 的前 n 个字符拼接到 str1
```

使用成员函数 strcat(…) 和 strncat(…)，可以将一个字符串拼接到另一个字符串的末尾。

【例 10-11】 程序 10-10 使用与拼接操作相关的两个函数。

程序 10-10　使用 strcat 和 strncat

```
1   /***************************************************************
2    * 使用 strcat 和 strncat 函数                                   *
3    * 将一个字符串拼接到另一个字符串的末尾                           *
4    ***************************************************************/
5   #include <cstring>
6   #include <iostream>
7   using namespace std;
8
9   int main ( )
10  {
11      // 使用 strcat 函数
12      char str1 [20] = "This is ";
13      const char*  str2 = "a string.";
14      strcat (str1, str2);
```

```
15    cout << "str1: " << str1 << endl;
16    // 使用 strncat 函数
17    char str3 [20] = "abcdefghijk";
18    const char* str4 = "ABCDEFGHIJK";
19    strncat (str3, str4, 4);
20    cout << "str3: " << str3 << endl;
21    return 0;
22  }
```

运行结果：
```
str1: This is a string.
str3: abcdefghijkABCD
```

词法分析

对字符串的常见操作之一是查找嵌入在字符串中的标记符。**字符串标记符**（string token）是由分隔符（如空格）分隔的子字符串。<cstring> 库定义了 strtok 函数。要在字符串中查找所有标记符，必须多次调用 strtok 函数。当 strtok 函数被调用时，将执行三个特定任务：

1）在字符串中搜索不在分隔符集中的第一个字符。从第一个参数指向的字符开始搜索。如果找到一个字符，会将指针 p 指向该字符。如果找不到，则什么也不做。

2）然后搜索分隔符集中字符第一次出现的位置。如果找到该字符，会将该字符更改为空字符。

3）返回在第一个任务中设置的指针。

在图 10-6 中，strtok 函数查找不在分隔符集中的第一个字符，并将 p 指向该字符。然后，查找分隔符集中的任意字符并将其更改为空字符。这意味着已经创建了一个从 p 到空字符的字符串。在第二次及以后的调用中，第一个参数是空指针，strtok 函数从上一次调用中创建的空字符开始搜索。当函数到达字符串末尾时，p 为空。

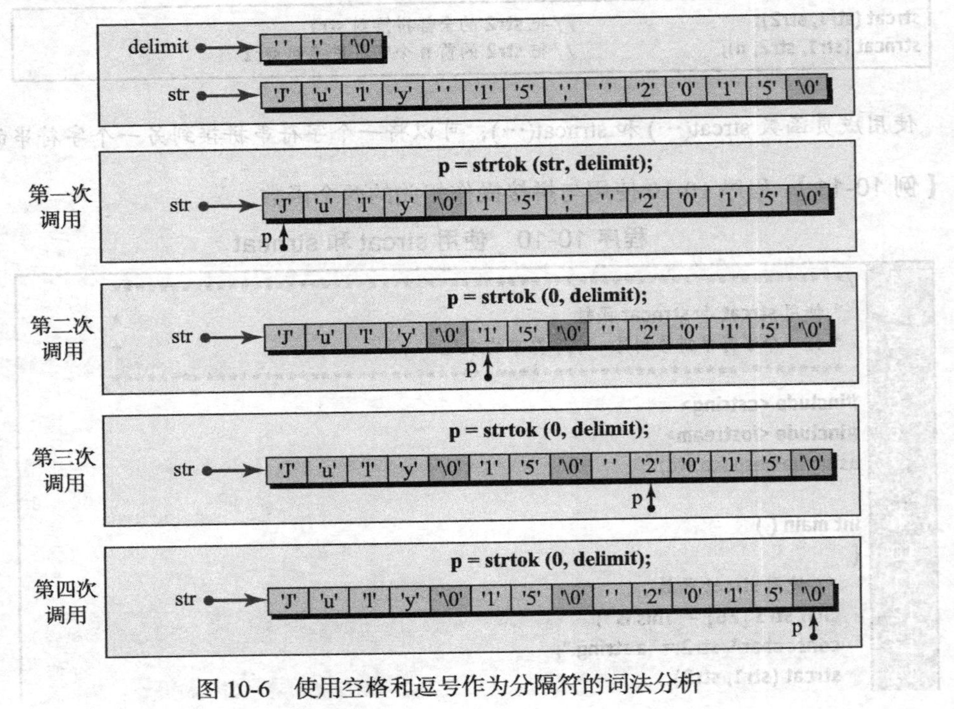

图 10-6　使用空格和逗号作为分隔符的词法分析

```
char* p = strtok (str, delimiter);        // 使用分隔符 delimiter 对字符串 str 进行词法分析
```

strtok(…) 函数可以用于使用分隔符字符将字符串拆分为标记符。

【例 10-12】 程序 10-11 使用 strtok(…) 函数将字符串拆分为标记符。程序多次循环调用 strtok，直到 p 指向一个空字符（原始字符串的结尾）。请注意，原始字符串在此过程中发生了更改，这意味着原始字符串不能是常量字符串。

请注意，程序 10-11 处理了其他两个条件。首先，如果字符串为空，则第 15 行中的 p 被设置为 null，从而 while 被跳过，这意味着不打印任何内容。其次，如果字符串仅由一个单词组成，则在第一次迭代中，第 18 行打印完唯一的单词后，p 指向一个空字符，循环终止。

程序 10-11 字符串的词法分析

```
1   /***************************************************************
2    * 本程序演示如何使用 strtok 函数                                *
3    * 提取一个日期字符串中的标记符                                  *
4    ***************************************************************/
5   #include <cstring>
6   #include <iostream>
7   using namespace std;
8
9   int main ( )
10  {
11    // 声明一个字符串和一个指针
12    char str [ ] = "July 15, 2015";
13    char* p;
14    // 使用 strtok 提取所有单词
15    p = strtok (str, ", ");  // 第一次调用
16    while (p)
17    {
18        cout << p << endl;
19        p = strtok (0, ", "); // 第二次、第三次、第四次调用
20    }
21    return 0;
22  }
```

```
运行结果：
July
15
2015
```

C 字符串存在的问题

虽然 C 字符串提供了一种处理字符串的方式，但是和 C++ 字符串相比，C 字符串更容易出错，缺乏健壮性。因此，我们建议尽可能使用 C++ 字符串。我们将在下一节中讨论 C++ 字符串。

10.2 C++ 字符串类

C++ 库提供了一个名为 string 的类。与 C 字符串的字符数组相比，string 类的对象通常

被称为 C++ 字符串对象。要使用这个类的对象和成员函数，我们必须在程序中包含头文件 <string>。

C++ 字符串库定义了类类型，而 C 字符串库定义了字符数组。使用 C 字符串时，必须创建一个字符数组并应用在 <cstring> 库中定义的函数。在使用 C++ 字符串时，我们可以构造一个字符串对象并应用在 <string> 库中预定义的成员函数。我们强烈建议尽可能多地使用 C++ 字符串。

为了使用 C++ 字符串，需要包含 <string> 头文件。

10.2.1　总体设计思路

为了更好地理解字符串类的工作方式，我们必须考虑设计人员在设计字符串类时使用的总体思路。字符串类具有私有数据成员和公共成员函数。用户调用公共成员函数来操作字符串对象。通常，数据成员包括指向字符数组的指针。其他数据成员用于存储有关字符数组的信息。数据成员通常在栈内存中创建，但字符数组本身在堆中分配，因为直到运行时才定义其大小。虽然 C++ 字符串可以被可视化为字符数组，但和前述的 C 字符串不一样，它不是以空字符结尾。图 10-7 显示了 C++ 字符串的总体概况。

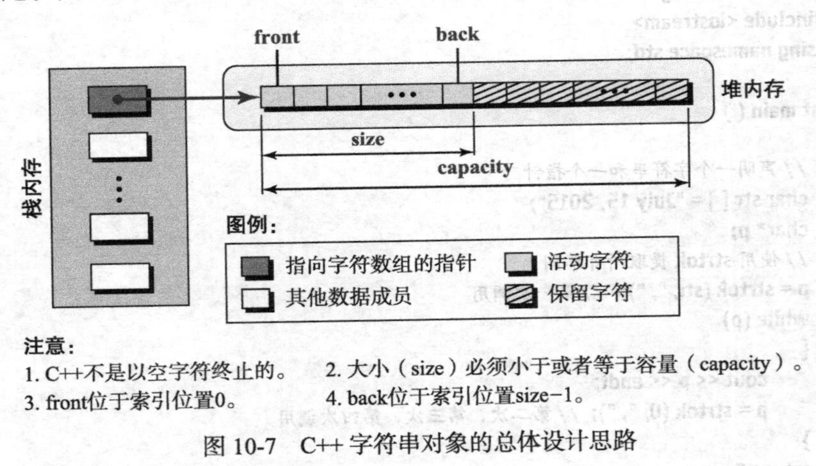

图 10-7　C++ 字符串对象的总体设计思路

C++ 字符串是字符数组，但它不是以空字符终止的。

在讨论 C++ 字符串类的每个成员函数之前，我们先讨论在库中使用的一个元数据类型和一个常量：size_type 和 npos。它们不是基本数据类型，而是使用基本类型在库中创建的类型定义。size_type 是无符号整数（unsigned int），它被定义为无符号以避免出现负整数，因为大小（size）必须始终为正。大小（size）有两个用途：作为字符数组的索引；包含一组字符的计数。在这两种情况下，size 值都不能为负数。常量 npos 的类型为 size_type，其值被设置为 −1。常量 npos 用来表示向前移动时已经经过了最后一个元素，或者向后移动时已经经过了第一个元素。换言之，npos 是一个不存在的索引。这两个标识符定义在 std::string 名称空间。在程序中使用这两个标识符时，必须使用名称空间来限定它们，如下所示：

```
string :: size-type length;
string :: npos;
```

10.2.2 C++ 字符串库

在我们讨论 C++ 字符串类库中的成员函数之前，必须先给出这些成员函数的函数原型（如表 10-2 所示）。

表 10-2 C++ 库成员节选

```
// 构造函数
string :: string ()
string :: string (size_type count, char c)
string :: string (const char* cstr)
string :: string (const char* cstr, size_type count)
// 析构函数
string :: ~string ()
// 拷贝构造函数
string :: string (const string& strg)
string :: string (const string& strg, size_type index, size_type length = npos)
// 有关大小和容量的操作
size_type string :: size ( ) const
size_type string :: max_size ( ) const
void string :: resize (size_type n, char c )
size_type string :: capacity ( ) const
void string :: reserve (size_type n = 0)
bool string :: empty ( ) const
// 输入和输出
istream& operator>> (istream& in, string& strg)
ostream& operator<< (ostream& out, const string& strg)
istream& getline (istream& in, string& strg)
istream& getline (istream& in, string& strg, char delimit)
// 获取给定位置的字符
const char& string :: operator[ ] (size_type pos) const
char& string :: operator[ ] (size_type pos)
const char& string :: at (size_type pos) const
char& string :: at (size_type pos)
// 获取给定开始位置和长度的子字符串
string string :: substr (size_type pos = 0, size_typ length = npos) const
// 查找给定字符的位置（正向查找和反向查找）
size_type string :: find (char c, size_type index = 0) const
size_type string :: rfind (char c, size_type index = npos) const
// 查找字符集中任意字符的位置（正向查找和反向查找）
size_type string :: find_first_of (const string& temp, size_type pos = 0)
size_type string :: find_last_of (const string& temp, size_type pos = npos)
// 查找字符集以外任意字符的位置（正向查找和反向查找）
size_type string :: find_first_not_of (const string& temp, size_type pos = 0)
size_type string :: find_last_not_of (const string& temp, size_type pos = npos)
```

（续）

```
// 比较两个字符串
int string :: compare (size_type pos1, size_type n1,const string strg2,
                       size_type pos2, size_type n2) const
int string :: compare (size_type pos1, size_type n1,
                       const char* cstr, size_type n2) const
// 两个字符串的逻辑比较（逻辑比较运算符包括 <、<=、>、>=、==、!=）
bool string :: operatorOper (const string strg1, const string strg2)
bool string :: operatorOper (const string strg1, const char* cstr)
bool string :: operatorOper (const char* cstr, const string strg1)
// 把一个字符附加到一个字符串的末尾
void string :: push_back (char c)
// 使用另一个字符串修改一个字符串（附加、插入、替换和赋值）
string&  string :: append (const string& temp)
string&  string :: insert (size_type pos, const string& temp)
string&  string :: replace (size_type pos, size_type n, const string& temp)
string&  string :: assign (size_type pos, size_type n, const string& temp)
// 清理和擦除字符串
void string :: clear ()
string& string :: erase (size_type pos = 0, size_type n = npos)
// 使用赋值运算符
string& string :: operator= (const string& strg)
string& string :: operator= (const char* cstr)
string& string :: operator= (char c)
// 使用复合赋值运算符（加法）
string& string :: operator+= (const string& strg)
string& string :: operator+= (const char* cstr)
string& string :: operator+= (char c)
// 使用加法运算符
string& string :: operator+ (const string& strg1, const string& strg2)
string& string :: operator+ (const string& strg1, const char* cstr2)
string& string :: operator+ (const char* cstr1, const string& strg2)
string& string :: operator+ (const string& strg1, char c)
// 转换为字符数组
const char* string :: data ( ) const
// 转换为 C 字符串
const char* string :: c_str () const
```

10.2.3 C++ 字符串定义的操作

本节简要描述表 10-2 中定义的操作。

构造函数

C++ 字符串定义了一个默认构造函数和三个参数构造函数。

默认构造函数。表 10-2 中所示的默认构造函数简单明了。它通过将图 10-7 中的指针数据成员设置为 0（空指针）来创建空字符串。以下代码片段显示了如何创建空字符串。

```
string strg;                          // 创建一个空字符串对象
```

参数构造函数。除了默认的构造函数之外，string 类还允许我们以三种不同的方式创建 string 对象，如表 10-2 所示。我们可以使用一组相同值的字符、字符串字面量、字符串字面量的一部分来创建字符串对象。以下代码片段显示了如何创建这些对象：

```
string strg1 (5 , 'a');               // 字符串 "aaaaa"
string strg2 ("hello");               // 字符串 "hello"
string strg3 ("hello", 2);            // 字符串 "he"
```

字符串 strg1 由五个相同的字符组成（此处的 size_type 定义长度）。字符串 strg2 由字符串字面量组成。在这种情况下，函数将字面量中的所有字符（末尾的空字符除外）复制到字符串对象。字符串 strg3 是字符串字面量的一部分。如果我们想使用 C 字符串对象的一部分（这里和后面其他成员函数中），我们必须从字符串的开头开始，因为指向字符串文本的指针是常量指针，不能移动。但是，我们可以定义应该复制的字符数。在这种情况下，我们只需要两个字符，这意味着只使用“he”来创建 C++ 字符串对象。

析构函数

字符串类的析构函数仅仅删除堆中创建的字符数组，并弹出栈中分配的所有数据成员。换言之，调用 delete 运算符释放分配的内存是由析构函数完成的，这有助于防止内存泄漏。

拷贝构造函数

string 类允许我们使用两个不同的拷贝构造函数：一个拷贝构造函数拷贝完整的现有对象，一个拷贝构造函数拷贝现有对象的一部分。

```
string strg (oldStrg);                        // 使用全部 oldStrg
string strg (oldstrg1, index, length);        // 使用部分 oldStrg
```

大小和容量

C++ 字符串对象使用堆中的字符数组。如果在操作期间必须减小数组的大小，则会使用成员函数 resize 更改数组大小。然而，如果在操作期间必须增大字符串的大小，则需要重新分配内存。必须在堆中创建更大的数组，复制现有元素的值，填充新元素，并回收原始内存。这些操作由后台的私有成员函数完成。但是，如果需要对大小进行多次小增量的更改，则此过程可能会为系统带来巨大的开销。为了避免这种开销，系统允许用户预留空间，这将导致创建的数组比实际所需的数组要大。

大小和最大字符数。存在两个函数用于返回字符串大小的值。size 函数返回字符串对象中当前的字符数。max_size 函数返回一个字符串对象可以拥有的最大字符数，它通常是一个与系统相关的非常大的数字。

```
size_type n = strg.size ();               // 获取大小
size_type n = strg.max_size ();           // 获取最大字符数
```

调整大小。函数 resize 更改字符串的大小。如果 n < size，则从字符串末尾删除字符，使字符串大小等于 n；如果 n > size，则将字符 c 的副本添加到字符串末尾，使字符串大小为 n。

```
strg.resize (n, 'c');                    // 调整大小，使用 'c' 填充剩余的字符串
```

容量和预留量。函数 capacity 返回字符数组的当前容量。如果我们没有预留量，则容量和大小是一样的。我们可以调用 reserve 函数使容量（capacity）大于大小（size）。

```
size_type n = strg.capacity ();                    // 获取容量
strg.reserve (n);                                  // 预留一个较大的数组
```

但是，也存在一些限制。如果函数的参数小于大小（size），则不会发生任何事情（容量（capacity）不能小于大小（size））。如果参数定义了一个小的增量，则系统可能会增加其值。

判断字符串是否为空。如果大小（size）为 0，则函数 empty 返回 true，否则返回 false。

```
bool fact = strg.empty();                          // 检查字符串是否为空
```

【例 10-13】 程序 10-12 使用有关大小、容量、是否为空、预留容量的函数。

程序 10-12　测试与大小和容量相关的函数

```
 1 /***************************************************************
 2  * 本程序先创建一个字符串对象，                                    *
 3  * 然后测试其大小，最大字符数、预留容量前后的容量                      *
 4  ***************************************************************/
 5 #include <string>
 6 #include <iostream>
 7 using namespace std;
 8 
 9 int main ( )
10 {
11   // 创建一个字符串对象
12   string strg ("Hello my friends");
13   // 测试大小、最大字符数和容量
14   cout << "Size: " <<  strg.size () << endl;
15   cout << "Maximum size: " << strg.max_size() << endl;
16   cout << "Capacity: " << strg.capacity() << endl;
17   cout << "Empty? " << boolalpha << strg.empty() << endl;
18   cout << endl;
19   // 预留容量后重新测试
20   strg.reserve (20);
21   cout << "Size: " <<  strg.size () << endl;
22   cout << "Maximum size: " << strg.max_size() << endl;
23   cout << "Capacity: " << strg.capacity() << endl;
24   cout << "Empty? " << boolalpha << strg.empty();
25   return 0;
26 }
```

```
运行结果：
Size: 16
Maximum size: 1073741820
Capacity: 16
Empty? false

Size: 16
Maximum size: 1073741820
Capacity: 32
Empty? false
```

在程序 10-12 中，我们尝试预留总共 20 个字符，但系统预留了 32 个字符（比当前大小多 16 个）。系统认为，如果我们已经创建了 16 个字符的原始字符串，那么额外的 4 个位置可能不够。

输入和输出

我们将在第 13 章和第 16 章中进一步讨论类对象的输入 / 输出。在本节中，我们将简要讨论如何输入字符串对象以及如何输出字符串对象。

输入 / 输出运算符。到目前为止，我们在基本数据类型上使用的输入和输出运算符是输入对象（istream）和输出对象（ostream）的成员函数。要输入或者输出的对象是参数。如果要输入或者输出字符串，则参数必须是 string 类的实例。

输入对象连接到键盘（cin）或者文件；输出对象连接到监视器（cout）或者文件。输出运算符的第二个参数是常量字符串，因为输出不应更改字符串对象。但输入运算符的第二个参数不是常量，因为从键盘或者文件读取的输入更改了字符串的内容。

输入运算符从输入流中逐字符读取数据。输入运算符需要知道什么时候应该停止读取。函数的设计者决定当遇到空白字符时将停止。这意味着，如果输入流中有一个空格或者有一个换行字符，则读到空白字符时将停止读取。

输出运算符将字符串对象从开始到结束都写入输出流。输出运算符没有输入运算符的限制。

```
cin >> strg;                         //读取若干字符到一个字符串对象
cout << strg;                        //输出一个字符串对象的字符
```

【例 10-14】 程序 10-13 演示了如何在一个字符串对象上使用输入 / 输出运算符。

程序 10-13 使用输入 / 输出运算符

```
1   /***************************************************************
2    * 本程序在一个字符串对象上使用输入 / 输出运算符                *
3    ***************************************************************/
4   #include <string>
5   #include <iostream>
6   using namespace std;
7
8   int main ( )
9   {
10    // 构造一个默认的字符串对象
11    string strg;
12    // 输入和输出 strg 对象的值
13    cout << "Input the string: " ;
14    cin >>  strg;
15    cout << strg << endl;
16    return 0;
17  }
```

运行结果：

```
Input the string: Hello
Hello
```

运行结果：

```
Input the string: Hi my friends
Hi
```

在第一次运行中，我们仅仅键入了字符串“hello”。所有字符都被读取并输出。在第二次运行中，我们键入了字符串“Hi my friends”，当遇到第一个空格时，读取停止。只有字符串“Hi”存储在字符串对象中，并被打印输出。

函数 getline。如前所述，输入运算符有预定义的分隔符，用于停止从输入流中读取数据。为了给用户更多的控制权，istream 对象包含一个名为 getline 的函数。getline 函数有两个版本：第一个版本使用 '\n' 作为分隔符，这意味着它可以读取整行；第二个版本允许用户定义自己的分隔符字符。

```
getline (in, strg);                    // 当读取到 '\n' 时停止
getline (in, strg, 'c');               // 当读取到字符 c 时停止
```

【例 10-15】 程序 10-14 演示了如何输入一行和多行字符串内容。

程序 10-14 使用 getline 输入

```
1   /***************************************************************
2    * 本程序在一个字符串对象上使用 getline 函数                      *
3    ***************************************************************/
4   #include <string>
5   #include <iostream>
6   using namespace std;
7
8   int main ( )
9   {
10      // 构造一个默认的字符串对象
11      string strg;
12      // 创建一个由一行内容构成的字符串
13      cout << "Enter a line of characters: " << endl;
14      getline (cin, strg);
15      cout << strg << endl << endl;
16      // 创建一个由多行内容构成的字符串
17      cout << "Enter lines of characters ended with $: " << endl;
18      getline (cin, strg, '$');
19      cout << strg;
20      return 0;
21  }
```

```
运行结果：
Enter a line of characters:
This is a line of text.
This is a line of text.

Enter lines of characters ended with $:
This is a multi-line set of
characters to be
stored in a string.$
This is a multi-line set of
characters to be
stored in a string.
```

请注意，输入显示为**灰色**。在第一部分，我们使用默认分隔符，即 '\n' 字符（回车键）。在第二部分，我们使用 '$' 字符作为分隔符。两个分隔符字符均未被打印。

访问字符

当一个字符串对象被实例化之后，如果我们知道某个字符的索引（它是在字符串中相对于零的位置），就可以访问这个字符，以实现检索或者更改。string 类为此提供了四个成员函数。前两个使用重载的下标运算符 [] 返回一个字符，作为右值或者左值。我们将在第 13 章中讨论运算符重载，目前，只需要了解运算符 [] 给出了一个函数的名称，允许我们像访问数组一样使用运算符 []。其他两个成员函数使用 at 函数来选择字符。

```
char c = strg [pos];                    // 字符 c 可以被更改
char c = strg.at (pos);                 // 字符 c 可以被更改
const char c = strg [pos];              // 字符 c 不能被更改
const char c = strg.at (pos);           // 字符 c 不能被更改
```

下标运算符不检查字符串的大小，如果下标超出范围，则可能会产生不可预测的结果并导致程序终止；at 函数检查大小，如果位置参数不在数组的范围内，则引发异常。我们将在第 14 章中学习如何处理异常。

访问字符的函数允许我们提供要返回的字符的位置。

【例 10-16】 程序 10-15 演示如何提取一个字符串中的单个字符。

程序 10-15　提取字符串的单个字符

```
1   /***************************************************************
2    * 本程序演示                                                   *
3    * 如何在一个字符串中提取单个字符                                 *
4    ***************************************************************/
5   #include <string>
6   #include <iostream>
7   using namespace std;
8
9   int main ( )
10  {
11    // 构造一个字符串对象
12    string strg ("A short string");
13    // 提取并打印索引 5 和 8 处的字符
14    cout << "Character at index 5: " << strg [5] << endl ;;
15    cout << "Character at index 8: " << strg.at(8) << endl;
16    return 0;
17  }
运行结果：
Character at index 5: r
Character at index 8: s
```

【例 10-17】 程序 10-16 演示如何将一行文本中的所有小写字符更改为大写字符。我们首先使用下标运算符（[]）来提取字符（作为右值）。然后，我们使用相同的运算符作为左值来存储字符的大写版本。

程序 10-16　把字符串中的所有字符转换为大写字符

```
1   /***************************************************************
2    * 本程序演示如何使用运算符 [] 作为左值和右值，                    *
```

```
3    * 把一行文本转换为大写字符                                              *
4    ***********************************************************************/
5  #include <string>
6  #include <iostream>
7
8  using namespace std;
9  int main ( )
10 {
11    string line;
12
13    cout << "Enter a line of text: " << endl;
14    getline (cin, line);
15    for (int i = 0; i < line.size(); i++)
16    {
17        line[i] = toupper (line[i]);
18    }
19    cout << line;
20    return 0;
21 }
```

运行结果：
```
Enter a line of text:
This is a line of text to be capitalized.
THIS IS A LINE OF TEXT TO BE CAPITALIZED.
```

【例 10-18】 程序 10-17 创建并测试一个将字符串中的字符反转的函数。该函数使用按引用传递来反转传递给它的同一字符串对象。在函数内部，我们使用拷贝构造函数创建一个临时字符串，然后反转该字符串。我们将在本章后面讨论的问题中使用此函数。

程序 10-17　反转字符串对象

```
1  /***********************************************************************
2   * 本程序使用一个函数来反转一个字符串对象                                  *
3   ***********************************************************************/
4  #include <string>
5  #include <iostream>
6  using namespace std;
7
8  void reverse (string& strg);    // 函数声明
9
10 int main ( )
11 {
12    // 声明字符串对象
13    string strg;
14    // 输入原始字符串对象并打印
15    cout << "Enter a string: ";
16    getline (cin, strg);
17    cout << "Original string: " << strg << endl;
18    // 反转字符串对象并打印
19    reverse (strg);
20    cout << "Reversed string: " << strg;
21    return 0;
22 }
```

```
23  /*************************************************************
24   * 该函数反转按引用传递给它的字符串                              *
25   *************************************************************/
26  void reverse (string& strg)
27  {
28    string temp (strg);
29    int size = strg.size () ;
30    for (int i = 0; i < size; i++)
31    {
32        strg [i] = temp [size - 1 - i];
33    }
34  }
运行结果:
Enter a string: Hello my friends.
Original string: Hello my friends.
Reversed string: .sdneirf ym olleH
```

提取子字符串

通过给定的第一个字符的索引和要提取的字符数（长度），可以从字符串中提取一个子字符串（一组连续字符）。由于只能将最左边的参数设置为默认值，如果只给定一个参数，则将其作为 pos。如果两个参数都省略，则返回整个字符串。注意，函数被定义为常量，这意味着宿主对象不能被更改。以下代码片段中 result 包含了在此处理过程中创建的字符串。

```
string result =  strg.substr (pos, n);          // result 包含 n 个字符
```

【例 10-19】 程序 10-18 演示了 substr 函数的两种用法。

程序 10-18 提取两个子字符串

```
1   /*************************************************************
2    * 本程序演示如何从一个字符串对象中                              *
3    * 提取两个子字符串                                            *
4    *************************************************************/
5   #include <string>
6   #include <iostream>
7   using namespace std;
8
9   int main ( )
10  {
11    // 构造一个字符串对象
12    string strg ("The C++ language is fun to work with.");
13    // 提取两个子字符串
14    cout << strg.substr(8) << endl ;
15    cout << strg.substr(4,12) << endl;
16    return 0;
17  }
运行结果:
language is fun to work with.
C++ language
```

查找字符

在 C++ 字符串中搜索的用途非常广泛，覆盖了多种情况。在特定搜索中，给定一个由

特定字符组成的搜索参数，我们希望在宿主字符串对象中找到该字符。但是，在宿主对象中可能存在多个搜索参数的副本。在正向搜索中，我们查找第一个副本；在反向搜索中，我们查找最后一个副本。请注意，搜索不会更改被搜索的对象，它只查找搜索参数的位置。如果我们想要更改特定位置的字符或者字符串，则必须使用前面讨论过的其他成员函数。

正向搜索或者反向搜索给定字符。可以使用两个成员函数（find 和 rfind）来查找字符，这两个函数分别用于正向搜索和反向搜索特定的字符，如图 10-8 所示。

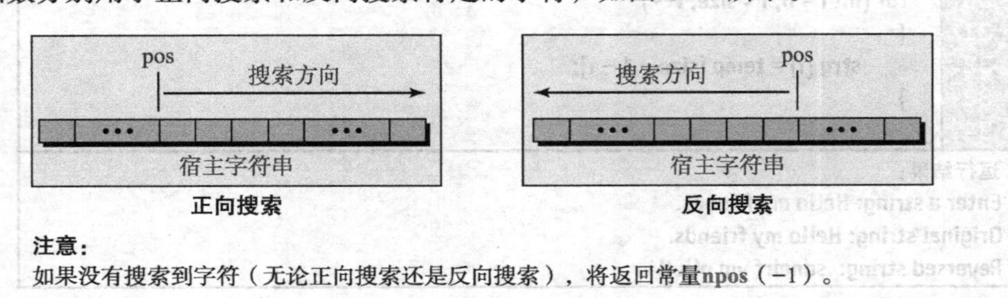

图 10-8　搜索字符串中的字符

表 10-2 显示了字符搜索函数的定义。搜索在宿主字符串中进行。第一个参数定义要查找的字符；第二个参数定义起始索引。如果缺少第二个参数，在正向搜索中默认为 0，在反向搜索中默认为 npos。如果搜索成功，则返回相应字符的索引，否则返回常量 npos（−1）。

警告：两个函数都返回一个无符号整数（size_type）。但是，当找不到字符时，两个函数都返回常量 npos，它是一个值为 −1 的整数。由于 −1 不能作为无符号整数返回，因此会包装该值并返回一个非常大的数字（最大整数 −1）。我们必须意识到这个问题，并在程序中正确地处理它。

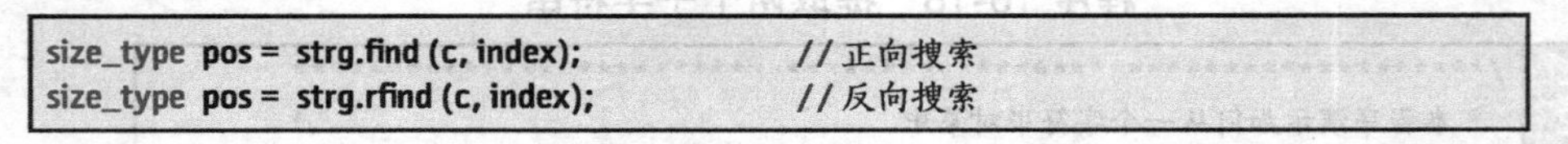

```
size_type pos = strg.find (c, index);          // 正向搜索
size_type pos = strg.rfind (c, index);         // 反向搜索
```

正向搜索或者反向搜索字符集中的任意字符。更有趣的搜索是查找字符集中的任意字符的情况。例如，假设我们想找到宿主字符串中的第一个元音。元音属于字符集“aeiou”。我们查找这些字符中的任何一个，然后返回正向搜索中第一个元音的索引或者反向搜索中最后一个元音的索引。另一种情况是查找不属于字符集中的字符。例如，我们希望在宿主字符串中找到第一个或者最后一个非元音字符。

C++ 包括各种各样的搜索字符集：C++ 字符串、C 字符串、C 字符串的一部分，甚至单个字符。为了使函数的格式更容易理解，我们创建了一个定义字符集的临时宿主对象，然后在搜索函数中使用该临时对象。图 10-9 显示了这种思想。存在四个搜索函数，可以与这四种字符集结合在一起使用。

- 函数 find_first_of 执行正向搜索，用于在字符串中查找匹配给定字符集中任意字符的第一个字符。如果搜索成功，则返回对应的索引，否则返回 npos。
- 函数 find_first_not_of 执行正向搜索，用于在字符串中查找不匹配给定字符集中任意字符的第一个字符。如果搜索成功，则返回对应的索引，否则返回 npos。
- 函数 find_last_of 执行反向搜索，用于在字符串中查找匹配给定字符集中任意字符的最后一个字符。如果搜索成功，则返回对应的索引，否则返回 npos。
- 函数 find_last_not_of 执行反向搜索，用于在字符串中查找不匹配给定字符集中任意字符的最后一个字符。如果搜索成功，则返回对应的索引，否则返回 npos。

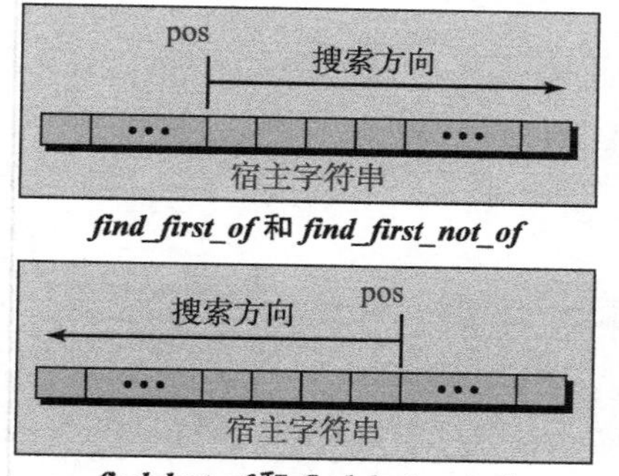

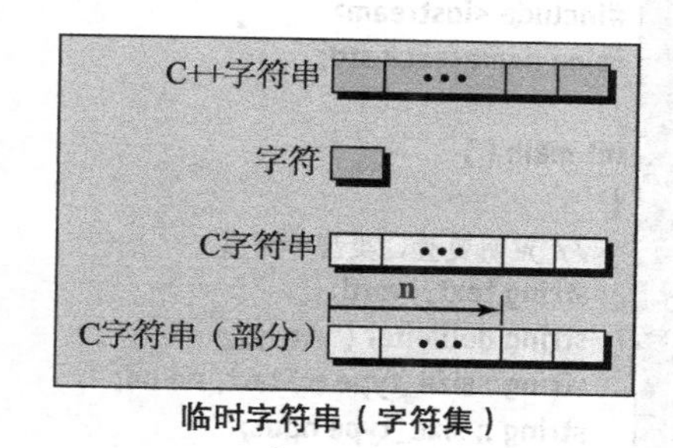

图 10-9　四个搜索函数的思想

以下代码片段演示如何使用这些搜索函数。注意，参数 set 是临时字符串，需要使用图 10-9 中定义的四种方法之一创建临时字符串，参数 index 定义了搜索的起点。

```
size_type pos = strg.find_first_of (set, index);        // 正向搜索
size_type pos = strg.find_last_of (set, index);         // 反向搜索
size_type pos = strg.find_first_not_of (set, index);    // 正向搜索
size_type pos = strg.find_last_not_of (set, index);     // 反向搜索
```

词法分析

我们可以使用搜索函数来查找子字符串的开头和结尾，然后从字符串中提取子字符串。例如，假设我们正在查找文本中的单词。文本中的单词通常由空格或者换行符（'\n'）分隔。如果我们创建由这两个字符作为单词分隔符的字符集，我们就可以从文本中提取单词。图 10-10 显示了我们如何实现这一点的策略。

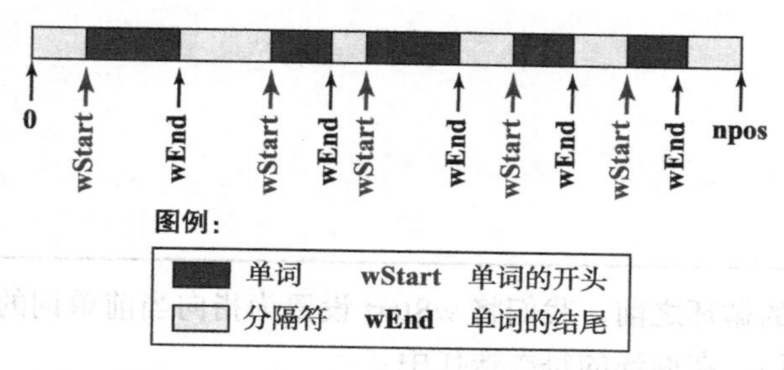

图 10-10　从一行文本中抽取单词的策略

程序 10-19 使用两个搜索函数和一个循环来提取单词。首先使用函数 find_first_not_of 来查找单词的开头（单词不包含分隔字符集中的字符），然后使用 find_first_of 来查找单词的结尾（单词后面属于分隔字符集的字符）。在本例中，分隔符由两个字符组成：空格符和换行符。分隔符也可以包含其他字符，比如逗号、分号等，但是为了保持程序简单，我们忽略了这些分隔符。

程序 10-19　从一行文本中提取单词

```
1   /***************************************************************
2    * 本程序使用字符串搜索函数                                     *
3    * 在一行文本中查找和提取单词                                   *
```

```
4     ****************************************************************/
5   #include <string>
6   #include <iostream>
7   using namespace std;
8
9   int main ( )
10  {
11    // 声明变量、类型和常量
12    string text, word;
13    string delimiter (" \n");
14    string:: size_type wStart, wEnd;
15    string :: size_type npos;
16    // 从键盘输入一行文本
17    cout << "Enter a line of text: " << endl ;
18    getline (cin, text);
19    // 搜索、查找和打印单词
20    cout << "Words in the text:" << endl;
21    wStart = text.find_first_not_of (delimiter, 0);
22    while (wStart < npos)
23    {
24          wEnd = text.find_first_of (delimiter, wStart);
25          cout <<  text.substr (wStart, wEnd – wStart) << endl;
26          wStart = text.find_first_not_of (delimiter, wEnd);
27    }
28    return 0;
29  }
```

```
运行结果:
Enter a line of text:
This is a line of text.
Words in the text:
This
is
a
line
of
text.
```

在程序10-19中的循环之前，我们将wStart设置为指向当前单词的开头，就是文本中的第一个单词（第21行）。在循环的每次迭代中：

1）首先将wEnd设置为当前单词的结尾（第24行）。

2）然后提取wStart和wEnd之间的子字符串（第25行）。

3）最后更新wStart的值，以指向下一次迭代的下一个单词的开头（第26行）。

当跳出循环时，所有的单词都被检索并打印出来了。请分析循环，如果在单词之间找到两个或者三个空格，会发生什么？

比较字符串

C++字符串为比较两个字符串提供了两种方式：整数型和布尔型。

结果为整数型的字符串比较。结果为整数型的字符串比较用于比较两个字符串，并返回三个整数值之一：当两个字符串相等时返回零；当第一个字符串大于第二个字符串时返回正数；当第一个字符串小于第二个字符串时返回负数。存在一些语法规则：第一个字符串是宿主字符串；字符串参数可以是 C++ 字符串或者 C 字符串；两者都可以是字符串或者子字符串。C++ 字符串的子字符串可以通过给定的开始索引和长度来定义；C 字符串的子字符串必须从第一个字符开始。图 10-11 显示了这种思想。

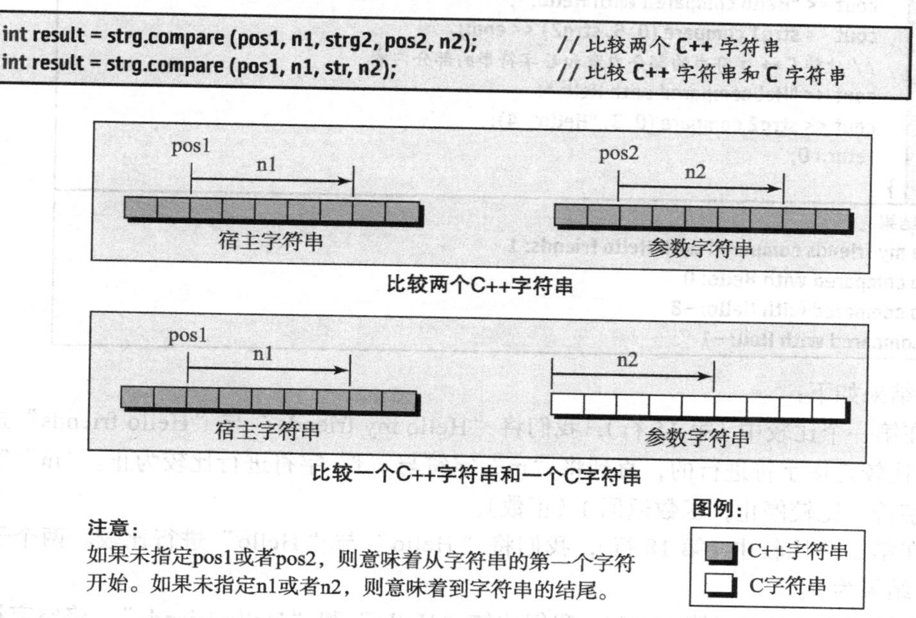

图 10-11 比较两个字符串

表 10-2 给出了两个用于比较两个字符串的成员函数。第一个函数将宿主字符串与另一个 C++ 字符串进行比较；第二个函数比较宿主字符串和 C 字符串。注意，C 字符串不使用 pos 参数，因为 C 字符串必须从头开始。

通常，pos1 和 n1 定义第一个字符串的开始索引和长度；pos2 和 n2 定义第二个字符串的开始索引和长度。图 10-11 显示了如何使用这些成员函数。

【例 10-20】 程序 10-20 比较两个字符串。

程序 10-20 结果为整数型的字符串比较

```
1  /***************************************************************
2   * 本程序用于测试结果为整数型的字符串比较                        *
3   ***************************************************************/
4  #include <string>
5  #include <iostream>
6  using namespace std;
7
8  int main ( )
9  {
10   // 声明两个 C++ 字符串
11   string strg1 ("Hello my friends");
12   string strg2 ("Hello friends");
13   // 比较两个 C++ 字符串
14   cout << strg1 << " compared with " << strg2 << ": ";
15   cout << strg1.compare (strg2) << endl;
16   // 比较两个 C++ 字符串的部分内容
```

```
17	    cout << "Hello compared with Hello: ";
18	    cout << strg1.compare( 0, 5, strg2, 0, 5) << endl;
19	    // 比较第一个 C++ 字符串的部分内容和 C 字符串
20	    cout << "Hello compared with Hello: ";
21	    cout << strg1.compare (0, 5, strg2) << endl;
22	    // 比较 C++ 字符串的部分内容和 C 字符串的部分内容
23	    cout << "Hel compared with Hell: ";
24	    cout << strg2.compare (0, 3, "Hello" ,4);
25	    return 0;
26	}
运行结果:
Hello my friends compared with Hello friends: 1
Hello compared with Hello: 0
Hello compared with Hello: -8
Hel compared with Hell: -1
```

比较结果如下:

1）在第一个比较中（第 15 行），我们将“Hello my friends”与“Hello friends”进行比较。这种比较是逐字符进行的，直到将“m”字符与“f”字符进行比较为止。“m”字符大于“f”字符，比较停止，函数返回 1（正数）。

2）在第二个比较中（第 18 行），我们将“Hello”与“Hello”进行比较。两个子字符串相等，结果为 0。

3）在第三个比较中（第 21 行），我们比较“Hello”和“Hello friends”。将空字符与空格进行比较。结果是一个负数（−8）。

4）在第四个比较中（第 24 行），我们将“Hel”与“Hell”进行比较。当我们到达第四个字符时，我们将空字符与“l”进行比较。结果是一个负数（−1）。

在结果为整数型的字符串比较中，必须注意，不要将整数结果转换为布尔类型，因为 1 和 −1 都转换为 true，这是不正确的。

结果为布尔型的字符串比较。结果为布尔型的字符串比较用于比较两个字符串，并返回布尔值（真或者假）。和结果为整数型的字符串比较一样，比较两个字符串。第一个字符串是宿主对象，第二个字符串是 C++ 字符串或者 C 字符串。与结果为整数型的字符串比较不同，我们不能比较两个子字符串。如果希望比较子字符串，则必须把子字符串转换为临时的 C++ 字符串，然后进行比较。

函数原型使用运算符重载，我们将在第 13 章中讨论。函数原型中的术语 oper 可以是关系运算符或者等性运算符之一（<，<=，>，>=，== 或者 !=）。我们可以使用这些运算符比较这两个字符串，就好像它们是基本类型一样。读者可能想知道我们是否可以比较两个 C 字符串。我们可以这样做，但 C 字符串被认为是基本类型，并且可以对 C 字符串应用普通的关系运算符和等性运算符（我们不需要 <string> 头）。注意，在逻辑比较中，两个字符串中的其中一个必须是 C++ 字符串。

以下代码片段演示了如何使用这些成员函数。

```
bool result = strg1 oper strg2;          // 比较两个 C++ 字符串
bool result = strg oper str;             // 比较 C++ 字符串和 C 字符串
bool result = str oper strg;             // 比较 C 字符串和 C++ 字符串
```

【例 10-21】 程序 10-21 比较字符串。

程序 10-21 使用逻辑运算符比较字符串

```
1  /*************************************************************
2   * 本程序用于测试使用逻辑运算符比较两个字符串                    *
3   *************************************************************/
4  #include <string>
5  #include <iostream>
6  using namespace std;
7
8  int main ( )
9  {
10    // 创建四个 C++ 字符串
11    string strg1;
12    string strg2 (5, 'a');
13    string strg3 ("Hello Friends");
14    string strg4 ("Hi People", 4);
15    // 使用六个逻辑运算符（关系运算符和等性运算符）比较字符串
16    cout << "strg1 < strg2 : " << boolalpha << (strg1 < strg2);
17    cout << endl;
18    cout << "strg2 >= strg3: " << boolalpha << (strg2 >= strg3);
19    cout << endl;
20    cout << "strg1 == strg2: " <<  boolalpha << (strg1 == strg2);
21    cout << endl;
22    cout << "Hi P != strg4: " << boolalpha << ("Hi P" != strg4);
23    return 0;
24 }
```

运行结果：

```
strg1 < strg2 : true
strg2 >= strg3: true
strg2 < strg3: false
strg1 == strg2: false
Hi P != strg4: false
```

【例 10-22】 如果一个字符串正读和反读内容相同，那么它就是回文（palindrome）。例如，“rotor”“dad”和“noon”都是回文。更复杂的回文例子包括“Madam, I'm Adam”和“Able was I ere I saw Elba”。通过将字符串和它的反转字符串进行比较，可以轻松地判断该字符串是否是回文。在某些情况下，在反转字符串之前，必须删除标点符号并使字符完全相同。程序 10-22 演示了如何检测回文字符串。

程序 10-22 测试回文

```
1  /*************************************************************
2   * 本程序用于测试一个输入字符串是否为回文                        *
3   *************************************************************/
4  #include <string>
5  #include <iostream>
6  using namespace std;
7
8  // 声明两个函数
9  void reverse (string& strg);
```

```
10  bool isPalindrome (string& strg);
11
12  int main ( )
13  {
14    // 构建默认的字符串对象
15    string strg;
16    // 输入
17    cout << "Enter a string: ";
18    getline (cin, strg);
19    // 检测回文
20    if (isPalindrome (strg))
21    {
22         cout << strg << " is a palindrome.";
23    }
24    else
25    {
26         cout << strg << " is not a palindrome.";
27    }
28    return 0;
29  }
30  /***********************************************************
31   * isPalindrome 函数调用 reverse 函数，                    *
32   * 比较其参数和反转后的参数                                 *
33   ***********************************************************/
34  bool isPalindrome (string& strg)
35  {
36    string temp (strg);
37    reverse (temp);
38    return (temp == strg);
39  }
40  /***********************************************************
41   * reverse 函数反转字符串参数                               *
42   ***********************************************************/
43  void reverse (string& strg)
44  {
45    string temp (strg);
46    int size = strg.size () ;
47    for (int i = 0; i < size; i++)
48    {
49         strg[i] = temp [size - 1 - i];
50    }
51  }
```

运行结果：

```
Enter a string: rotor
rotor is a palindrome.
```

运行结果：

```
Enter a string: mom
mom is a palindrome.
```

运行结果：

```
Enter a string: son
son is not a palindrome.
```

添加字符到字符串末尾（push_back）

我们经常需要在字符串中添加一个字符。字符串库定义了函数 push_back，用于在字符串末尾添加一个字符。库成员函数 push_back 的使用方法如下所示：

```
strg.push_back (c);                 // 把字符 c 添加到字符串 strg 的末尾
```

在本章的程序设计部分，我们将演示如何将字符添加到字符串前面。我们还将演示如何从字符串前面或者后面弹出一个字符。

字符串修改操作

C++ 字符串库为修改操作提供了一组不同的函数定义。但是，如果我们构造一个临时字符串作为参数，则成员函数的数量可以减少到四个。temp 字符串是我们要查找的字符集。

```
temp (strg);                  // 完整的 C++ 字符串
temp (strg, pos2, n2);        // 部分 C++ 字符串
temp (1, c);                  // 由一个字符构成的字符串
temp (cstr);                  // 完整的 C 字符串
temp (cstr, n);               // C 字符串的前 n 个字符
```

图 10-12 显示了四组函数如何利用临时字符串参数来修改宿主字符串。

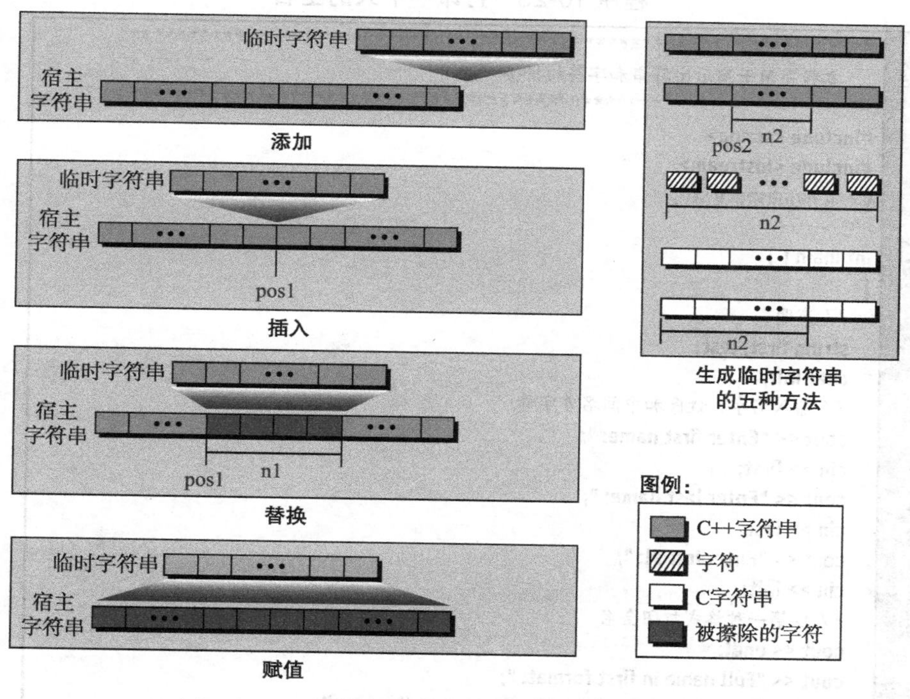

图 10-12　添加、插入、替换和赋值操作的通用思想

以下代码片段演示了如何使用这四种类型的函数。

```
strg.append (temp);               // 添加
strg.insert (pos1, temp);         // 插入
strg.replace (pos1, n1, temp);    // 替换
strg.assign (temp);               // 赋值
```

部分或者全部擦除

有两个函数可以完全或者部分擦除字符串中的字符而不破坏字符串。字符串对象仍然存在，但它的部分或者全部字符被擦除。

```
strg.clear ();                    // 擦除字符串中的所有字符
strg.erase (pos, n);              // 擦除字符串中的部分字符
```

重载的运算符

一些修改操作是使用重载的运算符来实现的。C++ 字符串重载了赋值运算符（=）、复合赋值运算符（+=）和加法运算符（+），如下所示。其中 temp、temp1 和 temp2 是 string 类的实例。

```
string strg = temp;               // 赋值
string strg +=  temp;             // 复合赋值
string strg =  temp1 + temp2;     // 加法
```

【例 10-23】 程序 10-23 演示了如何读取一个人的名字、姓氏和中间名缩写，并以两种格式创建其全名。

程序 10-23　打印一个人的全名

```
1   /***********************************************************************
2    * 本程序用于演示字符串和字符的拼接                                         *
3    ***********************************************************************/
4   #include <string>
5   #include <iostream>
6   using namespace std;
7
8   int main ( )
9   {
10      // 声明
11      string first, last;
12      char init;
13      // 输入名字、姓氏和中间名首字母
14      cout << "Enter first name: ";
15      cin >> first;
16      cout << "Enter last name: ";
17      cin >> last;
18      cout << "Enter initial: ";
19      cin >> init;
20      // 以第一种格式打印全名
21      cout << endl;
22      cout << "Full name in first format: ";
23      cout << first + " " + init + "." + " " + last << endl << endl;
24      // 以第二种格式打印全名
25      cout << "Full name in second format: ";
26      cout << last + ", " + first + " " + init + ".";
27      return 0;
28  }
```

```
运行结果：
Enter first name: John
Enter last name: Brown
```

```
Enter initial: A

Full name in first format: John A. Brown

Full name in second format: Brown, John A.
```

【例 10-24】 在本例中，我们将演示如何将文本左对齐的文件更改为文本右对齐的文件。对于这个问题有几种解决方法，这取决于我们有多少关于文件的信息。如果我们知道文件的最大行字符数，问题就很简单了。否则，我们需要找到文件的最大行字符数。为了找到文件最大的行字符数，我们可以使用两种方法。在第一种方法中，我们可以将输入文件中的各行内容读取到字符串数组中，找到最大行字符数，然后将数组写入输出文件。问题是我们需要在栈内存中创建一个数组。在第二种方法中，我们可以读取输入文件两次。第一次，我们只是查找最大的行字符数。然后，我们关闭文件并再次读取，以更改行并将其存储到输出文件中。程序 10-24 采用第二种方法。

程序 10-24 把右对齐的行写入文件

```
1   /***************************************************************
2    * 本程序读取左对齐的文件，                                      *
3    * 然后创建一个右对齐的文件                                      *
4    ***************************************************************/
5   #include <string>
6   #include <iostream>
7   #include <fstream>
8   #include <cassert>
9   using namespace std;
10
11  int main ()
12  {
13     // 声明两个文件和一个字符串对象
14     ifstream inputFile;
15     ofstream outputFile;
16     string line;
17     // 读取输入文件，仅查找最大行字符数
18     inputFile.open ("inFile.dat");
19     assert (inputFile);
20     int maxSize = 0;
21     while (!inputFile.eof())
22     {
23         getline (inputFile, line);
24         if (line.size() > maxSize)
25         {
26             maxSize = line.size();
27         }
28     }
29     inputFile.close ();
30     // 读取输入文件，创建输出文件
31     inputFile.open ("inFile.dat");
32     assert (inputFile);
33     outputFile.open ("outFile.dat");
34     assert (outputFile);
```

```
35    while (!inputFile.eof())
36    {
37        getline (inputFile, line);
38        string temp (maxSize - line.size() , ' ');
39        line.insert (0, temp);
40        line.append ("\n");
41        outputFile << line;
42    }
43    inputFile.close();
44    outputFile.close();
45    return 0;
46 }
```

下面并排显示了程序运行后输入文件和输出文件的内容。

输入文件的内容	输出文件的内容
This is a line. This is the second line. This is a new longer line. This is a shorter one. This is the longest line by far.	This is a line. This is the second line. This is a new longer line. This is a shorter one. This is the longest line by far.

转换

可以把 C++ 字符串对象转换为字符数组或者 C 字符串。

```
const char* arr = strg.data ();        // 转换为字符数组
const char* str = strg.c_str ();       // 转换为 C 字符串
```

10.3　程序设计

在本节中，我们首先使用字符串类中的成员函数创建四个自定义函数。然后我们使用这些函数来解决字符串处理中的一些经典问题。

10.3.1　四个自定义函数

C++ 库中字符串类定义了一组成员函数，但我们还可以根据需要创建自定义的函数。我们创建的函数不能成为类成员函数，但它们可以使用类类型作为参数。我们通过定义四个与 C++ 字符串类相关的新函数来演示这种能力。我们根据 C++ 字符串类中定义的成员函数来创建这些函数。为了在所有程序中使用自定义的函数，我们将它们收集到一个名为 customized.h 的头文件中，需要时包含该文件即可。我们将这四个函数分别命名为：pushBack、pushFront、popBack 和 popFront。在后面的章节中，我们将看到其他类库也定义了这四种函数（尽管名称不同），但是字符串类中只定义了一个函数，其名称为 push_back。图 10-13 显示了这四个自定义函数的行为。

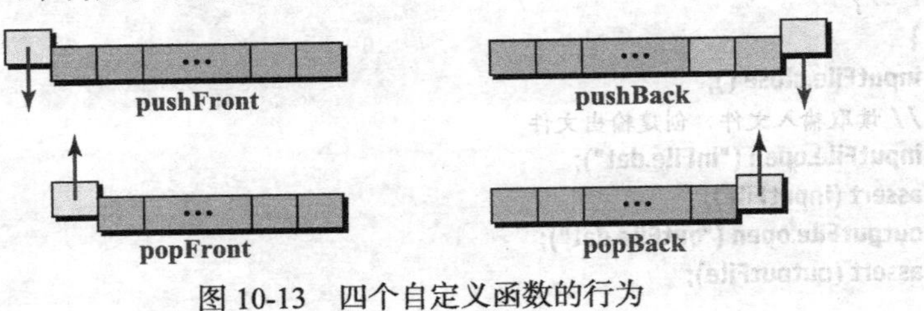

图 10-13　四个自定义函数的行为

程序 10-25 显示了创建四个自定义函数的头文件代码清单。程序 10-26 演示了如何使用普通的字符串测试这些自定义函数。

程序 10-25　四个自定义函数的头文件

```
1   /**********************************************************************
2    * 定义四个自定义函数的头文件。                                          *
3    * 函数 pushFront 在字符串的头部添加给定字符。                           *
4    * 函数 pushBack 在字符串的尾部添加给定字符。                            *
5    * 函数 popFront 删除字符串头部的字符。                                  *
6    * 函数 popBack 删除字符串尾部的字符                                     *
7    **********************************************************************/
8   #ifndef custom_H
9   #define custom_H
10  #include <iostream>
11  #include <string>
12  using namespace std;
13
14  // 函数 pushFront 的定义
15  void pushFront (string& strg, char c)
16  {
17    string temp (1, c);
18    strg.insert (0, temp);
19  }
20  // 函数 pushBack 的定义
21  void pushBack (string& strg, char c)
22  {
23    string temp (1, c);
24    strg.append (temp);
25  }
26  // 函数 popFront 的定义
27  char popFront (string& strg)
28  {
29    int index = 0;
30    char temp = strg [index];
31    strg.erase (index, 1);
32    return temp;
33  }
34  // 函数 popBack 的定义
35  char popBack (string& strg)
36  {
37    int index = strg.size () - 1;
38    char temp = strg [index];
39    strg.erase (index, 1);
40    return temp;
41  }
42  #endif
```

程序 10-26　测试四个自定义函数

```
1   /**********************************************************************
2    *本程序用于测试                                                        *
3    *在自定义头文件中定义的四个自定义函数                                  *
```

```
4    *************************************************************/
5    #include "customized.h"
6    #include <string>
7    #include <iostream>
8    using namespace std;
9
10   int main ( )
11   {
12     // 声明原始字符串
13     string strg ("abcdefgh");
14     // 测试 pushFront 函数
15     cout << "String before calling pushFront: " << strg << endl;
16     pushFront (strg, 'A');
17     cout << "String after calling pushFront: " << strg << endl;
18     cout << endl;
19     // 测试 pushBack 函数
20     cout << "String before calling pushBack: " << strg << endl;
21     pushBack (strg, 'Z');
22     cout << "String after calling pushBack: " << strg << endl;
23     cout << endl;
24     // 测试 popFront 函数
25     cout << "String before calling popFront: " << strg << endl;
26     char c1 = popFront (strg);
27     cout << "String after calling popFront: " << strg << endl;
28     cout << "The popped character: " << c1 << endl;
29     cout << endl;
30     // 测试 popBack 函数
31     cout << "String before calling popBack: " << strg << endl;
32     char c2 = popBack (strg);
33     cout << "String after calling popBack: " << strg << endl;
34     cout << "The popped character: " << c2 << endl;
35     cout << endl;
36     return 0;
37   }
```

运行结果：

```
String before calling pushFront: abcdefgh
String after calling pushFront: Aabcdefgh

String before calling pushBack: Aabcdefgh
String after calling pushBack: AabcdefghZ

String before calling popFront: AabcdefghZ
String after calling popFront: abcdefghZ
The popped character: A

String before calling popBack: abcdefghZ
String after calling popBack: abcdefgh
The popped character: Z
```

10.3.2 数值进制编码系统的转换

在计算机科学中，我们使用不同的数值进制编码系统（这些在附录 B 中描述）：二进制、八进制、十进制和十六进制。每个数值进制编码系统使用一组符号和一个基数。基数（base）定义系统中使用的符号总数。表 10-3 显示了我们在程序设计中使用的基数和符号。注意，符号 A、B、C、D、E、F 的值分别代表 10、11、12、13、14 和 15。

表 10-3 数值进制编码系统

编码系统	基数	符号
二进制	2	0, 1
八进制	8	0, 1, 2, 3, 4, 5, 6, 7
十进制	10	0, 1, 2, 3, 4, 5, 6, 7, 8, 9
十六进制	16	0, 1, 2, 3, 4, 5, 6, 7, 8, 9, A, B, C, D, E, F

尽管所有的数值进制编码系统都使用不同的符号表示整数，但十进制编码系统是程序中唯一可以直接用作整数的进制编码系统。其他进制编码系统中的数值必须用字符串表示。

> 在 C++ 中，十进制编码系统的数值作为整数使用，其他进制编码系统的数值作为字符串使用。

这意味着当我们使用不同的编码系统时，必须将字符串转换为整数，反之亦然。在本节中，我们给出了实现各种进制数值转换的总体设计思想。在学习本节其余部分之前，我们建议读者复习附录 B 的内容。

把其他进制转换为十进制

图 10-14 演示了如何把一个包含三个字符的字符串转换为一个整数。

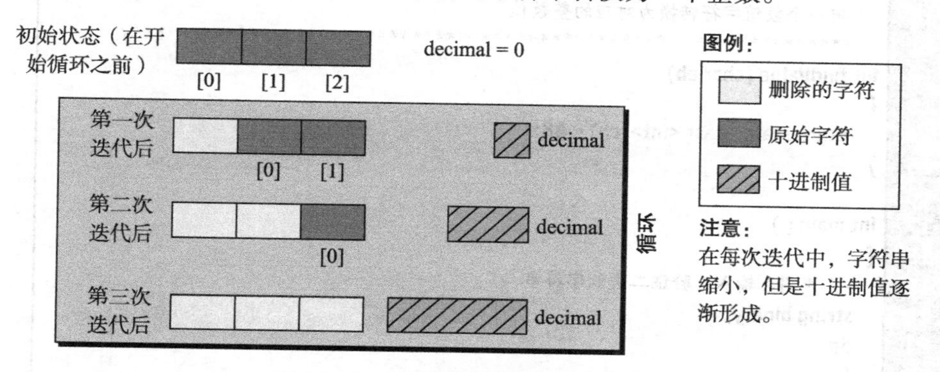

图 10-14 把其他进制转换为十进制

我们使用了一个由字符串大小控制的循环。当字符串为空时，我们退出循环。在开始循环之前，我们将变量 decimal 的值设置为 0。在每次迭代中，我们将 decimal 之前的值乘以字符串的基数。然后弹出字符串前面的字符，将其转换为十进制，并将其添加到 decimal 中。注意，在每次迭代中，decimal 的值都会增加，但字符串的大小会减小。当程序跳出循环时，decimal 的值被计算出来。算法如表 10-4 所示。

表 10-4 将字符串转换为十进制的算法

```
set base                                // 设置 base（基数）为 2、8 或 16
decimal = 0
input string
while (string not empty)
{
  decimal *= base;
```

（续）

ch = popFront (string)	// 使用前面开发的函数 popFront
decimal += findValue (ch)	// 使用函数把 ch 转换为值
}	
output decimal	

【例 10-25】 程序 10-27 将二进制字符串转换为十进制整数。在这种情况下，基数是 2。我们使用了之前开发的 popfront 函数（包含在头文件中）。我们还编写了一个名为 findValue 的简单函数，用于把提取的字符转换为对应的整数值。

程序 10-27　把二进制字符串转换为十进制整数

```
1   /***************************************************************
2    * 本程序把二进制字符串转换为十进制整数                          *
3    ***************************************************************/
4   #include "customized.h"
5   #include <string>
6   #include <iostream>
7   using namespace std;
8   /***************************************************************
9    * 该函数                                                        *
10   * 把一个数值字符转换为对应的整数值                              *
11   ***************************************************************/
12  int findValue (char ch)
13  {
14    return static_cast <int>(ch) – 48;
15  }
16
17  int main ( )
18  {
19    // 声明、输入、验证二进制字符串
20    string binary;
21    do
22    {
23        cout << "Enter binary string: ";
24        getline (cin, binary);
25    } while (binary.find_first_not_of ("01") < binary.size());
26    // 初始化并计算十进制整数值
27    int base = 2;
28    int decimal = 0;
29    while (!binary.empty())
30    {
31        decimal *= base;
32        char ch = popFront (binary);
33        decimal += findValue (ch);
34    }
35    cout << "Decimal value: " << decimal;
36    return 0;
37  }
```

运行结果：

```
Enter binary string: 11 101
```

Enter binary string: 11101 Decimal value: 29	
运行结果: Enter binary string: 1181 Enter binary string: 11111 Decimal value: 31	
运行结果: Enter binary string: 111000111 Decimal value: 455	

关于程序 10-27 有三个要点:

第一个要点在第 21 ～ 25 行中，我们处理一行表示二进制字符串的文本。字符串只能由字符“0”和“1”组成，这意味着在使用之前必须验证输入字符串。验证功能是通过我们在本章前面学习到的函数 find_first_not_of 来完成的。此函数查找不是“0”或者“1”的字符，并返回相应的索引。如果返回的索引小于字符串的大小，则意味着在字符串中找到了 0 或者 1 以外的字符，我们将继续循环以重新读取要转换的二进制字符串。

第二个要点是第 31 行。对于表 10-3 中给出的算法，每次进入循环时，必须将 decimal 值重置为 decimal 和基数值的乘积。

第三个要点是 popFront 函数。我们知道在每次迭代中，必须提取字符串中的下一个字符，这是由头文件中定义的 popFront 函数正确完成的。

从十进制转换为其他进制

图 10-15 显示了如何把十进制数值转换为包含三个字符的字符串。

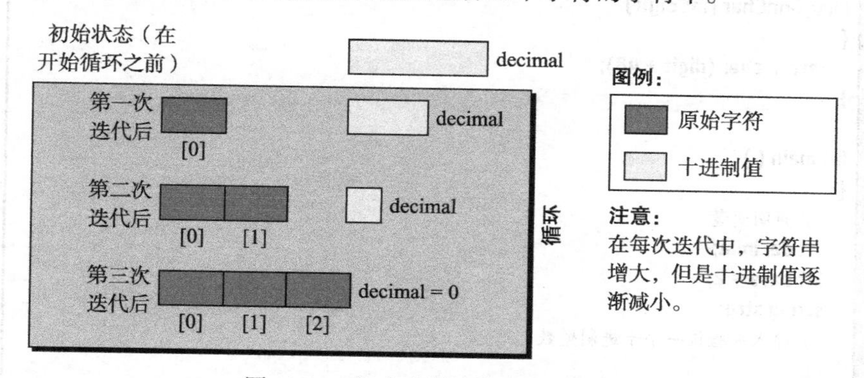

图 10-15　把十进制转换为其他进制

我们使用了一个由十进制数的值控制的循环。当值为 0 时，程序停止。在开始循环之前，字符串是空的。在每次迭代中，我们将十进制数之前的值除以基数，得到余数。然后，我们将结果值转换为一个字符，并将其添加到字符串的前面。注意，在每次迭代中，十进制的值都会减小，直到变为 0，但字符串的大小会增加。当跳出循环时，转换后的字符串就完成了。实现算法如表 10-5 所示。

表 10-5　将十进制转换为其他进制字符串的算法

set base input decimal while (decimal > 0)	// 设置 base（基数）为 2、8 或 16

（续）

{	
value = decimal % base;	
ch = findChar (value)	//使用函数把值转换为字符
string.pushFront (ch)	//使用前面开发的函数 pushFront
}	
output string	

【例 10-26】 程序 10-28 演示了如何将十进制整数转换为二进制字符串。

程序 10-28　把十进制整数转换为二进制字符串

```
1   /***************************************************************
2    * 本程序把十进制整数转换为二进制字符串                          *
3    ***************************************************************/
4   #include <string>
5   #include "customized.h"
6   #include <iostream>
7   using namespace std;
8
9   /***************************************************************
10   * 该函数使用 char 函数                                          *
11   * 把一个整数转换为一个字符                                      *
12   ***************************************************************/
13  char findChar (int digit)
14  {
15    return char (digit + 48);
16  }
17
18  int main ( )
19  {
20    // 声明变量
21    int decimal;
22    int base = 2;
23    string strg;
24    // 输入并验证一个十进制整数
25    do
26    {
27        cout << "Enter a positive decimal: " ;
28        cin >> decimal;
29    } while (decimal <= 0);
30    // 转换为二进制字符串
31    while (decimal > 0)
32    {
33        int digit = decimal % base;
34        char ch = findChar (digit);
35        pushFront (strg, ch);
36        decimal /= base;
37    }
38    // 输出二进制字符串
39    cout  << "Binary: " << strg;
```

```
40    return 0;
41 }
```

运行结果：

```
Enter a positive decimal: 35
Binary: 100011
```

运行结果：

```
Enter a positive decimal: 7
Binary: 111
```

运行结果：

```
Enter a positive decimal: 126
Binary: 1111110
```

输入的十进制值的验证很简单，通过 do-while 循环完成。为了在二进制字符串中插入字符，我们使用前文创建并被包含在头文件中的 pushFront 函数。

本章小结

C 字符串是一个以空字符结尾的字符数组，可以像数组一样创建。字面量字符串是一个常量值，由包含在两个双引号中的字符组成。可以使用提取运算符（>>）或者 getline 库函数输入 C 字符串。可以使用插入运算符（<<）输出 C 字符串。

C++ 字符串类包含在 <string> 头文件中。它是一个类，用于创建一个以非空字符结尾的字符数组，该字符数组在堆内存中创建。像其他类一样，字符串类也定义了数据成员和成员函数。

思考题

1. 以下代码片段使用了 C 字符串。修改代码，使用 C++ 字符串。

```
const char* str = "This is a string.";
cout << strlen (str) << endl;
```

2. 以下代码片段使用了 C 字符串。修改代码，使用 C++ 字符串。

```
const char* str1 = "This is a string.";
const char* str2 = "This is another one.";
cout << strcmp (str1, str2) << endl;
```

3. 以下代码片段使用了 C 字符串。使用 C++ 字符串重构代码，实现相同的功能。

```
char str1 [ ] = "This is the first string.";
const char* str2 = "Here is another one.";
strcpy (str1, str2);
cout << str1 << endl;
```

4. 以下代码片段使用了 C 字符串。使用 C++ 字符串重构代码，实现相同的功能。

```
char str1 [ ] = "This is the first string.";
const char* str2 = "Here is another one.";
strncpy (str1, str2, 4);
cout << str1 << endl;
```

5. 以下代码片段使用了 C 字符串。修改代码，使用 C++ 字符串。

```
char str1 [40] = "The time has come. ";
```

```
const char* str2 = "Are your ready?";
strcat (str1, str2);
cout << str1 << endl;
```

6. 以下代码片段使用了 C 字符串。修改代码，使用 C++ 字符串。

```
char str1 [40] = "The time has changed. ";
const char* str2 = "Do you know? My dear friend.";
strncat (str1, str2, 12);
cout << str1 << endl;
```

7. 以下代码片段使用了 C 字符串。修改代码，使用 C++ 字符串。

```
char str [ ] = "This is a long string.";
*strchr (str, 's') = 'S';
*strrchr (str, 's') = 'S';
cout << str << endl;
```

8. 以下代码片段使用了 C 字符串。修改代码，使用 C++ 字符串。

```
const char* str = "This is a long string.";
char* p = strstr (str, "is");
cout << *p << endl;
```

9. 以下代码片段使用了 C 字符串来删除字符串中的第一个字符。修改代码，使用 C++ 字符串。

```
const char* str = "ABCDEFGH";
str = str + 1;
cout << str << endl;
```

10. 以下代码片段使用了 C 字符串来删除字符串中除最后一个字符以外的所有字符。修改代码，使用 C++ 字符串。

```
const char* str = "ABCDEFGH";
str = str + strlen (str) – 1;
cout << str << endl;
```

11. 以下代码片段使用了 C++ 字符串。修改代码，使用 C 字符串。

```
string strg ("ABCDEFGH");
strg.push_back ('I');
cout << strg << endl;
```

12. 编写一个代码片段，将一个 C++ 字符串分割成两个大小相等的字符串。如果原始字符串的字符数是奇数，则代码在拆分前在字符串末尾添加一个空白字符。
13. 编写一个代码片段，提取字符串的前四个字符和最后四个字符。然后打印提取的字符串。

编程题

1. 编写一个函数，获取给定字符在 C++ 字符串中的计数。编写程序测试该函数。
2. 编写一个函数，把 C++ 字符串中的所有字符转换为大写。编写程序测试该函数。
3. 编写一个函数，从 C++ 字符串中移除给定字符的所有匹配项。编写程序测试该函数。
4. 编写一个函数，从 C++ 字符串中移除所有重复的字符（每个字符仅保留一个）。编写程序测试该函数。
5. 编写一个函数，给定两个字符串，创建另一个字符串，仅包含两个字符串中共同的字符。一种解决方案是先移除各字符串中重复的字符（第 4 题），然后再创建包含共同字符的字符串。

6. 编写一个函数，包括三个 C++ 字符串参数。在第一个参数中查找第二个参数（作为子字符串），然后替换为第三个参数。如果查找子字符串成功，则函数返回修改后的第一个参数，否则返回原始的第一个参数。编写程序测试该函数。
7. 编写一个函数，在第二个字符串的中间插入给定的字符串。如果第二个字符串包含的字符数为奇数，则插入前先重复第二个字符串的最后一个字符。编写程序测试该函数。
8. 修改程序 10-21，以处理复杂回文，例如“Madam, I'm Adam”或者“A man, a plan, a canal: Panama.”。
9. 编写一个程序，把一个正整数转换为对应的八进制字符串（即基数为 8）。例如，整数 123 将转换为字符串“2322”。提示：可以使用函数 toChar 把每位八进制数字转换为八进制字符，使用本章中开发的 pushFront 函数把每个八进制字符插入字符串中，使用 toString 函数重复调用其他两个函数直到全部转换完成为止。
10. 编写一个程序，把一个正整数转换为对应的十六进制字符串（即基数为 16）。例如，整数 23456 将转换为字符串“5BA0”（仅使用大写字母时）。提示：可以使用 toChar 函数把每位十六进制数字转换为十六进制字符，使用本章中开发的 pushFront 函数把每个十六进制字符插入字符串中，使用 toString 函数重复调用其他两个函数直到全部转换完成为止。
11. 编写一个程序，把一个八进制字符串转换为十进制整数。八进制字符串仅使用字符‘0’到‘7’，其基数是 8。提示：可以使用本章中开发的 popFront 函数从八进制字符串中提取下一个字符，使用 toInt 函数把一个八进制字符转换为八进制数字，使用 toDecimal 函数重复调用其他两个函数来完成转换任务。注意，必须验证八进制字符串，验证可以使用本章中讨论的 find_first_not_of 函数。
12. 编写一个程序，把一个十六进制字符串转换为十进制整数。十六进制字符串仅使用字符‘0’到‘9’和‘A’到‘F’，其基数是 16。提示：可以使用本章中开发的 popFront 函数从十六进制字符串中提取下一个字符，使用 toInt 函数把一个十六进制字符转换为十六进制数字，使用 toDecimal 函数重复调用其他两个函数来完成转换任务。注意，必须验证十六进制字符串，验证可以使用本章中讨论的 find_first_not_of 函数。
13. 在数据可以表示为不同形式的情况下（如编程题 9 到 11 所示），我们可以使用一个类。我们可以用最常见的形式（例如无符号整数）声明唯一的数据成员，然后使用 set 函数和 get 函数以四个不同的基（2、8、10 和 16）设置和获取字符串形式的数据值。为这个类提供接口文件、实现文件和应用程序文件。定义一些辅助函数，包括 toInt、toChar、pushFront、popFront 和 validate（用于验证传递给 set 函数的字符串）。
14. 创建一个名为 Address 的类，该类表示 Internet 地址。版本 4 中的 Internet 地址（IPv4 地址）是介于 0 和 4 294 967 295 之间的十进制值。换言之，IPv4 地址可以是连接到互联网上的超过 40 亿台计算机的地址。IP 地址的两种常见表示形式是二进制格式（即基数为 2）和点分十进制格式（即基数为 256），如下所示。二进制格式是由四组 8 位二进制组成的字符串。点分十进制格式是由四个十进制值（0 到 255）组成的字符串，由英文句点分隔。注意，在二进制和点分十进制格式中，所有元素（位或者十进制值）都必须存在，尽管有些可能是零。下面显示了一个 IPv4 地址的示例。

```
十进制格式: 71832456
二进制格式: 00000100 01001000 00010011 10001000
点分十进制格式: 4.72.19.136
```

第 11 章 | C++ Programming: An Object-Oriented Approach

类之间的关系

在面向对象的程序设计中，类通常不是独立地使用。类之间存在相互关系。一个程序通常使用若干类，这些类之间存在不同的关系。图 11-1 显示了我们在本章中讨论的关系分类。

学习目标

阅读并学习本章后，读者应该能够

- 定义类之间的继承关系，包括公共继承、受保护继承和私有继承。
- 了解如何为继承关系中的几个类编写类定义和成员函数定义。
- 把关联定义为类之间的关系，并定义其变体：聚合和组合。
- 了解如何为关联关系中的几个类编写类定义和成员函数定义。
- 定义类之间的依赖关系，以显示一个类可以使用另一个类的对象。
- 了解如何为依赖关系中的类编写类定义和成员函数定义。

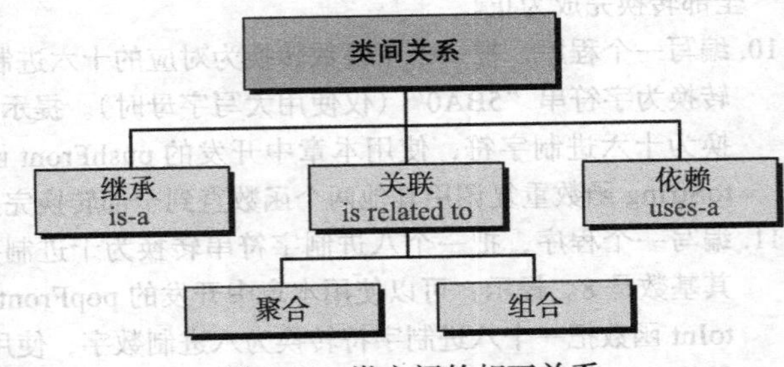

图 11-1 类之间的相互关系

11.1 继承关系

在面向对象程序设计中，**继承**（inheritance）是从一个更通用的概念派生出一个更具体的概念。这和现实生活中的概念一样。例如，生物学中动物的概念比马的概念更为普遍。我们可以给出动物的定义，然后添加定义内容来创建马的定义。从更具体到更一般存在一种关系（is-a）。所有的马都是动物，但并非所有的动物都是马。

11.1.1 总体思路

为了显示继承中类之间的关系，我们使用统一建模语言（Unified Modeling Language，UML）。UML 是一种以图形方式显示类和对象之间关系的语言。我们在附录 Q 中更详细地讨论 UML，但是在本章中使用了其中一些图表。在 UML 中，类显示为矩形框。继承关系由一条以空心三角形结尾的线表示，该三角形从更具体的类指向更一般的类。图 11-2 显示了三个继承关系。

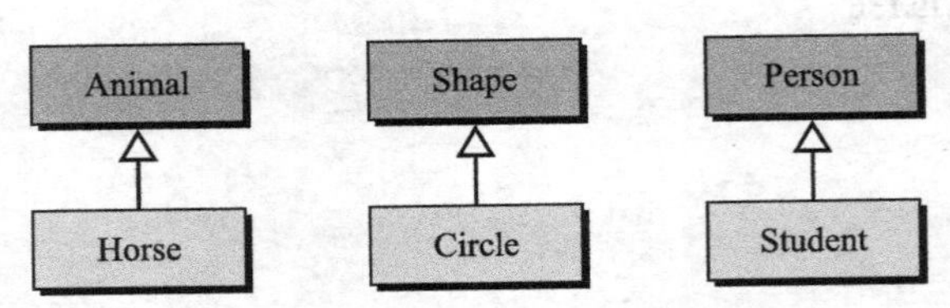

图 11-2 继承的 UML 图

在 C++ 中，最一般的类被称为**基类**（base class），而更具体的类被称为**派生类**（derived

class)。更一般的类也称为**超类**（superclass），更具体的类也称为**子类**（subclass）。

从 UML 图中，我们可以得出继承关系中对象集合之间的关系，如图 11-3 所示，这表明马的集合小于动物的集合。

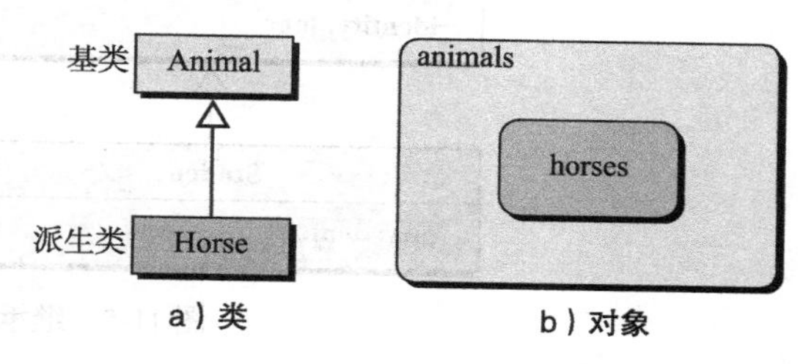

图 11-3 继承关系中的类和对象

一个特定的概念必须具有一般概念的特点，但特定的概念可以有更多的特点。换言之，马首先是动物，然后是马。这就是在 C++ 中派生类扩展其基类的原因。这里术语扩展（extend）表示派生类必须具有在基类中定义的所有数据成员和成员函数，但派生类可以添加内容。换言之，派生类继承了基类的所有数据成员和成员函数（需要重新定义的构造函数、析构函数和赋值运算符除外），它还可以创建新的数据成员和成员函数。稍后我们将讨论为什么不能继承构造函数、析构函数和赋值运算符。

派生类继承了基类中的所有成员（有些例外），它还可以添加新成员。

为了从基类创建派生类，在 C++ 中有三种选择：私有继承（private inheritance）、受保护继承（protected inheritance）和公共继承（public inheritance）。为了显示想要使用的继承类型，我们在类名后面插入一个冒号，后跟一个关键字（private、protected 或者 public）。图 11-4 显示了这三种类型的继承，其中 B 是基类，D 是派生类。

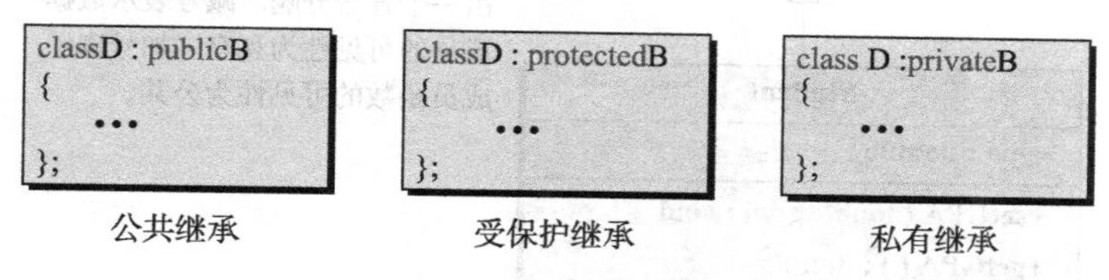

图 11-4 继承的三种类型

默认的继承类型是私有的。换言之，如果我们不指定类型（public、protected 或者 private），系统假定我们需要私有继承。因为私有继承（稍后我们将看到）很少使用，所以我们需要显式定义继承的类型。最常见的是公共继承。

11.1.2 公共继承

尽管默认的继承类型是私有继承，但到目前为止最常见的是公共继承。其他两种类型的继承很少使用。我们将在本章后面简要讨论其他两种类型的继承，但在本节中，我们将重点讨论公共继承。一些其他的面向对象程序设计语言，例如 Java，只有公共继承。

最常用的继承是公共继承。

【例 11-1】 在本例中，我们设计了两个类，Person 和 Student，其中 Student 类继承自 Person 类。我们知道学生是一个人。我们假设 Person 类只使用一个数据成员：identity（身份标识，如社会保险号）。我们还假设 Student 类需要两个数据成员：identity 和 gpa。但是，由于数据成员 identity 已经在 Person 类中定义了，因此不需要在 Student 类中定义 identity，因为它继承自 Person。图 11-5 显示了扩展的 UML 类图，它包括两个用来容纳数据成员的区域。

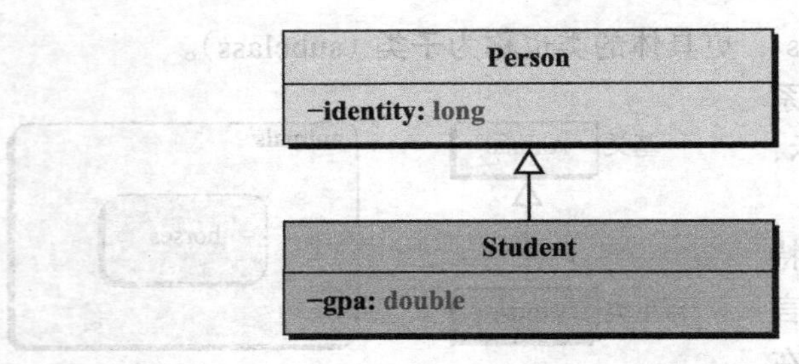

图 11-5　继承关系中的两个类

根据图 11-5，我们可以立即看出继承的优势。Student 类使用 Person 类的数据成员，并添加了自己的一个数据成员。

【例 11-2】 在本例中，我们将成员函数添加到两个类中。我们暂时忽略构造函数和析构函数，因为它们不能被继承；我们假设使用合成构造函数和合成析构函数。我们为每个类添加一个访问器函数和一个更改器函数。我们使用一个更为扩展的 UML 图，其中增加了另一个定义成员函数的区域，如图 11-6 所示。

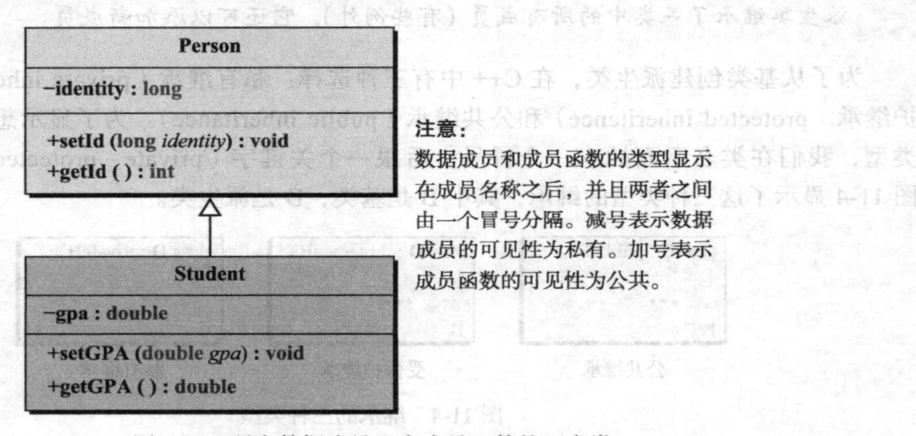

图 11-6　既有数据成员又有成员函数的两个类

根据图 11-6，我们可以看到另一个优势。Student 类的对象必须设置和获取 identity 值，但 identity 已经在 Person 类中定义，Student 类不需要设置或获取它。

【例 11-3】 程序 11-1 在一个简单的程序中演示了上述继承关系。稍后我们将演示如何使用继承来实现独立编译。

程序 11-1　公共继承

```
1   /***************************************************************
2    * 本程序演示如何让类 Student 继承类 Person，                  *
3    * 因为学生也是人                                              *
4    ***************************************************************/
5
6   #include <iostream>
7   #include <cassert>
8   #include <string>
9   using namespace std;
10
11  /***************************************************************
```

```
12    * Person 类的类定义                                                    *
13    ********************************************************************/
14  class Person
15  {
16    private:
17        long identity;
18    public:
19        void setId (long identity);
20        long getId( ) const;
21  };
22  /********************************************************************
23    * Person 类的函数 setId 的定义                                        *
24    ********************************************************************/
25  void Person :: setId (long id)
26  {
27    identity = id;
28    assert (identity >= 100000000 && identity <= 999999999) ;
29  }
30
31  /********************************************************************
32    * Person 类的函数 getId 的定义                                        *
33    ********************************************************************/
34  long Person :: getId () const
35  {
36    return identity;
37  }
38
39  /********************************************************************
40    * Student 类的类定义                                                  *
41    ********************************************************************/
42
43  class Student : public Person
44  {
45    private:
46        double gpa;
47    public:
48        void setGPA (double gpa);
49        double getGPA () const;
50  };
51
52  /********************************************************************
53    * Student 类的函数 setGPA 的定义                                      *
54    ********************************************************************/
55  void Student :: setGPA (double gp)
56  {
57    gpa = gp;
58    assert (gpa >=0 && gpa <= 4.0);
59  }
60
61  /********************************************************************
62    * Student 类的函数 getGPA 的定义                                      *
63    ********************************************************************/
```

```
64  double Student :: getGPA() const
65  {
66      return gpa;
67  }
68
69
70  /*******************************************************************
71   * 应用程序函数（main），使用上述两个类                                *
72   *******************************************************************/
73  int main ( )
74  {
75      // 实例化并使用 Person 对象
76      Person person;
77      person.setId (111111111L);
78      cout << "Person Information: " << endl;
79      cout << "Person's identity: " << person.getId ( );
80      cout << endl << endl;
81      // 实例化并使用 Student 对象
82      Student student;
83      student.setId (222222222L);
84      student.setGPA (3.9);
85      cout << "Student Information: " << endl;
86      cout << "Student's identity: " << student.getId() << endl;
87      cout << "Student's gpa: " << student.getGPA();
88      return 0;
89  }
```

```
运行结果：
Person Information:
Person's Identity: 111111111

Student Information:
Student's identity: 222222222
Student's gpa: 3.9
```

关于程序 11-1，有三个要点：

1）这个程序有两个类。我们实例化了 Person 类的一个对象，还实例化了 Student 类的一个对象。

2）虽然我们没有为 Student 类定义数据成员 identity，但该类的对象从 Person 类继承了 identity。

3）我们用 identity 而不是 name 来标识一个人。我们使用 identity 是因为 name 的类型是 string，它是一个类类型，如果我们使用它，实际上会向继承关系添加一个组合关系，但是我们还没有讨论组合关系。

私有数据成员

在深入探讨继承的主题之前，我们先讨论基类和派生类对象的内容。如图 11-7 所示，基类对象只有一个数据成员，但派生类对象有两个数据成员：一个是继承的，一个是创建的。

在公共继承中，基类中的私有数据成员被派生类继承，但它在隐私方面更进一步；基类中的私有数据成员在派生类中变得不可访问（有时称为隐藏）；必须通过基类自己的成员函数访问其私有数据成员（我们稍后将讨论）。

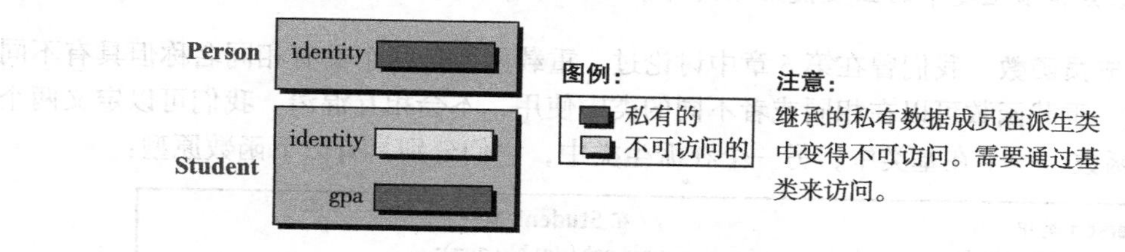

图 11-7　公共继承中的私有数据成员

> 基类中的私有成员成为派生类中不可访问（隐藏）的成员。

公共成员函数

图 11-8 将公共成员函数添加到图 11-7 中。我们可以看到，基类只有两个成员函数，而派生类有四个成员函数：两个是继承的，两个是在派生类中新定义的。

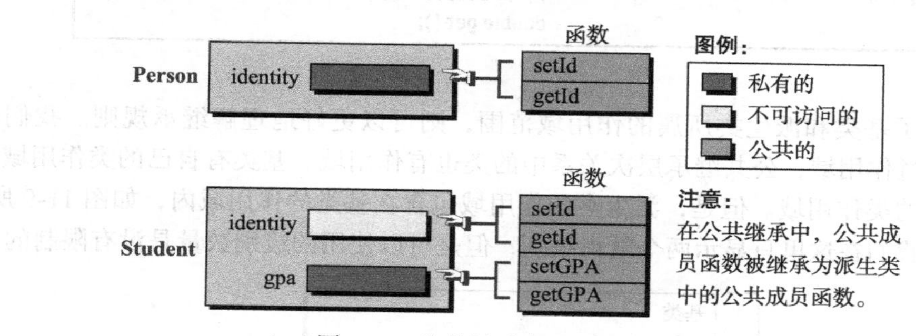

图 11-8　公共继承中的公共成员函数

> 基类中的公共成员成为派生类中的公共成员。

访问私有数据成员

下面显示了我们如何访问每个类中的私有数据成员：

1）在基类中，我们可以通过类中定义的公共成员函数（setId 和 getId）访问私有数据成员。

2）在派生类中，我们需要两组公共成员函数：

- 为了访问继承的私有数据成员（隐藏的），我们使用在基类中定义的公共成员函数（setId 和 getId）。
- 如果需要访问在派生类中定义的数据成员，我们将使用在派生类中定义的成员函数（setGPA 和 getGPA）。

在不同类中具有相同名称的函数

在前面的两个类 Person 和 Student 中，我们使用了两个成员函数（setId 和 setGPA）来设置数据成员，以及两个成员函数（getId 和 getGPA）来获取数据成员。我们可以使用两个同名函数（一个在基类中，另一个在派生类中）吗？换言之，我们可以在基类中定义一个名为 set 的函数，在派生类中定义另一个名为 set 的函数吗？类似地，我们可以在基类中定义一个名为 get 的函数，在派生类中定义另一个名为 get 的函数吗？这两个问题的答案都是肯定的，但我们需要使用两个不同的概念：重载函数和重写函数。

> 要对基类和派生类中的函数使用相同的名称，需要重载或者重写的成员函数。

重载的成员函数。我们曾在第 6 章中讨论过，重载函数是两个具有相同名称但具有不同签名的函数。重载函数可以在相同或者不同的类中使用，不会相互混淆。我们可以定义两个名为 set 的函数，一个在基类中，另一个在派生类中，它们分别具有以下函数原型：

```
//在 Person 类中                    //在 Student 类中
void set (long identity);          void set (double gpa);
```

由于签名是不同的，故可以在 Person 类中使用第一个函数，在 Student 类中使用第二个函数。

重写的成员函数。如果具有相同名称的两个函数的签名也相同，则被称为重写的成员函数，如下所示。

```
//在 Person 类中                    //在 Student 类中
long get ();                       double get ();
```

类作用域

如果了解了基类和派生类所属的作用域范围，则可以更好地理解继承规则。我们曾在第 6 章中讨论过作用域，公共继承层次关系中的类也有作用域。基类有自己的类作用域，派生类也有自己的类作用域。但是，派生类的作用域包含在基类的作用域内，如图 11-9 所示。请注意，尽管我们在这里只显示两个继承级别，但是可以使用的级别数量是没有限制的。

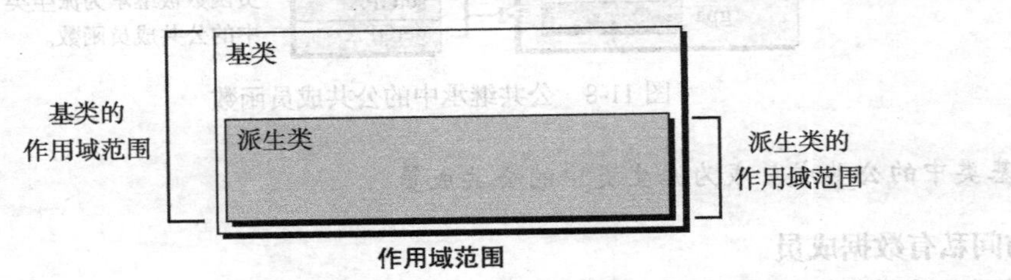

图 11-9　基类和派生类的作用域范围

在图 11-9 中，两个类的作用域表明，基类中定义的名称在派生类中也是可见的，但反过来却不成立。这个概念与函数中的块作用域一样。

系统如何区分调用的是哪个函数呢？必须牢记，我们不是通过名称来调用类中的成员函数，而是让实例对象调用适当的函数。换言之，如果 person 和 student 分别是 Person 和 Student 类的两个对象，则可以使用以下两条语句：

```
//使用 Person 的对象                //使用 Student 的对象
person.set (111111111L);           student.set (3.7);
person.get ();                     student.get ();
```

作用域告诉我们编译器如何根据以下规则调用函数：

1）编译器试图找到一个匹配的函数（使用名称和参数），该函数属于调用它的对象的类。

2）如果没有找到匹配的函数，编译器将搜索从超类继承的函数。

3）如果还找不到匹配项，则继续搜索，直至到达基类为止。

4）如果在所有类中都找不到匹配项，则产生编译错误。

委托授权（delegation of duty）

派生类中的重载或者重写成员函数可以通过调用相应的成员函数来将其操作的一部分委托给更高级别的类中的成员函数。使用 void 成员函数很容易实现这一点，带返回值的函数则稍微复杂些，因为必须返回两个值（这需要使用结构或者类对象）。在这种情况下，我们暂时专注讨论 void 成员函数。

例如，可以设计 Student 类中的成员函数 set：设置 Person 类的数据成员 identity 和自己类的数据成员 gpa。但是，如前所述，数据成员 identity 隐藏在 Student 类中，不能被 Student 类中的 set 函数访问，但是 Student 类中的 set 函数可以调用 Person 类的 set 函数。同样，我们可以在 Person 类中设计一个只打印数据成员 identity 值的打印函数 print。我们也可以在 Student 类中设计一个打印函数 print，调用 Person 类中的打印函数 print 来完成部分工作。下面是带委托的函数 set 和 print 的定义。

```
//使用 Person 的对象
void Person :: set (long id)
{
    identity = id;
}

void Person :: print ()
{
    cout << name;
}
```

```
//使用 Student 的对象
void Student :: set (long id, double gp)
{
    Person :: set (id); //委托
    gpa = gp;
}

void Student :: print ()
{
    Person :: print (); // 委托
    cout << gpa;
}
```

注意，要调用基类中的 set 函数或者 print 函数，必须使用类作用域运算符：Person::set(…)或者 Person::print(…)。

未继承的成员：调用

派生类中有五个未继承的成员函数：默认构造函数、参数构造函数、拷贝构造函数、析构函数和赋值运算符。我们将赋值运算符的讨论推迟到第 13 章，其他四个问题将在本章讨论。

构造函数和析构函数不被继承，因为派生类对象的数据成员自然多于相应的基类。派生类的构造函数必须构造更多的数据成员，派生类的析构函数必须销毁更多的数据成员。

> 构造函数、析构函数和赋值运算符未被继承，必须重新定义它们。

然而，我们在这里遇到了一个难题。派生类的构造函数无法初始化基类的数据成员，因为它们隐藏在派生类中。类似地，派生类的析构函数无法删除基类的数据成员，因为它们隐藏在派生类中。

如果派生类的构造函数在初始化时调用基类的构造函数，然后初始化派生类的数据成员，就可以解决构造函数问题。同样，如果派生类的析构函数首先删除派生类的数据成员，然后调用基类的析构函数，则可以解决析构函数问题，如图 11-10 所示。

注意，构造函数和析构函数中的活动顺序是相反的。因为析构函数是由系统而不是用户调用的，所以除非对象使用可能需要用户干预的指针或者文件，否则活动是在后台完成的。

【例 11-4】 表 11-1 并排显示了 Person 和 Student 类的默认构造函数、参数构造函数、拷贝构造函数和析构函数。请注意，派生类部分使用（调用）基类构造函数。在调用基类的构造函数之后，初始值设定项可以初始化派生类的私有成员。

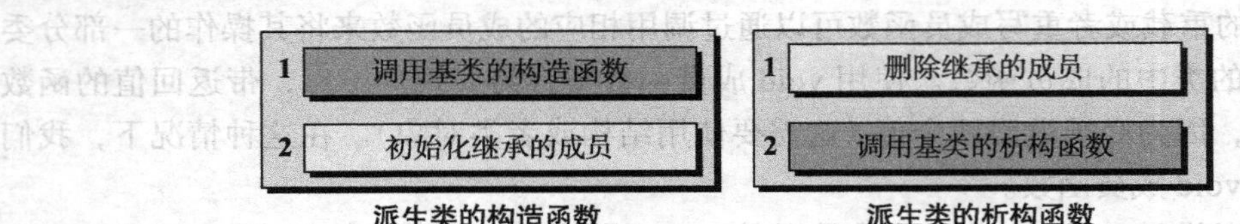

图 11-10　继承中的构造函数和析构函数

表 11-1　构造函数和析构函数的定义

基类	派生类

// 默认构造函数
Person :: Person ()
: identity (0)
{
}
	// 默认构造函数
	Student :: Student ()
	: Person (0), gpa (0.0)
	{
	}

// 参数构造函数
Person :: Person (long id)
: identity (id),
{
}
	// 参数构造函数
	Student :: Student (long id, double gp)
	: Person (id), gpa (gp)
	{
	}

// 拷贝构造函数
Person :: Person (const Person& obj)
: identity (obj.identity)
{
}
	// 拷贝构造函数
	Student :: Student (const Student& st)
	: Person (st) , gpa (std.gpa)
	{
	}

// 析构函数
Person :: ~Person ()
{
}
	// 析构函数
	Student :: ~Student ()
	{
	}

拷贝构造函数调用中的 st 引用了一个 Student 对象。读者可能疑惑如何将这个对象传递给 Person 的拷贝构造函数，因为后者需要一个 Person 对象类型的参数。如果我们知道每当使用需要基类对象的派生类的对象时，对象会被切片，属于派生类的数据成员被删除，那么这个问题就可以迎刃而解。换言之，在函数调用 Person(st) 中，st 只是从 Person 类继承的 Student 类的一部分。

图 11-11 显示了如何使用参数构造函数构造基类和派生类对象。对于基类（Person），我们必须初始化唯一的数据成员。在派生类（Student）中，我们可以通过调用基类的构造函数，然后初始化新的数据成员来构造继承的部分。基类的构造函数是公共成员，可以在派生类中访问。

委托和调用的比较

委托和调用是不同的概念，执行方式也不同。在委托中，派生的成员函数使用类解析运算符（::）将其部分职责委托给基类。在调用中，派生类的构造函数在初始化期间调用基类的构造函数，这不需要类解析运算符。

独立编译

我们曾经讨论过独立编译，但是在继承的情况下讨论其实现方式会非常有趣，如图 11-12 所示。

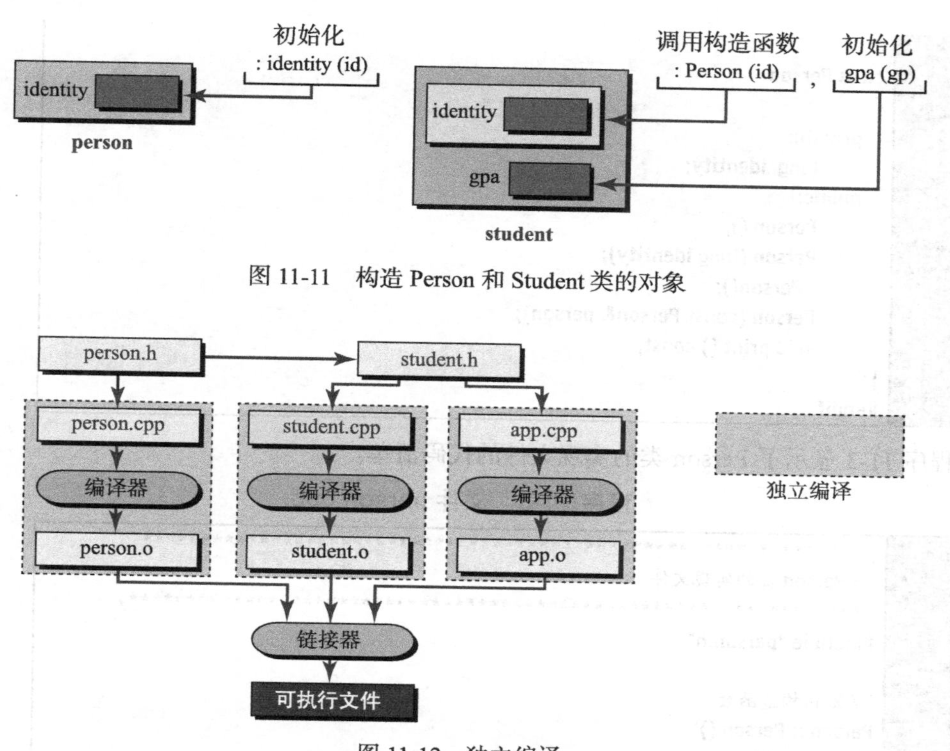

图 11-11 构造 Person 和 Student 类的对象

图 11-12 独立编译

图 11-12 包括几个要点。Person 类被编译并发送给任何只需要使用该类的用户。如果 Student 类可以访问 Person 类的接口文件，则 Person 类和 Student 类将被编译为两个不同的实体，因为 Student 类的接口文件必须包含 Person 类的接口文件。

使用独立编译的示例

接下来，我们为上述简单继承关系的示例（其中包含构造函数、析构函数和其他函数）提供接口文件、实现文件和应用程序文件。我们添加了额外的数据字段，删除了 set 函数和 get 函数，仅仅使用 print 函数来打印一个人或者一个学生的信息。

Person 类的接口文件和实现文件。我们首先给出 Person 类的接口文件和实现文件。这个类是独立的，可以与应用程序文件一起使用。程序 11-2 显示了 Person 类的接口文件的代码清单。

程序 11-2 文件 person.h

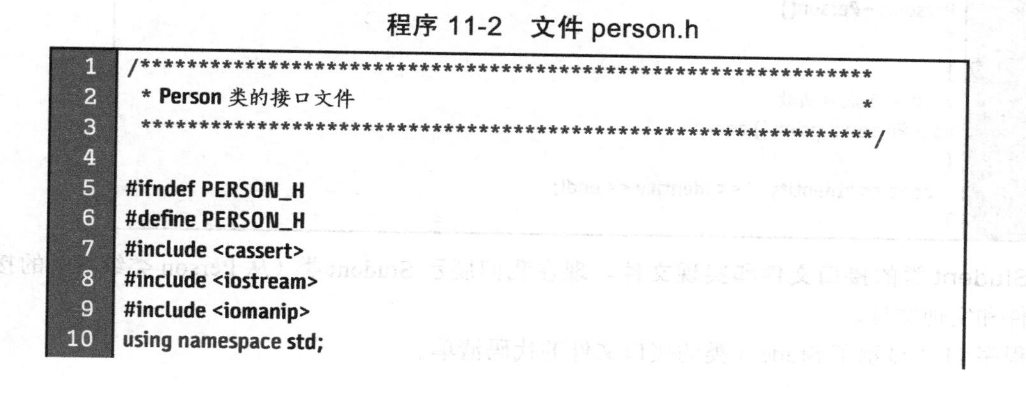
```
1   /*****************************************************************
2    * Person 类的接口文件                                            *
3    *****************************************************************/
4
5   #ifndef PERSON_H
6   #define PERSON_H
7   #include <cassert>
8   #include <iostream>
9   #include <iomanip>
10  using namespace std;
```

```
11
12  class Person
13  {
14      private:
15          long  identity;
16      public:
17          Person ();
18          Person (long identity);
19          ~Person();
20          Person (const Person& person);
21          void print () const;
22  };
23  #endif
```

程序 11-3 显示了 Person 类的实现文件的代码清单。

程序 11-3　文件 person.cpp

```
1   /***************************************************************
2    * Person 类的实现文件                                          *
3    ***************************************************************/
4   #include "person.h"
5
6   // 默认构造函数
7   Person :: Person ()
8   : identity (0)
9   {
10  }
11  // 参数构造函数
12  Person :: Person (long id)
13  : identity (id)
14  {
15      assert (identity >= 100000000 && identity <= 999999999);
16  }
17  // 拷贝构造函数
18  Person :: Person (const Person& person)
19  : identity (person.identity)
20  {
21  }
22  // 析构函数
23  Person:: ~Person()
24  {
25  }
26  // 访问器成员函数
27  void Person :: print () const
28  {
29      cout << "Identity: " << identity << endl;
30  }
```

Student 类的接口文件和实现文件。现在我们展示 Student 类（从 Person 类继承）的接口文件和实现文件。

程序 11-4 显示了 Student 类的接口文件的代码清单。

程序 11-4 文件 student.h

```
1  /***************************************************************
2   * Student 类的接口文件                                          *
3   ***************************************************************/
4  #ifndef STUDENT_H
5  #define STUDENT_H
6  #include "person.h"
7
8  class Student: public Person
9  {
10   private:
11       double gpa;
12   public:
13       Student ( );
14       Student (long identity, double gpa);
15       ~Student();
16       Student (const Student& student);
17       void print () const;
18 };
19 #endif
```

程序 11-5 显示了 Student 类的实现文件的代码清单。

程序 11-5 文件 student.cpp

```
1  /***************************************************************
2   * Student 类的实现文件                                          *
3   ***************************************************************/
4  #include "student.h"
5
6  // 默认构造函数
7  Student :: Student ()
8  : Person (), gpa (0.0)
9  {
10 }
11 // 参数构造函数
12 Student :: Student (long id, double gp)
13 : Person (id), gpa (gp)
14 {
15   assert (gpa >= 0.0 && gpa <= 4.0);
16 }
17 // 拷贝构造函数
18 Student :: Student (const Student& student)
19 : Person (student), gpa (student.gpa)
20 {
21 }
22 // 析构函数
23 Student :: ~Student()
24 {
25 }
26 // 访问器成员函数
27 void Student ::print () const
28 {
```

```
29     Person :: print ();
30     cout << "GPA: " << fixed << setprecision (2) << gpa << endl;
31 }
```

使用 Person 类和 Student 类的应用程序文件。程序 11-6 显示了一个简单的应用程序文件。客户端创建应用程序文件，使用 Person 类和 Student 类。

程序 11-6 文件 app.cpp

```
1  /*************************************************************
2   * 本应用程序测试 Person 类和 Student 类                        *
3   *************************************************************/
4  #include "student.h"
5
6  int main ( )
7  {
8     // 实例化并使用 Person 对象
9     Person person (111111111);
10    cout << "Information about person: " << endl;
11    person.print ();
12    cout << endl;
13    // 实例化并使用 Student 对象
14    Student student (222222222, 3.9);
15    cout << "Information about student: " << endl;
16    student.print ();
17    cout << endl;
18    return 0;
19 }
```

```
运行结果:
Information about person:
Identity: 111111111

Information about student:
Identity: 222222222
GPA: 3.90
```

11.1.3 有关公共继承的进一步讨论

在本节中，我们将讨论有关公共继承的其他问题。

受保护成员

到目前为止，我们只讨论了类中的私有成员和公共成员。C++ 还为成员定义了另一个修饰符：protected。**受保护成员**（protected member）在基类中（或者在没有派生的情况下）充当私有成员。当我们扩展一个类时，受保护成员的角色就会显现出来。在派生类和从派生类派生的所有类中都可以访问受保护成员。换言之，受保护成员像私有成员一样被继承，但它不会隐藏在派生类中。派生类中定义的函数可以轻松地访问受保护成员，而无须调用从基类继承的函数。图 11-13 显示了私有成员和受保护成员之间的差异。

图 11-13 显示了两种情况。在第一种情况下，我们只有一个私有数据成员和一个公共成员函数。当我们使用派生时，私有成员 x 是隐藏的。虽然 set(x) 是公共的，并且在派生类中可见，但是不能直接调用该函数，因为参数 x 不可见。我们需要创建另一个函数，使用另一

个虚拟变量 y 来访问 x，如下所示：

```
void Derived :: set (int y)
{
    Base :: set (y);           // 调用继承的函数
}
```

在继承中使用私有数据成员

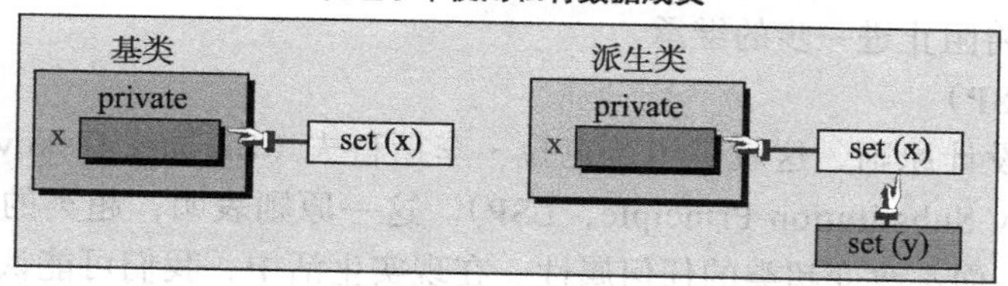

在继承中使用受保护数据成员

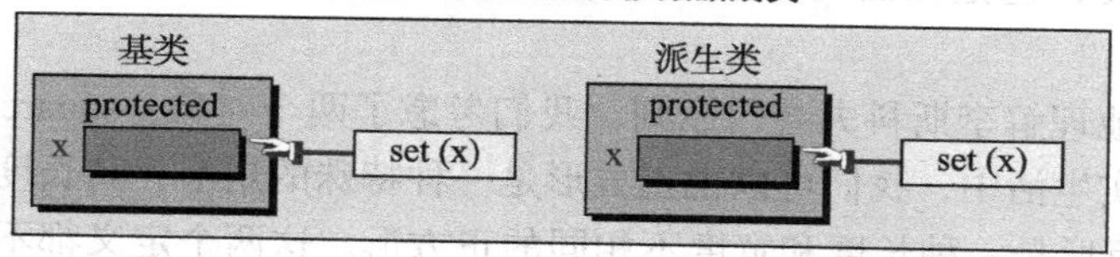

图 11-13　私有成员和受保护成员

如果我们已经将数据成员定义为受保护的，那么它在派生类中可见，我们可以直接使用 set(x)，而不必使用一个虚拟变量创建另一个函数。问题是我们应该采用哪种方法：使用私有数据成员还是受保护数据成员。接下来我们将讨论这两种方法的优缺点。

使用私有数据成员。使用私有数据成员强化了封装的概念。正如我们在前面的章节中讨论的，封装意味着隐藏类的数据成员。当我们使用私有数据成员时，数据对类外部的实体以及派生类都是隐藏的。另一方面，使用私有数据成员意味着在派生类中创建额外的代码。因此，私有数据成员的优点是更强的封装性，其缺点是在派生类中创建额外的代码。如果我们是基类的设计者，并且希望保护数据成员不被直接访问，那么应该使用私有数据成员。

使用受保护数据成员。使用受保护的数据成员可以简化派生类的编码。然而，它破坏了封装的思想。在后文中将看到，有时我们必须使用受保护的数据成员，因为如果使用私有数据成员，编码将变得非常复杂。

阻止继承

有时候我们可能希望阻止继承，例如，当我们定义了一个类，但作为设计者并不希望用户从这个类继承并创建派生类时。C++ 标准允许我们使用修饰符 final，如下所示。

```
class First final
{
    ...
}
```

我们还可以使用修饰符 final 在继承层次结构中的任何位置终止继承。例如，我们可能有一个从基类派生的类，但不希望层次结构继续继承下去，如下所示。

```
class First
{
```

```
    ...
}
class Second final : public First
{
    ...
}
```

Second 类继承 First 类，但继承在此被阻止。我们不希望有人试图从 Second 类中创建另一个子类。修饰符 final 将阻止进一步的继承。

李斯科夫替代原则（LSP）

有一种有时被忽略的设计原则，这就是由芭芭拉·李斯科夫（Barbara Liskov）提出的**李斯科夫替代原则**（Liskov Substitution Principle，LSP）。这一原则表明，超类的对象必须始终可由子类的对象替代，而不改变超类的任何属性。在现实生活中，我们可能认为一个对象是另一个对象的特殊种类，但是在面向对象的程序设计中，我们必须在编写代码之前检查 LSP。

【例 11-5】 为了更好地理解李斯科夫替代原则，我们考虑了两个对象：square（正方形）和 rectangle（矩形）。在现实生活中，我们可以说正方形是一种特殊的矩形，其长度和宽度是相同的。我们也可以说，矩形是一种长度和宽度不相同的正方形。这两个定义都不符合面向对象程序设计中的 is-a 关系。为了理解其原因，我们来考虑这两种方法，如图 11-14 所示。

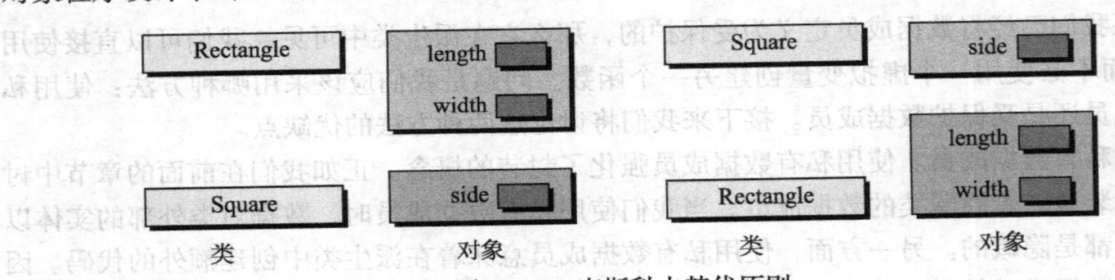

图 11-14　李斯科夫替代原则

在图 11-14 中，在左边的情况下，我们可以认为 Square 类可以从 Rectangle 类派生，因为正方形是一种特殊的矩形。但是，我们看到 Square 类的对象只有一个数据成员，而不是两个。Square 类型的对象无法封装 Rectangle 类型的对象，因此它不可被替代。

同样在图 11-14 中，在右边的情况下，Rectangle 类的对象比 Square 类的对象具有更多的数据成员，但这些成员都不能从 Square 类复制。换言之，Rectangle 类的对象与 Square 类的对象没有关系，因此不可被替代。

这并不意味着不能在继承层次结构中包含正方形类和矩形类。这意味着正方形类和矩形类不能互相继承，但它们可以在同一级别、从一个公共的基类继承。

继承树

在 C++ 中，可以形成**继承树**（inheritance tree）。例如，可以有两个从 Person 类继承的类：Student 和 Employee。很明显，学生是一个人，员工也是一个人。图 11-15 显示了这个继承树。在本章的后面部分，我们将定义继承自 Student（学生）类的其他类，如 Undergraduate（本科生）类和 Graduate（研究生）类。我们还定义了从 Employee（员

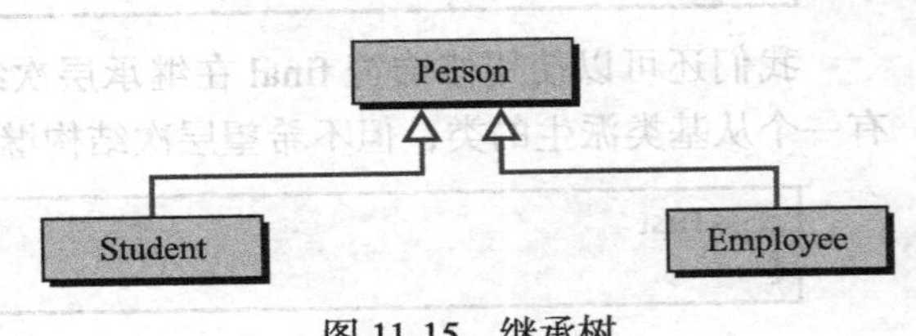

图 11-15　继承树

工）类继承的 Staff（职员）类和 Professor（教师）类。

【例 11-6】 在本例中，我们为图 11-15 所示的类创建接口文件、实现文件和应用程序文件。它们分别为程序 11-7 至程序 11-13。

程序 11-7 文件 person.h

```
1  /***************************************************************
2   * Person 类的接口文件                                          *
3   ***************************************************************/
4  #ifndef PERSON_H
5  #define PERSON_H
6  #include <iostream>
7  #include <string>
8  using namespace std;
9
10 // Person 类的定义
11 class Person
12 {
13   private:
14       string name;
15   public:
16       Person (string nme);
17       ~Person();
18       void print () const;
19 };
20 #endif
```

程序 11-8 文件 person.cpp

```
1  /***************************************************************
2   * Person 类的实现文件                                          *
3   ***************************************************************/
4  #include "person.h"
5
6  // Person 类的构造函数
7  Person :: Person (string nm)
8  :name (nm)
9  {
10 }
11 // Person 类的析构函数
12 Person :: ~Person()
13 {
14 }
15 // 成员函数 print 的定义
16 void Person :: print () const
17 {
18   cout << "Name: " << name << endl;
19 }
```

程序 11-9 文件 student.h

```
1  /***************************************************************
2   * Student 类的接口文件                                         *
```

```
3    ********************************************************************/
4  #ifndef STUDENT_H
5  #define STUDENT_H
6  #include "person.h"
7
8  // Student 类的定义
9  class Student : public Person
10 {
11   private:
12        string name;
13        double gpa;
14   public:
15        Student (string name, double gpa);
16        ~Student ( );
17        void print () const;
18 };
19 #endif
```

程序 11-10　文件 student.cpp

```
1  /*******************************************************************
2   * Student 类的实现文件                                              *
3   *******************************************************************/
4  #include "student.h"
5
6  // Student 类的构造函数
7  Student :: Student (string nm, double gp)
8  :Person (nm), gpa (gp)
9  {
10 }
11 // Student 类的析构函数
12 Student :: ~Student ()
13 {
14 }
15 // 成员函数 print 的定义
16 void Student :: print () const
17 {
18   Person :: print();
19   cout << "GPA: " << gpa << endl;
20 }
```

程序 11-11　文件 employee.h

```
1  /*******************************************************************
2   * Employee 类的接口文件                                             *
3   *******************************************************************/
4  #ifndef EMPLOYEE_H
5  #define EMPLOYEE_H
6  #include "person.h"
7
8  // Employee 类的定义
9  class Employee : public Person
10 {
```

```
11   private:
12       string name;
13       double salary;
14   public:
15       Employee (string name, double salary);
16       ~Employee ( );
17       void print () const;
18 };
19 #endif
```

程序 11-12 文件 employee.cpp

```
1  /*************************************************************
2   * Employee 类的实现文件                                      *
3   *************************************************************/
4  #include "employee.h"
5
6  // Employee 类的构造函数
7  Employee :: Employee (string nm, double sa)
8  :Person (nm), salary (sa)
9  {
10 }
11 // Employee 类的析构函数
12 Employee :: ~Employee()
13 {
14 }
15 // 成员函数 print 的定义
16 void Employee :: print () const
17 {
18    Person :: print();
19    cout << "Salary: " << salary << endl;
20 }
```

程序 11-13 文件 application.cpp

```
1  /*************************************************************
2   * 应用程序文件，使用前面创建的类                              *
3   *************************************************************/
4  #include "student.h"
5  #include "employee.h"
6
7  int main ()
8  {
9    // 实例化并使用 Person 类的对象
10   cout << "Person: " << endl;
11   Person person ("John");
12   person.print ();
13   cout << endl << endl;
14   // 实例化并使用 Student 类的对象
15   cout << "Student: " << endl;
16   Student student ("Mary", 3.9);
17   student.print ();
18   cout << endl << endl;
```

```
19    // 实例化并使用 Employee 类的对象
20    cout << "Employee: " << endl;
21    Employee employee ("Juan", 78000.00);
22    employee.print ();
23    cout << endl << endl;
24    return 0;
25  }
```

```
运行结果:
Person:
Name: John

Student:
Name: Mary
GPA: 3.9

Employee:
Name: Juan
Salary: 78000
```

11.1.4　继承的三种类型

虽然公共继承是迄今为止最常见的派生类型，但 C++ 允许我们使用其他两种派生类型：私有继承和受保护继承。在讨论这两种派生的应用之前，我们使用图 11-16 显示了每种派生中元素的状态。表中的术语 Hidden（隐藏）表示派生类的公共成员函数无法访问该成员。它必须由基类的公共成员函数访问。

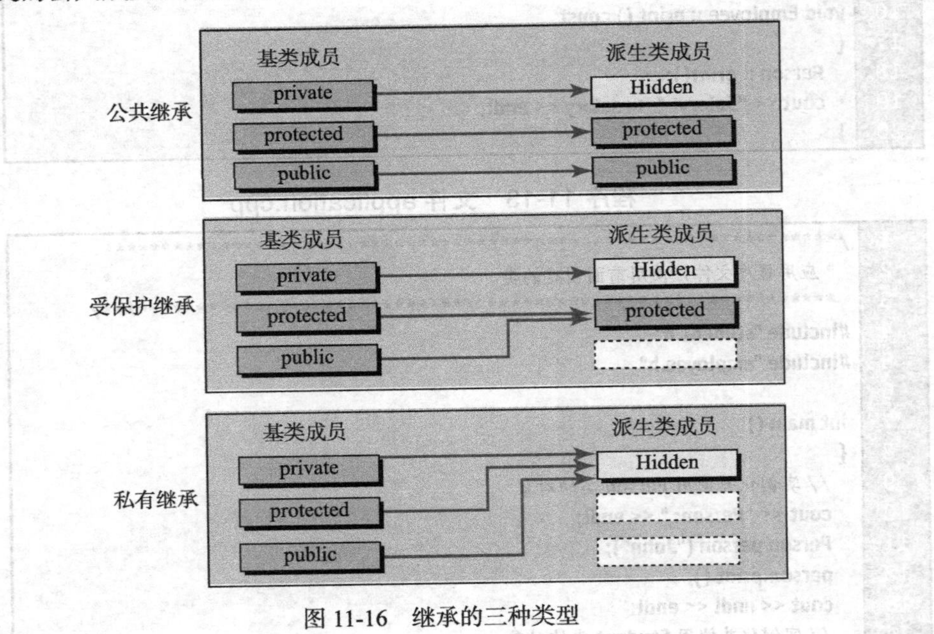

图 11-16　继承的三种类型

公共继承

公共继承（public inheritance）是我们大多数时候使用的继承类型。这种派生类型定义了基类对象和派生类对象之间的 is-a 关系，因为基类的公共接口成为派生类的公共接口。换言之，派生类的对象是基类的对象。注意，基类的私有成员在派生类中保持私有，并且不能通过派生类的公共函数访问基类的私有成员，必须通过基类的成员函数访问基类的私有成员。

受保护继承

受保护继承（protected inheritance）很少见，实际上从未使用过。

私有继承

私有继承（private inheritance）远没有公共继承普遍，但存在一些应用场景。在图 11-16 中，我们看到基类的公共成员成为派生类中的私有成员。这种属性允许继承实现（即代码重用）。假设我们已经为一个类编写了代码。现在我们想为另一个类编写代码，但两者之间没有 is-a 关系。但是，我们注意到，第一个类的一些公共成员可以帮助我们对新类的一些公共成员进行编码。我们在第二个类中私有继承第一个类，以使用第一个类的某些函数。以下是一个示例。我们不需要指定继承是私有的，因为私有继承是默认的。

```
void First :: functionFirst ( )
{
    ... ;
}
void Second :: functionSecond( )
{
    First : FunctionFirst()          // 使用 FunctionFirst 执行某些工作
    ... ;                            // 添加代码完成剩余的工作
}
```

在后续章节中讨论数据结构时，我们将看到私有继承的例子。例如，栈不是链表，但栈可以从链表继承代码。

私有继承不定义类型继承，它定义实现继承。

11.2 关联关系

并不是类之间的所有关系都可以定义为继承。在面向对象的程序设计中，存在相互间有其他关系的类。本节将讨论的第二种关系是**关联**（association）。如今，正在开发的程序更多地使用关联而不是继承。两个类之间的关联表示一种关系。例如，我们可以定义一个名为 Person 的类和另一个名为 Address 的类。Person 类的对象可能与 Address 类的对象相关：一个人居住在某个住所，而住所由某个人占用。Address 类不是从 Person 类继承的，反之也不存在继承关系。换言之，一个人不是一个地址，一个地址也不是一个人。我们不能将其中任何一个类定义为另一个类的子类。这种类型的关系在 UML 图中显示为两个类之间的实线，如图 11-17 所示。

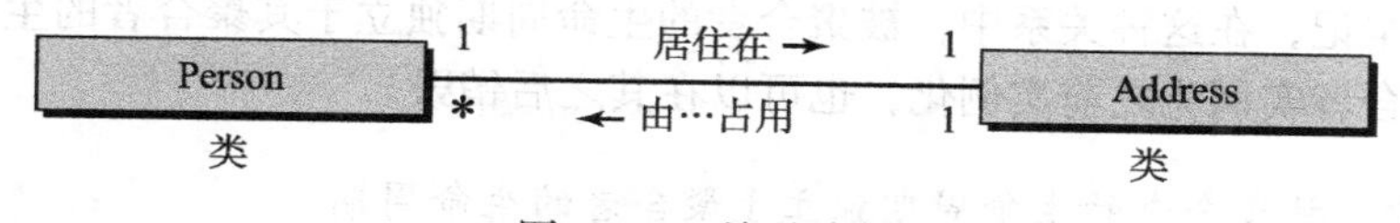

图 11-17　关联关系

关联图还显示了类之间的关系类型，这显示为指向对应类的箭头和文本。关联图中表示的另一条信息是**多重性**（multiplicity）。多重性定义参与关联的对象数。多重性显示在类旁边的线条的末尾。图 11-17 显示一个人只能有一个地址，但一个地址可以属于任何数量的居住者（所有居住在该地址的人）。表 11-2 显示了不同类型的多重性。

作为另一个示例，假设我们需要定义学生和他们所学课程之间的关联关系。我们可以定义两个类：Course（课程）和 Student（学生）。如果我们假设一个学生每学期可以修读五门课，一门课最多包含四十个学生，则可以在关联图中描述这些关系，如图 11-18 所示。

表 11-2　关联图中的多重性

关键字	说明
n	正好 *n* 个对象
*	任意对象个数，包括 0 个对象
0…1	0 个或者 1 个对象
n…*m*	*n* 到 *m* 个对象
n, *m*	*n* 个或者 *m* 个对象

Student　1　修读 →　0…5　Course
类　0…40　← 被…修读　1　类

图 11-18　学生和课程之间的关联关系

图 11-18 中的关联关系定义了多对多关系。表示多对多关系的关联不能直接被实现，因为它在程序中创建了无限多的对象（循环关系）。通常，这种类型的关联是以避免无穷大循环的方式来实现的。例如，一个 Student 对象可以有一个包含五门课程名称（不是完整的 Course 对象）的列表，一个 Course 对象可以有一个包含四十个学生姓名（不是完整的 Student 对象）的列表。在本章后面，我们将讨论一个实现这种关系的程序。

11.2.1　聚合关系

聚合（aggregation）是一种特殊的关联，其中的关系涉及所有权。换言之，聚合模拟了“has-a”的关系。其中一个类称为聚合者（aggregator），另一个类称为被聚合者（aggregatee）。聚合者类的对象包含被聚合者类的一个或者多个对象。图 11-19 显示了聚合的 UML 图。请注意，聚合的符号是放置在聚合者位置的空心菱形。

聚合关系是单向的，它不能是双向的，因为这样会创建无限多的对象。如图 11-19 所示，一个人可以有一个出生日期（Date 类的对象），但是一个 Date 对象可以与多个事件相关，而不仅仅是一个人的出生日期。

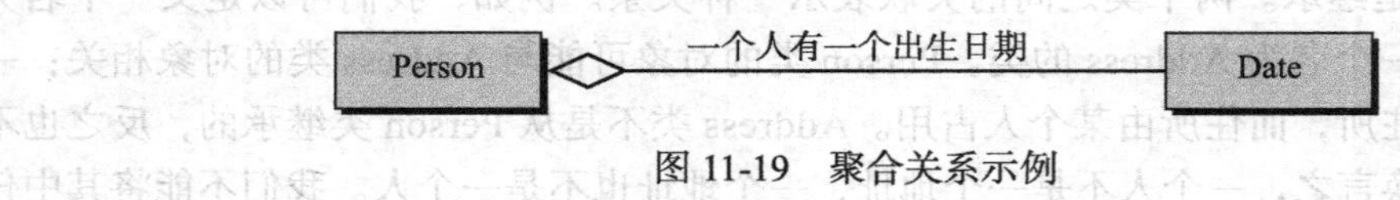

图 11-19　聚合关系示例

聚合是从聚合者到被聚合者的一对多关系。

我们必须牢记，在这种关系中，被聚合者的生命周期独立于其聚合者的生命周期。被聚合者可以在聚合者实例化之前实例化，也可以在其之后销毁。

在聚合中，被聚合者的生命周期独立于聚合者的生命周期。

【例 11-7】　在本例中，我们创建一个 Person 类和一个 Date 类。Date 类是独立的，可以用来表示任何事件。Person 类使用 Date 类的对象来定义一个人的生日。

Date 类。程序 11-14 显示了 Date 类的接口文件的代码清单。程序 11-15 显示了 Date 类的实现文件的代码清单。

程序 11-14 文件 date.h

```
1  /******************************************************************
2   * Date 类的接口文件                                              *
3   ******************************************************************/
4  #ifndef DATE_H
5  #define DATE_H
6  #include <iostream>
7  #include <cassert>
8  using namespace std;
9
10 class Date
11 {
12   private:
13       int month;
14       int day;
15       int year;
16   public:
17       Date (int month, int day, int year);
18       ~Date ();
19       void print() const;
20 };
21 #endif
```

程序 11-15 文件 date.cpp

```
1  /******************************************************************
2   * Date 类的实现文件                                              *
3   ******************************************************************/
4  #include "date.h"
5
6  // Date 类的参数构造函数
7  Date :: Date (int m, int d, int y)
8  : month (m), day (d), year (y)
9  {
10   if ((month < 1) || (month > 12))
11   {
12       cout << "Month is out of range. ";
13       assert (false);
14   }
15   int daysInMonths [13] = {0, 31, 28, 31, 30, 31, 30, 31,
16                                      31, 30, 31, 30 ,31};
17   if ((day < 1) || (day > daysInMonths [month]))
18   {
19       cout << "Day out of range! ";
20       assert (false);
21   }
22   if ((year < 1900) || (year > 2099))
23   {
24       cout << "Year out of range! " ;
25       assert (false);
```

```
26      }
27   }
28   //Date 类的析构函数
29   Date :: ~Date ()
30   {
31   }
32   // 成员函数 print 的定义
33   void Date :: print() const
34   {
35      cout << month << "/" << day <<  "/" << year << endl;
36   }
```

Person 类。程序 11-16 显示了 Person 类的接口文件的代码清单。

程序 11-16　文件 person.h

```
1    /***************************************************************
2     * Person 类的接口文件                                          *
3     ***************************************************************/
4    #ifndef PERSON_H
5    #define PERSON_H
6    #include "date.h"
7
8    // Person 类的定义
9    class Person
10   {
11     private:
12           long identity;
13           Date birthDate;
14     public:
15           Person (long identity, Date birthDate);
16           ~Person ( );
17           void print ( ) const;
18   };
19   #endif
```

程序 11-17 显示了聚合了 Date 类的 Person 类的实现文件的代码清单。

程序 11-17　文件 person.cpp

```
1    /***************************************************************
2     * Person 类的实现文件                                          *
3     ***************************************************************/
4    #include "person.h"
5
6    //构造函数
7    Person :: Person (long id, Date bd)
8    : identity (id), birthDate (bd)
9    {
10      assert (identity > 111111111 && identity < 999999999);
11   }
12   //析构函数
13   Person :: ~Person ( )
14   {
```

```
15 }
16 //打印函数
17 void Person :: print ( ) const
18 {
19     cout << "Person Identity: " << identity << endl;
20     cout << "Person date of birth: ";
21     birthDate.print ();
22     cout << endl << endl;
23 }
```

应用程序。程序 11-18 显示了使用 Person 类的应用程序的实现文件的代码清单。

程序 11-18 文件 app.cpp

```
1  /***************************************************************
2   * 本应用程序测试 Person 类                                      *
3   ***************************************************************/
4  #include "person.h"
5
6  int main ( )
7  {
8    //实例化
9    Date date1 (5, 6, 1980);
10   Person person1 (111111456, date1);
11   Date date2 (4, 23, 1978);
12   Person person2 (345332446, date2);
13   // 输出
14   person1.print ( );
15   person2.print ( );
16   return 0;
17 }
运行结果:
Person Identity: 111111456
Person date of birth: 5/6/1980

Person Identity: 345332446
Person date of birth: 4/23/1978
```

11.2.2 组合关系

组合（composition）是一种特殊的聚合，其中被包含者（containee）的生命周期取决于包含者（container）的生命周期。例如，一个人和他的姓名之间的关系就是一个组合的例子。姓名必须是人的名称才能存在。图 11-20 显示了员工和其姓名之间的关系。姓名本身由三个字符串对象组成。请注意，组合符号是放置在包含者一侧的实心菱形。Employee 和 Name 之间存在组合关系，Employee 是包含者。Name 与字符串之间存在组合关系，Name 是包含者。

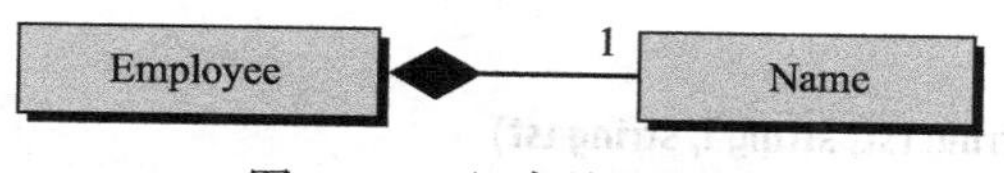

图 11-20 组合关系示例

聚合和组合之间的区别通常是概念上的，这取决于设计者如何考虑这种关系。例如，设计师可能认为一个姓名必须永远属于一个人，不能有自己的生命周期。另一位设计师可能认为，即使一个人死了，姓名也会继续存在。区别还取决于我们设计类的环境。例如，在汽车制造厂，汽车与其发动机之间的关系是组合关系，不安装在汽车上就不能使用发动机。在发动机工厂，每个发动机都有自己的生命周期。

与聚合关系一样，组合关系也是作为类实现的，其中包含者类具有被包含者类的数据成员（或者数据成员列表）。但是，被包含者对象是在包含者对象内创建的，它们不具有独立的生命周期。

【例 11-8】 我们创建一个 Employee 类，一个 Employee 对象有两个数据成员：salary 和 name。name 本身是一个类的对象，包含三个字段：名字、中间名缩写和姓氏。

Name 类。程序 11-19 显示了 Name 类的接口文件。

程序 11-19　文件 name.h

```
1  /***************************************************************
2   * Name 类的接口文件                                             *
3   ***************************************************************/
4
5  #ifndef NAME_H
6  #define NAME_H
7  #include <string>
8  #include <iostream>
9  #include <cassert>
10 using namespace std;
11
12 class Name
13 {
14   private:
15             string first;
16             string init;
17             string last;
18   public:
19        Name (string first, string init, string last);
20        ~Name ( );
21        void print ( ) const;
22 };
23 #endif
```

程序 11-20 显示了 Name 类的实现文件。

程序 11-20　文件 name.cpp

```
1  /***************************************************************
2   * Name 类的实现文件                                             *
3   ***************************************************************/
4  #include "name.h"
5
6  // 构造函数
7  Name :: Name (string fst, string i, string lst)
8  :first (fst), init (i), last (lst)
9  {
10   assert (init.size () == 1);
```

```
11    toupper (first[0]);
12    toupper (init [0]);
13    toupper (last[0]);
14  }
15  //析构函数
16  Name :: ~Name ( )
17  {
18  }
19  // 打印成员函数
20  void Name :: print ( ) const
21  {
22    cout << "Emplyee name: " << first << " " << init << ". ";
23    cout << last << endl;
24  }
```

Employee 类。程序 11-21 显示了 Employee 类的接口文件的代码清单。程序 11-22 显示了 Employee 类的实现文件的代码清单。

程序 11-21　文件 employee.h

```
1   /*******************************************************************
2    * Employee 类的接口文件                                           *
3    *******************************************************************/
4   #ifndef EMPLOYEE_H
5   #define EMPLOYEE_H
6   #include "name.h"
7
8   class Employee
9   {
10    private:
11        Name name;
12        double salary;
13    public:
14        Employee (string first, string init, string last,
15                                           double salary);
16        ~Employee ( );
17        void print ( ) const;
18  };
19  #endif
```

程序 11-22　文件 employee.cpp

```
1   /*******************************************************************
2    * Employee 类的实现文件                                           *
3    *******************************************************************/
4
5   #include "employee.h"
6
7   // 构造函数
8   Employee :: Employee (string fst, string i, string lst,
9                                          double sal)
10  : name (fst, i, lst), salary (sal)
11  {
```

```
12     assert (salary > 0.0 and salary < 100000.0);
13  }
14  // 析构函数
15  Employee :: ~Employee ( )
16  {
17  }
18  // 打印成员函数
19  void Employee :: print ( ) const
20  {
21     name.print();
22     cout << "Salary: " << salary << endl << endl;
23  }
```

应用程序。程序 11-23 显示了测试 Employee 类的应用程序文件的代码清单。

程序 11-23 文件 app.cpp

```
1   /***************************************************************
2    * 本应用程序测试 Employee 类                                  *
3    ***************************************************************/
4   #include "employee.h"
5
6   int main ( )
7   {
8      // 实例化
9      Employee employee1 ("Mary", "B", "White", 22120.00);
10     Employee employee2 ("William", "S", "Black", 46700.00);
11     Employee employee3 ("Ryan", "A", "Brown", 12500.00);
12     // 输出
13     employee1.print ( );
14     employee2.print ( );
15     employee3.print ( );
16     return 0;
17  }
```

```
运行结果:
Emplyee name: Mary B. White
Salary: 22120

Emplyee name: William S. Black
Salary: 46700

Emplyee name: Ryan A. Brown
Salary: 12500
```

11.3 依赖关系

我们可以在两个类之间定义的第三种关系是**依赖关系**（dependency）。依赖关系比继承关系或者关联关系弱。尽管存在几种依赖性定义，但我们使用的是最流行的定义。我们说依赖关系是“uses”关系的模型。如果类 A 以某种方式使用（use）类 B，则类 A 依赖于类 B。换言之，如果类 A 在不知道类 B 是否存在的情况下无法执行其完整任务，则类 A 依赖于类 B。这种情况发生的场合包括：

- 类 A 使用类型 B 的对象作为成员函数中的参数。
- 类 A 具有一个成员函数，该成员函数返回类型 B 的对象。
- 类 A 有一个成员函数，该函数有一个类型为 B 的局部变量。

11.3.1 UML 图

我们同时使用 UML 类图和 UML 序列图来显示依赖关系（参见附录 Q）。

UML 类图

尽管 UML 类图的依赖关系有很多不同版本，但是我们用虚线和箭头表示从依赖类到被依赖类的依赖关系。图 11-21 显示了一个依赖关系的示例，其中类 A 依赖于类 B。

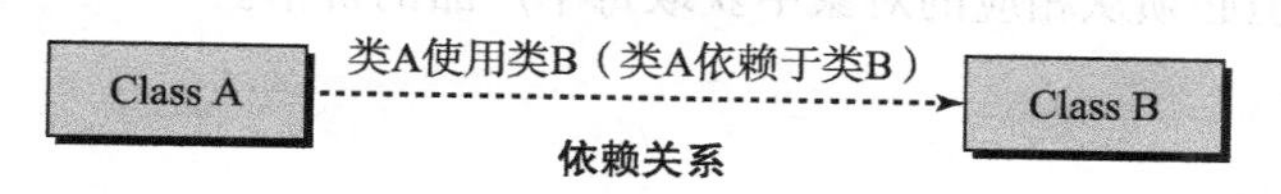

图 11-21 表示依赖关系的 UML 类图

UML 序列图

序列图用于显示对象之间的交互。main 函数和每个对象都有显示时间流逝的生命线。对象可以被实例化，并且对象的成员函数可以被调用。

作为一个简单的示例，我们有两个类：First 和 Second。类 First 有一个名为 fun() 的成员函数，用户不能直接在应用程序中调用该函数（出于某种原因，例如安全性）。我们想在类 Second 中使用另一个名为 funny() 的函数来调用类 First 中的函数 fun()。但是，我们必须把类 First 的一个实例传递给 funny()，以便用于调用 fun()（图 11-22）。

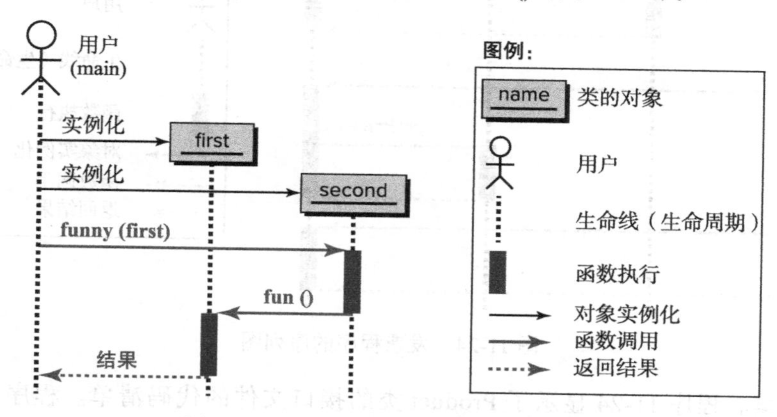

图 11-22 类 First 和类 Second 的序列图

在图 11-22 中，main 函数实例化了类 First 的一个对象和类 Second 的一个对象。main 函数调用类 Second 的对象的成员函数 funny(…)，并将类 First 的对象作为参数传递。然后，类 Second 的对象可以使用从 main 接收的对象的名称，调用类 First 的 fun() 函数。注意，First 对象和 Second 对象之间的关系是依赖关系。Second 对象在其成员函数 funny(…) 中使用类 First，并且 Second 对象调用了 First 对象的成员函数。

11.3.2 一个综合的示例

我们使用一个综合的示例来演示依赖关系的基本概念。假设我们要为销售的产品列表开具发票。我们有一个名为 Invoice 的类和一个名为 Product 的类。类 Invoice 使用类 Product 的实例作为其成员函数 add 的参数。图 11-23 显示了 UML 类图。

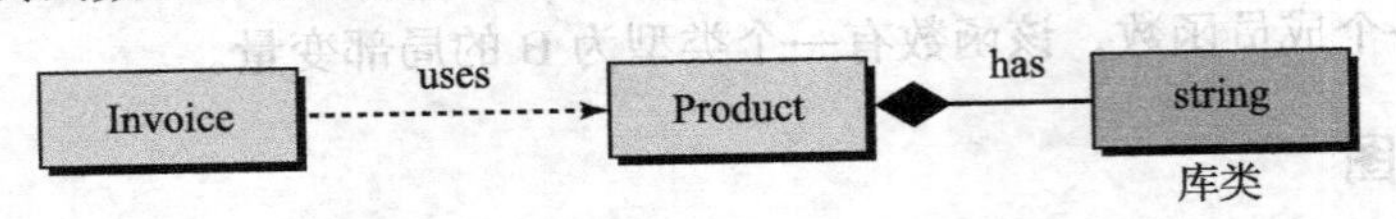

图 11-23 类 Invoice 和类 Product 的 UML 类图

我们在图 11-24 中显示了两个产品的序列图。注意，main 函数必须实例化两个 Product 类型的对象和一个 Invoice 类型的对象。然后，main 函数调用 Invoice 类中的 add 函数将产品添加到发票中，但必须从相应的对象中获取每个产品的价格。

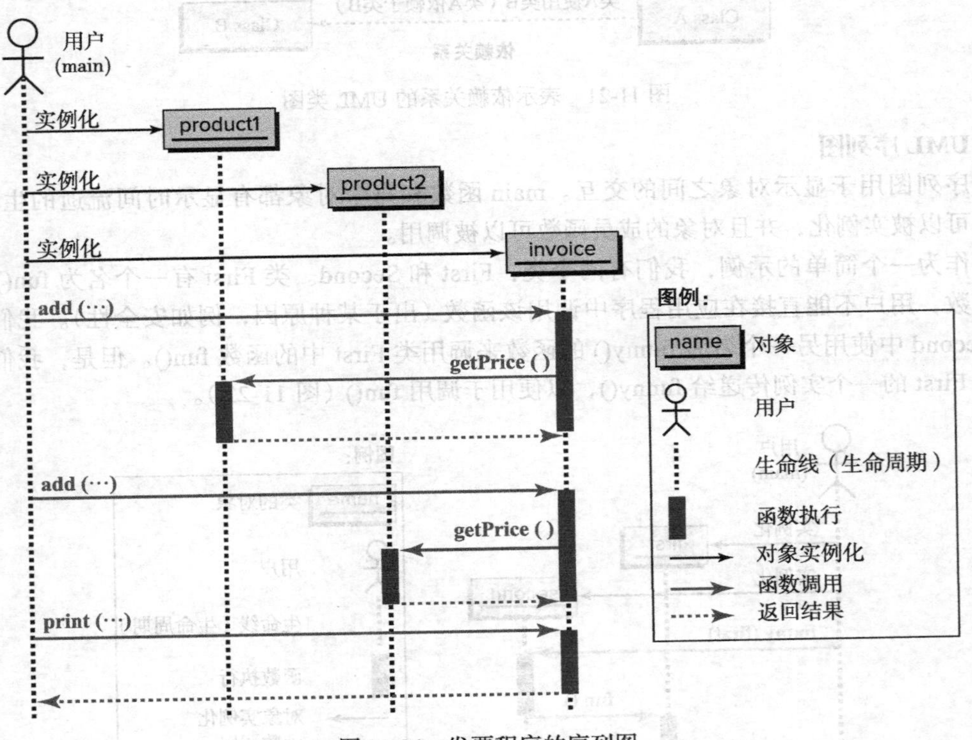

图 11-24 发票程序的序列图

Product 类。程序 11-24 显示了 Product 类的接口文件的代码清单。程序 11-25 显示了 Product 类的实现文件的代码清单。

程序 11-24 文件 product.h

```
1  /***************************************************************
2   * Product 类的接口文件                                         *
3   ***************************************************************/
4  #ifndef PRODUCT_H
5  #define PRODUCT_H
6  #include <string>
7  #include <iostream>
8  using namespace std;
9
```

```
10  class Product
11  {
12    private:
13          string name;
14          double unitPrice;
15    public:
16          Product (string name, double unitPrice);
17          ~Product ( );
18          double getPrice ( ) const;
19  };
20  #endif
```

程序 11-25　文件 product.cpp

```
1   /*****************************************************************
2    * Product 类的实现文件                                           *
3    *****************************************************************/
4   #include "product.h"
5
6   //构造函数
7   Product :: Product (string nm, double up)
8   : name (nm), unitPrice (up)
9   {
10  }
11  //析构函数
12  Product :: ~Product ( )
13  {
14  }
15  //成员函数 getPrice
16  double Product :: getPrice ( ) const
17  {
18    return unitPrice;
19  }
```

Invoice 类。程序 11-26 显示了 Invoice 类的接口文件的代码清单。程序 11-27 显示了 Invoice 类的实现文件的代码清单。

程序 11-26　文件 invoice.h

```
1   /*****************************************************************
2    * Invoice 类的接口文件                                           *
3    *****************************************************************/
4   #ifndef INVOICE_H
5   #define INVOICE_H
6   #include "product.h"
7
8   class Invoice
9   {
10    private:
11          int invoiceNumber;
12          double invoiceTotal;
13    public:
14          Invoice (int invoiceNumber);
```

```
15          ~Invoice ( );
16          void add (int quantity, Product product);
17          void print ( ) const;
18      };
19      #endif
```

程序 11-27　文件 invoice.cpp

```
1   /*****************************************************************
2    * Invoice 类的实现文件                                            *
3    *****************************************************************/
4   #include "invoice.h"
5   
6   // 构造函数
7   Invoice :: Invoice (int invNum)
8   : invoiceNumber (invNum), invoiceTotal (0.0)
9   {
10  }
11  // 析构函数
12  Invoice :: ~Invoice ( )
13  {
14  }
15  // 成员函数 add
16  void Invoice :: add (int quantity, Product product)
17  {
18    invoiceTotal += quantity * product.getPrice ();
19  }
20  // 成员函数 print
21  void Invoice :: print ( ) const
22  {
23    cout << "Invoice Number: " << invoiceNumber << endl;
24    cout << "Invoice Total: " << invoiceTotal << endl;
25  }
```

应用程序。程序 11-28 显示了测试 Invoice 类的应用程序文件的代码清单。

程序 11-28　文件 application.cpp

```
1   /*****************************************************************
2    * 本应用程序测试 Invoice 类                                       *
3    *****************************************************************/
4   #include "invoice.h"
5   
6   int main ( )
7   {
8     //实例化两个 Product 对象
9     Product product1 ("Table", 150.00);
10    Product product2 ("Chair", 80.00);
11    //为两个产品创建发票
12    Invoice invoice (1001);
13    invoice.add (1, product1);
14    invoice.add (6, product2);
15    invoice.print ();
16    return 0;
```

```
17 }
```

运行结果:

```
Invoice Number: 1001
Invoice Total: 630
```

11.4　程序设计

在本节中，我们将讨论两个项目。第一个演示如何创建词法分析器类，第二个演示如何模拟大学的注册过程。

11.4.1　词法分析器类

一个常见的问题是使用分隔符列表把一个字符串拆分为单词。例如，我们可能需要从文本中提取单词。文本中的单词由空格和换行符分隔。这种情况下，单词被称为标记符(token)，分隔单词的字符是分隔符。例如，下面的字符串中有七个标记符（单词）。

```
"This is a book about C++ language"
```

在 C++ 语言中，没有可以把字符串拆分为标记符的类，但是我们可以创建一个自定义类来实现该功能。字符串类包含的成员函数可用于查找在字符串中存在的字符，或者查找在字符串中不存在的字符。我们可以使用字符串库中的这些函数来创建一个词法分析类(Tokenizer)。

类之间的关系

在着手编写 Tokenizer 类之前，我们先展示用于描述类之间关系的 UML 类图。我们在该任务中仅仅使用了两个类：Tokenizer 类和 string 类。图 11-25 显示了这两个类之间的关系。

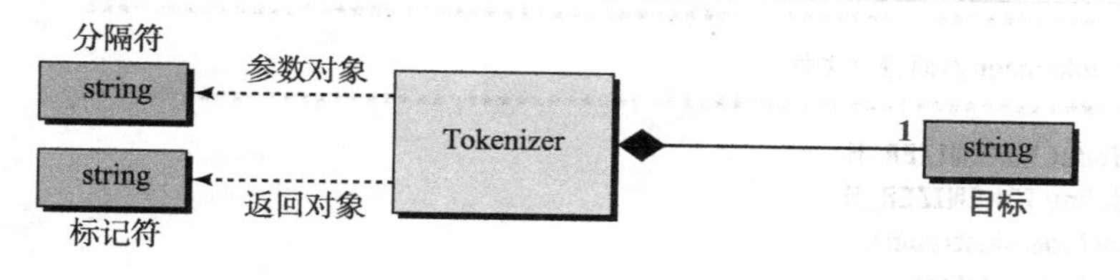

图 11-25　类之间的关系

在图 11-25 中，Tokenizer 类使用了 string 类 3 次。Tokenizer 类使用 string 类创建目标字符串（需要拆分的文本），创建分隔符字符串，以及最后创建标记符。一个字符串作为目标字符串，一个字符串作为分隔符字符串，但是创建了多个作为标记符的字符串。词法分析器对象和标记符对象之间的关系是依赖关系（标记符是 Tokenizer 类的函数返回的对象）。

序列图

图 11-26 是对象间相互作用的序列图。

在实例化 target 和 tokenizer 对象之后，程序使用一个循环从 target 对象中抽取标记符。用户调用 tokenizer 对象的 nextToken() 函数，而 nextToken() 函数则调用 target 对象的成员函数 find_first_not_of() 和 find_first_of()。然后 tokenizer 对象返回一个 token 对象实例给用户。注意，在序列图中没有包含 token 对象，因为它们仅仅作为返回的对象（依赖关系）。

程序

由于 string 类是 C++ 库中包含预定义公共接口的类，因此我们只需要创建一个类

Tokenizer。Tokenizer 的接口文件参见程序 11-29。Tokenizer 类仅包含两个数据成员：tagert 和 delim，其类型均为 string。然而，类返回类型为 string 的 token（标记符，第 21 行）。Tokenizer 类仅包含两个成员函数（构造函数和析构函数除外）：第一个函数检查在目标字符串中是否存在标记符，第二个函数返回目标字符串中的下一个标记符。

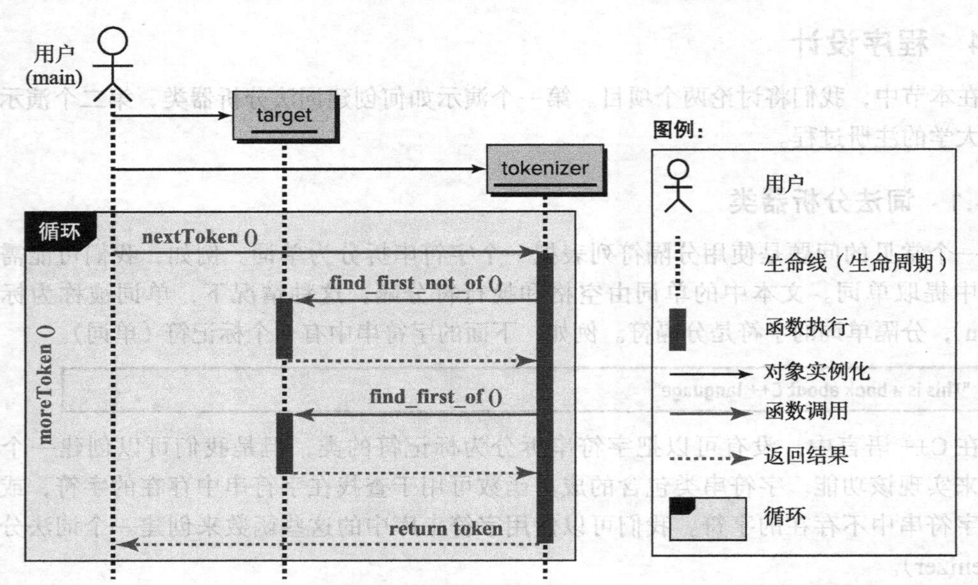

图 11-26　词法分析器设计的序列图

程序 11-29　文件 tokenizer.h

```
1   /***************************************************************
2    * Tokenizer 类的接口文件                                      *
3    ***************************************************************/
4   #ifndef TOKENIZER_H
5   #define TOKENIZER_H
6   #include <iostream>
7   #include <string>
8   using namespace std;
9
10  class Tokenizer
11  {
12    private:
13        string target;
14        string delim;
15        int begin;
16        int end;
17    public:
18        Tokenizer (const string& target, const string& delim);
19        ~Tokenizer ();
20        bool moreToken() const;
21        string nextToken();
22  };
23  #endif
```

我们还给出了实现 Tokenizer 类的实现文件，它包含四个函数：构造函数、析构函数、检查是否存在标记符的函数、返回下一个标记符的函数（程序 11-30）。

我们可以创建应用程序文件，创建并使用 Tokenizer 类。程序 11-31 是一个示例。

程序 11-30 文件 tokenizer.cpp

```
1  /*******************************************************************
2   * Tokenizer 类的实现文件                                          *
3   *******************************************************************/
4  #include "tokenizer.h"
5  
6  // 构造函数
7  Tokenizer :: Tokenizer (const string& tar, const string& del)
8  : target (tar), delim (del)
9  {
10   begin = target.find_first_not_of (delim, 0);
11   end = target.find_first_of (delim, begin);
12 }
13 // 析构函数
14 Tokenizer :: ~Tokenizer()
15 {
16 }
17 // 检查是否存在更多标记符
18 bool Tokenizer :: moreToken ( ) const
19 {
20   return (begin != - 1);
21 }
22 // 返回下一个标记符
23 string Tokenizer :: nextToken ( )
24 {
25   string token = target.substr (begin, end - begin);
26   begin = target.find_first_not_of (delim, end);
27   end = target.find_first_of(delim, begin);
28   return token;
29 }
```

程序 11-31 文件 app.cpp

```
1  /*******************************************************************
2   * 本应用程序测试 Tokenizer 类                                      *
3   *******************************************************************/
4  #include "tokenizer.h"
5  
6  int main ( )
7  {
8    // 需要拆分为标记符的目标字符串 target
9    string target ("This is the string to be tokenized. \n");
10   // 分隔字符串 delimit 定义分隔字符集
11   string delimit (" \n");   // 分隔符由 ' ' 和 '\n' 组成
12   // 实例化 Tokenizer 对象
13   Tokenizer tokenizer (target, delimit);
14   // 遍历目标字符串 target，查找所有标记符
15   while (tokenizer.moreToken ())
```

```
16   {
17       cout << tokenizer.nextToken () << endl;
18   }
19   return 0;
20 }
```

运行结果：

```
This
is
the
string
to
be
tokenize.
```

11.4.2 注册

在这一部分中，我们使用类之间的关联和依赖关系，为学院或者大学中的院系设计一个简单的注册系统。

UML 类图

我们使用了六个类，如图 11-27 所示。

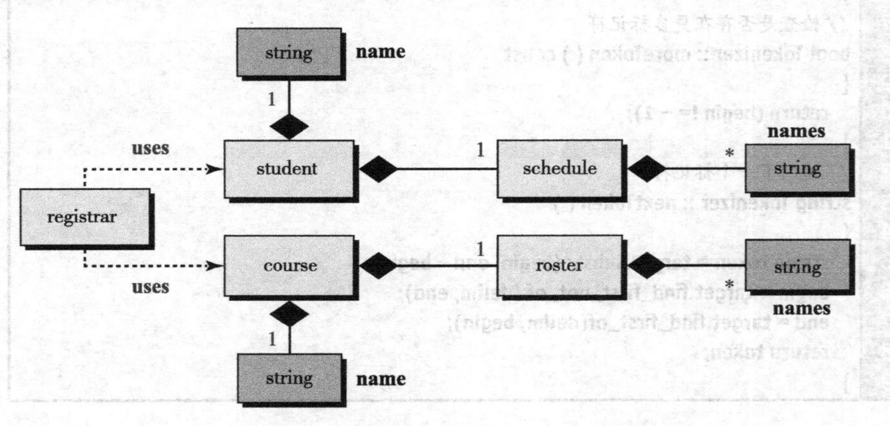

图 11-27　注册系统中各类之间的关系

每个 student（学生）对象包含一个 string 对象（作为学生姓名）和一个 schedule 对象。schedule（学生课表）对象包含一个字符串数组（课程名称）。每个 course（课程）对象包含一个 string 对象（作为课程名称）和一个 roster 对象。roster（课程花名册）对象包含一个字符串数组（学生姓名）。registrar 类仅使用 student 对象和 course 对象。

UML 序列图

在着手编写五个用户自定义类之前，我们先展示反映这些类之间交互关系的序列图（图 11-28）。注意，roster 对象由 course 对象创建，schedule 对象由 student 对象创建。registrar、course 和 student 对象由 main 函数创建。为了使 UML 序列图简单明了，我们仅显示一个学生和一门课程。

程序

CourseRoster（课程花名册）类。程序 11-32 是接口文件。程序 11-33 是实现文件。

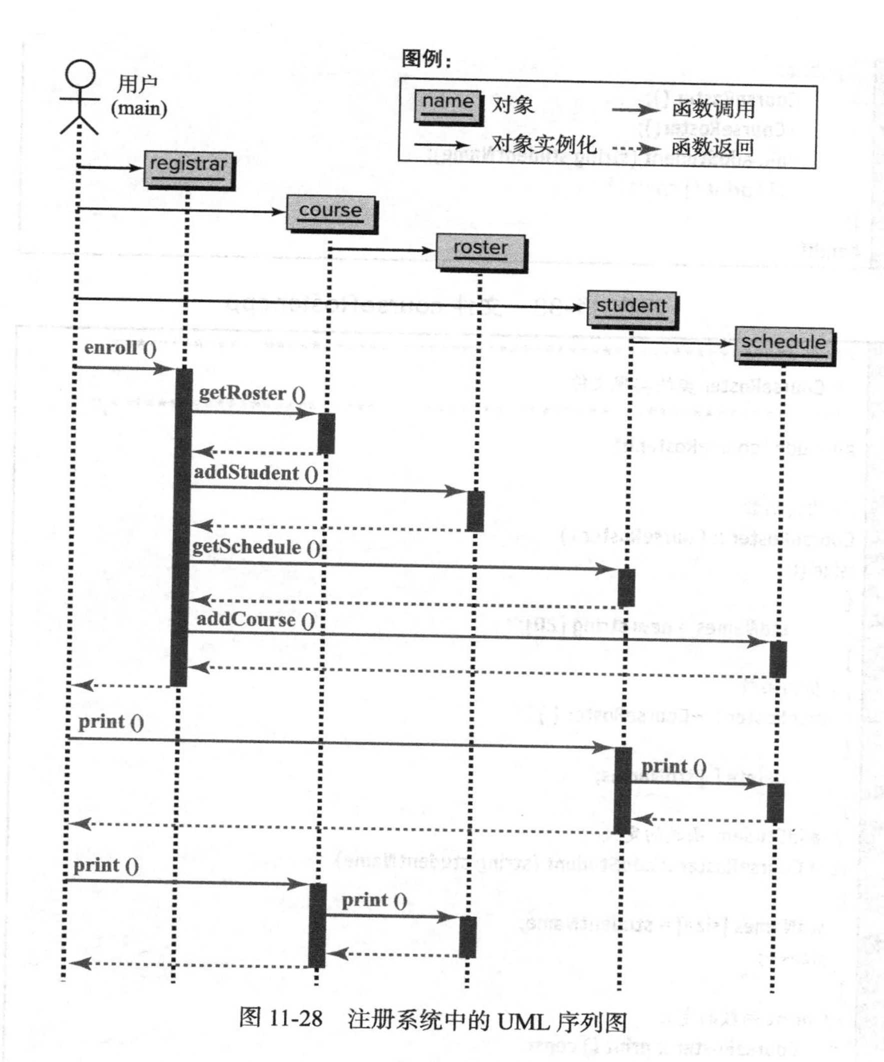

图 11-28 注册系统中的 UML 序列图

程序 11-32 文件 courseRoster.h

```
1   /***************************************************************
2    * CourseRoster 类的接口文件                                    *
3    ***************************************************************/
4   #ifndef COURSEROSTER_H
5   #define COURSEROSTER_H
6   #include <string>
7   #include <iostream>
8   #include <cassert>
9   using namespace std;
10
11  // 类定义
12  class CourseRoster
13  {
14    private:
15        int size;
16        string* stdNames;
```

```
17   public:
18         CourseRoster ();
19         ~CourseRoster();
20         void addStudent (string studentName);
21         void print () const;
22   };
23   #endif
```

程序 11-33　文件 courseRoster.cpp

```
1    /***************************************************************
2     * CourseRoster 类的实现文件                                     *
3     ***************************************************************/
4    #include "courseRoster.h"
5
6    //构造函数
7    CourseRoster :: CourseRoster ()
8    :size (0)
9    {
10         stdNames = new string [20];
11   }
12   //析构函数
13   CourseRoster :: ~CourseRoster ( )
14   {
15         delete [ ] stdNames;
16   }
17   //addStudent 函数的定义
18   void CourseRoster :: addStudent (string studentName)
19   {
20     stdNames [size] = studentName;
21     size++;
22   }
23   // print 函数的定义
24   void CourseRoster :: print () const
25   {
26     cout << "List of Students" << endl;
27     for (int i = 0; i < size; i++)
28     {
29         cout << stdNames[i] << endl;
30     }
31     cout << endl;
32   }
```

Course（课程）类。程序 11-34 是接口文件。程序 11-35 是实现文件。

程序 11-34　文件 course.h

```
1    /***************************************************************
2     * Course 类的接口文件                                           *
3     ***************************************************************/
4    #ifndef COURSE_H
5    #define COURSE_H
6    #include <cassert>
```

```
7   #include <string>
8   #include <iostream>
9   #include "courseRoster.h"
10  using namespace std;
11
12  // 类定义
13  class Course
14  {
15    private:
16          string name;
17          int units;
18          CourseRoster* roster;
19    public:
20          Course (string name, int units);
21          ~Course ();
22          string getName() const;
23          CourseRoster* getRoster () const;
24          void addStudent (string name);
25          void print () const;
26  };
27  #endif
```

程序 11-35 文件 course.cpp

```
1   /****************************************************************
2    * Course 类的实现文件                                          *
3    ****************************************************************/
4   #include "course.h"
5
6   // 构造函数
7   Course :: Course (string nm, int ut)
8   : name (nm), units (ut)
9   {
10    roster = new CourseRoster;
11  }
12  // 析构函数
13  Course :: ~Course ()
14  {
15  }
16  // getName 函数的定义
17  string Course :: getName() const
18  {
19    return name;
20  }
21  // addStudent 函数的定义
22  void Course :: addStudent (string name)
23  {
24    roster - >addStudent (name);
25  }
26  // getRoster 函数的定义
27  CourseRoster* Course :: getRoster () const
28  {
```

```
29    return roster;
30  }
31  // print 函数的定义
32  void Course :: print () const
33  {
34    cout << "Course Name: " << name << endl;
35    cout << "Number of Units: " << units << endl;
36    roster - > print ();
37  }
```

StudentSchedule（学生课表）类。程序 11-36 是接口文件。程序 11-37 是实现文件。

程序 11-36　文件 studentSchedule.h

```
1   /**************************************************************
2    * StudentSchedule 类的接口文件                                  *
3    **************************************************************/
4   #ifndef STUDENTSCHEDULE_H
5   #define STUDENTSCHEDULE_H
6   #include <string>
7   #include <iostream>
8   #include <cassert>
9   using namespace std;
10
11  // 类定义
12  class StudentSchedule
13  {
14    private:
15          int size;
16          string* courseNames;
17    public:
18          StudentSchedule ();
19          ~StudentSchedule();
20          void addCourse (string course);
21          void print () const;
22  };
23  #endif
```

程序 11-37　文件 studentSchedule.cpp

```
1   /**************************************************************
2    * StudentSchedule 类的实现文件                                  *
3    **************************************************************/
4   #include "studentSchedule.h"
5
6   // 构造函数
7   StudentSchedule :: StudentSchedule ()
8   :size (0)
9   {
10    courseNames = new string [5];
11  }
12  // 析构函数
13  StudentSchedule :: ~StudentSchedule ()
14  {
```

```
15      delete [ ] courseNames;
16  }
17  // addCourse 函数的定义
18  void StudentSchedule :: addCourse (string name)
19  {
20      courseNames [size] = name;
21      size++;
22  }
23  // print 函数的定义
24  void StudentSchedule :: print () const
25  {
26      cout << "List of Courses" << endl;
27      for (int i = 0; i < size; i++)
28      {
29          cout << courseNames[i] << endl;
30      }
31      cout << endl;
32  }
```

Student（学生）类。程序 11-38 是接口文件。程序 11-39 是实现文件。

程序 11-38　文件 student.h

```
1   /***************************************************************
2    * Student 类的接口文件                                          *
3    ***************************************************************/
4   #ifndef STUDENT_H
5   #define STUDENT_H
6   #include <cassert>
7   #include <string>
8   #include <iostream>
9   #include "studentSchedule.h"
10  using namespace std;
11
12  // 类定义
13  class Student
14  {
15    private:
16          string name;
17          StudentSchedule* schedule;
18    public:
19          Student (string name);
20          ~Student ();
21          string getName () const;
22          StudentSchedule* getSchedule () const;
23          void addCourse (string name);
24          void print () const;
25  };
26  #endif
```

程序 11-39　文件 student.cpp

```
1   /***************************************************************
2    * Student 类的实现文件                                          *
```

```
3    *****************************************************************/
4    #include "student.h"
5
6    // 构造函数
7    Student :: Student (string nm)
8    :name (nm)
9    {
10     schedule = new StudentSchedule;
11   }
12   // 析构函数
13   Student :: ~Student ()
14   {
15   }
16   // getName 函数的定义
17   string Student :: getName () const
18   {
19     return name;
20   }
21   // getSchedule 函数的定义
22   StudentSchedule* Student :: getSchedule () const
23   {
24     return schedule;
25   }
26   // addCourse 函数的定义
27   void Student :: addCourse (string name)
28   {
29     schedule -> addCourse (name);
30   }
31   // print 函数的定义
32   void Student :: print () const
33   {
34     cout << "Student name: " << name << endl;
35     schedule - >print ();
36   }
```

Registrar（注册器）类。程序 11-40 是接口文件。程序 11-41 是实现文件。

程序 11-40 文件 registrar.h

```
1    /*****************************************************************
2     * Registrar 类的接口文件                                            *
3     *****************************************************************/
4    #ifndef REGISTRAR_H
5    #define REGISTRAR_H
6    #include "course.h"
7    #include "student.h"
8
9    // 类定义
10   class Registrar
11   {
12     public:
13           Registrar ();
14           ~Registrar();
```

```
15        void enroll (Student student, Course course);
16    };
17    #endif
```

程序 11-41　文件 registrar.cpp

```
1   /*****************************************************************
2    * Registrar 类的实现文件                                          *
3    *****************************************************************/
4   #include "registrar.h"
5
6   // 构造函数
7   Registrar :: Registrar ()
8   {
9   }
10  // 析构函数
11  Registrar :: ~Registrar ()
12  {
13  }
14  // 注册函数 enroll
15  void Registrar :: enroll (Student student, Course course)
16  {
17     (course.getRoster ()) -> addStudent (student.getName());
18     (student.getSchedule()) -> addCourse (course.getName());
19  }
```

应用程序。程序 11-42 是应用程序文件。注意，应用程序创建一个注册器（Registrar）对象。在第 20 章中，我们将学习如何控制注册器对象，使其具有唯一性（singleton 模式），但目前我们假设只创建了一个注册器对象。然后程序 11-42 创建三个学生对象和三个课程对象。注册器对象负责将学生注册到他们想要的课程中。请注意，我们不会直接实例化 StudentSchedule（学生课表）对象或者 CourseRoster（课程花名册）对象。第一个对象在 Student 类中实例化，第二个对象在 Course 类中实例化。

程序 11-42　文件 application.cpp

```
1   /*****************************************************************
2    * 本应用程序测试注册系统中的类                                      *
3    *****************************************************************/
4   #include "registrar.h"
5
6   int main ( )
7   {
8      // 实例化一个 Registrar（注册器）对象
9      Registrar registrar;
10     // 实例化三个 Student（学生）对象
11     Student student1 ("John");
12     Student student2 ("Mary");
13     Student student3 ("Ann");
14     // 实例化三个 Course（课程）对象
15     Course course1 ("CIS101", 4);
16     Course course2 ("CIS102", 3);
17     Course course3 ("CIS103", 3);
```

```
18    // 让 Registrar 对象把学生注册到课程
19    registrar.enroll (student1, course1);
20    registrar.enroll (student1, course2);
21    registrar.enroll (student2, course1);
22    registrar.enroll (student2, course3);
23    registrar.enroll (student3, course1);
24    // 打印每个学生的信息
25    student1.print();
26    student2.print();
27    student3.print();
28    // 打印每门课程的信息
29    course1.print();
30    course2.print();
31    course3.print();
32    return 0;
33  }
```

运行结果：

```
Student name: John
List of Courses
CIS101
CIS102

Student name: Mary
List of Courses
CIS101
CIS103

Student name: Ann
List of Courses
CIS101

Course Name: CIS101
Number of Units: 4
List of Students
John
Mary
Ann

Course Name: CIS102
Number of Units: 3
List of Students
John

Course Name: CIS103
Number of Units: 3
List of Students
Mary
```

本章小结

在面向对象的程序设计中，继承是指从通用的概念派生出一个更具体的概念。更一般的类称为基类，更具体的类称为派生类。派生类继承除参数构造函数、默认构造函数、拷贝构造函数、析构函数和赋值运算符之外的所有基类成员。有三种继承类型：public、protected和private。默认类型是private，但最常见的是public。

两个类之间的关联显示了它们之间的关系。聚合是一种特殊的关联，在这种关联中，关系涉及所有权："has-a"关系。组合是一种特殊的聚合，其中被包容者的生命周期取决于包容者的生命周期。

依赖性是类之间的弱关系，它构建了"uses"(使用)关系模型。

思考题

1. 假设我们定义了以下两个类。使用委托为两个类编写set函数和print函数的定义。

```
//类 First 的声明
class First
{
    private:
        int a;
    public:
        void set (int a);
        void print () const;
};
//类 Second 的声明
class Second : public First
{
    private:
        int b;
    public:
        void set (int a, int b);
        void print () const;
};
```

2. 假设我们定义了以下两个类。不使用委托为两个类编写set函数和print函数的定义。

```
//类 First 的声明
class First
{
    protected:
        int a;
    public:
        void set (int a);
        void print () const;
};
//类 Second 的声明
class Second : public First
{
    private:
        int b;
    public:
        void set (int a, int b);
        void print () const;
};
```

3. 假设我们定义了如下所示的类First，如果类Second是类First的公共派生类，那么成员one、two和set的可访问性分别是什么？

```
class First
{
    private:
        int one;
    protected:
        int two;
    public:
        void set (int one, int two);
}
```

4. 重复第 3 题，但假设类 Second 是类 First 的受保护派生类。
5. 重复第 3 题，但假设类 Second 是类 First 的私有派生类。
6. 假设我们定义了如下所示的三个类。为所有三个类编写构造函数和打印函数的定义。

```
// 类 First 的定义
class First
{
    private:
        int a;
    public:
        First (int a);
        void print ( ) const;
};
// 类 Second 的定义
class Second : public First
{
    private:
        int b;
    public:
        Second (int a, int b);
        void print ( ) const;
};
// 类 Third 的定义
class Third : public Second
{
    private:
        int c;
    public:
        Third (int a, int b, int c);
        void print () const;
};
```

7. 假设我们定义了如下所示的三个类。为所有三个类编写构造函数和打印函数的定义。

```
// 类 First 的定义
class First
{
    private:
        int a;
    public:
        First (int a);
        void print ( ) const;
};
// 类 Second 的定义
class Second : public First
{
```

```
    private:
        int b;
    public:
        Second (int a, int b);
        void print ( ) const;
};
// 类 Third 的定义
class Third : public First
{
    private:
        int c;
    public:
        Third (int a, int c);
        void print () const;
};
```

8. 假设我们定义了如下所示的两个类。为这两个类编写构造函数和打印函数的定义。

```
// 类 First 的定义
class First
{
    private:
        int a;
        double b;
    public:
        First (int a, double b);
        void print ( ) const;
};
// 类 Second 的定义
class Second
{
    private:
        First f;
        char c;
    public:
        Second (First f, char c);
        void print ( ) const;
};
```

9. 绘制在程序 11-6 中 main 函数与 Person 对象和 Student 对象交互的 UML 序列图。
10. 绘制在程序 11-6 中 main 函数与 Person 对象和 Date 对象交互的 UML 序列图。注意，程序中包括两个 Person 类的实例对象和两个 Date 类的实例对象。
11. 绘制在程序 11-21 中各对象之间交互的 UML 序列图。

编程题

1. 设计一个名为 Square 的类，定义正方形的几何图形。类必须有一个名为 side 的数据成员，该数据成员定义正方形的边长。然后定义两个成员函数 getPeri 和 getArea，以获取正方形的周长和面积。接下来设计一个名为 Cube 的类，定义一个立方体，要求 Cube 类继承自 Square 类。类 Cube 不需要额外的数据成员，但需要成员函数 getArea 和 getVolume。为这两个类提供适当的构造函数和析构函数。
2. 设计一个名为 Rectangle（矩形）的类，该类包含两个私有数据成员：length（长度）和 width（宽度）。为该类定义构造函数和析构函数，并编写成员函数以获取矩形的周长和面积。然后定义一个名为 Cuboid（表示长方体）的类，该类继承自 Rectangle 类，并包含一个额外的数据成员：height（高度）。

然后为 Cuboid 类编写构造函数和析构函数，并编写成员函数来获取 Cuboid 对象的表面积和体积。

3. 创建一个名为 Sphere（球体）的类，继承自类 Circle（圆）。可以通过沿直径旋转一个圆来创建球体。
   a. 为名为 Circle 的类创建一个接口文件，其中包含一个私有数据成员：radius（半径）。为该类定义一个参数构造函数和一个析构函数，并编写成员函数获取圆的周长和面积，计算公式如下：

```
perimeter = 2 * π * radius                area = π * radius * radius
```

   b. 为名为 Sphere 的类创建一个接口文件。同样为该类定义一个参数构造函数和一个析构函数，并编写成员函数获取球体的表面积和体积。
   c. 使用以下计算公式为类 Sphere 定义一个实现文件，用于获取球的表面积和体积。在计算公式中，perimeter 是圆的周长，area 是圆的面积（参见 a 中的定义）。

```
surface = 2 * radius * perimeter                volume = (4 / 3) * radius * area
```

   d. 编写一个应用程序文件，测试类 Circle 和 Sphere。
4. 创建一个名为 Cylinder（圆柱体）的类，继承自类 Circle（圆）。可以通过在圆对象中增加一个高度来创建圆柱体。
   a. 为名为 Circle 的类设计一个接口文件（参见第 3 题）。
   b. 为名为 Cylinder 的类定义一个接口文件。同样为该类定义一个参数构造函数和一个析构函数，并编写成员函数用于获取圆柱体的表面积和体积。
   c. 使用以下计算公式，为类 Cylinder 定义一个实现文件，获取圆柱体的表面积和体积。在计算公式中，perimeter 是圆的周长，area 是圆的面积（参见 a 中的定义）。

```
surface = height * perimeter                volume = height * area
```

   d. 编写一个应用程序文件，测试类 Circle 和 Cylinder。
5. 创建一个简单的 Employee 类，该类支持两种类型的员工，即按月给薪员工和按小时给薪员工。所有员工都有四个数据成员：姓名、员工编号、出生日期和聘用日期。按月给薪员工有月工资和年度奖金（表示为 0% 到 10% 之间的百分比）。按小时给薪员工有小时工资和加班费（50% 到 100% 之间）。
6. 设计一个名为 Student（学生）的类，包含两个数据成员：name 和 gpa。然后定义一个名为 Course（课程）的类，其数据成员包含选修该课程的学生数，以及在堆内存中创建的学生数组。要求 Student 类包含一个用于打印有关 Student 对象的成员函数。要求 Course 类包含课程中注册的学生信息，并打印所有注册学生的信息。
7. 设计一个名为 Course（课程）的类，包含两个数据成员：name 和 units。然后设计一个名为 Student 的类，包含三个数据成员：name、gpa 和选修的课程列表 list。要求将列表实现为一个数组并保存在堆内存中。为 Course 类和 Student 类创建构造函数、析构函数以及操作所需的所有成员函数。在一个应用程序中测试这两个类。
8. 编写一个程序，模拟如下图所示的三个类。

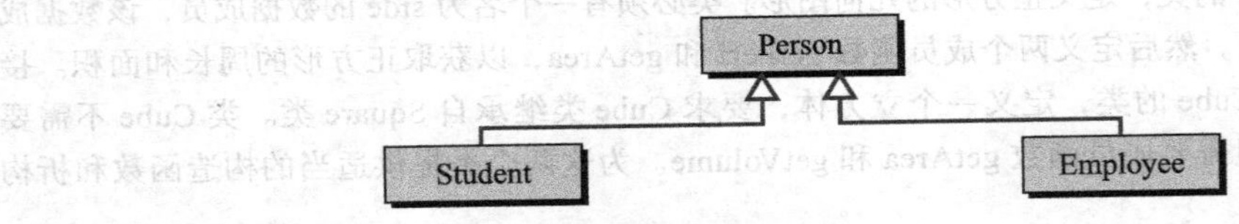

   a. 使用 name 作为 Person 类的唯一数据成员。
   b. 使用 name 和 gpa 作为 Student 类的数据成员。
   c. 使用 name 和 salary 作为 Employee 类的数据成员。
9. 编写一个程序，模拟如下图所示的七个类。前三个类（Person、Student 和 Employee）已经在第 8 题中定义。创建代码定义其他四个类。

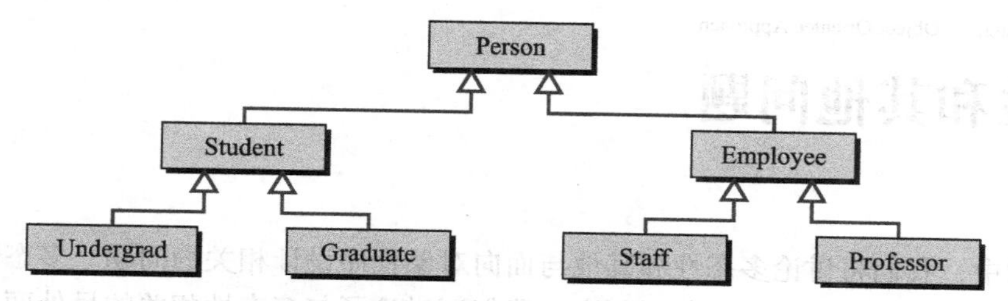

a. 使用 year（1、2、3、4）作为 Undergraduate 类的数据成员。

b. 使用 goal（master、phd）作为 Graduate 类的数据成员。

c. 使用 status（manager、nomanager）作为 Staff 类的成员。

d. 使用 status（part-time、full-time）作为 Professor 类的成员。

我们推荐使用 enum 结构来定义以上四种取值要求以避免有效值的检测。

10. 定义一个名为 Point 的类，表示一个坐标为 x 和 y 的点。然后编写成员函数，使用 Point 类获取两个点之间的距离。使用如下图所示的依赖关系。

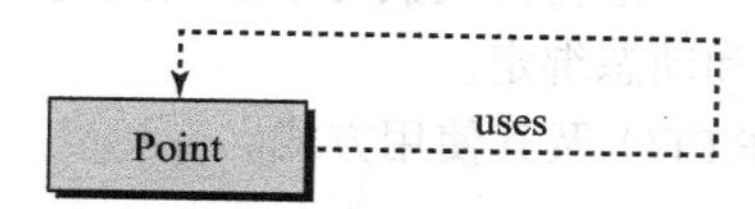

11. 重新设计在第 7 章定义的 Fraction 类，使得 Fraction 类依赖于其自身，如下图所示。支持四种操作：add（加）、subtract（减）、multiply（乘）和 divide（除）。

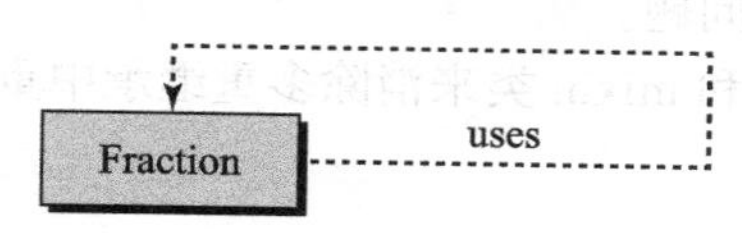

第 12 章

C++ Programming: An Object-Oriented Approach

多态性和其他问题

在本章中，我们将讨论多态性和其他与面向对象程序设计相关的问题。多态性使我们能够在不同的类中编写一个函数的若干版本。我们还讨论了与多态性相关的另外两个问题：抽象类和多重继承。

学习目标

阅读并学习本章后，读者应该能够

- 定义多态性并描述实现多态性的三个必要条件。
- 定义虚拟表以及理解虚拟表如何帮助系统决定在运行时使用哪个虚函数。
- 讨论虚析构函数以及为什么虚析构函数对于正确使用多态性是必需的。
- 区分继承中的静态绑定和动态绑定。
- 讨论运行时类型信息（RTTI）及其使用方式。
- 讨论抽象类及其使用。
- 讨论接口类及其使用。
- 讨论多重继承及其相关问题。
- 演示如何使用虚基继承和 mixin 类来消除多重继承中重复数据成员的问题。

12.1 多态性

在前面的章节中，我们提到了面向对象程序设计的主要支柱之一是**多态性**。多态性允许我们编写一个函数的若干版本，每个版本都在一个单独的类中。然后，当我们调用函数时，将执行适合被引用对象的版本。我们在日常语言中也看到了同样的概念。我们用一个动词（函数）来表示不同的操作。例如，我们说“打开”，意思是打开一扇门、打开一个罐子或者打开一本书，具体操作是由上下文决定的。类似地，在 C++ 中，我们可以调用一个名为 printArea 的函数来打印三角形的面积或者矩形的面积。

为了更好地理解多态性的概念，我们必须首先理解与插件兼容的对象的概念。与电气设备进行类比可能会有所帮助，如图 12-1 所示。

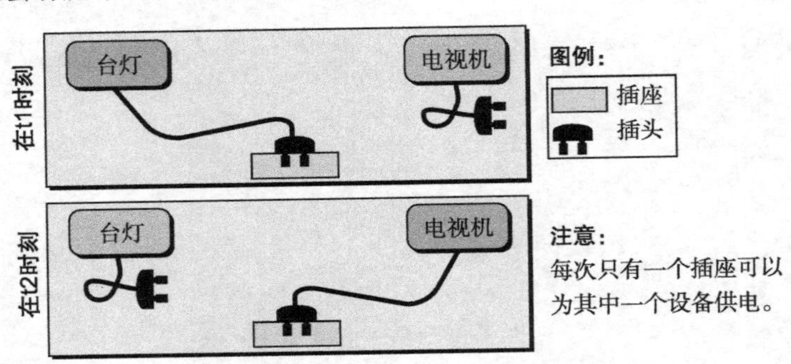

图 12-1　一个插座与多个插头兼容设备

假设我们有两个电气设备（例如台灯和电视机）。我们只有一个插座，每次只能为其中一个设备供电。每个设备都有一个可以插入唯一插座的插头。在t1时刻，我们把台灯插入插座。在t2时刻，我们拔下台灯，把电视机插到插座上。我们可以这样做，因为这两个设备是插头兼容的，它们的插头遵循相同的标准。关于插头兼容设备，有趣的一点是所有设备都从插座获得相同的东西（电力），但每个设备都执行不同的任务。

12.1.1 多态性的条件

我们通过将多态性与电气插头相比较来定义继承中多态性的条件。

指针或者引用

为了进行比较，我们需要一些能够扮演插座角色的东西，一旦创建了插座，就可以接收插头兼容的对象。在C++中，指针或者引用可以扮演这个角色。我们可以定义一个指向基类的指针（或者引用），然后我们可以让指针指向继承层次结构中的任何对象。因此，指针和引用变量有时被称为多态变量。

可交换对象

我们需要发挥设备作用的插头兼容对象。继承层次结构中的对象扮演此角色。

虚函数

我们需要一些东西来扮演电源的角色，提供给执行不同任务的不同设备。C++中的虚函数可以胜任此角色。虚函数通过关键字virtual进行修饰。例如，我们可以在所有类中都有一个print函数，所有类的打印函数名称都相同（print），但每个函数打印的内容都不同。

对于多态性，我们需要指针（或者引用）、可交换的对象以及虚函数。

【例12-1】 程序12-1演示了仅使用前两个条件的不完整的多态性程序。我们定义了两个类。然后我们创建了一个指针（模拟一个插座），该指针可以在不同的时间接收每个类的一个对象（插头兼容的对象）。为了简单起见，我们只为每个类定义一个公共成员函数，并让系统添加默认的构造函数。

程序12-1 一个不完整的多态性程序

```
1   /***************************************************************
2    * 一个简单的程序                                              *
3    * 演示多态性的前两个条件                                      *
4    ***************************************************************/
5   #include <iostream>
6   #include <string>
7   using namespace std;
8
9   // Base 类的定义，以及内联的 print 函数
10  class Base
11  {
12    public:
13        void print () const {cout << "In the Base" << endl;}
14  };
15  // Derived 类的定义，以及内联的 print 函数
16  class Derived : public Base
17  {
```

```
18  public:
19      void print () const {cout << "In the Derive" << endl;}
20  };
21
22  int main ( )
23  {
24      // 创建一个指向 Base 类的指针（模拟插座）
25      Base* ptr;
26      // 让 ptr 指向一个 Base 类的对象
27      ptr = new Base ();
28      ptr -> print();
29      delete ptr;
30      // 让 ptr 指向一个 Derived 类的对象
31      ptr = new Derived();
32      ptr -> print();
33      delete ptr;
34      return 0;
35  }
运行结果：
In the Base
In the Base
```

在第 27 行，ptr 指向 Base 类的对象，在第 28 行，我们调用在 Base 类中定义的函数。在第 31 行，我们将相同的指针指向 Derived 类的对象，在第 32 行，我们试图调用 Derived 类中定义的函数，但结果表明调用了 Base 类中定义的函数。程序实现了多态性的前两个条件，但第三个条件（虚函数）没有实现。打印函数不是虚函数。

结果应该是符合预期的，因为变量 ptr 被定义为指向 Base 的指针。它可以指向 Derived 类型的对象，因为 Derived 对象是 Base 对象（继承定义了 is-a 关系）。但是，当变量 ptr 想要调用 print 函数时，它仍然是指向 Base 的指针，因此它调用在 Base 类中定义的 print 函数。我们没有更改指针的类型，我们只是强制指针指向 Derived 类。

【例 12-2】 我们重写前面的例子，但是我们使 print 函数成为虚函数。结果是每次调用都能激活正确的函数，如程序 12-2 所示。在第 13 行和第 19 行中，我们将修饰符 virtual 添加到 print 函数。第 28 行为 Base 类调用适当的 print 函数，第 32 行为 Derived 类调用适当的 print 函数。

程序 12-2　一个正确的多态性程序，使用了所有三个条件

```
1   /***************************************************************
2    * 一个简单的程序                                              *
3    * 演示了如果满足所有三个必要条件，则多态性成立                *
4    ***************************************************************/
5   #include <iostream>
6   #include <string>
7   using namespace std;
8
9   // Base 类的定义，以及内联的 print 函数
10  class Base
11  {
12      public:
```

```
13        virtual void print () const {cout << "In the Base" << endl;}
14    };
15    // Derived 类的定义，以及内联的 print 函数
16    class Derived : public Base
17    {
18      public:
19        virtual void print () const {cout << "In the Derive" << endl;}
20    };
21
22    int main ( )
23    {
24      //创建一个指向 Base 类的指针（模拟插座）
25      Base* ptr;
26      //让 ptr 指向一个 Base 类的对象
27      ptr = new Base ();
28      ptr -> print();
29      delete ptr;
30      //让 ptr 指向一个 Derived 类的对象
31      ptr = new Derived();
32      ptr -> print();
33      delete ptr;
34      return 0;
35    }
```

```
运行结果：
In the Base
In the Derived
```

virtual 修饰符并不是必需的

虽然我们已经将 virtual 修饰符添加到两个 print 函数，但这并不是必需的。当一个函数被定义为虚函数时，类层次结构中具有相同签名的所有函数都会自动成为虚函数。

机制

为了理解虚函数如何参与多态行为，我们必须了解虚拟表（vtable）。在多态性中，系统为类层次结构中的每个类创建一个虚拟表。虚拟表中的每个条目都有一个指向相应虚函数的指针。在应用程序中创建的每个对象都将有一个额外的数据成员，该成员是指向对应的虚拟表的指针。在我们的简单程序中，有两个对象和两个虚拟表，每个虚拟表都有一个条目，如图 12-2 所示。

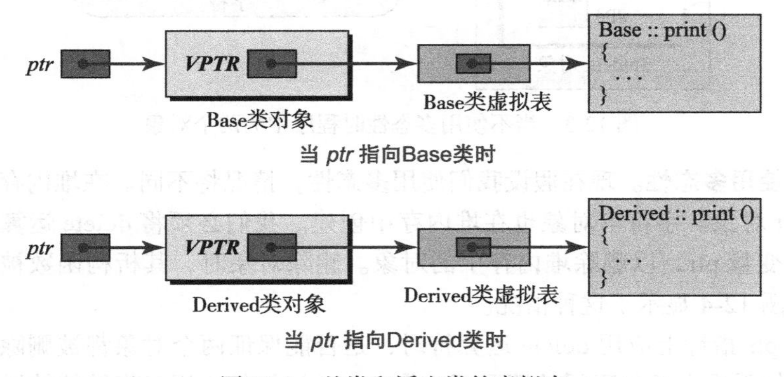

图 12-2　基类和派生类的虚拟表

当ptr指向Base对象时，将使用添加到Base类对象的VPTR指针，该指针指向Base类的vtable。在这种情况下，vtable只有一个条目，它调用Base类中唯一的虚函数。当ptr指向Derived类对象时，将使用Derived对象的vtable，并调用Derived类中定义的print函数。

12.1.2　构造函数和析构函数

在类层次结构中构造函数和析构函数同样是成员函数，虽然是特殊类型的成员函数。它们是否需要被声明为虚函数呢？下面我们分别讨论构造函数和析构函数的情况。

构造函数不能为虚函数

构造函数不能是虚函数，尽管构造函数是成员函数，但基类和派生类的构造函数名称是不同的（不同的签名）。

虚析构函数

尽管析构函数的名称在基类和派生类中有所不同，但通常不按其名称调用析构函数。当设计中任何地方有一个虚拟成员函数时，我们也应该使析构函数成为虚函数，以避免内存泄漏。为了理解这种问题，我们讨论了两种情况：当我们不使用多态性时；当我们使用多态性时。

情况1：不使用多态性。假定基类在堆内存中分配了一个字符串类型的数据成员。由于派生类继承了字符串数据成员，因此派生类在堆内存中也分配了一个数据成员。图12-3显示了我们不使用多态性的情况。我们创建了一个Person类和一个Student类。Person类有一个名为name的字符串类型的数据成员，它在堆内存中创建。Student类从Person类继承了name，但Student类还添加了另一个数据成员gpa。

在这种情况下，不存在内存泄漏。当程序终止时，会调用Person类和Student类的析构函数，这些析构函数会自动调用string类的析构函数，从而删除堆中分配的内存。

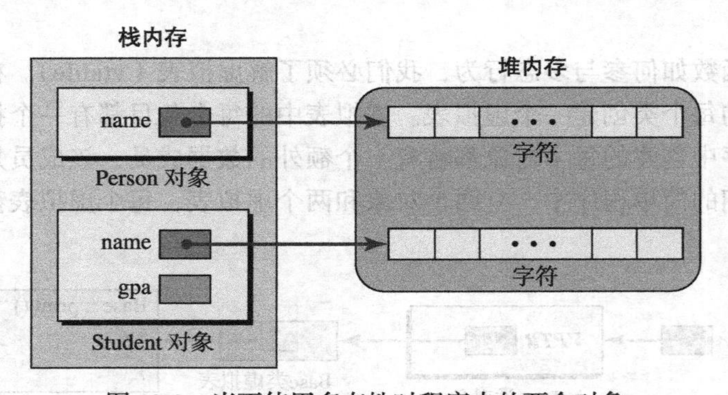

图12-3　当不使用多态性时程序中的两个对象

情况2：使用多态性。现在假设我们使用多态性，情况将不同。在堆内存中创建Person对象和Student对象。字符串对象也在堆内存中创建。我们必须将delete运算符应用于栈内存中的多态性变量ptr，以删除堆内存中的对象。删除对象时，其析构函数被调用，字符串对象被删除。图12-4显示了这种情况。

当我们在ptr指针上应用delete运算符时，是否能保证两个对象都被删除？为了回答这个问题，让我们看看在两种不同的情况下，将delete运算符应用于指针的结果。

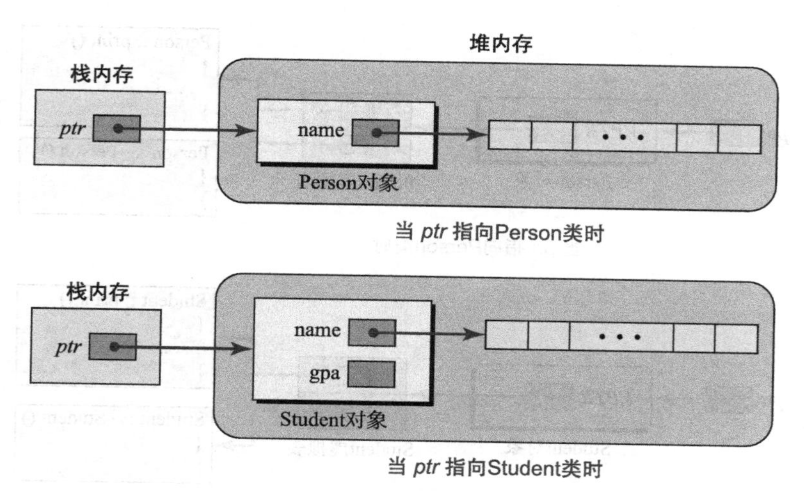

图 12-4 当使用多态性时程序中的两个对象

```
ptr = new Person (...);
...
delete ptr;          //删除 Person 对象，因为 ptr 的类型为 Person*
```

```
ptr = new Student (...);
...
delete ptr;          //不会删除 Student 对象，因为 ptr 的类型为 Person*
```

在第一种情况下，指针 ptr 是指向 Person 类型的指针，delete 运算符可以删除 Person 对象。当 Person 对象被删除时，Person 对象的析构函数被调用，而该析构函数又调用字符串类的析构函数。堆中创建的字符被取消分配。没有内存泄漏。

在第二种情况下，指针 ptr 仍然是指向 Person 类型的指针，这意味着指针 ptr 可以删除 Person 类的对象（但是不存在，因此不执行任何操作），但不能删除 Student 类的对象。如果 Student 类的对象未被删除，则其析构函数不会被调用，这意味着不会调用字符串类的析构函数，即不会释放堆内存中的字符串。结果导致了内存泄漏。请注意，当程序超出作用域时，Student 对象占用的内存最终将被释放，但指向字符的指针将成为悬空指针，系统永远无法删除字符串中为存储字符所创建的内存位置。

解决方案是使基类的析构函数为虚函数，从而自动使派生类的析构函数成为虚函数。在这种情况下，系统允许两个具有不同名称的不同成员函数是虚函数，它们都被添加到虚拟表中。图 12-5 显示了虚拟表中有两个条目，当删除堆中的一个对象时，程序知道应该调用哪个析构函数。

C++ 建议我们总是为多态性的基类定义一个显式析构函数，并使其成为虚函数。使用虚析构函数可以防止多态性中可能发生的内存泄漏。

【例 12-3】 本例演示如何使用多态性通过指向 Person 类的指针来打印 Student 类中的信息。

Person 类。 程序 12-3 显示了 Person 类的接口文件的代码清单。

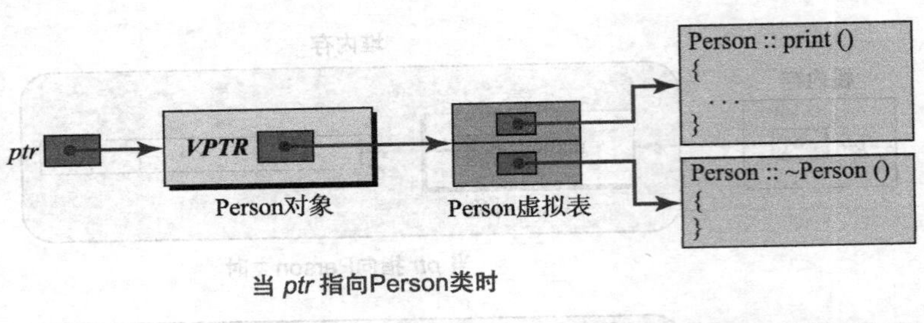

当 ptr 指向Person类时

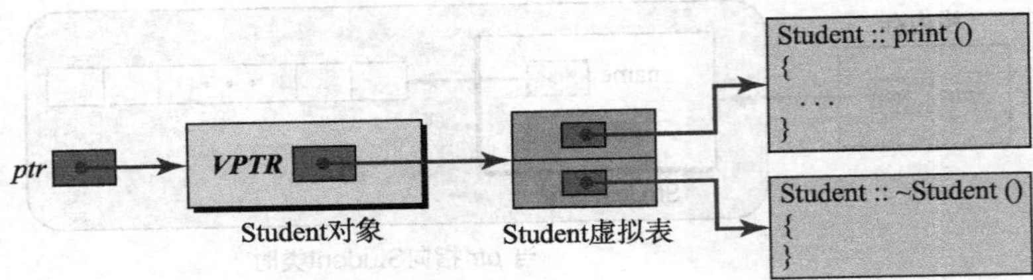

当 ptr 指向Student类时

图 12-5　当使用虚析构函数时的虚拟表

程序 12-3　文件 person.h

```
1   /*************************************************************
2    * Person 类的接口文件                                        *
3    *************************************************************/
4   #ifndef PERSON_H
5   #define PERSON_H
6   #include <iostream>
7   #include <string>
8   using namespace std;
9
10  class Person
11  {
12    private:
13          string name;
14    public:
15          Person (string name);
16          virtual ~Person ();
17          virtual void print () const;
18  };
19  #endif
```

程序 12-4 显示了 Person 类的实现文件的代码清单。

程序 12-4　文件 person.cpp

```
1   /*************************************************************
2    * Person 类的实现文件                                        *
3    *************************************************************/
4   #include "Person.h"
5
6   // Person 类的构造函数的定义
7   Person :: Person (string nm)
8   : name (nm)
9   {
```

```
10  }
11  // Person 类的析构函数的定义（虚函数）
12  Person :: ~Person ()
13  {
14  }
15  // print 函数的定义（虚函数）
16  void Person :: print () const
17  {
18     cout << "Name: " << name << endl;
19  }
```

Student 类。程序 12-5 显示了 Student 类的接口文件的代码清单。程序 12-6 显示了 Student 类的实现文件的代码清单。

程序 12-5　文件 student.h

```
1   /****************************************************************
2    * Student 类的接口文件                                         *
3    ****************************************************************/
4   #ifndef STUDENT_H
5   #define STUDENT_H
6   #include "person.h"
7
8   class Student: public Person
9   {
10    private:
11        double gpa;
12    public:
13        Student (string name, double gpa);
14        virtual void print () const;
15  };
16  #endif
```

程序 12-6　文件 student.cpp

```
1   /****************************************************************
2    * Student 类的实现文件                                         *
3    ****************************************************************/
4   #include "Student.h"
5
6   // Student 类的构造函数的定义
7   Student :: Student (string nm, double gp)
8   : Person (nm), gpa (gp)
9   {
10  }
11  // Student 类的 print 函数的定义（虚函数）
12  void Student :: print () const
13  {
14    Person :: print ();
15    cout << "GPA: " << gpa << endl;
16  }
```

应用程序。程序 12-7 显示了应用程序文件的代码清单。

程序 12-7 文件 app.cpp

```
1  /*************************************************************************
2   * 本应用程序测试 Person 类和 Student 类                                  *
3   *************************************************************************/
4  #include "Student.h"
5
6  int main ( )
7  {
8     // 创建一个指针 ptr，作为多态性变量
9     Person* ptr;
10    // 在堆内存中实例化 Person 对象
11    ptr = new Person ("Lucie");
12    cout << "Person Information";
13    ptr -> print();
14    cout << endl;
15    delete ptr;
16    // 在堆内存中实例化 Student 对象
17    ptr = new Student ("John", 3.9);
18    cout << "Student Information";
19    ptr -> print();
20    cout << endl;
21    delete ptr;
22    return 0;
23 }
```

运行结果：

```
Person Information
Name: Lucie
Student Information
Name: John
GPA: 3.9
```

更好地利用多态性

读者有可能注意到，在程序 12-3 中并不一定需要多态性。我们可以在不使用同一指针指向不同对象的情况下获得相同的结果。我们可以使用 person.print( ) 来代替 ptr－>print()，我们也可以使用 student.print() 来代替 ptr－>print()。然而，程序中显示了多态性的概念，只需要一个指针，就可以指向不同的对象。

更好的示例是必须使用多态性的时候。假设我们需要一个对象数组，数组中的所有元素都必须是相同的类型，这意味着如果对象是不同的类型，我们就不能使用对象数组。但是，我们可以使用指针数组，其中每个指针都可以指向基类的对象（前面例子中的 Person）。换言之，我们可以创建一个多态性变量数组而不是一个多态性变量。程序 12-8 与程序 12-7 基本相同，区别在于我们使用了指针数组。

程序 12-8 在多态性中使用一个指针数组

```
1  /*************************************************************************
2   * 修改应用程序文件                                                        *
3   * 使用指针数组演示多态性的实际用途                                        *
4   *************************************************************************/
5  #include "student.h"
6
```

```
7   int main ( )
8   {
9     //声明一个多态性变量（指针）的数组
10    Person* ptr [4];
11    //在堆内存中实例化四个对象
12    ptr[0] = new Student ("Joe", 3.7);
13    ptr[1] = new Student ("John", 3.9);
14    ptr[2] = new Person ("Bruce");
15    ptr[3] = new Person ("Sue");
16    //为每一个对象调用虚函数 print
17    for (int i = 0; i < 4; i++)
18    {
19        ptr[i] -> print ();
20        cout << endl;
21    }
22    //删除堆内存中的对象
23    for (int i = 0; i < 4; i++)
24    {
25        delete ptr [i];
26    }
27    return 0;
28  }
```

```
运行结果：
Name: Joe
GPA: 3.70

Name: John
GPA: 3.90

Name: Bruce

Name: Sue
```

程序 12-8 的虚拟表如图 12-6 所示，其中有四个对象，每个类型有两个虚函数。

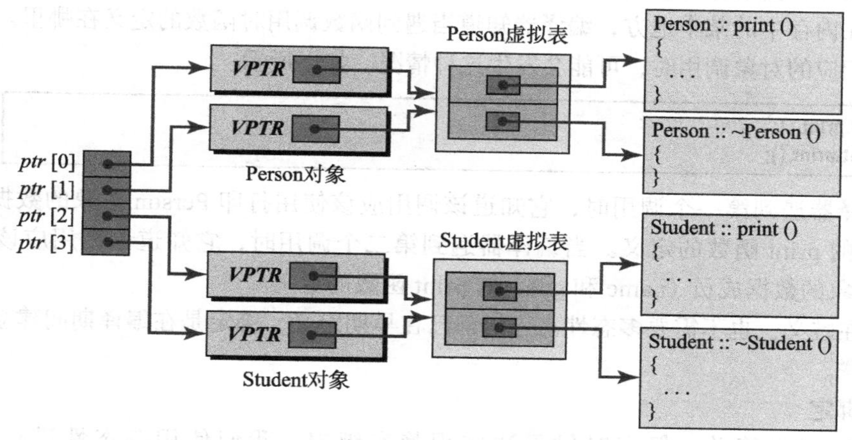

图 12-6　程序 12-8 中各对象的虚拟表

在图 12-6 中，每个类只有一个虚拟表，但是 Person 类型的两个对象都有一个指向 Person

虚拟表的 VPTR 指针，而 Student 类的两个对象都有一个指向 Student 虚拟表的 VPTR 指针。

其他语言中的多态性

如果读者了解其他面向对象的程序设计语言，你可能会想，为什么在其他语言中找不到上述所有条件，即使在这些程序设计语言的讨论中经常提到多态性。这三个条件总是存在于其他语言中，但它们可能是隐藏的。例如，在 Java 语言中（在 C++ 之后开发设计出来的程序设计语言），多态性的发生有以下三个条件：

1）Java 支持公共继承（事实上，Java 只有一种继承类型，即 public），这意味着我们可以通过继承来交换对象。

2）在 Java 中，虽然在用户级别中没有指针，但是类对象总是在堆内存中创建，并且有指向这些对象的变量（称为引用类型）。每个类对象都可以通过多态变量（插座）来访问。

3）Java 中的所有成员函数默认都是虚函数，这意味着多态函数存在的第三个条件也得到了满足。

12.1.3　绑定

绑定（binding）是一个与多态性有关的问题。虽然不需要额外的代码来确保绑定是正确的，但是我们应该理解绑定会发生什么现象。如前几章所述，函数分为两个实体：函数调用和函数定义。这里的绑定表示函数调用（例如 print()）和函数体（例如 void print {…}）之间的关联。

我们知道函数定义是在某个地方创建的，函数调用则发生在另一个地方。如果一个函数只有一个定义，就不会发生混淆。无论何时调用函数，都会执行相应的定义。

然而我们在继承中看到，可能存在两个具有相同签名的函数（重写函数）。这意味着函数只有一种调用形式，但可能有多个定义。如果一个 Person 对象调用了 print 函数，则执行一个定义；如果一个 Student 对象调用了 print 函数，则执行另一个定义。这里的绑定是指程序如何将函数调用绑定（关联）到函数定义。绑定分为两种情况：静态绑定和动态绑定。

静态绑定

静态绑定（static binding，有时称为编译时绑定或者早期绑定）一词出现在一个函数包含多个定义时，但编译器在编译程序时知道要使用哪个版本的定义。我们知道函数的每个定义都存储在内存中的某个地方，编译器知道当遇到函数调用时函数的定义在哪里。例如，当函数由其相应的对象调用时，可能会发生这种情况，如下所示。

```
person.print ();
student.print ();
```

当编译器遇到第一个调用时，它知道该调用应该使用打印 Person 对象的数据成员（只有 name）的 print 函数的定义。当编译器遇到第二个调用时，它知道该调用应该使用打印 Student 对象的数据成员（name 和 gpa）的 print 函数的定义。

不存在歧义，也不需要多态性。一切都已在早期定义。绑定是在编译期间建立的，编译器建立关联。

动态绑定

静态绑定很简单，但有时候无法实现静态绑定。我们使用多态性进行**动态绑定**（dynamic binding，也称为后期绑定或者运行时绑定），这意味着我们必须在运行时将调用绑定到相应定义。当编译期间对象未知时，需要进行动态绑定。例如，在程序执行期间，当不同的对象被插入一个多态变量中，并且程序需要调用适当的函数时，那么绑定必须在运行时完成。因此，我们需要一个虚函数来强制运行时系统创建一个虚拟表，显示哪个对象需要哪个函数，并将调用绑定到适当的函数。多态性与动态绑定紧密相连，因为我们希望能够执行适当的函数定义。

12.1.4 运行时类型信息

在处理类的层次结构时，有时我们需要知道正在处理的对象的类型，有时我们想要更改对象的类型。

使用 typeid 运算符

如果我们想在运行时获取对象的类型，可以使用 <typeinfo> 头文件来访问类 type_info 的对象（注意类名有下划线，但是头文件名没有）。类 type_info 没有构造函数、析构函数或者拷贝构造函数。创建类型为 type_info 的对象的唯一方法是使用名为 typeid 的重载运算符。我们可以通过将表达式传递给运算符 typeid（其求值结果为一个类型）来创建一个 type_info 对象。例如，typeid（5）、typeid（object_name）、typeid（6+2），等等。然后，我们可以使用 type_info 类中定义的四个成员函数或者运算符，如表 12-1 所示，其中 t1 和 t2 是 type_info 类的对象。

表 12-1 type_info 对象的操作

t1 == t2	// 如果 t1 和 t2 是同一类型，则返回 true
t1 != t2	// 如果 t1 和 t2 是不同类型，则返回 true
t1.name ( )	// 返回一个 C 字符串（t1 的名称）
t1.before (t2)	// 如果 t1 在 t2 前面（在继承关系中），则返回 true

程序 12-9 演示了如何获取两个对象的类型。

程序 12-9 测试 typeid 运算符

```
1  /***************************************************************
2   * 本应用程序使用 typeid 运算符获取类的名称                        *
3   ***************************************************************/
4  #include <iostream>
5  #include <string>
6  #include <typeinfo>
7  using namespace std;
8
9  class Animal {};
10 class Horse {};
11
12 int main ( )
13 {
14   Animal a;
15   Horse h;
16   cout << typeid(a).name() << endl;
17   cout << typeid(h).name();
18   return 0;
19 }
```

运行结果：

```
6Animal
5Horse
```

注意，类的名称前面是每种情况下的字符数（6和5）。

使用 dynamic_cast 运算符

我们已经看到，在多态性关系中，我们可以向上转换（upcast）一个指针，这意味着我们在派生类中创建一个指针，该指针指向基类，如下所示：

```
Person* ptr1 = new Student
```

在上述代码片段中，new 运算符返回的指针是指向 Student 对象的指针，但是我们将其分配给指向 Person 对象的指针（指针是向上转换的）。

C++ 还允许我们向下转换（downcast）一个指针，使指针指向层次较低的对象。这可以使用 dynamic_cast 运算符来完成，如下所示：

```
Student* ptr2 = dynamic_cast <Student*) (ptr1)
```

这种强制转换证明了 Student 类是从 Person 类派生的类，因为 ptr1 可以向下转换为 ptr2。一些程序员认为这是一种类型检查的形式。但是，我们不建议程序员使用 dynamic_cast 运算符进行类型检查，因为涉及大量系统开销。

12.2 其他问题

在本节中，我们将讨论抽象类和多重继承。我们介绍一个使用多层继承类的项目。

12.2.1 抽象类

到目前为止，我们设计的类称为具体类。**具体类**（concrete class）可以实例化并创建该类型的对象。当我们创建一组类时，有时会发现所有类都具有一组相同的行为列表。例如，假设我们定义了两个名为 Rectangle 和 Square 的类。这两个类至少有两个共同的行为：getArea() 和 getPerimeter()。我们如何强制这两个类的创建者（特别是当每个类由不同的实体创建时）为每个类提供两个成员函数的定义？我们知道，计算这些几何图形的面积和周长的公式是不同的，这意味着每个类必须创建自己的 getArea 和 getPerimeter 版本。

面向对象程序设计的解决方案是创建一个**抽象类**（abstract class），强制所有派生类的设计者将这两个定义添加到类中。具有一个抽象类的一组类必须具有纯虚函数的声明和定义。

抽象类是具有至少一个纯虚函数的类。

纯虚函数的声明

抽象类是具有至少一个纯虚函数的类。**纯虚函数**（pure virtual function）是一个虚函数，其声明被赋值为零，在抽象类中没有定义。下面显示了 Shape 类的两个纯虚成员函数：

```
virtual double getArea (0) = 0;
virtual double getPerimeter (0) = 0;
```

纯虚函数的定义

抽象类不定义其纯虚函数，但继承自抽象类的每个类必须提供每个纯虚函数的定义，或者将其声明为纯虚函数，以便在继承层次结构的下一个较低级别中定义。

无法实例化

我们无法从抽象类实例化对象，因为抽象类没有其纯虚函数的定义。对于要从某个类实例化的对象，该类必须具有所有成员函数的定义。这意味着，如果要为实例化定义具体类，则抽象类必须是多态继承的。

> 无法实例化抽象类，因为抽象类中没有纯虚成员函数的定义。

接口

抽象类可以同时具有虚函数和纯虚函数。然而，在某些情况下，我们可能需要为继承的类创建一个蓝图。我们可以定义一个全是纯虚函数的类。这个类有时被称为接口，我们不能从这个类中创建任何实现文件，只能创建接口文件。

> 接口是抽象类的一种特殊情况，其中所有成员函数都是纯虚函数。

Shape（形状）类

为了演示抽象类的使用，我们创建了五个具体类来表示形状。所有类都继承自抽象类Shape，如图 12-7 所示。

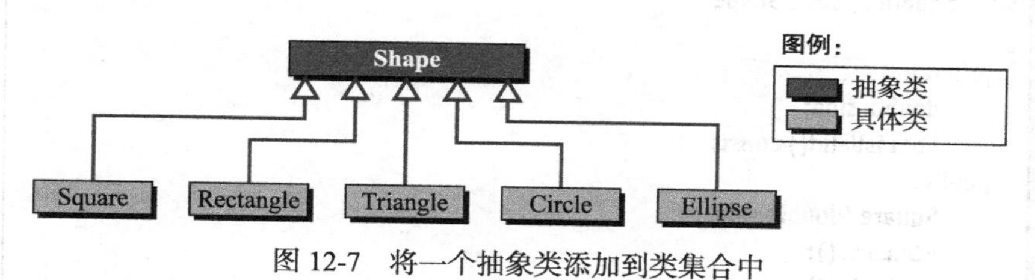

图 12-7　将一个抽象类添加到类集合中

程序 12-10 是 Shape 类的接口文件。

注意，这个程序中没有数据成员，只有纯虚成员函数。原因是我们不希望从此类实例化对象，其目的只是强制派生类实现纯虚成员函数。强制所有类实现的第一个纯虚成员函数是 isValid() 成员函数，它用于在对象构造期间验证其数据成员。强制每个派生类实现的另外两个纯虚成员函数是计算相应形状的面积和周长。

程序 12-10　文件 shape.h

```
1   /*****************************************************************
2    * 抽象类 Shape 的接口文件                                        *
3    *****************************************************************/
4   #ifndef SHAPE_H
5   #define SHAPE_H
6   #include <iostream>
7   #include <cassert>
8   #include <cmath>
9   using namespace std;
10
11  // 类定义
12  class Shape
13  {
14    protected:
15          virtual bool isValid () const = 0;
```

```
16    public:
17        virtual void print () const = 0 ;
18        virtual double getArea () const = 0 ;
19        virtual double getPerimeter () const = 0;
20  };
21  #endif
```

程序 12-11 是 Square 类的接口文件。它只有一个数据成员、一个用于验证唯一数据成员的私有成员函数、三个公共成员函数。

程序 12-11　文件 square.h

```
1   /***************************************************************
2    * Square 类的接口文件                                          *
3    ***************************************************************/
4   #ifndef SQUARE_H
5   #define SQUARE_H
6   #include "shape.h"
7   
8   // 类定义
9   class Square : public Shape
10  {
11    private:
12        double side;
13        bool isValid() const;
14    public:
15        Square (double side);
16        ~Square ();
17        void print() const;
18        double getArea () const;
19        double getPerimeter () const;
20  };
21  #endif
```

程序 12-12 是 Square 类的实现文件的代码清单。

程序 12-12　文件 square.cpp

```
1   /***************************************************************
2    * Square 类的实现文件                                          *
3    ***************************************************************/
4   #include "square.h"
5   
6   // 构造函数
7   Square :: Square (double s)
8   :side (s)
9   {
10    if (!isValid ())
11    {
12        cout << "Invalid square!";
13        assert (false);
14    }
15  }
16  // 析构函数
```

```
17  Square :: ~Square ()
18  {
19  }
20  // print 函数的定义
21  void Square :: print () const
22  {
23    cout << "Square of side " << side << endl;
24  }
25  //计算面积
26  double Square :: getArea () const
27  {
28    return (side * side);
29  }
30  //计算周长
31  double Square :: getPerimeter () const
32  {
33    return (4 * side);
34  }
35  //私有函数 isValid
36  bool Square :: isValid () const
37  {
38    return (side > 0.0);
39  }
```

程序 12-13 是 Rectangle 类的接口文件。它有两个数据成员、一个私有成员函数、三个公共成员函数。

程序 12-13　文件 rectangle.h

```
1   /***********************************************************************
2    * Rectangle 类的接口文件                                              *
3    ***********************************************************************/
4   #ifndef RECTANGLE_H
5   #define RECTANGLE_H
6   #include "shape.h"
7
8   //类定义
9   class Rectangle : public Shape
10  {
11    private:
12          double length;
13          double width;
14          bool isValid() const;
15    public:
16          Rectangle (double length, double width);
17          ~Rectangle ();
18          void print () const;
19          double getArea() const;
20          double getPerimeter() const;
21  };
22  #endif
```

程序 12-14 是 Rectangle 类的实现文件的代码清单。

程序 12-14　文件 rectangle.cpp

```
1   /**************************************************************
2    * Rectangle 类的实现文件                                        *
3    **************************************************************/
4   #include "rectangle.h"
5
6   // 构造函数
7   Rectangle :: Rectangle (double lg, double wd)
8   : length (lg), width (wd)
9   {
10    if (!isValid())
11    {
12        cout << "Invalid rectangle!";
13        assert (false);
14    }
15  }
16  // 析构函数
17  Rectangle :: ~Rectangle ()
18  {
19  }
20  // print 函数的定义
21  void Rectangle :: print () const
22  {
23    cout << "Rectangle of " << length << " X " << width << endl;
24  }
25  // 计算面积
26  double Rectangle :: getArea() const
27  {
28    return length * width;
29  }
30  // 计算周长
31  double Rectangle :: getPerimeter() const
32  {
33    return 2 * (length + width);
34  }
35  // 私有函数 isValid
36  bool Rectangle :: isValid () const
37  {
38    return (length > 0.0 && width > 0.0);
39  }
```

程序 12-15 是 Triangle 类的接口文件。它有三个数据成员、一个私有成员函数、三个公共成员函数。

程序 12-15　文件 triangle.h

```
1   /**************************************************************
2    * Triangle 类的接口文件                                         *
3    **************************************************************/
4   #ifndef TRIANGLE_H
5   #define TRIANGLE_H
6   #include "shape.h"
```

```
7
8  // 类定义
9  class Triangle : public Shape
10 {
11   private:
12        double side1;
13        double side2;
14        double side3;
15        bool isValid () const;
16   public:
17        Triangle (double side1, double side2, double side3);
18        ~Triangle ();
19        void print() const;
20        double getArea() const;
21        double getPerimeter() const;
22 };
23 #endif
```

程序 12-16 是 Triangle 类的实现文件的代码清单。

程序 12-16　文件 triangle.cpp

```
1  /*******************************************************************
2   * Triangle 类的实现文件                                          *
3   *******************************************************************/
4  #include "triangle.h"
5
6  // 构造函数
7  Triangle :: Triangle (double s1, double s2, double s3)
8  : side1(s1), side2(s2), side3 (s3)
9  {
10   if (!isValid())
11   {
12        cout << "Invalid triangle!";
13        assert (false);
14   }
15 }
16 // 析构函数
17 Triangle :: ~Triangle ()
18 {
19 }
20 // print 函数的定义
21 void Triangle :: print() const
22 {
23   cout << "Triangle of : " << side1 << " X " << side2 << " X ";
24   cout << side3 << endl;
25 }
26 // 计算面积
27 double Triangle :: getArea() const
28 {
29   double s = (side1 + side2 + side3) / 2;
30   return (sqrt (s * (s - side1) * (s - side2) * (s - side3)));
31 }
```

```
32 //计算周长
33 double Triangle :: getPerimeter() const
34 {
35   return (side1 +  side2 + side3);
36 }
37 //私有函数 isValid
38 bool Triangle :: isValid () const
39 {
40   bool fact1 = (side1 + side2) > side3;
41   bool fact2 = (side1 + side3) > side2;
42   bool fact3 = (side2 + side3) > side1;
43   return (fact1 && fact2 && fact3);
44 }
```

程序 12-17 是 Circle 类的接口文件。它只有一个数据成员（半径 radius）、一个私有成员函数、三个公共成员函数。

程序 12-17　文件 circle.h

```
1  /***************************************************************
2   * Circle 类的接口文件                                          *
3   ***************************************************************/
4  #ifndef CIRCLE_H
5  #define CIRCLE_H
6  #include "shape.h"
7
8  // 类定义
9  class Circle : public Shape
10 {
11   private:
12         double radius;
13         bool isValid () const;
14   public:
15         Circle (double radius);
16         ~Circle ();
17         void print() const;
18         double getArea() const;
19         double getPerimeter() const;
20 };
21 #endif
```

程序 12-18 是 Circle 类的实现文件的代码清单。

程序 12-18　文件 circle.cpp

```
1  /***************************************************************
2   * Circle 类的实现文件                                          *
3   ***************************************************************/
4  #include "circle.h"
5
6  // 构造函数
7  Circle :: Circle (double r)
8  : radius (r)
9  {
```

```
10   if (!isValid())
11   {
12       cout << "Invalid circle!";
13       assert (false);
14   }
15 }
16 // 析构函数
17 Circle :: ~Circle ()
18 {
19 }
20 // print 函数的定义
21 void Circle :: print() const
22 {
23   cout << "Circle of radius : " << radius << endl;
24 }
25 // 计算面积
26 double Circle :: getArea() const
27 {
28   return (3.14 * radius * radius);
29 }
30 // 计算周长
31 double Circle :: getPerimeter() const
32 {
33   return 2 * 3.14 * radius;
34 }
35 // 私有函数 isValid
36 bool Circle :: isValid () const
37 {
38   return (radius > 0);
39 }
```

程序 12-19 是 Ellipse 类的接口文件。它类似于 Circle 类，但需要两个数据成员（两个半径）、一个私有成员函数、三个公共成员函数。

程序 12-19　文件 ellipse.h

```
1  /****************************************************************
2   * Ellipse 类的接口文件                                          *
3   ****************************************************************/
4  #ifndef ELLIPSE_H
5  #define ELLIPSE_H
6  #include "shape.h"
7
8  // 类定义
9  class Ellipse : public Shape
10 {
11   private:
12       double radius1;
13       double radius2;
14       bool isValid () const;
15   public:
16       Ellipse (double radius1, double radius2);
17       ~Ellipse ();
```

```
18        void print() const;
19        double getArea () const;
20        double getPerimeter () const;
21  };
22  #endif
```

程序 12-20 是 Ellipse 类的实现文件的代码清单。

程序 12-20　文件 ellipse.cpp

```
1   /***************************************************************
2    * Ellipse 类的实现文件                                        *
3    ***************************************************************/
4   #include "ellipse.h"
5
6   //构造函数
7   Ellipse :: Ellipse (double r1, double r2)
8   : radius1 (r1), radius2 (r2)
9   {
10    if (!isValid())
11    {
12        cout << "Invalid ellipse!";
13        assert (false);
14    }
15  }
16  //析构函数
17  Ellipse :: ~Ellipse ()
18  {
19  }
20  //print 函数的定义
21  void Ellipse :: print() const
22  {
23    cout << "Ellipse of radii: " << radius1 << " X " <<;
24    cout << radius2 << endl;
25  }
26  //计算面积
27  double Ellipse :: getArea () const
28  {
29    return (3.14 * radius1 * radius2);
30  }
31  //计算周长
32  double Ellipse ::getPerimeter () const
33  {
34    double temp = (radius1 * radius1 + radius2 * radius2) / 2;
35    return (2 * 3.14 * temp);
36  }
37  //私有函数 isValid
38  bool Ellipse :: isValid () const
39  {
40    return (radius1 > 0 && radius2 > 0);
41  }
```

程序 12-21 是测试上述类的对象实例的应用程序。

程序 12-21　文件 app.cpp

```
1  /*****************************************************************
2   * 本应用程序测试上述所有的类                                      *
3   *****************************************************************/
4  #include "square.h"
5  #include "rectangle.h"
6  #include "triangle.h"
7  #include "circle.h"
8  #include "ellipse.h"
9
10 int main ( )
11 {
12    // 实例化并测试 Square 对象
13    cout << "Information about a square" << endl;
14    Square square (5);
15    square.print ();
16    cout << "area: " << square.getArea () << endl;
17    cout << "Perimeter: " << square.getPerimeter () << endl;
18    cout << endl;
19    // 实例化并测试 Rectangle 对象
20    cout << "Information about a rectangle" << endl;
21    Rectangle rectangle (5, 4);
22    rectangle.print ();
23    cout << "area: " << rectangle.getArea () << endl;
24    cout << "Perimeter: " << rectangle.getPerimeter () << endl;
25    cout << endl;
26    // 实例化并测试 Triangle 对象
27    cout << "Information about a triangle" << endl;
28    Triangle triangle (3, 4, 5);
29    triangle.print ();
30    cout << "area: " << triangle.getArea () << endl;
31    cout << "Perimeter: " << triangle.getPerimeter () << endl;
32    cout << endl;
33    // 实例化并测试 Circle 对象
34    cout << "Information about a circle" << endl;
35    Circle circle (5);
36    circle.print ();
37    cout << "area: " << circle.getArea () << endl;
38    cout << "Perimeter: " << circle.getPerimeter () << endl;
39    cout << endl;
40    // 实例化并测试 Ellipse 对象
41    cout << "Information about an ellipse" << endl;
42    Ellipse ellipse (5, 4);
43    ellipse.print ();
44    cout << "area: " << ellipse.getArea () << endl;;
45    cout << "Perimeter: " << ellipse.getPerimeter ()<< endl;
46    return 0;
47 }
```

运行结果：

```
Information about a square
Square of size 5
```

```
area: 25
Perimeter: 20

Information about a rectangle
Rectangle of 5 X 4
area: 20
Perimeter: 18

Information about a triangle
Triangle of : 3 X 4 X 5
area: 6
Perimeter: 12

Information about a circle
Circle of radius : 5
area: 78.5
Perimeter: 31.4

Information about an ellipse
Ellipse of radii: 5 X 4
area: 62.8
Perimeter 28.4339
```

12.2.2 多重继承

C++ 允许**多重继承**，即从多个类派生出一个类。作为一个简单的示例，我们可以设计一个名为 TA（助教）的类，它继承自两个类：Student（学生）和 Professor（教师）（图 12-8）。图 12-8 显示了 UML 图和每个类的对象。

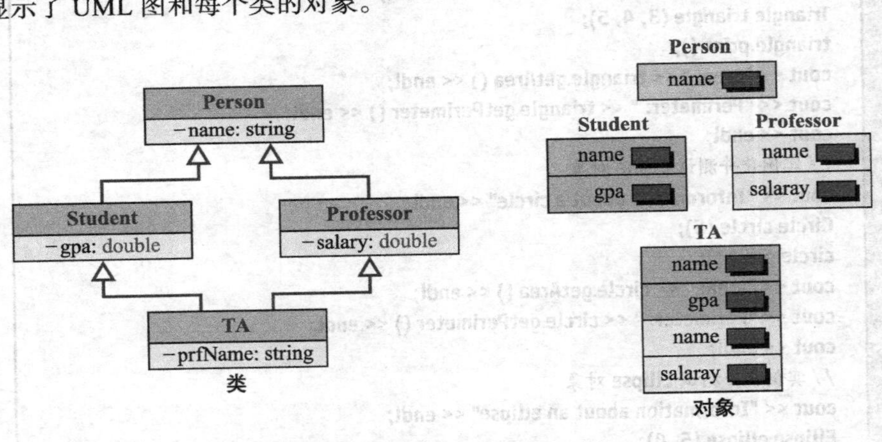

图 12-8 多重继承中的类和对象

不幸的是，当我们对这些类进行编码时，继承失败，因为 TA 类从 Student 类和 Professor 类继承数据成员 name。在 TA 类的对象中，我们有数据成员 name 的两个副本，这在 C++ 中是不允许的。

Person 类只定义了一个数据成员 name。此数据成员在 Student 对象和 Professor 对象中都被继承。由于 TA（助教）类继承了 Student 类和 Professor 类，因此数据成员 name 在 TA 类中重复了。在这种情况下，如果不删除重复的数据成员，则无法使用继承。

虚基继承

对于多重继承中重复的共享数据成员的问题的解决方案之一是使用**虚基继承**（virtual base inheritance）。在这种类型的继承中，两个类可以使用 virtual 关键字从公共基继承。图 12-9 显示了这种方法。

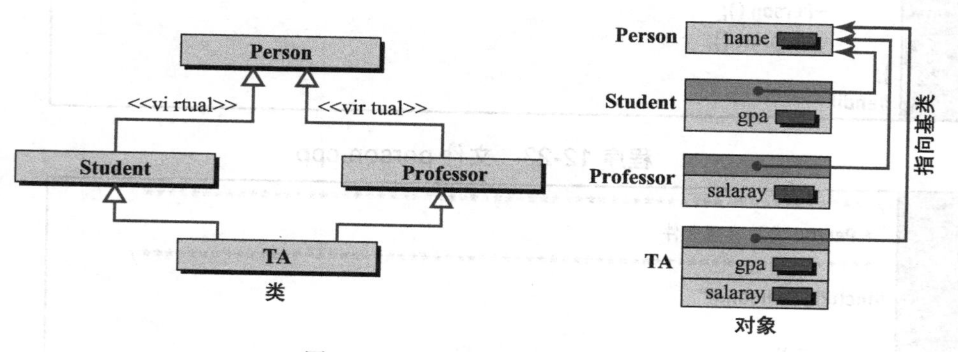

图 12-9　虚基继承中的类和对象

在这种情况下，存在以下四个类。

```
class Person {...};
class Student: virtual public Person {...};
class Professor: virtual public Person {...};
class TA: public Student, public Professor {...};
```

当我们使用虚基继承时，虚基类的对象不会存储在派生类的每个对象中。虚基类的对象单独存储，每个派生类都有指向此对象的指针。

在使用虚基技术时，我们必须避免使用委托（如第 11 章所述）。换言之，我们不能通过调用 Student 类和 Professor 类中相应的 print 函数来定义 TA 类中的 print 函数，因为数据成员 name 将被打印三次。解决这种困境的方法不止一种。我们建议的方法是设置共有的数据成员为受保护数据成员（protected），从而在所有的派生类中都可见，因此可以避免使用委托成员函数。

【例 12-4】 在本例中，我们给出了图 12-9 中类的接口文件、实现文件和应用程序文件。程序 12-22 显示了 Person 类的接口文件的代码清单。注意，数据成员 name 具有受保护的可访问性。

程序 12-23 显示了 Person 类的实现文件的代码清单。

程序 12-22　文件 person.h

```
1   /***************************************************************
2    * Person 类的接口文件                                           *
3    ***************************************************************/
4   #ifndef PERSON_H
5   #define PERSON_H
6   #include <iostream>
7   #include <cassert>
8   using namespace std;
9
10  class Person
```

```
11  {
12    protected:
13        string name; // 受保护的数据成员
14    public:
15        Person (string name);
16        ~Person ();
17        void print ();
18  };
19  #endif
```

程序 12-23 文件 person.cpp

```
1   /***************************************************************
2    * Person 类的实现文件                                          *
3    ***************************************************************/
4   #include "person.h"
5
6   // 构造函数
7   Person :: Person (string nm)
8   : name (nm)
9   {
10  }
11  // 析构函数
12  Person :: ~Person ()
13  {
14  }
15  // print 成员函数
16  void Person :: print ()
17  {
18    cout << "Person" << endl;
19    cout << "Name: " << name << endl << endl;
20  }
```

程序 12-24 显示了虚继承 Person 类的 Student 类的接口文件的代码清单。注意，数据成员 gpa 具有受保护的可访问性。

程序 12-25 显示了 Student 类的实现文件的代码清单。注意，print 函数没有调用 Person 类的 print 函数（没有使用委托）。

程序 12-24 文件 student.h

```
1   /***************************************************************
2    * Student 类的接口文件                                         *
3    ***************************************************************/
4   #ifndef STUDENT_H
5   #define STUDENT_H
6   #include "person.h"
7
8   class Student: virtual public Person   // 虚继承
9   {
10    protected:
11        double gpa; // 受保护的数据成员
12    public:
13        Student (string name, double gpa);
```

```
14        ~Student ();
15        void print ();
16  };
17  #endif
```

程序 12-25　文件 student.cpp

```
1   /*************************************************************
2    * Student 类的实现文件                                        *
3    *************************************************************/
4   #include "Student.h"
5
6   // 构造函数
7   Student :: Student (string name, double gp)
8   : Person (name), gpa (gp)
9   {
10    assert (gpa <= 4.0);
11  }
12  // 析构函数
13  Student :: ~Student()
14  {
15  }
16  // print 成员函数，使用受保护的数据成员（name）
17  void Student :: print ()
18  {
19    cout << "Student " << endl;
20    cout << "Name: " << name << " ";
21    cout << "GPA: " << gpa << endl << endl;
22  }
```

程序 12-26 显示了虚继承 Person 类的 Professor 类的接口文件的代码清单。注意，数据成员 salary 具有受保护的可访问性。

程序 12-27 显示了 Professor 类的实现文件的代码清单。注意，print 函数没有调用 Person 类的 print 函数（没有使用委托）。

程序 12-26　文件 professor.h

```
1   /*************************************************************
2    * Professor 类的接口文件                                      *
3    *************************************************************/
4   #ifndef PROFESSOR_H
5   #define PROFESSOR_H
6   #include "person.h"
7
8   class Professor: virtual public Person   // 虚继承
9   {
10    protected:
11        double salary; // 受保护的数据成员
12    public:
13        Professor (string name, double salary);
14        ~Professor ();
15        void print ();
```

```
16 };
17 #endif
```

程序 12-27 文件 professor.cpp

```
1  /***************************************************************
2   * Professor 类的实现文件                                        *
3   ***************************************************************/
4  #include "professor.h"
5
6  // 构造函数
7  Professor :: Professor (string nm, double sal)
8  : Person (nm), salary (sal)
9  {
10 }
11 // 析构函数
12 Professor :: ~Professor ()
13 {
14 }
15 // print 成员函数
16 void Professor :: print ()
17 {
18    cout << "Professor " << endl;
19    cout << "Name: " << name << "  ";
20    cout << "Salary: " << salary << endl << endl;
21 }
```

程序 12-28 显示了 TA 类的接口文件的代码清单。注意，该类没有数据成员。

程序 12-29 显示了 TA 类的实现文件的代码清单。注意，print 函数没有调用 Student 类的 print 函数，也没有调用 Professor 类的 print 函数（没有使用委托）。

程序 12-28 文件 ta.h

```
1  /***************************************************************
2   * TA 类的接口文件                                               *
3   ***************************************************************/
4  #ifndef TA_H
5  #define TA_H
6  #include "student.h"
7  #include "professor.h"
8
9  class TA: public Professor, public Student   // 双重继承
10 {
11    public:
12         TA (string name, double gpa, double sal);
13         ~TA ();
14         void print ();
15 };
16 #endif
```

程序 12-29 文件 ta.cpp

```
1  /***************************************************************
2   * TA 类的实现文件                                               *
```

```
3   ************************************************************************/
4   #include "ta.h"
5
6   // 构造函数
7   TA :: TA (string nm, double gp, double sal)
8   : Person (nm), Student (nm, gp), Professor (nm, sal)
9   {
10  }
11  // 析构函数
12  TA :: ~TA ()
13  {
14  }
15  // print 成员函数
16  void TA :: print ()
17  {
18    cout << "Teaching Assistance: " << endl;
19    cout << "Name: " << name << "  ";
20    cout << "GPA: " << gpa << "  ";
21    cout << "Salary: " << salary << endl << endl;
22  }
```

程序 12-30 显示了测试上述所有四个类的应用程序的代码清单。

程序 12-30 文件 application.cpp

```
1   /************************************************************************
2    * 本应用程序测试上述所有的四个类                                          *
3    *（Person、Student、Professor 和 TA）                                    *
4    ************************************************************************/
5   #include "ta.h"
6
7   int main ( )
8   {
9     // 测试 Person 类
10    Person person ("John");
11    person.print ();
12    // 测试 Student 类
13    Student student ("Anne", 3.9);
14    student.print ();
15    // 测试 Professor 类
16    Professor professor ("Lucie", 78000);
17    professor.print ();
18    // 测试 TA 类
19    TA ta ("George", 3.2, 20000);
20    ta.print ();
21    return 0;
22  }
运行结果：
Person
Name: John

Student
Name: Anne   GPA: 3.9
```

```
Professor
Name: Lucie   Salary: 78000

Teaching Assistance:
Name: George   GPA: 3.2   Salary: 20000
```

混入类（mixin class）

对于多重继承中的公共基类问题的另一个解决方案是使用 mixin 类。mixin 类永远不会被实例化（它包含一些纯虚函数），但 mixin 类可以向其他类添加数据成员。例如，我们可以将 Student 对象视为一个拥有额外数据成员（gpa）的 Person 对象。我们可以将 Professor 对象视为一个拥有额外数据成员（salary）的 Person 对象。我们还可以将 TA 对象视为一个拥有两个额外数据成员（gpa 和 salary）的 Person 对象。这些额外的数据成员可以通过 mixin 类的方式添加到这些类中，如图 12-10 所示。

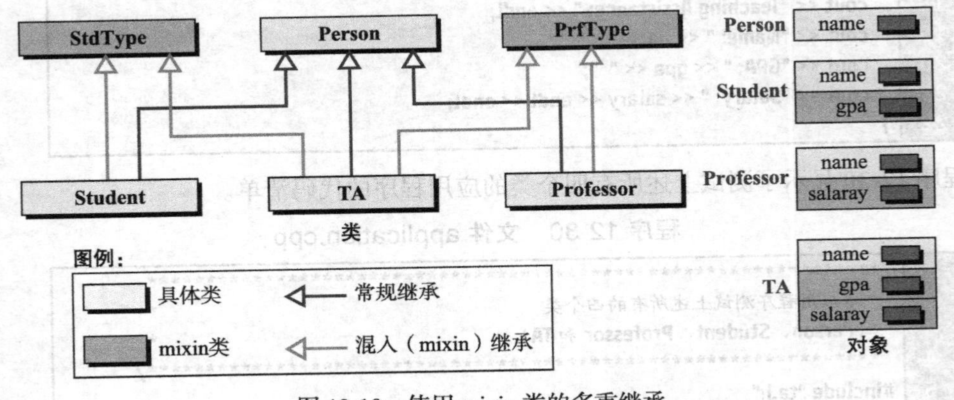

图 12-10　使用 mixin 类的多重继承

图 12-10 显示了一个学生、一个教师和一个助教都是 Person 类的对象实例。Student 对象有一个额外的资格（选修课程），Professor 对象有一个额外的资格（教授课程），TA 对象有两个资格（选修课程和教授课程）。注意，在这种情况下仍然存在多重继承（TA 类继承自 StdType 和 PrfType），但是没有公共的基类，StdType 和 PrfType 不会被实例化，因而不会产生我们提到的多重继承的问题。

在这里，可以使用一个帮助理解的类比。考虑一家卖一种冰激淋的冰激淋店，它允许顾客最多买两种配品。Person 对象没有任何配品，Student 对象或者 Professor 对象有一种配品，TA 对象则有两种配品。我们假设配品不能单独售卖，必须随冰激淋一起售卖。

【例 12-5】 在本例中，我们为图 12-10 所示的 mixin 类开发代码。我们给出了五个接口文件、三个实现文件和一个应用程序文件的代码。注意，mixin 类是抽象类（接口），没有实现。程序 12-31 显示了 StdType 类的接口文件的代码清单。

程序 12-31　文件 stdtype.h

```
1  /***************************************************************
2   * 抽象类 StdType 的接口文件                                      *
3   ***************************************************************/
4  #ifndef STDTYPE_H
5  #define STDTYPE_H
6  #include <iostream>
```

```
7   using namespace std;
8
9   class StdType
10  {
11    protected:
12        double gpa;
13    public:
14        virtual void printGPA( ) = 0 ;
15  };
16  #endif
```

程序 12-32 显示了 PrfType 类的接口文件的代码清单。

程序 12-32 文件 prftype.h

```
1   /****************************************************************
2    * 抽象类 PrfType 的接口文件                                   *
3    ****************************************************************/
4   #ifndef PRFTYPE_H
5   #define PRFTYPE_H
6   #include <iostream>
7   using namespace std;
8
9   class PrfType
10  {
11    protected:
12        double salary;
13    public:
14        virtual void printSalary () = 0;
15  };
16  #endif
```

程序 12-33 显示了 Person 类的接口文件的代码清单。

程序 12-33 文件 person.h

```
1   /****************************************************************
2    * 具体类 Person 的接口文件                                    *
3    ****************************************************************/
4   #ifndef PERSON_H
5   #define PERSON_H
6   #include <iostream>
7   #include <string>
8   #include <iomanip>
9   using namespace std;
10
11  class Person
12  {
13    private:
14        string name;
15    public:
16        Person (string name);
17        void print ();
18  };
19  #endif
```

程序 12-34 显示了 Student 类的接口文件的代码清单。注意，Student 类没有自己的数据成员，它仅继承 Person 类和 StdType 类的数据成员。

程序 12-34　文件 student.h

```
1  /*************************************************************************
2   * 具体类 Student 的接口文件                                              *
3   * 该类继承两个类：Person 和 StdType                                      *
4   *************************************************************************/
5  #ifndef STUDENT_H
6  #define STUDENT_H
7  #include "person.h"
8  #include "stdtype.h"
9
10 class Student: public Person, public StdType
11 {
12   public:
13       Student (string name, double gpa);
14       void printGPA();
15       void print();
16 };
17 #endif
```

程序 12-35 显示了 Professor 类的接口文件的代码清单。注意，Professor 类没有自己的数据成员，它仅继承 Person 类和 PrfType 类的数据成员。

程序 12-35　文件 professor.h

```
1  /*************************************************************************
2   * 具体类 Professor 的接口文件                                            *
3   * 该类继承两个类：Person 和 PrfType                                      *
4   *************************************************************************/
5  #ifndef PROFESSOR_H
6  #define PROFESSOR_H
7  #include "person.h"
8  #include "prftype.h"
9
10 class Professor : public Person, public PrfType
11 {
12   public:
13       Professor (string name, double salary);
14       void printSalary();
15       void print ();
16 };
17 #endif
```

程序 12-36 显示了 TA 类的接口文件的代码清单。注意，TA 类没有自己的数据成员，它仅继承 Person 类、StdType 类和 PrfType 类的数据成员。

程序 12-36　文件 ta.h

```
1  /*************************************************************************
2   * 具体类 TA 的接口文件                                                   *
3   * 该类继承三个类：Person、StdType 和 PrfType                             *
4   *************************************************************************/
```

```
5   #ifndef TA_H
6   #define TA_H
7   #include "person.h"
8   #include "stdtype.h"
9   #include "prftype.h"
10
11  class TA: public Person, public StdType, public PrfType
12  {
13    public:
14          TA (string name, double gpa, double salary);
15          void printGPA ();
16          void printSalary();
17          void print ();
18  };
19  #endif
```

程序 12-37 显示了 Person 类的实现文件的代码清单。

程序 12-37　文件 person.cpp

```
1   /*****************************************************************
2    * 具体类 Person 的实现文件                                        *
3    *****************************************************************/
4   #include "person.h"
5
6   // 构造函数
7   Person :: Person (string nm)
8   : name(nm)
9   {
10  }
11  // print 成员函数
12  void Person :: print ( )
13  {
14    cout << "Name: " << name << endl;
15  }
```

程序 12-38 显示了 Student 类的实现文件的代码清单。注意，Student 类继承了 StdType 类的受保护数据成员 gpa。该数据成员可以在 Student 类中访问但不能在该类中初始化，我们必须在构造函数中对其进行赋值。

程序 12-38　文件 student.cpp

```
1   /*****************************************************************
2    * 具体类 Student 的实现文件                                       *
3    *****************************************************************/
4   #include "student.h"
5
6   // 构造函数
7   Student :: Student (string na, double gp)
8   :Person (na)
9   {
10    gpa = gp;                          // 赋值，不是初始化
11  }
12  // printGPA 成员函数
```

```
13  void Student :: printGPA ()
14  {
15     cout << "GPA: " << fixed << setprecision (2) << gpa << endl;
16  }
17  // print 成员函数
18  void Student :: print ()
19  {
20     Person :: print();
21     printGPA ();
22  }
```

程序 12-39 显示了 Professor 类的实现文件的代码清单。注意，Professor 类继承了 PrfType 类的受保护数据成员 salary。该数据成员可以在 Professor 类中访问但不能在该类中初始化，我们必须在构造函数中对其进行赋值。

程序 12-39　文件 professor.cpp

```
1   /***************************************************************
2    * 具体类 Professor 的实现文件                                  *
3    ***************************************************************/
4   #include "professor.h"
5   
6   // 构造函数
7   Professor :: Professor (string nm, double sal)
8   : Person (nm)
9   {
10     salary = sal;                      // 赋值，不是初始化
11  }
12  // printSalary 成员函数
13  void Professor :: printSalary ()
14  {
15     cout << "Salary: ";
16     cout << fixed << setprecision (2) << salary << endl;
17  }
18  // 通用的 print 成员函数
19  void Professor :: print ()
20  {
21     Person :: print();
22     printSalary();
23  }
```

程序 12-40 显示了 TA 类的实现文件的代码清单。注意，TA 类继承了 PrfType 类的受保护数据成员 salary，并继承了 StdType 类的受保护数据成员 gpa。这些数据成员可以在 TA 类中访问但不能在该类中初始化。我们必须在构造函数中对它们进行赋值。

程序 12-40　文件 ta.cpp

```
1   /***************************************************************
2    * 具体类 TA 的实现文件                                         *
3    ***************************************************************/
4   #include "ta.h"
5   
6   // 构造函数
```

```
7   TA :: TA (string nm, double gp, double sal)
8   : Person (nm)
9   {
10     gpa = gp;                    // 赋值，不是初始化
11     salary = sal;                // 赋值，不是初始化
12  }
13  //打印 gpa 的成员函数
14  void TA :: printGPA ()
15  {
16     cout << "GPA: " << gpa << endl;
17  }
18  // 打印 salary 的成员函数
19  void TA :: printSalary ()
20  {
21     cout << "Salary: ";
22     cout << fixed << setprecision (2) << salary << endl;
23  }
24  // 通用的 print 成员函数
25  void TA :: print ()
26  {
27     Person :: print();
28     printGPA ();
29     printSalary();
30  }
```

程序 12-41 显示了测试上述 mixin 类的思想的简单应用程序的代码清单。用户定义应用程序。

程序 12-41 文件 application.cpp

```
1   /***************************************************************
2    * 本应用程序测试上述三个类                                      *
3    ***************************************************************/
4   #include "student.h"
5   #include "professor.h"
6   #include "ta.h"
7   
8   int main ( )
9   {
10     // 实例化四个对象
11     Person per ("John");
12     Student std ("Linda", 3.9);
13     Professor prf("George", 89000);
14     TA ta ("Lucien", 3.8, 23000);
15     // 打印有关人的信息
16     cout << "Information about person" << endl;
17     per.print();
18     cout << endl << endl;
19     // 打印有关学生的信息
20     cout << "Information about student" << endl;
21     std.print ();
22     cout << endl << endl;
23     // 打印有关教师的信息
```

```
24      cout << "Information about professor" << endl;
25      prf.print();
26      cout << endl << endl;
27      // 打印有关助教的信息
28      cout << "Information about teaching assistance " << endl;
29      ta.print();
30      cout << endl << endl;
31      return 0;
32  }
```

```
运行结果：
Information about person
Name: John

Information about student
Name: Linda
GPA: 3.90

Information about professor
Name: George
Salary: 89000.00

Information about teaching assistance
Name: Lucien
GPA: 3.80
Salary: 23000.00
```

本章小结

多态性允许我们编写一个函数的多个版本，每个版本都在一个单独的类中。然后，当我们调用函数时，将执行适合于被引用对象的版本。多态性需要满足三个条件：指针或者引用、继承层次结构和虚函数。与多态性相关的一个问题是绑定：包括静态绑定和动态绑定。

在 C++ 中，包括具体类和抽象类。具体类可以被实例化，可以创建其类型的对象实例。

抽象类可以用作具体类的共同基类。抽象类必须至少有一个声明赋值为 0 且没有定义的纯虚函数。我们不能从抽象类实例化一个对象，但抽象类可以被继承。接口是抽象类的一种特殊情况，其中所有成员函数都是纯虚函数。

思考题

1. 假设我们有一个名为 Base 的类，以及从 Base 继承的两个类：Derived1 和 Derived2。我们希望在多态关系中使用这些类。如果每个类都有一个虚打印函数和一个虚析构函数，请描述这些类的虚拟表中的条目。
2. 假设我们有以下两个类。演示如何使用栈内存以多态方式实例化这两个类的对象。

```
class First
{
    private:
        int fr;
    public:
        First (int fr);
        virtual ~First ();
```

```
class Second : public First
{
    private:
        int se;
    public:
        Second (int fr, int se);
        ~Second ();
```

```
        virtual void print () const;                    void print () const;
};                                                  };
```

3. 重复第 2 题，但在堆内存中创建对象。
4. 假设我们有一个名为 Base 的类和一个名为 Derived 的类。Base 类包含另一个名为 Base1 的类作为其私有数据成员。请绘制这些类的 UML 图。
5. 阅读以下代码片段。需要对函数 print() 附加什么修饰符，以便类 A 和类 B 可以支持多态性？

```
class A {...};
class B: public A {...};

int main()
{
        A* ptr
        ptr = new A ();
        ptr -> print ();
        ptr = new B ();
        ptr -> print ();
        return 0;
}
```

6. 假设我们有以下的类。在这种情况下，我们需要使用虚基还是 mixin 类？请说明理由。

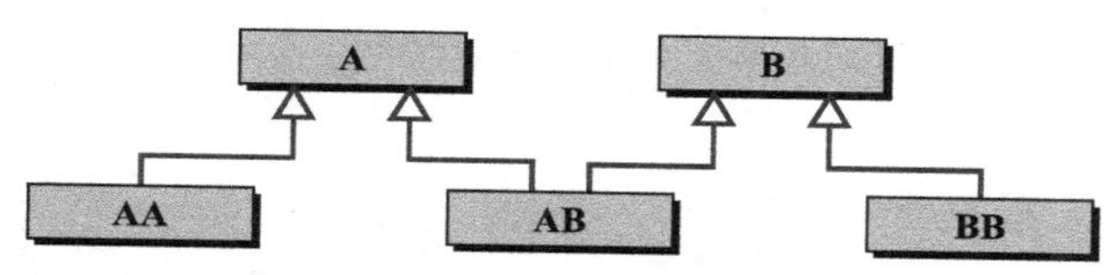

7. 假设我们有以下的类。在这种情况下，存在多重继承的问题吗？

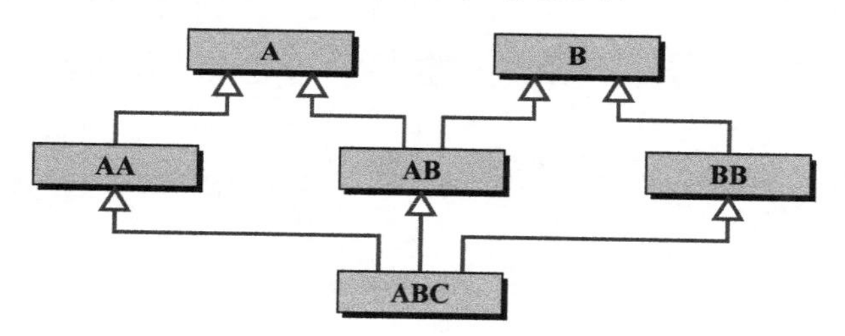

编程题

1. 修改程序 12-10 到程序 12-14，使其可以支持多态性。
2. 修改程序 12-25 到程序 12-35，使其可以支持多态性。
3. 假设我们在多态关系中有以下类：

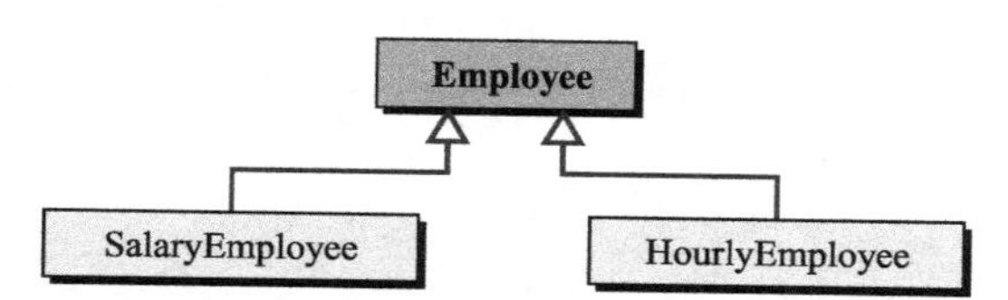

Employee（员工）类是抽象类。员工有名字、中间名缩写和姓氏。按月给薪员工每月领取固定工资。按小时给薪员工根据每月工作小时数和固定时薪领取工资。为所有三个类编写接口文件和实现文件，然后在应用程序文件中测试它们。
4. 修改第 3 题以添加另一个名为 SalaryHourlyEmployee 的类，该类的对象每月收到固定金额的工资，

如果每月工作超过180小时，则使用按小时给薪员工的相同比率获得额外工资。使用另外两个抽象类SalaryType和HourlyType，以避免多重继承的问题，如下所示。为所有类编写接口文件和实现文件，然后在应用程序文件中测试它们。

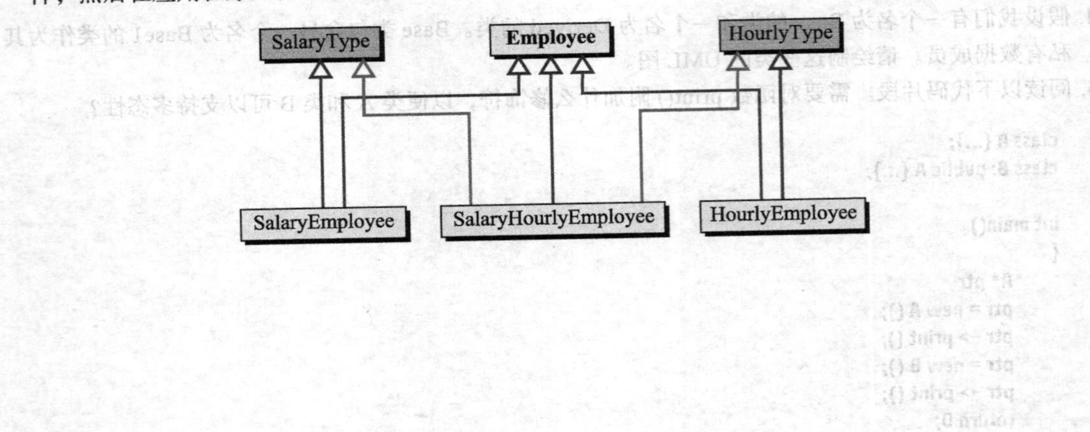

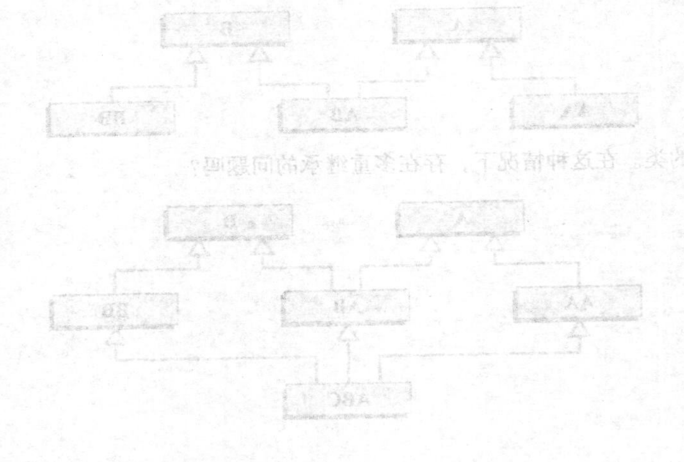

第 13 章

C++ Programming: An Object-Oriented Approach

运算符重载

在前面几章中，我们定义并使用了若干自定义类。如果我们要将用户定义的类型视为基本类型，则必须对这些自定义类型使用我们在基本类型上使用的相同操作。例如，我们知道当 x 和 y 都是整型或者浮点型的变量时，可以把这两个基本类型的对象相加，例如 $x + y$。如果两个对象都是 Fraction（分数）类型，我们应该也可以把它们相加，即 fract1 + fract2。当然，对用户定义类型的对象的操作必须是有意义的。从数学上我们知道两个分数可以相加，但另一方面，两笔贷款相加则没有什么意义。

学习目标

阅读并学习本章后，读者应该能够

- 解释在运算符重载中如何处理宿主对象、参数对象和返回对象。
- 区分可重载和不可重载的运算符。
- 了解运算符函数的作用以及如何将运算符函数用作成员函数和非成员函数。
- 演示如何将一元运算符重载为成员运算符函数，其中唯一的操作数将成为宿主对象。
- 演示如何将组合的二元运算符重载为成员运算符函数。
- 演示如何将二元运算符重载为友元函数。
- 演示如何使用构造函数把基本类型转换为用户自定义的类型。
- 演示如何使用转换函数把用户自定义的类型转换为基本类型。

13.1 对象的三种角色

用户自定义类型的对象可以在一个函数中扮演三种不同的角色：宿主对象、参数对象和返回对象。我们必须仔细研究与每种角色中的对象相关的问题，以了解编写函数或者重载运算符的过程。

13.1.1 宿主对象

为类定义非静态成员函数时，必须通过类的实例调用该函数。例如，类 Fun 可以包含一个非静态成员函数 functionOne，如表 13-1 所示。

表 13-1 宿主对象示例

```
void Fun :: functionOne (...)
{
   ...
}
int main ()
{
   fun1.functionOne (...);        // fun1 是受 functionOne 影响的对象
   fun2.functionOne (...);        // fun2 是受 functionOne 影响的对象
   fun3.functionOne (...);        // fun3 是受 functionOne 影响的对象
   ...
}
```

函数 functionOne 在第一次调用中作用于对象 fun1，在第二次调用中作用于对象 fun2，在第三次调用中作用于对象 fun3。在每种情况下，函数作用的对象称为**宿主对象**（host object）。在第一次调用中，宿主对象是 fun1；在第二次调用中，宿主对象是 fun2；在第三次调用中，宿主对象是 fun3。

如果你已经编写过实例成员函数，那么你可能想知道为什么在函数体中没有对宿主对象的引用。原因在于宿主对象（如上所述）一直在变化。第一次调用中是 fun1；第二次调用中是 fun2；第三次调用中是 fun3。这意味着函数的定义不能包含宿主对象的名称，因为名称不确定。解决方案是使用一个名为 this 的指针指向宿主对象（参见第 7 章中的讨论）。换言之，成员函数通过 **this 指针**间接访问宿主对象。在第一次调用中，系统使 this 指针指向 fun1；在第二次调用中 this 指针指向 fun2；在第三次调用中 this 指针指向 fun3。换言之，在每次调用中，宿主对象都是 *this 对象。

> 宿主对象就是 this 指针指向的对象。

【例 13-1】 假设 Fun 类仅包含一个名为 num 的整数成员函数。在表 13-2 中，左侧部分显示了我们通常如何定义 multiplyByTwo 函数，右侧部分显示了编译器如何将其转换为访问 this 指针指向的宿主对象。

表 13-2　成员函数的两种格式

//用户编写的格式	// 编译器转换后的格式

void Fun :: multiplyByTwo ()
{
　cout << num * 2;
}
	void Fun :: multiplyByTwo ()
	{
	　cout << (this -> num) * 2;
	}

我们可以直接使用右侧的代码来定义这个函数，但是按左侧所示的格式比较容易编码。换言之，宿主对象总是 this 指针指向的对象。我们可以使用 * 运算符（或者-> 运算符）访问宿主函数。

保护

某些成员函数的性质要求在宿主对象中进行更改，某些其他函数不应更改宿主对象。例如，如果我们要编写一个成员函数，用于输入数据成员的值，则必须更改宿主对象（更改器函数）。另一方面，如果我们要编写一个成员函数，输出数据成员的值，则不应该更改宿主对象（访问器函数）。为了防止成员函数更改宿主对象，必须使用 const 修饰符标识宿主对象为常量。由于宿主对象在成员函数中不可见，因此必须使用函数头末尾的 const 修饰符来提示编译器，我们不希望更改宿主对象。我们必须在函数声明和函数定义中都这样做，但是在这两种情况下函数调用是相同的。

> 如果函数不应该更改宿主对象，则宿主对象必须是常量，否则它必须是非常量。

【例 13-2】 表 13-3 显示了宿主对象的两种情况：一种是非常量，另一种是常量。

表 13-3　宿主对象的两种情况

// 宿主对象可以被更改	//宿主对象不能被更改
// 声明 void input (...);	// 声明 void output (...) const;

（续）

// 定义
void Fun :: input (...)
{
  ... ;
}
	// 定义
	void Fun :: output (...) const
	{
	  ... ;
	}

// 调用 fun1.input (...);	// 调用 fun1.output (...);

函数 input 应该为宿主对象的数据成员获取值，这意味着宿主对象将被该函数更改。另一方面，函数 output 应该打印宿主对象成员的副本，这意味着宿主对象不应该被更改。

13.1.2 参数对象

参数对象与宿主对象不同。宿主对象是成员函数的隐藏部分；参数对象则必须传递给成员函数。

传递参数的三种方式

正如我们在第 6 章中所讨论的，参数可以通过三种方式传递给函数：按值传递、按引用传递和按指针传递。

按值传递。第一种传递方式是**按值传递**（pass-by-value）。当参数是用户自定义类型时通常不采用按值传递方式，因为涉及调用拷贝构造函数，先复制对象，然后将其传递给函数。这种传递方式非常低效。

按引用传递。第二种传递方式是**按引用传递**（pass-by-reference）。按引用传递是我们在实践中遇到的最常见的传递方式。我们不复制对象，只是在函数头中定义一个别名，这样我们就可以访问对象。函数可以使用此名称访问原始对象并对其进行操作。按引用传递没有复制成本。

按指针传递。第三种传递方式是**按指针传递**（pass-by-pointer）。按指针传递并不常见，除非我们已经有一个指向对象的指针（例如，当对象在堆内存中创建时），可以将指针传递给函数。

> 将用户自定义的对象传递给函数的最常见方式是按引用传递。

保护

由于我们通常使用按引用传递的方式将用户自定义类型的对象传递给函数，因此函数可以更改原始对象。如果要防止这种更改（大多数情况下），应该在参数前面插入 const 修饰符，这意味着函数不能更改原始对象。但是，有时候函数需要更改参数（例如传递给输入函数的流）。在这种情况下，参数对象不应该是常量（见表 13-4）。

表 13-4　参数对象的两种类型

// 参数对象可以被更改	// 参数对象不能被更改

// 声明
void one (Type& para);
// 定义
void Fun :: one (Type& para)
{
  ... ;
}
	// 声明
	void two (const Type& para);
	// 定义
	void Fun :: two (const Type& para)
	{
	  ... ;
	}

（续）

// 调用 fun1.one (para);	// 调用 fun1.two (para);

13.1.3 返回对象

构造函数和析构函数不返回对象，它们创建或者销毁宿主对象。其他函数可以返回一个对象，可以定义一个返回对象实例的函数。

返回值的三种方式

我们可以通过三种方式从函数返回对象：按值返回、按引用返回和按指针返回。

按值返回。在**按值返回**（return-by-value）时，函数调用拷贝构造函数，创建对象的副本，并返回副本。众所周知，这是昂贵的返回方式，但是如果要返回的对象是在函数内部创建的，我们别无选择，因为在这种情况下，其他两种方法（按引用返回和按指针返回）不起作用。

> 如果返回的对象是在函数体中创建的，则必须使用按值返回。

按引用返回。按引用返回（return-by-reference）消除了复制的成本，但当对象是在函数体中创建的时，不能使用按引用返回。当函数终止时，在主体内创建的对象将被销毁，我们不能引用已销毁的对象。但是，当要返回的对象是按引用（或者按指针）传递的参数对象或者宿主对象时，我们可以按引用返回。即使函数终止，按引用（或者按指针）传递的原始对象也存在，因此可以引用。同样，当函数终止时，宿主对象也存在。

按指针返回。**按指针返回**（return-by-pointer）具有与按引用返回相同的限制和优势，但很少使用，除非原始对象是在堆内存中创建的。

> 如果返回的对象是作为参数传递给函数的，则必须使用按引用返回或者按指针返回。

保护

下一个问题是当返回的对象返回到应用程序后，是否允许或者防止被更改。答案取决于我们想要如何使用返回的对象。如果返回的对象仅用作右值，则应使用 const 常量修饰符保护其不被更改。另一方面，如果将返回的对象用作左值，则不应使用 const 常量修饰符。当返回的对象在这两种情况下都可以使用时，必须创建函数的两个版本：一个带有 const 常量修饰符，另一个不带 const 常量修饰符。

【例 13-3】 表 13-5 显示了使用两种方法按值返回的示例：按常量值返回和按值返回。

表 13-5　按值返回的两种类型

//对象可以被更改	//对象不能被更改

// 声明
Fun one (int value);
// 定义
Fun Fun :: one (int value)
{
  ...
  Fun fun (value);
  return fun;
	// 声明
	const Fun functOne (int value);
	// 定义
	const Fun Fun :: two (int value)
	{
	  ...
	  Fun fun (value);
	  return fun;

（续）

`}`
`// 调用`
`fun1.one (value) = ...;`
	`}`
	`// 调用`
	`fun1.two (value);`

【例 13-4】 表 13-6 显示了按引用返回宿主对象的两个示例。

表 13-6 按引用返回的两个示例

// 对象可以被更改	// 对象不能被更改

`// 声明`
`Fun& one ();`
`// 定义`
`Fun& Fun :: one()`
`{`
`  ...`
`  return *this;`
`}`
`// 调用`
`fun1.one () = ...;`
	`// 声明`
	`const Fun& two ();`
	`// 定义`
	`const Fun& Fun :: two ()`
	`{`
	`  ...`
	`  return *this;`
	`}`
	`// 调用`
	`fun1.two ();`

13.2 重载原理

运算符重载（operator overloading）是使用同一运算符的两个或者多个操作的定义。C++ 使用一组名为运算符的符号来处理基本数据类型（例如，整数和浮点数）。这些符号中的大多数都被重载以处理多种数据类型。例如，两个基本数据类型的加法运算符的符号（x + y）可以用于 int、long、long long、double 和 long double 类型的两个值的相加。这意味着 C++ 中下面的两个表达式使用相同的符号，但有两个不同的解释。第一个符号表示两个整数的相加；第二个符号表示两个实数的相加。

```
14 + 20                    14.21 + 20.45
```

如果我们理解了 C++ 如何使用两个不同的过程来把两个整数相加或者把两个浮点数相加，那么就可以更好地理解重载的含义。每个表达式中的加法符号表示不同的处理过程。在下面的示例中，编译器将第一个操作数转换为 14.0，然后执行加法操作：

```
14 + 20.35
```

C++ 语言中，符号 << 的使用更能表示两种不同的处理过程：符号 << 可以应用于整数类型，也可以应用于 ostream 类。以下两个表达式给出了该符号的两种不同的解释：

```
x << 5;                    cout << 5;
```

在左边的表达式中，符号 << 表示将整数对象中的二进制位向左移动五个位置（我们在附录 D 中讨论位操作）；在右边的表达式中，它表示将插入运算符应用于 cout 对象以打印基本数据类型 5 的值（cout 对象在类 ostream 中重载，我们将在本章后面讨论）。

重载是 C++ 语言的强大功能。重载允许用户重新定义用户自定义的数据类型的运算符，甚至可能执行新的处理。例如，我们可以重载加法符号（+）来把两个分数对象相加，而不是使用函数调用来执行相同的操作。

add (fr1, fr2)	fr1 + fr2

右边的表达式看起来更简洁、更自然。但是，我们必须仔细地重新定义分数对象的加法符号（+）的意义。

13.2.1　运算符的三种类别

可以将 C++ 中的运算符分为三类：不可重载、不推荐重载和可重载，如表 13-7 所示。

表 13-7　C++ 的运算符及其可重载性

运算符	元数	名称	可重载性
::	基本	作用域范围	不可重载
[]	后缀	数组下标	可重载
()	后缀	函数调用	可重载
.	后缀	成员选择	不可重载
->	后缀	成员选择	可重载
++	后缀	后缀递增	可重载
– –	后缀	后缀递减	可重载
++	前缀	前缀递增	可重载
– –	前缀	前缀递减	可重载
~	一元	按位非	可重载
!	一元	逻辑非	可重载
+	一元	正	可重载
–	一元	负	可重载
*	一元	解引用	可重载
&	一元	地址	不推荐重载
new	一元	分配对象	可重载
new []	一元	分配数组	可重载
delete	一元	删除对象	可重载
delete[]	一元	删除数组	可重载
type	一元	类型转换	可重载
.*	一元	指针成员选择	不可重载
->*	一元	指针成员选择	可重载
*	二元	乘	可重载
/	二元	除	可重载
%	二元	模（余数）	可重载
+	二元	加	可重载
–	二元	减	可重载
<<	二元	按位左移	可重载
>>	二元	按位右移	可重载
<	二元	小于	可重载
<=	二元	小于或者等于	可重载
>	二元	大于	可重载
>=	二元	大于或者等于	可重载
==	二元	等于	可重载
!=	二元	不等于	可重载
&	二元	按位与	不推荐重载
^	二元	按位异或	可重载
|	二元	按位或	不推荐重载
&&	二元	逻辑与	不推荐重载

（续）

运算符	元数	名称	可重载性
||	二元	逻辑或	不推荐重载
?:	二元	条件	不可重载
=	二元	简单赋值	可重载
oper=	二元	复合赋值	可重载
,	三元	逗号	不推荐重载

不可重载的运算符

C++ 语言不允许我们重载在表 13-7 中标记为“不可重载”的运算符。其背后的原因超出了本书的范围，但我们必须知道不能重载这些运算符。

不推荐重载的运算符

有六个运算符虽然可以重载，但是 C++ 强烈建议我们不要重载它们。不重载 &（地址）运算符的原因是 C++ 对这个运算符定义了一个非常特殊的含义，不能保证这个特殊的含义可由用户来实现。不建议重载其他五个运算符的原因是，这些运算符的内置版本按预先定义的顺序计算两个操作数（左操作数在前，右操作数在后）。成员函数不能保证此顺序。另外，第四个运算符和第五个运算符还有短路行为（参见第 4 章中的相关讨论），成员函数也不能保证短路行为。

可重载运算符

表 13-7 中定义的其余运算符可以重载，重载遵循 C++ 中定义的运算符的自然行为。我们不讨论所有的运算符，只讨论最常见的运算符。

13.2.2 重载的规则

在尝试为用户自定义的数据类型重载运算符之前，必须先了解重载的规则和限制：

- **优先级**。重载无法更改运算符的优先级。很好的例子是插入运算符（<<）和提取运算符（>>）。这些实际上是位运算符，具有非常低的优先级（2）。当我们为用户自定义的类使用运算符时，必须注意优先级列表中定义的运算符的优先级。
- **结合性**。重载不能更改运算符的结合性。大多数运算符具有从左到右的结合性，但也有一些具有从右到左的结合性。当我们重载一个运算符时，该运算符保持其结合性不变。
- **交换性**。重载不能改变运算符的交换性。例如，加法运算符（+）满足交换性：(a + b) 与（b + a）相同。而减法运算符不满足交换性：（a − b）与（b − a）不同。当我们重载运算符时，必须注意这条规则。
- **元数**。重载无法更改运算符的元数。如果原始运算符是一元运算符（只接收一个操作数），则重载的定义也必须是一元的。如果原始运算符是二元运算符（接收两个操作数），则重载的定义也必须是二元的。我们不需要担心唯一的三元运算符（?:），因为三元运算符不可重载。
- **不允许新的运算符**。我们不能发明运算符，我们只能重载现有的可重载运算符。例如，我们不能将符号 # 用作新的运算符，因为这个符号在 C++ 语言的运算符列表中没有定义。
- **不允许组合运算符**。我们不能组合两个运算符符号来创建新的运算符符号。例如，我们不能使用两个星号（**）的组合来定义一个新的操作符，它在某些其他语言中用于乘幂运算符。

13.2.3　运算符函数

要为用户自定义的数据类型重载运算符，必须编写**运算符函数**（operator function），该函数用作运算符。此函数的名称以保留字 operator 开始，后跟需要重载的运算符的符号。图 13-1 显示了运算符函数原型的一般形式。

返回类型 **需要重载的运算符的符号**（参数列表）
函数名称

图 13-1　运算符函数的格式

成员函数和非成员函数

大多数可重载运算符可以定义为成员函数或者非成员函数。少数可重载运算符只能作为成员函数重载；少数可重载运算符只能作为非成员函数重载。由于成员运算符函数和非成员运算符函数的语法不同，我们将分别进行讨论。

使用运算符或者运算符函数

重载之后，我们可以使用运算符本身或者运算符函数。例如，假设我们已经为分数类将一元负号运算符重载为成员函数。然后我们可以对分数对象应用该运算符，或者调用分数运算符函数，如下所示。左边的版本更简洁直观。运算符重载的全部目的是使用运算符本身来模拟内置类型的行为。

```
-fr //运算符                    fr.operator-( ) //运算符函数
```

13.3　重载为成员函数

尽管所有可重载运算符（插入运算符和提取运算符除外）都可以被重载为成员函数，但有些运算符更适合被重载为成员函数。

13.3.1　一元运算符

在一元运算符中，唯一的操作数成为运算符函数的宿主对象。不存在参数对象。这意味着我们应该只考虑两个对象：宿主对象和返回对象，如图 13-2 所示。

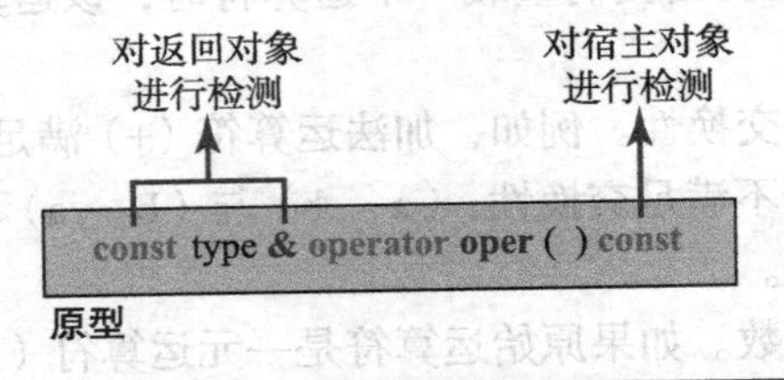
对返回对象
进行检测
对宿主对象
进行检测
const type & operator oper ( ) const
原型
1. 唯一的操作数是宿主对象。这个操作数可以是常量吗？
2.返回对象是结果。结果可以按引用返回吗？结果可以是常量吗？
检测列表

图 13-2　重载一元运算符的原则

正（+）运算符和负（−）运算符

我们首先讨论正（+）运算符和负（−）运算符，如图 13-3 所示。

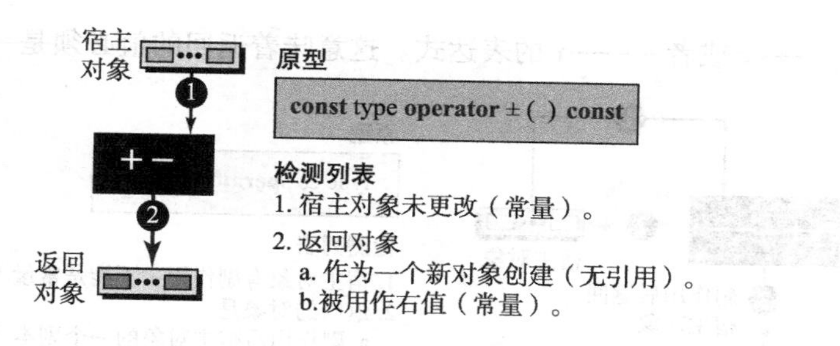

图 13-3　正运算符和负运算符的重载

示例

要定义类的正（+）运算符和负（−）运算符，必须首先定义这些运算符在应用于类的对象时的含义。尽管这两个运算符在应用于分数对象时有意义，但是在应用于贷款对象时却没有意义。根据这一准则，我们可以对分数类的正 / 负重载运算符进行声明和定义（分数类在第 7 章中定义），如表 13-8 所示。

表 13-8　一元运算符正（+）和负（−）

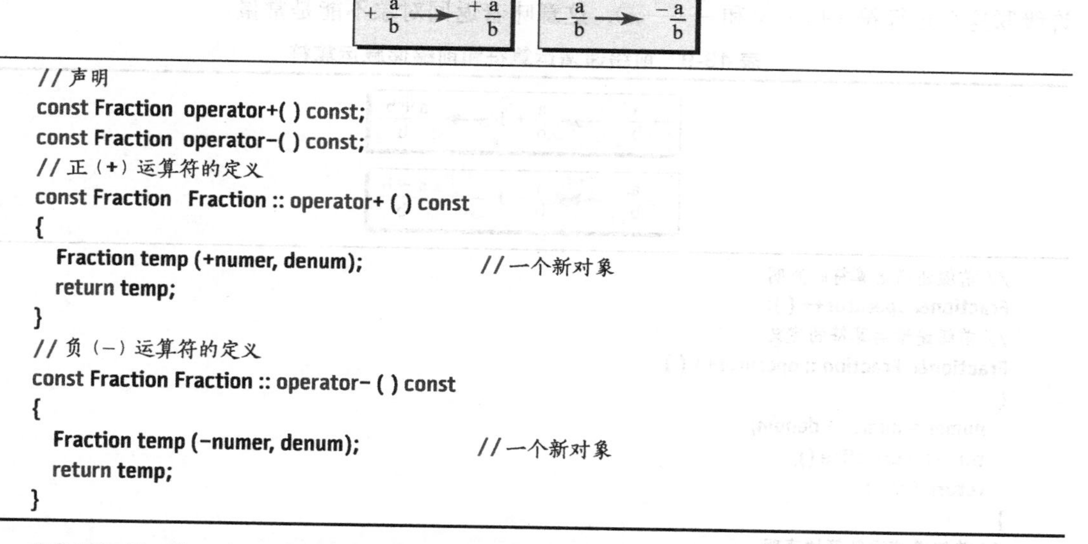

$+\frac{a}{b} \longrightarrow +\frac{a}{b}$　　$-\frac{a}{b} \longrightarrow -\frac{a}{b}$

```
// 声明
const Fraction  operator+( ) const;
const Fraction  operator-( ) const;
// 正（+）运算符的定义
const Fraction   Fraction :: operator+ ( ) const
{
   Fraction temp (+numer, denum);                    // 一个新对象
   return temp;
}
// 负（-）运算符的定义
const Fraction Fraction :: operator- ( ) const
{
   Fraction temp (-numer, denum);                    // 一个新对象
   return temp;
}
```

我们需要知道，在这种情况下没有参数对象。两个运算符都是一元运算符，宿主对象是唯一的操作数。宿主对象不应该被更改（没有副作用），返回的对象是修改后的宿主对象。

如表 13-8 所示，分数类的正（+）运算符和负（−）运算符的含义是将正号和负号应用于分子。我们可以在函数体中构造一个新的对象，并将这些运算符应用于宿主对象中表示分子的数据成员。然后，函数将临时分数对象作为常量值返回。我们将对象作为一个值返回，因为返回的对象是在函数体内部创建的，并且在函数终止时将不存在（我们不能将其按引用返回）。我们还使用 const 修饰返回对象，因为返回的对象只能用作右值（不能将其用作左值，例如在赋值运算符的左侧）。带有 const 常量修饰符的对象也不能级联在一起。

前缀递增运算符和前缀递减运算符

接下来我们将重载前缀递增运算符和前缀递减运算符。每个运算符都有一个改变其操作数的副作用；返回的对象是被改变对象的副本。C++ 允许级联这两个运算符，这意味着我们

可以编写诸如 ++++*x* 或者 −−−−*x* 的表达式。这意味着返回的值必须是一个左值（图 13-4）。

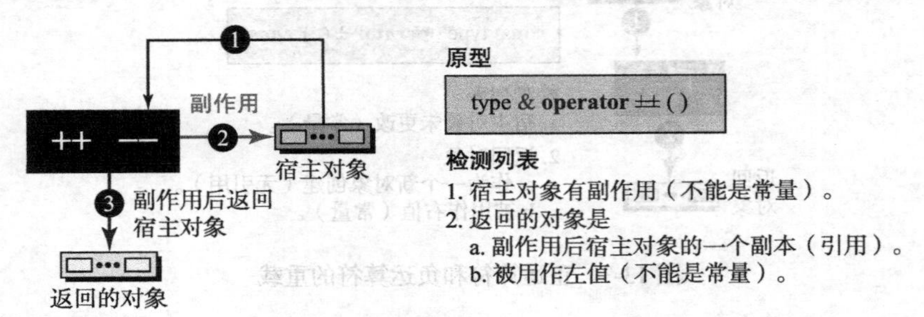

图 13-4　前缀递增运算符和前缀递减运算符的重载

根据图 13-4 中的检测列表，我们可以对分数类的前缀递增重载运算符和前缀递减重载运算符进行声明和定义，如表 13-9 所示。这两个运算符都要执行两个任务。运算符必须更改宿主对象，然后返回更改后的宿主对象（宿主对象不能是常量）。这意味着，我们不应该从宿主对象创建新对象，而是应该更改宿主对象，然后返回宿主对象。因为返回的对象是宿主对象的更改版本，所以我们可以通过引用而不是通过值返回（按值返回更昂贵）。C++ 允许级联这个运算符（++++x 和 −−−−x），这意味着返回对象不能是常量。

表 13-9　前缀递增运算符和前缀递减运算符

$$++\frac{a}{b} \longrightarrow \frac{a}{b}+1 \longrightarrow \frac{a+b}{b}$$

$$--\frac{a}{b} \longrightarrow \frac{a}{b}-1 \longrightarrow \frac{a-b}{b}$$

```
// 前缀递增运算符的声明
Fraction& operator++ ( );
// 前缀递增运算符的定义
Fraction&  Fraction :: operator++ ( )
{
   numer = numer + denom;
   this -> normalize ();
   return *this;
}
// 前缀递减运算符的声明
Fraction& operator-- ( );
// 前缀递减运算符的定义
Fraction&  Fraction :: operator-- ( )
{
   numer = numer - denom;
   this -> normalize ();
   return *this;
}
```

后缀递增运算符和后缀递减运算符

接下来我们将重载后缀递增运算符和后缀递减运算符。返回的对象是在副作用之前创建的（图 13-5）。

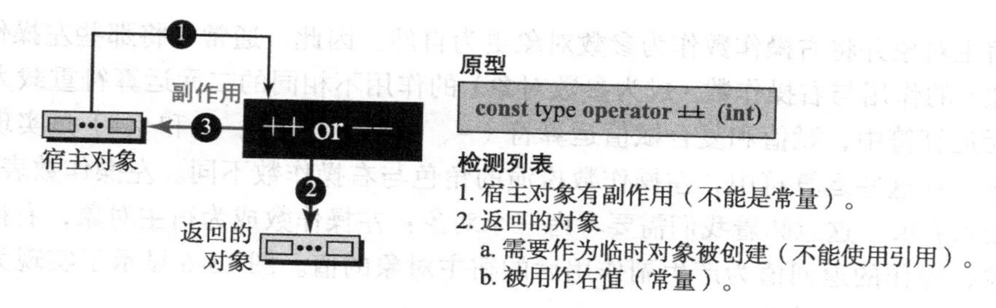

图 13-5　后缀递增运算符和后缀递减运算符的重载

由于在更改宿主对象之前无法返回它，因此需要从宿主对象中创建一个临时对象，将副作用应用于宿主对象，然后返回临时对象。哑整数参数（dummy integer parameter）创建一个唯一的签名来区分前缀运算符和后缀运算符的原型，这在程序中被忽略。

根据图 13-5 中的检测列表，我们可以对分数类的后缀递增重载运算符和后缀递减重载运算符进行声明和定义，如表 13-10 所示。返回的对象是在更改前创建的临时对象，这意味着我们不能通过引用返回该对象。返回的对象是常量，因为 C++ 不允许级联该操作，例如 x++++ 或者 x----。还要注意，由于前缀运算符和后缀运算符的副作用相同，我们可以调用后缀运算符来更改宿主对象。

表 13-10　后缀递增运算符和后缀递减运算符

$$\frac{a}{b}{++} \longrightarrow \frac{a}{b}+1 \longrightarrow \frac{a+b}{b}$$

$$\frac{a}{b}{--} \longrightarrow \frac{a}{b}-1 \longrightarrow \frac{a-b}{b}$$

```
//后缀递增运算符的声明
const Fraction operator++ (int);
//后缀递增运算符的定义
const Fraction  Fraction :: operator++ (int dummy)
{
   Fraction temp (numer, denom);
   ++(*this);
   return temp;
}
//后缀递减运算符的声明
const Fraction operator-- (int);
//后缀递减运算符的定义
const Fraction  Fraction :: operator-- (int dummy)
{
   Fraction temp (numer, denom);
   --(*this);
   return temp;
}
```

13.3.2　二元运算符

二元运算符包含两个操作数（左操作数和右操作数）。如果我们想将一个二元运算符重载为成员函数，我们必须将其中一个操作数作为宿主对象，将另一个作为参数对象。将左操作数作为宿主对象并将右操作数作为参数对象更为自然。因此，通常只将那些左操作数（成为宿主对象）的作用与右操作数（成为参数对象）的作用不相同的二元运算符重载为成员函数。在二元运算符中，赋值和复合赋值运算符（=、+=、-=、*=、/= 和 %=）是实现此目的的最佳候选。在这些运算符中，左操作数扮演的角色与右操作数不同。左操作数表示左值，右操作数表示右值。这意味着我们需要考虑三个对象：左操作数成为宿主对象，右操作数成为参数对象，操作的返回值为产生副作用后的宿主对象的值。图 13-6 显示了实现为成员函数的二元运算符的一般原型。

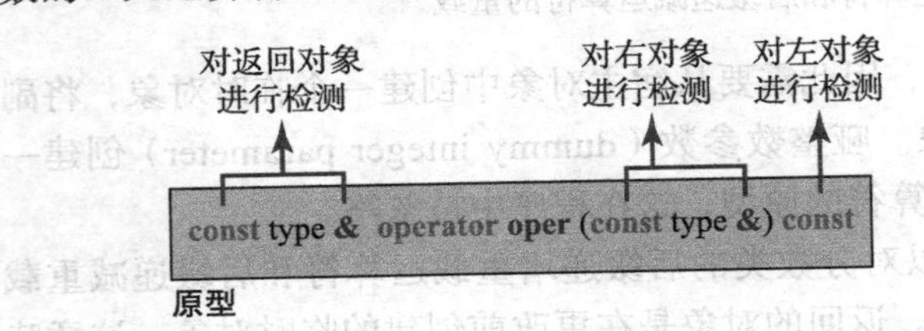

图 13-6　二元运算符的原理

赋值运算符

此类别中的第一个候选项是赋值运算符。这是一种不对称操作，其中左操作数和右操作数的性质不同。左操作数是一个接收操作副作用的左值对象；右操作数是一个不应在处理过程中更改的右值对象。在继续讨论此运算符之前，必须指出，要使用此运算符，左操作数和右操作数必须已经存在。换言之，此运算符不同于拷贝构造函数，后者从现有对象创建新对象。对于赋值运算符，两个对象都必须存在。我们只更改左对象，使左对象是右对象的精确副本。

> 在赋值运算符中，左对象（宿主）和右对象（参数）都必须已存在。

图 13-7 显示了简单的赋值操作符。首先，由于副作用，宿主对象（左操作数）不能是常量。其次，返回的对象不能是常量，因为我们可以级联该运算符（x = y = z）。

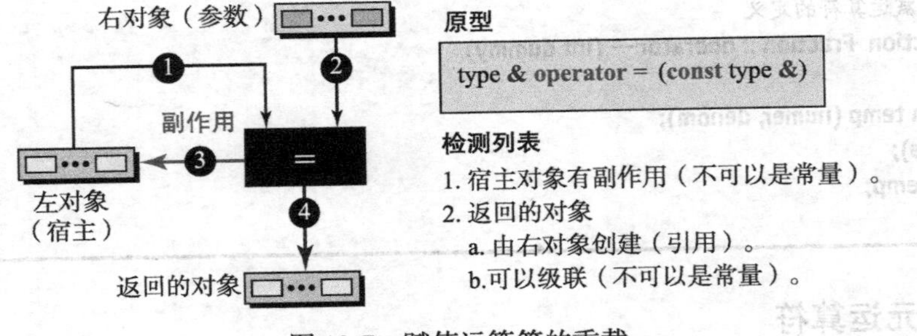

图 13-7　赋值运算符的重载

关于这个运算符，我们还必须提到另一个重要的问题。如果没有为我们的类定义赋值运算符，则系统将提供一个（合成赋值运算符）。但是，合成运算符可能不是我们需要的运算符。此外，重载运算符需要注意以下事项：

- 应该验证宿主对象和参数对象不是同一个对象（地址不同）。如果对象是在堆内存中创建的，这一点尤其重要。由于在复制参数对象的内容之前必须删除宿主对象，如果两个对象相同，则参数对象（与宿主对象的物理地址相同）也被删除了，从而没有了要复制的内容。
- 我们知道赋值对象是从右到左结合的。换言之，如果有 y = x，那么可以将操作结果分配给另一个对象 z = y = x，这被解释为 z = (y = x)。但是，C++ 要求 z 被看作对 y 的引用。这就是返回的对象必须通过引用返回的原因。

【例 13-5】 表 13-11 显示了分数类赋值运算符的声明和定义。

表 13-11　赋值运算符

$$\frac{a}{b} = \frac{c}{d} \longrightarrow \frac{c}{d}$$

```
//赋值运算符的声明
Fraction& operator= (const Fraction& right)
//赋值运算符的定义
Fraction& Fraction :: operator= (const Fraction& right)
{
  if (*this != right) //检查二者不是同一个对象
  {
      numer = right.numer;
      denom = right.denom;
  }
  return *this;
}
```

复合赋值运算符

我们可以重载的另一组运算符是复合赋值运算符（+=、-=、*=、/= 和 %=）。复合赋值的含义如下所示：

fract1 += fract2	表示	fract1 = fract1 + fract2
fract1 -= fract2	表示	fract1 = fract1 - fract2
fract1 *= fract2	表示	fract1 = fract1 * fract2
fract1 /= fract2	表示	fract1 = fract1 / fract2
fract1 %= fract2	表示	fract1 = fract1 % fract2

在上述赋值运算中，fract1 是宿主对象，fract2 是参数对象。这个过程与赋值运算符的过程相同，只是我们必须仔细定义宿主对象的分子和分母。返回的对象是更改后的宿主对象。

【例 13-6】 表 13-12 显示了分数类的复合赋值运算符的声明和定义。

表 13-12　重载复合赋值运算符

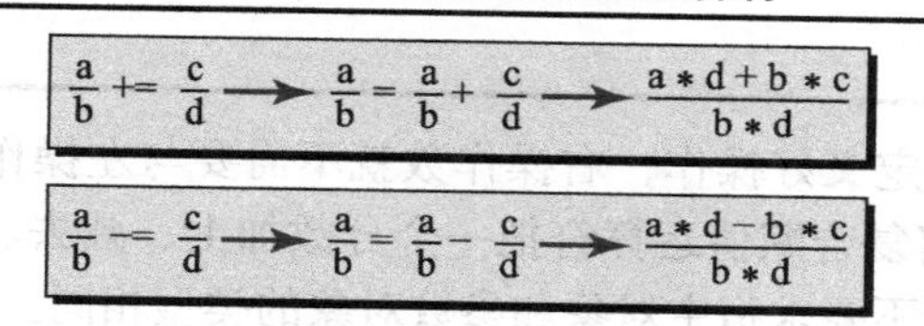

（续）

$$\frac{a}{b} *= \frac{c}{d} \longrightarrow \frac{a}{b} = \frac{a}{b} * \frac{c}{d} \longrightarrow \frac{a*c}{b*d}$$

$$\frac{a}{b} /= \frac{c}{d} \longrightarrow \frac{a}{b} = \frac{a}{b} / \frac{c}{d} \longrightarrow \frac{a*d}{b*c}$$

```
// 运算符 += 的声明
Fraction& operator+= (const Fraction& right)
// 运算符 += 的定义
Fraction& Fraction :: operator+= (const Fraction& right)
{
   numer  = numer * right.denom + denom * right.numer;
   denom = denom * right.denom;
   normalize ();
   return *this;
}
// 运算符 -= 的声明
Fraction& operator-= (const Fraction& right)
// 运算符 -= 的定义
Fraction& Fraction :: operator-= (const Fraction& right)
{
   numer = numer * right.denom - denom * right.numer;
   denom = denom * right.denom;
   normalize ();
   return *this;
}
// 运算符 *= 的声明
Fraction& operator*= (const Fraction& right)
// 运算符 *= 的定义
Fraction& Fraction :: operator*= (const Fraction& right)
{
   numer = numer * right.numer;
   denom = denom * right.denom;
   normalize ();
   return *this;
}
// 运算符 /= 的声明
Fraction& operator/= (const Fraction& right)
// 运算符 /= 的定义
Fraction& Fraction :: operator/= (const Fraction& right)
{
   numer = numer * right.denom;
   denom = denom * right.numer;
   normalize ();
   return *this;
}
```

在表13-12中，只要定义好操作，右操作数就不需要与左操作数具有相同的对象类型。例如，我们可以使用任何复合赋值运算符把一个分数加上、减去、乘以或者除以一个整数。操作定义良好。重载规则不要求宿主对象和参数对象的类型相同。

13.3.3 其他运算符

还有其他可以实现为成员函数的运算符。接下来我们将讨论其中的一些运算符。

间接运算符和成员选择运算符

在 C++ 中，我们经常使用间接运算符（*）或者成员选择运算符（->）来访问存储在栈内存或者堆内存中的对象或者对象的成员。在重载这两个运算符之前，我们必须注意有关这两个运算符的使用要点。

- 这两个运算符都是一元运算符，这意味着它们都只有一个操作数，该操作数必须是指向相应对象的指针。间接运算符返回操作数指向的对象；成员选择运算符将指针操作数更改为可指向相应对象的任何成员的指针并返回该成员。在第一种情况下，运算符位于操作数之前（前缀）；在第二种情况下，运算符位于操作数之后（后缀），如下所示：

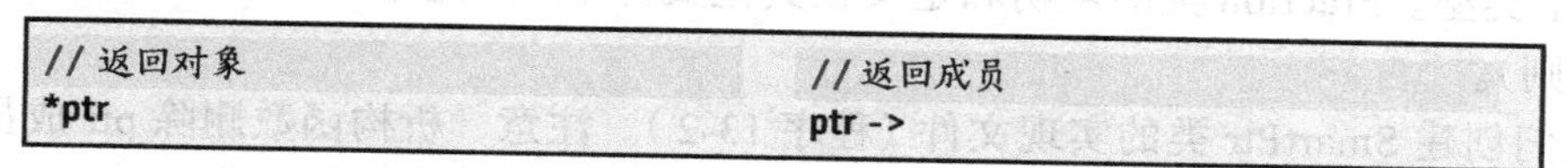

```
// 返回对象                    // 返回成员
*ptr                           ptr ->
```

- 如果我们想要重载这两个运算符，则需要一个充当指针的类，因为操作数（宿主对象）必须是指针。

智能指针（smart pointer）。当我们使用指针指向栈内存中的对象时，不需要重载间接运算符和成员选择运算符（* 和 ->）。我们可以使用指针（作为复合数据类型）指向我们的对象。当对象超出作用域范围时，它会自动从堆栈中弹出，并且没有内存泄漏。但是，当我们在堆内存中创建一个对象时，我们必须记住删除该对象以避免内存泄漏（参见第 9 章中的讨论）。为此，我们创建了一个**智能指针**，如图 13-8 所示。

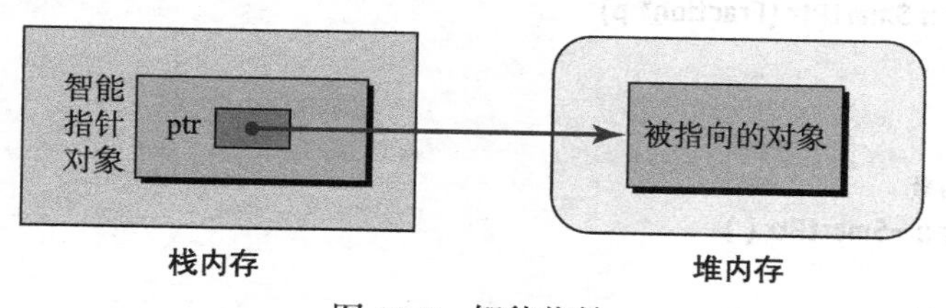

图 13-8 智能指针

智能指针对象是一个只有一个数据成员的对象：指向所需类型（例如分数）的指针。当调用此对象的构造函数时，将创建指向所需类型的指针。对于 SmartPtr 类，这两个操作符 * 和 -> 也被重载，从而可以指向所需的对象和所需对象的一个成员。

【例 13-7】 我们创建了一个名为 SmartPtr 的智能指针类，该类的接口文件显示在程序 13-1 中。

程序 13-1 文件 smartptr.h

```
1  /*****************************************************************
2   * SmartPtr 类的接口文件                                          *
3   *****************************************************************/
4  #ifndef SMARTPTR_H
5  #define SMARTPTR_H
6  #include <iostream>
7  using namespace std;
8
```

```
9   class Fraction;     // 前向声明
10  class SmartPtr
11  {
12    private:
13        Fraction* ptr;
14    public:
15        SmartPtr (Fraction* ptr);
16        ~SmartPtr ( );
17        Fraction& operator* ( ) const;
18        Fraction* operator->( ) const;
19  };
20  #endif
```

注意，我们在第9行中添加了一个前向声明，告诉头文件SmartPtr声明中的名称Fraction是一个类型。Fraction类的声明和定义在其他文件中（参见第7章中所讨论的关于独立编译的原则）。

接下来我们创建SmartPtr类的实现文件（程序13-2）。注意，析构函数删除ptr成员分配的内存（第14行）。

程序 13-2　文件 smartptr.cpp

```
1   /***************************************************************
2    * SmartPtr 类的实现文件                                        *
3    ***************************************************************/
4   #include "smartptr.h"
5
6   //构造函数
7   SmartPtr :: SmartPtr (Fraction* p)
8   : ptr (p)
9   {
10  }
11  //析构函数
12  SmartPtr :: ~SmartPtr ( )
13  {
14    delete ptr;
15  }
16  //重载间接运算符
17  Fraction& SmartPtr :: operator* ( ) const
18  {
19    return *ptr;
20  }
21  //重载箭头（成员选择）运算符
22  Fraction* SmartPtr :: operator-> ( ) const
23  {
24    return ptr;
25  }
```

接下来我们创建应用程序来测试SmartPtr类（程序13-3）。

程序 13-3　文件 app.cpp

```
1   /***************************************************************
2    * 本应用程序测试 SmartPtr 类                                    *
```

```
3   *************************************************************************/
4   #include "smartptr.h"
5   #include "fraction.h"
6
7   int main ( )
8   {
9       //创建一个智能指针对象
10      SmartPtr sp = new Fraction (2, 5);
11      //使用运算符 * 访问成员
12      cout << "Fraction: " << endl;
13      (*sp).print ();
14      cout << endl;
15      //使用运算符 -> 访问成员
16      cout << "Fraction: " << endl;
17      sp -> print ();
18      cout << endl;
19      return 0;
20  }
```

```
运行结果:
Fraction: 2 / 5
Fraction: 2 / 5
```

注意，我们可以在应用程序中使用以下语句在堆内存中创建分数，而不使用智能指针。

```
Fraction* ptr = new Fraction (2, 5);
(*ptr).print ();
ptr -> print ();
```

不同之处在于，堆内存中创建的分数对象不会自动删除，我们必须在应用程序中删除它，以避免可能的内存泄漏。智能指针则自动为我们完成这个任务。

下标运算符

另一个可以被重载为成员函数的运算符是下标运算符（[]）。但是，我们必须注意不要改变这个操作符的语义。下标运算符是一个二元运算符，其中左操作数是数组的名称，右操作数是定义数组中元素索引的类型。这意味着，只有当我们的类型是数组或者其行为类似于数组（例如字符串或者列表）时，才应该重载此运算符。

如果我们想正确地重载该运算符，我们需要两个版本。在第一个版本中，该运算符被重载为访问器函数；在第二个版本中，该运算符被重载为更改器函数。图 13-9 显示了下标运算符的设计。

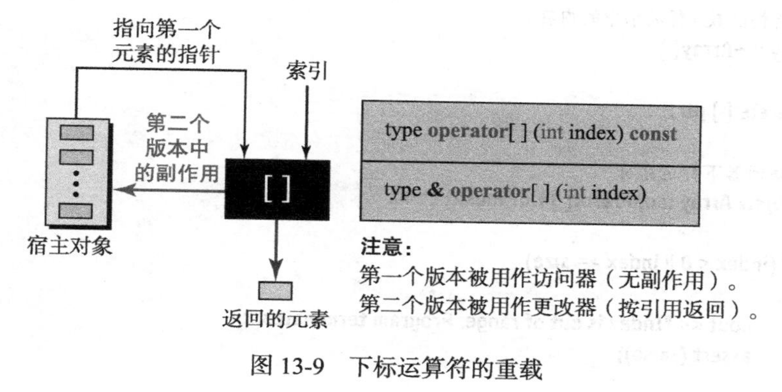

图 13-9　下标运算符的重载

假设我们需要一个名为 Array 的类，在该类中我们可以控制数组的大小，并且整个数组是在堆内存中创建的。程序 13-4 显示了其接口文件。

程序 13-4　文件 array.h

```
1  /***************************************************************
2   * Array 类的接口文件                                            *
3   ***************************************************************/
4  #ifndef ARRAY_H
5  #define ARRAY_H
6  #include <iostream>
7  #include <cassert>
8  using namespace std;
9  
10 class Array
11 {
12   private:
13       double* ptr;
14       int size;
15   public:
16       Array (int size); // 构造函数
17       ~Array( ); // 析构函数
18       double& operator[ ] (int index) const; // 访问器
19       double& operator[ ] (int index) ; // 更改器
20 };
21 #endif
```

程序 13-5 显示了 Array 类的实现文件的代码清单。

程序 13-5　文件 array.cpp

```
1  /***************************************************************
2   * Array 类的实现文件                                            *
3   ***************************************************************/
4  #include "array.h"
5  
6  // 构造函数（在堆中分配内存）
7  Array :: Array (int s)
8  :size (s)
9  {
10   ptr = new double [size];
11 }
12 // 析构函数（释放堆中的内存）
13 Array :: ~Array( )
14 {
15   delete [ ] ptr;
16 }
17 // 访问器下标运算符
18 double& Array :: operator[ ] (int index) const
19 {
20   if (index < 0 || index >= size)
21   {
22       cout << "Index is out of range. Program terminates.";
23       assert (false);
```

```
24   }
25   return ptr [index];
26 }
27 //更改器下标运算符
28 double& Array :: operator[ ] (int index)
29 {
30   if (index < 0 || index >= size)
31   {
32       cout << "Index is out of range. Program terminates.";
33       assert (false);
34   }
35   return ptr [index];
36 }
```

程序 13-6 是一个简单的应用程序。

程序 13-6　文件 app.cpp

```
1  /*****************************************************************
2   * 本应用程序测试 Array 类                                        *
3   *****************************************************************/
4  #include "array.h"
5
6  int main ( )
7  {
8    //实例化一个包含三个元素的数组对象
9    Array arr (3);
10   //使用更改器下标运算符 [] 存储值
11   arr[0] = 22.31;
12   arr[1] = 78.61;
13   arr[2] = 65.22;
14   //使用访问器下标运算符 [] 获取值
15   for (int i = 0; i < 3; i++)
16   {
17       cout << "Value of arr [" << i << "]: " << arr[i] << endl;
18   }
19   return 0;
20 }
```

```
运行结果:
Value of arr [0]: 22.31
Value of arr [1]: 78.61
Value of arr [2]: 65.22
```

函数调用运算符

另一个可以重载的一元运算符是**函数调用运算符**，如下所示：

```
name( 参数列表 )
```

函数调用运算符的重载允许我们创建一个函数对象（有时称为仿函数 functor）：一个能行使函数功能的类的对象实例。如果一个类重载了函数调用运算符，编译器允许我们像调用函数一样实例化这个类的对象。我们实际上是将一个函数封装在一个类中，其优点是可以在不同的应用程序中使用。我们可以实例化类的对象来调用嵌入在类中的函数。

函数对象和函数之间的区别在于函数对象可以保持其状态（函数没有状态）。例如，名为 smallest 的函数对象可以保存上一次调用中的最小值（作为数据成员）。这样，只要对象在作用域范围内，我们可以调用该函数对象任意多次，结果总是返回最小的值。

【例 13-8】 程序 13-7 显示了 Smallest 类的接口文件的代码清单。

程序 13-7　文件 smallest.h

```
1   /**********************************************************************
2    * Smallest 类的接口文件                                              *
3    **********************************************************************/
4   #ifndef SMALLEST_H
5   #define SMALLEST_H
6   #include <iostream>
7   using namespace std;
8
9   class Smallest
10  {
11    private:
12        int current;
13    public:
14        Smallest ( );
15        int operator ( ) (int next);  //函数调用运算符
16  };
17  #endif
```

程序 13-8 显示了 Smallest 类的实现文件的代码清单。

程序 13-8　文件 smallest.cpp

```
1   /**********************************************************************
2    * Smallest 类的实现文件                                              *
3    **********************************************************************/
4   #include "smallest.h"
5
6   // 构造函数
7   Smallest :: Smallest ( )
8   {
9     current = numeric_limits <int> :: max();
10  }
11  // 重载函数调用运算符
12  int Smallest :: operator() (int next)
13  {
14    if (next < current)
15    {
16        current  =  next;
17    }
18    return current;
19  }
```

程序 13-9 显示了用于测试函数对象的应用程序文件的代码清单。

程序 13-9　文件 app.cpp

```
1   /**********************************************************************
2    * 本应用程序测试 Smallest 类                                         *
```

```
3    ***********************************************************************/
4    #include "smallest.h"
5    #include <iostream>
6
7    int main ( )
8    {
9      //实例化一个 Smallest 对象
10     Smallest smallest;
11     //在对象上应用函数调用运算符
12     cout << "Smallest so far: " << smallest (5) << endl;
13     cout << "Smallest so far: " << smallest (9) << endl;
14     cout << "Smallest so far: " << smallest (4) << endl;
15     return 0;
16   }
运行结果:
Smallest so far: 5
Smallest so far: 5
Smallest so far: 4
```

在第 19 章中，当我们讨论 C++ 如何在标准模板库（STL）中使用函数对象时，将进一步地讨论函数对象（仿函数）。

关于函数对象，有趣的是 C++ 不限制函数调用中的参数数目。这意味着我们可以用多个参数来重载这个运算符。一种应用是使用函数对象来获取定义二维数组（甚至是具有更多维度的数组）的类中的元素。第一个参数可以定义第一个维度（行）的索引，第二个参数可以定义第二个维度（列）的索引。

13.4 重载为非成员函数

将二元运算符重载为成员函数时，其中一个操作数必须是宿主对象。如果每个操作数在操作中具有不同的角色时，这没有任何问题。但是，在一些运算符中，如（a + b）或者（a < b），两个操作数的作用是相同的，且两者都与结果无关。在这些情况下，最好使用非成员函数重载运算符。我们有两个选择：全局函数或者友元函数。

使用全局函数来重载二元运算符没有任何限制。尽管有许多人支持这一策略，但该方法存在一个缺点。运算符函数的定义代码会更冗长且更复杂，因为需要使用类的访问器函数和更改器函数来访问类的数据成员。我们不开发这些类型的函数。

C++ 允许函数被声明为类的友元函数。**友元函数**（friend function）没有宿主对象，但它被授予权限（友元），这样友元函数就可以在不调用公共成员函数的情况下访问类的私有数据成员和成员函数。我们使用友元函数来重载选定的运算符。

13.4.1 二元算术运算符

我们推迟到现在才讨论重载二元算术运算符，因为更适合将二元算术运算符重载为友元函数。图 13-10 显示了其设计方法。两个操作数必须事先存在。我们在函数内部创建一个新对象，并将其作为常量对象返回。我们可以将这两个操作数按引用

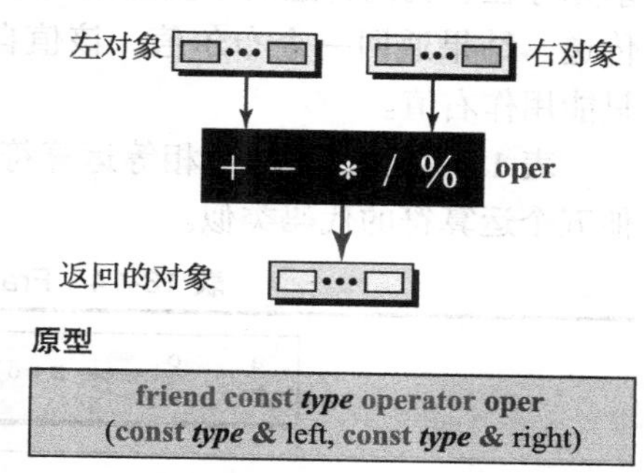

图 13-10 重载二元算术运算符

传递，但返回对象不能是引用类型，因为返回对象是在函数定义内创建的。

表 13-13 显示了分数类的算术运算符。请注意，模运算符（%）不能应用于 Fraction 类，因为此操作对分数没有意义。我们给出了第一个运算符的声明和定义，其余的类似。注意，仅在声明中需要指定修饰符 friend。

请注意，结果是右值，这意味着它们只能在需要右值时使用。但是，我们可以实例化一个对象，然后根据需要将结果分配给该对象（fract = fract1 + fract2）。我们必须确保重载了赋值运算符。

表 13-13　Fraction 类的二元算术运算符

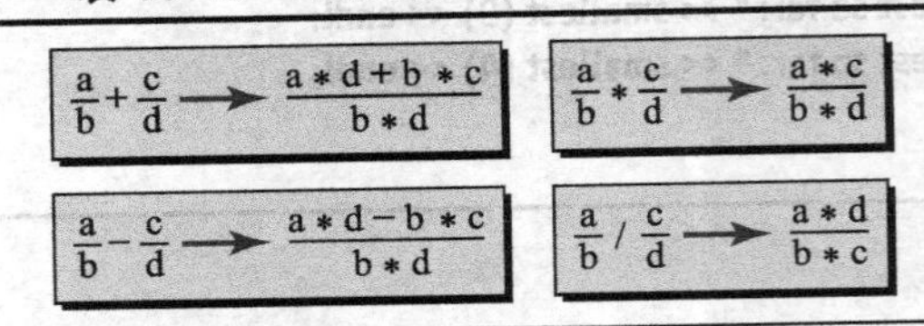


```
// 运算符 + 的声明
friend const Fraction operator+ (const Fraction& left,
                                 const Fraction& right);
// 运算符 + 的定义
const Fraction operator+ (const Fraction& left,
                          const Fraction& right)
{
    int newNumer = left.numer * right.denom + right.numer* left.denom;
    int newNumer = left.numer * right.denom + right.numer* left.denom;
    int newDenom = left.denom * right.denom;
    Fraction result (newNumer, newDenom);
    return result;
}
```

13.4.2　等性运算符和关系运算符

我们也可以使用友元函数重载两个等性运算符（== 和！ =）和四个关系运算符（<、<=、> 和 >=）。运算符函数的结构与算术运算符相同，但是返回值的类型为布尔型（true 或者 false）。

图 13-11 显示了设计方法。两个操作数必须事先存在。我们将这两个参数函数作为常量引用传递。结果返回一个布尔值，该值自动为常量，只能用作右值。

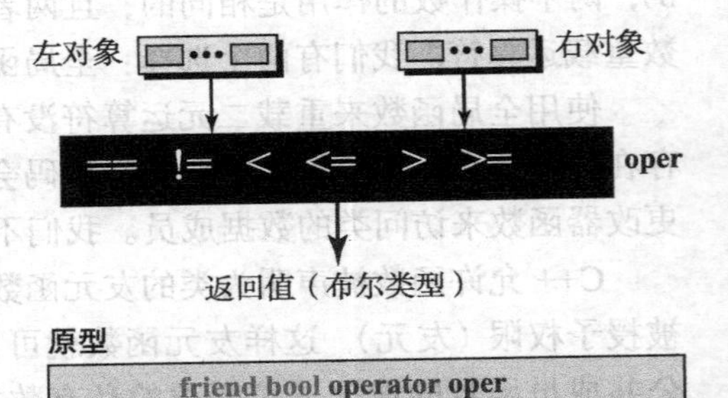

图 13-11　重载等性运算符和关系运算符

表 13-14 仅仅显示了相等运算符的代码。其他五个运算符的代码类似。

表 13-14　Fraction 类的等性运算符和关系运算符

a/b == c/d → a * d == b * c	a/b != c/d → a * d != b * c

（续）

$\frac{a}{b} < \frac{c}{d} \longrightarrow a*d<b*c$　　$\frac{a}{b} <= \frac{c}{d} \longrightarrow a*d<=b*c$

$\frac{a}{b} > \frac{c}{d} \longrightarrow a*d>b*c$　　$\frac{a}{b} >= \frac{c}{d} \longrightarrow a*d>=b*c$

```
// 运算符 == 的声明
friend bool operator== (const Fraction& left, const Fraction& right);
// 运算符 == 的定义
bool operator== (const Fraction& left, const Fraction& right)
{
   return (left.numer * right.denom == right.numer * left.denom) ;
}
```

13.4.3 提取运算符和插入运算符

基本类型的值可以使用提取运算符（>>）从输入流对象中提取，也可以使用插入运算符（<<）插入输出流中。我们可以为类类型重载这两个运算符，如图 13-12 所示。

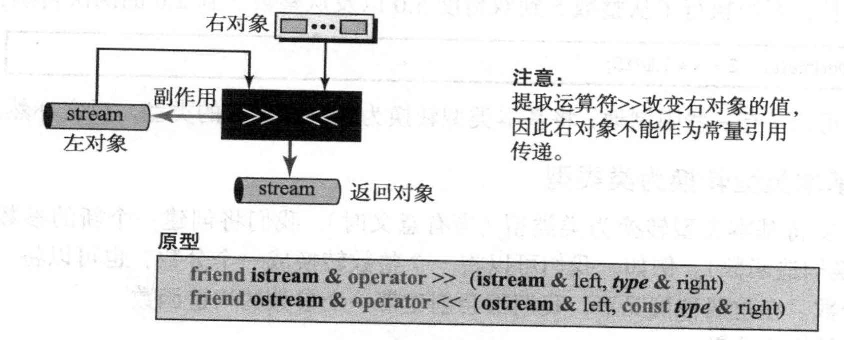

图 13-12　重载提取运算符和插入运算符

这些运算符都是二元运算符，但左操作数在提取运算符的情况下是 istream 类的对象，在插入运算符的情况下是 ostream 类的对象。换言之，在这两种情况下，左操作数都是库中类的对象，而不是我们要为这些运算符重载的类的对象。另一方面，右操作数是类对象。注意，如果我们想使用键盘作为输入流，输入流对象是 cin；如果我们想使用监视器作为输出流，输出流对象是 cout。否则，左参数应该是声明为 istream 或者 ostream 类型的文件。

【例 13-9】 在本例中，我们为 Fraction 类创建提取赋值运算符和插入赋值运算符。参见表 13-15。

表 13-15　重载提取运算符和插入运算符

```
// 运算符 >> 的声明
friend istream& operator >> (istream& left, Fraction& right) ;
// 运算符 >> 的定义
istream& operator >> (istream& left, Fraction& right)
{
   cout << "Enter the value of numerator: " ;
   left >> right.numer;
   istream << "Enter the value of denominator: " ;
```

（续）

```
    left >> right.denom;
    right.normalized( );
    return left ;
}
// 运算符 << 的声明
friend ostream& operator << (ostream& left, const Fraction& right) ;
// 运算符 << 的定义
ostream& operator << (ostream& left, const Fraction& right)
{
    left << right.numer << "/" << right.denom << endl;
    return left;
}
```

13.5 类型转换

当我们处理基本数据类型时，有时会在表达式中使用混合类型，并期望系统根据转换规则使它们成为相同的类型。例如，在以下表达式语句中，计算了一个半径为5的圆的周长。为了计算周长，系统执行了从整数5到双精度5.0以及从整数2到2.0的两次转换：

```
double perimeter = 2 * 5 * 3.1415;
```

我们也可以（有一些限制地）将基本类型转换为用户自定义的类型，反之亦然。

13.5.1 基本类型转换为类类型

如果需要将基本类型转换为类类型（当有意义时），我们将创建一个新的参数构造函数（也称为转换构造函数）。例如，我们可以将一个整数转换成一个分数，也可以将一个实数转换成一个分数。在这两种情况下，我们都会创建带一个参数的构造函数。

把整数转换为分数

这可以很容易地使用一个构造函数来完成，该构造函数接收一个被设置为分子的整数参数。分母被设置为1。

【例 13-10】 表13-16显示了如何将整数转换为分数对象。

表 13-16　把整数值转换为分数对象

```
// 声明
Fraction (int n);
// 定义
Fraction :: Fraction (int n)
: numer (n), denom (1)
{
}
```

我们可以把分数和整数相加，因为编译器在加法运算前将整数4转换为分数4/1。

```
Fraction fract1 (2, 5);
Fraction fract2 = fract1 + 4
```

把实数转换为分数

我们可以使用构造函数将实数转换为分数。分数可以是实数。例如，分数7/4是实数1.75。这意味着我们也可以将实数（如1.75）转换为其相应的分数表示。给定值1.75，我们可以将其重写为175/100，然后将其规范化为7/4。

表13-17显示了如何把实数转换为分数对象。构造函数接收成为分子的实数（值）。我们将分母设置为1，并不断地将分子和分母乘以10，直到分子的小数部分消失。分子则是该

值的整数转换。在标准化分子和分母后，即获得分数对象。

表 13-17 把实数值转换为分数对象

```
// 声明
Fraction (double value);
// 定义
Fraction :: Fraction (double value)
{
   denom = 1;
   while ((value – static_cast <int> (value)) > 0.0)
   {
        value *= 10.0;
        denom *= 10;
   }
   numer = static_cast <int> (value);
   normalize ();
}
```

13.5.2 类类型转换为基本类型

有时我们需要将类类型转换为基本类型。我们可以使用转换运算符来实现这一点，转换运算符是一种运算符函数，其中术语 operator 和运算符符号被替换为基本数据类型，如表 13-18 所示。

运算符的语法看起来有些奇怪（因为没有返回值），但我们必须将其视为 double 类型的构造函数。该构造函数将一个分数转换为一个双精度值。我们知道构造函数没有返回值，构造函数构造一个类型。

表 13-18 把类类型转换为基本类型

```
// 声明
operator double ();
// 定义
Fraction :: operator double ()
{
   double num = static_cast <double> (numer);
   return (num / denom);
}
```

13.6 设计类

在本节中，我们将创建三个类来说明重载运算符在某些应用程序中的重要性：使用重载运算符重新创建分数类；通过重载适当的运算符来定义日期类；设计和实现一个多项式类，可以用于计算机科学的许多领域。

13.6.1 带重载运算符的 Fraction 类

我们曾经在第 7 章中创建了一个 Fraction（分数）类。在本节中，我们将展示如何为此类重载多个运算符。

接口文件

程序 13-10 显示了 Fraction 类的接口文件的代码清单。

程序 13-10 文件 fraction.h

```
1  /*******************************************************************
2   * Fraction 类的接口文件                                               *
3   *******************************************************************/
4  # ifndef Fraction_H
5  # define Fraction_H
6  # include <iostream>
7  # include <cassert>
8  # include <iomanip>
9  # include <cmath>
10 using namespace std;
11
12 // Fraction 类的定义
13 class Fraction
14 {
15    private:
16       int numer;
17       int denom;
18       int gcd (int n, int m = 1);  // 辅助函数
19       void normalize ();    // 辅助函数
20    public:
21       Fraction (int numer, int denom);  // 参数构造函数
22       Fraction (double value);  // 参数构造函数
23       Fraction ();  // 默认构造函数
24       Fraction (const Fraction& fract);  // 拷贝构造函数
25       ~Fraction ();  // 析构函数
26       // 成员运算符
27       operator double ();  // 转换运算符
28       const Fraction  operator+( ) const;    // 一元正运算符
29       const Fraction  operator-( ) const;  // 一元负运算符
30       Fraction& operator++ ( );  // 前缀递增运算符
31       Fraction& operator--( );    // 前缀递减运算符
32       const Fraction operator++ (int); // 后缀递增运算符
33       const Fraction operator--(int);  // 后缀递减运算符
34       Fraction& operator= (const Fraction& right); // 赋值运算符
35       Fraction& operator+= (const Fraction& right); // 复合赋值运算符
36       Fraction& operator-= (const Fraction& right); // 复合赋值运算符
37       Fraction& operator*= (const Fraction& right); // 复合赋值运算符
38       Fraction& operator/= (const Fraction& right); // 复合赋值运算符
39       //友元算术运算符
40       friend const Fraction operator+
41             (const Fraction& left, const Fraction& right); //加法运算符
42       friend const Fraction operator-
43             (const Fraction& left, const Fraction& right); //减法运算符
44       friend const Fraction operator*
45             (const Fraction& left, const Fraction& right); //乘法运算符
46       friend const Fraction operator/
47             (const Fraction& left, const Fraction& right); //除法运算符
48       // 友元关系运算符
49       friend  bool  operator==
50             (const  Fraction& left, const Fraction& right);
51       friend  bool  operator!=
```

```
52              (const Fraction& left, const Fraction& right);
53          friend bool operator<
54              (const Fraction& left, const Fraction& right);
55          friend bool operator<=
56              (const Fraction& left, const Fraction& right);
57          friend bool operator>
58              (const Fraction& left, const Fraction& right);
59          friend bool operator>=
60              (const Fraction& left, const Fraction& right);
61          //提取运算符和插入运算符
62          friend istream& operator >> (istream& left, Fraction& right) ;
63          friend ostream& operator << (ostream& left, const Fraction& right) ;
64  };
65  #endif
```

实现文件

程序 13-11 显示了 Fraction 类的实现文件的代码清单。

程序 13-11　文件 fraction.cpp

```
1   /*********************************************************************
2     * Fraction 类的实现文件                                            *
3     *********************************************************************/
4   #include "fraction.h"
5
6   //参数构造函数
7   Fraction :: Fraction (int num, int den = 1)
8   : numer (num), denom (den)
9   {
10    normalize ();
11  }
12  //参数构造函数
13  Fraction :: Fraction (double value)
14  {
15    denom = 1;
16    while ((value - static_cast <int> (value)) > 0.0)
17    {
18        value *= 10.0;
19        denom *= 10;
20    }
21    numer = static_cast <int> (value);
22    normalize ();
23  }
24  //默认构造函数
25  Fraction :: Fraction ( )
26  : numer (0), denom (1)
27  {
28  }
29  //拷贝构造函数
30  Fraction :: Fraction (const Fraction& fract)
31  : numer (fract.numer), denom (fract.denom)
32  {
33  }
```

```
34  //析构函数
35  Fraction :: ~Fraction ()
36  {
37  }
38  //转换运算符
39  Fraction :: operator double ()
40  {
41     double num = static_cast <double> (numer);
42     return (num / denom);
43  }
44  //一元正运算符
45  const Fraction  Fraction :: operator+ ( ) const
46  {
47     Fraction temp (+numer, denom);
48     return temp;
49  }
50  //一元负运算符
51  const Fraction   Fraction :: operator- ( ) const
52  {
53     Fraction temp (-numer, denom);
54     return temp;
55  }
56  //前缀递增运算符
57  Fraction&  Fraction :: operator++ ( )
58  {
59     numer = numer + denom;
60     this -> normalize ();
61     return *this;
62  }
63  //前缀递减运算符
64  Fraction&  Fraction :: operator-- ( )
65  {
66     numer = numer - denom;
67     this -> normalize ();
68     return *this;
69  }
70  //后缀递增运算符
71  const  Fraction  Fraction :: operator++ (int)
72  {
73     Fraction temp (numer, denom);
74     ++(*this);
75     return temp;
76  }
77  //后缀递减运算符
78  const Fraction  Fraction :: operator-- (int)
79  {
80     Fraction temp (numer, denom);
81     --(*this);
82     return temp;
83  }
84  //赋值运算符
85  Fraction& Fraction :: operator= (const Fraction& right)
```

```
86  {
87      if (*this != right)
88      {
89          numer = right.numer;
90          denom = right.denom;
91      }
92      return *this;
93  }
94  // 复合赋值运算符（+=）
95  Fraction& Fraction :: operator+= (const Fraction& right)
96  {
97      numer  = numer * right.denom + denom * right.numer;
98      denom = denom * right.denom;
99      normalize ();
100     return *this;
101 }
102 // 复合赋值运算符（-=）
103 Fraction& Fraction :: operator-= (const Fraction& right)
104 {
105     numer = numer * right.denom - denom * right.numer;
106     denom = denom * right.denom;
107     normalize ();
108     return *this;
109 }
110 // 复合赋值运算符（*=）
111 Fraction& Fraction :: operator*= (const Fraction& right)
112 {
113     numer = numer * right.numer;
114     denom = denom * right.denom;
115     normalize ();
116     return *this;
117 }
118 // 复合赋值运算符（/=）
119 Fraction& Fraction :: operator/= (const Fraction& right)
120 {
121     numer = numer * right.denom;
122     denom = denom * right.numer;
123     normalize ();
124     return *this;
125 }
126 // 加法运算符（友元函数）
127 const Fraction operator+ (const Fraction& left, const Fraction& right)
128 {
129     int newNumer = left.numer * right.denom + right.numer * left.denom;
130     int newDenom = left.denom * right.denom;
131     Fraction result (newNumer, newDenom);
132     return result;
133 }
134 // 减法运算符（友元函数）
135 const Fraction operator- (const Fraction& left, const Fraction& right)
136 {
137     int newNumer = left.numer * right.denom - right.numer * left.denom;
```

```
138	   int newDenom = left.denom * right.denom;
139	   Fraction result (newNumer, newDenom);
140	   return result;
141	}
142	//乘法运算符（友元函数）
143	const Fraction operator* (const Fraction& left, const Fraction& right)
144	{
145	   int newNumer = left.numer * right.numer;
146	   int newDenom = left.denom * right.denom;
147	   Fraction result (newNumer, newDenom);
148	   return result;
149	}
150	//除法运算符（友元函数）
151	const Fraction operator/ (const Fraction& left, const Fraction& right)
152	{
153	   int newNumer = left.numer * right.denom;
154	   int newDenom = left.denom * right.numer;
155	   Fraction result (newNumer, newDenom);
156	   return result;
157	}
158	//相等运算符（友元函数）
159	bool operator== (const Fraction& left, const Fraction& right)
160	{
161	   return (left.numer * right.denom == right.numer * left.denom) ;
162	}
163	//不相等运算符（友元函数）
164	bool operator!= (const Fraction& left, const Fraction& right)
165	{
166	   return (left.numer * right.denom != right.numer * left.denom) ;
167	}
168	//小于运算符（友元函数）
169	bool operator< (const Fraction& left, const Fraction& right)
170	{
171	   return (left.numer * right.denom < right.numer * left.denom) ;
172	}
173	//小于或者等于运算符（友元函数）
174	bool operator<= (const Fraction& left, const Fraction& right)
175	{
176	   return (left.numer * right.denom <= right.numer * left.denom) ;
177	}
178	//大于运算符（友元函数）
179	bool operator> (const Fraction& left, const Fraction& right)
180	{
181	   return (left.numer * right.denom > right.numer * left.denom) ;
182	}
183	//大于或者等于运算符（友元函数）
184	bool operator>= (const Fraction& left, const Fraction& right)
185	{
186	   return (left.numer * right.denom >= right.numer * left.denom) ;
187	}
188	//提取运算符（友元函数）
189	istream&  operator >> (istream& left, Fraction& right)
```

```
190 {
191     cout << "Enter the value of numerator: " ;
192     left >> right.numer;
193     cout << "Enter the value of denominator: " ;
194     left >> right.denom;
195     right.normalize( );
196     return left ;
197 }
198 // 插入运算符（友元函数）
199 ostream&  operator << (ostream& left, const Fraction& right)
200 {
201     left << right.numer << "/" << right.denom ;
202     return left;
203 }
204 // 辅助函数（最大公因子）
205 int Fraction :: gcd (int n, int m)
206 {
207     int gcd = 1;
208     for (int k = 1; k <= n && k <= m; k++)
209     {
210         if (n % k == 0 && m % k == 0)
211         {
212             gcd = k;
213         }
214     }
215     return gcd;
216 }
217 // 辅助函数（规范化分数）
218 void Fraction :: normalize ()
219 {
220     if (denom == 0)
221     {
222         cout << "Invalid denomination in fraction. Need to quit." << endl;
223         assert (false);
224     }
225     if (denom < 0)
226     {
227         denom = -denom;
228         numer = -numer;
229     }
230     int divisor = gcd (abs (numer), abs (denom));
231     numer = numer / divisor;
232     denom = denom / divisor;
233 }
```

应用程序

程序 13-12 显示了测试 Fraction 类的操作的应用程序文件的代码清单。

程序 13-12　文件 app.cpp

```
1 /***************************************************************
2  * 本应用程序测试 Fraction 类                                    *
3  ***************************************************************/
```

```
4   # include "fraction.h"
5
6   int main ()
7   {
8     // 创建两个对象并测试正/负运算符
9     Fraction fract1 (2, 3);
10    Fraction fract2 (1, 2);
11    cout << "fract1: " << fract1 << endl;
12    cout << "fract2: " << fract2 << endl;
13    cout << "Result of +fract1: " << +fract1 << endl;
14    cout << "Result of -fract2: " << -fract2 << endl << endl;
15    // 创建四个对象并测试 ++ 和 -- 运算符
16    Fraction fract3 (3, 4);
17    Fraction fract4 (4, 5);
18    Fraction fract5 (5, 6);
19    Fraction fract6 (6, 7);
20    cout << "fract3: " << fract3 << endl;
21    cout << "fract4: " << fract4 << endl;
22    cout << "fract5: " << fract5 << endl;
23    cout << "fract6: " << fract6 << endl << endl;
24    ++fract3;
25    --fract4;
26    fract5++;
27    fract6--;
28    cout << "Result of ++fract3: " << fract3 << endl;
29    cout << "Result of --fract4: " << fract4 << endl;
30    cout << "Result of fract5++: " << fract5 << endl;
31    cout << "Result of fract6--: " << fract6 << endl << endl;
32    // 测试复合赋值运算符
33    Fraction fract7 (3, 5);
34    Fraction fract8 (4, 7);
35    Fraction fract9 (5, 8);
36    Fraction fract10 (7, 9);
37    fract3 += 2;
38    fract4 -= 3;
39    fract5 *= 4;
40    fract6 /= 5;
41    cout << "Result of fract7 += 2: " << fract7 << endl;
42    cout << "Result of fract8 -= 3: " << fract8 << endl;
43    cout << "Result of fract9 *= 4: " << fract9 << endl;
44    cout << "Result of fract10 /= 5: " << fract10 << endl << endl;
45    //创建两个新的对象，并测试友元算术运算符
46    Fraction fract11 (1, 2);
47    Fraction fract12 (3, 4);
48    cout << "fract11: " << fract11 << endl;
49    cout << "fract12: " << fract12 << endl;
50    cout << "fract11 + fract12 : " << fract11 + fract12 << endl;
51    cout << "fract11 - fract12 : " << fract11 - fract12 << endl;
52    cout << "fract11 * fract12 : " << fract11 * fract12 << endl;
53    cout << "fract11 / fract12 : " <<  fract11 /  fract12 << endl << endl;
54    // 创建两个新的对象，并测试关系运算符
55    Fraction fract13 (2, 3);
```

```
56    Fraction fract14 (1, 3);
57    cout << "fract13: " << fract13 << endl;
58    cout << "fract14: " << fract14 << endl;
59    cout << "fract13 == fract14: " << boolalpha;
60    cout << (fract13 == fract14) << endl;
61    cout << "fract13 != fract14: " << boolalpha;
62    cout << (fract13 != fract14) << endl;
63    cout << "fract13 >  fract14: " << boolalpha;
64    cout << (fract13 >  fract14) << endl;
65    cout << "fract13 <  fract14: " << boolalpha;
66    cout << (fract13 <  fract14) << endl << endl;
67    //使用转换构造函数创建两个新的对象
68    Fraction fract15 (5); // 把整数转换为分数
69    Fraction fract16 (23.45); // 把双精度数转换为分数
70    cout << "fract15: " << fract15 << endl;
71    cout << "fract16: " << fract16 << endl << endl;
72    //把分数对象转换为双精度数
73    Fraction fract17 (9, 13);
74    cout << "double value of fract17 (9, 13): ";
75    cout << setprecision (2) << fract17.operator double () << endl << endl;
76    //测试提取运算符
77    Fraction fract18;
78    cin >> fract18;
79    cout << "fract18: " << fract18 << endl;
80    return 0;
81 }
```

运行结果：

```
fract1: 2/3
fract2: 1/2
Result of +fract1: 2/3
Result of -fract2: -1/2

fract3: 3/4
fract4: 4/5
fract5: 5/6
fract6: 6/7

Result of ++fract3: 7/4
Result of --fract4: -1/5
Result of fract5++: 11/6
Result of fract6--: -1/7

Result of fract7+= 2: 13/5
Result of fract8 -= 3: -17/7
Result of fract9 *= 4: 5/2
Result of fract10 /= 5: 7/45

fract11: 1/2
fract12: 3/4
fract11 + fract12 : 5/4
fract11 - fract12 : -1/4
fract11 * fract12 : 3/8
```

```
fract11 / fract12 : 2/3

fract13: 2/3
fract14: 1/3
fract3 == fract14: false
fract13 != fract14: true
fract13 > fract14: true
fract13 < fract14: false

fract15: 5/1
fract16: 469/20

double value of fract17 (9, 13): 0.69

Enter the value of numerator: 6
Enter the value of denominator: 7
Fract18 : 6/7
```

13.6.2　Date 类

本节创建了一个新的类 Date，将日期表示为三个整数的组合：month（月）、day（日）和 year（年）（例如，2/5/2016）。虽然 C++ 中存在一个可以用于获取日期和时间的库类，但该类用于定义当前日期和时间。我们需要一个可以显示过去日期或者未来日期的类型。

日期定义从起始日（随不同的日历而变化）开始经过的天数。在我们的文化中，起始日以公历的第一天 1/1/1 为基准。但是，日期是根据月、日和年来定义的，而不是给出天数（接近 800000 的数值），月日年的方式使日期值不那么大。因为每个月的天数不是固定的，而且闰年有 366 天而不是 365 天，所以日期的计算十分复杂。

设计策略

为了克服这些困难，我们在设计 Date 类时采用了以下几种策略：

1）我们把日历的起始基准日期设置为 1/1/1900，而不是 1/1/1，因为在 16 世纪为了确定闰年对历法进行了更改。

2）为了遵循面向对象程序设计的原则，存储在一个对象中的各数据成员必须相互独立，不允许存在可以用其他数据成员计算出的冗余数据成员，因此仅储存 month、day 和 year。虽然我们需要通过从起始基准日期经过的总天数来计算日期，也需要确定是星期几，但是我们可以使用成员函数来计算额外的信息。

3）为了确定日期是星期几，我们使用一个名为 findTotalDays() 的函数来计算从起始基准日期开始的总天数。我们已知 1900 年 1 月的第一天是星期一。

4）我们使用递增运算符和递减运算符跳转到后一天或者前一天。但是，由于这可能会更改日期的月份或者年份，因此我们使用两个名为 plusReset() 和 minusReset() 的函数，在必要时调整新的日期。

5）为了计算加减若干天数后的新日期，我们将使用复合赋值运算符，但是为了使调整更容易，我们使用递增或者递减来定义函数。换言之，我们将日期递增 20 次，而不是给日期增加 20 天。

6）为了获取两个日期之间相差的天数，我们对每个日期应用函数 findTotalDays()，然后做减法。

不变式

我们必须首先考虑类不变式（参见我们在第 7 章中的讨论）。我们需要考虑的日期对象的不变式如下所示：

1）月份应该在 1 到 12 之间（1 月到 12 月）。

2）每个月的日期应在 1 和该月的天数（28、30、31）之间；如果月份是 2 月且年份是闰年，则应在 1 和 29 之间。

3）年份应该大于或者等于起始年份（我们为日历设置的起始基准日期的年份）。

程序文件

我们分别创建接口文件、实现文件和应用程序文件。

接口文件。程序 13-13 显示了 Date 类的接口文件的代码清单。

程序 13-13　文件 date.h

```
1   /***************************************************************
2    * 接口文件 date.h 定义了类 Date                                 *
3    ***************************************************************/
4   #ifndef DATE_H
5   #define DATE_H
6   #include <iostream>
7   #include <cmath>
8   #include <cassert>
9   #include <string>
10  using namespace std;
11
12  class Date
13  {
14    private:
15          // 实例数据成员
16          int month;
17          int day;
18          int year;
19          // 静态数据成员和成员函数
20          static const int startWeekDay;
21          static const int startYear;
22          static const int daysInMonths [ ];
23          static const string daysOfWeek [ ];
24          static const string monthsOfYear [ ];
25          static bool isLeap (int year);
26          // 私有辅助函数
27          bool isValid () const;
28          string findWeekDay ( );
29          int findTotalDays( ) const;
30          void plusReset ( );
31          void minusReset ( );
32    public:
33          // 构造函数和析构函数
34          Date (int month, int day, int year);
35          Date ( );
36          ~Date ( );
37          // 成员运算符函数
```

```
38        Date& operator++( );
39        Date& operator--( );
40        Date operator++ (int);
41        Date operator--(int);
42        Date& operator+= (int days);
43        Date& operator-= (int days);
44        bool operator== (const Date& right ) const;
45        bool operator!= (const Date& right ) const;
46        Date& operator= (const Date& right );
47        // 友元运算符函数
48        friend int operator- (const Date& date1, const Date& date2);
49        friend ostream& operator<< (ostream& output , const Date& date);
50  };
51  #endif
```

实现文件。我们给出了接口文件中定义的所有成员函数的实现代码，我们还初始化了接口文件中声明的静态数据成员。程序 13-14 显示了 Date 类的实现文件的代码清单。

程序 13-14　文件 date.cpp

```
1   /*****************************************************************
2    * 实现文件 date.cpp                                             *
3    * 为 Date 类定义了实例数据成员函数和辅助函数                    *
4    *****************************************************************/
5   #include "date.h"
6
7   // 参数构造函数
8   Date :: Date (int m, int d, int y )
9   : month (m), day (d), year (y)
10  {
11    if (!isValid ())
12    {
13        cout << "Date is not valid; program terminates!" << endl;
14        assert (false);
15    }
16  }
17  // 默认构造函数
18  Date :: Date ()
19  : month (1), day(1), year(1900)
20  {
21  }
22  // 析构函数
23  Date :: ~Date ()
24  {
25  }
26  // 前缀递增运算符
27  Date&  Date :: operator++()
28  {
29    day++;
30    plusReset ();
31    return *this;
32  }
33  // 前缀递减运算符
```

```
34  Date& Date :: operator--( )
35  {
36     day--;
37     minusReset ();
38     return *this;
39  }
40  //后缀递增运算符
41  Date Date :: operator++ (int)
42  {
43     Date temp (month, day, year);
44     ++(*this);
45     return temp;
46  }
47  //后缀递减运算符
48  Date Date :: operator--(int)
49  {
50     Date temp (month, day, year);
51     --(*this);
52     return temp;
53  }
54  //复合赋值运算符（+=）
55  Date& Date :: operator+= (int days)
56  {
57     for (int i = 1; i <= days; i++)
58     {
59          ++(*this);
60     }
61     return *this;
62  }
63  //复合赋值运算符（-=）
64  Date& Date :: operator-= (int days)
65  {
66     for (int i = days; i >= 1; i--)
67     {
68          --(*this);
69     }
70     return *this;
71  }
72  // 运算符（==）
73  bool Date :: operator== (const Date& right ) const
74  {
75     bool fact1 = (month == right.month);
76     bool fact2 = (day == right.day);
77     bool fact3 = (year == right.year);
78     return (fact1 && fact2 && fact3);
79  }
80  // 运算符（!=）
81  bool Date :: operator!= (const Date& right ) const
82  {
83     return !(*this == right);
84  }
85  // 赋值运算符（=）
```

```
86  Date& Date :: operator= (const Date& right )
87  {
88      if (*this != right)       // 检查是否为自我赋值
89      {
90          month = right.month;
91          day = right.day;
92          year = right.year;
93      }
94      return *this;
95  }
96  // 减法运算符（友元函数）
97  int operator-(const Date& date1, const Date& date2)
98  {
99      return (date1.findTotalDays() - date2.findTotalDays());
100 }
101 // 输出运算符（友元函数）
102 ostream& operator<< (ostream& output , const Date& date)
103 {
104     cout << Date :: daysOfWeek [(date.findTotalDays()
105                                  + Date :: startWeekDay)% 7] << "   ";
106     cout << Date :: monthsOfYear [date.month] << " ";
107     cout << date.day << " ";
108     cout << date.year << endl;
109 }
110 // 用于验证的私有函数
111 bool Date :: isValid () const
112 {
113     bool validMonth = (month >= 1) && (month <=12);
114     bool validYear = (year >= startYear);
115     bool validDay = (day >= 1) && (day <= (Date:: daysInMonths[month]
116                          + (isLeap (year) && month == 2)));
117     return (validMonth && validYear && validDay);
118 }
119 // 用于递增后调整日期的私有函数
120 void Date :: plusReset ()
121 {
122     bool extraDay = (isLeap (year) && month == 2);
123     if (day > daysInMonths[month] + extraDay)
124     {
125         day = 1;
126         month++;
127     }
128     if (month > 12)
129     {
130         month = 1;
131         year++;
132     }
133 }
134 // 用于递减后调整日期的私有函数
135 void Date :: minusReset ()
136 {
137     if (day < 1)
```

```
138    {
139        month--;
140        if (month < 1)
141        {
142            month = 12;
143            year--;
144        }
145        bool extraDay = isLeap (year) && (month == 2);
146        day = daysInMonths[month] + extraDay ;
147    }
148 }
149 // 用于获取总天数的私有函数
150 int Date :: findTotalDays() const
151 {
152     int totalDays = 0;
153     int currentYear = startYear;
154     while (year > currentYear)
155     {
156         totalDays += 365 + isLeap(currentYear);
157         currentYear++;
158     }
159     int currentMonth = 1;
160     while (month > currentMonth)
161     {
162         totalDays += daysInMonths [currentMonth];
163         if (currentMonth == 2)
164         {
165             totalDays += isLeap(year);
166         }
167         currentMonth++;
168     }
169     totalDays += day - 1;
170     return totalDays;
171 }
172 // 初始化静态数据成员
173 const int Date :: startWeekDay = 1;
174 const int Date :: startYear = 1900;
175 const int Date :: daysInMonths [ ] = {0, 31, 28, 31, 30, 31,
176                                       30, 31, 31, 30, 31, 30, 31};
177 const string Date :: daysOfWeek [ ] = {"Sun", "Mon", "Tue", "Wed",
178                                        "Thr", "Fri", "Sat"};
179 const string Date :: monthsOfYear [ ] = {"", "Jan", "Feb", "Mar", "Apr",
180        "May", "Jun", "Jul", "Aug", "Sep", "Oct", "Nov", "Dec"};
181 // 静态成员函数的定义
182 bool Date :: isLeap (int year)
183 {
184     return (year % 400 == 00) || ((year % 4 == 0)&& (year % 100 != 0));
185 }
```

应用程序文件。应用程序文件必须创建 Date 类型的对象并对其进行操作。用户需要获得公共接口，以便能够将其包含在程序中；用户还需要实现文件的编译版本，以便能够创建自己的应用程序文件的版本。程序 13-15 是一个示例。

程序 13-15　文件 app.cpp

```
1  /****************************************************************
2   * 本应用程序文件 date.cpp 使用 Date 对象                          *
3   ****************************************************************/
4  #include "date.h"
5
6  int main ( )
7  {
8    // 创建两个日期对象并打印
9    Date date1 (2, 8, 2014);
10   Date date2 (10, 15, 1944);
11   cout << "date1: " << date1;
12   cout << "date2: " << date2;
13   // 创建另外两个日期对象，递增日期，并打印
14   Date date3 = date1;
15   Date date4 = date2;
16   date3++;
17   date4++;
18   cout << "date3: " << date3;
19   cout << "date4: " << date4;
20   // 将前面的日期增加若干天数和减去若干天数
21   date3 += 20;
22   date4 -= 130;
23   cout << "date3 after change: " << date3;
24   cout << "date4 after change: " << date4;
25   // 计算并打印出两个日期相差的天数
26   cout <<"Difference between date3 and date4: "
27                                  << date3 - date4 << " days.";
28   return 0;
29 }
```

```
运行结果:
date1: Sat   Feb 8 2014
date2: Sun   Oct 15 1944
date3: Sun   Feb 9 2014
date4: Mon   Oct 16 1944
date3 after change: Sat   Mar 1 2014
date4 after change: Thr   Jun 8 1944
Difference between date3 and date4: 25468 days.
```

13.6.3　多项式

多项式是计算机科学中众多不同领域（例如，网络和网络安全）的基本数学结构之一。具有一个变量的多项式可以定义如下，其中 n 是指数，a 是项的系数。

$$a_n x^n + a_{n-1} x^{n-1} + \ldots + a_2 x^2 + a_1 x + a_0$$

多项式的表示方法

为了对多项式进行数学运算，我们对每个项使用两个值。例如，项 $3.0x^5$ 的系数为 3.0，指数为 5。最大的指数称为多项式的次数（degree）。在我们的实现中，我们假设系数是 double 类型，指数是 unsigned int 类型。在这些假设下，我们可以用至少三种方法在计算机

中存储多项式。

在第一种方法中，我们可以定义任意大小的数组来处理多项式的最大次数，并让指数作为数组的索引，系数则对应数组元素的值。这种方法不是很有效。如果我们有多个多项式，它们的次数不同，则需要为每个多项式创建相同的大数组。

第二种方法非常有效，我们使用了一个包含两个数据成员的结构：系数和次数。我们只存储那些系数不是 0.0 的项。这个实现可以使用一个链表（在第 18 章中讨论）来完成，其中每项都通过一个指针链接到下一个项。

我们在本章中使用的第三种方法比第一种方法效率高，但比第二种方法效率低。第三种方法可以实现为一个数组，其中数组大小是相应多项式的次数加 1。例如，如果我们将 3 次多项式乘以 5 次多项式，则分别创建三个大小为 4、6 和 16 的数组，而不是三个大小为 16 的数组（第一种情况）。数组的索引定义了指数。可能存在一些系数为 0 的项，但每个对象都有自己的大小。图 13-13 中，堆内存位于栈内存的左侧，指数最高的项位于左侧，指数最低的项位于右侧。注意，多项式的次数是 5，数组的大小是 6，因为次数从 0 到 5。

多项式：$4.0x^5 + 2.1x^4 + 3.6x^2 + 1.0$

4.0	2.1	0.0	3.6	0.0	1.0
5	4	3	2	1	0

堆　ptr　栈

图 13-13　5 次多项式的表示方法

运算操作

我们为多项式类定义了五种数学运算：加、减、乘、除和求余。假设我们有以下两个多项式（Poly1 和 Poly2）：

Poly1 = $+4.00x^5 +2.00x^3 +5.00x^2 +1.00x^1 +4.00$
Poly2 = $+2.00x^2 + 6.00$

数学上为这些多项式定义了以下五种运算操作：

Poly1 + Poly2 = $+4.00x^5 +2.00x^3 +7.00x^2 + 1.00x^1 + 10.00$
Poly1 - Poly2 = $+4.00x^5 +2.00x^3 +3.00x^2 +1.00x^1 -2.00$
Poly1 * Poly2 = $+8.00x^7 +28.00x^5 + 10.00x^4 + 14.00x^3 +38.00x^2 +6.00x^1 +24.00$
Poly1 / Poly2 = $+2.00x^3 -5.00x^1 +2.50$
Poly1 % Poly2 = $+31.00x^1 -11.00$

多项式类

我们基于前面的讨论定义了一个多项式类 Poly。

接口文件。程序 13-16 显示了 Poly 类的接口文件。我们定义了三个构造函数（第三个是除法中使用的单项式构造函数）、一个析构函数、一个拷贝构造函数和一个赋值运算符。我们为多项式类重载了加法、减法、乘法、除法和余数运算符。我们还重载了提取运算符。我们使用一个成员函数来输入构成多项式的系数。函数 max 用于求两个多项式的最大次数。

程序 13-16　文件 poly.h

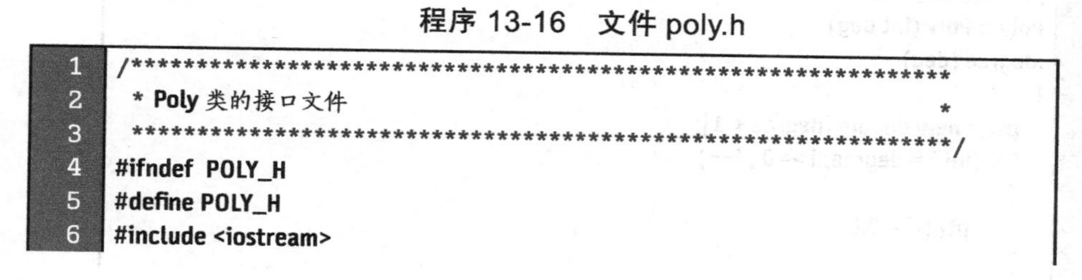

```
1  /***************************************************************
2   * Poly 类的接口文件                                           *
3   ***************************************************************/
4  #ifndef POLY_H
5  #define POLY_H
6  #include <iostream>
```

```
7   #include <string>
8   #include <cassert>
9   #include <iomanip>
10  using namespace std;
11
12  // Poly 类的定义
13  class Poly
14  {
15    private:
16        int degree;
17        double* ptr;
18    public:
19        Poly ();
20        Poly (int degree);
21        Poly (int degree, double coef); // 单项式
22        ~Poly ();
23        Poly (const Poly& origin);
24        Poly& operator= (const Poly& right );
25        void fill ();
26        int max (int x, int y);
27        friend const Poly operator+ (const Poly& left, const Poly& right);
28        friend const Poly operator- (const Poly& left, const Poly& right);
29        friend const Poly operator* (const Poly& left, const Poly& right);
30        friend const Poly operator/ (const Poly& left, const Poly& right);
31        friend const Poly operator% (const Poly& left, const Poly& right);
32        friend ostream& operator << (ostream& left, const Poly& poly);
33  };
34  #endif
```

实现文件。程序 13-17 显示了 Poly 类中声明的函数的实现代码。

程序 13-17　文件 poly.cpp

```
1   /***************************************************************
2    * Poly 类的实现文件                                              *
3    ***************************************************************/
4   #include "poly.h"
5
6   // 默认构造函数
7   Poly :: Poly ()
8   :degree (0)
9   {
10    ptr = 0;
11  }
12  // 参数构造函数
13  Poly :: Poly (int deg)
14  :degree (deg)
15  {
16    ptr = new double [degree + 1];
17    for (int i = degree; i >=0 ; i--)
18    {
19        ptr [i] = 0.0;
20    }
```

```
21  }
22  //创建包含一个项的对象的构造函数
23  Poly :: Poly (int deg, double coef)
24  :degree (deg)
25  {
26      ptr = new double [degree + 1];
27      for (int i = degree; i >=0 ; i--)
28      {
29          ptr [i] = 0.0;
30      }
31      ptr [degree] = coef;
32  }
33  // 析构函数
34  Poly :: ~Poly ()
35  {
36      delete [ ] ptr;
37  }
38  // 拷贝构造函数
39  Poly :: Poly (const Poly& origin)
40  {
41      ptr = new double [degree + 1];
42      for (int i = origin.degree ; i >= 0 ; i--)
43      {
44          ptr [i] = origin.ptr [i];
45      }
46  }
47  //赋值运算符（=）
48  Poly& Poly :: operator= (const Poly& right )
49  {
50      this -> degree = right.degree;
51      this -> ptr = new double [degree + 1];
52      for (int i = right.degree; i >= 0; i--)
53      {
54          (this -> ptr) [i] = right.ptr [i];
55      }
56      return *this;
57  }
58  //加法运算符（+）
59  const Poly operator+ (const Poly& left, const Poly& right)
60  {
61      Poly result (max (left.degree , right.degree));
62      for (int i = result.degree; i >= 0; i--)
63      {
64          if (i <= left.degree && i <= right.degree)
65          {
66              result.ptr [i] = left.ptr [i] + right.ptr[i];
67          }
68          else if (i <= left.degree && i > right.degree)
69          {
70              result.ptr [i] = left.ptr [i];
71          }
72          else
```

```
73          {
74                  result.ptr [i] = right.ptr [i];
75          }
76      }
77      return result;
78  }
79  // 减法运算符 (-)
80  const Poly operator- (const Poly& left, const Poly& right)
81  {
82      Poly result (max (left.degree , right.degree));
83      for (int i = result.degree; i >= 0; i--)
84      {
85          if (i <= left.degree && i <= right.degree)
86          {
87                  result.ptr [i] = left.ptr [i] - right.ptr[i];
88          }
89          else if (i <= left.degree && i > right.degree)
90          {
91                  result.ptr [i] = left.ptr [i];
92          }
93          else
94          {
95                  result.ptr [i] = -right.ptr [i];
96          }
97      }
98      return result;
99  }
100 // 乘法运算符（*）
101 const Poly operator* (const Poly& left, const Poly& right)
102 {
103     int degree = left.degree + right.degree;
104     Poly result (degree);
105     for (int i = result.degree ; i >= 0; i--)
106     {
107         result.ptr [i] = 0;
108     }
109     for (int i = left.degree; i >= 0; i--)
110     {
111         for (int j = right.degree; j >= 0; j--)
112         {
113                 result.ptr [i + j] += (left.ptr [i] * right.ptr [j]);
114         }
115     }
116     return result;
117 }
118 // 除法运算符（/）
119 const Poly operator/ (const Poly& left, const Poly& right)
120 {
121     Poly result (left.degree - right.degree);
122     Poly temp (left.degree);
123     temp = left;
124     int i = temp.degree;
```

```
125     int j = right.degree;
126     int k = i - j;
127     while (i >= j)
128     {
129         double coef = temp.ptr [i] / right.ptr [j];
130         Poly poly (k , coef);
131         temp = temp - ( poly * right);
132         result = result + poly;
133         i--;
134         k = i - j;
135     }
136     return result;
137 }
138 // 余数运算符（%）
139 const Poly operator% (const Poly& left, const Poly& right)
140 {
141     Poly result (left.degree - right.degree - 1);
142     Poly temp (left.degree);
143     temp = left;
144     result = temp - (temp /right) * right;
145     return result;
146 }
147 // 插入运算符（<<）（输出）
148 ostream& operator<< (ostream& output, const Poly& poly)
149 {
150     string sign;
151     for (int i = poly.degree; i >= 0 ; i--)
152     {
153         if (poly.ptr[i] > 0.0 || poly.ptr[i] < 0.0)
154         {
155             cout << fixed << showpos << setprecision (2);
156             cout << poly.ptr[i];
157             cout << noshowpos;
158             if (i != 0)
159             {
160                 cout << "x^";
161                 cout << i;
162             }
163             cout << " " ;
164         }
165     }
166     cout << endl;
167     return output;
168 }
169 // 辅助函数
170 int max (int x, int y)
171 {
172     if (x >= y)
173     {
174         return x;
175     }
176     return y;
```

```
177 }
178 //用于填充多项式系数的函数
179 void Poly :: fill ()
180 {
181   for (int i = degree; i >= 0 ; i--)
182   {
183       cout << "Enter coefficient for exponent " << i << ": " ;
184       cin >> ptr[i];
185   }
186   cout << endl;
187 }
```

应用程序文件。程序 13-18 显示了一个用于测试我们为多项式定义的运算的应用程序。

程序 13-18 文件 app.cpp

```
1  /***************************************************************
2   * 本应用程序用于测试 Poly 类                                    *
3   ***************************************************************/
4  #include "poly.h"
5
6  int main ()
7  {
8    //创建并填充两个多项式类
9    Poly poly1 (5);
10   poly1.fill ();
11   Poly poly2 (2);
12   poly2.fill ();
13   //打印两个多项式的值
14   cout << "Printing the first two polynomials: " << endl;
15   cout << "Poly1 : " << poly1 << endl;
16   cout << "Poly2 : " << poly2 << endl;
17   //在创建的两个多项式上应用五种运算
18   Poly poly3 = poly1 + poly2;
19   Poly poly4 = poly1 - poly2;
20   Poly poly5 = poly1 * poly2;
21   Poly poly6 = poly1 / poly2;
22   Poly poly7 = poly1 % poly2;
23   // 打印运算的结果
24   cout << "Printing the results of operations: " << endl;
25   cout << "Poly1 + Poly2: " << poly3 << endl;
26   cout << "Poly1 - Poly2: " << poly4 << endl
27   cout << "Poly1 * Poly2: " << poly5 << endl;
28   cout << "Poly1 / Poly2: " << poly6 << endl;
29   cout << "Poly1 % Poly2: " << poly7 << endl;
30   return 0;
31 }
```

```
运行结果:
Enter coefficient for degree 5: 4
Enter coefficient for degree 4: 0
Enter coefficient for degree 3: 2
Enter coefficient for degree 2: 5
Enter coefficient for degree 1: 1
```

```
Enter coefficient for degree 0: 4

Enter coefficient for degree 2: 2
Enter coefficient for degree 1: 0
Enter coefficient for degree 0: 6

Printing the first two polynomials:
Poly1 : +4.00x^5 +2.00x^3 +5.00x^2 +1.00x^1 +4.00
Poly2 : +2.00x^2 +6.00

Printing the result of operations:
Poly1 +  Poly2: +4.00x^5 +2.00x^3 +7.00x^2 +1.00x^1 +10.00
Poly1 - Poly2: +4.00x^5 +2.00x^3 +3.00x^2 +1.00x^1 -2.00
Poly1 *  Poly2: +8.00x^7 +28.00x^5 +10.00x^4 +14.00x^3 +38.00x^2 +6.00x^1 +24.00
Poly1  /  Poly2: +2.00x^3 -5.00x^1 +2.50
Poly1 % Poly2: +31.00x^1 -11.00
```

请注意，将商（Poly1 / Poly2）乘以除数（Poly2），然后将结果加上余数（Poly1 % Poly2），我们就可以得到被除数（Poly1）。

本章小结

用户自定义类型的对象在一个函数中可以扮演三个不同的角色：宿主对象、参数对象和返回对象。非静态成员函数需要通过类的实例（被称为宿主对象）来调用。可以使用按值传递、按引用传递或按指针传递将参数对象传递给成员函数。成员函数也可以返回对象。

C++ 中的运算符可以分为三类：不可重载运算符、不推荐重载运算符和可重载运算符。必须注意，不能改变要重载的运算符的优先级、结合性、交换性和元数。还需要注意，不能创建新的运算符符号或者组合现有运算符的符号来创建新的运算符符号。要为用户自定义的数据类型重载运算符，我们需要编写运算符函数，该函数充当运算符。

我们演示了如何将一些运算符重载为成员函数。要将运算符重载为非成员函数，我们有两个选择：重载运算符为全局函数或者友元函数。本书采用第二个选项。友元函数是从类中授予友元访问权限的函数，它可以访问类的私有数据成员。

思考题

1. 给定名为 Fun 的类，指出以下一元运算符函数原型中存在的错误：

```
Fun operator+ ( );
```

2. 给定名为 Fun 的类，指出以下一元运算符函数原型中存在的错误：

```
Fun operator? (int x, int y, int z)
```

3. 给定名为 Fun 的类，指出以下二元运算符函数原型中存在的错误：

```
void operator+ (const Fun& fun, const Fun& fun);
```

4. 给定名为 Fun 的类，指出以下运算符重载原型中存在的错误：

```
double operator [ ] (int x, int y);
```

5. 给定名为 Fun 的类，指出以下运算符重载原型中存在的错误：

```
void operator( ) (int x, int y);
```

6. 给定名为 Fun 的类，以下两个重载运算符原型有什么区别？它们可以在 Fun 类中同时存在吗？

```
Fun operator += (const Fun& fun);
Fun operator += (int x);
```

7. 假设 fun1 和 fun2 是类 Fun 的实例对象。请为以下操作编写运算符函数声明。
   a. fun1 += fun2
   b. fun2 = fun1
   c. fun1++
   d. fun1 (a)
8. 编写函数，重载以下类的复合赋值运算符（+=）。

```
class Sample
{
    private:
        int value;
    public:
        Sample (int value);
        ~Sample();
        ...
}
```

编程题

1. 创建一个名为 Time 的类。Time 类以小时、分钟和秒定义时间点，并指定 AM 或者 PM。小时可以在 1 到 12 之间，分钟可以在 1 到 59 之间，秒也可以在 1 到 59 之间。同时定义一个值为 0 或者 1 的整数来表示 AM 或者 PM。创建默认构造函数，将时间设置为午夜。重载前缀递增运算符（++），把时间增加一秒钟（表示时钟的每一个节拍）。重载 operator() 运算符以查找从午夜开始的持续时间（秒）。重载运算符 +=，增加一段时间（以秒为单位），以获取新的时间点。
2. 设计一个表示复数的 Complex 类。数学中复数的定义为 $x + iy$，其中 x 定义复数的实部，y 是复数的虚部。字母 i 代表−1 的平方根（也就是说 i^2 等于−1）。编写运算符函数，为类 Complex 重载运算符：+=、−=、*=、/= 和 <<。请注意，两个复数之间存在以下关系：

$$(x_1 + iy_1) + (x_2 + iy_2) = (x_1 + x_2) + i\,(y_1 + y_2)$$
$$(x_1 + iy_1) - (x_2 + iy_2) = (x_1 - x_2) + i\,(y_1 - y_1)$$
$$(x_1 + iy_1) * (x_2 + iy_2) = (x_1x_2 - y_1y_2) + i\,(x_1y_2 + x_2y_1)$$
$$(x_1 + iy_1) / (x_2 + iy_2) = ((x_1x_2 + y_1y_2) + i\,(-x_1y_2 + x_2y_1)) / \text{denominator}$$
$$\text{denominator} = x_2^2 + y_2^2$$

3. 重新设计 Complex 类（参见第 2 题），但只为 Complex 类重载运算符：+、−、*、/ 和 <<。
4. 声明和定义表示一组整数的类 Set（集合）。集合是没有重复项、未排序的数据集合。类 Set 应该只包含私有的数据成员：一个指向动态分配的整数数组的指针和一个保存集合大小的整数。下面显示了要为集合定义的运算符。

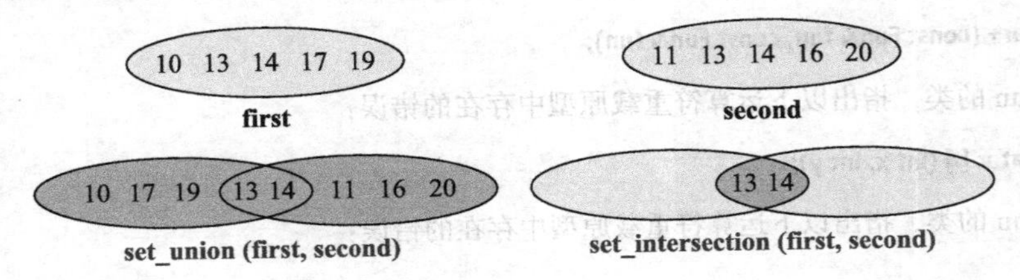

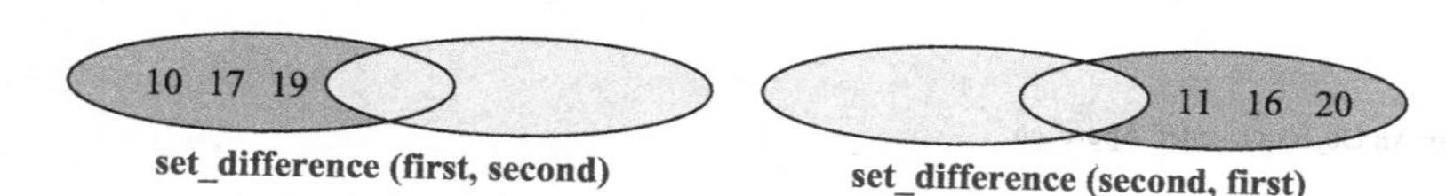

你的解决方案必须包括接口文件、实现文件和应用程序文件。要求包含以下成员方法：

a. 创建空集合的构造函数。

b. 析构函数。

c. 向集合中添加元素的函数（重载 += 运算符）。

d. 从集合中移除元素的函数（重载 -= 运算符）。

e. 用于获取两个集合的交集的二元友元函数（重载运算符 *）。

f. 用于获取两个集合的并集的二元友元函数（重载运算符 +）。

g. 用于获取两个集合的差集的二元友元函数（重载运算符 -）。

h. 打印集合内容的函数。

5. 创建一个表示货币估值器的类 Money（美元和美分的组合）。重载二元加法运算符计算两个货币值对象之和，以获得新值。重载二元减法运算符，从较大的值中减去一个货币值，以获得一个新值。另外，重载 *= 运算符以将货币值乘以一个整数，重载 /= 运算符以将货币值除以一个整数。
6. 重新设计类 Money，仅重载关系运算符（<、>、== 和 !=），以比较两个 Money 对象。
7. 在第 9 章中，我们设计了一个 Pascal 类来定义给定 n 的所有项的系数值。另一种方法是重载函数调用运算符来计算任意项的系数。例如，可以调用 operator()(5, 3)，以获取 x^5y^3 的系数。基于以下公式，可以获取一个项的系数：

x^iy^j 的系数 = factorial (*i* + *j*) / (factorial (*i*) * factorial (*j*))

8. 设计一个 BigInteger 类。BigInteger 类处理以数字字符串形式给出的任意大小的整数，如“345672134579098765”。使用以下准则：

a. 使用我们在第 10 章中为字符串定义的 popBack 函数和 pushFront 函数，来提取大整数最右边的数字，或者在大整数的左边插入一个数字。

b. 重载运算符 <<，以打印大整数。

c. 重载运算符 +，以实现两个大整数的相加。注意，我们在数学中把两个数相加时，可能有进位（数字 1）。例如，将整数 7 和 8 相加得到 5，并将进位添加到下一列。

d. 重载运算符–，以从一个整数中减去另一个整数。注意，当我们从一个数字中减去另一个数字时，可能有借位（数字 1）。例如，从 5 中减去整数 7 将得到 8，并在下一列中使用借位。请注意，如果第一个整数小于第二个整数，则减法的结果必须取补，结果为负整数。

e. 将前导零添加到整数以使其大小在加法和减法中相同，并在打印结果时删除前导零。

第 14 章

C++ Programming: An Object-Oriented Approach

异常处理

在本章中，我们将重点讨论异常处理。异常处理是 C++ 错误处理的术语。我们首先定义异常处理的概念，然后展示如何使用类来实现异常处理。我们还将讨论可以在程序中使用的库异常类。

学习目标

阅读并学习本章后，读者应该能够

- 讨论错误处理的传统方法。
- 讨论在函数中使用 try-catch 块和 throw 语句的三种模式处理异常。
- 讨论定义函数可以引发的异常类型的异常规范，包括任何异常、预定义的异常和无异常。
- 讨论栈展开过程及其对捕获异常的影响。
- 讨论类中的异常，以及如何使用名为 function-try 块的 try-catch 块版本在构造函数中处理异常。
- 强调我们应该避免析构函数中的异常，因为它们破坏了栈展开过程。
- 讨论 C++ 中标准异常类的总体布局和格式，并讨论它们的公共接口及其用途。
- 演示如何创建从标准异常类继承的异常类。

14.1 概述

当我们编写源代码时，在编译过程中可能出现一些错误。然而，许多新手程序员会认为，如果一个程序编译成功，一切就没有问题。另一方面，经验丰富的程序员在用预先确定的一组测试数据测试程序之前，并不会对程序感到满意。即使通过了一个完整的计划测试阶段，程序运行时也会偶尔出现错误。这样的错误被称为异常，因为它很罕见。本章的主题是如何处理这些异常。

14.1.1 错误处理的传统方法

运行时错误表示程序有问题。如果没有语法错误，程序将被编译，但如果产生一个阻止程序继续运行的运行时错误，结果必须中止程序。用于处理运行时错误的常见方法有好几种，我们在这里讨论其中四种方法。

让运行时环境中止程序

第一种方法是什么都不做，当出现异常时让程序终止。

【例 14-1】 在本例中，我们编写了一个简单的程序，该程序从用户那里获取两个整数，将第一个整数除以第二个整数，然后打印结果。程序最多重复五次计算，如程序 14-1 所示。

程序 14-1 两个整数相除

```
1   /***************************************************************
2    * 本程序演示被零除将终止程序运行                               *
```

```
3    *******************************************************************/
4  #include <iostream>
5  using namespace std;
6
7  int main ()
8  {
9     int num1, num2, result;
10    for (int i = 0; i < 5; i++)
11    {
12        cout << "Enter an integer: ";
13        cin >> num1;
14        cout << "Enter another integer: ";
15        cin >> num2;
16        result = num1 / num2;  // 此语句可能导致异常
17        cout << "The result of division is: " << result << endl;
18    }
19    return 0;
20 }
```

运行结果：

```
Enter an integer: 12
Enter another integer: 5
The result of division is: 2
Enter an integer: 10
Enter another integer: 3
The result of division is: 3
Enter an integer: 6
Enter another integer: 0
```

程序编译时没有错误，因为没有语法错误。当我们运行程序时，循环应该重复五次，读取五对整数，将第一个整数除以第二个整数，然后打印结果。在第一次迭代和第二次迭代中一切顺利，但是在第三次迭代中，程序中止（并没有错误消息），因为我们为第二个整数输入了 0，而 C++ 运行时系统不允许除以 0。循环过早终止，程序中止。

请求运行时环境中止程序

第二种方法是测试可能发生的错误。这种方法的优点是可以打印一条消息来解释是什么原因导致程序中止。

【例 14-2】 在本例中，我们重复了上一个程序，但是在每次迭代中检查除数的值，如果除数为零，则打印一条消息并中止程序（参见程序 14-2）。

程序 14-2 强制终止除法运算

```
1  /*******************************************************************
2   * 本程序演示终止程序运行并打印错误信息                              *
3   *******************************************************************/
4  #include <iostream>
5  #include <cassert>
6  using namespace std;
7
8  int main ()
9  {
10    int num1, num2, result;
11    for (int i = 0; i < 5; i++)
```

```
12  {
13      cout << "Enter an integer: ";
14      cin >> num1;
15      cout << "Enter another integer: ";
16      cin >> num2;
17      if (num2 == 0)
18      {
19          cout << "No division by zero!. Program is aborted." << endl;
20          assert (false);
21      }
22      result = num1 / num2;
23      cout << "The result of division is: " << result << endl;
24  }
25  return 0;
26  }
```

```
运行结果：
Enter an integer: 8
Enter another integer: 3
The result of division is: 2
Enter an integer: 9
Enter another integer: 6
The result of division is: 1
Enter an integer: 7
Enter another integer: 0
No division by zero. Program is aborted.
```

经过两次迭代，程序被中止，但是我们打印了一条消息来解释发生了什么。同样程序再次被运行时环境中止，但我们告诉运行时系统使用 assert 宏中止程序。

使用错误检查

第三种方法是在每次迭代中检查第二个数值的值，如果是零，则跳过除法。这比前两种方法更可取，因为我们可以跳过第二个数值为零的情况，并继续处理其他情况。

【例 14-3】 因为我们知道如果除数为零，程序将中止，所以我们可以使用 if-else 语句，并且仅当除数不为零时才执行除法运算。程序 14-3 显示了这种方法。

程序 14-3 一个传统的错误检查程序

```
1   /*****************************************************************
2    * 本程序使用传统的错误检查方法                                    *
3    * 防止程序被终止                                                  *
4    *****************************************************************/
5   #include <iostream>
6   using namespace std;
7
8   int main ( )
9   {
10    int num1, num2, result;
11    for (int i = 0; i < 4; i++)
12    {
13        cout << "Enter an integer: ";
14        cin >> num1;
15        cout << "Enter another integer: ";
```

```
16          cin >> num2;
17          if (num2 == 0)
18          {
19                  cout << "Division cannot be done in this case." << endl;
20          }
21          else
22          {
23                  result = num1 / num2;
24                  cout << "The result of division is: " << result << endl;
25          }
26      }
27      return 0;
28  }
```

```
运行结果：
Enter an integer: 8
Enter another integer: 2
The result of division is: 4
Enter an integer: 7
Enter another integer: 1
The result of division is: 7
Enter an integer: 8
Enter another integer: 0
Division cannot be done in this case.
Enter an integer: 9
Enter another integer: 7
The result of division is: 1
```

计算在程序的第 17 ～ 25 行中完成。如果计算无法完成，则跳过。循环将继续处理其余的数据。

使用函数返回值进行错误检查

第四种方法涉及函数。在 C++ 程序中执行语句总是发生在函数中，而不是在函数之外。我们可以在函数外部（在全局区域）初始化变量，但初始化不是执行语句。如果初始化有问题，则会由编译器检测到。由于执行语句总是发生在一个函数中，我们可以说运行时错误总是发生在一个函数中。以前在面向过程的程序设计中，通常在函数中执行每个计算，并通过检查函数的返回值来检查错误。一个示例是 main 函数，如果没有问题则返回 0，如果存在问题，则返回其他值。以前的程序员使用这种思想设计了一些程序，在程序中如果出现了错误，则函数返回一个特定的值。

【例 14-4】 在本例中，我们重写了前一个程序，使用函数的返回值进行错误检查，如程序 14-4 所示。

程序 14-4　使用函数返回值进行错误检查

```
1  /***************************************************************
2   * 本程序使用函数返回值                                          *
3   * 表示发生了运行时错误                                          *
4   ***************************************************************/
5  #include <iostream>
6  using namespace std;
7
8  // 函数声明
```

```
9   int quotient (int first, int second);
10
11  int main ()
12  {
13     int num1, num2, result;
14     for (int i = 0; i < 3; i++)
15     {
16         cout << "Enter an integer: ";
17         cin >> num1;
18         cout << "Enter another integer: ";
19         cin >> num2;
20         result = quotient (num1, num2);
21         if (result == -1)
22         {
23             cout << "Error, division by zero." << endl;
24         }
25         else
26         {
27             cout << "Result of division is: " << result << endl;
28         }
29     }
30     return 0;
31  }
32
33  // 函数定义
34  int quotient (int first, int second)
35  {
36     if (second == 0)
37     {
38         return -1;
39     }
40     return (first / second);
41  }
```

```
运行结果：
Enter an integer: 6
Enter another integer: 5
The result of division is: 1
Enter an integer: 7
Enter another integer: 0
Error, division by zero.
Enter an integer: 8
Enter another integer: 2
The result of division is: 4
```

注意，当除数为零时，程序将打印一条错误消息，而不是打印计算结果。

传统方法的问题

让我们分析这些方法，以了解为什么我们需要一个新的方法（异常处理）。

1）第一种方法最糟糕。我们让程序在没有任何警告的情况下中止运行。

2）第二种方法稍好。程序仍将中止运行，但用户将得到通知。

3）第三种方法比前两种方法都好，因为会导致程序中止的整数对将被忽略，而程序将继续处理其余的数据。这种方法的问题在于，用于处理错误的代码与程序的功能代码混合在一起。换言之，这里的问题是耦合。错误处理代码与执行任务的代码紧密地耦合在一起，以至于很难区分它们。

4）第四种方法最好，但不能适用于所有情况。此外，模块化程序设计的原理规定函数的返回值只能用于一个目的，而不能用于两个目的。在该例中，一个值（-1）用于报告错误，其他的值则用于返回计算结果。

14.1.2 异常处理的方法

为了避免前面提到的四个问题，C++ 开发了异常处理方法。在这种方法中，运行时系统检测到错误，但不会中止程序。异常处理方法允许程序处理错误。

> *在异常处理方法中，会检测到运行时错误，但程序会处理该错误，仅在必要时中止程序。*

当使用这种方法时，我们仍然需要添加额外的代码，但是用于错误处理的代码与执行程序逻辑的代码之间不存在耦合。检测和处理错误的代码必须遵循一种标准模式，每个 C++ 程序员都应该掌握如何处理异常。

try-catch 语句块

C++ 中的异常处理方法使用了所谓的 try-catch 语句块。try-catch 语句块由两个子句组成。第一个子句称为 **try 子句**，包含可能导致程序中止的代码。运行时环境尝试执行代码。如果代码成功执行，则程序流将继续。如果代码无法执行，系统将抛出异常（基本类型或者类的对象），但不会中止程序。第二个子句称为 **catch 子句**，它允许程序处理异常并在可能的情况下继续处理程序的其余部分。

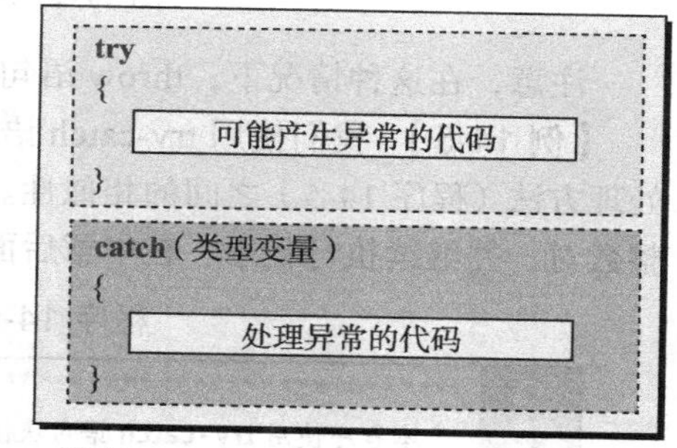

注意：
try子句检测错误的可能性并抛出异常对象。catch子句处理异常以防止程序中止。两条语句必须依次放置，中间不能有其他任何代码，它们属于同一个语句块。

图 14-1 简单的 try-catch 语句块

图 14-1 显示了简单的 try-catch 语句块。在本章后面，我们将看到 try-catch 语句块可以有多个 catch 子句。

图 14-2 显示了 throw 操作符的两个版本。第一个用于抛出异常；第二个用于重新抛出异常（我们将在本章后面讨论第二个版本）。

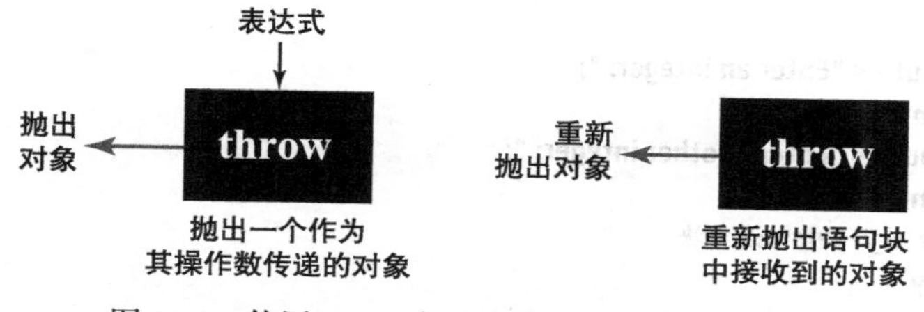

图 14-2 使用 throw 表达式抛出一个对象（异常）

注意，图 14-2 所示的表达式没有返回值。只为了其副作用被调用，并将抛出一个异常。通常在表达式后添加分号来将其更改为表达式语句。

三种模式

异常处理方法通常使用以下三种模式。

第一种模式。在第一个模式中，try-catch 语句块完全包含在一个函数中。如果抛出异常，则忽略 try 子句中 throw 语句之后的其余部分，并将控制转移到 catch 子句。除非在 catch 子句中中止程序，否则函数将在 catch 子句之后继续执行。图 14-3 显示了这种模式。

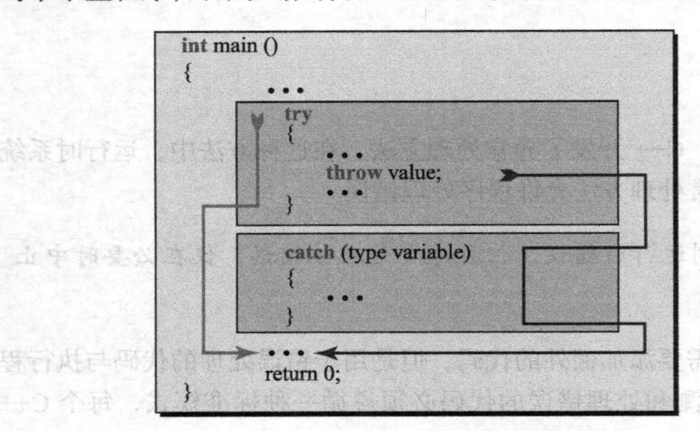

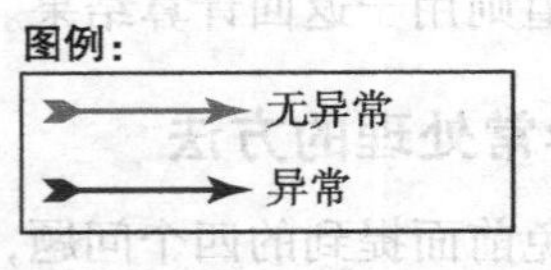

注意：
throw语句直接包含在try语句块中。

图 14-3　调用函数中的 try-catch 语句块

注意，在这种情况下，throw 语句包含在 try 子句中。这种模式并不常见。

【例 14-5】我们使用 try-catch 语句块重写程序 14-3，以比较传统错误检查方法和异常处理方法（程序 14-5）之间的相似性。注意，在本例中，异常处理意味着只忽略引起异常的整数对，并继续执行程序。在本章后面，我们将看到异常处理可能需要其他操作。

程序 14-5　使用 try-catch 语句块

```
1   /***************************************************************
2    * 本程序使用 try-catch 语句块检测错误，                          *
3    * 抛出可被程序捕获和处理的异常                                    *
4    ***************************************************************/
5   #include <iostream>
6   using namespace std;
7
8   int main ()
9   {
10      int num1, num2, result;
11      for (int i = 0; i < 3; i++)
12      {
13          cout << "Enter an integer: ";
14          cin >> num1;
15          cout << "Enter another integer: ";
16          cin >> num2;
17          // try-catch 语句块
18          try
19          {
20              if (num2 == 0)
21              {
22                  throw 0;        //抛出一个整数类型的对象
23              }
```

```
24          result = num1 /num2;
25          cout << "The result is: " << result << endl;
26      }
27      catch (int x)
28      {
29          cout << "Division by zero cannot be performed." << endl;
30      }
31  }
32  return 0;
33 }
```

运行结果：

```
Enter an integer: 6
Enter another integer: 5
The result is: 1
Enter an integer: 7
Enter another integer: 0
Division by zero cannot be performed.
Enter an integer: 8
Enter another integer: 2
The result is: 4
```

在程序 14-5 中，try-catch 语句块是第 18 ～ 30 行。有几个问题需要解释。try 子句包含我们怀疑会产生运行时错误的代码。我们提供逻辑（在本例中是一个决策分支语句）来抛出异常。如果抛出异常，则控制权转移到 catch 子句。如果发生这种情况，异常必须是内置类型或者用户自定义类型的对象。我们决定在本例中抛出 int 类型的对象（值为 0）。如果没有抛出异常，则执行第 24 行和第 25 行，并忽略 catch 子句。还要注意，catch 子句看起来像带异常对象类型的参数的函数，但它不是一个函数。catch 子句是 try-catch 语句块中的子句，它的参数（x）是整数类型的变量。程序输出与程序 14-4 一致。

第二种模式。在第二种模式中，try-catch 语句块仍然在 main 函数中，但在 try 子句中调用的另一个函数中会抛出异常。本例中的 throw 语句位于被调用函数中。当抛出异常时，被调用函数中的其余代码将被忽略，程序流将转移到主调函数中的 catch 语句块。图 14-4 显示了这种模式。

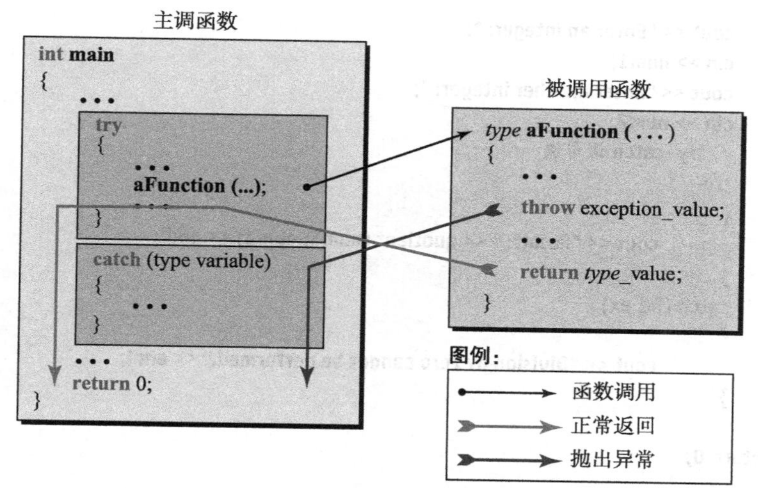

图 14-4　第二种模式（在被调用函数中抛出异常）

第二种模式是首选模式，因为这种模式遵循划分任务的结构化程序设计原则。操作的任务是在被调用函数中定义的，如果有问题，被调用函数将抛出异常。异常处理的任务在主调函数中。

当我们将 throw 语句放入一个函数中时，就意味着在被调用的函数中创建了两个不同的返回点。如果没有抛出异常，被调用函数将在到达 return 语句时返回，程序流返回到主调函数。在这种情况下，catch 子句被忽略。如果抛出异常，被调用函数将提前终止，程序流将返回到主调函数中的 catch 子句。

图 14-4 展示了使用异常处理的另一个优势。我们不必从函数中返回两种值，一种用于成功，另一种用于错误。在发生错误的情况下，将抛出异常并终止函数。

此模式中的 throw 语句是否也包含在 try 子句中？答案是肯定的，但包含是间接的。try 子句包含被调用的函数，被调用的函数包含 throw 语句。

【例 14-6】 在本例中，我们使用第二种模式来重写前一个程序。计算商的任务属于一个名为 quotient 的函数，如果有问题，该函数将抛出 int 类型的异常。main 函数只负责调用函数 quotient，并在引发异常时捕获和处理异常。注意，在这种情况下，捕获异常仅意味着忽略整数对并继续下一个整数对的处理。程序 14-6 显示了该情况。

程序 14-6 捕获由函数抛出的异常

```
/***************************************************************
 * 本程序使用 try-catch 语句块                                    *
 * 捕获由一个函数抛出的异常                                        *
 ***************************************************************/
#include <iostream>
using namespace std;

int quotient (int first, int second);     // 函数声明

int main ()
{
   int num1, num2, result;
   for (int i = 0; i < 3; i++)
   {
        cout << "Enter an integer: ";
        cin >> num1;
        cout << "Enter another integer: ";
        cin >> num2;
        //try-catch 语句块
        try
        {
              cout << "Result: " << quotient (num1, num2) << endl;
        }
        catch (int ex)
        {
              cout << "Division by zero cannot be performed." << endl;
        }
   }
   return 0;
}
```

```
31  // 函数定义
32  int quotient (int first, int second)
33  {
34      if (second == 0)
35      {
36          throw 0;
37      }
38      return first / second;
39  }
```

运行结果：

```
Enter an integer: 12
Enter another integer: 4
Result: 3
Enter an integer: 16
Enter another integer: 0
Division by zero cannot be performed.
Enter an integer: 7
Enter another integer: 2
Result: 3
```

此模式允许函数向调用方抛出异常。换言之，在被调用函数和主调函数之间划分出抛出异常和处理异常的责任。我们创建了一个函数，我们不需要公开函数的代码，我们只需要告诉用户函数的功能（公共接口），并告诉用户函数可能抛出的异常。用户的责任是捕获和处理异常。其优点在库函数的情况下是显而易见的。

异常处理方法的优点之一是我们可以设计能够抛出异常的函数。调用方负责处理异常。

第三种模式。有时我们需要在被调用函数中有一个 try-catch 语句块。当被调用的函数属于独立实体（例如类中的成员函数）时，可能会发生这种情况。当一个函数被另一个函数调用时，如果捕获并处理了一个异常，存在两种情况。

1）被调用函数可以继续执行剩余的代码，并在完成后返回到主调函数。这是第一种模式，在这种模式中，被调用函数与异常处理不相关。

2）被调用函数无法继续执行剩余的代码。控制必须返回到主调函数。否则，程序必须中止。这是第三种模式。我们需要在这两个函数中都有 try-catch 语句块。catch 子句重新抛出异常给主调函数，以便主调函数捕获该异常。图 14-5 显示了此模式。

throw 语句的位置

throw 语句必须直接或者间接包含在 try 子句中才能捕获异常。当 throw 语句显式地位于 try 子句中时，属于直接包含（第一种模式）；当 throw 语句位于 try 子句中调用的函数中时，属于间接包含（第二种模式）。我们讨论的第三种模式是一种特殊情况，其中有两条 throw 语句。图 14-6 显示了直接 try 子句和间接 try 子句之间的区别。

隐藏的 throw 语句

有时我们会看到一个 try-catch 语句块，但看不到 throw 语句。当我们使用一个定义不可见的预定义函数或者库函数时，就会发生这种情况。我们只在 try 子句中看到函数调用，但函数定义包含一条 throw 语句。例如，一些字符串成员函数（例如 at() 函数）在索引超出范围时抛出异常。当我们在程序中调用此函数时，可能需要将函数调用包含在一个 try-catch

语句块中，以捕获异常并防止程序中止。

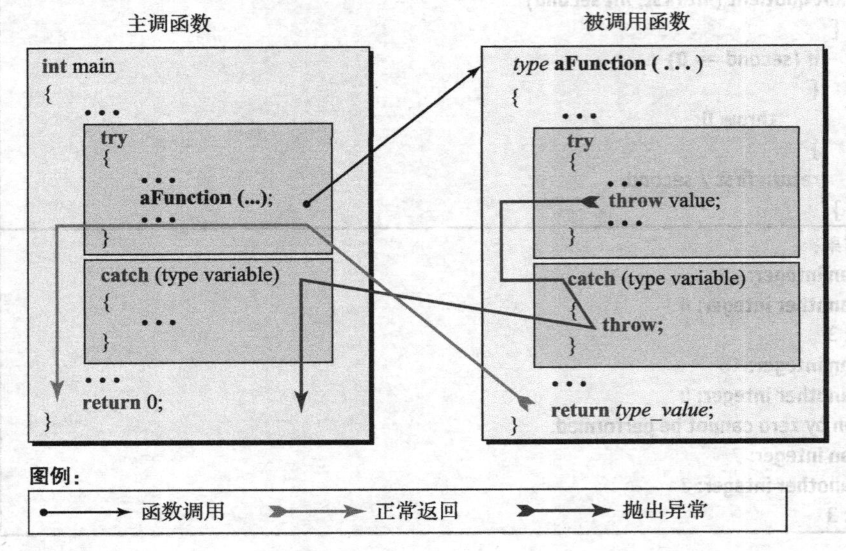

图 14-5　第三种模式（主调函数和被调用函数中都有 try-catch 语句块）

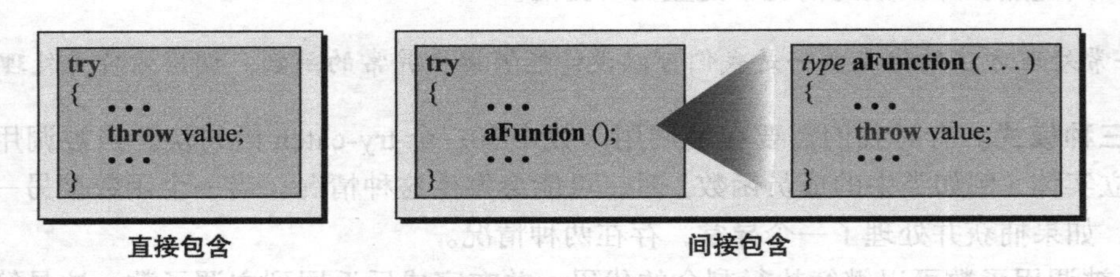

图 14-6　throw 语句的直接包含和间接包含

多条 catch 子句

图 14-7 显示了一个函数可能抛出两种不同类型异常的情况。由于每种类型的处理可能不同，我们可以有两种类型的 catch 子句。

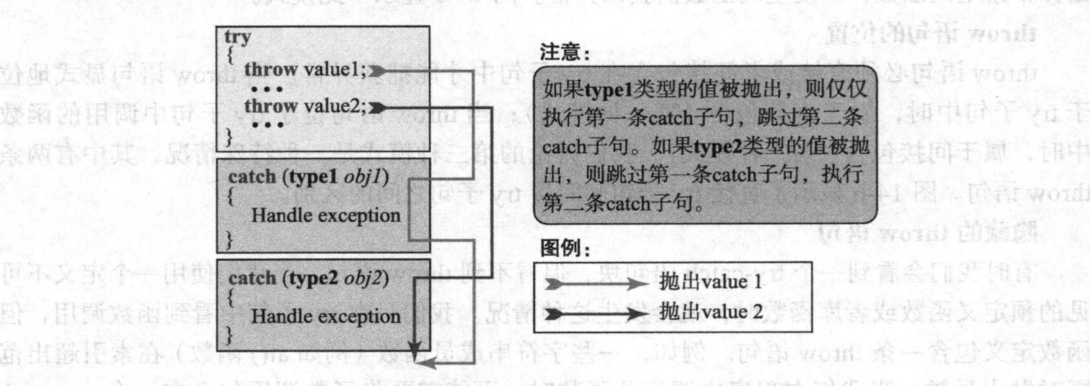

图 14-7　具有多条 catch 子句的 try-catch 语句块

通用 catch 子句

catch 子句只能捕获其参数类型的异常。整数类型的 catch 子句无法捕获抛出的浮点类型的异常。如果要捕获任意类型的异常，可以使用省略号作为 catch 子句中的参数。在这种情况下，如果需要以不同的方式处理特定的异常类型，则将通用（任意类型）catch 子句编码为最后一个 catch 子句。下面显示了这样一个 try-catch 语句块。

```
try
{
  ...
}
catch (int x) //特定类型的 catch 子句
{
  ...
}
catch (...) //省略号表示任意类型的异常
{
  ...
}
```

异常传播

函数可能会抛出异常，但这并不意味着必须在抛出异常的同一个函数中捕获和处理该异常。如果未在抛出异常的位置捕获和处理异常，该异常将被自动传播到函数调用层次结构中的上一个函数，直到路径中的某个函数捕获并处理该异常为止。最后一个可以捕获并处理被抛出异常的函数是 main 函数。如果异常没有被 main 函数捕获和处理，它将被传播到运行时系统，运行时系统捕获异常并导致 main 函数中止，从而中止整个程序。此过程称为**异常传播**（exception propagation）。图 14-8 显示了三个函数中的函数调用路径和异常传播路径：函数 main 由运行时系统调用，函数 first 由函数 main 调用，函数 second 由函数 first 调用。

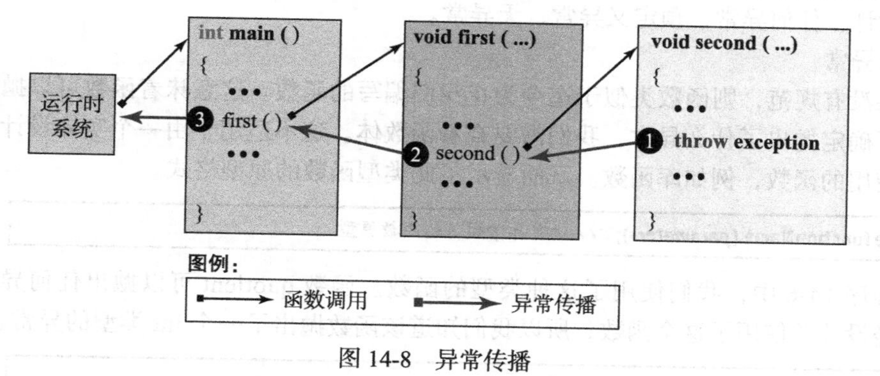

图 14-8 异常传播

如果在函数 second 中抛出了一个异常（显式或者隐式），但未被捕获。则异常将被传播到函数 first，依旧没有被捕获。异常最终被传播到运行时系统，在那里异常被捕获，程序被终止。如果任何函数（second、first 或 main）使用 try-catch 语句块并处理传播的异常，则异常不会到达运行时系统，程序也不会中止。

注意：函数 second 在抛出异常的位置终止；函数 first 在调用函数 second 的位置终止；main 函数则在调用函数 first 的位置终止。这意味着在到达运行时系统之前，在向后路径的某个地方必须有 try-catch 语句块来处理异常。这正是我们在程序 14-6 中的处理方式：异常是在函数 quotient 中抛出的，但是在异常到达运行时系统（这会导致程序中止）之前，我们在 main 函数中捕获并处理异常。

重新抛出异常

在第三种模式中，我们简要讨论了重新抛出一个异常。catch 子句有可能无法或者不希望处理部分或者全部异常。在这种情况下，可以将异常重新抛给上一级的函数。这可能发生在处理异常之前，我们需要释放内存、关闭网络连接或者执行其他维护活动时。我们执行维护活动，然后重新抛出由主调函数处理的异常。要重新抛出异常，我们使用不带任何操作数的 throw 运算符。下面是重新抛出异常的示例。

```
try
{
  ...
}
catch (type variable)
{
  ... //某些处理工作
  throw; // 重新抛出异常给主调函数
}
```

14.1.3 异常规范

当我们为自己的程序编写函数时，我们知道可以在函数中抛出什么类型的异常，并且可以在主调函数中创建适当的 try-catch 语句块。但是，如果我们编写一个函数供其他人使用，我们通常只向用户提供该函数的签名作为公共接口。换言之，用户不知道函数体的内容以及可能抛出哪些异常。在这种情况下，建议在函数头中添加表达式，以告诉用户如果在主调函数中捕获异常，要使用哪种类型的抛出对象。为此，我们说一个函数可以用三种方法中的一种进行设计：任何异常、预定义异常、无异常。

任何异常

如果没有规范，则函数类似于迄今为止我们编写的函数。这意味着函数可以抛出任何异常。为了确定抛出了什么异常，我们需要查看函数体。这不适用于由一个实体设计并由另一个实体使用的函数，例如库函数。下面显示了此类型函数的原型格式。

```
type functionName (parameters);  // 没有异常规范的函数原型
```

在程序 14-6 中，我们使用了这种类型的函数。函数 quotient 可以抛出任何异常。因为我们已经设计并使用了这个函数，所以我们知道该函数抛出了一个 int 类型的异常。

```
int quotient (int first, int second);  // 函数声明
```

预定义异常

如果函数的设计者和用户不同，我们必须在函数头部定义函数抛出的异常（在声明中复制这部分内容）。此类型函数的语法如下所示。函数头部定义了可能从函数中抛出的所有异常对象的类型。

```
type functionName (parameters) throw (type1, type2, ..., typen);
```

我们可以添加一个预定义的规范来显示函数 quotient 可能会抛出一个 int 类型的异常，

如下所示：

```
int quotient (int first, int second) throw (int);  // 函数声明
```

无异常

第三种可能性是向用户声明此函数不引发异常，这意味着用户不需要使用 try-catch 语句块。此类函数的语法如下所示。函数头不使用 throw 关键字，但括号为空。

```
type functionName (parameters) throw ();
```

以下是本规范的一个示例。如果一个函数应该打印两个整数，那么就没有错误条件，不需要进行异常处理。

```
int print (int first, int second) throw ();  // 函数声明
```

14.1.4 栈展开

异常处理中最重要的概念之一是**栈展开**。栈展开与管理分配给程序的内存密切相关。

程序内存

在第 9 章中，我们讨论了当 C++ 程序开始运行时，运行时系统为程序指定了四个内存区域：代码内存、静态内存、栈内存和堆内存。代码内存（程序内存）保存在程序运行期间执行的程序指令。静态内存存储静态变量和全局变量的值。这些值与局部变量分开。栈内存是一个后进先出的内存，类似于我们在餐馆看到的托盘栈。最后一个被推入栈中的托盘是第一个被弹出的托盘。栈内存负责跟踪每个函数的三种信息：参数值、局部变量值和代码内存中主调函数的返回地址。堆内存是程序用来存储可能超过函数生存期的信息的可用内存。

压入和弹出栈内存

异常处理基于栈内存的行为。每个程序由一组相互调用的函数组成。在函数调用期间，有关主调函数的数据（例如参数、局部变量和返回地址）被压入栈内存中。当函数正常或者异常终止（通过抛出异常）时，将使用返回地址。图 14-9 显示了在函数调用期间数据如何被压入栈中。

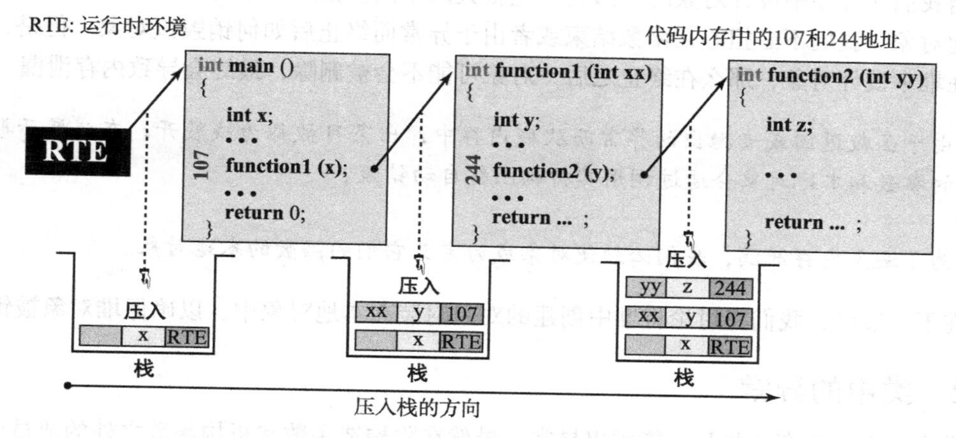

图 14-9　将函数信息压入栈中

在图 14-9 所示的例子中，main 函数被压入栈中，除了局部变量（x）之外没有其他参数，

在 RTE（运行时环境）中调用 main 函数的指令地址被压入栈中。当 main 函数调用 function1 时，function1 的参数（xx）、局部变量（y）和函数调用指令的地址（107）被压入栈中。同样，当 function1 调用 function2 时，function2 的参数（yy）、局部变量（z）和函数调用的地址（244）都存储在栈中。现在栈有三个条目，因为有三个函数调用。注意，每个参数条目可以有零个或者多个参数，每个局部变量记录可以有零个或者多个变量。

图 14-10 显示了在从每个函数返回期间如何从栈中弹出信息。这被称为栈展开。

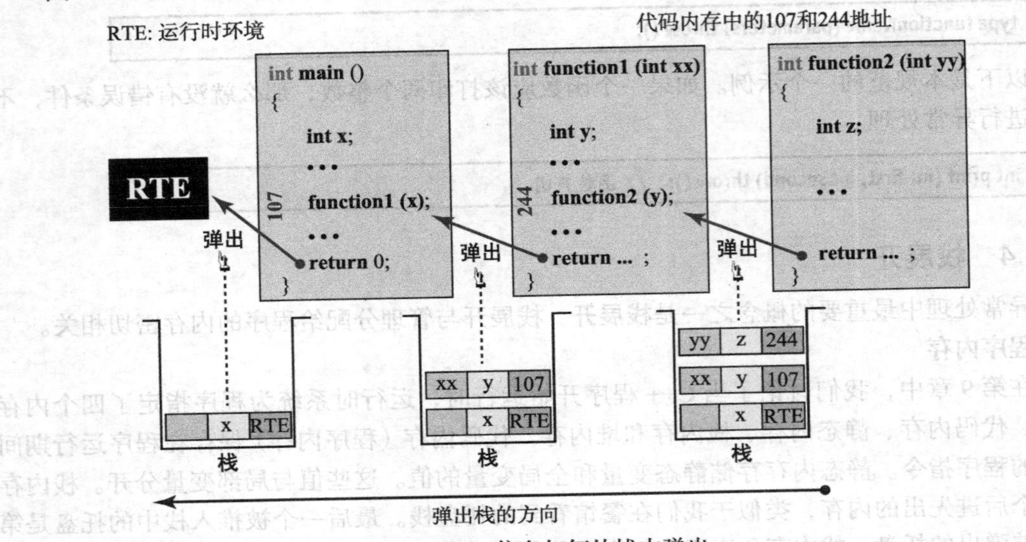

图 14-10　信息如何从栈中弹出

当抛出异常并且控制权返回主调函数时，也会发生栈展开。换言之，当抛出异常时，正在运行的函数会立即弹出堆栈上的最后一个条目而不是等待函数结束运行。

栈展开的效果

栈展开最重要的效果是，当函数返回或者终止时，从栈中弹出相应的条目，并销毁参数和本地对象。如果参数或者对象是类的实例，则会自动调用它们的析构函数进行销毁。

当我们在程序中设计对象时，这会产生很大的不同。如果一个对象被创建为一个参数或者本地对象，我们不必担心在函数结束或者由于异常而终止后如何销毁该对象。但是，如果我们在堆中设计对象，那么在终止之后，对象可能不会被删除，最终会导致内存泄漏。

由于函数返回或者抛出的异常而从栈内存中弹出条目被称为栈展开。在栈展开期间，函数的参数和本地对象会通过调用其析构函数自动销毁。

为了避免内存泄漏，我们必须使对象成为定义它们的函数的本地对象。

在下一节中，我们将讨论将堆中创建的对象封装在本地对象中，以确保堆对象被销毁。

14.2　类中的异常

类中定义的所有函数都可能抛出异常。虽然在除构造函数和析构函数之外的成员函数中处理异常的方式与在独立函数中的处理方式相同，但我们必须注意构造函数和析构函数中的异常。

14.2.1 构造函数中的异常

构造函数是一个函数，但它不同于常规函数，因为构造函数是为创建和初始化对象而设计的。在我们学习如何在构造函数中捕获异常之前，先看看在构造函数中抛出异常时会发生什么。我们考虑两种情况：在第一种情况下，对象完全在栈内存中创建；在第二种情况下，对象部分在堆内存中创建。

在栈内存中创建对象

我们考虑两种情况：在第一种情况下，不会在构造函数中抛出异常；在第二种情况下，会在构造函数中抛出异常。我们假设有一个名为 Integer 的类，它存储一个整数。这种类型的类被称为封装类，在面向对象的程序设计中很常见。

1）在构造函数中不抛出异常。使用栈内存来存储一个对象时，当调用构造函数时，对象会在栈的顶部创建，当析构函数完成执行时，对象会被销毁。调用构造函数将在栈顶部创建对象；构造函数的执行将初始化对象并分配资源（如果有）。另一方面，当调用析构函数时，在析构函数执行期间资源被释放，然后对象被弹出并销毁。换言之，对象在构造函数执行之前就已经存在了，对象在析构函数执行之后就不再存在了，如图 14-11 所示。将对象压入栈是第一步，弹出是最后一步。

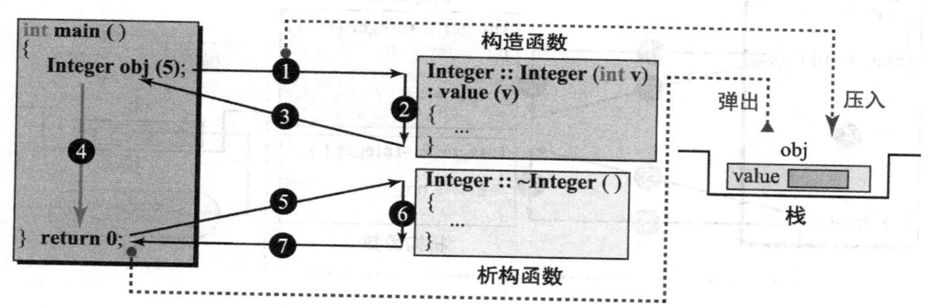

注意：

1. 当调用构造函数时，对象被创建并且被压入栈。
2. 构造函数的执行将初始化对象并且分配资源（如果有）。
3. 初始化后的对象被返回main函数备用。
4. main函数使用对象。
5. 当对象超出作用域范围时，析构函数被调用。
6. 析构函数释放资源（如果有）。
7. 对象从栈中弹出。

图 14-11　当无异常抛出时的构造函数和析构函数

2）在构造函数中抛出异常。假设在构造函数体中抛出了异常。C++ 是这样设计的，如果一个类的构造函数不能完全完成它的任务，则该对象的析构函数将永远不会被调用。图 14-12 显示了与 1 相同的场景，但是构造函数抛出了一个异常，程序终止。栈中分配的所有内存被释放。

在堆内存中对象的部分创建

假设 Integer 对象具有指向堆内存中创建的整数的指针。我们再次讨论两种情况。

1）在构造函数中不抛出异常。图 14-13 显示了构造函数中没有抛出异常的情况。我们假设构造函数在初始化期间在堆内存中创建对象，然后在主体执行期间在对象中存储值。当整数对象超出作用域范围时，将调用析构函数并释放内存。释放内存后，整数对象将从栈内存中弹出。

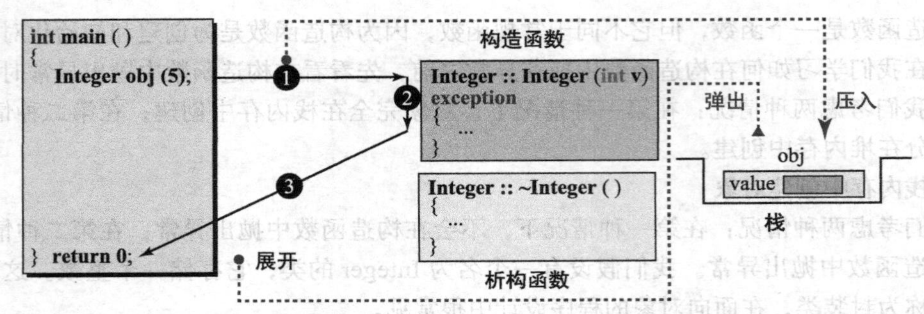

注意：
1. 当调用构造函数时，对象被创建并且被压入栈。
2. 在构造函数初始化（或者分配资源）过程中引发了异常。
3. 抛出异常，终止构造函数的执行。永远不会调用析构函数，但是栈展开弹出对象，此时销毁对象。

图 14-12　当在构造函数中抛出异常

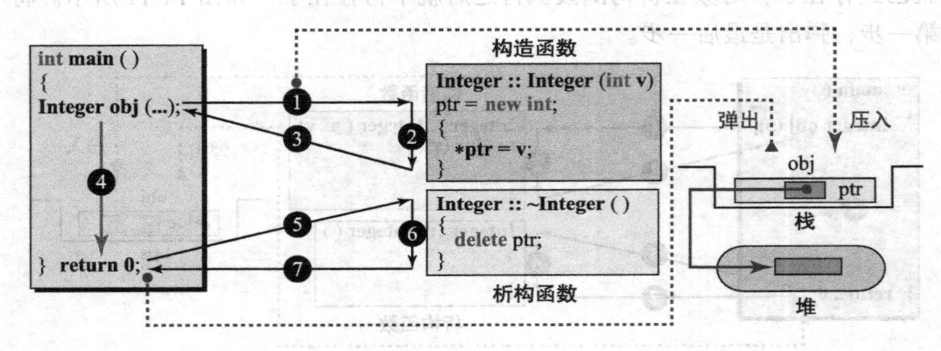

注意：
1. 当调用构造函数时，对象被创建并且被压入栈中。
2. 在构造函数体内执行初始化操作时，将在堆中分配内存，并且值被存储在堆内存中。
3. 构造函数终止，对象返回到main函数。
4. main函数使用对象。
5. 当对象超出作用域范围时，将调用析构函数。
6. 析构函数删除堆中的整数。
7. 析构函数返回后对象超出作用域范围，对象被弹出。

图 14-13　当无异常抛出时的构造函数和析构函数

2）在构造函数中抛出异常。假设当值存储在堆中创建的变量中时，在构造函数体中抛出异常。在这种情况下，构造函数将在未完成其任务的情况下终止。由于对象没有完全构造，因此不会调用析构函数，这意味着永远不会删除堆中分配的内存。结果可能产生内存泄漏。虽然整数对象在栈展开期间从栈中弹出，但不会调用析构函数，也不会释放堆中的内存（图 14-14）。

这里存在的问题是，我们将两个相关任务（分配内存和释放内存）分离开来（通过将数据存储在分配的内存中的另一个任务）。由于第三个任务失败（抛出异常），构造函数已分配内存，但其任务尚未完成。这意味着不会调用析构函数来释放内存。分配内存和释放内存的两个相关任务是部分完成的。内存已分配，但未释放。结果可能产生内存泄漏。

解决方案是将内存的分配和释放结合到一个原子任务中，让一个对象同时负责内存的分配和释放。这就是我们在第 13 章中讨论的智能指针的用武之地。我们需要一个智能指针来处理内存分配和取消分配，而不必担心将数据存储在分配的内存中。

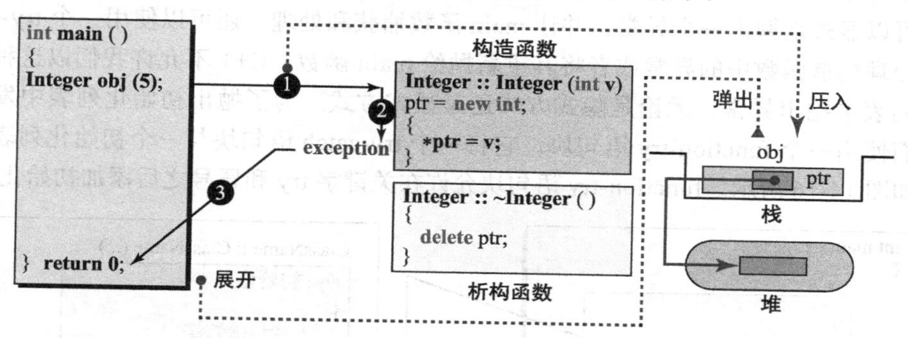

注意：
1. 当调用构造函数时，对象被创建并且被压入栈中。
2. 在堆中地址分配已完成，但是当构造函数需要将值存储在堆中分配的对象中时发生了异常。
3. 抛出异常，main函数终止。在栈展开期间从栈中弹出整数对象，但是因为没有调用析构函数，堆中的内存并未释放。

图 14-14　当有异常抛出时的构造函数和析构函数

使用智能指针进行内存管理。为了把内存的分配和内存的释放紧密联系在一起，我们使用智能指针代替原始指针。智能指针在调用其构造函数时分配内存，在调用其析构函数时释放内存。然而，构造函数没有其他的职责，这意味着当分配结束时，构造任务就完成了，如果抛出任何异常，则栈展开将调用析构函数。换言之，因为智能指针的构造函数只有一个任务，所以总是能保证调用其析构函数，不存在无法释放内存的问题。图 14-15 显示了相同的场景，但使用智能指针代替原始指针。读者可能想知道如果堆中的内存分配失败时会发生什么，答案是没有分配内存，故不用担心。

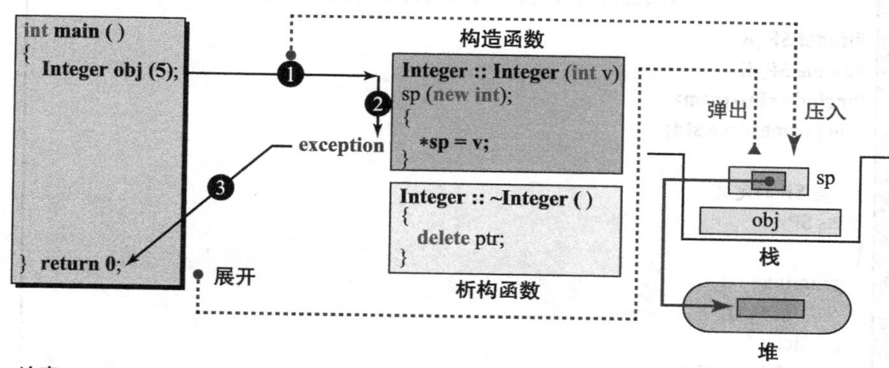

注意：
1. 在栈中创建一个整数对象，继而创建一个智能指针对象，并将其压入栈中。
2. 在构造函数中，智能指针对象在堆中创建一个存储位置，但是在填充存储位置期间抛出异常，这就意味着整数类的构造函数还未完全构造好对象。
3. main函数终止，在栈展开期间从栈中弹出智能指针对象。因为智能指针对象是完全构造的，所以会调用智能指针对象的析构函数，以删除堆中的内存。不会调用整数对象的析构函数，但是不会产生内存泄漏，因为智能指针对象释放了内存。

图 14-15　使用智能指针

构造函数中的 try-catch 语句块：function-try 语句块

迄今为止所讨论的示例中，异常在构造函数体中被隐式地抛出，并传播到 main 函数。我们还可以显式地抛出一个异常，并让 main 函数捕获和处理。还可以使用一个 try-catch 语句块来处理构造函数中的异常或者将其重新抛给 main 函数。C++ 不允许我们以这种方式在初始化列表中抛出异常，无论是隐式方式还是显式方式。为了抛出初始化列表中发生的异常，我们使用一个 function-try 语句块，它将一个 try-catch 语句块与一个初始化列表组合在一起，如图 14-16 所示。function-try 语句块允许在关键字 try 和冒号之后添加初始化。

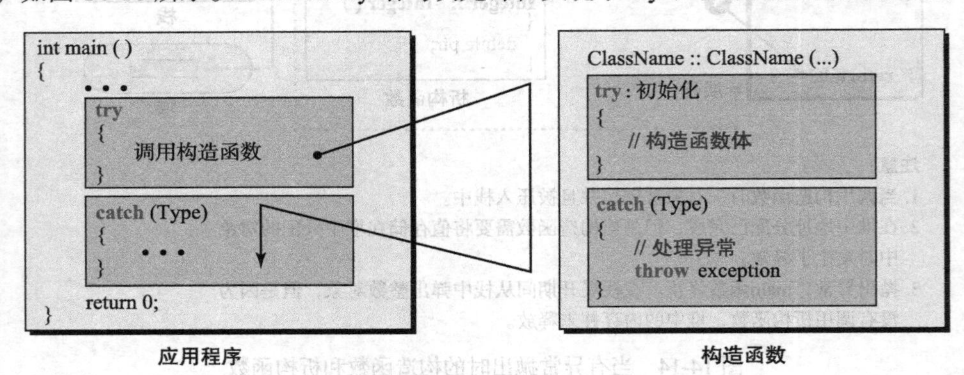

图 14-16　使用 function-try 语句块

一个示例

我们给出了一个 Integer（整数）类的简单示例。程序 14-7 显示了 Integer 类中使用的智能指针（SP）类的接口文件。

程序 14-7　文件 sp.h

```
1	/***************************************************************
2	 * SP 类的接口文件                                              *
3	 ***************************************************************/
4	#ifndef SP_H
5	#define SP_H
6	#include <iostream>
7	using namespace std;
8	
9	// 类 SP 的定义
10	class SP
11	{
12	  private:
13	      int* ptr;
14	  public:
15	      SP (int* ptr);
16	      ~SP ( );
17	      int& operator* ( ) const;
18	      int* operator-> ( ) const;
19	};
20	#endif
```

程序 14-8 显示了 SP 类的实现文件。

程序 14-8 文件 sp.cpp

```
1  /*****************************************************************
2   * SP 类的实现文件                                                *
3   *****************************************************************/
4  #include "sp.h"
5
6  // 构造函数
7  SP :: SP (int* p)
8  : ptr (p)
9  {
10 }
11 // 析构函数
12 SP :: ~SP ()
13 {
14   delete ptr;
15 }
16 // 重载运算符（*）
17 int& SP :: operator* ( ) const
18 {
19   return *ptr;
20 }
21 // 重载运算符（->）
22 int* SP :: operator-> ( ) const
23 {
24   return ptr;
25 }
```

程序 14-9 显示了 Integer 类的接口文件。

程序 14-9 文件 integer.h

```
1  /*****************************************************************
2   * Integer 类的接口文件                                           *
3   *****************************************************************/
4  #ifndef INTEGER_H
5  #define INTEGER_H
6  #include "sp.h"
7
8  // 类 Integer 的定义
9  class Integer
10 {
11   private:
12       SP sp;
13   public:
14       Integer (int value);
15       ~Integer ();
16       int getValue();
17 };
18 #endif
```

程序 14-10 显示了 Integer 类的实现文件。

程序 14-10 文件 integer.cpp

```
/***************************************************************
 * Integer 类的实现文件                                        *
 ***************************************************************/
#include "integer.h"

// 构造函数，使用 function-try 语句块
Integer :: Integer (int v)
try: sp (new int)
{
  *sp = v;
}
catch (...)
{
  throw;
}
// 析构函数
Integer :: ~Integer ()
{
}
// 访问器函数
int Integer :: getValue()
{
  return *sp;
}
```

程序 14-11 显示了应用程序文件。

程序 14-11 文件 app.cpp

```
/***************************************************************
 * 本应用程序测试 Integer 类                                   *
 ***************************************************************/
#include "integer.h"

int main ( )
{
  for (int i = 0; i < 1000000; i++)
  {
        try
        {
                Integer integer (i);
                cout << integer.getValue() << endl;
        }
        catch (...)
        {
                cout << "Exception is thrown" << endl;
        }
  }
  return 0;
}
```

运行结果：

```
0
```

```
1
2
3
4
5
6
...
87245
Exception is thrown
87247
...
92101
Exception is thrown
92103
...
99999
```

为了节省篇幅，我们没有显示所有的输出，但是我们可以看到整数 87246 和 92102 引发了异常。

14.2.2 析构函数中的异常

析构函数在栈展开期间被调用。如果析构函数因抛出异常而中断，则停止栈展开过程。由于这个原因，如果在析构函数中抛出任何异常，则 C++ 调用一个名为 terminator 的全局函数，终止整个程序。

必须避免在析构函数中引发异常。

14.3 标准异常类

本章前几节讨论的异常涉及基本数据类型的异常。C++ 定义了一组标准异常类，用于其类库，如图 14-17 所示。C++ 包含从一个名为 exception 的类派生的标准**异常类**（exception class）。

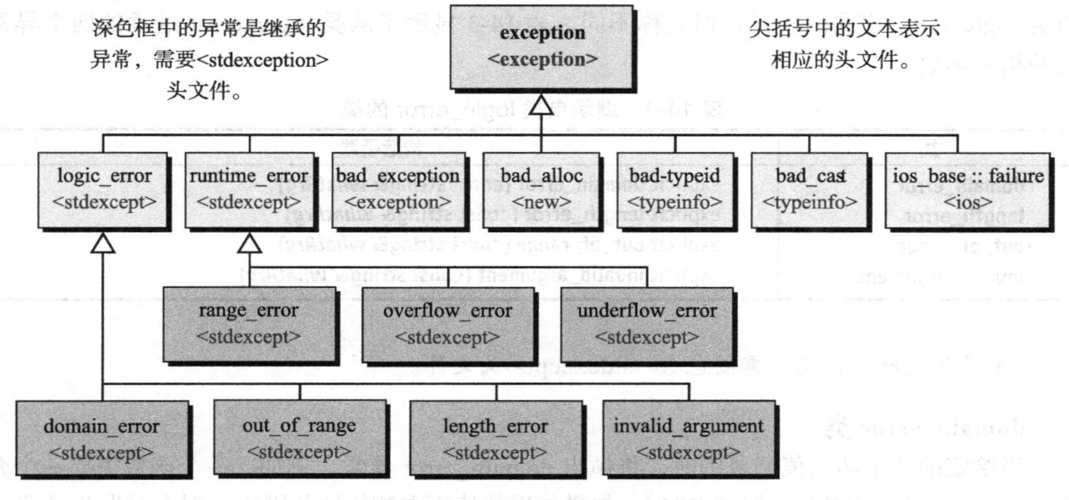

图 14-17 标准异常类的层次结构

仔细研究一下图 14-17 中所示的类的目的和格式，可以帮助我们了解当库函数抛出异常时如何进行处理，并使用这些类或者继承于这些类来创建自定义的异常类。

图 14-17 中层次结构的顶部是在 <exception> 头文件中定义的 exception 类。表 14-1 显示了 exception 类的公共接口。

表 14-1　类 exception 的公共接口

```
exception () throw ()  // 构造函数
exception (const exception&) throw () // 拷贝构造函数
exception&  operator = (const exception&) throw () // 赋值运算符
virtual ~exception () throw ()  // 析构函数
virtual const char* what () const throw ()  // 成员函数
```

类 exception 中定义的所有函数都有异常规范，该规范定义了在函数中不会引发异常。成员函数 what 是一个虚函数，它返回一个 C 字符串，该字符串描述所发生的错误。派生类提供函数 what 的实现。

14.3.1　逻辑错误

如图 14-17 所示，C++ 定义了一个名为 logic_error（逻辑错误）的类。这是其他四个与逻辑错误相关的类的基类。逻辑错误是与函数先决条件相关的错误，这些错误在编译时无法被检测。

> 逻辑错误与函数先决条件相关。

表 14-2 显示了类 logic_error 的公共接口。这些类还从类 exception 继承了 what 成员函数。

表 14-2　类 logic_error 的公共接口

```
explicit logic_error (const  string& whatArg) // 构造函数
virtual const char* what () const throw ()  // 成员函数
```

从类 logic_error 派生出四个类：domain_error（域错误）、length_error（长度错误）、out_of_range（越界错误）和 invalid_argument（无效参数错误）。所有这些类的构造函数都具有与类 logic_error 相同的模式，但名称不同。表 14-3 显示了从类 logic_error 继承的四个异常类的构造函数。

表 14-3　继承自类 logic_error 的类

类	构造函数
domain_error	explicit domain_error (const string& *whatArg*)
length_error	explicit length_error (const string& *whatArg*)
out_of_range	explicit out_of_range (const string& *whatArg*)
invalid_argument	explicit invalid_argument (const string& *whatArg*)

> 要使用逻辑错误类，需要包含 <stdexcept> 头文件。

domain_error 类

当给定的数据超出值域范围时，将抛出 domain_error 异常。例如，一个函数需要一个介于 0.0 和 4.0 之间的参数（例如 GPA），如果传递的值不在该值域范围中，则会引发此异常。

length_error 类

如果对象的长度大于或者小于预定义的长度，则引发 length_error 异常。例如，如果字符串的大小超过成员函数 max_size 返回的值，则字符串类将引发异常。如果数组的大小超过了其预定义的大小，我们可以使用这个类抛出异常。

out_of_range 类

如果索引超出库中某个类的范围，则抛出 out_of_range 异常。例如，字符串类有一个名为 at(…) 的函数，该函数返回其参数索引处的值。如果传递的整数超出了当前字符串对象的索引范围，则会引发此类型的异常。如果数组的索引超出范围，我们可以使用这个类抛出异常。

invalid_argument 类

当存在逻辑错误，但异常的性质与之前定义的三个类都不匹配时，通常会发生 invalid_argument 异常。例如，当我们有一个位集，其中每个位的值应该是 0 或者 1。

14.3.2 运行时错误

如图 14-17 所示，C++ 定义了一个名为 runtime_error（运行时错误）的类。**运行时错误**通常是与一个函数的后置条件相关，例如向上溢出、向下溢出、返回值越界等。

> 要使用运行时错误类，需要包含 <stdexcept> 头文件。

表 14-4 显示了类 runtime_error 的公共接口，同样继承了类 exception 的 what 成员函数。

表 14-4　类 runtime_error 的公共接口

```
explicit runtime_error (const string& whatArg) // 构造函数
virtual const char* what () const throw () // 成员函数
```

从类 runtime_error 派生出三个类：underflow_error（向下溢出错误）、overflow_error（向上溢出错误）和 range_error（范围错误）（参见图 14-17）。所有这些类的构造函数都具有与类 runtime_error 相同的模式，但名称不同。所有类都继承了 what 函数。表 14-5 显示了从类 runtime_error 继承的三个异常类的构造函数。

> 运行时错误与函数的后置条件相关。

表 14-5　继承自类 runtime_error 的类

类	构造函数
underflow_error	explicit underflow_error (const string& *whatArg*)
overflow_error	explicit overflow_error (const string& *whatArg*)
range_error	explicit range_error (const string& *whatArg*)

underflow_error 类

我们在第 3 章中讨论了向下溢出的概念。在算术运算中存在向下溢出的情况。但是，这种类型的错误通常不会被任何算术运算符定义。underflow_error 类可以用于在用户自定义的函数中引发异常。

overflow_error 类

我们在第 3 章中还讨论了向上溢出的概念。在算术运算中也存在向上溢出的情况。但是，这种类型的错误通常不会被任何算术运算符定义。overflow_error 类可以用于在用户自定义的函数中引发异常。

range_error 类

range_error 异常的设计目的是当函数的结果超出预定义的范围时引发异常（与 out_of_range 比较，out_of_range 与函数参数范围内的错误相关）。<cmath> 头文件中的预定义数学函数不会引发这些错误，但由其他来源设计的函数可能会引发这些错误。

14.3.3 其他五个类

如图 14-17 所示，还有五个直接从 exception 类派生的类。这些类与 exception 类具有相同的成员函数，但构造函数、拷贝构造函数、赋值运算符和析构函数具有相应类的名称。在这五个函数中，at() 函数都是相同的。

bad_exception（错误异常）类

如果函数有异常规范，则不会抛出异常。但是，如果函数中发生了一些事情，则会抛出异常。这时函数会抛出一个 bad_exception 类对象。

bad_alloc（错误内存分配）类

头文件 <new> 定义了与动态内存分配相关的类型和函数。如果无法分配请求的内存，则 new 运算符将抛出 bad_alloc 类的对象。

bad_typeid（错误类型）类

当我们定义一个无法实现的类型时，会引发 bad_typeid 异常。例如，如果我们试图对空指针（*P）解引用，将引发此类型的异常。

bad_cast（错误类型转换）类

当动态类型转换失败时，会抛出 bad_cast 类的对象。

failure（失败）类

头文件 <ios> 定义了一个类 failure，可以用作所有输入 / 输出类中抛出异常的基类。请注意，类的名称是 failure，但类的作用域是 ios_base。构造函数中唯一的参数用于定制 at() 函数显示的消息（具体请参见表 14-6）。

表 14-6　类 failure 的公共接口

```
explicit failure (const string& mesg)  // 构造函数
virtual ~failure ()  // 析构函数
virtual const char* what () const  throw ()  // 成员函数
```

在 C++ 中，输入 / 输出类在语言支持的异常处理之前定义。但是，在异常类被添加到语言之后，failure 类被添加到异常类的层次结构中。

输入 / 输出类（如前所述）保存状态，以显示在相应的流中是否发生了错误。为了在输入 / 输出操作中使用异常，C++ 添加了两个名为 exception 的函数，如表 14-7 所示。

表 14-7　输入 / 输出异常类的公共接口

```
void exception (iostate flags) // 设置触发异常的标志
iostate exception() const  // 返回异常的候选标志
```

表 14-7 中显示的第一个函数签名定义了哪些标志被触发时将抛出 failure 类型的异常。第二个函数签名返回为抛出异常定义的标志。

14.3.4 使用标准异常类

我们无须创建自己的异常对象，可以使用 <exception> 或者 <stdexcept> 头文件中定义的标准异常类的对象。

【例 14-7】 程序 14-12 显示了如何使用 invalid_argument 类来处理程序 14-6 中定义的商问题。

程序 14-12 使用 invalid_argument 类

```
/***************************************************************
 * 本程序演示如何使用 invalid_argument 类的对象                *
 * 检测函数中的除零异常                                        *
 ***************************************************************/
#include <stdexcept>
#include <iostream>
using namespace std;

// 函数声明
int quotient (int first, int second);

int main ()
{
  int num1, num2, result;
  for (int i = 0; i < 3; i++)
  {
     cout << "Enter an integer: ";
     cin >> num1;
     cout << "Enter another integer: ";
     cin >> num2;
     // try-catch 语句块
     try
     {
        cout << "Result of division: " << quotient (num1, num2);
        cout << endl;
     }
     catch (invalid_argument ex)
     {
        cout << ex.what () << endl;
     }
  }
  return 0;
}
// 函数定义
int quotient (int first, int second)
{
  if (second == 0)
  {
     throw invalid_argument ("Error! Divide by zero!");
  }
  return first / second;
}
```

```
运行结果：
Enter an integer: 20
Enter another integer: 4
Result of division:  5
Enter an integer: 12
Enter another integer: 0
Error! Divide by zero!
Enter an integer: 14
Enter another integer: 5
Result of division:  2
```

本章小结

我们讨论了传统错误处理的四种方法。第一种方法是什么都不做，当出现错误时让程序在没有任何警告的情况下中止。第二种方法是测试可能的错误，并在出现错误时打印消息。第三种方法是使用错误检查。第四种方法是使用函数返回值进行错误检查。

为了避免这四种方法存在的问题，C++ 开发了异常处理方法。在异常处理方法中，运行时系统检测到错误，但不会中止程序，它允许程序处理错误。异常处理方法使用 try-catch 语句块。我们讨论了 try-catch 语句块的三种模式。在第一种模式中，try-catch 语句块完全位于一个函数中。在第二种模式中，try-catch 语句块位于一个函数中，但在 try 子句中调用的函数抛出异常。在第三种模式中，try-catch 语句块位于被调用的函数中。

我们需要注意构造函数和析构函数中的异常。如果一个对象不能完全由构造函数创建完成，则不会调用相应的析构函数。如果对象全部或者部分在堆内存中创建，这可能会造成严重的问题。在这种情况下抛出异常可能会导致内存泄漏。为了将内存的分配和内存的释放紧密联系在一起，我们使用智能指针代替原始指针。智能指针在其构造函数被调用时分配内存，在其析构函数被调用时释放内存。标准异常类是在 C++ 库中定义的，并且是从 exception 类派生的异常类。标准异常类分为逻辑错误和运行时错误。

思考题

1. 运行以下简短程序时会发生什么？

```
#include <iostream>
using namespace std;

int main ()
{
    int value = 30;
    if (value > 20) throw value;
    cout << value;
    return 0;
}
```

2. 尝试编译以下简短程序时会发生什么？

```
#include <iostream>
using namespace std;

int main ()
```

```
{
    int value = 30;
    try
    {
        if (value > 30) throw value;
    }
    cout << value;
    return 0;
}
```

3. 运行以下简短程序时会发生什么?

```
#include <iostream>
using namespace std;

int main ()
{
    int value = 30;
    try
    {
        if (value < 20) throw value;
    }
    catch (int value)
    {
        cout << "In the catch clause." << endl;
    }
    cout << value << endl;
    return 0;
}
```

4. 运行以下简短程序时会发生什么?

```
#include <iostream>
using namespace std;

void fun (int x )
{
    if (x < 10) throw 10.0;
}

int main ()
{
    try
    {
        fun (5);
    }
    catch (int value)
    {
        cout << value << endl;
    }
    return 0;
}
```

5. 指出以下函数定义中的错误。

```
void fun (int x) throw ()
{
```

```
    if (x < 10) throw 10.0;
}
```

6. 指出以下函数定义中的错误。

```
void fun (double x) throw (double)
{
    if (x < 10.0) throw 10.0;
}
```

7. 以下程序的输出结果是什么？

```
#include <iostream>
using namespace std;

void fun (int x ) throw (int)
{
    if (x > 1000) throw 10000;
}

int main ()
{
    try
    {
        fun (1002 );
    }
    catch (int value)
    {
        cout << value << endl;
    }
    return 0;
}
```

8. 以下程序的输出结果是什么？

```
#include <iostream>
using namespace std;

void second (int x ) throw (int)
{
    if (x > 1000) throw x;
}

void first (int x )
{
    try
    {
        second (1200);
    }
    catch (...)
    {
        throw x * 10;
    }
}

int main ()
{
```

```
    try
    {
        first (10);
    }
    catch (int value)
    {
        cout << value << endl;
    }
    return 0;
}
```

编程题

1. 我们可以使用类的对象，而不是使用基本数据类型作为要抛出的异常类型。重写程序 14-5，并使用名为 DivByZero 的类来抛出异常对象而不是整数。类 DivByZero 不需要任何数据成员或者成员函数。
2. 重写第 1 题的程序，但为 DivByZero 类定义构造函数和成员函数 what()。
3. 重写第 2 题的程序，使用从标准异常类 invalid_argument 继承的 DivByZero 类，invalid_argument 适用于此目的。当我们把两个整数相除时，可以调用库运算符 (/)，其中第二个参数是一个不能为 0 的整数。
4. 编写一个程序，提示用户以小时、分钟和秒的形式输入时间。然后程序以秒为单位计算时间。程序必须抛出三个不同的异常对象：HExcept（如果小时为负）、MExcept（如果分钟不在 0 到 59 之间）和 SExcept（如果秒不在 0 到 59 之间）。创建三个从标准异常类 out_of_range 继承的异常类。使用多态性（按引用捕获）和一个 catch 子句来处理所有情况。
5. 当函数 at 尝试访问不在范围内的字符时，<string> 库抛出类型为 out of_range 的异常。编写一个程序，创建由英文大写字母组成的字符串，并打印给定索引处的字符。如果用户输入的索引小于 1 或者大于 26，请使用 try-catch 语句块捕获错误。注意，在这种情况下，我们不需要 throw 语句，因为库抛出了一个异常。我们只需要捕获异常即可。

第 15 章

C++ Programming: An Object-Oriented Approach

泛型编程：模板

在本章中，我们将重点讨论泛化（generalization），这意味着要编写一个可以在若干特殊情况下使用的通用程序。C++ 把这个过程称为模板编程（template programming）。我们首先讨论如何编写泛型函数（称为函数模板），然后讨论泛型类（称为类模板）。

学习目标

阅读并学习本章后，读者应该能够

- 讨论函数模板，并将其作为创建一组具有相同代码逻辑但其代码可以应用于不同数据类型的函数的工具。
- 给出函数模板的语法，并演示编译器如何在编译期间生成一组非模板函数。
- 讨论函数模板中的几个问题，包括非模板参数、显式类型确定、预定义操作、特殊化和重载。
- 讨论如何为模板函数创建单独的接口文件和应用程序文件。
- 讨论类模板，并将其作为创建一组类的工具，这些类只在它们所包含的数据成员的类型上有所不同。
- 讨论编译程序时涉及类模板的两种方法：包含方法和单独编译。
- 讨论模板友元函数和继承类中模板的概念。

15.1 函数模板

不管使用哪种语言编写程序，有时我们需要将相同的代码应用于不同的数据类型。例如，可能有一个函数查找两个整数类型中的较小者，另一个函数查找两个浮点类型中的较小者，等等。首先，我们应该考虑代码（程序逻辑），然后考虑要使用的数据类型。我们可以把这两个任务分开。我们可以编写一个程序来解决一般数据类型的问题，然后将程序应用到我们需要的特定数据类型。这被称为泛型编程或者模板编程。

C++ 中的函数是在零个或者多个对象上应用操作并创建零个或者多个对象的实体。使用**函数模板**（function template），可以在编写程序时定义操作，在编译程序时定义数据类型。换言之，我们可以定义函数族，其中每个函数都有一个或者多个不同的数据类型。

15.1.1 使用函数族

如果我们不使用模板和泛型编程，则必须定义函数族。假设我们需要比较并查找程序中各种数据类型的较小数据项，例如，假设我们需要找到两个字符、两个整数和两个浮点数之间的较小值。由于数据类型不同，如果没有模板，我们需要编写三个函数，如程序 15-1 所示。注意，在 ASCII 中，大写字母在小写字母之前，这意味着 'B' 小于 'a'。

程序 15-1 使用三个重载函数

```
1  /*********************************************************************
2   *本程序用于查找三种不同数据类型的数据中的较小者                          *
```

```
3     ***************************************************************/
4  #include <iostream>
5  using namespace std;
6
7  //查找两个字符中较小者的函数
8  char smaller (char first, char second)
9  {
10    if (first < second)
11    {
12        return first;
13    }
14    return second;
15 }
16 // 查找两个整数中较小者的函数
17 int smaller (int first, int second)
18 {
19    if (first < second)
20    {
21        return first;
22    }
23    return second;
24 }
25 //查找两个双精度数中较小者的函数
26 double smaller (double first, double second)
27 {
28    if (first < second)
29    {
30        return first;
31    }
32    return second;
33 }
34
35 int main ()
36 {
37    cout << "Smaller of 'a' and 'B': " << smaller ('a', 'B') << endl;
38    cout << "Smaller of 12 and 15: " << smaller (12, 15) << endl;
39    cout << "Smaller of 44.2 and 33.1: " << smaller (44.2, 33.1) << endl;
40    return 0;
41 }
```

运行结果：

```
Smaller of 'a' and 'B': B
Smaller of 12 and 15: 12
Smaller of 44.2 and 33.1: 33.1
```

如果我们使用模板，只需编写一个函数，就可以避免编写三个类似的函数。

15.1.2 使用函数模板

我们首先给出要定义的函数模板的语法，然后给出实例化的概念，即编译器处理实例化的方式。

语法

为了创建一个模板函数，我们可以为每个泛型类型使用一个占位符。表 15-1 显示了泛型函数的一般语法，其中 T、U、…、Z 在调用函数时被实际类型替换。

表 15-1　模板函数的语法

```
template <typename T, typename U, ..., typename Z>
T functionName  (U first, ... Z last)
{
   ...
}
```

如表 15-1 所示，模板头包含关键字 template，后面是两个尖括号内的符号类型列表。模板函数头遵循函数头的规则，除了模板头中声明的类型是符号类型。虽然允许多个泛型类型，但具有两个以上泛型类型的函数模板却很少见。一些旧代码使用术语 class 而不是 typename。我们使用关键字 typename，因为它是当前的标准，并在 C++ 库中使用。

我们使用函数模板把程序 15-1 更改为程序 15-2。在程序 15-2 中，我们只有一个泛型类型，但是该类型被使用了三次：两次作为参数，一次作为返回类型。在这个程序中，参数的类型和返回值的类型是相同的。

我们可以看到，程序 15-2 的结果与程序 15-1 的结果相同。我们只编写了一个模板函数而不是三个重载函数，从而节省了代码。注意，我们只使用了一个类型名 T，它用于定义两个参数和返回值的类型。我们这样做是因为两个参数的类型和返回值的类型是相同的。

程序 15-2　使用一个函数模板

```
1  /****************************************************************
2    * 本程序使用模板函数                                            *
3    * 查找不同数据类型的两个值中的较小者                            *
4    ****************************************************************/
5  #include <iostream>
6  using namespace std;
7  
8  // 模板函数的定义
9  template <typename T>
10 T smaller (T first, T second)
11 {
12    if (first < second)
13    {
14         return first;
15    }
16    return second;
17 }
18 
19 int main ( )
20 {
21    cout << "Smaller of a and B: " << smaller ('a', 'B') << endl;
22    cout << "Smaller of 12 and 15: " << smaller (12, 15) << endl;
23    cout << "Smaller of 44.2 and 33.1: " << smaller (44.2, 33.1) << endl;
24    return 0;
25 }
```

运行结果：

```
Smaller of a and B: B
Smaller of 12 and 15: 12
Smaller of 44.2 and 33.1: 33.1
```

程序 15-3 展示了如何创建一个泛型交换函数来交换两个整数值、两个双精度值等。我们把该函数命名为 exchange，以避免与名为 swap 的库函数混淆。

程序 15-3 交换两个值

```
1  /**************************************************************
2   * 本程序使用模板函数交换两个值                                *
3   **************************************************************/
4  #include <iostream>
5  using namespace std;
6
7  // 模板函数的定义
8  template <typename T>
9  void exchange (T& first, T& second)
10 {
11    T temp = first;
12    first = second;
13    second = temp;
14 }
15
16 int main ()
17 {
18    // 交换两个整数
19    int integer1 = 5;
20    int integer2 = 70;
21    exchange (integer1, integer2);
22    cout << "After swapping 5 and 70: ";
23    cout << integer1 << " " << integer2 << endl;
24    // 交换两个双精度浮点数
25    double double1 = 101.5;
26    double double2 = 402.7;
27    exchange (double1, double2);
28    cout << "After swapping 101.5 and 402.7: ";
29    cout << double1 << " " << double2 << endl;
30    return 0;
31 }
```

```
运行结果：
After swapping 5 and 70: 70 5
After swapping 101.5 and 402.7: 402.7 101.5
```

模板实例化

使用模板函数将非模板函数定义的创建推迟到编译时。这意味着在编译涉及函数模板的程序时，编译器会根据函数调用的需要创建函数的任意多个版本。在程序 15-2 中，编译器创建了名为 smaller 的函数的三个版本，用它们的参数处理三个函数调用，如图 15-1 所示。此过程称为模板实例化（template instantiation），但不应与根据类型的对象实例化混淆。

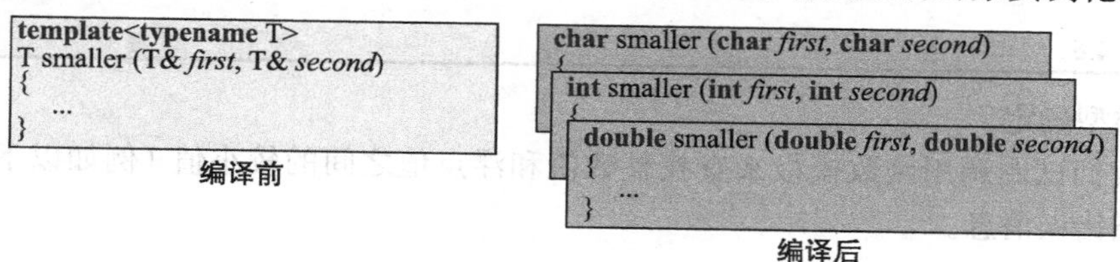

图 15-1 将一个函数模板实例化为若干函数

15.1.3 其他函数模板版本

除前一节中讨论的基本函数模板语法外，还存在几个其他函数模板版本。

非类型模板参数

有时我们需要在函数模板中定义一个值而不是一个类型。换句话说，对于我们想要使用的所有模板函数，参数的类型可能是相同的。在这种情况下，我们可以定义一个模板，但不能具体定义其类型。

假设我们需要一个函数来打印任意数组的元素，而不管元素的类型和数组的大小。我们知道每个元素的类型在不同的数组之间可能有所不同，但是数组大小总是一个整数（或者无符号整数）类型。我们使用程序 15-4 来实现。我们有两个模板参数：T 和 N。参数 T 可以是任何类型；参数 N 是非类型（类型预先定义为整数）。

程序 15-4　打印数组

```
1  /***************************************************************
2   * 本程序使用模板函数                                          *
3   * 打印任意类型的任意数组的元素                                *
4   ***************************************************************/
5  #include <iostream>
6  using namespace std;
7
8  // 模板函数 print 的定义
9  template <typename T, int N>
10 void print (T (&array) [N])
11 {
12   for (int i = 0; i < N ; i++)
13   {
14        cout << array [i] << " ";
15   }
16   cout << endl;
17 }
18
19 int main ()
20 {
21   // 创建两个数组
22   int arr1 [4] = {7, 3, 5, 1};
23   double arr2 [3] = {7.5, 6.1, 4.6};
24   // 调用模板函数
25   print (arr1);
26   print (arr2);
27   return 0;
28 }
```

运行结果：

```
7 3 5 1
7.5 6.1 4.6
```

显式类型确定

如果我们试图调用函数模板来查找整数值和浮点值之间的较小值（例如以下语句），则会得到一条错误消息。

```
cout << smaller (23, 67.2); // 错误！对于同一模板类型 T 使用两种不同的数据类型
```

换言之，我们传递给编译器的第一个参数（T 类型）为整数值 23，第二个参数（T 类型）为浮点值 67.2。T 类型是一个模板类型，它所有值的类型必须相同。如果我们在调用期间定义显式类型转换，就可以避免前面的错误。这是通过定义尖括号内的类型来完成的，如下所示：

```
cout << smaller <double> (23, 67.2); //23 将转换为 23.0
```

我们告诉编译器，我们要使用 T 的值是 double 类型的 smaller 程序版本。然后编译器创建该版本，并在调用 smaller 之前将 23 转换为 23.0。

预定义操作

我们可以使用函数 smaller 比较两个整数、两个双精度数或者两个字符，因为小于运算符（<）是为这些类型定义的。这意味着我们可以为任何类型使用这个函数模板，只要该类型定义了小于（<）的重载运算符。例如，我们知道库字符串类已经重载了这个运算符。我们可以告诉编译器用字符串替换 T，然后调用该函数。这意味着以下语句有效，结果是字符串“Bi”。

```
cout << smaller ("Hello" , "Bi"); // 结果是 "Bi"
```

C 字符串类型和 Rectangle（长方形）类型没有重载该运算符，如果重载的话，我们将得到一个编译错误。解决方案是接下来讨论的模板特化。

模板特化

C 样式字符串（C-style string）并没有定义小于运算符（<）。这意味着我们不能使用模板函数 smaller 来查找两个 C 样式字符串中的较小者。但是，有一种解决方案，称为**模板特化**（template specialization）。我们可以定义另一个具有特定类型（而不是模板类型）的函数，如程序 15-5 所示。关于这个程序有两个要点。

1）我们必须在头部之前使用 template<>，以显示这是前面定义的模板函数的特化。

2）当使用一种特殊的类型替换 T 时要特别小心。C 样式字符串的类型为 const char*，这意味着每次使用 T 时，都必须将其替换为 const char*。

程序 15-5 函数模板特化

```
1  /***************************************************************
2   * 使用模板特化的模板函数                                        *
3   ***************************************************************/
4  #include <iostream>
5  #include <string>
6  #include <cstring>
7  using namespace std;
8
9  // 模板函数的定义
10 template <typename T>
11 T smaller (const T& first, const T& second)
12 {
13   if (first < second)
14   {
```

```
15          return first;
16      }
17      return second;
18  }
19  // 模板函数的特化
20  template <>
21  const char* smaller (const (const char*) & first, const (const char*) & second)
22  {
23      if (strcmp (first, second ) < 0)
24      {
25          return first;
26      }
27      return second;
28  }
29
30  int main ( )
31  {
32      // 使用两个字符串对象调用模板函数
33      string str1 = "Hello";
34      string str2 = "Hi";
35      cout << "Smaller (Hello ,  Hi): " << smaller (str1, str2) << endl;
36      //使用两个 C 字符串对象调用模板函数
37      const char* s1 = "Bye";
38      const char* s2 = "Bye Bye";
39      cout << "Smaller (Bye, Bye Bye)" << smaller (s1, s2) << endl;
40      return 0;
41  }
运行结果：
Smaller (Hello , Hi): Hello
Smaller (Bye, Bye Bye): Bye
```

函数模板的重载

我们在第 7 章中讨论了一般函数（非模板）的重载。我们可以对函数模板应用相同的概念。我们可以重载一个函数模板，来拥有多个同名但签名不同的函数。通常，模板类型相同，但参数的数量不同。

例如，我们重载模板函数 smaller 以接收两个或者三个参数（我们称之为 smallest，因为该函数使用了两个以上的参数）。程序 15-6 显示了代码清单。注意，我们已经根据第一个函数定义了第二个函数。这就是第二个函数较短的原因。

程序 15-6　重载函数 smaller 的程序

```
1   /***************************************************************
2    *模板函数 smaller 的重载版本                                      *
3    ***************************************************************/
4   #include <iostream>
5   using namespace std;
6
7   // 带两个参数的模板函数
8   template <typename T>
9   T smallest (const T& first, const T& second)
10  {
```

```
11      if (first < second)
12      {
13          return first;
14      }
15      return second;
16  }
17  //带三个参数的模板函数
18  template <typename T>
19  T smallest (const T& first, const T& second, const T& third)
20  {
21      return smallest (smallest (first, second), third);
22  }
23
24  int main ( )
25  {
26      //调用带三个整数参数的重载版本
27      cout << "Smallest of 17, 12, and 27 is ";
28      cout << smallest (17, 12, 27)  << endl;
29      //调用带三个双精度数参数的重载版本
30      cout << "Smallest of 8.5, 4.1, and 19.75 is ";
31      cout << smallest (8.5, 4.1, 19.75) << endl;
32      return 0;
33  }
运行结果：
Smallest of 17, 12, and 27 is 12
Smallest of 8.5, 4.1, and 19.75 is 4.1
```

15.1.4 接口文件和应用程序文件

模板函数的定义可以放在接口文件中，该头文件可以包含在应用程序文件中。这意味着我们可以为模板函数编写一个定义，然后在不同的程序中使用该定义。在这种情况下，接口文件必须包括函数的定义，而不仅仅是声明。程序 15-7 展示了我们如何创建一个接口文件 smaller.h，并将函数模板的定义放入其中。任何程序都可以包含该接口并使用该函数（程序 15-8）。

程序 15-7 函数模板的定义

```
1   /*****************************************************************
2    * 名为 smaller 的函数模板的接口文件                                  *
3    *****************************************************************/
4   #ifndef SMALLER_H
5   #define SMALLER_H
6   #include <iostream>
7   using namespace std;
8
9   template <typename T>
10  T smaller (const T& first, const T& second)
11  {
12      if (first < second)
13      {
14          return first;
```

```
15    }
16    return second;
17  }
18  #endif
```

程序 15-8　使用函数模板

```
1   /*****************************************************************
2    * 测试函数模板的应用程序文件                                      *
3    *****************************************************************/
4   #include "smaller.h"
5
6   int main ( )
7   {
8     cout << "Smaller of 'a' and 'B': " << smaller ('a', 'B') << endl;
9     cout << "Smaller of 12 and 15: " << smaller (12, 15) << endl;
10    cout << "Smaller of 44.2 and 33.1: " << smaller (44.2, 33.1) << endl;
11    return 0;
12  }
运行结果:
Smaller of a and B: B
Smaller of 12 and 15: 12
Smaller of 44.2 and 33.1: 33.1
```

15.2　类模板

类模板（class template）的概念使得 C++ 语言非常强大。在前几章中，我们了解到类是数据成员和成员函数的组合。我们可以创建一个具有数据类型的类和另一个具有相同功能但具有不同数据类型的类。在这些情况下，我们可以使用类模板。模板在 C++ 库中被使用，如 string 类和 stream 类。模板也用于标准模板库 STL 中（我们将在第 19 章中学习）。要创建类模板，我们必须使数据成员和成员函数都是通用的。

15.2.1　接口和实现

我们知道，类有接口和实现。当我们需要创建一个类模板时，我们必须在接口和实现中都使用泛型参数。

接口

类的接口必须为使用参数化类型的数据成员和成员函数定义类型名（typename）。表 15-2 显示了一个简单类模板的语法。我们只使用一个数据成员、一个默认构造函数和两个成员函数。在这种情况下，构造函数不使用数据成员。访问器函数返回 T 类型的值。更改器函数带一个 T 类型的参数。

表 15-2　一个简单类模板的语法

```
template <typename T>
class Name
{
  private:
      T data;
```

（续）

```
public:
    Name ();                      // 默认构造函数
    T get () const;               // 访问器函数
    void set (T data);            // 更改器函数
};
```

实现

在实现中，我们必须为每个使用泛型类型的成员函数指定类型名（typename）。表 15-3 显示了我们在表 15-2 中定义的简单类的实现语法。

表 15-3　在表 15-2 中定义的类的实现

```
//函数 get 的实现
template <typename T>
T name <T > :: get () const
{
  return data;
}
//函数 set 的实现
template <typename T>
void name < T > :: set (T  d )
{
  data = d;
}
```

请注意，解析运算符（::）之前使用的类的名称应该是 Name<T>，而不仅仅是 Name。

为了专注于语法，我们定义了一个非常简单的名为 Fun 的类，该类只有一个数据成员（可以是 int、double、char 甚至 string 类型）。我们想展示语法在定义类的接口、实现和应用程序文件中的实际使用方法。我们将在本章后面介绍更复杂的类。在本章的末尾，我们将创建一个更相关的类（一个泛型数组），以更好地显示 C++ 中模板的实用性。

程序 15-9 演示了类 Fun 的接口文件。注意，每次需要定义一个类型时（第 13 行），我们必须使用类型名（本例中的 T）。同时，在类定义的前面必须使用 template <typename T> 的声明。

程序 15-9　类 Fun 的接口文件

```
1   /***************************************************************
2    *名为 Fun 的类的接口文件                                        *
3    ***************************************************************/
4   #ifndef FUN_H
5   #define FUN_H
6   #include <iostream>
7   using namespace std;
8
9   template <typename T>
10  class Fun
11  {
12    private:
13        T data;
14    public:
```

```
15        Fun (T data);
16        ~Fun ();
17        T get () const;
18        void set (T data);
19    };
20    #endif
```

程序 15-10 演示了类 Fun 的实现文件。注意，每个函数定义都必须使用带有 template <typename T> 的声明的模板函数。每次需要类型声明时，我们使用 T 作为泛型类型。

程序 15-10　类 Fun 的实现文件

```
1    /*************************************************************
2     *模板类 Fun 的实现文件                                        *
3     *************************************************************/
4    #ifndef FUN_CPP
5    #define FUN_CPP
6    #include "fun.h"
7
8    //构造函数
9    template <typename T>
10   Fun <T> :: Fun (T d)
11   : data (d)
12   {
13   }
14   //析构函数
15   template <typename T>
16   Fun <T> :: ~Fun ()
17   {
18   }
19   //访问器函数
20   template <typename T>
21   T Fun <T> :: get () const
22   {
23       return data;
24   }
25   //更改器函数
26   template <typename T>
27   void Fun <T> :: set (T d)
28   {
29       data = d;
30   }
31   #endif
```

在非模板类和模板类的实现中，我们注意到的一个非常重要的区别是编译器在编译应用程序文件时需要看到模板函数的参数化版本。换言之，应用程序文件（稍后定义）需要将实现文件作为头文件，这意味着我们需要将宏 ifndef、define 和 endif 添加到实现文件中。

程序 15-11 显示了使用模板类 Fun 的应用程序。注意，我们将 fun.cpp 作为头文件来包含以帮助编译器创建类的不同版本。还要注意，要实例化模板类（第 9 ～ 12 行），必须定义替换 typename T 的实际类型。

Fun1 类的数据成员的实际类型是 int，Fun2 的数据成员的实际类型是 double，Fun3 的数据成员的实际类型是 char，Fun4 的数据成员的实际类型是 string。

程序 15-11 使用类 Fun 的应用程序文件

```
/*****************************************************************
 * 用于测试模板类 Fun 的应用程序文件                                   *
 *****************************************************************/
#include "fun.cpp"

int main ( )
{
  //使用四种不同的数据类型实例化四个类
  Fun <int> Fun1 (23);
  Fun <double> Fun2 (12.7);
  Fun <char> Fun3 ('A');
  Fun <string> Fun4 ("Hello");
  //显示每个类的数据值
  cout << "Fun1: " << Fun1.get() << endl;
  cout << "Fun2: " << Fun2.get() << endl;
  cout << "Fun3: " << Fun3.get() << endl;
  cout << "Fun4: " << Fun4.get() << endl;
  //设置两个类的数据值
  Fun1.set(47);
  Fun3.set ('B');
  //显示新设置的数据值
  cout << "Fun1 after set: " << Fun1.get() << endl;
  cout << "Fun3 after set: " << Fun3.get() << endl;
  return 0;
}
```

运行结果：

```
Fun1: 23
Fun2: 12.7
Fun3: A
Fun4: Hello
Fun1 after set: 47
Fun3 after set: B
```

15.2.2 编译

当项目中有多个函数模板或者类模板时，我们可以编译和链接与程序相关的不同文件。基本上有两种方法。

包含方法

第一种方法称为包含，如图 15-2 所示。在这种方法中，我们将声明和定义放在头文件中，然后将头文件包含在应用程序中。当然，一般建议使用单独的头文件，一个用于声明，一个用于定义。这两个文件的分离使我们能够快速地切换到其他方法。可以将声明放在（.h）文件中，将定义放在（.cpp）文件中。我们不单独编译定义文件，只是将其包含在应用程序文件中，并且只编译应

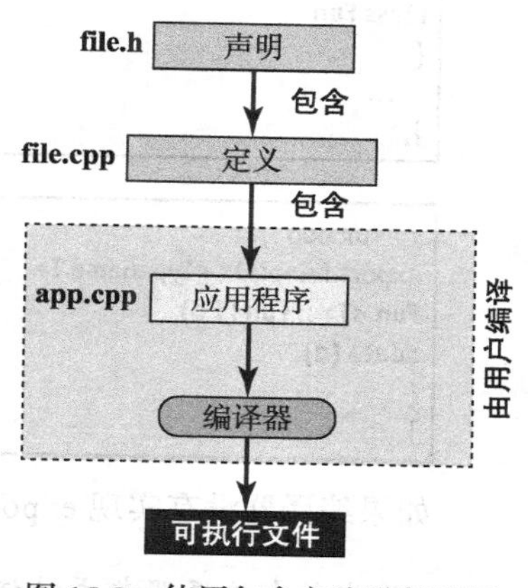

图 15-2 使用包含方法进行编译

用程序文件。注意声明文件必须包含在定义文件中，定义文件必须包含在应用程序文件中。我们只编译应用程序文件。

独立编译

在前面基于面向对象程序设计而编写的所有程序都使用了在第 7 章中讨论的独立编译概念。在这个模型中，存在三个不同的文件：接口文件、实现文件和应用程序文件，如图 15-3 所示。正如我们在第 7 章中讨论的那样，独立编译具有对最终用户隐藏实现的优势，从而促进了封装的概念。

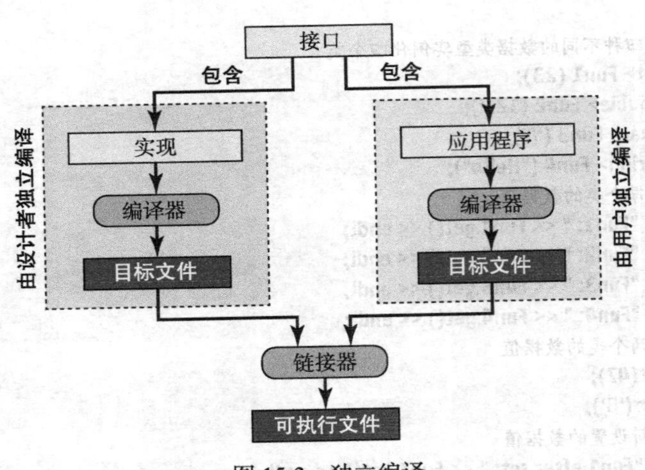

图 15-3　独立编译

虽然这种方法适用于非模板函数和类，但是如果程序中包含模板函数或者模板类，则需要加以修改。问题是实现文件不能独立编译，因为编译器需要应用程序文件来确定编译中必须使用每个模板函数或者模板类的哪些实例。

C++ 标准已经找到了解决方案。我们必须使用关键字 export 导出每个模板声明或者定义。我们在每个类型名前面插入关键字 export。例如，在类 Fun 的定义中，我们添加这个关键字，如下所示。对于每个成员函数（如以下的构造函数），我们也需要添加关键字 export：

```
// fun.h
export template <typename T>
class Fun
{
   ...
};
```

```
// fun.cpp
export template <typename T>
Fun <T> :: Fun (T d)
: data (d)
{
}
```

如果编译器没有实现 export 关键字，则必须使用包含方法。

并非所有编译器都支持使用 export 关键字进行独立编译。

示例

在本例中，我们创建一个类模板来模拟栈。如前所述，栈是一种结构，其中最后一个被压入栈的项将是第一个从栈弹出的项，如图 15-4 所示。

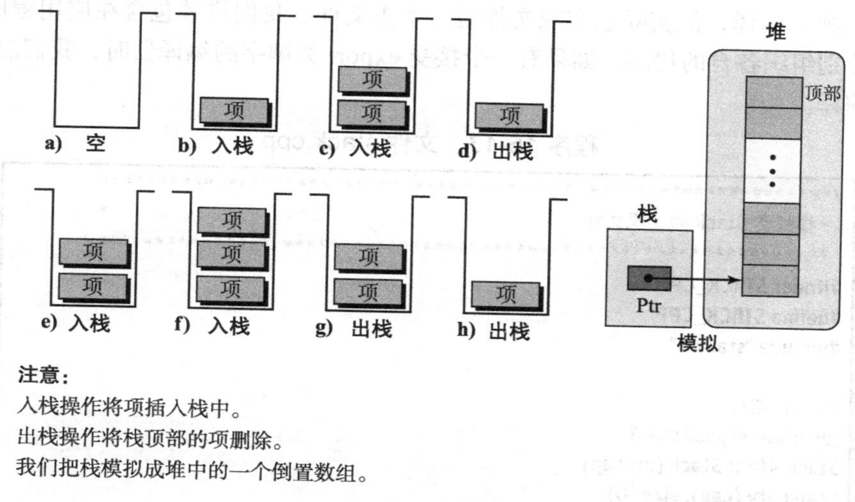

图 15-4 栈的操作

为了模拟栈，我们使用 T 类型的数组，但是我们只提供可以在数组末尾插入元素（入栈）和在数组末尾删除元素（出栈）的操作。当一项被弹出时，它的值可以被检索。程序 15-12 显示了接口文件。

在这个程序中，我们将栈定义为一个数组（第 14 行），但是为数组定义的操作只允许压入和弹出顶部元素。不允许访问或者更改其他元素。

程序 15-12 文件 stack.h

```
1   /*******************************************************************
2    * 模板类 Stack 的接口文件                                          *
3    *******************************************************************/
4   #ifndef STACK_H
5   #define STACK_H
6   #include <iostream>
7   #include <cassert>
8   using namespace std;
9
10  template <typename T>
11  class Stack
12  {
13    private:
14        T* ptr;
15        int capacity;
16        int size;
17    public:
18        Stack (int capacity);
19        ~Stack ();
20        void push (const T& element);
21        T pop ();
```

```
22 };
23 #endif
```

程序 15-13 显示了实现文件。注意，我们将实现文件定义为单独的文件，表面上好像我们意图使用独立编译，但实际上实现文件是一个头文件。我们将其包含在应用程序文件中。两个文件分别组织各自的代码，如果有一个接受 export 关键字的编译器时，我们总是可以切换到独立编译。

程序 15-13　文件 stack.cpp

```
1  /******************************************************************
2   *模板类 Stack 的实现文件                                          *
3   *****************************************************************/
4  #ifndef STACK_CPP
5  #define STACK_CPP
6  #include "stack.h"
7  
8  //构造函数
9  template <typename T>
10 Stack <T> :: Stack (int cap)
11 : capacity (cap), size (0)
12 {
13   ptr = new T [capacity];
14 }
15 //析构函数
16 template <typename T>
17 Stack <T> :: ~Stack ()
18 {
19   delete [ ] ptr;
20 }
21 //入栈函数 push
22 template <typename T>
23 void Stack <T> :: push (const T& element)
24 {
25   if (size < capacity)
26   {
27         ptr[size] = element;
28         size++;
29   }
30   else
31   {
32         cout << "Cannot push; stack is full." << endl;
33         assert (false);
34   }
35 }
36 //出栈函数 pop
37 template <typename T>
38 T Stack <T> :: pop ()
39 {
40   if (size > 0)
41   {
42         T temp = ptr [size - 1];
```

```
43          size--;
44          return temp;
45      }
46      else
47      {
48          cout << "Cannot pop; stack is empty." << endl;
49          assert (false);
50      }
51  }
52  #endif
```

程序 15-14 显示了应用程序文件。我们使用一个整数替换类型名（typename），以实例化应用程序文件，其中的整数也可以是 double、char 甚至是用户自定义的类型。

程序 15-14　用于实例化栈类 Stack 的应用程序文件

```
1   /*********************************************************************
2    * 测试使用整数的栈类 Stack 的应用程序文件                              *
3    *********************************************************************/
4   #include "stack.cpp"
5
6   int main ()
7   {
8     Stack <int> stack (10);
9     stack.push (5);
10    stack.push (6);
11    stack.push (7);
12    stack.push (3);
13    cout << stack.pop () << endl;
14    cout << stack.pop ();
15    return 0;
16  }
```

```
运行结果：
3
7
```

15.2.3　其他问题

我们在这里简要讨论其他问题。在后续章节中，我们将更详细地探讨这些问题。

友元函数

模板类的声明可以包含友元函数。实现友元函数有三种方法：模板类可以包含一个非模板函数作为友元函数；模板类可以包含一个模板函数作为友元函数；模板类也可以包含一个特化的模板函数作为友元函数。

别名

有时可以使用关键字 typedef 为模板类定义别名。这允许我们使用别名作为程序中类的完整定义。例如，对于先前定义的栈模板类 Stack，我们可以定义以下别名：

```
typedef stack <int> iStack;
typedef stack <double> dStack;
typedef stack <string, int> siStack;
```

然后我们可以在程序中使用这些类型定义，如下所示：

```
iStack s1;
dStack s2;
siStack s3;
```

15.2.4 继承

我们可以从一个模板类或者非模板类派生出另一个模板类。例如，假设我们将类 First 定义为模板类：

```
template < typename T >
class First
{
...
}
```

然后我们可以定义另一个类 Second，它公开继承类 First，如下所示：

```
template < typename T >
class Second : public First <T>
{
  ...
}
```

继承的模板类没有什么特别之处，而且也不是很常见。故本书将不展开讨论。

15.2.5 回顾

我们已经讨论了一些实际上是模板类的类。在本节中，我们将这些类视为泛型类。

字符串类

在第 10 章中，我们讨论了字符串类 string。string 类是名为 basic_string 的模板类的特化。模板类 basic_string 的总体设计如下：

```
template < typename charT >
class basic_string
{
  ...
};
```

C++ 类库定义了这个类的两个特化：一个用于 char 类型，另一个用于 wchar_t 类型，如下所示。在附录 A 中，我们使用第一个特定化；第二个特定化与之类似。

```
typedef basic_string <char> string;
typedef basic_string <wchar_t> wstring;
```

输入 / 输出类

另一组实际上是泛型类的类是输入 / 输出类。我们之前讨论过的所有输入类实际上都是模板类的特化。例如，istream 类是 basic_istream 类的一个特化，如下所示：

```
typedef basic_istream <char> istream;
typedef basic_istream <wchar_t> wistream;
```

本章小结

函数模板允许我们定义一个函数，但类型的定义要推迟到编译时确定。编译程序时，编译器根据函数调用时使用的不同类型，创建不同类型的函数版本。函数模板的语法包含保留关键字 template 和在尖括号中的保留关键字 typename。为了便于共享，函数模板被放置在一个接口文件中，该文件被包含在正在编译的程序中。函数模板中的类型确定有三种规则：我们不能混合类型；如果需要混合类型，则辅助类型必须显式类型化；函数中的操作（例如比较）必须对指定的类型有效，如果不是，我们可以使用函数模板的特化，也就是说，我们可以定义另一个具有特定类型的函数来处理特定情况。只要签名不同，我们就可以重载模板函数。

通过在接口和实现中使用泛型参数，可以创建类模板。类的接口必须为使用参数化类型的数据成员和成员函数定义类型名。在实现中，必须为涉及泛型类型的每个成员函数使用该类型名。但是，编译器在编译应用程序文件时必须看到模板函数的参数化版本，这意味着实现文件必须作为头文件被包含。虽然可以使用包含方法或者通过独立编译来包含实现文件，但是并不是所有的编译器都提供独立编译。只要用不同的标识符指定每个类型名，类模板就可以包含多个类型。

思考题

1. 以下函数的多少个版本在编译期间被实例化了？

```
template <typename T>
void fun (T x)
{
    ...
}
int main ()
{
    fun (7);
    fun (12.5);
    fun ("Hello");
    return 0;
}
```

2. 以下函数的多少个版本在编译期间被实例化了？

```
template <typename T>
void fun (T x)
{
    ...
}
int main ()
{
    Sample (7);
    Sample (9) ;
    return 0;
}
```

3. 编写一个函数模板，计算两个数值（例如整数、长整数、双精度数或者长双精度数）的平均值。
4. 重载第 3 题中编写的函数模板，计算三个数值的平均值。
5. 编写一个名为 Pair 的 struct 模板，创建一个包含两个任意类型的数据项（例如，一个整数和一个双精度数、一个双精度数和一个字符，等等）的结构。

编程题

1. 编写一个模板函数，获取任何类型的数组中最小元素的索引。使用三个类型分别为 int、double 和 char 的数组测试该函数。然后打印最小元素的值。
2. 如果我们能在数组中找到最小的元素，则可以使用选择排序（selection sort）算法对数组进行排序。在该算法中，我们在数组中找到最小的元素，并将其与第一个元素交换。然后在数组的其余部分找到最小的，并将其与第二个元素交换。重复上述过程，直到数组完全排序。编写一个程序，对三个分别为 int、double 和 char 类型的数组进行排序。
3. 编写一个模板函数以搜索数组中给定的值。用两个 int 和 char 类型的数组测试该函数。请注意，搜索双精度值可能无法给出正确的结果，因为没有为浮点值定义相等运算符。
4. 定义一个函数，反转任何类型的数组中元素的顺序。用整数、双精度数、字符和字符串类型的数组测试程序。使用辅助函数 swap 交换任意两个元素。使用辅助函数 print 打印交换前后的数组的内容。
5. 创建一个模板类 Array，用于处理堆内存中任何类型和任何大小的对象数组。定义成员函数 add 以将元素添加到数组末尾。定义函数 print 以打印数组中的所有元素。使用 int、double 和 char 类型的数组测试程序。
6. 队列（queue）是先进先出的结构。队列的一个例子是排队等候服务的人。我们可以使用一个数组来实现一个队列，在该数组的末尾插入项（入队），然后从数组的前面删除项（出队）。创建一个模板类和一个任何类型的队列。然后在两个单独的应用程序文件中测试队列：一个作为整数队列，另一个作为字符串队列。添加适当的 try-catch 语句块以捕获向满队列中加入项，或者从空队列中删除项时产生的异常。

输入 / 输出流

C++ 将输入 / 输出操作视为被称为流的库类。在本章中，我们首先介绍数据的来源和接收器。然后我们讨论流，这些流是将我们的程序连接到这些数据源和数据接收器的类。我们将讨论控制台流、文件流和字符串流。最后，我们讨论与流相关的数据格式。

学习目标

阅读并学习本章后，读者应该能够

- 讨论使用流对象实现程序和数据源或者数据接收器之间的通信。
- 讨论流类的层次结构。
- 强调作为层次结构中所有类的虚基的类 ios 不能实例化，但 ios 定义了整个层次结构中所有类继承的数据成员。
- 讨论控制台流（istream、ostream 和 iostream 类），并强调用户不能从这些类实例化对象。
- 显示 C++ 已经从类 istream 实例化了一个对象，并从类 ostream 实例化了三个对象，并将这些对象存储在 <iostream> 头文件中。
- 显示我们可以使用在控制台流中定义的成员函数从键盘读取数据并写入监视器。
- 讨论文件流（ifstream、ofstream 和 fstream）的三个类，这些类用于连接作为数据源或者数据接收器的文件。
- 显示文件流能够实现文本和二进制的输入 / 输出。
- 演示如何按顺序和随机访问数据源或者数据接收器。
- 讨论字符串流（istringstream、ostringstream 和 stringstream）可以作为应用程序和字符串类之间的适配器使用。
- 演示如何使用 ios 类中定义的标志和字段直接格式化数据，以及如何使用标准的或者自定义的操作符进行格式化。

16.1 概述

当我们运行程序时，数据必须存储在内存中。存储在内存中用于处理的数据来自外部数据源，并进入外部数据接收器。这意味着我们必须考虑数据的来源和接收器。数据源可以是键盘、文件或者字符串。数据接收器可以是监视器、文件或者字符串。图 16-1 显示了输入 / 输出数据和程序之间的关系。

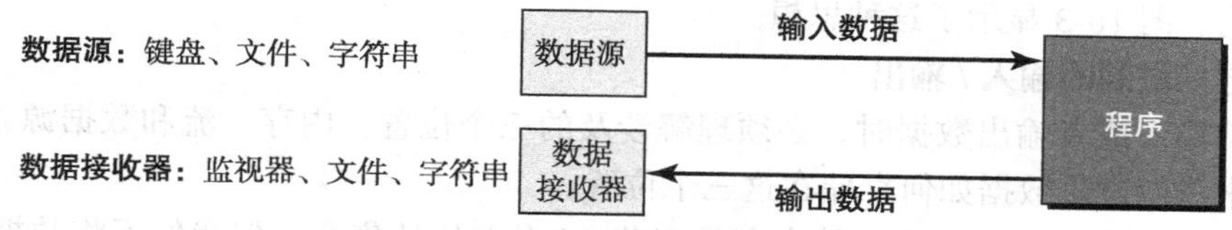

图 16-1 一个与数据源和数据接收器相关的程序

迄今为止，我们已经使用键盘（作为数据源）和监视器（作为数据接收器）来处理数据。

键盘是临时数据源，监视器是临时数据接收器。当我们运行程序时，输入数据必须再次输入，输出数据将在监视器上重新创建。

另一方面，文件是数据的永久来源和接收器。文件可以被保存。文件还可以物理地转移到另一台计算机上，可以反复使用。

作为内部数据源或者数据接收器的字符串是为特殊目的而设计的，我们将在本章后面讨论。

数据源或者数据接收器可以是控制台、文件或者字符串。

16.1.1　流

数据源和数据接收器不能直接连接到程序。在它们和程序之间需要一个中介器来控制数据流，同时在读取或者写入时解释数据。我们有输入流和输出流。输入流位于数据源和程序之间；输出流位于程序和数据接收器之间。

图 16-2 显示了我们通常对**流**（stream）的看法：流入或者流出程序的字节序列。字节从输入流中提取并插入输出流中。输入 / 输出流可以是双向的，但这在图中没有显示。

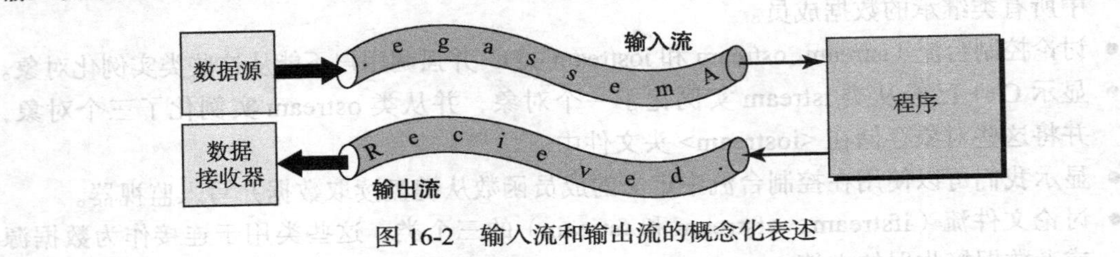

图 16-2　输入流和输出流的概念化表述

16.1.2　数据表示

虽然我们可以从数据源接收数据，并且可以将数据以文本或者二进制格式发送到数据接收器，但是数据接收器和数据源是计算机系统的一部分，并且被组织为字节序列（8 位块）。我们只能将一个字节序列发送到数据接收器，或者从数据源接收一个字节序列。了解这一事实可以更容易地处理不同的数据源和数据接收器。

数据接收器只能接收字节序列；数据源只能发送字节序列。

流的角色

C++ 标准创建输入 / 输出流的原因之一在于，将从数据源接收到的数据字节更改为程序所需的数据类型，并将程序发送到数据接收器的数据类型更改为数据接收器可以存储的字节。换言之，流的主要任务可以被认为是将字节转换为数据类型，以及将数据类型转换为字节。当然，流还有其他的工作，例如控制字节在计算机中的流入和流出，我们将在本章中逐渐展开讨论。图 16-3 显示了这种思想。

文本与二进制的输入 / 输出

当我们输入或者输出数据时，必须理解涉及的三个位置：内存、流和数据源 / 数据接收器。接下来我们讨论数据如何存储在这三个位置。

内存中的数据存储。计算机的内存是字节（8 位）块的集合，但我们不将数据存储为单个字节，我们将数据存储为一组连续的字节。例如，无论变量的值是多少，int 类型的每个数据对象都占用 4 个字节的内存。换言之，值为 23 的小整数和值为 4294967294 的大整数都占用4个字节的内存或者32位的二进制数据。

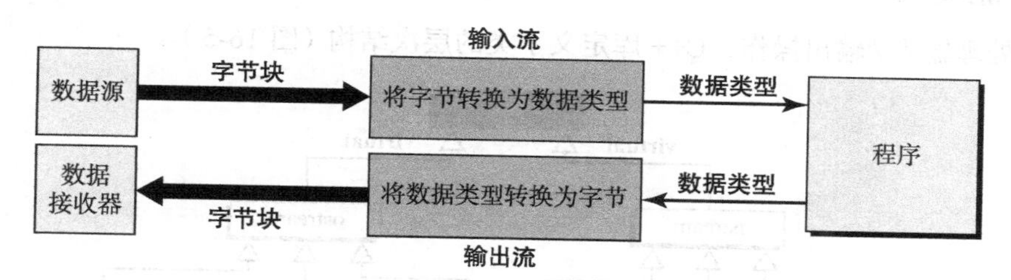

图16-3　流作为转化器的主要角色

> 内存中的数据以二进制形式存储。

数据源或者数据接收器中的数据存储。数据源和数据接收器也将数据存储在一个字节块中，但数据可以是二进制或者文本形式。如果数据项存储为二进制，则数据类型定义字节数；如果数据项存储为文本，则数据值定义字节数。例如，整数23可以存储为4个字节（二进制），也可以存储为两个字符"2"和"3"（文本）。整数4 294 967 294可以以二进制形式存储为4个字节，或者以文本形式存储为10个字节。

问题是我们需要使用哪种存储方法。以下内容可能有助于我们做出决定。

- 作为数据源的键盘只能接受文本形式的数据。类似地，作为数据接收器的监视器只将数据显示为一组字符。
- 作为数据源或者数据接收器的文件可以处理文本和二进制的输入/输出。问题是我们如何将数据存储在一个文件中：以二进制形式还是文本的形式。如果我们想在一个文件中存储一个对象，则应该使用二进制格式。
- 作为数据源或者数据接收器的字符串必须作为文本进行输入/输出，字符串是字符的集合（我们将在本章后面更详细地讨论这一点）。

在流缓冲区中存储数据。流将数据批量发送到数据接收器，或者从数据源批量接收数据。这意味着流缓冲区应该与它所连接的数据接收器或者数据源完全匹配。换言之，如果数据源或者数据接收器使用文本数据，则流必须使用文本数据；如果数据源或者数据接收器使用二进制数据，则流必须使用二进制数据。稍后我们将看到，默认情况下，流被设计为文本模式下的操作。如果希望它们以二进制模式操作，则流对象必须被实例化为二进制。这里存在一个问题：如果内存是二进制的，而流是文本或者二进制的（基于它所连接的数据源或者数据接收器），那么如何解决这种差异？这是使用流的主要目的之一。流总是以二进制形式接收内存中的数据或者将数据传递到内存，但它需要转换数据以满足数据源或者数据接收器的需求。图16-4显示了这一职责。

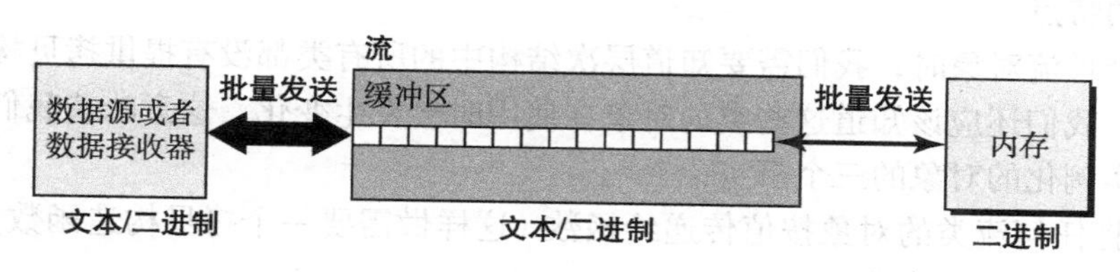

注意：
流负责将数据从文本转换为二进制，反之亦然。

图16-4　流的职责

16.1.3 流类

为了处理输入 / 输出操作，C++ 库定义了类的层次结构（图 16-5）。

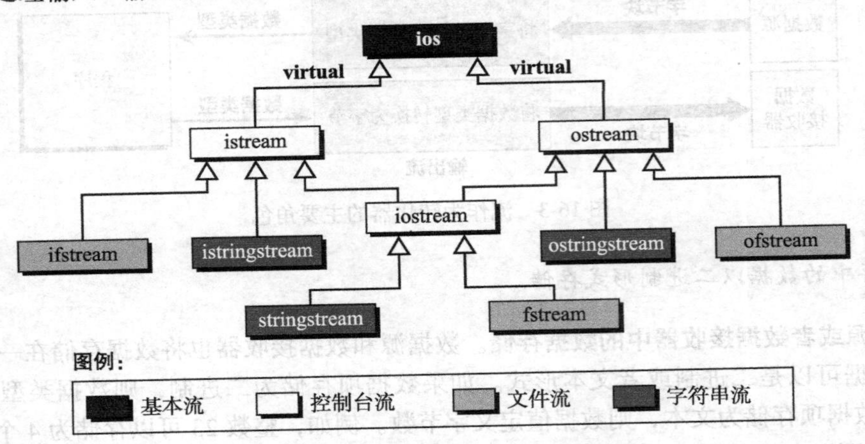

图 16-5 C++ 中的基本流层次结构

ios 类

在层次结构的顶部是 ios 类，它充当虚基类。ios 类是所有输入 / 输出类的基类。它定义由所有输入 / 输出流类继承的数据成员和成员函数。由于 ios 类不会被实例化，因此不使用其数据成员和成员函数。ios 类的数据成员和成员函数被其他流类使用。

ios 类不能被实例化。

其他类

为了方便起见，我们将 istream、ostream 和 iostream 称为控制台类（它们用于将程序连接到控制台），将 ifstream、ofstream 和 fstream 称为文件流（它们用于将程序连接到文件），并将 istringstream、ostringstream 和 stringstream 称为字符串流。

使用流

要使用任何流，我们必须考虑表 16-1 中所示的五个操作步骤：创建适当的流对象；将流对象连接到数据源或者数据接收器；从数据源读取数据或者将数据写入数据接收器；将数据源或者数据接收器与流对象断开连接；销毁流对象。

表 16-1 使用流的五个步骤

1）创建流对象
2）将流对象连接到数据源或者数据接收器
3）从流中读取或者写入数据
4）断开数据源或者数据接收器
5）销毁流对象

流对象的特点

当我们处理流对象时，我们需要知道层次结构中的所有类都没有提供拷贝构造函数或者赋值运算符。我们还应该知道这些类的对象在使用时会发生变化。这意味着我们应该记住关于从这些类实例化的对象的三个事实：

- 不能将任何流类的对象按值传递给函数（这样做需要一个拷贝构造函数）。传递这些对象必须按引用传递。
- 不能从函数按值返回任何流类的对象（这样做需要一个拷贝构造函数）。返回必须按引用返回。

- 不能使用常量修饰符修饰传递或者返回的任何流类的对象。这些对象的性质要求能修改。

16.2 控制台流

在本节中，我们将讨论控制台流：istream、ostream 和 iostream。我们可以使用前两个，但 iostream 类不能实例化，它被定义为层次结构中其他类的超类，我们将在本章后面讨论。

16.2.1 控制台对象

我们首先讨论从两个控制台流类创建的对象。

istream 对象：cin

istream 类定义了一个类，允许我们从键盘读取数据到程序中。我们不能实例化这个类。但是，系统已经创建了此类的一个名为 cin 的实例对象，该对象存储在 <iostream> 头文件中。系统还将 cin 对象连接到键盘，如图 16-6 所示。

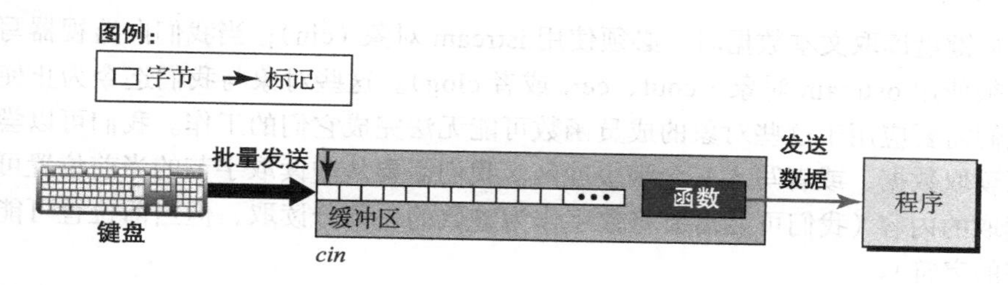

图 16-6　与程序和键盘相关的 cin 对象

当程序超出作用域范围时，对象 cin 将被销毁，并自动与键盘断开连接。这意味着表 16-1 中定义的五个任务中有四个是由系统自动为我们完成的，我们只是使用这个类中定义的成员函数从键盘读取数据。

> 系统从 istream 类创建了一个名为 cin 的对象，并将其存储在 <iostream> 头文件中。

ostream 对象：cout、cerr 和 clog

ostream 类定义了一个类，允许我们将程序中的数据写入监视器。我们不能实例化这个类。但是，系统已经创建了这个类的三个对象（名为 cout、cerr 和 clog），这些对象存储在 <iostream> 头文件中。系统还将这三个对象都连接到监视器。当我们的程序超出作用域范围时，这些对象会被自动销毁，并自动与监视器断开连接。这意味着表 16-1 中定义的五个任务中有四个是由系统自动为我们完成的。我们只需使用此类中定义的成员函数将数据写入监视器。图 16-7 显示了这些对象与监视器和程序的连接。

关于这些对象需要注意以下要点。首先，cout 绑定到 cin 对象，这意味着每次我们想要通过 cin 对象输入数据时，cout 对象都会被刷新（清空）。其次，cerr 和 clog 都被设计为向控制台发送错误。不同之处在于，cerr 对象在每次操作后立即刷新其内容，clog 对象收集错误消息，当程序终止或者被显式刷新时才刷新其内容。

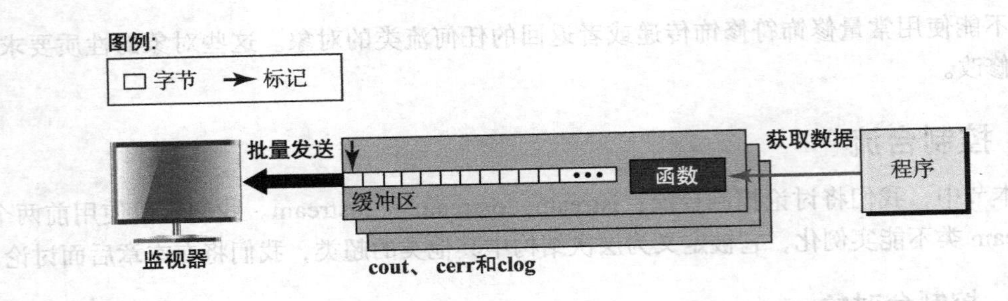

注意：
有三个同步对象。
当监视器连接到流时，标记位于缓冲区的最开始位置。
一次写入操作后，标记向右移动。

图 16-7　与程序和监视器相关的 cout、cerr 和 clog 对象

16.2.2　流状态

当我们从键盘读取文本数据时，必须使用 istream 对象（cin）；当我们向监视器写入文本数据时，必须使用 ostream 对象（cout、cerr 或者 clog）。这些对象与我们迄今为止使用的对象不同。我们需要应用于这些对象的成员函数可能无法完成它们的工作。我们可以尝试从一个空缓冲区读取数据，或者写入一个满缓冲区。我们需要从中读取字节的当前位置可能不包含所需要读取的内容（我们可能需要将数字作为整数的一部分读取，但当前位置可能包含一个不是数字的字符）。

基于上述原因，流必须跟踪其状态。**流状态**（stream state）存储在表示流对象状态的数据成员中。流还需要成员函数来测试其状态。由于流的状态对于控制台流、文件流和字符串流是相同的，因此状态数据成员和相应的成员函数在 ios 类中定义，ios 类由所有流类继承。

> 与流状态相关的数据成员和成员函数在 ios 类中定义，但由所有流类继承。

状态数据成员

ios 流定义了一个名为 iostate（输入 / 输出状态）的类型。这种类型的实现依赖于系统，但我们知道该类型定义了三个常量：eofbit、failbit 和 badbit。我们可以将 iostate 数据成员看作由三个位组成的位集，如图 16-8 所示。

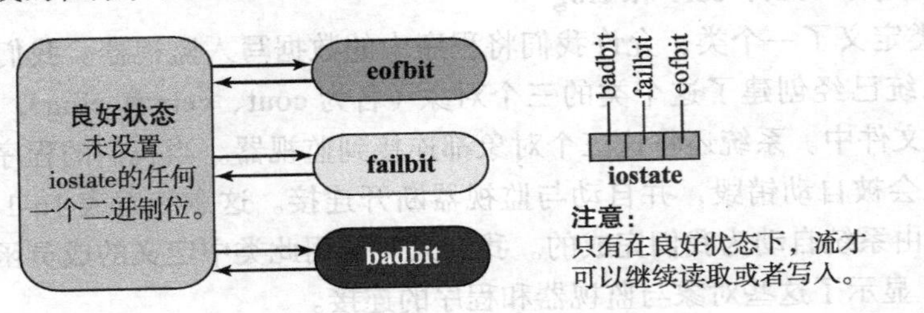

图 16-8　流的状态

表 16-2 显示了 iostate 类型中定义的二进制位。请注意，eofbit 不适用于输出流。

表 16-2　iostate 使用的常量值

常量	输入流	输出流
ios :: eofbit	没有更多的字符供提取	不适用

（续）

常量	输入流	输出流
ios :: failbit	无效的读取操作	无效的写入操作
ios :: badbit	流数据完整性丢失	流数据完整性丢失
ios :: goodbit	一切正常	一切正常

当我们尝试提取流缓冲区中不存在的字节（流位置已到达缓冲区末尾，并且没有可读取的字符）时，会设置 eofbit 标志位。当我们试图提取一个与我们想要的数据不匹配的字节时，就会设置 failbit 标志位。例如，如果我们试图读取一个整数，我们必须提取数字（或者定义一组数字结尾的空白），而不是任何其他字符。请注意，如果设置了两个标志位（eofbit 或者 failbit）中的一个，则会自动设置 badbit 标志位。当流的完整性丢失时，例如当内存不足时，当存在转换问题时，当抛出异常时等，就会设置 badbit 标志位。如前所述，没有消息就是好消息：当三个二进制位都没有被设置时，流处于良好状态，可以使用。

与状态相关的成员函数

ios 类还提供用于检查状态的成员函数。类库中定义了几个成员函数，其中大多数依赖于系统。例如，存在一个名为 rdstate 的成员函数，它返回 iostate 数据成员的当前二进制位的值。我们可以使用表 16-3 中定义的成员函数来检查输入 / 输出状态。

表 16-3　检查流状态的成员函数

函数	返回值
bool eof ()	如果设置了 eofbit 标志位，结果为 true；否则，结果为 false
bool fail ()	如果设置了 failbit 或者 badbit 标志位，结果为 true；否则，结果为 false
bool bad ()	如果设置了 badbit 标志位，结果为 true；否则，结果为 false
bool good ()	如果流状态良好，结果为 true；否则，结果为 false
clear ();	清除所有的标志位（设置为 0）
operator void* ();	返回一个指针；如果不为空，则解释为 true
const bool operator! ();	如果设置了 badbit 或者 failbit 标志位，则返回 true

表 16-3 中的前四个函数非常简单，它们返回 iostate 数据成员中相应标志位的状态。第五个函数（clear）将所有三个标志位重置为零（假）。有趣的是第六个函数和第七个函数。第六个函数是转换构造函数（参见第 6 章中讨论的重载运算符）。例如，我们定义了一个转换构造函数，将 Fraction（分数）对象的实例转换为双精度值。转换构造函数将宿主对象（流的实例）转换为指向 void 的指针（而不是布尔值），因为我们知道指向 void 的指针可以在任何表达式中使用。当指针为空时，其值为 0，在布尔表达式中被解释为假；不为空时，在布尔表达式中被解释为真。最后一个函数是一个重载运算符，如果设置了 failbit 或者 badbit，则返回 true。它不检查 eofbit。最后两个函数非常有用，因为我们可以将它们应用于流对象（不带任何运算符），或者将它们应用于操作返回的对象。请注意，istream 和 ostream 中的大多数操作（包括提取操作符和插入操作符）都返回 stream 类型的对象。

程序 16-1 显示了我们讨论的一些要点。

程序 16-1　测试 cin 状态

```
1  /***************************************************************
2   *本程序演示如何测试流的状态                                    *
3   ***************************************************************/
```

```
4  #include <iostream>
5  using namespace std;
6
7  int main ()
8  {
9      int n;
10     cout << "Enter a line of integers and eof at the end: " << endl;
11     while (cin >> n)
12     {
13         cout << n * 2 << " ";
14     }
15     return 0;
16 }
```

运行结果：

```
Enter a line of integers and eof at the end:
14 24 11 78 19 32 ^Z
28 48 22 156 38 64
```

在第 11 行中，我们使用了表达式 cin>>n。虽然我们没有正式讨论运算符 >>，但我们在前文中使用过这个运算符。提取运算符 >> 从连接到键盘的 cin 对象中读取一个整数。稍后我们将看到该运算符将一个整数读入其参数，并返回对 cin 对象的引用。while 语句中的表达式要求一个布尔值，但语句（cin>>n）返回的是 cin 对象的实例。必须进行类型转换才能将 cin 对象更改为布尔值。这是通过调用 operator void*() 来完成的，如果上一个操作成功，则返回一个指针（true），如果不成功，则返回一个空指针（false）。这意味着循环持续执行，直到设置了三个标志位中的一个。当我们在键盘上键入 ^Z 时，这意味着我们终止向 cin 缓冲区输入数据，并且 eofbit 被设置。

16.2.3　输入 / 输出

由于 istream 和 ostream 类的对象已经被实例化并连接到控制台，因此我们的唯一任务是使用这两个流类中定义的成员函数从缓冲区读取数据或者将数据写入缓冲区。如前所述，连接到唯一 istream 对象的键盘和连接到三个 ostream 对象的监视器只能接收文本数据。这意味着 istream 和 ostream 对象中的缓冲区只存储一组单独解释为字符的 8 位字节。由于计算机的内存以二进制形式存储数据，因此 ostream 对象的成员函数必须将二进制数据转换为单个字符。同样地，istream 对象的成员函数必须读取一组字符并将其解释为二进制数据。

字符成员函数

istream 和 ostream 类提供了若干字符成员函数和重载的运算符，用于从键盘读取字符，以及将字符写入监视器。

读取和写入单个字符。表 16-4 显示了从键盘读取单个字符的两个成员函数，以及向监视器写入单个字符的一个成员函数。istream 类提供了两个版本的 get 函数。第一个 get 函数返回缓冲区中字符的整数（ASCII）值，该字符作为整数（转换为 4 字节）存储在内存中。由于此函数不返回对 istream 对象的引用，因此无法在调用该函数时测试流的状态。第二个 get 函数读取存储在其参数中的

表 16-4　读取和写入单个字符

istream 类	ostream 类
int get()	无
istream& get (char& c)	ostream& put (char& c)

一个字符。然而，ostream 对象只提供一个 put 函数，它将其参数中定义的一个字符的副本写入缓冲区。

程序 16-2 显示了如何使用第一个 get 函数读取并打印五个字符的 ASCII 值。请注意，循环由 i 的值控制，而不是由流的状态控制。

请注意，我们键入由五个字符组成的一组数据并存储在 cin 对象缓冲区中。程序逐个读取这五个字符，并使用插入运算符（<<）将它们发送到监视器。

程序 16-2　测试第一个 get 函数

```
1   /***************************************************************
2    * 本程序用于读取若干字符并打印它们的 ASCII 值                    *
3    ***************************************************************/
4   #include <iostream>
5   using namespace std;
6
7   int main ()
8   {
9     int x;
10    cout << "Enter five characters (no spaces): ";
11    for (int i = 0; i < 5; i++)
12    {
13      x = cin.get ();
14      cout << x << " ";
15    }
16    return 0;
17  }
运行结果:
Enter five characters (no spaces): ABCDE
65 66 67 68 69
```

接下来，我们使用 istream 类中的 get 函数和 ostream 类中的 put 函数，将每个单词的第一个字母变为大写（参见程序 16-3）。

注意，在程序 16-3 中，我们定义了两个变量：当前字符 c 和前一个字符 pre。如果前一个字符是空格或者换行符（'\n'），则将当前字符转换为大写。为了将第一个单词的首字母大写，我们将 pre 设置为换行符。在每次迭代之后，我们将当前字符存储为前一个字符 pre。

程序 16-3　测试 get 函数和 put 函数

```
1   /***************************************************************
2    * 本程序把每个单词的首字母转换为大写                              *
3    ***************************************************************/
4   #include <iostream>
5   using namespace std;
6
7   int main ()
8   {
9     char c;
10    cout << "Enter a multi-line text and EOF as the last line." << endl;
11    char pre = '\n';
12    while (cin.get(c))
13    {
```

```
14        if (pre == ' ' || pre == '\n' )
15        {
16            cout.put (toupper (c));
17        }
18        else
19        {
20            cout.put (c);
21        }
22        pre = c;
23    }
24    return 0;
25 }
```

运行结果：

```
Enter a multi-line text and EOF as the last line.
This is the text that we want to capitalize
This Is The Text That We Want To Capitalize
each word.
Each Word.
^Z
```

读取一个C字符串。istream类提供了两个成员函数，用于读取一组字符并从中创建一个C字符串。第一个函数get从键盘读取 $n-1$ 个字符，并将其存储为C字符串（在数组末尾添加一个空字符）。第二个函数getline读取 $n-1$ 个字符或者由分隔符（默认为'\n'）终止的一组字符，以先到者为准。请注意，分隔符将从流中删除，并不会添加到字符数组中。这些函数的原型如表16-5所示。

表16-5　读取一个C字符串

istream类	ostream类
istream& get (char* s, int n)	
istream& getline (char* s, int n, char d ='\n')	

注意，在表中显示的两种情况下，当读取了n个字符或者达到分隔符时，将设置eofbit标志位，并且不能从流中读取更多的字符。

我们在程序16-4中测试了第二个函数。我们创建了一个可以容纳80个字符的数组来接收一行内容。然后我们键入一行文本并输入回车键。程序根据输入行的内容创建了一个字符串（不带'\n'字符），并在末尾添加了一个空字符。

程序16-4　测试getline函数

```
1  /***************************************************************
2   *本程序用于测试getline函数                                      *
3   ***************************************************************/
4  #include <iostream>
5  using namespace std;
6
7  int main ()
8  {
9     char str2 [80];
10    cin.getline (str2, 80, '\n');
11    cout << str2;
12    return 0;
```

```
13 }
```

运行结果：

```
This is a line to make a string out of it.
This is a line to make a string out of it.
```

基本数据类型的输入 / 输出

istream 和 ostream 类基本上是字符流类，但这并不意味着我们不能从输入流中提取基本类型的值，也不意味着我们不能将基本数据类型的值插入输出流中。基本数据类型的输入和输出也被视为文本输入 / 输出，因为函数提取字符并将它们组合在一起以获得基本数据类型，或者函数从基本数据类型创建字符并将它们插入输出流中。这些工作是由两个称为插入运算符（<<）和提取运算符（>>）的重载运算符完成的。表 16-6 显示了这两个运算符的一般声明，其中类型可以是 bool、char、short（有符号和无符号）、int（有符号和无符号）、long（有符号和无符号）、float、double、long double 或者 void*。

表 16-6　提取和插入一个基本数据类型的数据

istream& operator >> (*type& x*);	// 提取下一个数据项
ostream& operator << (*type& x*);	// 插入下一个数据项

对于基本数据类型，提取运算符和插入运算符的行为是不同的。我们简单回顾一下这些操作。

提取基本数据类型。当遇到类似“cin >> 变量”的语句时，基本数据类型的提取运算符使用以下四个规则。提取运算符（>>）逐个读取字符，直到找到不属于相应类型语法的字符：

1）如果正在提取布尔值，则必须找到可以解释为 0 或者 1 的值。

2）如果正在提取字符值，则读取输入流中的下一个字符（包括空白字符）并将其存储在相应的变量中。

3）如果正在提取整数值，则读取数字，直到遇到非数字字符。然后生成已提取字符的整数值。未提取的字符仍保留在流中。

4）如果正在提取浮点值，则读取的数字之间只有一个小数点。如果遇到第二个小数点或者非数字字符，它将停止读取并生成浮点值。

插入基本数据类型。基本数据类型可以使用插入运算符（<<）插入流中。该值被视为一组字符并插入流中。例如，整数 124 被作为三个字符（1、2 和 4）插入流中。

C 类型字符串的输入 / 输出

当我们使用 cin 对象输入 C 字符串时，必须注意不要超过字符数组的大小，否则，会发生运行时错误，程序将被中止。输出 C 字符串时不存在问题。

C++ 字符串的输入 / 输出

提取运算符和插入运算符为 string 类进行了重载，它们可以按照定义的方式使用。但是，当遇到第一个空白字符时，输入停止。如果要读取空白字符（整行），必须使用 <string> 头文件中定义的 getline 函数（全局函数）。函数签名如下：

```
istream& getline (istream& in, string& str);
```

其他成员函数

istream 和 ostream 类中定义了其他成员函数，但这些类的实例化对象不会使用这些成员函数。这些成员函数由层次结构中的其他流类继承，我们将在下一节中讨论这些类。

16.3　文件流

在上一节中，我们使用键盘和监视器作为数据源和数据接收器，但键盘和监视器上的数据是临时的。程序终止后，数据将不存在，不能再次使用。如果需要，必须重新创建数据。文件是数据的永久源或者接收器。创建后，文件可以被其他程序使用。文件还可以保存以备将来使用。

要使用文件，必须使用在 <fstream> 头文件中定义的文件流类。这样做允许我们将文件连接到程序，以便从文件读取数据或者写数据到文件中。

> 要使用文件流，必须包含 <fstream> 头文件。

如图 16-5 所示，文件流类由三个类组成：ifstream、ofstream 和 fstream。这些类用于从文件输入数据，或者将数据输出到文件，或者输入和输出的是同一个文件。它们分别从 istream、ostream 和 iostream 类继承，并继承了这些类的所有成员函数。文件流类定义了一些新的函数，主要用于实例化和打开文件。

> 控制台流中定义的所有成员函数都在文件流中被继承。

16.3.1　文件输入 / 输出

数据源或者数据接收器必须始终是一个连接到流对象的实体。对于文件输入 / 输出，数据源是文件，数据接收器是文件。文件是驻留在磁盘（辅助存储器）上的实体。换言之，我们有文件和必须连接到文件的流对象，如图 16-9 所示。这意味着我们必须构造一个流，将该流连接到相应的文件，从该文件读取输入或者写数据到该文件，断开该文件，并销毁该流。

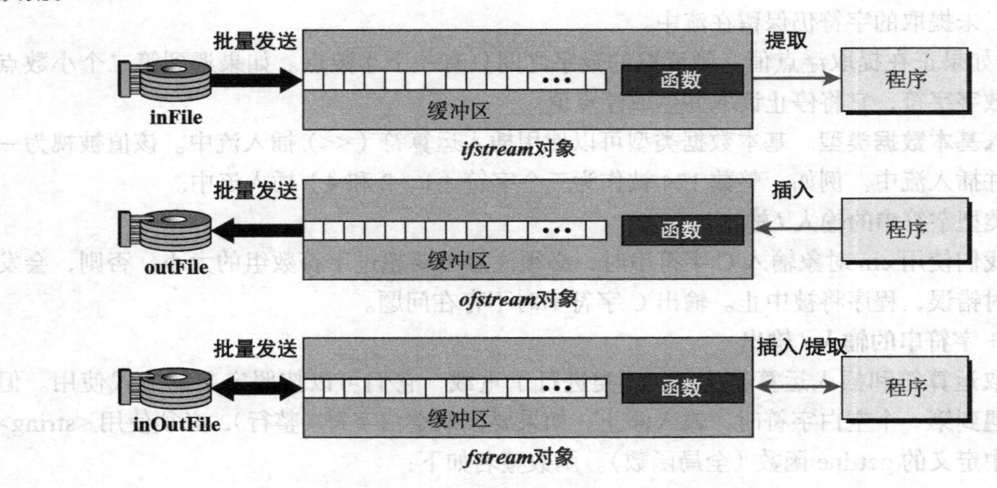

图 16-9　将程序连接到文件的文件流对象

构建流

我们可以实例化三个流中的任何一个，如下所示。为此，我们必须在程序中包含

<fstream> 头文件。

```
ifstream inStream;
ofstream outStream;
fstream inOutStream;
```

连接到文件（打开文件）

流的实例化只创建该类型的对象。如果要读取或者写入文件，必须将实例化的流连接到相应的文件。此步骤也称为打开文件，因为该步骤是使用相应流类中名为 open 的成员函数完成的。open 函数使用文件名（C 字符串）作为第一个参数，使用文件打开模式作为第二个参数。目前我们暂时忽略打开模式，并使用默认模式。参数 inFile、outFile 和 inOutFile 是文件名。

```
inStream.open (const char* inFile, ...);
outStream.open (const char* outFile, ...);
inOutStream.open (const char* inOutFile, ...);
```

读取和写入

在实例化流并打开文件之后，我们可以从为输入而打开的文件中读取数据，或者写数据到为输出而打开的文件中，或者读取并写入的是为读/写而打开的文件。没有为文件流定义新的读写成员函数，这些流继承了那些为控制台流定义的成员函数，如前所述。

断开文件连接（关闭文件）

当我们处理完这些文件后，必须使用文件流中定义的成员函数 close 关闭文件。由于流一次只能连接到一个文件，因此成员函数 close 没有参数。

```
inStream.close ();
outStream.close ();
inOutStream.close ();
```

销毁流对象

当流对象超出作用域范围时，此步骤自动完成。

测试是否成功打开文件

我们知道文件是程序的外部实体。成员函数 open 连接由参数指定名称的文件。如果操作系统没有打开或者无法打开文件，会发生什么情况？例如，输入文件可能已被删除或者损坏，或者由于磁盘已满而无法打开输出文件。三个文件流类都提供了一个函数 is-open，来测试文件是否已被成功打开并连接到该流。测试结果是布尔值 true 或者 false。

```
inStream.is-open ()
outStream.is-open ()
inOutStream.is-open ()
```

两个示例

程序 16-5 演示了如何使用上面描述的步骤来实例化流、打开文件并测试文件是否被打开、写入文件以及关闭文件（流对象的销毁在后台完成）。

程序 16-5 测试输出数据到一个文件的五个步骤

```
1   /***************************************************************
2    * 本程序用于创建输出文件并写内容到文件中                        *
```

```
3   ***************************************************************/
4   #include <iostream>
5   #include <fstream>
6   #include <cassert>
7   using namespace std;
8
9   int main ()
10  {
11    // 实例化一个 ofstream 对象
12    ofstream outStrm;
13    // 创建一个文件并连接到 ofstream 对象
14    outStrm.open ("integerFile");
15    if (!outStrm.is_open())
16    {
17        cout << "integerFile cannot be opened!";
18        assert (false);
19    }
20    // 使用重载的插入运算符写内容到文件中
21    for (int i = 1; i <= 10; i++)
22    {
23        outStrm << i * 10 << " ";
24    }
25    // 关闭文件
26    outStrm.close ();
27    // 执行 return 语句后，ofstream 对象自动被销毁
28    return 0;
29  }
```

在程序 16-6 中，我们实例化一个 ifstream 类并将其连接到在程序 16-5 中创建的 integerFile。然后我们逐个读取整数并显示这些整数。注意，我们使用两个流：连接到 integerFile 的 ifstream 和自动连接到监视器的 ostream。如前所述，ostream 由系统自动实例化（cout 对象）。

程序 16-6　从现有文件中读取数据

```
1   /***************************************************************
2    * 本程序用于从现有文件中读取数据                               *
3    * 并显示到监视器上                                             *
4    ***************************************************************/
5   #include <iostream>
6   #include <fstream>
7   #include <cassert>
8   using namespace std;
9
10  int main ()
11  {
12    int data;
13    // 实例化一个 ifstream 对象
14    ifstream inStrm;
15    // 连接一个现有文件到 ifstream 对象
16    inStrm.open ("integerFile");
17    if (!inStrm.is_open())
```

```
18        {
19            cout << "integerFile cannot be opened!";
20            assert (false);
21        }
22        //从 ifstream 对象读取数据，并写入 cout 对象
23        for (int i = 1; i <= 10; i++)
24        {
25            inStrm >> data ;
26            cout << data << " " ;
27        }
28        //断开文件 integerFile 和 ifstream 的连接
29        inStrm.close ();
30        //执行 return 语句后，ifstream 对象自动被销毁
31        return 0;
32    }
运行结果：
10 20 30 40 50 60 70 80 90 100
```

当程序 16-5 运行时，不会在监视器上输出任何内容，但我们可以使用文本编辑器检查名为 integerFile 的文件（数据文件）的内容，以确认该文件是否包含以下内容行。

```
10 20 30 40 50 60 70 80 90 100
```

注意，这次我们可以看到监视器上打印的整数。我们使用提取运算符从 integerFile 中读取数据。此运算符跳过空白字符并读取数据。每次读取之后，我们将数据写入 cout 对象（监视器）。

16.3.2 文件打开模式

在程序 16-5 和程序 16-6 中，我们忽略了文件打开模式。在这两种情况下（连接到 ofstream 和连接到 ifstream 时），我们都使用了 integerFile 的默认打开模式。文件打开模式是在 ios 类中定义的类型，由流类层次结构中的所有类继承。文件打开模式不适用于控制台流或者字符串流，因为我们不需要打开这些流，文件打开模式只适用于文件流。文件打开模式定义为一个类型，但其实现依赖于系统。大多数情况下，文件打开模式是作为一个位字段（见附录 E）实现的，其中包含 6 个二进制位，这些二进制位可以独立使用，或者与其他二进制位结合使用。如表 16-7 所示。

表 16-7 文件流的文件打开模式

openmode: in / out / app / ate / trunc / binary	
ios :: in	打开文件用于输入
ios :: out	打开文件用于输出；文件内容被清空
ios :: app	打开文件用于输出；写入文件末尾（添加内容）
ios :: ate	打开文件后立即移动到文件末尾
ios :: trunc	把文件长度截断为 0
ios :: binary	以二进制模式读 / 写文件（默认是文本模式）

其中一些模式可以组合，如图 16-10 所示。二进制模式可以与所有其他模式组合（图中未显示）。

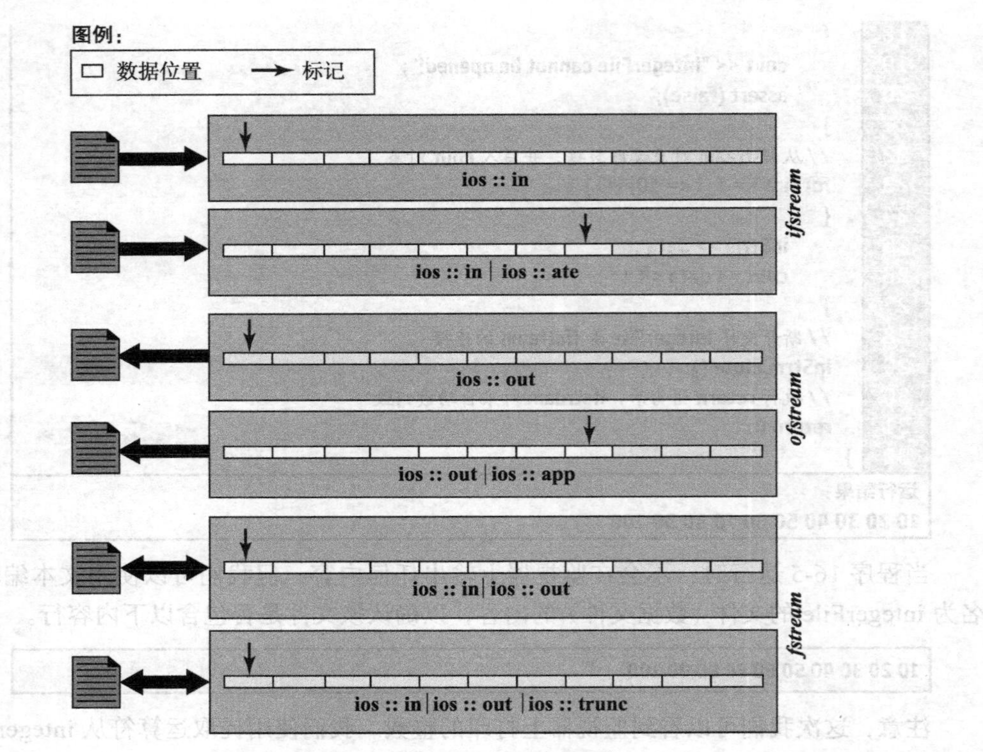

图 16-10 文件打开模式的常用组合

用于输入的文件打开模式

通常，对输入文件我们使用以下两种模式之一。第一种模式（ios::in）打开文件用于输入，并将标记放在缓冲区的第一个字节。这允许我们从第一个字节开始读取文件。每次读取后，标记移动到下一个字节，直至到达一个空位置。在这种情况下，eofbit 标志位被设置，将无法再读取文件。

第二种模式（ios :: in | ios :: ate），打开文件用于读取，但标记的位置在最后一个字节之后。这意味着已经设置了 eofbit 标志位，我们无法读取。然而，第二种模式有其他应用，例如通过向文件头移动标记来反向读取字节。另一种应用是获取文件的大小，稍后我们将演示。

我们可以使用第一种模式（ios :: in）读取名为 file1 的文件的内容，其内容如下所示：

```
This is the file that we want to open
and read its contents.
```

程序 16-7 逐字符读取文件，并在监视器上输出字符，直到 eofbit 标志位被设置。

程序 16-7 使用第一种输入模式来读取一个文件

```
1  /***************************************************************
2   *本程序用于打开文件，逐字符读取文件内容，                        *
3   *并把每个字符输出到监视器                                      *
4   ***************************************************************/
5  #include <iostream>
6  #include <fstream>
```

```
7    #include <cassert>
8    using namespace std;
9
10   int main ()
11   {
12     //变量声明
13     char ch;
14     //实例化一个 ifstream 对象
15     ifstream istrm ;
16     //打开 file1，并测试它是否被正确打开
17     istrm.open ("file1", ios :: in);
18     if (!istrm.is_open())
19     {
20         cout << "file1 cannot be opened!" << endl;
21         assert (false);
22     }
23     //逐字符读取文件 file1 并输出到监视器
24     while (istrm.get (ch))
25     {
26         cout.put(ch);
27     }
28     //关闭流
29     istrm.close ();
30     return 0;
31   }
运行结果：
This is the file that we want to open
and read its contents.
```

用于输出的文件打开模式

如图 16-10 所示，我们可以使用两种输出模式来写入文件。在第一种模式（ios :: out）中，我们打开文件用于输出。如果文件包含任何数据，则这些数据将被删除。在第二种模式（ios :: out | ios :: app）中，我们打开文件并在文件末尾写入数据（添加数据）。保留现有数据。

我们创建一个名为 file2 的文件，并将 file1 的内容复制到其中。注意，我们使用打开模式（ios :: in）打开 file1 进行输入，这意味着标记被设置在缓冲区的第一个字节上。我们使用打开模式（ios :: out）打开 file2，这意味着缓冲区被清空，标记被设置为第一个字节。在程序 16-8 中，我们从 file1 中一次读取一个字符，然后将其写入 file2。两个文件中的标记按顺序移动。

程序 16-8　使用输入和输出模式实现文件拷贝

```
1    /***************************************************************
2     * 本程序用于打开文件，逐字符读取文件内容，                    *
3     * 并把每个字符写入另一个文件                                  *
4     ***************************************************************/
5    include <iostream>
6    #include <fstream>
7    #include <cassert>
8    using namespace std;
9
```

```
10  int main ()
11  {
12      //变量声明
13      char ch;
14      //实例化一个 ifstream 对象和一个 ofstream 对象
15      ifstream istr;
16      ofstream ostr;
17      //打开 file1 和 file2，并测试它们是否被正确打开
18      istr.open ("file1", ios :: in);
19      if (!istr.is_open())
20      {
21          cout << "file1 cannot be opened!" << endl;
22          assert (false);
23      }
24      ostr.open ("file2", ios :: out);
25      if (!ostr.is_open())
26      {
27          cout << "file2 cannot be opened!" << endl;
28          assert (false);
29      }
30      //逐字符读取文件 file1 的内容并写入文件 file2 中
31      while (istr.get (ch))
32      {
33          ostr.put(ch);
34      }
35      //关闭文件 file1 和 file2
36      istr.close ();
37      ostr.close ();
38      return 0;
39  }
```

如果比较程序 16-7 和程序 16-8，我们会发现两个程序都在做同样的事情：复制。但存在差异。程序 16-7 将 file1 的内容复制到监视器；程序 16-8 将 file1 的内容复制到 file2。在第一个程序中，我们只有一个要实例化的流（cout 已经被系统实例化了）；在第二个程序中，我们必须实例化两个流。在第一个程序中，我们只能打开一个文件（cout 已经被系统打开了）；在第二个程序中，我们必须打开两个文件。在第一个程序中，我们只关闭一个文件（cout 是系统自动关闭的）；在第二个程序中，我们必须关闭这两个文件。在第一个程序中，在监视器上创建的副本是临时的（但是我们在运行程序时可以看到数据）；在第二个程序中，我们什么也看不到，但是我们可以打开 file2，以查看 file2 是 file1 的精确副本。

打开模式（ios :: out | ios :: app）可以用于打开现有文件进行内容添加。文件的内容不受影响，我们在文件末尾添加数据。例如，假设我们决定在 file1 末尾添加日期。如程序 16-9 所示，我们只需打开文件，并将日期作为 C 字符串添加到文件末尾。

程序 16-9 添加数据到一个文件中

```
1   /**************************************************************
2    * 本程序用于打开文件，把当前日期                                *
3    * 添加到文件末尾，然后关闭文件                                  *
4    **************************************************************/
5   #include <iostream>
```

```
6   #include <fstream>
7   #include <cassert>
8   using namespace std;
9
10  int main ()
11  {
12    //实例化一个 ofstream 对象
13    ofstream ostr;
14    //打开 file1，并连接到 ofstream 对象
15    ostr.open ("file1", ios :: out | ios :: app);
16    if (!ostr.is_open())
17    {
18        cout << "file1 cannot be opened!";
19        assert (false);
20    }
21    //把 C 字符串格式的日期添加到文件 file1 末尾
22    ostr << "\nOctober 15, 2016.";
23    //关闭文件
24    ostr.close ();
25    return 0;
26  }
```

当我们使用文本编辑器打开文件 file1 时，我们看到日期已添加到文件末尾（在新行中），如下所示：

```
This is the file that we want to open
and read its contents.
October 15, 2015.
```

用于输入/输出的文件打开模式

如果使用模式（ios :: in | ios :: out）将文件连接到 fstream 对象，则可以打开文件进行输入/输出。假设我们有一个整数文件。我们决定在文件的末尾加上整数的和。我们可以打开文件进行读写（ios :: in | ios :: out）。请注意，这种情况与前面的示例不同。我们必须逐个读取每个整数（输入），然后在文件末尾写入整数之和（输出）。变更前文件的内容如下所示：

```
12 14 17 20 21 25 32 27 56 18
```

程序 16-10 显示了解决方案。

程序 16-10　打开文件同时用于输入和输出

```
1   /**************************************************************
2    *本程序用于打开文件，计算其内容之和，                           *
3    *并把结果添加到文件末尾                                         *
4    **************************************************************/
5   #include <iostream>
6   #include <fstream>
7   #include <cassert>
8   using namespace std;
9
10  int main ()
11  {
12    //实例化一个 fstream 对象
```

```
13    fstream fstr;
14    //打开 intFile，并连接到 fstream 对象
15    fstr.open ("intFile", ios :: in | ios :: out);
16    if (!fstr.is_open())
17    {
18        cout << "intFile cannot be opened!";
19        assert (false);
20    }
21    //读取所有的整数，计算累计和，直到检测到文件末尾
22    int num;
23    int sum = 0;
24    while (fstr >> num)
25    {
26        sum += num;
27    }
28    //清除文件，然后添加信息和累计和到文件末尾
29    fstr.clear ();
30    fstr << "\nThe sum of the numbers is: ";
31    fstr << sum;
32    //关闭流
33    fstr.close ();
34    return 0;
35 }
```

当循环（程序 16-10 中的第 24 行到第 27 行）终止时，eofbit 标志位被设置，我们不能继续使用流。在第 29 行中，我们清除流，然后将消息和累计和写到文件的末尾。结果是一个包含以下内容的文件：

```
12 14 17 20 21 25 32 27 56 18
The sum of the numbers is: 242
```

【例 16-1】 图 16-10 中的最后一个文件模式（ios :: in | ios :: out | ios :: trunc）允许我们打开文件，截断其内容，并向其写入新数据。这与模式（ios :: out）相同，但有一个区别。在带 trunc 的模式中，我们可以创建一个新文件；在只有 out 的模式中，我们只能删除一个旧文件的内容，并用新数据替换它，这是在我们想要保留文件名但包含新内容时使用的。

其他打开模式

我们没有使用表 16-7 中列出的最后三种模式，但在本章后面我们将使用这三种模式。

16.3.3 其他成员函数

类 istream 还定义了其他成员函数，但这些成员函数在 ifstream 类中有更多的应用（表 16-8）。

表 16-8　文件流中使用的其他成员函数

int gcount () const	// 统计字符个数
istream& unget ()	// 把最后一个字符放回流中
istream& putback (char c)	// 同 unget
int peak ()	// 查看内容但不提取
istream& ignore (int n = 1, int d = eof)	// 忽略若干字符

函数 gcount 给出最后一次输入中提取的字符数。unget 函数将从流中提取的最后一个字

符放回流中。函数 putback 与 unget 的作用相同。函数 peak 查看下一个字符的值而不从流中删除该字符。函数 ignore 跳过若干字符而不提取这些字符。我们演示如何使用 unget 函数，其余函数留作练习。

假设我们有一个只包含文本和整数的文件，如下所示：

```
We have 7, 12, 23, and 442 in this file.
```

我们想从这个文件中提取整数并忽略文本。程序 16-11 显示了如何使用 unget 函数来实现这一功能。

程序 16-11 逐字符浏览整个文件。如果一个字符不是数字字符，则被丢弃。如果字符是一个数字，则将其放回流中，以便再次提取为整数。

程序 16-11 使用 unget 函数

```
1  /****************************************************************
2   * 本程序用于从同时包含整数和字符的文件中                        *
3   * 仅读取整数内容                                                *
4   ****************************************************************/
5  #include <iostream>
6  #include <fstream>
7  #include <cassert>
8  using namespace std;
9
10 int main ()
11 {
12    //实例化一个 ifstream 对象并连接到一个文件
13    ifstream ifstr;
14    ifstr.open ("mixedFile" , ios :: in);
15    if (!ifstr.is_open())
16    {
17        cout << "The file mixedFile cannot be opened for reading!";
18        assert (false);
19    }
20    //仅读取整数，通过把数字字符放回流中并重新读取的方法
21    char ch;
22    int n;
23    while (ifstr.get (ch))
24    {
25        if (ch >= '0' && ch <= '9')
26        {
27            ifstr.unget();
28            ifstr >> n;
29            cout << n << " ";
30        }
31    }
32    //关闭文件
33    ifstr.close ();
34    return 0;
35 }
运行结果:
7 12 23 442
```

16.3.4　顺序访问与随机访问

文件是 8 位字节的顺序集合。我们不直接读取或者写入文件，我们读取或者写入流对象中的缓冲区。在读取过程中，系统以块移动的方式将文件内容复制到缓冲区；在写入过程中，缓冲区的内容以块移动的方式从缓冲区移动到文件。将字节从文件传输到缓冲区或者从缓冲区传输到文件是我们无法控制的。

当我们谈论**顺序访问**（sequential access）或者**随机访问**（random access）时，并不意味着文件是按顺序或者随机排列的（文件始终是字节的顺序集合），我们指的是如何访问流对象中的缓冲区：顺序访问或者随机访问。如果我们考虑缓冲区，就会发现它就像一个数组，我们总是可以按顺序（逐个元素）或者随机（我们想要的任何元素）访问数组。事实上，流中的缓冲区是由元素组成的，元素的索引从 0 到 $n-1$，其中 n 是系统定义的缓冲区大小。

ifstream 中的缓冲区具有指向下一个要读取的字节的标记。ofstream 中的缓冲区具有指向下一个要写入的字节的标记。fstream 中的缓冲区只有一个读取或者写入标记。

> 每个流只有一个标记，用于指示是否应读取或者写入下一个字节。

顺序访问

迄今为止，我们所展示的所有示例都使用顺序访问。在顺序访问中，流标记的移动由读/写函数控制。打开文件时，标记从缓冲区的开头处（索引为 0）开始。每次读取或者写入时，标记都会向缓冲区的末尾移动。标记指向下一个要读取或者写入的字节，但每次读取或者写入的字节数取决于数据类型。如果我们正在读取或者写入字符，则一次移动一个字节；如果我们正在读取或者写入格式化数据（基本类型或者类类型），则每次读取或者写入的移动可以是若干字节。

随机访问

使用控制台系统读取或者写入只能按顺序进行，但是当我们使用文件流时，可以使用随机访问。在随机访问中，我们可以从任何位置读取数据，也可以将数据写入任何位置。我们只需将流标记移动到相应的字符。istream 和 ostream 类提供了成员函数，允许我们查找标记的位置并将标记移动到所需的位置。表 16-9 显示了用于此目的的六个成员函数。

表 16-9　用于随机访问的成员函数

ios :: beg	ios :: cur	ios :: end

dir的值

输入	输出
int tellg (); istream& seekg (int *pos*); istream& seekg (int *off*, ios :: dir);	int tellp (); ostream& seekp (int *pos*); ostream& seekp (int *off*, ios :: dir);

查找当前位置的索引。每个类别中的第一个函数（tellg 或者 tellp）返回标记指向的当前字节的索引。虽然只有一个标记，但是当我们使用 istream 对象时，必须使用 tellg（g 代表 get），当我们使用 ostream 对象时，必须使用 tellp（p 代表 put）。

程序 16-12 显示了我们如何在一个文件中打印标记的位置和相应字符的值，文件中只有一个单词（“Hello !”）。

程序 16-12 打印字符的位置和值

```
/***************************************************************
 *本程序用于打印文件中字符的位置和值                                *
 ***************************************************************/
#include <iostream>
#include <fstream>
using namespace std;

int main ()
{
  // 变量的声明
  char ch;
  int n;
  // 实例化流对象，并打开文件
  ifstream istr;
  istr.open ("sample", ios :: in);
  // 获取字符和它们的位置
  n = istr.tellg ();
  while (istr.get(ch))
  {
      cout << n << " " << ch << endl;
      n = istr.tellg ();
  }
  // 关闭文件
  istr.close ();
  return 0;
}
```

运行结果：

```
0 H
1 e
2 l
3 l
4 o
5 !
```

移动标记。使用表 16-9 中的其他四个成员函数，我们可以移动标记指向下一个要读取或者写入的字节。移动可以是绝对的或者相对的。如果我们知道要移动到的字节的索引，我们可以使用两个成员函数和一个参数：seekg（location）和 seekp（location）。如果要将标记移动到相对于缓冲区的开始、当前或者结束位置的另一个位置，可以使用相对成员函数 seekg(off, dir) 或 seekp(off, dir)，其中偏移量（off）是正整数或者负整数，dir 是开始位置：cur 表示当前位置，beg 表示缓冲区的开始，end 表示缓冲区的结束。

假设我们有一个简单的文件，其内容如下所示：

```
There are wonderful things to do in life.
```

我们希望更改文件的内容，使每个单词都单独在一行上。程序 16-13 完成该工作。请注意，我们一次读取一个字符（顺序访问）。然后我们检查字符的值。如果是空格字符，我们将其更改为回车符（'\r'），而不是换行符（'\n'），因为换行符通常由两个字符组成，将替换空格和下一个单词的第一个字符。在读取一个字符并发现它是一个空格字符之后，我们必须替换这个空格字符。但是，标记已经前进到下一个单词的开头。在用回车符替换空格字符之前，必须将标记向后移动。我们使用 seekp 函数将标记移回一个位置。

程序 16-13 把空格字符替换为回车符

```
/*****************************************************************
 *本程序用于把文件中的每个单词放置在单独一行                       *
 *****************************************************************/
#include <iostream>
#include <fstream>
using namespace std;

int main ()
{
  fstream fstr;
  fstr.open ("file3" , ios :: in | ios :: out);

  char ch;
  while (fstr.get(ch))
  {
      if (isspace (ch))
      {
            fstr.seekp (-1 , ios :: cur);
            fstr.put ('\r');
      }
  }
  fstr.close ();
  return 0;
}
```

运行结果：
```
There
are
wonderful
things
to
do
in
life.
```

作为另一个例子，如果使用 ate（在末尾，at the end）打开模式打开文件，我们可以得到文件的大小（以字节为单位）。ate 打开模式将标记放在文件中最后一个字符之后。然后，我们可以使用成员函数 tellg，以获取最后一个字符后面的字符索引，如程序 16-14 所示。

下面显示了我们使用的文件的内容。注意，当我们检查 tellg 返回的值时，它给出了文件中句点后面字符的索引（44），但是由于索引从 0 开始而不是 1，我们知道文件中有 44 个字符，索引从 0 到 43。

```
This is the file whose size we want to find.
```

程序 16-14 获取文件的大小

```
/*****************************************************************
 *本程序用于获取文件的长度                                         *
```

```
3    ***********************************************************************/
4    #include <iostream>
5    #include <fstream>
6    using namespace std;
7
8    int main ()
9    {
10       //实例化流对象，并连接到文件
11       ifstream ifstr;
12       ifstr.open ("file4" , ios :: in | ios :: ate);
13       //查找最后一个字符之后的标记的值
14       cout << "File size: " << ifstr.tellg ();
15       //关闭文件
16       ifstr.close ();
17       return 0;
18   }
```

运行结果：

```
File size: 44
```

16.3.5 二进制输入 / 输出

控制台流不能用于二进制数据，因为数据源和数据接收器（键盘和监视器）都无法处理二进制数据。另一方面，连接到文件流的文件可以用于输入和输出二进制数据，其中 8 个二进制位被视为一个字节，而不与任何字符代码（ASCII 或者 Unicode）相关。当我们希望输入或者输出应解释为字节的数据时，这一点很重要。例如，我们可以编写一个程序来读取存储了音频 / 视频信息的文件。视频文件中的每个图片元素（像素）可以是 8 个二进制位，即可以存储为一个字节。我们可能还需要将用户自定义类型的对象存储为一组字节。在这些情况下，我们可以将数据以字节序列存储在文件中。为此，应以二进制模式打开文件。

文件流能够保存文本和二进制数据。

正如我们在本章开头所讨论的，数据存储在内存、流和文件中的 8 位字节中。区别在于我们如何解释存储在流和文件中的数据（它们是彼此的映像）。在文本输入 / 输出中，值转换为字节；在二进制输入 / 输出中，二进制位的确切模式被转换为字节。我们将展示一个大整数 1 464 740 402 是如何以文本和二进制的形式存储在内存、流和文件中的（图 16-11）。

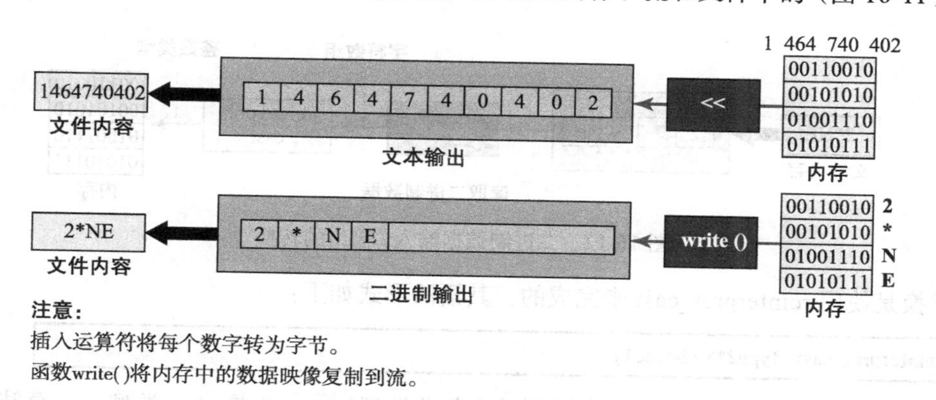

图 16-11　文本输出和二进制输出的区别

图16-11显示了在文本输出中，我们必须将整数的每个数字作为字符（文本）存储在流和文件中。在二进制输出中，我们只使用4个字节（内存中的数据映像）。很明显，在文本输出的情况下，文件可以通过文本编辑器打开，我们可以看到整数的值；在二进制输入的情况下，我们可以看到字符2*NE，这绝对不是存储在内存中的整数的值。当整数很大时，我们可以从图中看到二进制输出的好处。在文本输出中，文件存储10个字节；在二进制输出中，文件只存储4个字节。

成员函数

为了读写二进制数据，库定义了三个成员函数，如表16-10所示。这些成员函数是在istream和ostream类中定义的，但它们不在这两个类的对象中使用。ifstream和ofstream类继承并使用这些成员函数。

表16-10　读取和写入的成员函数

输入	输出
ifstream& read (char* s, int n) ofstream& readsome (char* s, int n)	ifstream& write (char* s, int n)

在表16-10中，函数read从流缓冲区最多读取n个字符，并填充名为s的字符数组。如果流中可用的字符数小于数组的大小，则eofbit标志位被设置。为了防止这种情况发生，第二个函数（readsome）读取字符，直到没有可用的字符或者读取n个字符为止。两个函数都不会向数组中添加空字符，这意味着它们不能用作C字符串。write函数将字符数组中的n个字符精确写入流缓冲区。

基本类型转换

读者可能已经注意到，这三个输入/输出函数实际上从流缓冲区读取字符，或者向流缓冲区写入字符，存储在内存中的字节与存储在流缓冲区中的字节之间不会发生转换。我们需要一个转换机制来将整数1 464 740 402转换为4个字符的数组（2，*，N，E），反之亦然（图16-12）。

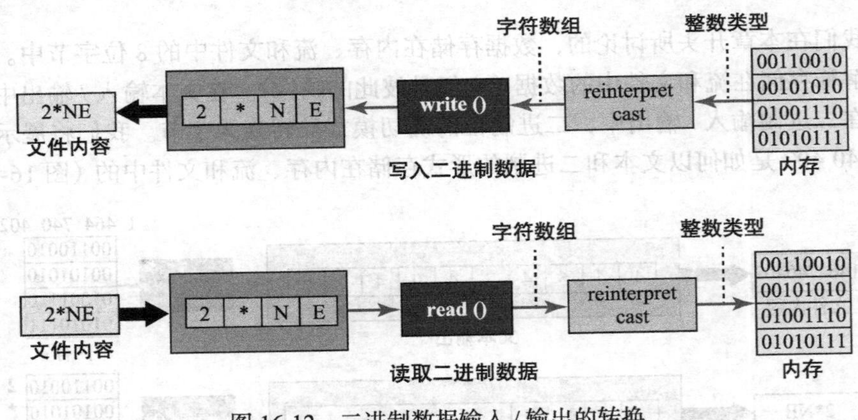

图16-12　二进制数据输入/输出的转换

转换是使用reinterpret_cast来完成的，其语法形式如下：

```
reinterpret_cast <type2*> (&type1)
```

在转换运算符中，type1是基本类型（或者类类型），type2是char类型。这意味着我们需要重写读写函数，如下所示：

```
ifstream& read (reinterpret_cast <char*> (&type), sizeof (type))
ifstream& readsome (reinterpret_cast <char*> (&type), sizeof (type))
ofstream& write (reinterpret_cast <char*> (&type), sizeof (type))
```

程序 16-15 演示了如何将 int 和 double 的值写入文件并读取它们，以确保正确存储了这些值。注意，我们使用 ofstream 对象创建一个新文件，然后使用 ifstream 对象检查该文件：在本例中，我们先写入数据，然后读取数据。

程序 16-15　写入和读取二进制数据

```
1   /***************************************************************
2    * 本程序演示如何将二进制数据写入文件                              *
3    * 并从文件中读取二进制数据                                        *
4    ***************************************************************/
5   #include <iostream>
6   #include <string>
7   #include <fstream>
8   #include <cassert>
9   using namespace std;
10
11  int main ()
12  {
13    int int1 = 12325;
14    double double1 = 45.78;
15    //创建一个新文件用于输出，并写入两个不同数据类型的值
16    ofstream strmOut ("Sample", ios :: out | ios :: binary);
17    if (!strmOut.is_open())
18    {
19        cout << "The file Sample cannot be opened for writing!";
20        assert (false);
21    }
22    strmOut.write (reinterpret_cast <char*> (&int1), sizeof (int));
23    strmOut.write (reinterpret_cast <char*> (&double1), sizeof (double));
24    strmOut.close ();
25
26    int int2;
27    double double2;
28    //打开同一文件用于输入，并读取两个不同数据类型的值
29    ifstream strmIn ("Sample", ios :: in | ios :: binary);
30    if (!strmIn.is_open())
31    {
32        cout << "The file Sample cannot be opened for reading!";
33        assert (false);
34    }
35    strmIn.read (reinterpret_cast <char*> (&int2) , sizeof (int));
36    strmIn.read (reinterpret_cast <char*> (&double2) , sizeof (double));
37    strmIn.close ();
38    //测试存储数据类型的值
39    cout << "Value of int2: " << int2 << endl;
40    cout << "Value of double2: " << double2 << endl;
41    return 0;
```

```
42  } // main 结束
```

运行结果：
```
Value of int2: 12325
Value of double2: 45.78
```

在程序16-15的第22行中，我们获取int类型变量的地址，将其解释为指向字符的指针，并将其内容作为字符写入文件。我们在第23行中做了同样的操作，但是变量的类型是double。

在第35行中，我们获取int类型变量的地址，将内存中保留的位置解释为指向字符的指针，并将从文件中提取的字符存储在其中。我们在第36行中也这么做了，但所在位置的类型为double。

注意，在这两种情况下（读取和写入），我们都将基本数据类型的地址解释为指向字符的指针。

用户自定义对象的转换

由于基本数据类型的对象以位（或者字节）序列的形式存储在内存中，因此可以执行转换。我们只需要更改读/写函数来访问类变量，而不是基本数据类型的变量。

读取数据时，我们需要一个空对象（从默认构造函数创建），用文件中的字符填充。写入数据时，需要将填充了内容的对象（由参数构造函数创建）更改为字符并写入文件，如下所示：

```
istream& read (reinterpret_cast <char*> (&object) , sizeof (class))
istream& write (reinterpret_cast <char*> (&object), sizeof (class))
```

我们可以创建一个表示学生信息（身份、姓名和GPA）的类，并将该类的实例写入一个文件，然后从该文件中读取实例。接下来我们将展示接口文件、实现文件和应用程序文件。程序16-16显示了接口文件。

程序16-16　文件student.h

```
1   /***************************************************************
2    * 本程序定义 Student 类的接口文件，                                *
3    * 该类将用于存储学生信息到二进制文件中                              *
4    ***************************************************************/
5   #ifndef STUDEN_H
6   #define STUDEN_H
7   #include <iostream>
8   #include <fstream>
9   #include <cassert>
10  #include <iomanip>
11  #include <cstring>
12  #include <string>
13  using namespace std;
14  
15  class Student
16  {
17      private:
18          int stdId;
19          char stdName [20];
20          double stdGpa;
```

```
21  public:
22      Student (int, const string&, double);
23      Student ();
24      ~Student ();
25      int getId() const;
26      string getName() const;
27      double getGpa () const;
28  };
29  #endif
```

程序 16-17 显示了 Student 类的实现文件。注意，我们使用访问器函数来获取数据成员的值，但是我们也可以为该类重载插入运算符。

程序 16-17　文件 student.cpp

```
1   /*****************************************************************
2    * Student 类的实现文件                                           *
3    *****************************************************************/
4   #include "student.h"
5   
6   // 参数构造函数
7   Student :: Student (int id, const string& name, double gpa)
8   : stdId (id), stdGpa (gpa)
9   {
10  strcpy (stdName, name.c_str() );
11    if (stdId < 1 || stdId > 99)
12    {
13        cout << "Identity is out of range. Program aborted.";
14        assert (false);
15    }
16    if (stdGpa < 0.0 || stdGpa > 4.0)
17    {
18        cout << "The gpa value is out of range. Program aborted.";
19        assert (false);
20    }
21  }
22  // 默认构造函数
23  Student :: Student ()
24  {
25  }
26  // 析构函数
27  Student :: ~Student ()
28  {
29  }
30  // 访问器函数
31  int Student :: getId() const
32  {
33    return stdId;
34  }
35  // 访问器函数
36  string Student :: getName() const
37  {
38    return stdName;
```

```
39  }
40  //访问器函数
41  double Student :: getGpa () const
42  {
43      return stdGpa;
44  }
```

程序16-18显示了用于将对象写入文件和从文件读取对象的应用程序文件。我们首先将几个对象更改为二进制数据并将它们写入文件。然后我们打开同一个文件，逐个读取学生对象。

程序 16-18　文件 app.cpp

```
1   /*************************************************************
2    * 本应用程序用于把学生记录写入一个二进制文件，                    *
3    * 然后依次读取这些记录                                        *
4    *************************************************************/
5   #include "student.h"
6
7   int main ()
8   {
9       // 打开文件 File.dat，用于二进制数据输出
10      fstream stdStrm1;
11      stdStrm1.open ("File.dat", ios :: binary | ios :: out);
12      if (!stdStrm1.is_open())
13      {
14          cout << "File.dat cannot be opened for writing!";
15          assert (false);
16      }
17      // 实例化五个对象
18      Student std1 (1 , "John", 3.91);
19      Student std2 (2 , "Mary", 3.82);
20      Student std3 (3 , "Lucie", 4.00);
21      Student std4 (4 , "Edward", 3.71);
22      Student std5 (5 , "Richard", 3.85);
23      // 把五个对象写入二进制文件，然后关闭文件
24      stdStrm1.write(reinterpret_cast <char*> (&std1), sizeof (Student));
25      stdStrm1.write(reinterpret_cast <char*> (&std2), sizeof (Student));
26      stdStrm1.write(reinterpret_cast <char*> (&std3), sizeof (Student));
27      stdStrm1.write(reinterpret_cast <char*> (&std4), sizeof (Student));
28      stdStrm1.write(reinterpret_cast <char*> (&std5), sizeof (Student));
29      stdStrm1.close ();
30      // 打开文件 File.dat，用于输入
31      fstream stdStrm2;
32      stdStrm2.open ("File.dat", ios :: binary | ios :: in);
33      if (!stdStrm2.is_open())
34      {
35          cout << "File.dat cannot be opened for reading!";
36          assert (false);
37      }
38      // 读取 Student 对象，显示它们，然后关闭文件 File.dat
39      cout << left << setw (4) << "ID" << " ";
40      cout << setw(15) << left << "Name" << " ";
```

```
41      cout << setw (4) << "GPA" << endl;
42      Student std;
43      for (int i = 0; i < 5; i++)
44      {
45          stdStrm2.read(reinterpret_cast <char*> (&std), sizeof (Student));
46          cout << setw (4) << std.getId() << " ";
47          cout << setw(15) << left << std.getName() <<" ";
48          cout << fixed << setw (4) << setprecision (2) << std.getGpa ();
49          cout << endl;
50      }
51      stdStrm2.close();
52      return 0;
53  }
```

运行结果：

```
ID   Name        GPA
1    John        3.91
2    Mary        3.82
3    Lucie       4.00
4    Edward      3.71
5    Richard     3.85
```

随机访问

我们还可以随机读取或者写入二进制文件。换言之，我们不需要读取所有学生的记录，而是可以使用学生的 ID 来读取特定学生的记录。然而，这涉及两个问题。

首先，我们必须确保文件中存储的所有对象的大小相同，这样我们就可以使用 seekg() 和 seekp() 函数。对于只有基本数据类型的数据成员的对象，每个对象的大小是其数据成员大小的总和。对于 Student 类，表示姓名的数组的大小也是固定的。

其次，根据给定的信息，我们必须采取措施来找到对象的位置。例如，我们可以使用上一个示例中学生的 ID，使用以下方法查找 Student 对象的位置：

```
seekg ((id - 1) * sizeof (Student))
```

但是，如果 ID 不是从 1 开始，则必须将 ID 转换为学生记录编号。

16.4 字符串流

字符串流使用三个类：istringstream、ostringstream 和 stringstream。这些类用于从字符串输入数据或者将数据输出到字符串。它们分别继承自 istream、ostream 和 iostream。除了少数例外，在超类中讨论的所有成员函数都可以在这些类中使用，并且可以添加更多的成员函数。字符串流类在 <sstream> 头文件中定义，使用字符串流类时，必须将其包含在程序中。

要使用字符串流，我们需要包含 <sstream> 头文件。

字符串流的源或者目标与控制台流或者文件流的源或者目标不同。在控制台流中，源或者目标是程序外部的物理设备。在文件流中，源或者目标是一个文件，一个存在于程序外部的实体。在字符串流中，源或者目标实际上是程序内部的字符串。换言之，我们从程序中的现有字符串中读取数据，我们将数据写入程序中的字符串。为了清晰起见，在本节中使用的图中，我们将数据源或者数据接收器显示为程序外部的字符串，但必须记住它是程序的一部分。

我们无法打开或者关闭字符串流，因为连接到这些流的数据源或者数据接收器实体不是外部的，它们是在程序中创建和销毁的。因此，字符串流没有 open() 或者 close() 函数。

> 字符串流没有定义 open() 和 close() 函数。

16.4.1 实例化

与文件流类似，每个字符串流都是使用构造函数实例化的。我们在表 16-11 中显示了 istringstream、ostringstream 和 stringstream 类的构造函数，其中使用了打开模式的默认值。

表 16-11　三个字符串流类的构造函数

```
istringstream (const string strg, ios :: openmode mod = ios :: in);
ostringstream (const string strg, ios :: openmode mod = ios :: out);
stringstream (const string strg, ios :: openmode mod = ios :: in |ios :: out );
```

每个构造函数实例化 istringstream、ostringstream 或 stringstream 类型的对象，并将该对象连接到第一个参数定义的字符串对象。请注意，必须包含一个字符串对象，但它可以是一个空字符串。

图 16-13 以图形方式显示了三个字符串流对象。

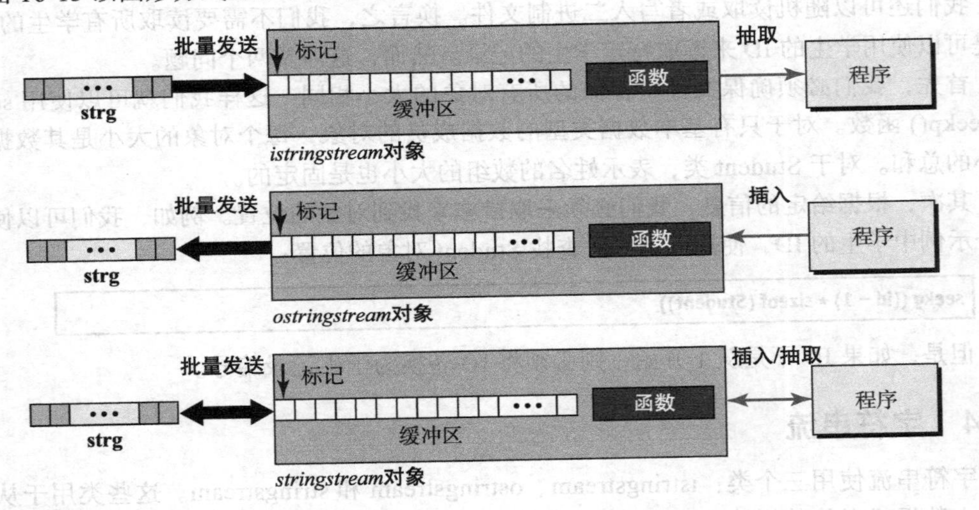

图 16-13　连接字符串和程序的字符串流对象

成员函数

类 istringstream 继承了在超类 istream 中定义的用于顺序和随机访问的所有成员函数（函数 open 和 close 除外）。ostringstream 同样如此，但它继承自超类 ostream。stringstream 类也如此，但它继承自超类 iostream。

此外，每个类都提供名为 str 的新成员函数的两个版本。第一个版本替换连接到流的字符串；第二个版本返回连接到流的字符串的副本。表 16-12 显示了这些函数的原型。

表 16-12　函数 str 的两个版本

void str (string *strg*);	// 把参数连接到宿主对象
string str () const;	// 返回连接到宿主对象的字符串

程序 16-19 是一个简单的字符串流示例。在程序的第一部分中，我们创建一个与字符串连接的 istringstream 对象，并打印该字符串。然后更改连接到相同 istringstream 的字符串并打印该字符串。在程序的第二部分中，我们对 ostringstream 进行了同样的操作处理。

程序 16-19　测试字符串流类

```
1  /***************************************************************
2   * 本程序演示如何使用字符串流类                                  *
3   ***************************************************************/
4  #include <iostream>
5  #include <string>
6  #include <sstream>
7  using namespace std;
8  
9  int main ()
10 {
11   // 使用 istringstream 对象
12   istringstream iss ("Hello friends!");
13   cout << iss.str () << endl;
14   iss.str ("Hello world!");
15   cout << iss.str () << endl << endl;
16   // 使用 ostringstream 对象
17   ostringstream oss ("Bye friends!");
18   cout << oss.str () << endl;
19   oss.str ("Bye world!");
20   cout << oss.str () << endl;
21   return 0;
22 }
```

```
运行结果：
Hello friends!
Hello world!

Bye friends!
Bye world!
```

16.4.2 应用：适配器

字符串流类最常见的应用是充当字符串类的适配器。我们知道，string 类没有构造函数来包装基本类型（char 除外）的值并从中生成字符串，也没有一个成员函数来展开一个字符串并从中取出基本数据类型的值。

在程序设计中，有时我们需要将基本数据类型转换为字符串，以及将字符串转换为基本数据类型。为此，我们使用 stringstream 类的对象作为适配器。

在封包过程中，我们将基本数据类型插入 ostringstream 对象中，然后将 ostringstream 对象更改为字符串对象。在解包过程中，我们将字符串更改为 istringstream 对象，并从 istringstream 对象中提取基本数据类型。这两种操作如图 16-14 所示。

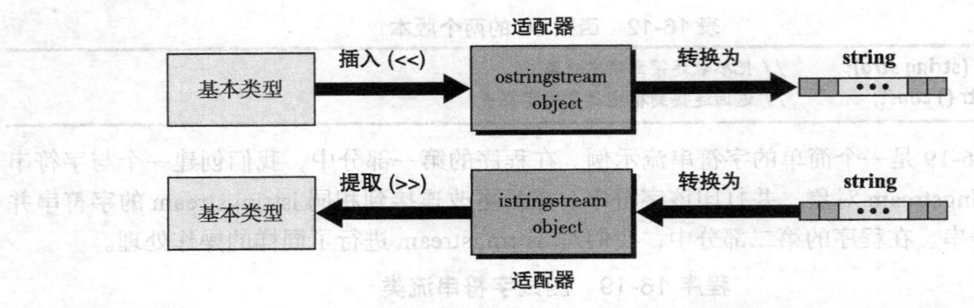

图 16-14　stringstream 类的对象作为适配器

我们可以创建一个函数模板来将任何基本数据类型转换为字符串。我们还可以创建一个函数模板来返回嵌入在字符串中的基本数据类型。我们在名为 convert.h 的头文件中包含这两个模板函数，如程序 16-20 所示。注意，在第 23 行中，我们必须使用函数模板特化（参见第 15 章中的讨论），因为 toData 模板函数只使用模板类型作为返回的类型。如果没有函数模板特化，模板函数就无法检测返回类型。

程序 16-20　文件 convert.h

```
1  /*****************************************************************
2   * 本头文件定义了两个模板函数。                                     *
3   * 第一个模板函数把基本数据类型插入字符串。                         *
4   * 第二个模板函数提取嵌入在字符串中的基本数据类型                   *
5   *****************************************************************/
6  #ifndef CONVERT_H
7  #define CONVERT_H
8  #include <iostream>
9  #include <string>
10 #include <sstream>
11 using namespace std;
12
13 // 函数 toString 把任何数据类型转换为字符串
14 template <typename T>
15 string toString (T data)
16 {
17   ostringstream oss ("");
18   oss << data;
19   return oss.str ();
20 }
21 // 函数 toData 提取嵌入在字符串中的数据
22 template <typename T>
23 T toData (string strg)
24 {
25   T data;
26   istringstream iss (strg);
27   iss >> data;
28   return data;
29 }
30 #endif
```

程序 16-21 演示了 convert.h 模板函数的使用。

程序 16-21　文件 app.cpp

```
1  /***************************************************************
2   *一个简单的应用程序，                                            *
3   *使用 toString 和 toData 模板函数                                 *
4   ***************************************************************/
5  #include "convert.h"
6
7  int main ( )
8  {
9    // 把整数 12 转换为字符串
10   string strg = toString (12);
11   cout << "String: " << strg << endl;
12   // 把字符串"15.67"转换为双精度浮点数
13   double data = toData <double> ("15.67");
14   cout << "Data: " << data;
15   return 0;
16 }
```

```
运行结果：
String: 12
Data: 15.67
```

16.5　格式化数据

在第 3 章中，我们讨论了操作符，并使用了其中的一些操作符。操作符使用 ios 类中定义的格式化数据成员（格式化标志、格式化字段和格式化变量）。由于所有流类都继承自 ios 类，因此它们都继承了这些数据成员和相应的成员函数，以对字段进行设置和取消设置并将值存储在变量中。为了创建自定义操作符，我们必须了解如何直接使用这些标志（flag）、字段（field）和变量（variable）。

16.5.1　直接使用标志、字段和变量

了解如何使用标志、字段和变量将有助于我们在本章后面创建定制的操作符。

格式化标志

如表 16-13 所示，ios 类定义了一个名为 fmtflags 的类型，它可以采用七个值中的任意组合。每个二进制位可以被设置（值 1）或者取消设置（值 0）。这些标志是独立的，我们可以使用这些标志中的一个或者多个来格式化数据。

表 16-13　类型 fmtflags 中的值

fmtflags: boolalph | skipws | showbase | showpoint | showpos | unitbuf | uppercase

标志	说明
ios :: boolalpha	显示布尔值表示（true 或者 false）
ios :: skipws	跳过空白字符（输入时默认设置）
ios :: showbase	当打印整数值时不显示数值的基数

（续）

ios :: showpoint	显示浮点数值的小数点
ios :: showpos	显示正数值的正号（加号 +）
ios :: unitbuf	每次操作后清空流
ios :: uppercase	显示十六进制的 A ～ F 和科学计数法的 E（大写）

我们可以使用 setf() 或者 unsetf() 函数设置或者取消设置这些标志位，如表 16-14 所示。

表 16-14　设置和取消设置格式化标志位的成员函数

```
fmtflags ios :: setf (flag)          // 设置对应的标志
fmtflags ios :: unsetf (flag)        // 取消对应标志的设置
```

格式化字段

类 ios 还定义了三个字段用于格式化数据。每个字段有两个或者三个域，可以单独设置或者取消设置。一次只能设置一个二进制位。表 16-15 显示了该类别中定义的三个字段。

表 16-15　按组设置的格式化字段

图例：
■ 默认域

basefield：**dec**　hex　oct
floatfield：fixed　scientific
adjustfield：**right**　left　internal

标志	值	说明
ios :: basefield	ios :: **dec** ios :: hex ios :: oct	设置整数值为十进制 设置整数值为十六进制 设置整数值为八进制
ios :: floatfield	ios :: fixed ios :: scientific	以固定格式显示浮点数值 以科学计数法格式显示浮点数值
ios :: adjustfield	ios :: **right** ios :: left ios :: internal	在字段中数据值右对齐 在字段中数据值左对齐 在符号后添加填充字符

每组中的字段域不是独立的，一次只能设置其中一个字段域。默认设置以深色显示，但请注意 floatfield 没有默认值（如果未显式选择，则由系统选择最佳值）。

这些字段域的设置与 fmtflags 使用的方法不同，因为它保证每个类别中一次只设置一个值（表 16-16）。

表 16-16　设置和取消设置字段的成员函数

```
fmtField ios :: setf (addingField, field)     // 设置对应字段
fmtField ios :: unsetf (field)                // 取消对应字段的设置
```

要设置字段，必须使用两个参数。第二个参数（字段）取消字段中所有域的设置；第一个参数随后设置所需域。

格式化变量

表 16-17 中所示的格式化变量是 width（int 类型）、precision（int 类型）和 fill（char 类型）。

这些格式化变量用于为一个值定义预留多少位置，在小数点后预留多少位置，以及应该使用什么字符来填充未使用的位置。有六个成员函数用于设置和取消设置这些标志，如

表 16-17 所示。请注意，如果未设置填充字段，则填充字符默认为空格。

表 16-17 设置和取消设置变量字段的成员函数

width [int] precision [int] fill [char]

成员函数	说明
int ios :: width (int n)	设置值使用的宽度
int ios :: width ()	重置 width 为 0
int ios :: precision (int n)	设置小数点后的精度
int ios :: precision ()	重置 precision 为 0
int ios :: fill (char c)	设置用于填充的字符类型
int ios :: fill ()	重置 fill 字段

程序 16-22 显示了我们如何直接格式化三个分别为布尔类型、整数类型和浮点类型的数据项。本程序有两个目标：展示可以使用格式化标志、字段和变量格式化数据；展示这个过程的冗长性和复杂性。稍后我们将展示使用操作符也可以实现相同的目标。

程序 16-22 打印三个数据项

```
1  /***************************************************************
2   *本程序演示如何使用格式化标志、字段和变量，                    *
3   *格式化三种数据类型                                            *
4   ***************************************************************/
5  #include <iostream>
6  using namespace std;
7
8  int main ()
9  {
10   //声明和初始化三个变量
11   bool b = true;
12   int i = 12000;
13   double d = 12467.372;
14   //打印值
15   cout << "Printing without using formatting" << endl;
16   cout << "Value of b: " << b << endl;
17   cout << "Value of i: " << i << endl;
18   cout << "Value of d: " << d << endl << endl;
19   //格式化布尔数据并重新打印
20   cout << "Formatting the Boolean data" << endl;
21   cout.setf (ios :: boolalpha);
22   cout << b << endl << endl;
23   //格式化整数数据并重新打印
24   cout << "Formatting the integer data type" << endl;
25   cout.setf (ios :: showbase);
26   cout.setf (ios :: uppercase);
27   cout.setf (ios :: hex, ios :: basefield);
28   cout.setf (ios :: right, ios :: adjustfield);
29   cout.width (16);
30   cout.fill ('*');
31   cout << i << endl << endl;
32   //格式化浮点数数据并重新打印
```

```
33    cout << "Formatting the floating-point data type" << endl;
34    cout.setf (ios :: showpoint);
35    cout.setf (ios :: right, ios :: adjustfield);
36    cout.setf (ios :: fixed, ios :: floatfield);
37    cout.width (16);
38    cout.precision (2);
39    cout.fill ('*');
40    cout << d << endl;
41    return 0;
42  }
```

运行结果：

```
Printing without using formatting
Value of b: 1
Value of i: 12000
Value of d: 12467.4

Formatting the Boolean data
true

Formatting the integer data type
**********0X2EE0

Formatting the floating-point data type
********12467.37
```

16.5.2　预定义操作符

在本节中，我们将讨论在本书中多次使用的标准操作符。我们的目标是更深入地探讨操作符，并研究操作符是如何被编写来实现其目标的。这样将有助于我们编写自己的操作符。

与格式化标志有关的操作符

在表 16-13 中，我们列出了 7 个格式化标志。我们在表 16-18 中再次列出了这 7 个格式化标志。在这 7 个标志中，我们正好有 14 个格式化标志操作符（设置和取消设置分别对应一个操作符）。黑体的是默认值。

表 16-18　使用格式化标志的操作符

标志	操作符	效果	输入	输出
boolalpha	**noboolalpha**	将布尔值显示为 0/1	√	√
	boolalpha	将布尔值显示为 false/true	√	√
skipws	noskipws	输入时不跳过空白字符	√	
	skipws	输入时跳过空白字符	√	
showbase	**noshowbase**	不显示进制的基		√
	showbase	显示进制的基		√
showpoint	**noshowpoint**	不显示小数点		√
	showpoint	显示小数点		√
showpos	**noshowpos**	不显示正号（+）		√
	showpos	显示正号（+）		√
unitbuf	**nounitbuf**	不刷新输出		√
	unitbuf	写入后刷新输出		√

（续）

标志	操作符	效果	输入	输出
uppercase	**nouppercase**	不显示大写字符		√
	uppercase	显示大写字符		√

与格式化字段有关的操作符

在表 16-15 中，我们列出了 3 个格式化字段。我们正好有 8 个与这些字段有关的操作符。黑体的是默认值（如表 16-19 所示）。

表 16-19 使用格式化字段的操作符

标志	操作符	效果	输入	输出
basefield	**dec**	把整数显示为十进制	√	√
	hex	把整数显示为十六进制	√	√
	oct	把整数显示为八进制	√	√
floatfield	fixed	以固定格式显示数据类型		√
	scientific	以科学计数法格式显示数据类型		√
adjustfield	**right**	在字段中数据右对齐		√
	left	在字段中数据左对齐		√
	internal	在字段中数据两端对齐		√

与格式化变量有关的操作符

在表 16-17 中，我们讨论了设置变量字段的成员函数。有 3 个操作符看起来像一个带参数的函数（表 16-20）。

表 16-20 使用变量字段的操作符

标志	操作符	效果	输入	输出
width	setw (n)	设置字段宽度为 n 个字符	√	√
precision	setprecision (n)	设置精度为 n 个字符		√
fill	setfill (c)	使用字符 c 作为填充字符	√	√

与标志和字段无关的操作符

有 4 个操作符与标志或者字段无关。它们在流上执行特定的操作（表 16-21）。

表 16-21 不使用标志或字段的操作符

操作符	效果	输入	输出
ws	提取空白字符	√	
endl	插入新行并刷新缓冲区		√
ends	插入字符串结束字符		√
flush	刷新流缓冲区		√

我们重复程序 16-22，但使用操作符代替格式化标志、字段和变量，以实现与程序 16-22 相同的目标（程序 16-23）。注意，我们需要为与格式化变量相关的操作符包含头文件 <iomanip>。

程序 16-23 使用操作符实现格式化

```
1  /***************************************************************
2   * 本程序演示如何使用预定义操作符，                                 *
3   * 以实现与使用格式化标志、字段和变量相同的目标                      *
```

```
4      ****************************************************************/
5   #include <iostream>
6   #include <iomanip>
7   using namespace std;
8
9   int main ()
10  {
11    //声明和初始化三个变量
12    bool b = true;
13    int i = 12000;
14    double d = 12467.372;
15    //打印值
16    cout << "Printing without using formatting" << endl;
17    cout << "Value of b: " << b << endl;
18    cout << "Value of i: " << i << endl;
19    cout << "Value of d: " << d << endl << endl;
20    //格式化布尔数据并重新打印
21    cout << "Formatting the Boolean data" << endl;
22    cout.setf (ios :: boolalpha);
23    cout << boolalpha << b << endl << endl;
24    //格式化整数数据并重新打印
25    cout << "Formatting the integer data type" << endl;
26    cout << showbase << uppercase << hex << right
27     << setw (16) << setfill ('*') << i << endl << endl;
28    //格式化浮点数数据并重新打印
29    cout << "Formatting the floating-point data type" << endl;
30    cout << showpoint << right << fixed << setw (16)
31     << setprecision (2) << setfill ('*') << d << endl << endl;
32    return 0;
33  }
```

运行结果：

```
Printing without using formatting
Value of b: 1
Value of i: 12000
Value of d: 12467.4

Formatting the Boolean data
true

Formatting the integer data type
***********0X2EE0
Formatting the floating-point data type
*********12467.37
```

16.5.3 操作符定义

如果要创建新的操作符，必须首先了解如何定义预定义的操作符。

不带参数的操作符

理解如何实现不带参数的操作符（表16-18、表16-19和表16-21）并不困难。系统有两个重载运算符，带一个接收指向函数的指针的参数，如下所示：

```
istream& istream :: operator >> (istream& (*pf) (istream&));
ostream& ostream :: operator << (ostream& (*pf) (ostream&));
```

第一个重载运算符使用一个指向函数的指针，该函数引用 istream 对象并返回对 istream 对象的引用（用于级联）。第二个重载运算符使用一个指向函数的指针，该函数引用一个 ostream 对象并返回一个对 ostream 对象的引用（用于级联）。

每个不带参数的操作符都可以作为函数实现，如下所示。函数名作为指向函数的指针传递给前面重载的运算符。

```
istream& name (itream& is)          ostream& name (ostream& os)
{                                   {
  action;                             action;
  return is;                          return os;
}                                   }
```

可以使用接收流参数并返回流的函数创建不带参数的操作符。

通过模拟实现操作符 boolalpha 和 noboolalpha，可以帮助我们理解如何编写新的操作符。我们把要创建的操作符称为 alpha 和 noalpha。程序 16-24 定义并使用了这两个模拟的操作符。

程序 16-24　实现两个定制的操作符的程序

```
1   /*****************************************************************
2    *本程序用于模拟实现操作符 boolalpha 和 noboolalpha              *
3    *****************************************************************/
4   #include <iostream>
5   using namespace std;
6
7   // 定义一个名为 alpha 的函数
8   ostream& alpha (ostream& os)
9   {
10    os.setf (ios :: boolalpha) ;
11    return os;
12  }
13  // 定义一个名为 noalpha 的函数
14  ostream& noalpha (ostream& os)
15  {
16    os.unsetf (ios :: boolalpha) ;
17    return os;
18  }
19
20  int main ()
21  {
22    //声明和初始化两个布尔变量
23    bool b1 = false;
24    bool b2 = true;
25    //使用操作符 alpha 和 noalpha 打印变量值
26    cout << alpha << b1 << " " << b2 << endl;
27    cout << noalpha << b1 << " " << b2 << endl;
28    return 0;
```

```
29 }
```

运行结果：

```
false true
0 1
```

了解如何模拟系统操作符有助于我们创建自己的操作符。假设我们想要创建一个名为 currency 的操作符（不带参数），它用于带 $ 打印值，$ 后跟前导星号以填充字段的开头，小数点后跟两位小数。程序 16-25 显示了这样一个定制的操作符。

程序 16-25　创建一个定制的操作符并测试

```
1  /***************************************************************
2   * 本程序使用一个定制的操作符                                      *
3   ***************************************************************/
4  #include <iostream>
5  #include <iomanip>
6  using namespace std;
7
8  ostream& currency (ostream& stream)
9  {
10   cout << '$';
11   stream.precision (2);
12   stream.fill ('*');
13   stream.setf(ios:: fixed, ios:: floatfield);
14   return stream;
15 }
16
17 int main ( )
18 {
19   cout << currency << setw (12) << 12325.45 << endl;
20   cout << currency << setw (12) << 0.36 << endl;
21   return 0;
22 }
```

运行结果：

```
$****12325.45
$********0.36
```

带参数的操作符

模拟带参数的操作符稍微有些复杂。指向函数的指针不能有参数。换言之，我们不能有一个指向带有参数的函数的指针，例如 setw(4)，指针只能是函数名 setw。这意味着我们必须改变我们的战略。我们可以将 setw(4) 视为对仅具有一个整数类型数据成员的类的构造函数的调用，而不是指向函数的指针。换句话说，以下语句：

```
cout << setw (4);
```

可以解释为带两个操作数的运算符。第一个操作数是类型 ostream 的一个实例；第二个操作数是类型 setw 的一个实例。接下来我们按以下步骤创建一个带一个参数的操作符。

1）我们需要创建一个类，其名称与操作符的名称相同，并且其中带一个数据成员，其类型与操作符的参数的类型相同。

2）对于第一个参数为 stream 类型（ostream 或者 istream）且第二个参数与步骤 1 中的

类相同的类，我们需要重载该类的插入或者提取运算符。

3）在重载运算符的函数体中，我们对操作进行编码，以达到操作符的目的。

通过定义包含一个数据成员的类，以及重载类的插入或者提取运算符，可以创建带一个参数的运算符。

接下来模拟实现操作符 setw(n)，但我们称之为 length(n)，以作为一个新的操作符。程序 16-26 显示了类 length（我们采用小写字母作为类的名称，和操作符保持一致）的接口文件。

程序 16-26 接口文件（length.h）

```
1  /****************************************************************
2   * 名为 length 的类的接口文件                                    *
3   ****************************************************************/
4  #ifndef LENGTH_H
5  #define LENGTH_H
6  #include <iostream>
7  using namespace std;
8
9  class length
10 {
11   private:
12     int n;
13   public:
14     length (int n);
15     friend ostream& operator << (ostream& stream, const length& len );
16 };
17 #endif
```

我们还实现了类 length，如程序 16-27 中所示。注意，我们希望运算符 << 在类 length 中用数据成员 n 的值设置宽度格式化变量。

程序 16-27 实现文件（length.cpp）

```
1  /****************************************************************
2   * 类 length 的实现文件                                          *
3   ****************************************************************/
4  #include "length.h"
5
6  // length 成员函数（参数构造函数）的定义
7  length :: length (int n1)
8  : n (n1)
9  {
10 }
11 // 重载运算符 <<
12 ostream& operator << (ostream& stream, const length& len )
13 {
14   stream.width (len.n);
15   return stream;
16 }
```

我们编写了程序 16-28 来应用这个新的操作符并对其进行测试。

程序 16-28 应用程序文件（application.cpp）

```
1   /*****************************************************************
2    * 本应用程序使用操作符 length(n)                                   *
3    *****************************************************************/
4   #include "length.h"
5
6   int main ()
7   {
8     cout << length (10) << 123 << endl;
9     cout << length (20) << 234 << endl;
10    return 0;
11  }
运行结果：
       123
                 234
```

16.6 程序设计

在本节中，我们设计了两个程序来演示输入/输出流的应用。

16.6.1 合并两个已排序文件

我们要对两个整数文件创建一个合并排序。假设有两个整数文件，其中已对整数进行排序（从最小到最大排列）。我们想合并这两个文件来创建一个新的文件，其中整数仍然是有序的，并且保留重复的数值。图 16-15 显示了 infile1、infile2 和 outfile 的内容。

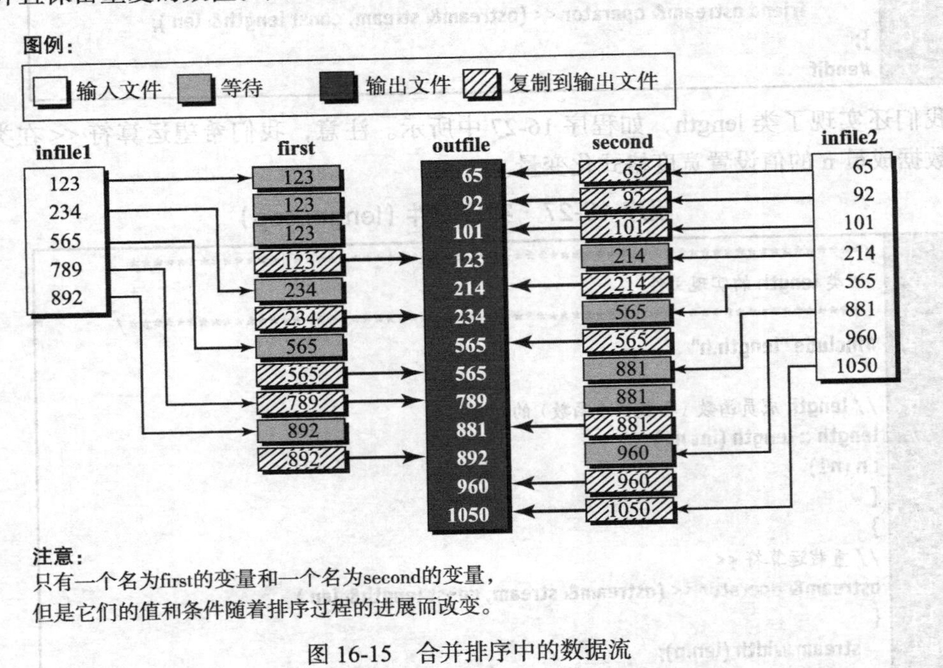

图 16-15 合并排序中的数据流

合并这些文件的过程需要 13 次迭代。在每次迭代中，变量 first 或者变量 second 的内容（但不是两者同时）被移动到输出文件。在移到输出文件之前，一些整数在相应的变量中会保留几次迭代。注意，比较和将整数移动到输出文件的决策总是基于变量 first 和 second 的内容。

程序 16-29 显示了代码清单。在第 14 行中，我们声明了两个变量：first 和 second，用于在处理过程中保存第一个和第二个输入文件中的整数。在第 15 行中，我们创建了一个 sentinel 变量（一个非常大的整数，参见前面章节中的介绍）。

程序 16-29　应用于两个文件的合并排序过程

```
1   /*****************************************************************
2    * 本程序用于合并两个已排序的整数文件，                            *
3    * 并创建一个新的有序的整数文件                                    *
4    *****************************************************************/
5   #include <iostream>
6   #include <fstream>
7   #include <assert.h>
8   #include <limits>
9   using namespace std;
10
11  int main ()
12  {
13    // 声明和初始化
14    int first, second;
15    int sentinel = numeric_limits <int> :: max();
16    // 实例化流，并打开文件
17    ifstream strm1 ("infile1");
18    ifstream strm2 ("infile2");
19    ofstream strm3 ("outfile");
20    if (!strm1.is_open())
21    {
22         cout << "Error opening infile1!" << endl;
23         assert (false);
24    }
25    if (!strm2.is_open())
26    {
27         cout << "Error opening infile2!" << endl;
28         assert (false);
29    }
30    if (!strm3.is_open())
31    {
32         cout << "Error opening outfile!" << endl;
33         assert (false);
34    }
35    // 处理
36    strm1 >> first;
37    strm2 >> second;
38    while (strm1 || strm2)
39    {
40        if (first <= second)
41        {
42              strm3 << first << " ";
43              strm1 >> first;
44              if (!strm1)
```

```
45             {
46                 first = sentinel;
47             }
48         }
49         else
50         {
51             strm3 << second << " ";
52             strm2 >> second;
53             if (!strm2)
54             {
55                 second = sentinel;
56             }
57         }
58     }
59     //关闭文件
60     strm1.close();
61     strm2.close();
62     strm3.close();
63     return 0;
64 }
```

16.6.2 对称密码

使用文件的信息传输必须以安全的方式处理。为此，常用的技术之一是密码学。**密码学**（cryptography）是一个起源于希腊语的词，意思是“秘密写作”，它涉及两种不同的机制：对称密钥密码学和非对称密钥密码学。密码学使用密钥将称为明文（plaintext）的原始消息转换为称为密文（ciphertext）的秘密消息。

密码学分为两大类：对称密码学和非对称密码学。我们在本节中讨论并使用第一种对称密码学的示例；在下一节中将使用第二种非对称密码学的示例。

在对称密钥加密中，我们使用相同的密钥进行加密和解密。密钥可以用于双向通信，这就是它被称为对称的原因。图16-16显示了对称密钥密码背后的一般概念。

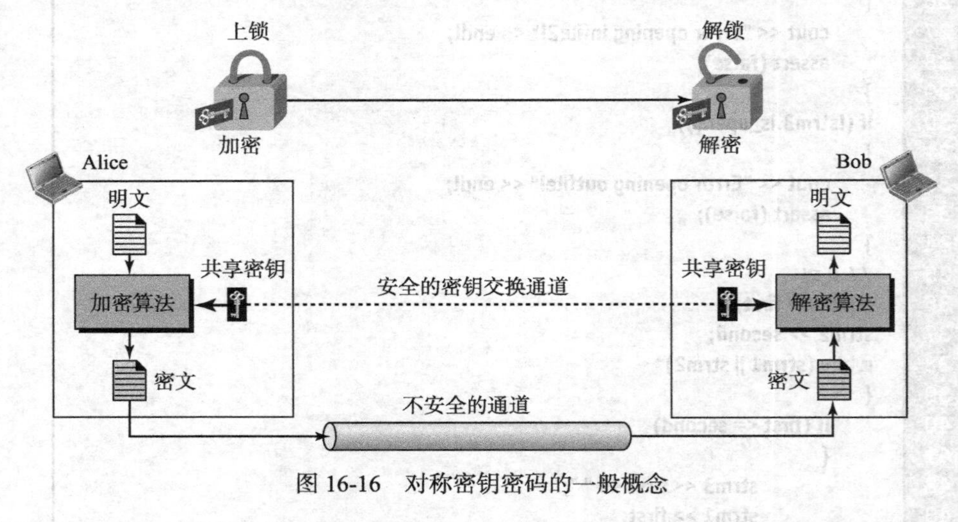

图16-16　对称密钥密码的一般概念

在图16-16中，一个实体Alice通过一个不安全的通道向另一个实体Bob发送一条消息，假设对手无法通过简单地窃听通道来理解消息的内容。

Alice 给 Bob 的原始消息是明文，通过通道发送的消息是密文。为了从明文创建密文，Alice 使用加密算法（encryption algorithm）和共享密钥（shared secret key）。

对称密钥密码也称为密钥密码。

为了从密文创建明文，Bob 使用解密算法（decryption algorithm）和相同的密钥。我们将加密和解密算法称为加密解密器（cipher）。密钥（key）是作为算法的加密解密器使用的一个值或者一组值。

对称密钥密码学可以分为传统密码学和现代密码学。传统密码学是简单的、面向字符的密码学，根据当今的标准，这种密码学是不安全的。现代密码学是复杂的、面向二进制位的、更安全的密码学。我们的程序使用其中一种传统密码学方法。现代密码学非常复杂，如 DES（数据加密标准），需要高水平的密码学和数论知识（参见 *Cryptography and Network Security 2008*，Behrouz A.Forouzan，McGraw Hill 出版）。

传统的对称加密解密器之一被称为单字母加密解密器（monoalphabetic cipher），其中明文中的一个字符（或者符号）在密文中总是被更改为相同的字符（或者符号），而不管它在文本中的位置如何。例如，如果算法将明文中的字母 A 改为字母 D，则每个字母 A 都改为字母 D。换言之，明文中的字母与密文之间的关系是一对一的。一种常见的单字母加密解密器是使用二维表进行加密和解密。图 16-17 显示了这样一个映射的例子，程序 16-30 演示了它的应用。

明文	a	b	c	d	e	f	g	h	i	j	k	l	m	n	o	p	q	r	s	t	u	v	w	x	y	z
密文	N	O	A	T	R	B	E	C	F	U	X	D	Q	G	Y	L	K	H	V	I	J	M	P	Z	S	W

图 16-17 一个单字母加密解密器的密钥示例

在这种情况下，明文和密文中的每个字符的密钥是不同的。如果明文中的字符是“a”，则密文中对应的字符将是“N”，依此类推。注意，我们用小写字母显示明本，用大写字母显示密文。

我们创建了一个名为 MonoAlpha 的类，它包含密钥（一个静态二维数组）。该类包含两个公共成员函数，即 encrypt（加密）和 decrypt（解密）。还有两个私有成员函数，用于将明文中的一个字符转换为密文，以及将密文中的一个字符转换为明文。它们由公共成员函数 encrypt（加密）和 decrypt（解密）调用。

程序 16-30 文件 monoalpha.h，MonoAlpha 类的接口文件

```
1  /***************************************************************
2   * MonoAlpha 类的接口文件                                      *
3   ***************************************************************/
4  #ifndef MONOALPHA_H
5  #define MONOALPHA_H
6  #include <iostream>
7  using namespace std;
8
9  class MonoAlpha
10 {
11   private:
```

```
12        static const char key [ ][2];
13        char searchEncrypt (char c);
14        char searchDecrypt (char c);
15    public:
16        MonoAlpha ();
17        ~MonoAlpha ();
18        void encrypt (const char* plainFile, const char* cipherFile);
19        void decrypt (const char* cipherFile, const char* plainFile);
20    };
21    #endif
```

实现文件。实现文件基于接口文件。encrypt 函数从明文文件中逐个提取字符。然后，该函数调用 searchEncrypt 函数，根据密钥表将每个字符转换为新字符。每个字符转换后，将插入密文文件中。decrypt 函数执行相反过程。程序 16-31 显示了实现的代码清单。

程序 16-31　文件 monoalpha.cpp，MonoAlpha 类的实现文件

```
1   /***************************************************************
2    * MonoAlpha 类的实现文件                                          *
3    ***************************************************************/
4   #include "monoalpha.h"
5   #include <fstream>
6   
7   // 构造函数
8   MonoAlpha :: MonoAlpha ()
9   {
10  }
11  // 析构函数
12  MonoAlpha :: ~MonoAlpha ()
13  {
14  }
15  // 公共成员函数
16  void MonoAlpha :: encrypt (const char* plainFile, const char* cipherFile)
17  {
18     ifstream istrm (plainFile, ios :: in);
19     ofstream ostrm (cipherFile, ios :: out);
20     char c1, c2;
21     while (istrm.get (c1))
22     {
23          c2 = searchEncrypt (c1);
24          ostrm.put(c2);
25     }
26     istrm.close ();
27     ostrm.close ();
28  }
29  // 公共成员函数
30  void MonoAlpha :: decrypt (const char* cipherFile, const char* plainFile)
31  {
32     ifstream istrm (cipherFile, ios :: in);
33     ofstream ostrm (plainFile, ios :: out);
34     char c1, c2;
35     while (istrm.get (c1))
```

```
36      {
37          c2 = searchDecrypt (c1);
38          ostrm.put(c2);
39      }
40      istrm.close ();
41      ostrm.close ();
42  }
43  // 私有成员函数
44  char MonoAlpha :: searchEncrypt (char c)
45  {
46      int i = 0;
47      while (true)
48      {
49          if (key[i][0] == c)
50          {
51              return key[i][1] ;
52          }
53          i++;
54      }
55  }
56  // 私有成员函数
57  char MonoAlpha :: searchDecrypt (char c)
58  {
59      int i = 0;
60      while (true)
61      {
62          if (key[i][1] == c)
63          {
64              return key[i][0];
65          }
66          i++;
67      }
68  }
69  // 静态密钥数组的定义
70  const char MonoAlpha :: key [ ][2] = {{'a', 'N'}, {'b', 'N'},
71          {'c', 'A'}, {'d', 'T'}, {'e', 'R'}, {'f', 'B'}, {'g', 'E'}, {'h', 'C'},
72          {'i', 'F'}, {'j', 'U'}, {'k', 'X'}, {'l', 'D'}, {'m', 'Q'}, {'n', 'G'},
73          {'o', 'Y'}, {'p', 'L'}, {'q', 'K'}, {'r', 'H'}, {'s', 'V'}, {'t', 'I'},
74          {'u', 'J'}, {'v', 'M'}, {'w', 'P'}, {'x', 'Z'}, {'y', 'S'}, {'z', 'W'} };
```

加密端的应用程序文件。我们需要两个应用程序文件，一个在加密端，另一个在解密端。然而，Alice 和 Bob 需要共享接口和实现文件，但对其他人隐藏。程序 16-32 显示了加密端的应用程序。Alice 需要把明文转换为密文，然后发送给 Bob。请注意，我们在明文中只使用了小写字母，在密文中只使用了大写字母，以使转换更容易，代码更安全。另外，应该避免使用标点符号，因为它为黑客提供了更多的线索。

程序 16-32 文件 app1.cpp，在发送端使用

```
1  /****************************************************************
2   * 用于加密消息的应用程序文件                                    *
3   ****************************************************************/
4  #include "monoalpha.h"
```

```
5
6  int main ()
7  {
8     MonoAlpha monoalpha;
9     monoalpha.encrypt ("plainFile", "cipherFile");
10    return 0;
11 }
```

明文文件的内容：
thisisthefiletoencrypt
密文文件的内容：
ICFVFVICRBFDRIYRGAHSLI

解密端的应用程序文件。程序16-33显示了解密端的应用程序。Bob需要转换从Alice收到的密文以创建明文。

单字母加密解密器程序仅用于教育目的，在实际系统中使用并不安全。

程序16-33　文件app2.cpp，在接收端使用

```
1  /*******************************************************************
2   * 用于解密消息的应用程序文件                                        *
3   *******************************************************************/
4  #include "monoalpha.h"
5
6  int main ( )
7  {
8     MonoAlpha monoalpha;
9     monoalpha.decrypt ("cipherFile", "plainFile");
10    return 0;
11 }
```

明文文件的内容：
ICFVFVICRBFDRIYRGAHSLI
密文文件的内容：
thisisthefiletoencrypt

本章小结

数据源或者数据接收器不能直接连接到程序。我们需要一个位于数据源/数据接收器和程序之间的中介，以控制数据流，并在读取或者写入时解释数据元素。这是使用流来完成的。使用流的主要目的之一是将存储在数据接收器或者数据源中的数据格式转换为内存中的格式。C++定义了一个流层次结构，其中ios类是所有输入/输出类的基类。

控制台流由三个类组成：istream、ostream和iostream。我们可以使用前两个，但iostream类不能实例化。

我们还可以使用文件流读取和写入文件。文件流由三个类组成：ifstream、ofstream和fstream。要使用它们，我们必须构造流，将文件连接到流（打开文件），进行输入或者输出，断开文件与流的连接（关闭文件），并销毁流（自动完成）。

C++库定义了三个类来输入或者输出字符串。字符串流类最常见的应用是充当字符串类的适配器。

输入/输出中的一个问题是格式化数据。ios类定义了用于格式化数据的标志、字段和变量。它们可以直接使用，也可以在操作符的名义下使用，操作符是使用它们的成员函数或者类。除了预定义的操作符之外，我们还可以定义自己的操作符。

思考题

1. 编写一个函数 openInput，以输入模式打开一个文件。文件名作为参数传递。
2. 编写一个函数 openOutput，以输出模式打开一个文件。文件名作为参数传递。
3. 编写一个函数，作为参数给定文件名和文件中字符的位置，返回文件中指定位置的字符。
4. 编写一个函数，作为参数给定文件名、文件中字符的位置、替换的新字符，替换文件中的字符。
5. 编写一个函数，拷贝一个文件中的内容到一个新文件。
6. 编写一个函数，比较两个文件，如果内容相同，则返回 true，否则返回 false。
7. 编写一个函数，把一个文件中的内容附加到另一个文件的末尾。
8. 编写一个函数，给定一个文件，把奇数位置的字符拷贝到第二个文件，把偶数位置的字符拷贝到第三个文件。
9. 编写一个函数，将一个包含若干字符的文件作为参数，在文件末尾的一个空白字符之后附加文件中字符的个数。
10. 编写一个函数，读取一个字符数组中的项，并创建包含这些字符的文件。
11. 演示如何读取一行包含 id（整数）、name（字符串）和 gpa（双精度数）的学生数据记录，并把读取的数据存储到相应的变量中。
12. 演示如何组合 id（整数）、name（字符串）和 gpa（双精度数）三个变量的值，并把它们存储在单个字符串中。

编程题

1. 编写一个程序，读取文本文件并将文件中的每个字符都改为大写。让用户以 C++ 字符串的形式输入文件名（记住将文件名改为 C 字符串以打开文件）。使用包含以下内容的文件测试程序：

```
This is a file of characters to be changed.
```

2. 使用文本编辑器创建由一个空格字符分隔的整数文件，如下所示。然后使用程序查找文件中是否存在给定的整数。

```
14 17 24 32 11 72 43 88 99
```

3. 如果文件中的元素已经排序（按顺序排列），则可以更快地搜索目标。使用文本编辑器创建由一个空格字符分隔的整数文件，如下所示。然后使用程序查找文件中是否存在给定的整数。请注意，如果文件很大，搜索时间会有巨大差异。我们用一个小文件进行测试。

```
14 17 24 32 48 52 64 74 81 92
```

4. 我们可以从随机文件中删除数据元素，因为内存中连接的流实际上是一个数组。我们将某数据项后面的其余数据项向流的开始位置移动一个位置，从而擦除该数据项。但是，移动数据项会在末尾创建一个重复的数据项。在数组中，可以减小数组大小以避免使用该副本。对于流，我们必须将最后一个元素更改为空元素。我们知道可以随机访问一个字符文件，因为字符的大小是固定的。假设你有以下字符文件：

```
ABCDEFGHIJKLMNOPQRSTUVWXYZ
```

编写一个程序，删除文件中给定位置（0 到 25）的字符。记住将结尾的重复字符更改为空字符。

5. 可以使用索引随机访问数组元素。然而，数组不是永久的。当程序终止时，数组的内容将被销毁。我们可以使用随机访问将数组模拟为二进制文件。创建两个程序。在第一个程序中，在一个文件中存储10个不同的双精度值。在第二个程序中，随机地检索第一个程序存储在文件中的一些值。注意，我们使用两个应用程序来说明在第一个程序终止后文件存在，并且可以被第二个程序访问。
6. 存在一个问题："我们如何创建一个字符串（例如名称）的随机访问文件？"这不能通过简单地使用string类来完成，因为string对象的大小不是预先定义的。在随机访问文件中，所有项的大小必须相同。解决方案是创建一个新类，该类只有一个string类型的数据成员。我们知道，当C++创建用户自定义类型的对象时，它会填充对象以使它们具有相同的大小。编写一个程序将一些名称存储在一个二进制文件中，然后随机访问其中的一个。
7. 创建存储在文件中并且可以随机访问的信息记录的一种方法是从中创建一个字符串。我们可以使用toString函数将可变大小和不同数据类型的记录更改为字符串，并添加填充以使所有字符串的大小相同。编写一个程序，创建五个学生的记录，其中每个记录由三部分组成：id（0到100之间的整数）、name（不同大小的字符串）和gpa（小于或者等于4.0的双精度值）。使字符串的大小相同（18个字符），并在每个记录的末尾添加一个换行符，以便使用getline函数读取每行。编写一个程序，在一个文件中存储五条记录。编写另一个程序来访问这些记录。
8. 编写一个更新银行客户的二进制文件的程序。数据由账号（整数）和账户余额（双精度值）组成。编写两个应用程序文件。在第一个应用程序中，为五个客户创建具有账号（从1000开始）和账户余额的二进制文件。在第二个应用程序中，对某些客户的账户余额进行更改（存款或者取款）（更新文件），并打印更新后的文件内容。

第 17 章

C++ Programming: An Object-Oriented Approach

递　归

我们知道计算机程序涉及迭代（循环）。在这一章中，我们介绍一种新的方法，叫作递归（recursion）。递归可以使某些类别的程序更容易实现。特别地，我们将展示如何使用递归来解决一些排序和搜索问题。最后，我们将展示如何使用递归来解决一个经典问题：汉诺塔。

学习目标

阅读并学习本章后，读者应该能够

- 讨论递归的概念，并将其与循环进行比较。
- 区分不返回值的递归函数和返回值的递归函数。
- 演示如何将某些循环函数更改为递归函数。
- 区分尾部递归和非尾部递归，以及如何将第二种递归更改为第一种递归以提高效率。
- 讨论如何使用辅助函数来提高某些递归函数的效率。
- 讨论列表排序，并演示如何编写高效的递归排序函数，如快速排序。
- 讨论搜索并演示如何编写递归搜索函数，如二分查找。
- 将汉诺塔作为经典的递归算法进行讨论。

17.1　概述

在前面的章节中，我们学习了编写函数来解决问题。当函数要求重复某项工作时，我们使用迭代（循环）。在本章中，我们将展示还可以使用递归来解决重复性问题。

17.1.1　循环与递归

在简单的迭代中，我们使用计数器重复一个任务 n 次；在递归中，函数只执行一次任务，然后调用自己 $n-1$ 次以实现相同的目标。不返回值的递归函数与返回值的递归函数的方法有区别，因此我们将分别加以讨论。

不返回值的递归函数

假设我们需要在一行上打印 n 个星号，n 的值是已知的。我们可以使用迭代或者递归的解决方案，如下所示。我们为迭代方案选择了 while 循环，因为 while 循环使两个函数之间的比较更容易。在两个函数中，我们只有一个变量 n。

```
// 迭代方法
void line (int n)
{
  while (n >= 1)
  {
     cout << "*";
     n--;
  }
  return;
}
```

```
// 递归方法
void line (int n)
{
  if (n < 1)
  {
     return;
  }
  cout << "*";
  line (n - 1);
}
```

迭代版本中存在一个隐式条件。如果 n 小于 1，则永远不会进入 while 循环并且函数返回。在每次迭代中，我们将 n 的值减少 1，直到 n 的值小于 1，然后从函数返回。

在递归版本中，当 n 小于 1 时，我们使用显式条件从函数返回。但是，我们不重复循环 n−1 次，而是使用参数 n−1 再次调用该函数。

如果我们比较这两种方法，就会发现两种方法都在做同样的事情。需要隐式条件或者显式条件从函数返回。在迭代版本中，我们使用语句 n−−，减少变量 n 的值；在递归版本中，我们用一个减少的参数（n−1）再次调用该函数。

在迭代版本中，语句（cout<<"*"）被调用 n 次，在循环的每次迭代中调用一次。在递归版本中，在对函数的每次调用中执行相同的语句。

在迭代版本中，只有一次函数调用；在递归版本中，有 n−1 次函数调用。在除最后一次以外的每次调用中，我们打印一个星号；在最后一次调用中，我们不打印任何内容。前几次调用称为一般情况（general case），最后一次调用称为基本情况（base case）。**一般情况**与那些执行某项操作的调用有关，基本情况与终止递归的条件有关。

图 17-1 图形化地显示了迭代和递归调用。

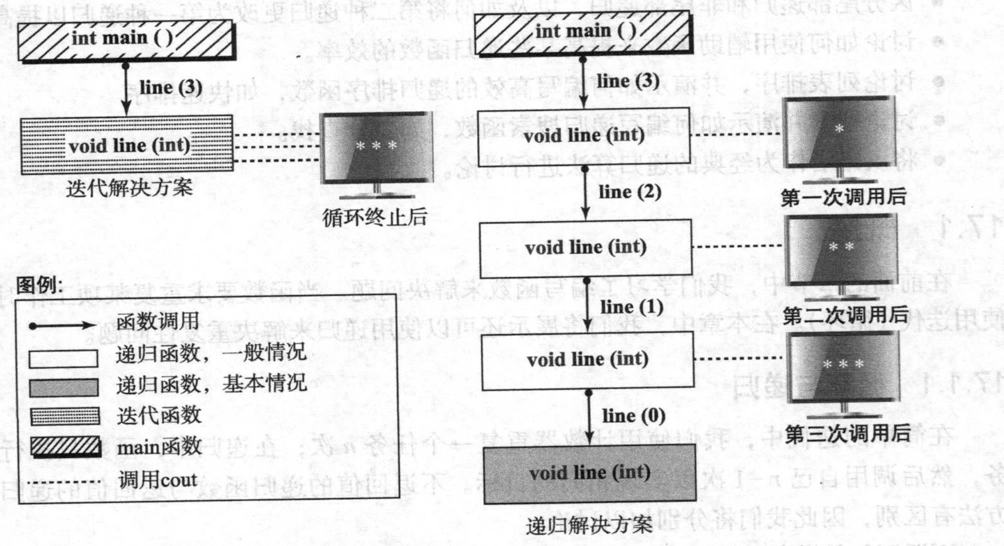

图 17-1　不返回值的函数的迭代调用和递归调用比较

返回值的递归调用

假设我们需要计算从 0 到 *n* 的所有数值的和（通常称为求和）。我们可以编写两个函数：一个迭代函数、一个递归函数。但是，函数不是 void 函数，每个函数都必须返回求和的值。

我们并排显示这两个函数，以比较迭代版本和递归版本之间的区别。同样，我们使用 while 循环，这会使得比较更容易。

```
// 迭代方法
int sum (int n)
{
   int result = 0;
   while (n >= 0)
   {
      result += n;
```

```
// 递归方法
int sum (int n)
{
   if (n <= 0)
   {
      return 0;
   }
```

```
    n--;                                    return sum (n - 1) + n;
  }                                       }
  return result;
}
```

在这两种情况下，我们都是求数值累加之和，即从 n 累加到 0。换言之，我们要求 sum = n+n−1+n−2+ ⋯ +1+0。在迭代的情况下，如果 n<0（终止情况），函数将跳过循环。在递归的情况下，函数显式地返回。这意味着条件（n<=0）是两个函数的基本情况或者终止情况。迭代函数在每次迭代中减少 n 的值，递归函数使用 n−1 调用相同函数。图 17-2 图形化地显示了每个函数的行为。

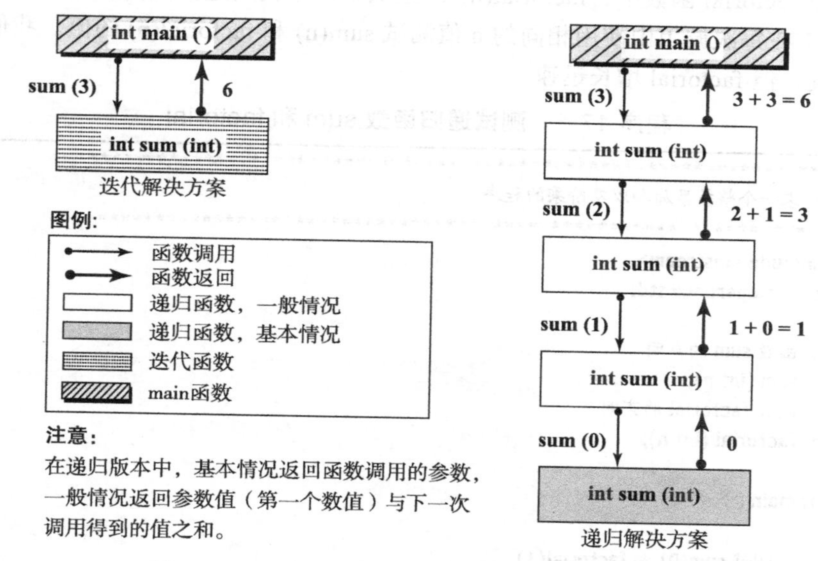

图 17-2 返回值的函数的迭代调用和递归调用比较

比较

当我们比较递归函数 line（图 17-1）和递归函数 sum（图 17-2）时，可以推导出不返回值的递归函数和返回值的递归函数之间的差异。在不返回值的递归函数的情况下，将连续调用一般情况，直至达到基本情况，一般情况不需要保存任何信息。在返回值的递归函数的情况下，调用一般情况，直至达到基本情况。在下一次调用返回之前，每个一般情况都必须保存一些信息（n 的值）。然后，基本情况将 sum(0) 的值返回给前一个一般情况，进而将 sum(1) 的值返回给前一个一般情况，依此类推。

17.1.2 递归算法

在本节中，我们将讨论简单的递归算法。

求和与阶乘

我们讨论了如何递归地求和的问题。求和函数 sum(n) 的孪生兄弟是阶乘 factorial(n) 函数（也称为乘积）。在求和函数中，我们将从 n 到 0 的数值相加；在阶乘函数中，我们将从 n 到 1 的数值相乘。

```
// 求和函数                                  // 阶乘函数
int sum (int n)                              int factorial (int n)
```

```
{                                   {
  if (n =< 0)                         if (n =< 1)
  {                                   {
     return 0;                           return 1;
  }                                   }
  return sum (n - 1) + n;             return factorial (n - 1) * n;
}                                   }
```

注意，sum(0) 是 0，factorial(1) 是 1。在 sum 函数中，我们使用加法运算符；在 factorial 函数中，我们使用乘法运算符。注意，在 sum 函数中，sum(n−1) 是从下一个调用返回的值；在 factorial 函数中，factorial(n−1) 是从下一个调用返回的值。

程序 17-1 在每次调用中使用相同的 n 值测试 sum(n) 和 factorial(n) 的值。我们可以看到 sum 增长缓慢，但 factorial 增长迅速。

程序 17-1 测试递归函数 sum 和 factorial

```
1   /*************************************************************
2    * 求一个整数累加和以及阶乘的程序                                *
3    *************************************************************/
4   #include <iostream>
5   using namespace std;
6
7   // 函数 sum 的声明
8   int sum (int n);
9   // 函数 factorial 的声明
10  int factorial (int n);
11
12  int main ( )
13  {
14    // 测试 sum(0) 和 factorial(1)
15    cout << "sum (0) = " << sum (0) << endl ;
16    cout << "factorial (1) = " << factorial (1) << endl << endl;
17    // 测试 sum(3) 和 factorial(3)
18    cout << "sum (3) = " << sum (3) << endl;
19    cout << "factorial (3) = " << factorial (3) << endl << endl;
20    // 测试 sum(7) 和 factorial(7)
21    cout << "sum (7) = " << sum (7) << endl;
22    cout << "factorial (7) = " << factorial (7);
23    return 0;
24  }
25  // 递归函数 sum(n) 的定义
26  int sum (int n)
27  {
28    if (n <= 0)
29    {
30       return 0;
31    }
32    return n + sum (n - 1);
33  }
34  // 递归函数 factorial(n) 的定义
35  int factorial (int n)
36  {
```

```
37    if (n <= 1)
38    {
39        return 1;
40    }
41    return n * factorial (n - 1);
42 }
```

```
运行结果:
sum (0) = 0
factorial (1) = 1

sum (3) = 6
factorial (3) = 6

sum (7) = 28
factorial (7) = 5040
```

最大公约数

数学和计算机科学领域中经常需要的一个函数是求两个正整数的最大公约数 gcd（greatest common divisor）。如果 x % y=0，则整数 y 是 x 的除数。两个正整数可以有多个公约数，但只有一个最大公约数。

例如，12 的除数是 1、2、3、4、6、12，140 的除数是 1、2、4、7、10、14、20、28、35、70 和 140，12 和 140 的公约数是 1、2、4。然而，最大的公约数是 4。如图 17-3 所示。

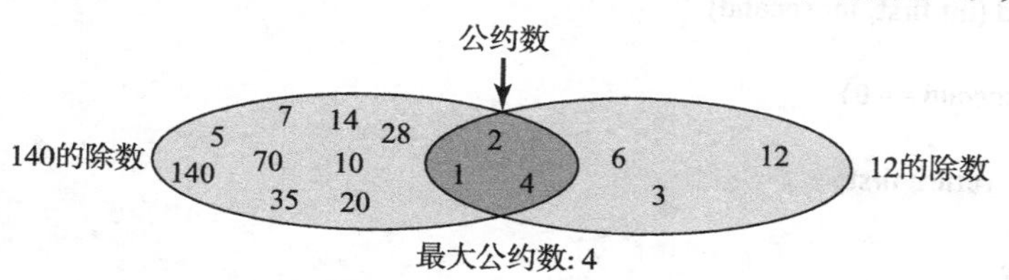

图 17-3　12 和 140 的最大公约数

两千多年前，一位名叫欧几里得的数学家开发了一种算法，可以递归地查找两个正整数之间最大的公约数。表 17-1 显示了欧几里得算法的基本情况和一般情况。

表 17-1　欧几里得算法的基本情况和一般情况

基本情况	一般情况
gcd (*x*, 0) = *x*	gcd (*x*, *y*) = gcd (*y*, *x* % *y*)

图 17-4 显示了递归函数 gcd 如何调用另一个版本，直到找到两个参数的最大公约数。

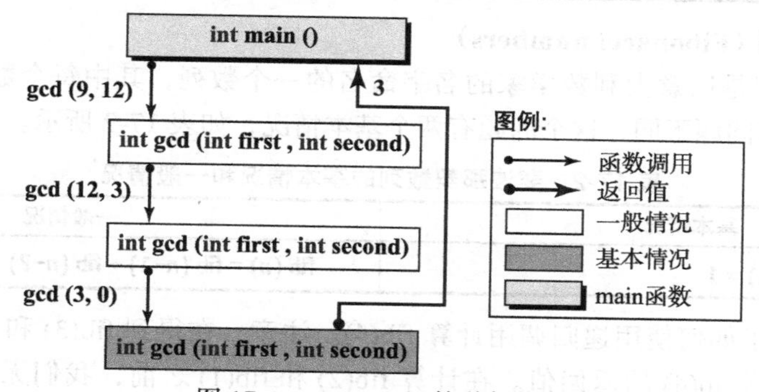

图 17-4　gcd (9, 12) 的递归调用

程序 17-2 演示了如何使用递归函数查找若干整数对的最大公约数。

程序 17-2　最大公约数

```
1  /**************************************************************
2   * 查找两个整数的最大公约数的程序                              *
3   **************************************************************/
4  #include <iostream>
5  using namespace std;
6  
7  // 函数 gcd 的声明
8  int gcd (int first, int second);
9  
10 int main ()
11 {
12   // 检查五个整数对的最大公约数
13   cout << "gcd (8, 6) = " << gcd (8, 6) << endl;
14   cout << "gcd (9, 12) = " << gcd (9, 12) << endl;
15   cout << "gcd (7, 11) = " << gcd (7, 11) << endl;
16   cout << "gcd (21, 35) = " << gcd (21, 35) << endl;
17   cout << "gcd (140, 12) = " << gcd (140, 12);
18   return 0;
19 }
20 // 最大公约数递归函数的定义
21 int gcd (int first, int second)
22 {
23   if (second == 0)
24   {
25       return first;
26   }
27   else
28   {
29       return gcd (second, first % second);
30   }
31 }
```

```
运行结果:
gcd (8, 6) = 2
gcd (9, 12) = 3
gcd (7, 11) = 1
gcd (21, 35) = 7
gcd (140, 12) = 4
```

斐波那契数列（Fibonacci numbers）

斐波那契数列是以意大利数学家的名字命名的一个数列，其中每个数都是前两个数的和。与前面的递归问题不同，这个问题有两个基本情况，如表 17-2 所示。

表 17-2　斐波那契数列的基本情况和一般情况

基本情况	一般情况
fib (0) = 0, fib (1) = 1	fib (*n*) = fib (*n*−1) + fib (*n*−2)

图 17-5 显示了如何使用递归调用计算 fib(4)。注意，在得到 fib(3) 和 fib(2) 的返回值之前，我们无法计算 fib(4) 的返回值。在计算 fib(2) 和 fib(1) 之前，我们无法计算 fib(3)。在

计算 fib(1) 和 fib(0) 之前，我们无法计算 fib(2)。最后，我们计算出 fib(0) 和 fib(1) 的结果，它们是基本情况。

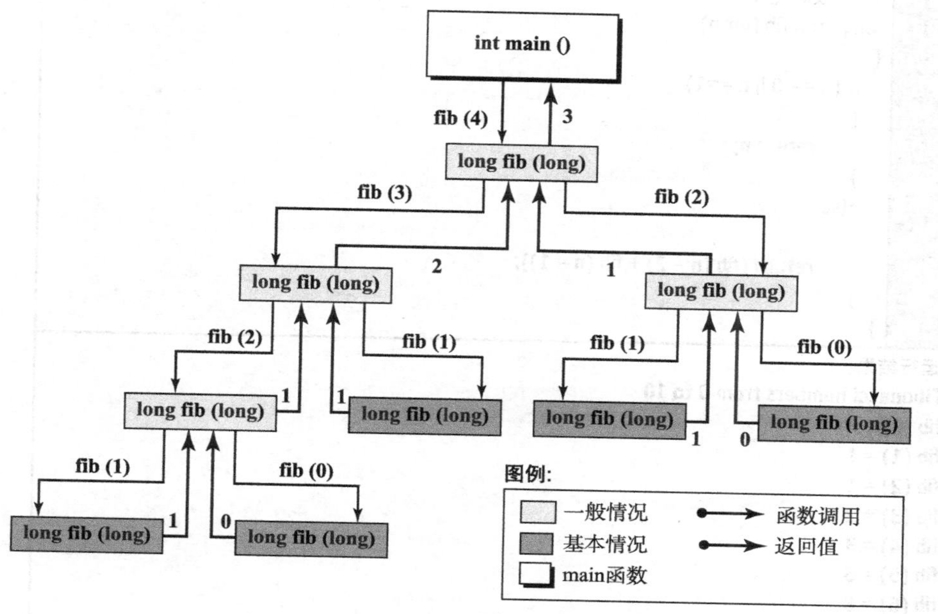

图 17-5　斐波那契数 fib(4) 的递归跟踪

根据表 17-2 中斐波那契数列的定义，我们编写了一个递归函数来计算其中的任何一个数，如程序 17-3 所示。

程序 17-3　斐波那契数列的递归解决方案

```
1  /**************************************************************
2   * 计算斐波那契数的程序                                        *
3   **************************************************************/
4  #include <iostream>
5  using namespace std;
6
7  // 函数的声明
8  long long fib (int n);
9
10 int main ( )
11 {
12   // 测试 0 到 10 的斐波那契数
13   cout << "Fibonacci numbers from 0 to 10" << endl;
14   for (int i = 0; i <= 10; i++)
15   {
16       cout << "fib (" << i << ") = " << fib(i) << endl;
17   }
18   cout << endl;
19   // 测试 35 和 36 的斐波那契数
20   cout << "Fibonacci numbers of 35 and 36" << endl;
21   cout << "fib (35) = " << fib(36) << endl;
22   cout << "fib (36) = " << fib(36) << endl;
```

```
23  return 0;
24  }
25  //函数的定义
26  long long fib (int n)
27  {
28    if (n == 0 || n ==1)
29    {
30          return n;
31    }
32    else
33    {
34          return (fib (n - 2) + fib (n - 1));
35    }
36  }
```

```
运行结果：
Fibonacci numbers from 0 to 10
fib (0) = 0
fib (1) = 1
fib (2) = 1
fib (3) = 2
fib (4) = 3
fib (5) = 5
fib (6) = 8
fib (7) = 13
fib (8) = 21
fib (9) = 34
fib (10) = 55

Fibonacci numbers of 35 and 36
fib (35) = 9227465
fib (36) = 14930352
```

结果表明，每个斐波那契数都是前两个数的和。

反转字符串

另一个可以递归解决的问题是反转字符串。我们可以定义基本情况和一般情况，如表 17-3 所示。换言之，如果字符串由零个或者一个字符组成，则反向字符串就是其本身。否则，我们需要找到除第一个字符以外的子字符串的反转字符串，然后将它与只包含第一个字符的子字符串连接起来。图 17-6 显示了递归反转字符串的方法。

表 17-3　反转字符串的基本情况和一般情况

基本情况	一般情况
if (length <= 1), return strg	return reverse (substr (1, length – 1)) + subtr (0, 1)

每个一般情况调用另一个一般情况，传递的参数为子字符串（第一个字符除外）。在一个子字符串中保存第一个字符，直到返回反转的字符串。然后将这两个字符串连接起来，并将结果返回给主调函数。换言之，给定字符串“ABCD”，第一次调用将“A”作为一个子字符串，并将“BCD”传递给下一次调用。当返回字符串“DCB”时，将其与字符串“A”连接起来以创建“DCBA”。

程序 17-4 显示了这个问题的递归解决方案。

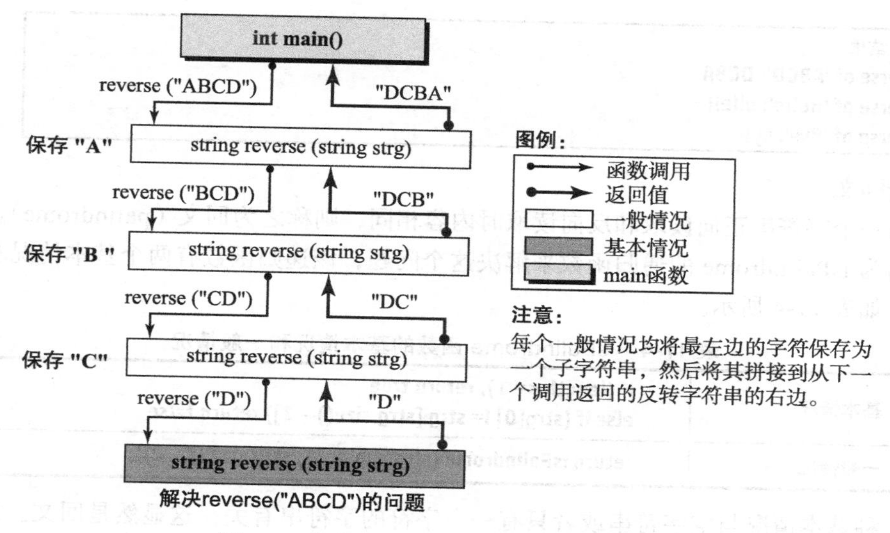

图 17-6 调用 reverse("ABCD") 涉及的步骤

程序 17-4 使用递归方法反转字符串

```
1  /***************************************************************
2   * 用递归方法反转字符串的程序                                    *
3   ***************************************************************/
4  #include <iostream>
5  #include <string>
6  using namespace std;
7
8  // 递归函数的声明
9  string reverse (string str);
10
11 int main ( )
12 {
13    // 使用一些字符串调用递归函数
14    cout << "Reverse of 'ABCD': " << reverse ("ABCD") << endl;
15    cout << "Reverse of 'Hello': " << reverse ("Hello") << endl;
16    cout << "Reverse of 'Bye': " << reverse ("Bye") << endl;
17    return 0;
18 }
19 // 递归函数的定义
20 string reverse (string str)
21 {
22   if (str.length () <= 1)
23   {
24       return str;
25   }
26   else
27   {
28       return reverse (str.substr (1, str.length() - 1)) + str.substr (0, 1);
29   }
30 }
```

```
运行结果:
Reverse of 'ABCD': DCBA
Reverse of 'Hello': olleH
Reverse of 'Bye': eyB
```

检查回文

如果一个字符串正向读取和反向读取时内容相同，则称之为回文（palindrome）。我们可以使用名为 isPalindrome 的递归函数来解决这个问题，该递归函数有两个基本情况和一个一般情况，如表 17-4 所示。

表 17-4　isPalindrome 函数的基本情况和一般情况

基本情况	if (length <= 1), return true else if (strg[0] != strg [strg.size() - 1]) return false
一般情况	return isPalindrome (strg.substr (1, strg.size () - 2));

第一种基本情况与空字符串或者只有一个字符的字符串有关，这显然是回文。第二种基本情况与第一个字符和最后一个字符不同的字符串有关，这显然不是回文。

图 17-7 显示了如何使用基本情况和一般情况来判断字符串是否为回文。

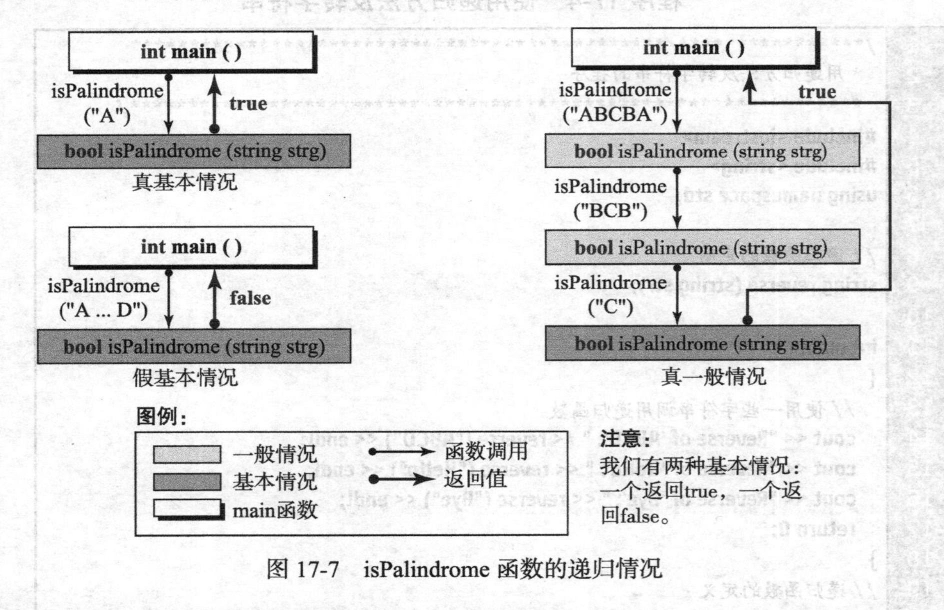

图 17-7　isPalindrome 函数的递归情况

程序 17-5 显示了递归函数 isPalindrome 的实现代码。

程序 17-5　使用递归函数 isPalindrome

```
1  /***************************************************************
2   * 检查一个字符串是否为回文的程序                                *
3   ***************************************************************/
4  #include <iostream>
5  #include <string>
6  using namespace std;
7
8  // 递归函数的声明
9  bool isPalindrome (string strg);
```

```
10
11  int main ( )
12  {
13    // 实例化一些字符串
14    string strg1 ("");
15    string strg2 ("rotor");
16    string strg4 ("hello");
17    // 检查是否为回文
18    cout << boolalpha;
19    cout << "Is '' a palindrome? " << isPalindrome (strg1) << endl;
20    cout << "Is 'rotor' a palindrome? " << isPalindrome (strg2) << endl;
21    cout << "Is 'hello' a palindrome? " << isPalindrome (strg3);
22    return 0;
23  }
24  // 递归函数的定义
25  bool isPalindrome(string strg)
26  {
27    if (strg.size () <= 1)
28    {
29        return true;
30    }
31    else if (strg[0] != strg [strg.size() - 1])
32    {
33        return false;
34    }
35    return isPalindrome (strg.substr (1, strg.size () - 2));
36  }
运行结果:
Is '' a palindrome? true    // 空字符串
Is 'rotor' a palindrome? true
Is 'hello' a palindrome? false
```

17.1.3　尾部递归函数和非尾部递归函数

在前面几节讨论的例子中，我们遇到了两种递归类型：尾部递归和非尾部递归（图 17-8）。

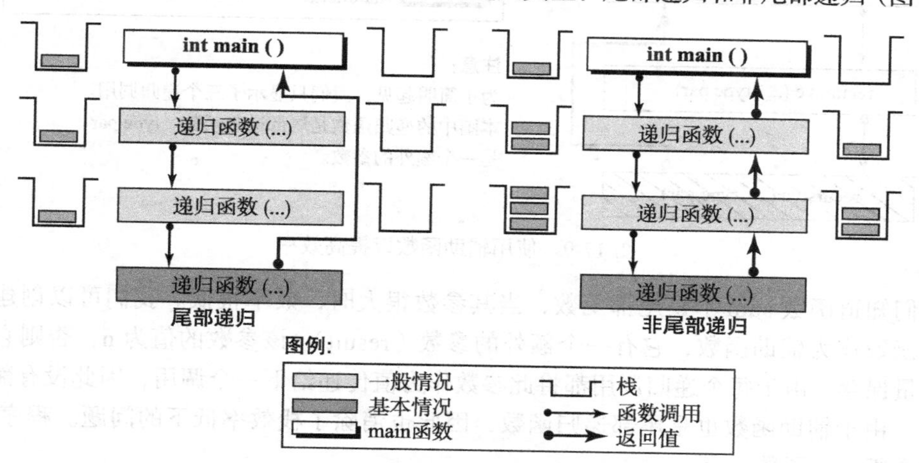

图 17-8　尾部递归函数和非尾部递归函数的区别

尾部递归（tail recursion）

在**尾部递归**中，每个一般情况在调用下一个一般情况或者基本情况后终止。换句话说，一般情况的职责仅仅是调用下一个一般函数或者基本函数。基本情况的职责是将整个操作的结果返回给 main 函数。在尾部递归中，保存下一个函数调用记录的栈只有一条记录，因为当调用下一个函数时，当前函数终止，其记录从栈中弹出。迄今为止，我们已经涉及的尾部递归函数的例子是 gcd 和 isPalindrome。注意，如果我们设计了一个 void 递归函数，则它是尾部递归函数，在这个函数中，即使是基本情况也不会向 main 函数返回任何内容。

非尾部递归（nontail recursion）

在**非尾部递归**中，在下一个调用返回之前，一般情况调用的职责不会终止。然后，当前函数将它所保存的信息与下一个调用返回的信息结合起来，并将其传递给上一个调用。最后的信息由第一个一般情况返回给 main 函数。在非尾部递归中，栈保存下一个递归调用的记录，直到调用返回为止。我们讨论的非尾部递归函数的例子包括 sum、factorial、fibonacci 和 reverse。

17.1.4　辅助函数

递归函数可能效率低下。特别地，非尾部递归函数之所以效率低下有两个原因。首先，每个一般情况必须保存信息，直到下一个调用的结果返回。其次，栈最终会保存许多记录，这可能需要大量的内存。为了提高效率，我们可以使用辅助函数。**辅助函数**（helper function）是一个尾部递归函数，其参数比非尾部递归函数多。我们可以使用非递归函数来调用辅助函数。图 17-9 显示了这种设计思想。

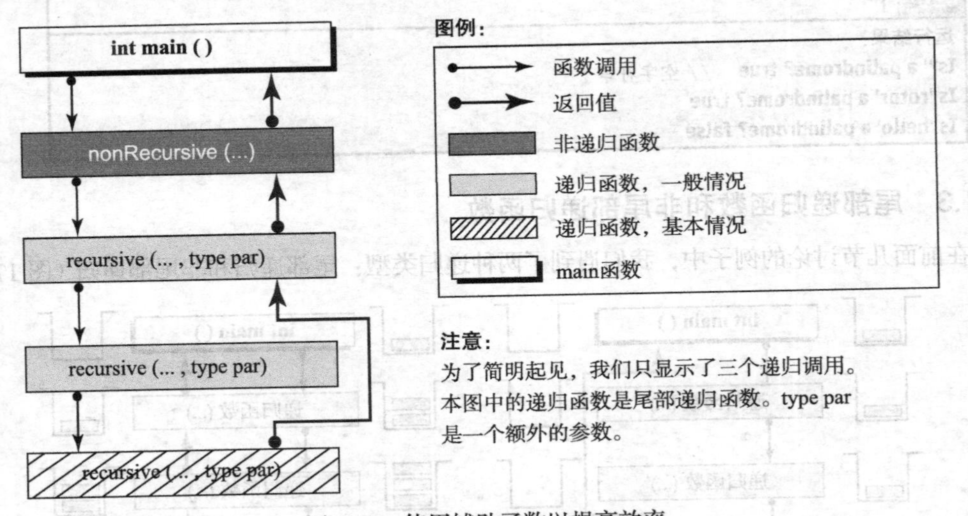

图 17-9　使用辅助函数以提高效率

我们知道函数 sum 不是尾部函数，当其参数很大时，效率很低。我们可以创建一个尾部递归函数作为辅助函数，它有一个额外的参数（result），该参数的值为 n，否则它将作为一个变量保存。由于每个递归调用都将此参数作为值传递给下一个调用，因此没有额外的内存使用。由于辅助函数也是尾部递归函数，因此也消除了栈效率低下的问题。程序 17-6 显示了这个版本的函数 sum。

程序 17-6　使用辅助函数的 sum 函数

```
1  /*************************************************************
2   * 本程序使用一个辅助函数创建一个尾部递归函数,                      *
3   * 该函数比非尾部递归函数高效                                      *
4   *************************************************************/
5  #include <iostream>
6  using namespace std;
7
8  // 函数的声明
9  int sum (int n);
10 int sum (int n, int result);
11
12 int main ( )
13 {
14   // 调用非递归函数四次
15   cout << "Sum (0) = " << sum (0) << endl;
16   cout << "Sum (1) = " << sum (1) << endl;
17   cout << "Sum (3) = " << sum (3) << endl;
18   cout << "Sum (7) = " << sum (7);
19   return 0;
20 }
21 // 非递归函数的定义
22 int sum (int n)
23 {
24   return sum (n, 0);
25 }
26 // 递归函数的定义
27 int sum (int n, int result)
28 {
29   if (n == 0)
30   {
31       return result;
32   }
33   return sum (n - 1, n + result);
34 }
```

```
运行结果:
Sum(0) = 0
Sum(1) = 1
Sum(3) = 6
Sum(7) = 28
```

作为另一个示例，我们使用一个辅助函数重新设计 isPalindrome 函数（程序 17-7）。请注意，isPalindrome 已经是一个尾部递归函数，但是重新设计该函数可以消除每次调用函数时创建子字符串的低效率问题。

程序 17-7　使用辅助函数的 isPalindrome 函数

```
1  /*************************************************************
2   *检查一个字符串是否为回文的程序,                                 *
3   *使用一个辅助函数以避免创建子字符串对象                           *
4   *************************************************************/
```

```
5  #include <string>
6  #include <iostream>
7  using namespace std;
8
9  // 函数的声明
10 bool isPalindrome (const string& strg);
11 bool isPalindrome (const string& strg, int left, int right);
12
13 int main ( )
14 {
15    // 检查字符串是否为回文
16    cout << boolalpha;
17    cout << "Is 'rotor" a palindrome? " << isPalindrome ("rotor") << endl;
18    cout << "Is 'madam' a palindrome? "  << isPalindrome ("madam") << endl;
19    cout << "Is 'Hello' a palindrome? "  << isPalindrome ("Hello");
20    return 0;
21 }
22 // 非递归函数 isPalindrome 的定义
23 bool isPalindrome(const string& strg)
24 {
25    return isPalindrome (strg, 0, strg.size () – 1);
26 }
27 // 递归函数 isPalindrome 的定义
28 bool isPalindrome(const string& strg, int left, int right)
29 {
30    if (right <= left)
31    {
32         return true;
33    }
34    else if (strg [left] != strg [right])
35    {
36         return false;
37    }
38    return isPalindrome (strg, left + 1 , right – 1);
39 }
```

运行结果：

```
Is 'rotor' a palindrome? true
Is 'madam' a palindrome? true
Is 'hello' a palindrome? false
```

17.2 递归排序和查找

在计算机科学中，我们经常需要排序和搜索列表。对于排好序的列表，搜索速度非常快。

17.2.1 快速排序

排序是指重新排列列表（例如数组）中的元素，使其值按顺序排列。我们在本节中介绍的排序算法称为快速排序（quicksort，一种递归算法）。这是一种在大多数算法库中使用的快速有效的算法。快速排序算法使用一种称为分区（partition）的非递归算法，我们将在下面讨论。

分区算法

分区算法围绕一个中心点（pivot，数组的一个元素）重新排列数组，以便所有大于或者等于中心点的元素移动到中心点之后，所有小于中心点的元素移动到中心点之前。通常选择第一个元素作为中心点。

以下是分区算法的实现代码。

```
int partition (int arr[], int i, int j) //i 和 j 是分区索引
{
   int p = i; // p 是中心点 pivot
   while (i < j) //外循环
   {
        while (arr [j] > arr[p] ) // 第一个内循环
        {
                j--;
        }
        swap (arr[j], arr[p]);
        p = j;
        j--;
        while (arr [i] <  arr [p] ) // 第二个内循环
        {
                i++;
        }
        swap (arr[i], arr [p]);
        p = i;
        i++;
   }
   return p;
}
```

在外循环的每次迭代中，我们执行以下两组操作：

- 只要 arr[j] 大于 arr[p]，我们就向左移动 j。然后交换 arr[j] 和 arr[p] 的值并设置 p = j，再将 j 向左移动一个元素。
- 只要 arr[i] 小于 arr[p]，我们就向右移动 i。然后交换 arr[i] 和 arr[p] 的值并设置 p = i，再将 i 向右移动一个元素。

将中心点 p 设置为第一个元素，但它将移动到数组中的最终位置。注意，在第二次迭代之后，i 移动到 j 的右边，外部循环终止。

图 17-10 显示了一个由 10 个元素组成的数组，以及在外循环的两次迭代期间发生的变化。

快速排序算法

快速排序算法是一种无返回值（void）的递归算法，这意味着它不会向其调用方返回任何内容。快速排序算法从左到右递归地向下移动，直到到达一个空的分区数组。图 17-11 显示了快速排序算法背后的思想。

程序 17-8 演示了一个由 10 个元素组成的小数组的快速排序。

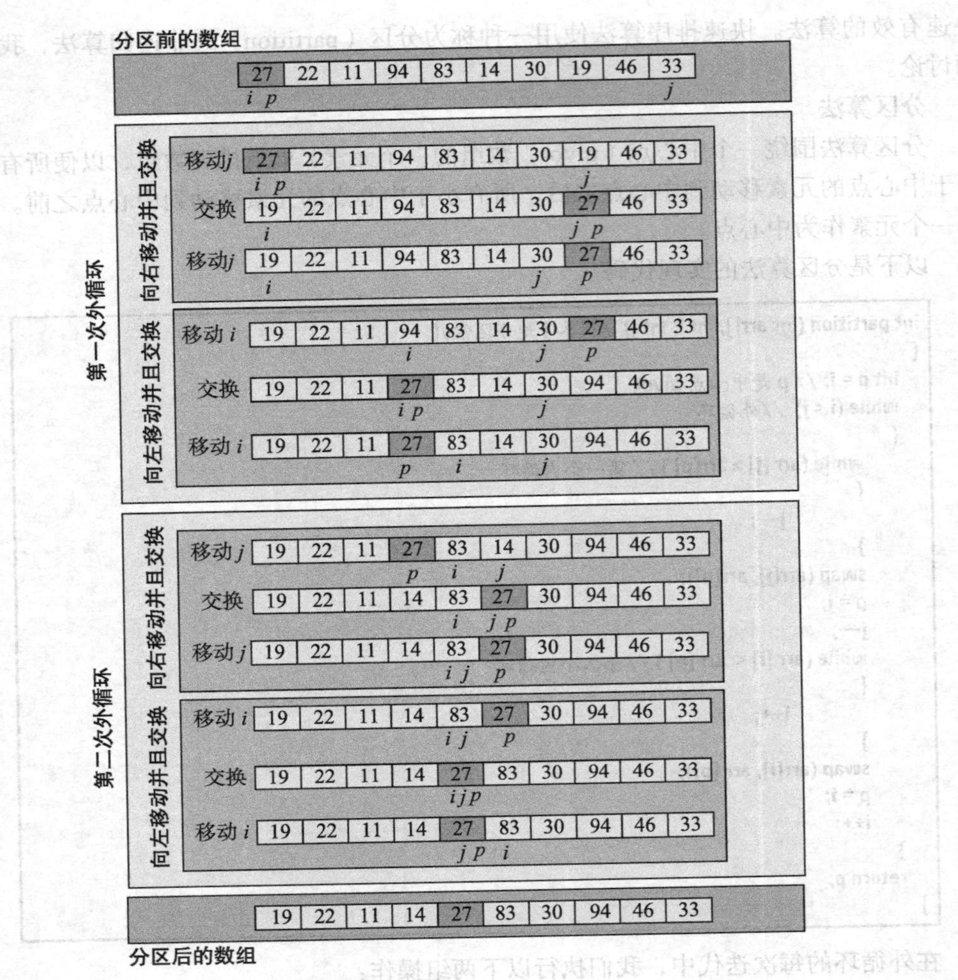

图 17-10　对数组应用分区算法

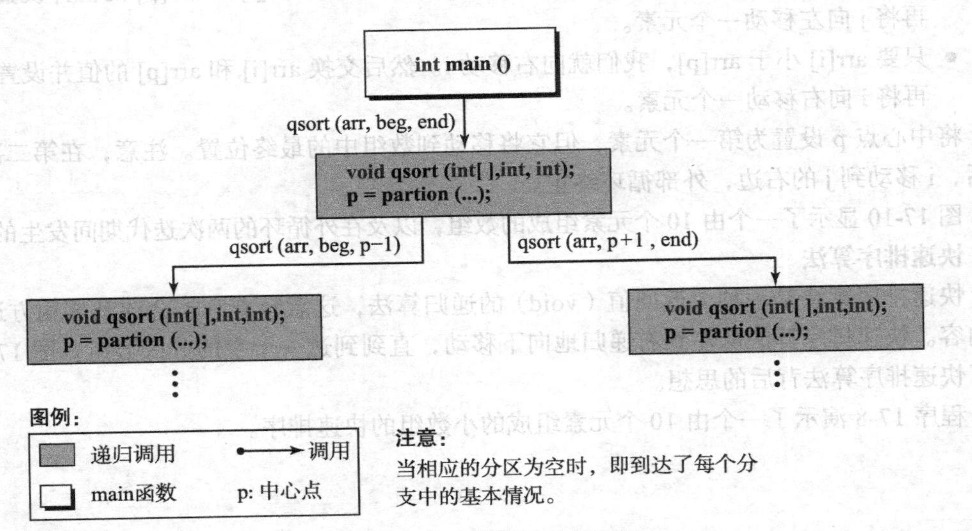

图 17-11　快速排序算法的思想

程序 17-8 快速排序程序

```
1  /**************************************************************
2   * 使用快速排序算法排序一个数组的程序,                              *
3   * 通过递归调用分区算法                                            *
4   **************************************************************/
5  #include <iostream>
6  using namespace std;
7
8  // 函数的声明
9  void swap (int& x, int& y);
10 void print (int array[], int size);
11 int partition (int arr[], int beg, int end);
12 void quickSort (int arr[], int beg, int end);
13
14 int main ()
15 {
16   // 声明一个未排序的数组
17   int array [10] = {27, 22, 11, 94, 83, 14, 30, 19, 46, 33};
18   // 打印未排序的数组
19   cout << "Original array: " << endl;
20   print (array, 10);
21   // 调用 quickSort 函数
22   quickSort (array, 0, 9);
23   // 打印排序后的数组
24   cout << "Sorted array: " << endl;
25   print (array, 10);
26   return 0;
27 }
28 // 交换函数
29 void swap (int& x, int& y)
30 {
31   int temp = x;
32   x = y;
33   y = temp;
34 }
35 // 打印数组的函数
36 void print (int array[], int size)
37   {
38   for (int i = 0; i < size; i++)
39   {
40       cout << array [i] << "  ";
41   }
42   cout << endl;
43 }
44 // 分区函数
45 int partition (int arr[], int beg, int end)
46 {
47   int p = beg ; // 初始化中心点
48   int i = beg;   // 初始化 i
49   int j = end;   // 初始化 j
50
```

```
51     while (i < j)
52     {
53         // 把 j 向左移动
54         while (arr [j] > arr[p] )
55         {
56             j--;
57         }
58         swap (arr[j], arr[p]);
59         p = j;
60         // 把 i 向右移动
61         while (arr [i] <  arr [p] )
62         {
63             i++;
64         }
65         swap (arr[i], arr [p]);
66         p = i;
67     }
68     return p;
69 }
70 // 快速排序函数
71 void quickSort (int arr[], int beg, int end)
72 {
73     if (beg >= end || beg < 0)
74     {
75         return;
76     }
77     int pivot = partition (arr, beg, end);
78     quickSort (arr, beg, pivot - 1);
79     quickSort (arr, pivot + 1, end);
80 }
```

```
运行结果：
Original array:
27  22  11  94  83  14  30  19  46  33
Sorted array:
11  14  19  22  27  30  33  46  83  94
```

17.2.2　二分查找法

我们常常需要搜索一个数组来找到某个元素的位置。搜索已排序数组要比搜索未排序数组容易得多。我们知道当物品整齐排列时，可以更容易地找到它们。例如，如果电话簿中的条目没有排序，则很难在其中搜索条目。**二分查找**（binary search，又称折半查找）算法被设计用于在已排序的数组中查找值。二分查找法从中间元素开始查找。对于要查找的值，存在三种情况：

- 如果该值等于中间元素，则停止搜索。已找到索引位置。
- 如果该值大于中间元素，则搜索将在数组的右半部分继续。
- 如果该值小于中间元素，则搜索将在数组的左半部分继续。

图 17-12 显示了一个二分查找法的示例。我们在 10 个元素的已排序数组中搜索值 30。三次尝试后，返回值的索引位置 5。

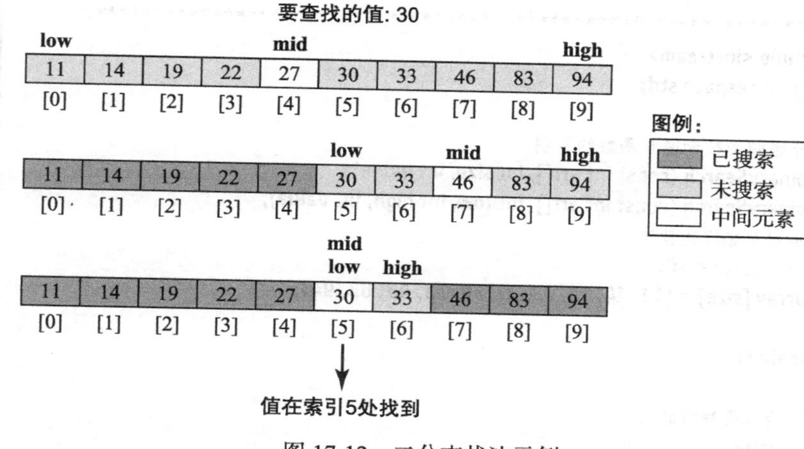

图 17-12　二分查找法示例

递归二分查找法

图 17-13 显示了使用辅助函数的递归二分查找法。

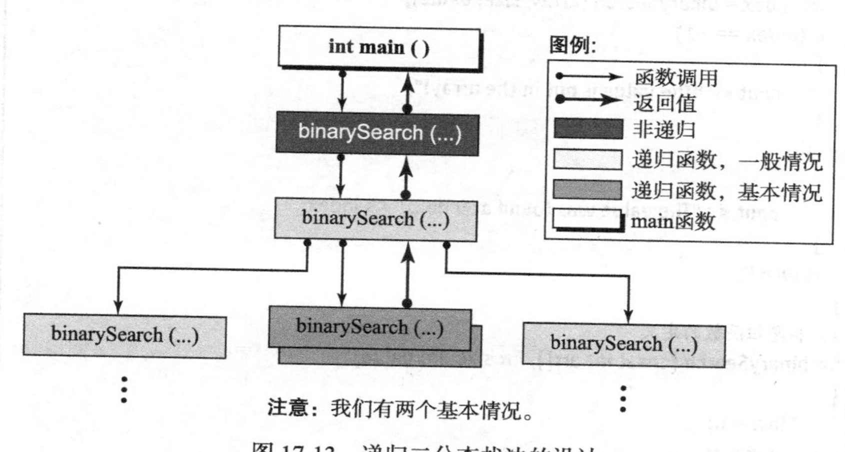

图 17-13　递归二分查找法的设计

如图 17-13 所示，我们有两个基本情况。在第一个中找不到值，在第二个中找到值。

在传统的二分查找法中，应用程序使用数组名称、大小和要查找的值调用搜索函数（如 binarySearch(arr, size, value)）。如果要查找的值不是中间元素，则必须再次调用，但新调用必须只搜索数组的左侧或者右侧部分。如果要搜索左侧部分，则不需要更改数组指针，但必须更改大小。如果要搜索右侧部分，则必须将数组指针移动到中间元素之后并调整大小。使用不移动数组指针但在每次调用中都使用数组的一部分（从 low 到 high）的辅助函数既方便又高效。辅助递归调用的格式为 binarySearch(arr, low, high, value)。在每次调用中，只需更改 low 或者 high 的值，arr 指针和搜索参数 value 保持不变。

程序 17-9 演示了如何使用辅助函数实现递归二分查找法。

程序 17-9　递归二分查找法

```
1  /*****************************************************************
2   *使用递归二分查找法，                                           *
3   *在一个数组中查找一个值                                         *
```

```
4    ***********************************************************************/
5    #include <iostream>
6    using namespace std;
7    
8    // 非递归函数和递归函数的声明
9    int binarySearch (const int arr[], int size, int value);
10   int binarySearch (const int arr[], int low, int high, int value);
11   // 声明要查找的数组
12   const int size = 10;
13   int array [size] = {11, 14, 19, 22, 27, 30, 33, 46, 63, 94};
14   
15   int main ()
16   {
17     // 输入要查找的值
18     int value;
19     cout << "Enter the value to be found: ";
20     cin >> value;
21     // 调用非递归搜索函数
22     int index = binarySearch (array, size, value);
23     if (index == -1)
24     {
25         cout << "The value is not in the array!";
26     }
27     else
28     {
29         cout << "The value was found at index: " << index;
30     }
31     return 0;
32   }
33   // 非递归函数的定义
34   int binarySearch (const int arr[], int size, int value)
35   {
36     int low = 0;
37     int high = size - 1;
38     return binarySearch (arr, low, high, value);
39   }
40   // 递归函数的定义
41   int binarySearch (const int arr[], int low, int high, int value)
42   {
43     int mid = (low + high) / 2;
44     if (low > high)
45     {
46         return -1;
47     }
48     else if (value == arr [mid])
49     {
50         return mid;
51     }
52     else if (value < arr[mid])
53     {
54         return binarySearch (arr, low, mid - 1, value);
55     }
```

```
56  else
57  {
58      return binarySearch (arr, mid + 1, high,  value);
59  }
60 }
```

运行结果：
Enter the value to be found: 11
The value was found at index: 0

运行结果：
Enter the value to be found: 27
The value was found at index: 4

运行结果：
Enter the value to be found: 94
The value was found at index: 9

运行结果：
Enter the value to be found: 10
The value is not in the array!

运行结果：
Enter the value to be found: 95
The value is not in the array!

17.2.3 汉诺塔

本节讨论一个更经典的递归问题：汉诺塔。通常使用递归方法求解。在汉诺塔问题中，有三个柱子（塔）。在第一个柱子上堆叠了一叠大小不同的圆盘（最小的放在顶部）。图 17-14 显示了四个圆盘的情况。

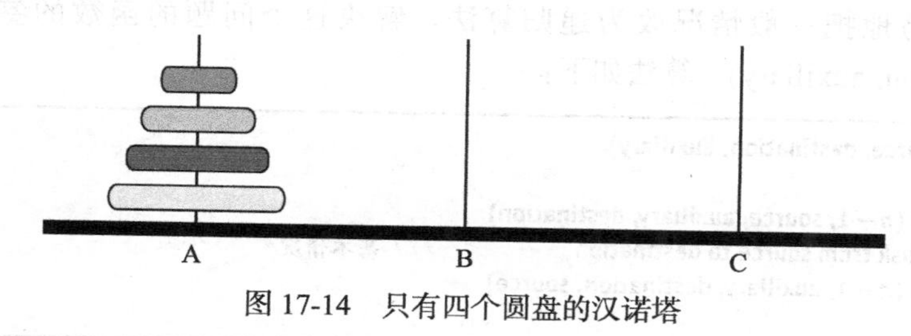

图 17-14　只有四个圆盘的汉诺塔

我们必须把圆盘从第一个柱子（A）移到第三个柱子（C）。第二个柱子（B）用于在移动过程中临时存放圆盘。移动圆盘的任务使用以下规则：

- 一次只能移动一个圆盘。
- 较大的圆盘不能堆叠在较小的圆盘之上。
- 临时柱子（B）可以用于临时放置圆盘。

这个问题之所以有趣有两个原因。第一，递归解决方案比迭代解决方案更容易编码。第二，这个问题的求解模式不同于我们前面讨论的简单示例。

基本情况

我们首先找到最简单情况（基本情况）的解决方案。我们假设只有一个圆盘。解决方案很简单：我们将圆盘从源柱子（A 或者 B）移动到目标柱子（C），不需要辅助柱子。

一般情况

现在我们讨论如果有 n 个圆盘（一般情况），如何解决这个问题。我们可以分三步完成：

- 我们必须将柱子 A 上的前 n–1 个圆盘移动到柱子 B，以便可以执行下一步移动操作。
- 我们将最后一个圆盘从源柱子 A 移动到目标柱子 C（基本情况）。
- 我们需要将柱子 B 上的 n–1 个圆盘（临时存放在柱子 B 上的所有圆盘）移动到柱子 C。

为了充分理解这个案例，我们将以图形方式显示移动步骤。图 17-15 显示了四个圆盘的情况。

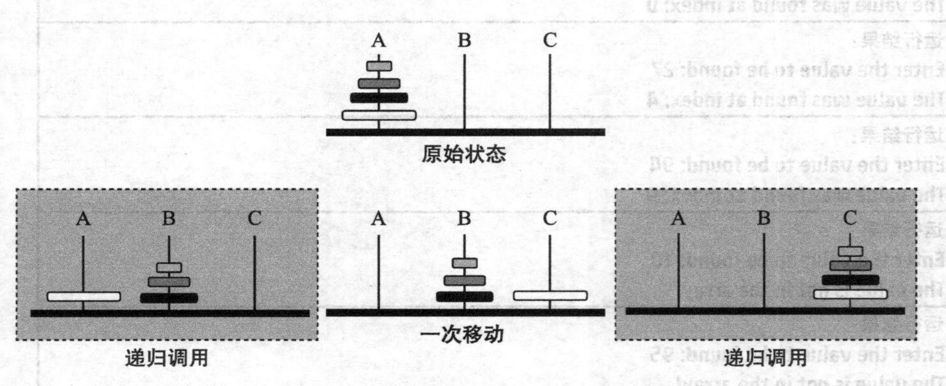

图 17-15　汉诺塔问题中的基本情况和递归情况

原始调用函数了解到有四个圆盘，源柱子是 A，目标柱子是 C，辅助柱子是 B。在第一次递归调用中，我们假定源柱子是 A，目标柱子是 B，辅助柱子是 C。在第二次递归调用中，我们假定源柱子是 B，目标柱子是 C，辅助柱子是 A。

算法

可以很容易地把一般情况改为递归算法。解决这个问题的函数的签名是 towers(n, source, destination, auxiliary)。算法如下：

```
towers (n, source, destination, auxiliary)
{
   Call towers (n – 1, source, auxiliary, destination)
   Move one disk from source to destination                // 基本情况
   Call towers (n – 1, auxiliary, destination, source)
}
```

程序

使用上面的算法，我们编写了程序 17-10 来解决任意数量的圆盘的汉诺塔问题。注意，为了清晰起见，我们使用一个单独的函数来处理基本情况。

程序 17-10　汉诺塔

```
1   /***************************************************************
2    *本程序用于求解汉诺塔问题                                      *
3    ***************************************************************/
4   #include <iostream>
5   using namespace std;
6
7   //函数声明
8   void towers (int, char, char, char);
9   void moveOneDisk (char, char);
```

```
10
11  int main ()
12  {
13     //变量声明
14     int n;
15     //输入
16     do
17     {
18          cout << "Enter number of disks (1 to 4): ";
19          cin >> n;
20     } while ((n < 1) || (n > 4));
21     //函数调用
22     towers (n, 'A', 'C', 'B');
23  }
24
25  // 函数 towers 的定义
26  void towers (int num, char source, char dest, char aux)
27  {
28     if (num == 1)
29     {
30          moveOneDisk (source, dest);
31     }
32     else
33     {
34          towers (num - 1, source, aux, dest);
35          moveOneDisk (source, dest);
36          towers (num - 1, aux, dest, source);
37     }
38  }
39  // 函数 moveOneDisk 的定义
40  void moveOneDisk (char start, char end)
41  {
42     cout << "Move top disk from " << start << " to " << end << endl;
43  }
```

运行结果：

```
Enter number of disks (1 to 4): 1
Move top disk from A to C
```

运行结果：

```
Enter number of disks (1 to 4): 2
Move top disk from A to B
Move top disk from A to C
Move top disk from B to C
```

运行结果：

```
Enter number of disks (1 to 4): 3
Move top disk from A to C
Move top disk from A to B
Move top disk from C to B
Move top disk from A to C
Move top disk from B to A
Move top disk from B to C
Move top disk from A to C
```

第一次运行非常简单：我们只有一个圆盘，将它从A移动到C。第二次运行也很简单：我们将顶部圆盘从A移动到B（辅助），将第二个圆盘移动到目的地C，然后将第一个圆盘从B移动到C。第三次运行则需要七次移动。前三步将前两个圆盘从A移动到B（辅助）。第四步将最后一个圆盘从A移动到C。最后三步将两个圆盘从B移动到C（使用A作为辅助）。图17-16显示了三个圆盘的原始状态和七个移动步骤。

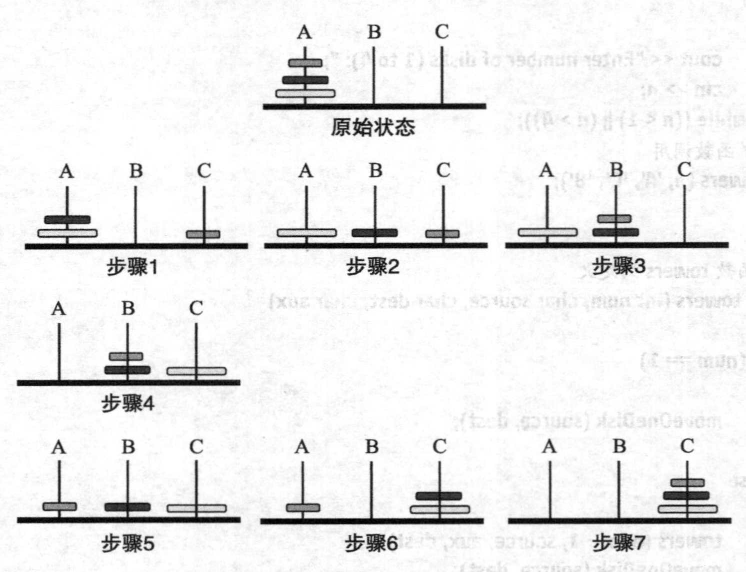

图 17-16　移动三个圆盘的步骤

17.3　程序设计

在这一节中，我们开发了两个问题的递归解决方案。对于这两个问题，递归解决方案比迭代解决方案更简短、更优雅。

17.3.1　字符串排列

字符串排列是查找字符串中所有可能的字符排列。给定一个 n 个字符的字符串，有 $n!$（阶乘）种排列。例如，当 $n=3$ 且给定字符串为“abc”时，我们可以得到以下六个字符串：

```
"abc"    "acb"    "bac"    "bca"    "cab"    "cba"
```

任何长度的字符串排列都可以通过迭代和递归的组合来解决。

迭代

在这个问题的所有递归解决方案中，我们必须先选择一个字符，然后再排列其余的字符。例如，在三字符“abc”的排列中，我们先选择字符“a”作为第一个字符，再将排列应用于其余字符。然后我们需要选择字符“b”并排列其余字符，依此类推。第一个字符的选择可以使用迭代来完成，它首先将第一个字符与字符串中的一个字符交换，然后调用递归解决方案，如下所示：

```
for (int i = 0; i < str.length () ; i++)
{
```

```
    swap (str [0], str[i]);
    permute (...);
}
```

递归

现在我们可以定义完整的递归函数来排列一个字符串。递归排列函数有两个参数：左参数是要排列的字符串；右参数是部分排列的字符串。递归函数从左字符串中删除一个字符，并将其添加到右字符串的末尾，然后再次进行排列。当左字符串为空时，将调用基本情况。以下是递归排列函数的实现代码：

```
void permute (string str, string p)
{
   if (str.length () == 0)
   {
       cout << p;
   }
   else
   {
       for (int i = 0; i < str.length(), i++)
       {
              swap (str[0], str[1]);
              permute ((str.substr (1, str.length() - 1), p + str.substr (0, 1);
       }
   }
}
```

增加一个非递归函数

我们用上面开发的递归函数作为辅助函数，并使用只有一个参数（要排列的字符串）的非递归函数。换句话说，我们有一个非递归函数，它调用带两个参数的递归函数，如下所示。注意，第二个参数是一个空字符串。

```
void permute (string s)
{
   permute (s, "");
}
```

这样，用户只能调用非递归函数 permute(str) 来显示字符串的所有排列。

程序 17-11 显示了任意字符串的排列程序。

程序 17-11　字符串的排列

```
1   /***************************************************************
2    * 本程序用于创建给定字符串的所有排列                          *
3    ***************************************************************/
4   #include <iostream>
5   #include <string>
6   using namespace std;
7
8   // 函数声明
9   void permute (string);
10  void permute (string, string);
11  void swap (char&, char&);
```

```
12
13  int main ()
14  {
15     // 排列字符串"xy"
16     cout << "Permutation of xy: ";
17     permute ("xy");
18     cout << endl;
19     // 排列字符串"abc"
20     cout << "Permutation of abc: ";
21     permute ("abc");
22     cout << endl;
23  }
24  // 非递归函数的定义
25  void permute (string s)
26  {
27     permute (s, "");
28  }
29  // 递归（辅助）函数 permute 的定义
30  void permute (string str, string p)
31  {
32     if (str.length () == 0)
33     {
34          cout << p << " ";
35     }
36     else
37     {
38          for (int i = 0; i < str.length (); i++)
39          {
40                 swap (str[0], str[i]);
41                 permute (str.substr (1, str.length() - 1), p + str.substr (0, 1)) ;
42          }
43     }
44  }
45  // 交换函数 swap 的定义
46  void swap (char& c1, char& c2)
47  {
48     char temp = c1;
49     c1 = c2;
50     c2 = temp;
51  }
```

运行结果：
Permutation of xy: xy yx
Permutation of abc: abc acb bac bca cab cba

17.3.2　素数

数学家把正整数分成三组：

- 整数 1
- 素数（也称为质数）
- 合数

整数 1 既不是素数也不是合数。素数可以被 1 及其本身整除。合数可以被 1、本身和其他整数整除。换言之，整数 1 只有一个除数（本身），一个素数有两个除数（整数 1 和本身），一个合数有两个以上的除数（整数 1、本身和其他小于自身的整数）。我们可以看到最小的素数是整数 2。小于 10 的素数包括：2、3、5 和 7。

素数判断

给定一个整数 *n*，我们如何才能判断它是否是素数？判别规则如下：如果一个整数不能被范围（2, …, floor(sqrt(n))）中的任何素数整除，则该整数是素数。

【例 17-1】 我们可以使用上面的规则来检查 97 是否为素数。floor(sqrt(97)) = 9。我们只需要检查所有小于或者等于 9 的素数，即 2、3、5 和 7。因为这些整数中没有一个可以整除 97，所以整数 97 是素数。

【例 17-2】 我们可以使用上面的规则来检查 301 是否为素数。floor(sqrt(301)) =17。我们只需要检查所有小于或者等于 17 的素数，即 2、3、5、7、11、13 和 17。因为这些整数中 7 可以整除 301（301 / 7 = 43），所以整数 301 不是素数。

递归函数

基于以上的观察，我们可以编写一个递归函数来判断一个整数是否是素数。我们可以不断地检查这个数是否可以被它的素数除数整除。然而，这意味着我们必须先创建一个素数除数列表，这也意味着我们已经知道如何判断一个整数是否为素数。我们可以放宽这个条件，检查给定的整数是否可以被范围（2, …, floor(sqrt(n))）中的所有整数整除。

以下代码片段是用于判断给定整数是否为素数的递归函数。它有两个基本情况和一个一般情况（递归情况）。第一个基本情况返回 false，如果数值可以被第一个参数整除，则终止函数。第二个基本情况返回 true，在检查了所有小于（floor(sqrt(num))）的整数并且未找到除数时终止函数。一般情况使用范围中下一个可能的除数再次调用该函数。

```
bool isPrime (int div, int num)
{
   if (num % div == 0)
   {
       return false;
   }
   else if (div >= floor (sqrt (num)))
   {
       return true;
   }
   return isPrime (div + 1, num);
}
```

增加一个非递归函数

读者可能已经注意到，前面的递归函数没有正确地处理两个特殊情况。首先，如果 num 为 1，它错误地返回 true。其次，如果 num 为 2，它错误地返回 false。这两种情况不能包括在递归函数中，但可以在调用递归函数的非递归函数中进行测试，如下所示：

```
bool isPrime (int num)
{
   if (num  <= 1)
   {
       return false;
```

```
    }
    else if (num == 2)
    {
        return true;
    }
    return isPrime (2, num);
}
```

注意，在处理完这两个特殊情况后，非递归函数调用递归函数（第一个参数为 2，从 2 开始递归处理）。

程序 17-12 显示了使用两个函数的完整程序。

程序 17-12　测试一个整数是否为素数

```
1   /**************************************************************
2    * 本程序用于测试一个整数是否为素数                              *
3    **************************************************************/
4   #include <iostream>
5   #include <cmath>
6   using namespace std;
7
8   // 非递归函数和递归函数的声明
9   bool isPrime (int num);
10  bool isPrime (int div, int num);
11
12  int main ( )
13  {
14    // 测试一些整数是否为素数
15    cout << "Is 1  prime?  " << boolalpha << isPrime(1)   << endl;
16    cout << "Is 2  prime?  " << boolalpha << isPrime(2)   << endl;
17    cout << "Is 7  prime?  " << boolalpha << isPrime(7)   << endl;
18    cout << "Is 21 prime?  " << boolalpha << isPrime(21)  << endl;
19    cout << "Is 59 prime?  " << boolalpha << isPrime(59)  << endl;
20    cout << "Is 97 prime?  " << boolalpha << isPrime(97)  << endl;
21    cout << "Is 301 prime? " << boolalpha << isPrime(301) << endl;
22    return 0;
23  }
24  // 非递归函数的定义，调用递归函数
25  bool isPrime (int num)
26  {
27    if (num  <= 1)
28    {
29        return false;
30    }
31    else if (num == 2)
32    {
33        return true;
34    }
35    return isPrime (2, num);
36  }
37  // 递归（辅助）函数的定义
38  bool isPrime (int div, int num)
```

```
39 {
40     if (num % div == 0)
41     {
42         return false;
43     }
44     else if (div >= floor (sqrt (num)))
45     {
46         return true;
47     }
48     return isPrime (div + 1, num);
49 }
```

运行结果：

```
Is 1  prime? false
Is 2  prime? true
Is 7  prime? true
Is 21 prime? false
Is 59 prime? true
Is 97 prime? true
Is 301 prime? false
```

本章小结

我们可以使用循环或者递归来解决迭代问题。在循环中，我们让函数重复自身操作，直到出现终止条件。在递归中，我们让一个函数调用自己，直到出现相同的终止条件。

我们有两种类型的递归：尾部递归和非尾部递归。尾部递归通常比非尾部递归更有效率，并且使用的计算机资源更少。通过定义更多参数，我们可以将非尾部递归更改为尾部递归。

我们使用递归的两个领域是排序和搜索一个列表，比如一个数组。

汉诺塔是一个经典的递归算法，它使用了递归的许多方面知识。

思考题

1. 根据以下代码，fun('G') 的打印结果是什么？解释该函数的作用。

```
void fun (char c)
{
    if ( c < 'A' || c > 'Z')
    {
        return;
    }
    cout << c << " ";
    fun (c + 1);
}
```

2. 根据以下代码，fun(5, 12) 和 fun(12, 5) 的返回结果是什么？解释该函数的作用。

```
int fun (int n, int m)
{
    if ( n == m)
    {
        return 0;
```

```
    }
    else if ( n > m)
    {
        return 1;
    }
    else
    {
        return fun (m, n);
    }
}
```

3. 根据以下代码，fun(5) 的打印结果是什么？解释该函数的作用。

```
void fun (int n)
{
    if (n < 0)
    {
        return;
    }
    cout << n << " ";
    fun (n - 1);
}
```

4. 根据以下代码，fun(5) 的返回结果是什么？解释该函数的作用。

```
double fun (int n)
{
    if (n <= 1)
    {
        return 0;
    }
    return fun (n - 1) + 1.0 / n;
}
```

5. 根据以下代码，fun(4) 的返回结果是什么？解释该函数的作用。

```
double fun (int n)
{
    if (n <= 1)
    {
        return 0;
    }
    return fun (n - 1) + (double) n / (n + 2);
}
```

6. 根据以下代码，fun(4) 的返回结果是什么？解释该函数的作用。

```
double fun (int n)
{
    if (n <= 1)
    {
        return 0;
    }
    return fun (n - 1) + double (n) / (2 * n - 1);
}
```

7. 根据以下代码，fun(4) 的返回结果是什么？解释该函数的作用。

```
double fun (int n)
```

```
{
    if (n <= 1)
    {
        return 0;
    }
    return fun (n - 1) + double (n) / (3 * n - 1);
}
```

8. 根据以下代码，fun(164) 的打印结果是什么？解释该函数的作用。

```
int fun (int n)
{
    if (n < 10)
    {
        return n
    }
    else
    {
        cout << (n % 10);
        return fun (n / 10);
    }
}
```

9. 解释以下函数的作用，以及当 n 为 3 并且 m 为 4 时其返回的内容。

```
int fun (int n, int m)
{
    if (n == 0 || m == 0)
    {
        return 1;
    }
    else
    {
        return fun (n, m - 1) + fun (n - 1, m);
    }
}
```

10. 根据以下代码，fun(19) 的打印结果是什么？解释该函数的作用。

```
void fun (int n)
{
    if (n > 0)
    {
        fun (n / 2);
        cout << n % 2;
    }
}
```

编程题

1. 编写 factorial 函数的尾部递归版本，并添加一个非递归函数来调用它。在程序中测试该函数。
2. 编写一个递归函数，一次查找 *n* 个对象的排列数 *k*，然后编写一个程序对其进行测试。公式如下：

```
P (n, k) = factorial (n) / factorial (n - k)
```

3. 编写一个递归函数，一次查找 *n* 个对象的组合数 *k*，然后编写一个程序对其进行测试。公式如下：

```
C (n, k) = factorial (n) / ((factorial (n - k) * (factorial (k))
```

4. 编写一个尾部递归函数，该函数反转其参数中的数字。例如，给定整数 12789，函数返回 98721。然后使用非递归函数来调用它。在程序中测试该函数。
5. 编写一个尾部递归函数，将十进制整数转换为二进制字符串，然后使用非递归函数来调用它。例如，十进制 78 将被转换为“1001110”。在程序中测试该函数。
6. 编写一个尾部递归函数，将十进制整数转换为十六进制字符串，并使用非递归函数来调用它。例如，十进制 78 将被转换为“4E”。在程序中测试该函数。
7. 编写一个尾部递归函数，查找整数数组中的最小整数。然后使用非递归函数来调用它。在程序中至少用三个包含 10 个整数的数组来测试该函数。注意，当我们在数组中查找最小的元素时，我们必须跟踪最小的元素和要检查的下一个索引。这意味着辅助函数必须带两个额外的参数。
8. 编写一个程序，找出给定整数的所有因子。因子是小于或者等于给定整数的数，给定整数除以该数时余数为零。请注意，任何整数至少有两个因子：1 和它本身。用几个整数的因子来测试你编写的函数，并在程序中把结果制成表格。编写一个递归（辅助）函数来查找两个数的因子。然后使用非递归函数来调用它。
9. 编写一个尾部递归函数和相应的非递归函数，测试一个整数是否为素数。记住，一个素数只能被 1 和它本身整除，但是我们可以检查从 1 到该数平方根的数值，以确保它是素数。记住，1 不是素数。数学家把正整数分成三组：合数、素数和 1。
10. 编写一个程序，找出给定整数的所有素数因子。素数因子是一个整数，它是一个因子，也是一个素数。素数因子在安全和密码学领域有许多应用。在编写程序时，可以使用第 8 题和第 9 题中提出的设计方法。
11. 编写一个递归函数，计算并打印一个字符串中给定字符的个数。在程序中测试该函数。